International Seminar on Agriculture, Biodiversity, Food Security and Health

IOP Conference Series: Earth and Environmental Science
Volume 883

Ambon, Indonesia
10 December 2020

Editors:

Esther Kembauw
Asri S. Mahulette
Natelda R. Timisela
Febby J. Polnaya
Robbi Rahim

ISBN: 978-1-7138-4865-3
ISSN: 1755-1307

Printed from e-media with permission by:

Curran Associates, Inc.
57 Morehouse Lane
Red Hook, NY 12571

CURRAN ASSOCIATES INC. proceedings.com

Some format issues inherent in the e-media version may also appear in this print version.

Institute of Physics
Dirac House, Temple Back
Bristol BS1 6BE UK

Phone: 44 1 17 929 7481
Fax: 44 1 17 920 0979

techtracking@iop.org

Additional copies of this publication are available from:

Curran Associates, Inc.
57 Morehouse Lane
Red Hook, NY 12571 USA
Phone: 845-758-0400
Fax: 845-758-2633
Email: curran@proceedings.com
Web: www.proceedings.com

TABLE OF CONTENTS

Preface

Peer review declaration

AGRICULTURE

The properties of rhizobacteria from tomato rhizosphere as biocontrol and biofertilizer 1
A M Kalay, H Kesaulya, A Talahaturuson, R Osok

Essential oil components of forest clove variants from Ambon Island, Maluku 9
A S Mahulette, J Riry, H Kesaulya, E Kembauw, I J Lawalata, A Y Wattimena, M H Makaruku, A Alfian

Inorganic fertilizers efficiency with using the liquid organic fertilizer to increase the cabbage yield (Brassica oleracea var. capitata L.).. 16
A E Marpaung, B Karo, S Barus

Green economic based on low-carbon development on small islands.. 22
A Kastanya, C C V Suhendy, D V Pattimahu, Iskar

Income analysis of coconut farming with land diversification in North Minahasa Regency 27
J N K Dumais, R Kaunang, J Lumingkewas, Y Rori

The Strategy of guava agribusiness development in Kendal Regency using a business model canvas............. 35
A Khamdi, W Roessali, Mukson

Isolation and identification of fenobucarb degrading bacteria from Pangalengan farm land 44
A Akhdiya, R A Sanjaya, Wartono

Food consumption patterns among university students in Indonesia during the transition period in new Normal Era of Covid-19 Pandemic.. 50
A E Yunianto, K Kristiandi, M Darawati, T H Doloksaribu, I Anggraeni, M Pasambuna

Species of pandan (Pandanus sp) in Gorom Island, East Seram Regency.. 56
A Sahupala, T E Siahaya, B B Seipala, L. Siahaya, L. Pelupessy, Y. D. Komul

Estimation rumen degradable protein of local feeds in dairy cattle using in sacco method 68
A Rosmalia, I G Permana, Despal, R Zahera

The effect of light and gibberellic acid concentrations on breaking dormancy of potato micro tuber 74
A K Karjadi, N Waluyo

The evaluation of protease enzyme effectivenes in broiler chicken diet containing jack bean seed (Canavalia ensiformis) with different protein level toward internal organ size.. 79
B P Mahardhika, M Ridla, R Mutia, D N Adli

Effort to reduce ammonia gas in the broiler chicken excreta with the addition of probiotic as substitute for antibiotic growth promoter 86
B P Mahardhika, R Mutia, M Ridla

Salt-induced growth promotion in rice varieties during nursery 93
B Kurniasih, N Arini, D Alvioliana, R I Nisa, R A Wulandari

Application of Drip Irrigation on Various Planting Media and Cow Urine in the Main Media of Preliminary Oil Palm Nursery 102
C I Wahyudin, A S Mahulette, V L Tanasale, D A Marasabessy, N Goo, W D Mariati, Y Asmi, R A Rifqah, Michala

The Effect of polyethylene glycol concentration on some varieties of kenaf (Hibiscus cannabinus L.) in enhancing the germination viability 106
C I Wahyudin, A S Mahulette, J I Nendissa, M H Makaruku, W D Mariati, A Haitami, Desrihastuti, Febrianti, Arby'In Pratiwi

Introduction to a systematic review and meta-analyses in Indonesia nutrition poultry: case study in probiotic 111
D N Adli, O Sjofjan, A Jayanegara, B P Mahardhika

Optimization of gamma ray irradiation dose on strawberry plantlets 116
D Saptadi, H Arisah, D Agisimanto

Physical and mechanical properties of cement-bonded particleboard made from squander of paper 121
R S Maail, J J Fransz, J Titarsole

Embryo-gynogenic induction of Indonesia shallot (Allium cepa L. Aggregatum group) cultivars using three different protocol unpollinated flower cultures 128
E S Lestari, Sulastriningsih, D C Prayantini, A Purwantoro, E Sulistyaningsih

Mycorrhizal fungi Glomus spp. propagation in zeolite enriched with mycorrhiza helper bacteria for controlling nematode in coffee 133
I N Asyiah, R Hindersah, R Harni, B N Fitriatin, W Anggraeni

Effects of physical and chemical treatments on seed germination and soybean seed-borne fungi 138
E P Ramdan, A Y Perkasa, T K K Azmi, Aisyah, R Kurniasih, P I Kanny, Risnawati, P Asnur

Resistance varieties and pattern of disease progress of rust (Pucciana horiana p. henn) in Chrysanthemum 144
Y A Bety, R Pangestuti

Improving seed germination and seedling growth of true seed shallot (TSS) using plant growth regulator seed priming 153
R Pangestuti, E Sulistyaningsih, B Kurniasih, R H Murti

The diversity of bird species based on the altitude of the protected forest area in Sirimau Mountain in Soya Village - Ambon City 163
C K Pattinasarany, L Latupapua, A Sanduan, Y Th Latupapua, F F Tetelay, F Soselisa

Characteristics of yellow sweet potato flakes (Ipomoea batatas L.) with the addition of moringa leaf flour (Moringa oleifera) 169
G H Augustyn, V N Lawalata, S G Sipahelut

Activity test of Bacillus Spp against bacterial wilt (R. solanacearum) on tomatoes by in vitro 177
G N C Tuhumury, J V Hasinu, H Kesaulya

Characterization of plant growth promoting rhizobacteria of maize 183
H Kesaulya, A Talahaturuson, A M Kalay, E Matatula, I J Lawalatta, M L Hehanussa, S J Nendissa

Edible insects: Alternative protein for sustainable food and nutritional security 195
H C D Tuhumury

Scoring system as an alternative audit method in food safety management system certification body............ 203
H D Wahyuni, S Nurjanah, W P Rahayu

Indigenous maize seed storage using buffalo dung ash practiced by smallholder farmers in Kisar Island 213
H Jesajas, E Kembauw, M J Matatula, A D Tagueha, I J Liur

Sustainable development strategy for agroforestry 219
I S Ruhimat, A Widiyanto

Soybean varieties tolerance to intercropping with maize 229
I Permanasari, E Sulistyaningsih, B Kurniasih, D Indradewa

Farmers empowerment level analysis in farming during the Covid-19 pandemic and its impact on farm income.. 237
I P N Damanik, M E Tahitu, M Turukay, F P Adam

Contribution of home-garden farming to household income and its sustainability in Yogyakarta City, Indonesia.. 244
Irham, D S Gusfarina, A W Widada, A Nurhayati

Level of escherichia coli contamination of broiler chicken meat in Ambon City Market.............................. 255
I J Liur, M Veerman

Livelihood strategy of coastal households during covid-19 pandemic: case study in Wermaktian District, Tanimbar Islands Regency 260
J F Sopamena, L O Kakisina, A E Pattiselanno

Analysis of macro nutrient content in biofouling waste organic fertilizer pearl oyster (Pinctada maxima L.) 266
J I Nendissa, M H Makaruku, V L Tanasale, A K Kilkoda, J Taribuka

Potential of Bacillus spp as a biocontrol agent against Ralstonia bacterial wilt in bananas 270
J V Hasinu, G N C Tuhumury, Henry Kesaulya

Digital extension and the development of agricultural performance in Indonesia 276
M A Nasir, Ismiasih, Jamhari

Response of soybean (Glycine max L.) that was applied by various liquid organic fertilizer in climate change at acid soil... 281
Jamilah, Rapialdi, M Ernita

Accelerated adoption of sugar palm farming technology to supports sustainable resource utilization in North Sulawesi 292
J B M Rawung, J G Kindangen, R Indrasti, A Gaffar

Breed availability analysis of local beef cattle in Seram Utara Timur Seti District Maluku Tengah Regency.. 298
J F Salamena, Rajab, B J Papilaya, R Sarfan

Study of dryland cultivation farming technology to support sustainable food independence on small islands.. 303
J Riry, A S Mahulette, A M Tapotubun, W A Riry

Food security among pandemic covid 19 313
J F Sahusilawane, A M Sahusilawane

Nutmeg Cultivation Intensity (Myristica fragrans Houtt) in Banda District 317
A Y Wattimena, M H Makaruku, E Kembauw, A S Mahulette

Implementation study of Good Agricultural Practices (GAP) of Red Fruit (Pandanus conoideus Lamk.) 322
M H Makaruku, A Y Wattimena, A S Mahulette, E Kembauw

Factors affecting the capacity of farming rice farmers in managing rice field in the District Tangerang 327
Khaerul Saleh, Suherman

Impact of salicylic acid and biosilica application on plant growth of shallot under water deficit 336
L D Indarwati, E Sulistyaningsih, B Kurniasih

The characteristic anda feasibility of Banda'S nutmeg agro-industry in Banda Island of Maluku Province 345
M Lawalata, N R Timisela, M Turukay, E D Leatemia, J M Luhukay

The effect of organic fertilizers on growth several varieties of soybeans 355
M Rizwan, M Dalimunthe, I A Pasaribu, H Satriawan

Response of growth and yield of two soybean varieties to seed pre-treatment applications 364
N Herawati, A R Aisah, I Mardian, B N Hidayah, B T R Erawati

Drought stress and K effects on contents of soil water and organic matter, CEC, exchangeable-K, yield and water productivity of sweet corn on Inceptisols 370
N Nurlaeny, D Herdiyantoro, R M Putra, N Ratuliami, W N Pratiwi, F S Nurfadilah

Effect of urea in steamed sago waste on rumen fermentation parameters in vitro tested 379
N C Tiven, T M Simanjorang

Consumption pattern and hope food pattern in Salahutu District, Maluku Tengah Regency 385
N R Timisela, W Girsang, L Tupamahu

Test for the antibacterial inhibition of kaffir lime leaf (Citrus hysteric D.C) extract against pathogen bacteria in improving food safety 395
S J Nendissa, D M Nendissa

Centella asiatica tendril growth of Samosir – Indonesia accession 401
N S Vinolina, R Sigalingging

Evaluation of effectiveness various seed processing combination technique to produce True Shallot Seed (TSS) with good quality 408
N Waluyo, A Rahayu, R Rosliani, T Wikan, R Gaswanto

The effect of replacing maize with re-binding banana hump flour on the meat quality, footpad color, and feather or hybrid duck 417
O Sjofjan, D N Adli, R H Syahputra

Characterization of pectin from tongka langit banana peels with various extraction temperature 421
P Picauly, G Tetelepta

The impact of el niño southern oscillation and covid-19 on the rice price dynamics in Indonesia: the vector error correction model approach 428
P Hasudungan, I Irham, A W Utami

Carrying capacity and environmental capacity of water resources analysis in Maluku Province of Indonesia 438
P J Kunu, S Talakua, Y Pesulima, L Uyara, S Laimeheriwa, R M Osok

Characteristics of chemical properties of fruit flour of mangrove (Bruguiera gymnorrhiza Lamk.) with lower cyanide content........ 451
F J Polnaya, V J Thenu, S Palijama, R Breemer

Effect of dietary palm kernel meal on laying hens 459
R D D Prasetya, M Rahmadani, Nahrowi, A Jayanegara

Vegetation composition in gold post-mining land of PT. Sultra Utama Nickel, Bombana District 464
F D Tuheteru, Husna, Basrudin, A Arif, Albasri, I Mustafa

Adaptation test of five introduced cherry tomato lines in Lembang highland........ 471
Redy Gaswanto

Organic matter and mixed biofertilizer for plant growth and yield of shallot grown in fertile soil 478
R Hindersah, A M Kalay, I A Ngabalin, M Jamlean, A Talahaturuson

Meat price volatility as implications for food security in Indonesia........ 485
R P Destiarni, A S Jamil, F Septya

Income adaptation of farmers as long covid-19 pandemy on sustainable ub forest management: a case from Indonesia........ 492
A Kustanti

The feasibility of environmentally friendly and conventional shallot farming in Bantul Regency 501
D R Kamardiani, M Fauzan, K S Riansyah

Contribution of agrosilvopastoral pattern to sustainable communities economic improvement (study in Namaa Hamlet, Pelauw Village, Haruku Island Sub-District, Central Maluku Regency) 507
M Sahureka, M Hadijah, H Marasabessy

Correlation of normalized difference water index and baseflow index in small island watershed landscapes........ 515
B Latuamury, M Talaohu, F Sahusilawane, W N Imlabla

Physiological characters of four lowland chilli varieties (Capsicum annum L.) with root cutting........ 526
S Kusumaningrum, E Sulistyaningsih, R Harimurti, K Dewi

Development of home-based ready-to-use supplementary food (RUSF-HB) to overcome nutrition-related problems among children under five during the covid-19 pandemic........ 535
S Sastrawan, M Menap, L Sulaiman, H Hendrayani

In vitro propagation of six selected sugarcane mutant clones through leaf explants........ 544
R Purnamaningsih, D Sukmadjaja, S Suhesti, S Rahayu

Utilization of herbal supplements on feed on growth and survival rate of catfish (Clarias sp.) 552
S N Pratama, R C Mukti

Collaborative and sustainable management strategy of the Wallacea Key biodiversity area (KBA) in Kasa Island 560
D V Pattimahu, A Kastanya

Study of changes in carbon stock in primary and secondary forest in protection forest of Soya Village, Ambon City 569
L Siahaya, L Pelupessy, Y D Komul, A Sahupala, T E Siahaya, B B Seipala

The impact of climate change on the resin productivity of dammar tree (Agathis alba) in Inamosol subdistrict, West Seram district, Maluku-Indonesia 577
J M Matinahoru

Institutional Characteristics of Supporting Potato Farming in South Sulawesi 582
R Indrasti, J B M Rawung, N R Sudolar, S Handoko, Nurjanani

Organoleptic characteristics of nutmeg tea with variations in the roast time 588
S Palijama, L O Kakisina, N R Timisela, J M Luhukay, R Breemer, S Laterissa, E Kaplale, F J Polnaya

Corn farming analysis in Babar Island, District of Babar Islands, Southwest Maluku Regency 593
S F W Thenu

Moler disease and cultivation practiced by shallot farmers in Brebes Central Java 598
Supriyadi, Supyani, Sh. Poromarto, Hadiwiyono

Performance of native chicken fed on ration containing fermented sago waste 603
T N Ralahalu, Ch C E Latupeirissa, S Ch H Hehanussa, S. Fredriksz

Identification of endophytic and rhizosphere bacteria in rice (Oryza sativa L.) in the experimental field at Payakumbuh State Agriculture Polytechnic, West Sumatra, Indonesia 610
Y Sondang, R Siregar, K Anty

The physicochemical characteristics of gandaria (Bouea macrophylla) leather with sugar concentration treatment 618
V N Lawalata

The utilization of agro-industrial waste for soil amendment and liquid biofertilizer mixed bacterial antagonist in eggplant production 622
S Boonraeng, N Punyoyai

In-situ kuskus conservation strategy in Wakasihu Village, West Leihitu District, Central Maluku District 629
Y Th Latupapua, F F Tetelay, F Soselisa, C K Pattinasarany, L Latupapua, A Sanduan

The influence of agricultural diversification on community income in East Taniwel District 642
La Ega, Muspida, E Kembauw, S J Ferdinandus, L J Sinay, M Sangadji, M Faisal, H S E S Aponno, M Wenno, E Regar, R Apituley

Moringa oleifera milk powder as a supplementary food for malnutrition children (SUSUKE) 650
S W Djoko, E Kembauw, I B D Kapelle

Clove processing as a source of increasing business income in Ambon City 653
E Kembauw, A S Mahulette, A P Kakisina, M T F Tuhumury, M C B Umanailo, M G I Kembauw

Author Index

IOP Conf. Series: Earth and Environmental Science **883** (2021) 011001 doi:10.1088/1755-1315/883/1/011001

Preface International Seminar on Agriculture, Biodiversity, Food Security, And Health (ABFSH) 2020

Agricultural Faculty of Pattimura University organized International Seminar on Agriculture, Biodiversity, Food Security, And Health (ABFSH) 2020 on 10 December 2020 in Ambon, Maluku, Indonesia. The theme of our international seminar is: "Sustainable Natural Resources And Food Production Development During Covid 19 Pandemi".

The main objective of this agricultural seminar is not only to generate information on the availability of food and raw materials, but also to grow healthy and well-nourished people. One of the important tasks besides ensuring the availability of sufficient and healthy food for the world's population in a sustainable manner so that all people can lead healthy and productive lives, which is more important in this era of COVID-19. With the world's population increasing in the future, the link between agriculture, biodiversity, food security and sustainability to support health and well-being needs to be addressed.

Writing scientific articles is a very important activity for a professional educator. This activity is not only necessary in order to obtain credit points for promotion or for certification purposes through portfolios, but with this international activity it has become a forum for building partnerships with outside partners, both main speakers and fellow lecturers and researchers from various other institutions and universities in Indonesia.

The international seminar was attended by 95 speakers from various institutions, Pattimura University - Ambon, and also other universities in Indonesia. This seminar can be held thanks to the help of various parties. For this reason, on this occasion allow us on behalf of the 2020 International Agricultural Seminar Committee to thank the Pattimura University Chancellor and his staff, the Dean of the Faculty of Agriculture and his staff, as well as various other parties who have participated to assist in the implementation of the seminar which we cannot mention one by one. We also convey our highest appreciation to all the committees who have worked hard for the success of this seminar. Last but not the least, we are thankful to IOP EES for producing the proceeding.

Editors

1. Dr. Esther Kembauw, SP., M.Si
2. Dr. Asri S. Mahulette, SP., MP
3. Dr. Natelda. R. Timisela, SP., MP
4. Dr. Febby. J. Polnaya, SP., MP
5. Dr. Robbi Rahim, M.Kom

Invited Speaker

1. Supot Boonraeng (Faculty of Agriculture Technology Chiang Mai Rajabhat University – Thailand)
2. Abu Hasan B Abdul Jalli (Agronomist, Landscape and Beescape Consultant, Meliponiculturist and Meliponary Consultant) Malaysia Kelulut Academy
3. Dr. Hellen C. D. Tuhumury, SP., M.FoodSc (Faculty of Agriculture Pattimura University Maluku - Indonesia)

Organizing Institutions

- Agricultural Faculty of Pattimura University, Ambon, Indonesia

IOP Conf. Series: Earth and Environmental Science **883** (2021) 011002 doi:10.1088/1755-1315/883/1/011002

Peer review declaration

All papers published in this volume of IOP Conference Series: Earth and Environmental Science have been peer reviewed through processes administered by the Editors. Reviews were conducted by expert referees to the professional and scientific standards expected of a proceedings journal published by IOP Publishing.

- **Type of peer review:**
  a. All papers submitted were checked their similarity index using Turnitin.
  b. Papers have similarity index more than 25% were rejected.
  c. After that, the format of the papers was checked. Papers that did not follow the IOP template will be sent back to the authors for revision.
  d. All papers passed the similarity index and have already follow the IOP template will be sent to the reviewers.
  e. Reviewers reviewed the papers and give comments to the papers.
  f. Scientific committee decided if the papers: accepted, accepted with major revision, accepted with minor revision, or rejected.
  g. Author revised their papers and send back their revision

- **Conference submission management system:**
  All papers from the conference submitted to using JOTForm Submission System

- **Number of submissions received:**
  143 Submission were received

- **Number of submissions sent for review:**
  143 Submission was sent for review, and the rest was rejected because out of scope, low-quality articles and high similarity

- **Number of submissions accepted:**
  100 Paper accepted based on IOP EES Scope and quality

- **Acceptance Rate (Number of Submissions Accepted / Number of Submissions Received X 100):**
  (100/143) x 100 = 69.93 %

- **Average number of reviews per paper:**
  2 Times

- **Total number of reviewers involved:**

  21 Reviewer Involved + 4 Editors

- **Any additional info on review process:**

  NONE

- **Contact person for queries:**
  Dr. Esther Kembauw, SP., M.Si
  Pattimura University
  Email: abfshconf@konferensi.id

IOP Conf. Series: Earth and Environmental Science **883** (2021) 012001 doi:10.1088/1755-1315/883/1/012001

The properties of rhizobacteria from tomato rhizosphere as biocontrol and biofertilizer

A M Kalay[1,*], H Kesaulya[1], A Talahaturuson[1] and R Osok[1]

[1] Faculty of Agriculture, University of Pattimura, Jl. Ir. M. Putuhena, Poka Campus, Ambon 97233, Indonesia

* E-mail: marthinkalay@gmail.com

Abstract. Low soil fertility and the presence of pathogen are the limiting factors in horticultural crops cultivation in Ambon City. This study aimed to isolate plant growth promoting rhizobacteria from tomato rhizosphere as a potential biocontrol, biostimulant and biofertilizer. The research consisted of two stages, namely (1) isolation of bacteria from tomato rhizosphere and purification of the isolates, and (2) characterization of colony morphology, gram reaction, hypersensitivity test and antagonist test to soil pathogens, phosphate dissolution, and phytohormones indole acetic acid (IAA) production. Forty-two isolates were found from tomato rhizosfer, 36 isolates suppressed growth of *Rhizoctonia solani* while 6 isolates increased the growth of those pathogen. The growth of *Phytophthora* sp. pathogen was decreased by 35 bacterial isolates and increased by 7 bacterial isolates. All isolates produce IAA but only 13 isolates are enabled to dissolve inorganic phosphate. Based on bacterial properties, the isolates of TT-22, TT-21, TT-12, TT-11, WT-11 and WT-332 are the promising isolates for developing the biological agent as pathogen control and biofertilizer.

1. Introduction

The presence of pathogens and low soil fertility are a major problem in optimizing horticultural and food crops production, and therefore is a major constraint in maintaining food availability in Ambon city. Horticultural and food crops are cultivated on entisol and inceptisol soils which often contain 1.67% C-organic; 0.12% N Total; 13.92 C/N Ratio; 9.08 mg / 100 g P_2O_5; 35 mg / 100 g K_20; 7.86 ppm P total; 0.44 cmol / kg K total; 5.86 pH (H_2O) and 4.5 pH (KCl), and they are classified as soil with low nutrient oil borne [1]. In addition, the lack of using organic and inorganic fertilizers, and the presence of soil borne pathogens in the nursery phase to production has been exacerbated this condition, as indicated by the presence of damping off and leaf blight diseases such as *Rhizoctonia solani*, *Sclerotium rolfsii*, and *Fusarium oxysporum* [2,3].

Organic materials such as compost and biological fertilizers can be used as a source of nutrients for plants and to improve soil physical and biological [4,5]. Biological fertilizers are inoculants with beneficial living organisms active ingredients which can be used to improve crop yields productivity and quality by increasing the availability of nutrients in the soil [6].

Microbes that play a role in spurring plant growth of plants are known as Plant Growth Promoting Rhizobacteria (PGPR). In general, the function of RPTT in increasing plant growth is characterized by the following mechanisms: (1) growth promoters (biostimulants) by synthesizing and regulating the concentration of various growth regulators or phytohormones such as indol acetic acid (IAA), gibberellin, cytokinins, and ethylene in the root environment; (2) a nutrient provider (biofertilizer) by tethering N2 from the air asymbiosis and dissolving P nutrients bound in the soil; and (3) controlling soil-derived pathogens (bioprotectant) by producing various anti-pathogenic compounds or metabolites such as siderophore, β-1,3-glucanase, chitinase, antibiotics, and cyanide [7,8,9].

Indigenic bacteria from the genus *Azotobacter*, *Acinetobacter Azospilium*, *Bacillus* and *Pseudomenas* are types of microorganisms found in soil and plants, especially at the base of the stems and roots. These four genera of bacteria have been widely used as Plant Growth Promoting Rhizobacteria (PGPR) to increase plant growth through nitrogen fixation, phosphate dissolving, and phytohormone production [6,9,10]; and to reduce the intensity of disease attacks caused by soil pathogens [2,11,12], and they are more effective if applied to their natural environmental conditions. This indicates that the environmental conditions play important roles in nutrient supply through N tethering, phosphate dissolving, and production of phytohormones as well as producing enzymes to suppress plant pathogens and disease.

This study aimed to isolate plant growth promoting rhizobacteria from tomato as they are potentally used as biocontrol, biostimulant and biofertilizer

2. Methods

This research was conducted from June to October 2019. Bacteria were isolated from tomato rhizosphere collected from three vegetable cultivated fields in Ambon City, namely Hutumuri Village, Waiheru Village, and Tawiri Village. Research activities included isolation, selection and purification of isolates; isolate characterization; and testing the ability of the isolates to stimulate the growth of tomato seedlings.

2.1. Isolation of Bacteria Tomato Rhizosphere

100 g of soil was taken from the rhizosphere of the fruiting tomato plant. Five very healthy plants were selected from each of the sampling locations. 500 g of soil was mixed and 10g was taken and added to 90 ml of sterile water in erlemeyer and was shaked using a hotplate stirrer for 15 minutes. A total of 1 ml of this solution was taken to 9 ml test tube and shaked using a vortex for 1 minute until a 10-5 dilution series was obtained. The 0.5 ml was taken using pipette and distributed in petridis on a petri dish (Ø 9 cm), and added with nutrient broth (NB) media while shaking it with a rotary petridis. The sample was then incubated at room temperature for 24-48 hours. Single colonies of growing bacteria were separated based on colony color and shape, grown and purified using NA media. Pure colonies were maintained on NA media in test tubes (slanted media) to be used as stock cultures.

2.2. Characterics Bacterial Isolates

Bacterial characterization was determined based on motile, size, shape, periphery, and colony color on nutrient agar (NA) media; gram reaction test; phosphate dissolution test; Indole Acetic Acid (IAA) phytohormone test; hypersensitivity test; and antagonist test.

2.2.1. Gram Reaction Test

Stored-bacterial colonies on NA media in a test tube were transferred to a glass object using a loop needle, added with two drops of 3% KOH solution and mixed for 5-10 seconds. If the sample looks slimy and unbroken when lifted up, it shows a positive reaction and is classified as gram-negative, conversely, if the sample is not slimy and removeable, it shows a negative reaction and is classified as gram-positive [13,14].

2.2.2. Phosphate Dissolution Test

Phosphate dissolution was tested using the Pikovskaya method [9,15]. The 72 hour old suspension isolated-bacteria on NA media was taken aseptically 1 ose and innoculated on solid Phikovkaya media containing tricalcium phosphate (Ca_3PO_4), then incubated for 72 hours at room temperature (27-28 ^{0}C). Phosphate dissolving activity is indicated by a clear zone around the bacterial colony. The phosphate dissolution index (IP) is measured as follows: IP = (clear zone diameter - colony diameter) / (colony diameter)

2.2.3. IAA phytohormone test

The cultured-bacterial on NB media containing 10 ml tryptophan was centrifuged at 3,000 rpm for 30 minutes. The 2 ml supernatant from centrifugation was taken and added with 2 drops of ortho phosphate solution and 4 ml of Salkowsky's reagent (50 ml 35% H_2SO_4 + 1 ml M FeCl3), incubated for 20 hours at room temperature without light. The sample was measured by spectrophotometry at wavelength (λ) = 353 nm. The resulted-absorbance were then entered into the standard curve equation of IAA 0-60 ppm to obtain the final concentration of IAA [16].

2.2.4. Hypersensitivity Test
The isolated-bacteria obtained after screening were carried out by the HR (Hyhipersensitive Reaction) test on tobacco plants and tomato seeds. Isolates were grown on Lurial Bertani (LB) media for 3 days while shaking with a shaker at a speed of 150 rpm at room temperature. A total of 1 ml isolated-bacterial suspension with a cell density of 10^7 cfu / ml was injected into tobacco leaves and observed for 5 days. The present of pathogenic bacteria is indicated by necrotic symptoms.
Hypersensitivity test was carried out as follows: 50 tomato seeds were immersed in the bacterial isolate suspension with a cell density of 10^7 cfu/ml for 5 and then planted in sterile soil in the seed boxes. Observations were made from the seeds germinating until 10 days old.

2.2.5. Antagonist Test
The isolated-bacteria obtained after hypersensitivity test were tested for antagonism against the fungal pathogens *Rhizoctonia* sp and *Phytopthora* sp. The antagonism testing of each isolated-bacteria against pathogens was conducted by using the dual culture method on PDA media. The antagonism of each isolate against pathogens was calculated by measuring the area of the pathogenic colony as control and compared to the growing area of the pathogenic colony with isolated-bacteria using the formula: $PIRG = \frac{l1-l2}{l1} x100\%$, where l_1 = area of pathogenic colonies without bacteria (control) and l_2 = area of pathogenic colonies with bacteria [17].

3. Results and discussion
The screening of the isolated-bacteria from tomato rhizosphere results in 42 isolates, and most of the bacterial colonies are white, circular shape, flat elevation, and entire margin (Table 1). Thirty three of them are motile, 26 are gram positive, 16 isolates are gram negative, while 13 isolates can be active as phosphate solvents, however all isolates can produce the IAA hormone and negative hypersensitivity test (Table 2).

Table 1. Morphology of bacteria isolated from tomato rhizosphere

No	Isolate		Morfologi Koloni			
	Code	Origin	Color	Shape	Elevation	Margin
1	WT-332	Waiheru	Yellow	Circular	Umbonate	Entire
2	WT-321	Waiheru	White	Irregular	Convex	Lombate
3	WT-31	Waiheru	White	Circular	Flat	Entire
4	WT-15	Waiheru	Bright yellow	Circular	Flat	Entire
5	WT-11	Waiheru	White	Circular	Umbonate	Lombate
6	WT-16	Waiheru	White	Circular	Flat	Entire
7	WT-13	Waiheru	White	Irregular	Umbonate	Lombate
8	WT-14	Waiheru	White	Circular	Flat	Entire
9	WT-12	Waiheru	Bright yellow	Circular	Flat	Entire
10	WT-33	Waiheru	Bright yellow	Irregular	Flat	Lombate
11	WT-331	Waiheru	White	Irregular	Umbonate	Lombate
12	WT-322	Waiheru	White	Circular	Flat	Entire
13	HT-11	Hutumuri	White	Irregular	Umbonate	Lombate
14	HT-322	Hutumuri	White	Irregular	Umbonate	Lombate
15	HT-22	Hutumuri	White	Circular	Flat	Entire
16	HT-331	Hutumuri	Bright yellow	Circular	Flat	Entire
17	HT-321	Hutumuri	White	Irregular	Convex	Lombate

18	HT-121	Hutumuri	White	Circular	Flat	Entire
19	HT-32	Hutumuri	White	Irregular	Convex	Lombate
20	HT-222	Hutumuri	White	Circular	Flat	Entire
21	HT-221	Hutumuri	Bright yellow	Circular	Flat	Entire
22	HT-21	Hutumuri	White	Circular	Flat	Entire
23	HT-31	Hutumuri	White	Circular	Flat	Entire
24	HT-132	Hutumuri	Bright yellow	Circular	Flat	Entire
25	HT-13	Hutumuri	White	Circular	Flat	Entire
26	HT-33	Hutumuri	Yellow	Irregular	Convex	Lombate
27	HT-131	Hutumuri	White	Circular	Flat	Entire
28	TT-13	Tawiri	White	Circular	Flat	Entire
29	TT-211	Tawiri	White	Irregular	Umbonate	Lombate
30	TT-32	Tawiri	White	Circular	Flat	Entire
31	TT-131	Tawiri	White	Circular	Flat	Entire
32	TT-14	Tawiri	White	Circular	Flat	Entire
33	TT-11	Tawiri	White	Circular	Flat	Entire
34	TT-12	Tawiri	White	Circular	Flat	Entire
35	TT-22	Tawiri	White	Circular	Flat	Entire
36	TT-232	Tawiri	Yellow	Circular	Flat	Entire
37	TT-222	Tawiri	White	Irregular	Convex	Lombate
38	TT-221	Tawiri	White	Irregular	Umbonate	Lombate
39	TT-212	Tawiri	White	Irregular	Umbonate	Lombate
40	TT-31	Tawiri	White	Irregular	Flat	Lombate
41	TT-33	Tawiri	White	Circular	Flat	Entire
42	TT-21	Tawiri	White	Irregular	Flat	Lombate

Table 2. Motile test, gram reaction, phosphate dissolving ability, production of IAA phytohormones, and hypersensitivity tests on tobacco and tomato seeds

No	Kode	Motile	Gram reaction	phosphate dissolving ability	IAA consentration (ppm)	hypersensitivity tests	
						on tobacco leaf	on tomato seeds
1	WT-332	+	+	1,15	1,000	-	-
2	WT-321	+	+	-	0,843	-	-
3	WT-31	+	+	-	0,865	-	-
4	WT-15	+	+	-	0,973	-	-
5	WT-11	+	+	1,20	0,865	-	-
6	WT-16	-	+	-	0,834	-	-
7	WT-13	-	+	-	0,816	-	-
8	WT-14	+	+	-	0,987	-	-
9	WT-12	+	+	-	1,000	-	-
10	WT-33	+	+	1,23	1,031	-	-
11	WT-331	-	+	-	0,861	-	-
12	WT-322	-	-	-	0,901	-	-
13	HT-11	+	+	-	0,883	-	-
14	HT-322	+	+	-	0,847	-	-
15	HT-22	+	-	1,22	1,175	-	-
16	HT-331	+	+	-	0,829	-	-
17	HT-321	-	+	-	0,924	-	-
18	HT-121	+	-	1,90	0,996	-	-
19	HT-32	+	+	-	0,861	-	-
20	HT-222	-	-	-	1,211	-	-
21	HT-221	+	-	-	1,391	-	-

22	HT-21	-	+	-	0,879	-	-
23	HT-31	+	-	-	0,825	-	-
24	HT-132	+	+	-	0,865	-	-
25	HT-13	+	+	-	0,982	-	-
26	HT-33	+	+	-	1,000	-	-
27	HT-131	+	+	-	0,870	-	-
28	TT-13	-	-	-	1,013	-	-
29	TT-211	+	+	-	0,973	-	-
30	TT-32	+	-	-	1,067	-	-
31	TT-131	+	-	1,22	1,031	-	-
32	TT-14	+	-	1,13	0,919	-	-
33	TT-11	+	-	1,17	1,049	-	-
34	TT-12	+	-	1,19	1,054	-	-
35	TT-22	+	+	1,25	0,829	-	-
36	TT-232	-	-	1,31	0,852	-	-
37	TT-222	+	-	-	0,847	-	-
38	TT-221	+	+	-	0,996	-	-
39	TT-212	+	+	-	0,987	-	-
40	TT-31	+	-	-	1,000	-	-
41	TT-33	+	-	1,27	0,870	-	-
42	TT-21	+	+	1,17	0,838	-	-

Forty two isolated- bacteria were tested for pathogenicity against the pathogens of the fungi *Rizoctonia solani* and *Phytopthora* sp. The result shows that not all isolates can suppress pathogen development, but some can increase pathogen growth. Tables 3 shows that HT-11, HT-32, HT-21, TT-222, TT-212, and TT-33 isolates can help increase the growth of the pathogen *R. solani*, while isolates WT-31, WT-15, WT-16, WT- 12, HT-22, HT-13, and TT-13, can increase the growth of pathogens *Phytopthora* sp.

Tabel 3. Antagonist test of isolated bacterial to suppress in vitro growth of pathogens *Rhizoctonia solani* and *Phytopthora* sp.

No	Isolate code	Percentage of suppress growth of pathogen	
		Rhizoctonia solani	*Phytopthora* sp
1.	WT-332	62,57	86,67
2.	WT-321	47,93	86,67
3.	WT-31	11,66	-(31,81)
4.	WT-15	56,77	-(45,17)
5.	WT-11	61,77	86,67
6.	WT-16	33,68	-(8,70)
7.	WT-13	62,41	86,67
8.	WT-14	21,45	86,67
9.	WT-12	71,28	-(27,19)
10.	WT-33	28,25	86,67
11.	WT-331	59,31	86,67
12.	WT-322	21,45	83,54
13.	HT-11	-(25,21)	77,47
14.	HT-322	51,82	86,67

15.	HT-22	15,07	-(12,11)
16.	HT-331	46,60	86,67
17.	HT-321	4,91	61,98
18.	HT-121	17,46	86,67
19.	HT-32	-(24,91)	86,67
20.	HT-222	46,41	83,54
21.	HT-221	48,31	86,67
22.	HT-21	-(2,10)	37,74
23.	HT-31	31,75	83,54
24.	HT-132	46,80	86,67
25.	HT-13	63,69	-(30,88)
26.	HT-33	13,62	83,54
27.	HT-131	32,18	86,67
28.	TT-13	46,41	-(19,09)
29.	TT-211	18,41	86,67
30.	TT-32	15,07	83,54
31.	TT-131	52,72	86,67
32.	TT-14	25,12	86,67
33.	TT-11	33,47	86,67
34.	TT-12	41,12	83,54
35.	TT-22	52,90	86,67
36.	TT-232	28,47	86,67
37.	TT-222	-(5,30)	86,67
38.	TT-221	58,13	86,67
39.	TT-212	-(8,27)	86,67
40.	TT-31	8,94	80,09
41.	TT-33	-(4,23)	86,67
42.	TT-21	49,99	86,67

The data in Table 2 and Table 3 is used as references for selecting potential isolates to be used as plant biostimulants, nutrient providers (biofertilizers), and plant disease control (biocontrol / bioprotectant). The ability of isolates to stimulate plant growth and yield is known by testing the ability of isolates to produce phytohormones, including Indol Acetic Acid IAA) or Indol Acetic Acid. Table 2 shows that all isolates have ability to produce IAA phytohormones. IAA is the active form of the auxin hormone [18], it plays a role in increasing plant growth and yield [19,20]. The main function of auxin hormone in plants is to promote the elongation of developing buds, affect stem and root growth, while zeatin is a natural cytokinin that can affect root growth and diffentiation, promote cell division and general growth, promote germination and delay aging [21]. The action of this hormone is synergistic with gibberellin and cytokinins, therefore it can also regulate physiological processes, such as growth, division and differentiation of cells and protein synthesis [22].

The phosphate dissolution test was carried out to ensure the ability of the isolates to dissolve the phosphate in vitro. The results shows that of the 42 isolates, 13 isolates have the ability as a phosphate solvent, namely isolates WT-332, WT-11, WT-33, HT-22, HT-121, TT-33, TT-21, TT-131, TT-14, TT-11, TT-12, TT-22, and TT-232 (Table 2). These results indicates that if the lilate can be used as an agent in the process of making biological fertilizers. Phosphate is a macro nutrient that plays a major role in increasing plant growth and yield. In the soil, phosphate is found as inorganic phosphate such as calcium phosphorus (Ca-P, aluminum phosphate (Al-P) and iron phosphate (Fe-P), and as organic phosphates such as phytin compounds, nucleic acids, and phospholipids [23,24]. Inorganic phosphates are difficult to dissolve in water. The presence of microbes through biological fertilizers will increase the availability of phosphate in plants. The activities of phosphate solubilizing microbia will produce organic acids including lactic, citric, succinic, glutamate, oxalate, malate, glyoxalate, and fumarate. The increasing of these acids can decrease soil pH and followed by the dissolution of the phosphate bound by Ca, Al [15,25].

In vitro hypersensitivity and antagonistic testing against pathogens *Phytopthora* sp and *Rizoctonia solani* will further strengthen the selection of isolates that have the potential to be developed as biological fertilizers. Although an isolate can produce phytohormones and dissolve phosphate but it has potency to be a pathogen or as a trigger of pathogens development, therefore this isolate cannot be recommended as biological fertilizer. Hypersensitivity test results also show that all isolates have no potential to be pathogens (Table 2). However, the results of invitro antagonistic testing against pathogens *Phytopthora* sp and *Rizoctonia solani* show that six isolates (HT-11, HT-32, HT-21, TT-222, TT-212, and TT-33) can help increase the growth of the pathogen R. solani, while seven isolates (WT-31, WT-15, WT-16, WT-12, HT-22, HT-13, and TT-13) can increase the growth of the pathogen *Phytopthora* sp (Table 3).

4. Conclusion

A total of 42 bacteria are isolated from the tomato rhizosphere. All isolates produce IAA but only 13 isolates are able to dissolve inorganic phosphate. Among all isolates, 36 isolates can suppress the growth of the pathogen *Rhizoctonia solani*, and 6 isolates can increase the growth of the pathogen. 35 isolates could suppress the growth of the pathogen *Phytophthora* sp, and 7 isolates can increase the growth of the pathogen. Based on the characteristic of bacteria, isolates TT-22, TT-21, TT-12, TT-11, WT-11 and WT-332 are promising isolates to be developed as biological agents both as pathogen controllers and as biological fertilizers to increase plant growth.

5. References

[1] Hindersah R, Hermawan W, Mutiarawati T and Kalay A M 2015 *Penguatan kemandirian sayuran di Teluk Ambon. Laporan Program Pengabdian Kepada Masyarakat.* Universitas Padjadjaran. Bandung.
[2] Kalay A M, Hindersah R, Talahaturuson A and Latupapua A I 2017 *IJSER* **8** 1288-1292.
[3] Hindersah R, Kalay A M, Risamasu R and Kurniati N 2019 *Budidaya sayuran di Kota Ambon: Pertanian Perkotaan. Penerbit. LoGoz*, Bandung.
[4] Yuwono N W 2009 *Jurnal Ilmu Tanah dan Lingkungan* **9** 137-141
[5] Kasno A. 2006 *J. Tanah Tropika* **14** *111-118.*
[6] Simanungkalit R D M, Suriadikarta D A, Saraswati R, Setyorini D and Hartatik W 2006 *Pupuk organik dan pupuk hayati: Organic fertilizer and biofertilizer.* Balai Besar Litbang Sumberdaya Lahan Pertanian, Badan Penelitian dan Pengembangan Pertanian.
[7] Tentua M 2006 *Plant growth promoting rhizobacteria:* Prospect for increasing nutrient acquisition and disease control. Available: http://www.umanitoba.ca/afs/agronomists-conf/2003/pdf/tentuarhizobacteria.pdf.
[8] Cattelan A J, Hartel P G and Fuhrmann J J 1999 *Soil Sci.Soc.Am.J.* **63** 1.670-1.680.
[9] Purwaningsih S 2012 *J.Tek.Ling* **13** 101-108.
[10] Rini C R and Sulochana K K 2007 *Journal of Tropical Agriculture* **45** 24-32.
[11] Chauhan S, Wadhwa K, Vasudeva M and Narula N 2012 *Archives of Agronomy and Soil Science* **58** 1365-1385.
[12] Bhosale H J, Kadam T A and Bobade A R 2013 *Journal of Environmental Biology* **34** 177-182.
[13] Schaad N W, Jones J B and Chun W 2001 *Laboratory guide for identification of plant pathogenic bacteria.* Third Edition. APPS Press, American Phytopathological Society, St. Paul Minnesota.
[14] Goszczynska T, Serfontein J J and Serfontein S 2000 *Introduction to practical phytobacteriology. A manual for phytobacteriology.* First Edition. ARC – Plant Protection Research Institute Pretoria, South Africa.

[15] Suba-Rao N S 1994 *Mikroorganiisme tanah dan pertumbuhan tanaman*. Edisi ke dua. Terjemahan Herawati Susilo. UI Press.
[16] Patten C L and Glick B R 2002 *Applied and Environmental Microbiology* **68** 3795-3801.
[17] Gaigole A H, Wagh G N and Khadse A C 2011 *Asiatic J. Biot. Resour* **4** 461-465.
[18] Wedhastri S 2002 *Jurnal Ilmu Tanah dan Lingkungan* **3** 45-51.
[19] Oedjijono, Lestanto U W, Nasution E M, Bondansari 2012 *Pengaruh Azospirillum spp. terhadap pertumbuhan tanaman jagung (Zea mays L.) dan kemampuan beberapa isolat dalam menghasilkan IAA. Prosiding Seminar Nasional "Pengembangan Sumber Daya Pedesaan dan Kearifan Lokal Berkelanjutan II*" Purwokerto, 27-28 Nopember 2012. Hal: 156-163.
[20] Akbari Gh, Abbas, Arab S M, Alikhani A H, and Allahdadi Arzanesh M H 2007 *World J. Agric. Sci.* **3** 523-529.
[21] Salisbury F B and Ross C W 1992 *Plant physiology*. Belmont, Calif.: Wadsworth Pub. Co.
[22] Darnel J, Lodish H and Baltimore H 1986 *Molecular Cell Biology*. Scientific American Books, Inc. New York.
[23] Damanik M M B, Bactiar E H, Fausi, Sarifudin F and Hamidah H 2011 *Kesuburan tanah dan pemupukan*. USU-Press, Medan.
[24] Rosmarkum A dan Yuwono, N W 2002 *Ilmu kesurudan tanah*. Penerbit Kanisium. Yogyakarta.
[25] Alexander M 1978 *Introduction to soil Microbiology*. 2 nd Ed, Willey Eastern Limited. New Delhi.

Essential oil components of forest clove variants from Ambon Island, Maluku

A S Mahulette[1,*], J Riry[1], H Kesaulya[2], E Kembauw[3], I J Lawalata[1], A Y Wattimena[1], M H Makaruku[1], A Alfian[4]

[1] Agrotechnology Study Program, Faculty of Agriculture, Universitas Pattimura, Ambon, Indonesia
[2] Plant Breeding Study Program, Faculty of Agriculture, Universitas Pattimura, Ambon, Indonesia
[3] Agribusiness Study Program, Faculty of Agriculture, Universitas Pattimura, Ambon, Indonesia
[4] Faculty of Health Agriculture and Marine Science, Universitas Muhammadiyah Palopo, South Sulawesi, Indonesia

* E-mail: mahulette_07@yahoo.co.id

Abstract. Forest clove is native to Maluku from the non-aromatic class. Forest cloves have several variants, including the accession group with a small morphological size in the population. So far, the complete information on essential components contained in this accession group is not studied. The research aims to determine the forest clove accession group's total essential oils with small morphological sizes originating from its distribution area on Ambon Island, Maluku. The analyses were performed on flowers, flower stalks, and leaves using Gas Chromatography-Mass Spectrometry (GC-MS). The analysis results identified 11 components in the flower section, 14 components on the flower stalk, and 13 components on the leaf. The main components analysis of the heat map profile revealed that the small morphological-sized forest clove accession groups' main components were germacrene-D, α-cubebene, eugenol, δ-cadinene, α-copaene, methyl eugenol.

1. Introduction

Forest cloves are classified as non-aromatic and endemic cloves from the Maluku Islands [1]–[5]. A previous study conducted by [3] identified three forest clove accessions in Maluku's distribution area. The forest clove variants found in the last research consisted of 3 accessions, namely forest clove accessions with large, medium, small leaf, flower, fruit, and seed morphological sizes [4], [3]. Characterization of the essential components of forest clove accession groups with large morphological sizes has been carried out [1]; however, the accessions groups with medium and small morphological sizes have not been informed. The study of clove essential components is important, considering that essential oil is currently very commercial in the trading market. It is widely used in the biopharmaceutical sector as an antioxidant, antibacterial, antifungal, antivirulence, and antiseptic [6], [7].

Forest clove with small morphological size on Ambon Island is found in Hitulama and Hitumesing villages, Central Maluku Regency [3], [8]. This accession grows together with other forest clove accession in one population. Its existence can be distinguished based on the appearance of plant morphology, which is somewhat different from the two other forest clove accession groups [3], [4]. So far, Physico-chemical studies of the essential oils of the forest clove accession group of small morphological sizes have not been reported. However, complete information on the essential components of this accession group is not yet available. The forest clove accession group with small morphological size was reported in the previous study to have low oil content, namely 3% on flowers, 0.77% on flower stalks, and 0.93% on leaves [2]. Also, it was reported that this accession group had not met the standard of oil trade because it had low eugenol levels [2].

Further research is needed on the potential utilization of other essential components. This study aims to inform the small morphological-sized forest clove accession group's complete essential components using Gas Chromatography-Mass Spectrometry (GC-MS). This information is vital to inform the potential of the essential components in this forest clove accession group to help develop these commodities towards biopharmaceuticals.

2. Methods

Samples for oil distillation of the forest clove accession group of small morphological sizes were taken from forest cloves more than 15 years old on Ambon Island (03°36'17.3 "E - 128°11'15.3" S, at an altitude of 179 masl). Samples were taken from 10 clove trees from the forest clove accession group with small morphological sizes, and then they were composited. Samples for distillation consisted of 3 kg of dried clove flowers with 14% moisture content, 5 kg of dried flower stalks, and 6 kg of withered leaves using a ventilated drying oven at 30 ° C. The handling of material samples to be distilled refers to the procedure [6]. Oil distillation was carried out at The Spice and Medicinal Crops Research Institute (BALITTRO), Bogor, Indonesia, using the steam-hydro distillation method for 8 hours. The distilled oil is then analyzed for its essential components using Gas Chromatography-Mass Spectrometry (GC-MS) at the Regional Health Laboratory of DKI Jakarta Province. Gas Chromatography-Mass Spectrometry (GC-MS) analysis using the Agilent Technologies 7890 Gas Chromatograph. The specifications for the tools used consist of columns in the form of HP Ultra; column length 30 x 0.25 (mm) I.D x 0.25 (μm) film thickness; mobile phase rate 1.2 μ / min; pressure 12 kPa; injector split ratio 8: 1; injector temperature 250 °C. The GC-MS chromatogram peak was then compared with the mass spectra of the NIST 2005 v.2.0 and Wiley 7 library 2003 [9], [10] to determine the component names.

The GC-MS results were presented qualitatively, including component name, component class, retention time (RT), quality (Q), and concentration (%). The main components are displayed in a heat map profile using the R Stat 3.1.0 software with the metabolomics package.

3. Results and discussion

3.1. The morphology of the small forest clove accession

The small morphological size of the forest clove accession had morphological differences with the other two forest clove accession groups in Ambon Island's distribution population. Forest cloves of small morphological size have smaller flower, fruit, and seed sizes in their population (Figure 1). The results of previous studies on forest cloves conducted by [1], [2], [4] indicated that small morphological size forest cloves had the smallest morphological sizes of leaves, flowers, fruits, and seeds compared to other forest clove accessions. According to [4], forest clove accessions of small morphological size have leaves length and width of ± 15.17 cm and 6.99 cm, respectively. The leaf morphology size is different from other forest clove accessions, which can reach ± 20.36 cm in length and 9.86 cm in width [1], [2], [4]. Apart from leaf size, it was also reported that there were differences in the morphology of flowers, fruits, and seeds. Small morphological groups of forest cloves were reported to have flower bud lengths and flower buds diameter of ± 2.35 cm and ± 0.57 cm, respectively, length and diameter of ripe fruit ± 2.75 cm and 1.40 cm, respectively, length and diameter of seed ± 1.97 cm and ± 0.90 cm, respectively [4].

3.2. Essential oils of small morphological size forest clove accession

GC-MS analysis identified 11 components in flower, 14 components on the flower stalk, and 13 components in leaves. Most of the components consisted of sesquiterpenoid fractions (Table 1). The forest clove accession group with a small morphological size had the highest concentration of volatile components in flowers, namely α-cubebene (27.83%), then eugenol (12.90%), germacrene-D (11.26%), δ-cadinene (8.58%), 2'- 3 ', 4' trimethocyacetophenone (8.44%); on flower stalks in the form of eugenol (24.26%), δ-cadinene (12.60%), methyleugenol (8.91%), 2'-3 ', 4' trimethocyacetophenone (8.79%); and the leaves are eugenol (22.83%), then followed by α-cubebene (17.62%), α-copaene (13.00), δ-cadinene (7.03%), caryophyllene (6.90%), methyleugenol (6.35%).

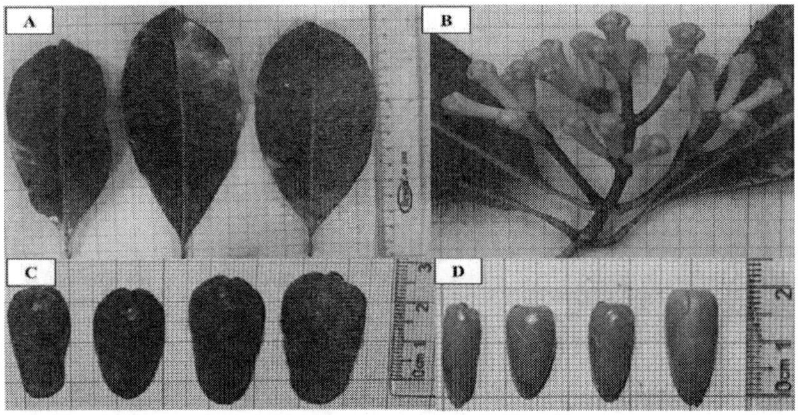

Figure 1. Morphology of leaves (A), flowers (B), fruit (C), seeds (D) of small morphological size of forest clove

The number of components contained in the small morphological size of the forest clove accession group of the flower, flower stalks, and leaf parts obtained in the study showed similarities to the number of components in the large morphological size forest clove accession group obtained in the previous study on Ambon Island by [1]. Previous studies on the large morphological-sized forest clove accession group consisted of 13 components in flowers, 11 components on flower stalks, and six components on leaves. This difference can be seen in the amount and concentration of content in part being analyzed. According to [11], [12], variations in the number of components can occur in the same species depending on environmental factors, plant genetics, and the part being analyzed. According to [1], [2], [13], forest cloves are classified as non-aromatic cloves, which contain up to 14 essential components, which is different from aromatic cloves, which only have six essential components. Even though there were more essential components, the eugenol content in the forest clove accession group of small morphological sizes in this study was at a low concentration (<25%). The low level of eugenol is in line with the results obtained in previous studies on the large morphological size forest clove accession group in its distribution area in Maluku [1], [2]. According to [2], [14], forest cloves are classified as non-aromatic wild type cloves which have low eugenol content in all parts of the plant.

Eugenol is a phenylpropanoid which is pale yellow (pale yellow), where the fraction will give the cloves a spicy aroma [15], [16]. The low levels of eugenol contained in forest cloves showed a difference with the aromatic clove group, which had the highest eugenol content in all parts of the plant, namely around 70-90% [13], [17], [18]. The difference in eugenol levels is due to forest cloves belonging to the wild type of non-aromatic cloves, which differ morphologically and metabolically from aromatic cloves[1], [3], [13], [19]. Although low eugenol levels were found, the small morphological size of forest cloves showed various variations in components and concentrations in the flower, flower stalks, and leaves.

IOP Conf. Series: Earth and Environmental Science **883** (2021) 012002 doi:10.1088/1755-1315/883/1/012002

3.3. The essential oil of forest clove accession groups with small morphology

bud oil					stem oil					leaf oil				
Components	class	RT	Q	cons. (%)	Components	Class	RT	Q	Cons. (%)	Components	Class	RT	Q	Cons. (%)
α-cubebene	S	21.187	99	27.83	Eugenol	PP	50.064	98	24.26	Eugenol	PP	50.059	98	22.83
Eugenol	PP	50.059	98	12.90	δ-cadinene	S	36.859	99	12.60	α-cubebene	S	21.149	99	17.62
Germacrene-D	S	34.097	98	11.26	Methyleugenol	PP	47.773	98	8.91	α-copaene	S	23.006	99	13.00
δ-cadinene	S	36.816	98	8.58	2'-3',4' Trimetho cyacetophenone	S	56.578	87	8.79	δ-cadinene	S	36.854	99	7.03
2'-3',4'Trimethocyacetophenone	S	56.573	90	8.44	α-cubebene	S	21.125	99	6.74	Caryophyllene	S	28.202	99	6.90
Caryophyllene	S	28.225	99	5.87	Methyl 4,47-trimethyl-4,7-dihydroindan-6-carboxylate	A	56.407	83	6.53	Methyleugenol	PP	47.768	98	6.35
α-copaene	S	23.044	99	17.49	α-copaene	S	22.982	99	6.28	3,4-Dihydro-1-hydroxy-5,8-dimethoxy-1,3-dimethyl-1H-2-benzopyran	A	56.402	78	6.09
Geraniol	S	40.473	87	1.40	Germacrene-D	S	34.063	98	5.79	Germacrene-D	S	34.054	99	5.71
Cadina-1,4-diene	S	38.068	97	1.36	Naphtalene, 1,2,3,4,4a,7-hexahydro-1,6-dimethyl-4-(1-methylethyl)	A	31.740	89	5.41	2',3',4' Trimethocyacetophenone	S	56.569	83	4.35
γ-Muurolene	S	33.216	99	1.28	α-cadinol	S	51.211	95	4.65	1S, Cis-Calamenene	S	40.464	96	3.24
Methyleugenol	PP	47.764	98	1.08	Caryophyllene	S	28.230	99	3.86	Naphtalene, 1,2,3,4,4a,7-hexahydro-1,6-dimethyl-4-(1-methylethyl)	A	48.792	70	1.98
					γ-Muurolene	S	33.197	99	2.40	α-cadinol	S	51.197	98	1.25
					Copaene	S	50.726	95	1.23	Humelene	S	31.897	78	1.18
					Cadina-1,4-diene	S	38.044	97	1.02					

Note: Data uses the 15-year-old clove accession group during the harvest season in March 2018; retention time (RT); quality (Q); sesquiterpenoid (S), phenylpropanoid (P), aliphatic (A), monoterpenoid (M)

3.4. The main components of the forest clove accession group of small morphological size

The heat map profile of the essential components of the forest clove accession group of small morphological sizes in this study showed six main components: germacrene-D, δ-cadinene, eugenol, α-cubebene, α-copaene, methyleugenol (Figure 2). These components are found in almost all parts of the flower, flower stalk, and leaf. The results showed similarities with the main component in the large morphological-sized forest clove accession group previously formulated by [1]. The similarity between the main components in the forest clove accession group obtained in the previous study was due to the similarity of plant species. According to [20], [21], the same species will show the similarity of the components contained due to genetic proximity.

The results obtained in the forest clove accession group of small morphological size and previous studies on the large morphological size of forest clove accessions showed that forest cloves were classified as wild non-aromatic type cloves with almost the same main volatile components. These results differ from the aromatic group of clove's main components, where the main components are eugenol, caryophyllene, and eugenol acetate with high concentrations [13], [17]. The difference in main components between forest cloves and aromatic cloves is more due to differences in plant species [1], [3].

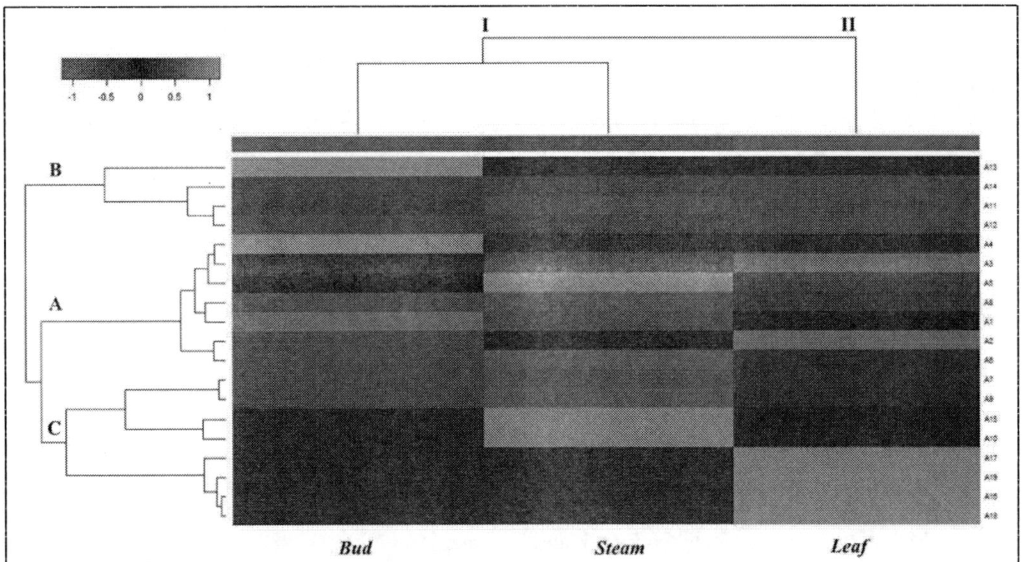

Figure 2. Heat map profile of the grouping of essential components of the flower, flower stalk (I), and leaf (II) on forest cloves of small morphological size; A1-A19 (types of essential components); A (main components), B and C (minor components).

Main components are depicted on the peak of the GC-MS chromatogram (Figure 3), where germacrene-D has a peak at m / z 204.0 (M +), δ-cadinene (CAS) at m / z 204.0 (M +), eugenol at m / z namely 164 (M +), α-cubebene at m / z 204.0 (M +), α-copaene at m / z 204.0 (M +), methyleugenol 178.0 (M +).

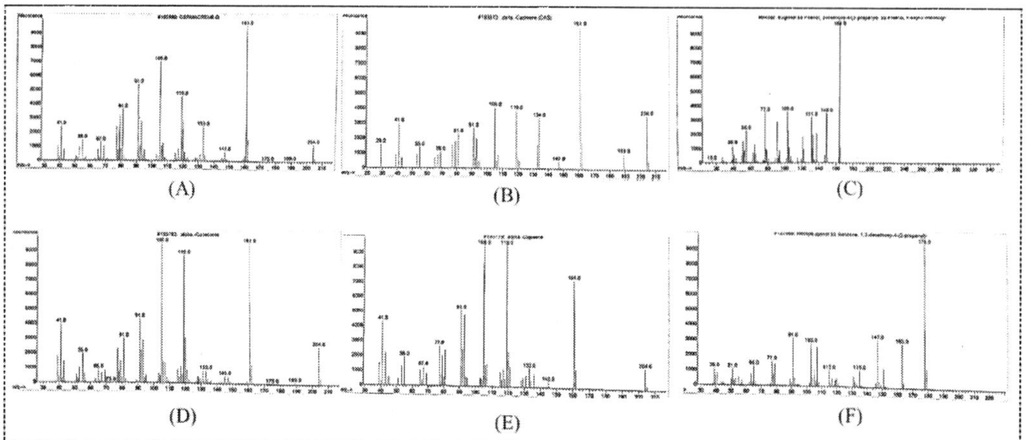

Figure 3. The chromatogram for main components of forest cloves of small morphological size; germacrene-D (A), δ-cadinene (CAS) (B), eugenol (C), α-cubebene (D), α-copaene (E), methyleugenol (F).

4. Conclusion

The forest clove accession group of small morphological size had the same main component as the previous study in the forest clove accession group's large morphological size. However, the differences were found in the concentrations contained in the flower, flower stalks, and leaves. The main components obtained in the small morphological size forest clove accession group consisted of 6 components: germacrene-D, δ-cadinene, eugenol, α-cubebene, α-copaene, and methyleugenol.

5. Acknowledgment

The author would like to thank the Indonesian Education Fund Management Agency (LPDP) for funding this research.

References

[1] Mahulette A S, Yahya S, and Wachjar A 2019 The physicochemical components and characteristic from essential oils of forest cloves *Syzygium aromaticum* (Myrtaceae) in Maluku Province, Indonesia *Plant Arch.* **19** 466–472.

[2] Mahulette A S, Hariyadi, Yahya S, and Wachjar A 2020 Physico-chemical properties of clove oil from three forest clove accession groups in Maluku in *1st International Conference on Sustainable Plantation (1st ICSP 2019)* **418** (2020) 012028 1–8.

[3] Mahulette A S, Hariyadi, Yahya S, Wachjar A, and Marzuki I 2019 Morpho-agronomical diversity of forest clove in Moluccas, Indonesia *Hayati J. Biosci.* **26** 156–162.

[4] Mahulette A S, Hariyadi, Yahya S, Wachjar A, and Alfian A 2019 Morphological traits of Maluku native forest clove (*Syzygium aromaticum* L. Merr & Perry.) *J. Trop. Crop Sci.* **6** 105–111, 2019.

[5] Alfian A, Mahulette A S, Zainal M, Hardin, and Bahrun A H 2019 Morphological character of raja clove (*Syzygium aromaticum* L . Merr & Perry .) native from Ambon Island in *The First International Conference of Interdisiplinary Research on Green Environment Approach for Sustainable Development (ICROEST 2019)* **343** (2019) 012150 1–4.

[6] Towaha J 2012 The benefits of cloves eugenol in various industries in Indonesia *J. Perspekt.* **11** 79–90.

[7] Mahulette A S, Alfian A, Zainal M, Nendissa J I, Wattimena A Y, Tanasale V L, Makaruku M H and Laisina J K J.2020 Growth of forest clove seedlings at different concentrations of

paclobutrazol in *The 2nd International Conference of Interdisciplinary Research on Green Environmental Approach for Sustainable Development* **575** (2020) 012081 1–7.

[8] Hariyadi, Mahulette A S, Yahya S, and Wachjar A 2019 Measuring the potential of biomass, carbon storage, and carbon sink of forest cloves in *The 1st International Seminar on Natural Resources and Environmental Management (ISenREM 2019)* **399** (2019) 012063 1–9.

[9] Wenqiang G, Shufen L, Ruixiang Y, Shaokun T, and Can Q 2007 Comparison of essential oils of clove buds extracted with supercritical carbon dioxide and other three traditional extraction methods *Food Chem.* **101** 1558–1564.

[10] Hossain M A, Al-hashmi R A, Weli A M, Al-riyami Q, and Al-Sabahib J N 2012 Constituents of the essential oil from different brands of *Syzigium caryophyllatum* L by gas chromatography-mass spectrometry *Asian Pac. J. Trop. Biomed.* **2** 1446–1449.

[11] Figueiredo A C, Barroso J G, Pedro L G, and Scheffer J J C 2008 Factors affecting secondary metabolite production in plants : volatile components and essential oils *Flavour Fragr. J.* **23** 213–226.

[12] Amini F, Asghari G R, Talebi S M, Askary M, and Shahbazi M. 2016 Effect of environmental factors on the compounds of the essential oil of Lippia citriodora *Biologija* **62** 194–201.

[13] Hariyadi, Mahulette A S, Yahya S, and Wachjar A 2020 Morphological characters and essential oil constituents extracted of two clove varieties (*Syzygium aromaticum* (L.) Merr. & L. M. Perry.) from Ambon Island, Indonesia *Plant Arch.* **20** 2208–2214.

[14] Koerniati S 1997 *Keanekaragaman Plasma Nutfah Cengkih dan Pelestariannya. (Monograf tanaman cengkih no. 2)*. Bogor (ID): Balittro, 1997.

[15] Barceloux D G 2008 *Medical Toxicology of Natural Substances: Foods, Fungi, Medicinal Herbs, Plants and Venomous Animals*. Hoboken, New Jersey: John Wiley & Sons.

[16] Kaufman T S 2015 The multiple faces of Eugenol. A versatile starting material and building block for organic and bio-organic synthesis and a convenient precursor toward bio-based fine chemicals *J. Braz. Chem. Soc.* **26** 1055–1085.

[17] Sohilait H J 2015 Chemical composition of the essential oils in Eugenia caryophylata, Thunb from Amboina Island *Sci. J. Chem.* **3** 95–99.

[18] Uddin M A, Shahinuzzaman M, Sohel Rana M, and Yaakob Z 2017 Study of chemical composition and medicinal properties of volatile oil from clove buds (*Eugenia caryophyllus*) *Int. J. Pharm. Sci. Res.* **8** 895–899.

[19] Hariyadi, Mahulette A S, Yahya S, and Wachjar A 2020 Agro-morphologies and physicochemical properties of flower bud , stem and leaf oils in two clove varieties (*Syzygium aromaticum* L . Merr . and Perry .) originated from Ambon island *Chiang Mai Univ. J. Nat. Sci.* **19** 516–530.

[20] Toncer O, Karaman S, Kizil S, and Diraz E 2009 Changes in essential oil composition of oregano (*Origanum onites* L .) due to diurnal variations at different development stages*Not. Bot. Hort. Agrobot. Cluj-Napoca* **37** 1842–4309.

[21] Oniga I, Vlase L, Toiu A, Benedec D, and Duda M 2010 Evaluation of phenolic acid derivatives and essential oil content in some *Melissa officinalis* L . varieties *Farmacia* **58** 764–769.

Inorganic fertilizers efficiency with using the liquid organic fertilizer to increase the cabbage yield (*Brassica oleracea* var. capitata L.)

A E Marpaung[1,*], B Karo[1] and S Barus[1]

[1] Research and Assessment Installation of Agricultural Technology Berastagi (Indonesia Vegetables Research Institute) Jl. Raya Medan-Berastagi Km. 60 Berastagi, 22156

* E-mail: agustinamarpaung@yahoo.com

Abstract. Using organic materials as fertilizer has contributed a lot to protecting the environment and the future of human life through sustainable agriculture. Local resources are used in such a way that synthetic nutrients can be reduced as low as possible. This study aims to determine the effect of liquid organic fertilizer (LOF) on cabbage yield with inorganic fertilizers' efficiency. The research was conducted in Berastagi experimental farm, Karo regency, with the soil type and altitude of 1,340 m asl, which began from July - September 2018. The design used was a randomized block design consisting of 16 treatments with three replication. The treatments tested were: A. Without inorganic fertilizer + LOF 1, B. 25% inorganic fertilizer + LOF 1, C. 50% inorganic fertilizer + LOF 1, D. 75% inorganic fertilizer + LOF 1, E. 100% inorganic fertilizer + LOF 1, F. Without inorganic fertilizer + LOF 2, G. 25% inorganic fertilizer + LOF 2, H. 50% inorganic fertilizer + LOF 2, I. 75% inorganic fertilizer + LOF 2, J. 100% Inorganic Fertilizer + LOF 2, K. Without inorganic fertilizers + LOF 3, L. 25% inorganic Fertilizer + LOF 3, M. 50% inorganic Fertilizer + LOF 3, N. 75% inorganic Fertilizer + LOF 3, O. 100% Inorganic Fertilizer + LOF 3, P. 100% Inorganic Fertilizer + without LOF (Control). The results showed that, in general, the application of liquid organic fertilizer for rabbit urine, fish fertilizer, and fish teillation could stimulate the growth and yield of cabbage better than inorganic fertilizers. Applicationthe liquid organic fertilizer can increase theheight growth of cabbage was 0.67 - 0.88%, plant diameter was 0.33 - 2.56%, crop weight per plant was 5.14 - 5.84%, production per plot was 2.96 – 9.78%, and the crop ratio of 3.36 - 10.84% compared to the use of inorganic fertilizers.

1. Introduction

Generally, farmers in Indonesia cultivate a lot of cabbage in the highlands. Cabbage is a vegetable rich in vitamins such as vitamin A 200 IU, B 20 IU, and C 120 IU, which are very important for health. The need for vegetables is increasing with the increasing population. Therefore, vegetables, especially cabbage, need to be increased to provide these needs [1],[2].

Currently, cabbage vegetable farmers in general still use fertilization with synthetic chemical fertilizers. The continued use of synthetic chemical fertilizers has had a negative impact on the soil and the environment. The negative impact that arises is damaging the soil structure (physical) and the environment. The soil becomes hard in the dry season and sticky in the rainy season with decreased soil porosity. Synthetic chemical fertilizers do not have properties that can directly improve soil's physical and biological functions[3],[4]. Based on the previous, it is necessary to have organic fertilizers that can increase cabbage production, such as synthetic chemical fertilizers.

Increasing vegetable productivity can be done by handling proper cultivation, one of which is fertilization. Fertilization can be done using synthetic chemical fertilizers and organic fertilizers [5]. Organic fertilizers have an essential role in improving the soil's physical, chemical, and biological properties. Although the nutrient content in organic fertilizers is relatively low, soil chemical properties far exceed synthetic chemical fertilizers [6].

One of the organic fertilizers is in the form of a liquid, one of the essential components in organic agriculture, which contains many macros, micro, hormone, and amino acids needed by plants. Liquid organic fertilizers have several benefits, including reducing the use of chemical fertilizers [7]. Liquid organic fertilizers can also increase the growth and production of mustard plants [8][9], tomatoes [10][11], soybeans [12], sweet corn [13], scallions [14]. The provision of liquid organic fertilizer from kirinyu and rabbit manure can increase cabbage growth and production[15]. Provision of liquid organic fertilizer from rabbit manure as the raw material can increase tuber weight per plant (49.21%), output per plot (48.35%), and tuber length (12.83%) compared without LOF on carrots [16].

Liquid organic fertilizer, which is used as the primary raw material, uses local natural ingredients around the agricultural location, such as rabbit manure, fish fertilizer, fish teillation, and coconut water. Rabbit urine has higher N and mucosa than other animals, which can bind available nutrients [17]. Fertilizers made from fish raw as a nutrient source can also induce Actinomycetes and Rhizobacteria groups that play a role in producing growth hormones around plant roots [18].Coconut water can replace synthetic root stimulants as a growth regulator, where the results of [19] research resulted in faster sprouting time, shoot length, number of leaves, length, and high root wet weight in tin seeds.

This research aims to determine the effect of giving liquid organic fertilizer on cabbage yield with inorganic fertilizers' efficiency.

2. Methods

The research was conducted at Berastagi experimental farm, Dolat Rayat District, Karo District, with andisol soil type, the altitude of 1,340 m asl, the temperature of 22-28^{0}C, humidity 80-90%, and rainfall 2,500 mm / year, and starting in July until September 2018.

The research used non-factorial Randomized Block Design (RBD) with three replications. The treatments tested were: A. Without inorganic fertilizer + liquid organic fertilizer (LOF) 1, A. Without inorganic fertilizer + LOF 1, B. 25% inorganic fertilizer + LOF 1, C. 50% inorganic fertilizer + LOF 1, D. 75% inorganic fertilizer + LOF 1, E. 100% inorganic fertilizer + LOF 1, F. Without inorganic fertilizer + LOF 2, G. 25% inorganic fertilizer + LOF 2, H. 50% inorganic fertilizer + LOF 2, I. 75% inorganic fertilizer + LOF 2, J. 100% Inorganic Fertilizer + LOF 2, K. Without inorganic fertilizers + LOF 3, L. 25% inorganic Fertilizer + LOF 3, M. 50% inorganic Fertilizer + LOF 3, N. 75% inorganic Fertilizer + LOF 3, O. 100% Inorganic Fertilizer + LOF 3, P. 100% Inorganic Fertilizer + without LOF (Control)

Making Liquid Organic Fertilizer

- LOF 1 (LOF rabbit urine) = EM0 + Coconut water + Rabbit urine (2,5: 5: 4,5)
- LOF 2 (LOF fish fertilizer) = EM0 + Coconut water + fish fertilizer (2.5: 5: 2)
- LOF 3 (LOF of fish teillation) = EM0 + Coconut water + fish teillation(2,5: 5: 2)

All ingredients are cooked to a boil and thenfermented for one month.

Experimental plots were made with a size of 1 m x 5 m; the distance between treatment plots was 0.75 m, and the distance between replicates was 1 m. The total population of cabbage consists of 24 plants. Then given manure according to the tested treatment with a dose of 300 g per plant and given at planting. Watering is done when it does not rain. Weeding depends on the growth of weeds in the field. According to the type tested with a concentration of 20 ml LOF/liter of water, the next fertilization of liquid organic fertilizer is given by spraying 1 x 1 week, starting at the ageof 2 weeks after the plant (WAP) until the plant at 10 WAP. Pest and disease control is carried out every 1 x 7 days using insecticides with active ingredients. Sipermetrin 50 g / l, Piridaben 135 g / l, and Spinetoram 120 g / l, with a concentration of 0.5 - 1.0 cc / l water and 2 g chlorothalonil fungicide. / l water and azoxystrobin 200 g / l + difenoconazole 125 g / l 0.5 - 1 ml / l water. Harvesting is carried out at the age of 3 months after planting. The variables observed were as follows: plant height and diameter at 1 and 2 months after plant, leaf width at twomonths after plant, crop weight per plant,production per plot, and crop ratio.

The observed data were analyzed with the F test and continued with the HSD average difference test at the 5% level.

3. Results and discussion

3.1. Plant height, plant diameter, and leaf width

Based on the results of the analysis of variance, it showed that the fertilization treatment had no significant effect on the growth of plant height, plant width, and cabbage leaf width at the age of 1 and 2 months after planting (Table 1).

Table 1. Effect of treatments on plant height, plant diameter, and leaf width

Varieties	Plant Height (cm)		Plant Diameter (cm)		Leaf Width (cm)
	1 MAP	2 MAP	1 MAP	2 MAP	
A. 0% + LOF1	31.08 a	37.25 a	55.38 a	65.25 a	30.25 a
B. 25% + LOF1	30.08 a	36.08 a	54.50 a	65.88 a	27.50 a
C. 50% + LOF1	31.42 a	35.33 a	53.00 a	66.04 a	27.83 a
D. 75% + LOF1	30.58 a	36.67 a	54.88 a	64.58 a	28.50 a
E. 100% + LOF1	28.54 a	32.58 a	51.46 a	61.96 a	28.67 a
F. 0% + LOF2	28.33 a	37.33 a	51.42 a	63.13 a	28.67 a
G. 25% + LOF2	31.33 a	35.33 a	53.25 a	66.79 a	27.67 a
H. 50% + LOF2	31.92 a	34.58 a	55.54 a	64.79 a	26.92 a
I. 75% + LOF2	31.46 a	36.50 a	54.13 a	67.17 a	29.58 a
J. 100% + LOF2	31.38 a	36.00 a	56.83 a	64.67 a	27.50 a
K. 0% + LOF3	29.42 a	36.00 a	50.88 a	63.79 a	27.50 a
L. 25% + LOF3	29.83 a	34.00 a	52.79 a	62.88 a	28.42 a
M. 50% + LOF3	30.92 a	35.58 a	51.71 a	64.04 a	29.33 a
N. 75% + LOF3	31.08 a	35.17 a	53.46 a	65.00 a	27.83 a
O. 100% + LOF3	31.63 a	36.00 a	53.33 a	65.04 a	28.92 a
P. 100% + 0	30.46 a	37.00 a	53.25 a	63.58 a	30.25 a
CV (%)	5,94	5,92	6,65	4,57	7,11

Note: Mean followed by the same letters on the same column is not significantly different according to honestly significantly different test at 5% level
MAP = Month after Planted
LOF 1 (LOF rabbit urine), LOF 2 (LOF fish fertilizer), LOF 3 (LOF of fish teillation)

Cabbage growth showed no significant difference between inorganic fertilization treatment with the reduced inorganic fertilizationaddition of liquid organic fertilizer and only LOF application. Application the LOF can increase the height growth of cabbage was 0.67 - 0.88% and plant diameter were 0.33 - 2.56%compared to inorganic fertilizers. The highest gain of cabbage was found in the treatment of liquid organic fertilizer 1 (rabbit urine) and liquid organic fertilizer 2 (fish fertilizer), each 37.25 cm and 37.33 cm. This shows that liquid organic fertilizer can be used by cabbage for growth. This is consistent with the research results of [20] that the rabbit manure liquid organic fertilizer can support cabbage plants' vegetative growth. Likewise, the application of fish fertilizer can also increase cabbage growth [21].

Plant diameter at the age of 2 months after plant, application of the liquid organic fertilizer rabbit manure and fish teillation without inorganic fertilizers produced a higher plant diameter, each 65.25 63.79 cm. Likewise, with leaf width, the treatment liquid organic fertilizer of rabbit urinewithout inorganicfertilizers resulted inthe most sweeping leaves than other treatments. This is consistent with the research results of [22], that rabbit urine can also support the height growth of shallot plants.

3.2. Crop weight per plant, production, and crop ratio

The application of liquid organic fertilizer resulted in cabbage crop weight per plant, production, and crop ratio, which were not significantly different from the application of 100% chemical fertilizers (Table 2). This shows that using liquid organic fertilizers can increase cabbage production even though chemical fertilizers are reduced and even without being used.

Table 2. Effect of treatments on crop weight per plant, production, and crop ratio

Treatments	Crop Weight (g)	Production (kg/5 m^2)	Crop Ratio (%)
A. 0% + LOF1	1.36 a	32.73 a	53.68 a
B. 25% + LOF1	1.47 a	31.60 a	55.15 a
C. 50% + LOF1	1.38 a	32.50 a	49.96 a
D. 75% + LOF1	1.52 a	36.20 a	52.46 a
E. 100% + LOF1	1.28 a	29.67 a	49.91 a
F. 0% + LOF2	1.26 a	30.43 a	51.23 a
G. 25% + LOF2	1.37 a	34.57 a	50.32 a
H. 50% + LOF2	1.42 a	34.10 a	53.03 a
I. 75% + LOF2	1.38 a	33.47 a	52.25 a
J. 100% + LOF2	1.41 a	34.67 a	52.21 a
K. 0% + LOF3	1.37 a	31.20 a	55.53 a
L. 25% + LOF3	1.35 a	31.63 a	52.38 a
M. 50% + LOF3	1.35 a	31.47 a	54.55 a
N. 75% + LOF3	1.31 a	29.23 a	51.11 a
O. 100% + LOF3	1.46 a	34.57 a	54.84 a
P. 100% + 0	1.29 a	29.53 a	49.51 a
CV (%)	13,88	13,15	6,36

Note: Mean followed by the same letters on the same column is not significantly different according to honestly significantly different test at 5% level
LOF = Liquid organic fertilizer
LOF 1 (LOF rabbit urine), LOF 2 (LOF fish fertilizer), LOF 3 (LOF of fish teillation)

In general, the cabbage crop weight per plant, production per plot, and crop ratio in the use of liquid organic fertilizers resulted have higher values than the use of 100% chemical fertilizers, although no significant. The liquid organic fertilizer can increase the crop weight per plant was 5.14 - 5.84%, production per plot was 2.96 – 9.78%, and the crop ratio of 3.36 - 10.84% inorganic fertilizers.The highest weight of cabbage crops per plant and production per plot was found in the treatment of using 75% inorganic fertilizer + LOF 1 (rabbit urine), each 1.52 kg, and 35.20 kg per plot. Meanwhile, the highest yield ratio resulted in using LOF 3 (fish teillation) without inorganic fertilizers, 55.53%. This is following the research results by [7], which shows that using liquid organic fertilizers can reduce the use of chemical fertilizers on chicory and increase growth and production. The obtained efficiency of urea fertilizer was 50% and 75 ml/L of liquid organic fertilizer treatment was able to increase the growth and production of chicory, which was higher than the 100% chemical fertilizer treatment although it was not significantly different.

4. Conclusion

In general, the application of liquid organic fertilizer for rabbit urine, fish fertilizer, and fish teillation could stimulate the growth and yield of cabbage better than inorganic fertilizers. Application the liquid organic fertilizer can increase the height growth of cabbage was 0.67 - 0.88%, plant diameter was 0.33 - 2.56%, crop weight per plant was 5.14 - 5.84%, production per plot was 2.96 – 9.78%, and the crop ratio of 3.36 - 10.84% compared to the use of inorganicfertilizers.

References

[1] Kumarawati NPM, Supartha IW and Yuliadhi KA 2013 Struktur Komunitas dan Serangan Hama-Hama Penting Tanaman Kubis (*Brassica oleracea* L.) *E-Jurnal Agroekoteknologi Tropika* **2** 252 ISSN: 2301- 6515https://ojs.unud.ac.id/index.php/JAT/article/view/7014

[2] Mujib A, Syabana MA and Hastuti D 2014 Uji Efektivitas Larutan Pestisida Nabati terhadap Hama Ulat Krop (*Crocidolomia pavonana* L.) pada Tanaman Kubis (*Brassica oleraceae*) *Jurnal Ilmu Pertanian dan Perikanan* **3** 67 ISSN 2302-6308 http://umbidharma.org/jipp

[3] Mutryarny E, Endriani and Letari SU 2014 Pemanfaatan Urine Kelinci untuk Meningkatkan Pertumbuhan dan Produksi Tanaman Sawi (*Brassica juncea* L) Varietas Tosakan *Jurnal Ilmiah Pertanian* **11** 23 **DOI:** https://doi.org/10.31849/jip.v11i2.1246

[4] Dewanto FG, Londok JJMR, Tuturoong RAV and Kaunang WB 2013 Pengaruh pemupukan kimia sintetis dan organik terhadap produksi tanaman jagung sebagai sumber pakan *Jurnal Zootek* **32** 1 ISSN 0852-2626DOI: https://doi.org/10.35792/zot.32.5.2013.982

[5] Utomo M, Sabrina T, Sudarsono, Lumbanraja J, Rusman B and Wawan 2016 *Ilmu Tanah: Dasar-dasar dan Pengelolaan* Kencana, Prenada Media Group, Jakarta, 433.

[6] Hartatik W, Husnain and Widowati LR 2015 Peranan Pupuk Organik dalam Peningkatan Produktivitas Tanah dan Tanaman *Jurnal Sumberdaya Lahan* **9** 107 DOI: http://dx.doi.org/10.21082/jsdl.v9n2.2015.%25p

[7] Sopha GS and Uhan TS 2013 Application of liquid organic fertilizer from city waste on reduce urea application on chinese mustard (*Brassica juncea* L) cultivation *AAB Bioflux* **5** 39 http://www.aab.bioflux.com.ro

[8] Manullang GS, Rahmi A and Astuti P 2014 Pengaruh jenis dan konsentrasi pupuk organik cair terhadap pertumbuhan dan hasil tanaman sawi (*Brassica juncea* L.*)* varietas tosakan *Jurnal AGRIFOR* **XIII** 33 ISSN : 1412 – 6885 DOI: https://doi.org/10.31293/af.v13i1.545

[9] ArinongA R and Lasiwua CD 2011 Aplikasi pupuk organik cair terhadap pertumbuhan dan produksi tanaman sawi *Jurnal Agrisistem* **7** 47 ISSN 1858.

[10] Rehatta H, Mahulete A and Pelu AM 2014 Pengaruh konsentrasi pupuk organik cair bioliz dan pemangkasan tunas air/wiwilan terhadap pertumbuhan dan produksi tanaman tomat (*Lycopersicon esculentum* Miller) *Jurnal Budidaya Pertanian* **10** 88.

[11]Marliah A, Hayati M and Muliansyah I 2012 Pemanfaatan pupuk organik cair terhadappertumbuhan dan hasil beberapa varietas tomat (*Lycopersicum esculentum* L.) *Jurnal Agrista* **16** 122.

[12]Hamzah S 2014 Pupuk Organik Cair dan Pupuk Kandang Ayam berpengaruh kepada pertumbuhan dan produksi kedelai (*Glycine max* L.) *Agrium* **1** 228.

[13] Syofia I, Munar A and Sofian M 2014 Pengaruh pupuk organik cair terhadap pertumbuhan dan hasil dua varietas tanaman jagung manis (*Zea Mays Saccharata* Sturt) *Agrium* **18** 208.

[14] Marpaung AE, Karo B and Dinata K 2016 Pemanfaatan pupuk organik cair (LOF) dari limbah pertanian asal sumber daya alami lokal pada budidaya sayuran bawang daun (*Allium fistulosum*L.*) Prosiding Seminar Nasional* Inovasi Teknologi Pertanian Modern Mendukung Pembangunan Pertanian Berkelanjutan, Balai Pengkajian Teknologi Pertanian, Balitbangtan, Bengkulu 316.

[15] Marpaung AE 2017 Pemanfaatan Jenis dan Dosis Pupuk Organik Cair (LOF) untuk Meningkatkan Pertumbuhan dan Hasil Sayuran Kubis *Jurnal Agroteknosains* **01** 117 ISSN: 2598-6228 DOI: http://dx.doi.org/10.36764/ja.v1i2.39

[16] Marpaung AE, Karo B and Sinaga R 2017 Pemanfaatan Kotoran Kelinci sebagai Pupuk Organik Cair (LOF) untuk Meningkatkan Pertumbuhan dan Hasil Wortel (*Daucus carota*) Varietas Lokal *Prosiding Seminar Nasional* Peran Teknologi Agronomi dalam Mempercepat Penciptaan dan Hilirisasi Inovasi Pertanian, Perhimpunan Agronomi Indonesia, Bogor 413.

[17] Simorangkir CA, Supriyanto A, Murdiono WE and Nihayati E 2017 Pemberian urine kelinci (*Leporidae*) dan KNO_3 pada pertumbuhan dan hasil tanaman stroberi (*Fragaria* sp) *Jurnal Produksi Tanaman* **5** 782.

[18] Zahroh F, Kusrinah and Setyawati SM 2018 Perbandingan Variasi Konsentrasi Pupuk Organik Cair dari Limbah Ikan Terhadap Pertumbuhan Tanaman Cabai Merah (*Capsicum annum* L.) *Al-Hayat: Journal of Biology and Applied Biology* **1** 50 DOI : 10.21580/ah.v1i1.2687

[19] Marpaung AE and Hutabarat RC 2015 Respons Jenis Perangsang Tumbuh Berbahan Alami dan Asal Setek Batang Terhadap Pertumbuhan Bibit Tin (*Ficus carica* L.) *J. Hort.* **25** 37 DOI: http://dx.doi.org/10.21082/jhort.v25n1.2015.p37-43

[20] Marpaung AE, Udiarto BK, Lukman L and Hardiyanto 2018 Potensi Pemanfaatan Formulasi Pupuk Organik Sumber Daya Lokal untuk Budidaya Kubis *J.Hort.* **28** 191 DOI: http://dx.doi.org/10.21082/jhort.v28n2.2018.p191-200

[21] Karo B, Marpaung AE and Hidayat T 2016 Pemanfaatan urin sapidan kelinci sebagai pupuk organik cair dalam peningkatan pertumbuhan dan produksi bawang daun (*Allium fistulosum* L) *Prosiding Seminar Nasional* Inovasi Teknologi Pertanian Modern Mendukung Pembangunan Pertanian Berkelanjutan, Balai Pengkajian Teknologi Pertanian, Balitbangtan, Bengkulu 323.

[22] Simamora ALB, Simanungkalit T and Ginting J 2014 Respons Pertumbuhan dan Produksi Bawang Merah (*Allium ascalonicum* L.) Terhadap Pemberian Vermikompos dan Urine Kelinci *J. Agroekoteknologi* **2** 533.

Green economic based on low-carbon development on small islands

A Kastanya[1,*], C C V Suhendy[1], D V Pattimahu[1] and Iskar[1]

[1] Forestry Department, Pattimura University, Jl. Ir. M. Putuhena, Kampus Poka, Ambon, Indonesia

* E-mail: aguskastanya@yahoo.com

Abstract. Global warming and climate change are having an impact on human and the planet. This occurs worldwide due to an exploitative economy, which does not consider the degradation of natural resources and the environment. Damage to terrestrial ecosystems in small island areas has a direct impact on the destruction of marine ecosystems, such as coral reefs, seagrasses, and another biodiversity. The most current devastating impact is the emergence of the "Covid-19 Pandemic", which shows that so far humans have adopted an individualistic way of life that ignores their environment. This analysis is carried out through a literature review of the research results and scientific discussions that have developed so far. The results of the study show that if the economy continues to run as it is now, the Earth will continue to be in crises such as floods, extreme weather, rising sea levels, food shortage, and outbreaks of more viruses. Changes in paradigm and human behaviour are needed, not only from the government as the main actor of development but also support from the entire community. Science and innovation that are currently developing a change in development towards the ecological-economy, environmental economics, which is currently known as the "Green Economy Model" is already covered by controlling global warming and climate change. The transformation from an Exploitative Economy to a Green Economy has developed in the concept of Low Carbon Development and Build Back Better as part of the Sustainable Development Goals (SDGs), which needs to be formed in an integrated manner as a concept for Sustainable Small Islands Development and bring welfare for the community.
Keywords: Green economy, low-carbon development, small islands

1. Introduction

Global warming and climate change are already impacting humans, biodiversity and this planet, and are occurring around the world as a result of the world's exploitative economy. An economy that does not consider the degradation of natural resources and the environment [9]. As a result, Indonesia is faced with an environmental crisis that must be resolved immediately due to global warming and climate change, such as air pollution, clean water crisis, forest destruction, soil pollution, soil erosion, garbage, floods, landslides, river pollution, and damage to other terrestrial ecosystems. A direct impact on the destruction of marine ecosystems, coral reefs, seagrass, and biodiversity. The environmental crisis's impact on natural resources, human life, economy and food [16, 9]. Furthermore, according to Czikkely *et al.* [12] , there are ten main problems due to environmental damage on the planet earth, which are: population explosion, climate change, degradation of biodiversity, disruption of the nitrogen and phosphorus cycle, water supply, ocean acidity, pollution, the damage of the ozone layer, overfishing and deforestation. A very devastating impact today is the emergence of a new virus that threatens the world today, called "Covid-19". Through COVID-19, Planet Earth provides a lesson that so far, humans have adopted an individualistic way of life that ignores their environment. If the economy continues as it is today, the Earth will continue to be in crisis after crisis, such as floods, extreme weather, rising sea levels, failure of food supplies, and more viruses.

Indonesia, an archipelago country that is very vulnerable to environmental damage, global warming and climate change, has committed to reducing gas emissions by 29% on its own and can be increased to 41% with foreign assistance from Business As Usual (BAU) by 2030. Greenhouse gas (GHG) reduction in the green economy concept, must be applied in all development sectors, especially in the land sector. UNFCCC decisions on REDD + (reducing emissions from deforestation and degradation) that have been planned and developed to become the main direction in the concept of development in Indonesia [8, 2, 4]. Centralised and exploitative economic development in Indonesia has resulted in damage to the mainland natural resources of small islands, coasts and the sea which has caused poverty for most of the community and has even triggered the development of new microorganisms such as the corona virus which is endangering the world today. Despite this predetermined commitment, it has not been fundamentally implemented in national and regional development planning, everything is still sectoral, especially now that everything is still concentrated on handling Covid. Therefore, the concept of renewal of development after Covid-19 must undergo a fundamental renewal.

Changes in paradigm and human behaviour are needed, especially the government and the main development actors supported by the whole society. This is not easy because the concept of sustainability has been developed for a long time since the world meeting in Rio in 1992, which resulted in various conventions but is still difficult to implement. Due to the economic growth as well as science and innovation which are far developed, it is urgent to have changes such as ecological-economy, environmental economics, which can be packaged in green economy lessons that have been covered by controlling global warming and climate change. The foundation that is packaged in current development is Low Carbon Development which is part of the Sustainable Development Goals (SDGs). The 17 integrated development goals have answered the goals of environmental sustainability and welfare which have a positive impact on the restoration of the planet earth. Because of the concept of sustainability and scientific development, the impact of global warming and climate change requires a change to a green economy, especially for the small island region of Maluku [9].

Environmental economy is a part of the economy that deals with the efficient use of resources. Because the environment provides direct value and the raw materials intended for economic activity, the environment and the economy are interdependent. For that reason, the way an economy is managed can have an impact on the environment which, in turn, might affect economic well-being and performance. One of the most famous critics of traditional economic thought on the environment is Herman Daly. In his first book, Steady-State Economics, Daly suggests, "enough is best", arguing that economic growth leads to environmental degradation and an unequal distribution of wealth. He emphasized that the economy is part of our environment, which is limited. Hence the idea that a steady-state economy is one in which there is an optimal level of population and economic activity that leads to sustainability. Daly calls for qualitative improvement in people's living - development - without lasting growth. Today, many of his ideas are linked to the concept of sustainable development. "(Environmental Literacy Council, 2015).

2. Characteristics of Small Islands and Climate Change Threats

Small islands have limited resources, for this reason, land use must be carried out efficiently to meet people's needs for water, food, building materials and a decent quality of life, and to maintain the functioning of the natural systems on which these depend.

Small islands are vulnerable to the threat of climate change. The problem that is often found on small islands in Maluku is a damage of coastal resources which results in reduced productive fisheries. Coral reefs are destroyed due to construction, dredging, pollution, silting and dynamization or fish poisoning.

The damage of mangrove forests has resulted in the unavailability of barriers to protect the area in the event of a tsunami or storm. In addition, if the mangrove forest to the seagrass beds is damaged, the aquatic animal life in it will also become extinct.

The most widespread environmental problem, affecting most of the small islands, is the unsafe disposal of domestic wastewater, particularly human waste and municipal waste. Thus, water pollution can become very serious problem, especially those related to clean water supply (rivers, groundwater and even rainwater catchments) as well as pollution of coastal waters, coral reefs and lagoons which are important for tourism, recreation and fisheries. This pollution can pose a very serious risk to human health.

Due to the narrow space of land on small islands it is imperative that comprehensive planning and proper land allocation for the most appropriate use or combination of land uses is required. The land is a limited and valuable commodity on an island. If the population is growing too fast, there will also be an expansion of population settlements which are generally located in coastal areas. This is certainly vulnerable if there is an increase in sea level or a tsunami due to climate change. In such condition, it is necessary to pay attention to the existence of the inhabitants of an island since not all of the small islands in Maluku are inhabited but many are uninhabited.

Land resources, which are the basis of agriculture, must be limited to the situation in small islands. The problem of erosion and loss of soil fertility on a narrow land needs to be considered so that it can be prevented for the sake of the sustainability of life on this small island.

When there is high rainfall in small and short watersheds with little water storage capacity, the water will quickly flow into the sea and very little can be stored in the ground. Such conditions during the dry season can cause serious water shortages that hinder development, and can create serious public health problems.

Damaged forest cover certainly causes a lot of water flow to stop flowing in the dry season. Thus water is often the most limiting factor in the development of small islands. The smaller the island, the more difficult the problem of solid waste disposal will be. Domestic waste dumps are often located in coastal swamps, and take land from other important uses.

Nature conservation measures are particularly important on islands where isolation has allowed the evolution of unique flora and fauna with a large number of endemic species, while small population sizes increase their vulnerability.

Demands to increase the local population on limited land resources will make the area difficult to reach by natural areas even if the land tenure situation allows such action. Stable habitat destruction and competition and predation by introduced species will increase pressure on native species.

The situation on many islands is of critical importance as the area of undisturbed natural habitat decreases. As a result, endemic species will be threatened with extinction. In this regard, the determination of several Essential Ecosystem Areas (KEE) in Maluku is urgently needed at this time.

One of the difficulties in developing infrastructure on small islands is the limited supply of sand and gravel without causing serious environmental problems. Removal of sand from the beach causes beach erosion and loss of the beach. This situation will certainly be more vulnerable if it is affected by climate change. Dredging of coral and sand from coastal waters destroys productive fisheries resources. No less important is the existence of onshore mining, such as what happened on Romang Island. This can affect the area available for agriculture and land use by leaving useless pits and mines.

If inhabitants of small islands wish to ensure a satisfactory environmental future, they must take measures for erosion prevention that are good at their resource base and necessary to stabilize population growth within the carrying capacity of their islands. Efforts to restore damaged resources, and to achieve comprehensive resource management and resource development activities, are focused on the critical coastal zone (which in islands can cover most or all of the land area). Small islands require a unique form of development that is adapted to environmental limitations and attracts many of the traditional residents who inhabit the islands. By considering the characteristics of small islands like that and the threat of climate change, both the Regional and Central Governments need to formulate appropriate development strategies in small islands which are very different from the large islands (continental).

3. Impact of the Covid-19 Pandemic

In 2019 Indonesia was ranked 62 in the world and fifth in the Southeast Asia Food Sustainability Index (FSI) in 2018, Indonesia scored 59.1 behind Ethiopia's 68.5 Food Loss and Waste (FLW)

Indonesia of around 300 kg/capita/year and number two in the world after Saudi Arabia. Governance crisis. The crisis of environment and natural resources is a crisis of governance. Governance crisis is a failure to regulate the actions of the state, market and community actors with interest in resources.

Sir David Attenborough, British television broadcaster and naturalist, already working for 60 years said at the World Economic Forum in Davos, that man has a terrifying destruction mechanism that can be annihilated living things and their environment unconsciously. World leaders reacted greatly slow to the climate crisis. This is because since the Industrial Revolution, the relationship between an urban society with the universe is getting away, so humans do not realize the impact their actions have on global ecosystems.

Through COVID-19, Planet Earth provides lessons that have been human adopt an individualistic way of life that ignores the fate of his environment. If the economy continues as it is now, the Earth will continue to create crisis after crisis, such as floods, extreme weather, rising sea levels, failure of food supplies, and outbreaks of more viruses.

A review in the scientific magazine Nature concluded, although the area with High biodiversity can be a source of seeds for new diseases, increasingly there is a lot of evidence that the loss of biodiversity can increase disease transmission. Scientists suggest preserving the area with endemic biodiversity, which only exists in certain geographic areas so that the prevalence of infectious diseases can be reduced.

4. Low Carbon Development on Small Islands

The commitment of the Ministry of National Development Planning / Bappenas to develop a policy "Low Carbon Development Planning (PPRK)". This policy is formulated to achieve inclusive economic growth through efforts to equalize development and alleviate poverty, while maintaining environmental quality and the availability of natural resources. Low Carbon Development supports the achievement of the 13th Sustainable Development Goals (SDGs) which is about Climate Action. Climate change must be integrated into the national development program. In this case, the Ministry of National Development Planning / Bappenas has a strategic role in ensuring the mainstreaming of climate change issues into national development plans.

In addition, several government efforts in collaboration with some other parties to support low carbon development in Maluku Province include:

- Adaptation activities carried out in collaboration with APIK USAID, and APIKI Maluku with the support of the Maluku Regional Government centred on Haruku Island, Aru Islands and Ambon, including the designation of several Climate Villages in Maluku.
- Preparation of a Road Map and Regional Action Plan for Maluku Climate Change Mitigation and Adaptation which will be implemented in the Maluku Development Planning, so that each regional apparatus can works based on the two documents.
- Mitigation activities on Saparua Island regarding the development of organic agriculture and handling of climate change with a survey of the biomass potential of organic agricultural crops and mangrove forests which was carried out between Yayasan Tiara Pusaka (YTP) Maluku with APIKI Maluku.
- Preparation of a document on the Proposal for the Establishment of a KEE on Buano Island by the Maluku Provincial Forestry Service, APIKI Maluku and other stakeholders.
- Research support such as the role of ants and birds on the small islands of Maluku as bioindicators of climate change.
- APIKI Maluku collaborates with BAPPEDA Maluku Province for the implementation of mangrove restoration in the Aru Islands by utilizing funds from the Green Climate Funds (GCF).
- Changes in the paradigm of government and society from exploitative development to a green economy through the integration of all planning documents (RTRW, RZWP3K), RPJPJP, RPJMD, KLHS, ROAD MAP MAPI, RAD MAPI) based on the conservation of biological diversity to achieve welfare and environmental sustainability.
- The transformation from an Exploitative Economy to a Green Economy has developed in the concept of Low Carbon Development and Build Back Better as part of the Sustainable Development Goals

(SDGs), which needs to be formed in an integrated manner as a concept for Sustainable Small Islands Development and bring welfare for the community.

References

[1] Anonymous 2015 National Forest Reference Emission Level for Deforestation and Forest Degradation in the Context of Decision 1/CP.16 para 70 UNFCCC - Encourages Developing Country Parties to Contribute to Mitigation Actions in the Forest Sector (Jakarta: DG Climate Change, Ministry of Environment and Forestry)

[2] Anonymous 2016 Perubahan Iklim, Perjanjian Paris, dan Nationally Determined Contribution (Jakarta: DG Climate Change, Ministry of Environment and Forestry)

[3] Burung Indonesia 2014 Critical Ecosystem Fund, Ecosystem Profile Wallacea Biodiversity Hotspot for Submission to the CEPF Donor Council

[4] Gregorio M, Nurrochmat D R, Paavolaa J, Sari I M, Fatorellia L, Pramova E, Locatelli B, Brockhaus M and Kusumadewi SD 2016 Climate policy integration in the land-use sector: Mitigation, adaptation and sustainable development linkages *J. Environmental Science & Policy* **67** 35-43

[5] Harbi J, Erbaugh J T, Sidiq M, Haasler B and Nurrochmat DR 2018 Making a bridge between livelihoods and forest conservation: lessons from Non-Timber Forest Products' utilization in South Sumatera, *Indonesia J. Forest Policy Econ* **94** 1-10

[6] Kastanya A 2002 Pengelolaan Hutan Produksi Lestari Berbasis Gugus Pulau di Maluku (Yogyakarta: Universitas Gadjah Mada)

[7] Kastanya A 2012 A Community and ecosystem-based sustainable forest management model for small islands in Maluku *J. Agricultural Science and Technology* **2** (4) 418

[8] Kastanya A 2014 Rencana Pengelolaan Multi LanskapPulau-Pulau Kecil Berbasis Ekonomi Hijau (Green Economics) di Kepulauan Maluku. Pidato Pengukuhan Guru Besar Tetap Bidang Ilmu Perencanaan dan Ekonomi Sumber Daya Hutan Jurusan Kehutanan Fakultas Pertanian (Ambon: Universitas Pattimura)

[9] Kastanya A 2019 Multi-landscape forest management of small islands in the Moluccas based on green economy. UFROBOGOR. IOP Conf. Series: Earth and Environmental Science 285 (2019) 012011. IOP Publishing doi:10.1088/1755-1315/285/1/012011.

[10] Indrawan M, Primack R B and Supriatna J 2007 Biologi Konservasi (Jakarta: Yayasan Obor Indonesia)

[11] Monk K A, de Fretes Y and Lilley G R 1997 The Ecology of Nusa Tenggara and Maluku. The Ecology Indonesia Series, Volume V (Oxford: Oxford University Press)

[12] Czikkely. M; Ymeri. P; Hoang N H; Herczeg. B; Horváth B and Neubauer E, 2018. Actual Issues of Environmental economics. Reviewed by Dr. habil Csaba Fogarassy. Szent István University, Faculty of Economics and Social Sciences Institute of Regional Economics and Rural Development. Climate Change Economics Research Centre. ISBN 978-963-269-770-3 Gödöllő, Hungary.

[13] Nurrochmat D R, Nugroho I A, Hardjanto, Purwadianto A, Maryudi A and Erbaugh J T 2017 Shifting contestation into cooperation: Strategy to incorporate different interest of actors in medicinal plants in Meru Betiri National Park, Indonesia *J. Forest Policy Econ* **83** 162– 168

[14] Sahide M A K, Nurrochmat D R, Giessen L 2015 The regime complex for tropical rainforest transformation: analysing the relevance of multiple global and regional land use regimes in Indonesia *J. Land Use Policy* **47**408–425

[15] Sukwika T, Darusman D, Kusmana C, Nurrochmat D R 2016 Evaluating the level of sustainability of privately managed forest in Bogor, Indonesia *J. Biodiversitas* **17** (1) 241-248

[16] Satria Arif, 2020. Pembangunan Berkelanjutan 2045, Tanggapan Atas Orasi 90 Tahun Prof Emil Salim. Prof. Arif Satria Rektor IPB bahan presentase Webinar KEHATI, 18 Juni 2020

IOP Conf. Series: Earth and Environmental Science **883** (2021) 012005 doi:10.1088/1755-1315/883/1/012005

Income analysis of coconut farming with land diversification in North Minahasa Regency

J N K Dumais[1,*], R Kaunang[1], J Lumingkewas[1] and Y Rori[1]

[1] Department of Social Economics / Agribusiness, Sam Ratulangi University Jln. Kampus Unsrat Kleak

* E-mail: joachimdumais@yahoo.com

Abstract. North Minahasa Regency is the second-largest coconut plant center in North Sulawesi, with a coconut plantation area of 44,886 ha in 2018 (North Sulawesi Province in Figures, 2018). The plantation of coconut in North Minahasa Regency is a plantation of the people traditionally managed. Less development of rejuvenation coconut plantation causes an increase in the proportion of coconut plantation that its age relatively old has been cut down because it is not productive anymore. Cutting down is causing an increase in unproductive land under the coconut tree. This land can be used to grow other plants such as corn, papaya, chilli, ginger, banana and economic plants that will increase the income of coconut's farmer. The objective of this research is to: 1) assess and analyze the existing condition of diversification and revenue land farming coconut exists in the District of Minahasa Utara; 2) calculate and compare the income per hectare of any diversification of land farming coconuts in the District of Minahasa Utara, and 3) produce a recommendation policy that is associated with alternative types of diversification of land farming coconuts in the District of Minahasa Utara. Research result showed five types of income diversification of land, namely crop monoculture of coconut processed into copra. Crop monoculture of coconut processed into coconut grain, coconut processed into coconut grain with corn, coconut into coconut grain with papaya, coconut processed into coconut grain with chilli. Farming that gives the highest income is coconut processed into coconut grain with papaya.

1. Introduction

Plantations commodity is one of the commodity agriculture that has a high opportunity to develop to increase the state's income and increase farmers' income. One of the commodity plantations that most cultivated by farmers is coconut. Coconut plant has a high economic value because almost all parts of the plants have benefits. The most important part of this plant is the fruit. Coconut meat can be consumed directly as food or used as raw material for processing various fruit coconut products [1].

Coconut plants are not efficient in land use when cultivated in monoculture. The low price of coconut selling at the level of the farmer makes income farmers not feasible. On the other hand, the coconut plant's level of productivity is also relatively low due to lack of maintenance. To increase the income of coconut's farmer, one of the efforts that can be done is diversification in the diversification of plants' diversification to efficiently use land that can increase the household economic security and sustainability efforts [2]. Some advantages of the effort crop diversification horizontally between the coconut plant are the production of a coconut plant and the increased income of farmers [3]. Many farmers take advantage of the rest of the land to grow to share the kind of plant food to planting diversified horizontally.

North Sulawesi has long been recognized as one of the main coconut producing areas in Indonesia. Most of the farmers had me take advantage of the rest of the land for planting some type of food. Planting pattern of cropping diversification under a tree of coconut or in between plant-ing coconut highly profitable, by for land in the bottom of the coconut tree can be planted with other crops [4]. In managing the farming, farmers look for profit, where costs are incurred to produce the maximal output and increase revenue.

North Minahasa Regency is the second-largest coconut plant center in North Sulawesi, with a coconut plantation area in 2018 of 44,886 ha [5]. Plantation of coconut in North Minahasa Regency is the farming with various crops worth economical as corn, papaya, chilli, ginger and plant. The objective of this research is to: 1) assess and analyze the conditions the existing condition of diversification and revenue land farming coconut exist in North Minahasa Regency; 2) to calculate and compare the income per hectare of each type of diversification of coconut farming land in North Minahasa Regency, 3) to produce the policy associated with alternative types of diversification of land farming coconuts in the North Minahasa Regency.

2. Methods

2.1. Time and Place of Research
This research was conducted for 12 months in North Minahasa Regency. The location's selection is made intentionally based on considerations rural village location that produces the most extensive crop types production.

2.2. Data Collection Method
The types of data to be collected in this study include primary data and secondary data. Primary data obtained through the giving of a questionnaire to the respondent who is a coconut's farmer with the interview directly to farmers coconuts. Secondary data obtained from agencies related to research such as the Central Bureau of Statistics, Department of Agriculture Regional North Minahasa Regency dan literature research, and the books are related.

2.3. Sampling method
Respondents were used in the study of this includes farmers coconut in the North Minahasa Regency. In purposive sampling, subdistrict with the largest crop production under the coconut in North Minahasa Regency. Six combinations of coconut plants, namely copra monoculture, coconut grain monoculture, diversification of copra with maize, diversification of coconut grains with corn, diversification of coconut grain with papaya and diversification of coconut grain with chili.

2.4. Variable measurement
1. A general overview of respondents: Age (year), an education level (Elementary School, Junior High School, Senior High School, College), the duration of potato farming (years)
2. Basic variables
   a. Land area, which is a land area planted with potatoes (ha)
   b. The Status and ownership of land is proprietary or not proprietary
   c. The production cost of the cost of farmers incurred during the potato production process for one harvest
      - Fixed costs; Tax (Rp/yr), cost of depreciation of tools (Rp),
      - Variable cost: Labor cost (Rp/HOK), Seed (Rp/kg), Fertilizer (Rp/kg) Pesticides, (Rp/kg), Transportation (Rp/day).
   d. Total production (kg or grains/ha)
   e. The price of coconut or copra and crop under the coconut at the level of farmer (Rp/kg or grains)

3. Results and discussion

This research was conducted in North Minahasa Regency. The subdistrict chosen as the sample is the District Kauditan as a subdistrict producer of coconut grains with crops between corn and chilli pepper. The District Kalawat as a producer of copra with crop between corn and the District Dimembe as a coconut grain producer to plant between papaya.

3.1. Land tenure status

Research results showed that coconut farmers' land ownership status in North Minahasa Regency is their land, not rent or profit-sharing.

3.1.1. Type and area of land for cultivated coconut

There are five types of farmer coconuts grown by farmers coconut, as shown in Table 4.

Table 1. Types of coconut farming being cultivated

No.	Type
1	Coconut monoculture plants that are processed into copra
2	Coconut monoculture plants that are processed into coconut seeds
3	Plant of coconut that is processed into coconut grains with crops between corn
4	Plant of coconut that is processed into coconut grain to plant between papaya
5	Plant of coconut that is processed into coconut grain to plant between cayenne pepper

In the general area of land greatly affects the number of products produced by so also affects farmers' income. Based on the research results, extensive land earned by farmers coconut varies between 0.5 to 7 hectares with an average of 1.6 Hectares to plant monocultures of copra and plants between corn, papaya, and chilli pepper varies between 0.25 to 1 Hectares.

3.2. Production, cost of production, revenue and income farmers copra per hectares per planting season

The average production costs of copra farmers can be seen in Table 5.

Table 2. Production Costs of Copra Farmers per hectare

No.	Type	Cost (Rp/Ha)
1	Fixed Costs	
	a. Tax	118,868
	b. Shrinkage of tools	10,335
2	Variable Costs	
	a. Copra processing :	
	Weeding	300,000
	Climb	520,755
	Collecting	333,585
	Peel off	190,189
	Fumigation / packaging	719,811
	Total Cost	2,193,542

The average taxes were incurred by farmers copra is quite different. The tax per hectare is Rp. 118,868 per planting season. The highest cost is the fumigation, which is Rp. 719,811. The second-largest cost is the climbing fee, which is Rp. 520,755 per hectare per planting season.

Table 3. Average production, cost of production, revenue and income of copra's farmer per hectare per planting season

No.	Type	Amount	Unit	Price (Rp/Unit)	Total (Rp/Ha
1	Production	576	kg		
2	Price			8,000	
3	Reception				4,608,000
4	Cost				2,193,542
5	Income				2,414,458

Table 6 showed that the cost of copra production per hectare per planting season is Rp. 2,193,542, the revenue of Rp. 4,608,000 and the income of Rp. 2,414,458.

3.3. Production, cost of production, revenue and income coconut grain's farmer per hectare

Table 7 showed that the tax per hectare is Rp.170,323 per planting season. The most significant expense is climbing—namely Rp.606,452. The second-largest cost is the collecting cost, which is Rp. 381,935 per hectare per planting season.

Table 4. Production Costs of Coconut Grain's Farmers per hectare

No.	Type	Cost (Rp/Ha)
1	Fixed Costs	
	a. Tax	170,323
	b. Tool's Depreciation	5,395
2	Variable Costs	
	a. Coconut grain processing :	
	Weeding	300,000
	Climb	606,452
	Collecting	381,935
	Peeling	162,581
	Total Cost	1,626,685

Table 5. Average production, cost of production, revenue and coconut grain's income per hectare per season planting

No.	Type	Amount	Unit	Price (Rp/Item)	Total (Rp/Ha)
1	Production	2,624	grain		
2	Price			2,000	
3	Reception				5,248,387
4	Cost				1,626,685
5	Income				3,621,702

Table 8 showed that the cost of coconut grains per hectare per planting season is Rp. 1,626,685, the revenue of Rp. 5,248,387 and the income of Rp. 3,621,702 .

3.4. Production, cost of production, revenue and income coconut grain with corn's farmer per hectare

Table 6. Average production costs of coconut grain with corn's farmers per hectare

No.	Type	Cost (Rp/Ha)
1	Fixed Costs	
	a. Tax	171,707
	b. Shrinkage of tools	44,232
2	Variable Costs	
	a. Coconut grain processing :	
	Weeding	280,000
	Climb	490,732
	Collecting	317,659
	Peeling	245,854
	b.Corn farming	
	Land Processing	930,000
	Weeding	168,000
	Fertilization	168,000
	Harvest	2,010,000
	Post harvest	1,800,000
	Purchase of fertilizer	1,249,850
	Total Cost	7,876,033

Table 9 showed that the farm coconut grains with corn, the cost of which the biggest is the cost of maize crop farming amounting to Rp. 1.24985 million per hectare per planting season, then the cost of corn post-harvest amounting to Rp. 1,800,000 per hectare per planting season. Table 10 shows that the income of coconut grain per hectare per season Planting is amounting to Rp. 610 732, while the income of corn is amounting to Rp. 2,794,220, so that the total income is Rp. 3,404,951.

Table 7. Production, cost of production, revenue and income coconut grain with corn's farmer per hectare

No.	Description	Amount Coconut Grain	Unit	amount Corn	Unit	Price (Rp / unit)	Total (Rp / Ha)
1	Production	972	grain	1,680	kg		
2	Price of coconut grain					2,000	
3	Prices of corn					3,500	
4	Acceptance of coconut grains						1,944,976
5	Cost of coconut grain						1,334,244
6	Coconut grain income						610,732
7	Receipt of corn						5,880,000
8	Corn costs						3,085,780
9	Corn income						2,794,220
10	Total Income						3,404,951

Table 10 showed that the income of coconut grain per hectare per season Planting is amounting to Rp. 610 732, while the income of corn is amounting to Rp. 2,794,220, so that the total income is Rp. 3,404,951.

Table 8. Cost of production coconut grain with papaya's farmer per hectare

No.	Type	Cost (Rp/Ha)
1	Fixed Costs	
	a. Tax	65,789
	b. Shrinkage of tools	39,789
2	Variable Costs	
	a. Coconut grain processing :	
	Weeding	280,000
	Climb	501,053
	Collecting	322,526
	Peeling	265,263
	b.Papaya farming	
	Land Processing	418,033
	Weeding	319,672
	Fertilization	137,705
	Harvest	137,705
	Transport	8,508,197
	Purchase fertilizer	321,311
	Total Cost	11,317,044

Table 11 showed that the farm coconut grains with papaya, the cost of which the biggest is the cost of transporting farm papaya, amounted to Rp. 8,508,197 per hectare per season planting, then the cost of climbing coconut amounted to Rp. 503,053 per hectare per planting season.

3.5. Production, cost of production, revenue and income coconut grain with papaya's farmer per hectare

Table 12 shows that the income of coconut grain per hectare per season Planting is amounting to Rp. 533 684, while revenue papaya is amounting to Rp. 16,253,770, so that the total income is Rp. 16,789,455.

Table 9. Cost of production of coconut grain with papaya's farmer per hectare per planting season

No.	Description	Amount Coconut Grain	Unit	Amount Papaya	Unit	Price (Rp / unit)	Total (Rp / Ha)
1	Production	951	grain	11,287	Fruit		
2	Price of coconut grain					2,000	
3	Papaya price					2,000	
4	Acceptance of coconut grains						1,902,000
5	Cost of coconut grain						1,366,316
6	Coconut grain income						535,684
7	Papaya reception						22,573,770
8	Papaya costs						6,320,000
9	Papaya income						16,253,770
10	Total Income						16,789,455

3.6. Production, cost of production, revenue and income coconut grain with chili's farmer per hectare

Table 10. Cost of production of coconut grain with chili's farmer per hectare per planting season

No.	Type	Cost (Rp / Ha)
1	Fixed Costs	
	a. Tax	112,118
	b. Tool's depreciation	6,855
2	Variable Costs	
	a. Coconut grain processing :	
	Weeding	300,000
	Climb	507,368
	Collecting	328,000
	Peeling	265,263
	b. Chili farming	
	Land Processing	2,410,714
	Weeding	1,150,000
	Fertilization	910,714
	Harvest	1,110,714
	Transport	2,400,000
	Purchase fertilizer	2,272,321
	Total Cost	11,774,070

Table 13 shows that the farm coconut grains with chilli, the cost of which the biggest is the cost of processing land chilli farming amounting to Rp. 2,410,714 per hectare per season planting, then the cost of transporting chilli amounting to Rp. 2,400,000 per hectare per growing season.

Table 11. Cost of production of coconut grain with chili's farmer per hectare per planting season

No.	Description	Amount Coconut Grain	Unit	Amount Chili	Unit	Price (Rp / unit)	Total (Rp / Ha)
1	Production	1,046	grain	707.4	Kg		
2	Price of coconut grain					1,800	
3	Price of chili					20,000	
4	Revenue of coconut grains						1,883,368
5	Cost of coconut grain						828,402
6	Coconut grain income						1,054,967
7	Chili Revenue						14,147,368
8	Chili Cost						3,379,464
9	Chili Income						10,767,904
10	Total Income						11,822,871

Table 14 showed that the income of coconut grain per hectare per season planting is amounting to Rp. 1,054,967, while chilli revenue is amounting to Rp. 10,767,904, so that the total income is Rp. 11,822,871.

4. Conclusion

In North Minahasa Regency, there are five types of income diversification of land, namely crop monoculture of coconut that is processed into copra, coconut grains. Plant of coconut that is processed into coconut grains with crops between corn. Plant of coconut that is processed into coconut grain to plant between papaya. Plant of coconut that is processed into coconut grain to plant between chilli. Farming that produces the highest income is coconut plant processed into coconut grain planted with papaya.

References

[1] Sukamto 2001 *Upaya Meningkatkan Produksi Kelapa* (Jakarta: PT. Penebar Swadaya)

[2] Mahmud Z 2003 *Pemberdayaan Petani Kelapa dengan Sistem Usahatani Kelapa Terpadu* (Proc Konferensi Nasional Kelapa V Tembilahan 22-24 Oktober 2002) pp 115-124

[3] Toha H M 2002 *Padi Gogo Sebagai Tanaman Sela Perkebunan dan HTI Muda* (Bogor: Badan Penelitian dan Pengembangan Pertanian Balai Penelitian Tanaman Padi)

[4] Kadekoh 2007 *Optimalisasi Pemanfaatan Lahan Kering Berkelanjutan* (Proc Seminar Nasional Pengembangan Inovasi Lahan Marginal 2007) pp 27-33

[5] Badan Pusat Statistik Provinsi Sulawesi Utara, 2019, *Provinsi Sulawesi Utara dalam Angka* (Manado:BPS Provinsi Sulut)

The Strategy of guava agribusiness development in Kendal Regency using a business model canvas

A Khamdi[1,*], W Roessali[1], Mukson[1]

[1] Department of Agribusinesses, Diponegoro University, Jl. Prof. Soedarto No. 50275, Tembalang, Semarang, Indonesia

* E-mail: alikhamdi2013@gmail.com;wroessali@gmail.com; mukson.fapetundip@gmail.com

Abstract. Kendal is the area in Central Java with the most red guava fruit production, according to the Central Bureau of Statistics Republic of Indonesia. The production of red guava in Kendal Regency increased from 97,050 to 222,548 quintals between 2014 and 2016. In 2017, output fell to 137,671 quintals; however, in 2018, production rose to 159,079 quintals, still falling short of the previous year's total. This study used a business model canvas (BMC) methodology and SWOT analysis to map the existing red guava agribusiness business model and create a development plan for it. The interpretation of the SWOT diagram shows that guava farmers in four districts of Kendal Regency are in quadrant I, which indicates that they are in growth conditions with good opportunities and strengths to support the expansion of a red guava business. BMC demonstrates the need for additional consumer segments in order to reduce wholesaler dependence with several strategies. 1) Expansion of the trading network (2) Increasing key resources, (3) Adding key activity, such as regulating the harvest period for red guava and post-harvest processing. (4) Value proposition, by improving the quality of red guava

1. Introduction

Indonesia as an agricultural country has abundant agricultural products which serve as a source of income for the Indonesian people. The plantation and agricultural industries provided a source of livelihood for many Indonesians. In addition, the agricultural and plantation sectors provide both markets and unprocessed materials for the industrial sector's output, as well as foreign exchange for development. [1]

According to the Central Statistics Agency (BPS), the agricultural sector contributed 13.41 percent of GDP on a constant 2010 basis in 2018, accounting for 31.86 percent of the 124.54 million national workforces. Agriculture, forestry, and fisheries are the second-largest contributors to GDP, with a total of 1,785,881 billion rupiah. Meanwhile, the manufacturing sector accounts for the first amount, which is about 2,739,415 billion rupiah. According to data from 2014 to 2017, Indonesia's agriculture, forestry, and fisheries sectors contributed more to the country's GDP. [2]

Bananas (7.16 million tons), mangoes (2.20 million tons), siam/tangerines (2.16 million tons), pineapples (1.79 million tons), and salak (0.95 million tons) are the top five annual fruit production commodities, according to 2017 statistical results. In addition to the five fruits listed above, guava has developed into an excellent commodity that is processed in large quantities. Guava production rose to 230,697 tons in 2018, up from 200,495 tons in 2017. [3]

With an output of 637,159 quintals, or 27.6% of national production, Central Java became the first province to contribute to national guava production in 2018. In Central Java, the Kendal Regency contributes the most to guava production. The production of guava in Kendal Regency increased from 97,050 quintals in 2014 to 222,548 quintals in 2016. In 2017, production fell to 137,671 quintals; however, in 2018, production rose to 159,079 quintals. [4]

The sustainability of the red guava agribusiness in Kendal district must be maintained and developed in order for farmers to reap the benefits of this agribusiness. Based on production fluctuations and previous studies, it appears that little has been done to improve the red guava agribusiness. The Business Model Canvas is one of the methods we can use to grow red guava agribusiness.

The Business Model Canvas is a strategic tool used to describe a business model and describe the procedure of how the organizations create, deliver, and capture value. The Business Model Canvas helps us to see more accurately the current or future business conditions. Changing a complex business concept to a simple one that is displayed on a single sheet of canvas containing nine key elements that are well integrated, which includes both internal and external strategic analysis. [5]

2. Methods

This study used a concurrent embedded design model that merged qualitative analysis methods as primary methods with quantitative methods to help and improve results.[6]. From June to July 2020, it was conducted in four districts: Patean, Pageruyung, Sukorejo, and Plantungan.

Interviews were used as the primary data in this analysis, with supporting data and a questionnaire serving as secondary data. Customer segments, value propositions, channels, customer relationships, revenue streams, key resources, key activities, and cost structure are the nine elements of the Business Model Canvas (BMC). Then, in every aspect, it was analyzed using a SWOT analysis. [7]

3. Results and discussion

Sukorejo, Pageruyung, Patean, and Plantungan Districts produce the most red guava in Kendal. This district is ideal for growing red guava because of its location. Pageruyung sub-district has a total area of 51.44 km2, of which 82.07 percent is used for agricultural land (Ricefield land and dry land), while the remaining 17.93 percent is used for state forests and grounds, and the area of agricultural land used as a Red guava farm covering 16 ha with 10,000 red guava trees producing 9,500 quintals.

Pageruyun's land elevation ranges from 250 to 600 meters above sea level. Sukorejo produces the most quintals, with 352 ha of red guava farming land and 220,000 trees producing 187,000 quintals. Patean has a total of 19,778 red guava trees on 25 hectares of red guava farming land, with a production yield of 24,401 quintals. Meanwhile, the Plantungan has 23.5 hectares of red guava trees, growing 20,860 quintals of red guava on 13.7 hectares of red guava agricultural property.

3.1 Current Business Model Condition

The state of the red guava agribusiness Busines Model Canvas in terms of the identification of the nine blocks is as can be seen in Figure 1, based on the results of interviews with both the Regional Government Department of Agriculture and Farmers:

<table>
<tr><th>Key Partners</th><th>Key Activities</th><th>Value Proporsitions</th><th>Customer Relationships</th><th>Customer Segment</th></tr>
<tr><td rowspan="3">1. Wholesalers
2. Plant seeds supllier
3. Fertilizer supplier
4. Workers
5. Land owner</td><td>1. Plantation
2. Fertilization
3. Plant Pest Controlling
4. Covering
5. Harvesting</td><td rowspan="3">1. Quantity
2. Quality
3. Adjust price</td><td>1. Kinship
2. Adjust price (cheaper)
3. Guarantee
4. Capital</td><td rowspan="3">1. Personal
2. Collector
3. Wholesalers</td></tr>
<tr><th>Key Resources</th><th>Channels</th></tr>
<tr><td>1. Land
2. Good Seed</td><td>1. Direct selling
2. Partner up</td></tr>
</table>

	3. Human resources 4. Finance			
Cost Struktur			***Revenue Streams***	
1. Workers salary 2. Fertilizer cost 3. Covering cost 4. Land rent ost			1. Selling	

Figure 1. Identification of the BMC Red Guava Farmers

3.1.1 Internal factors identification

1. ***Strenght***
   a. Availability of agricultural infrastructure
   b. The large production of quantity
   c. Well-built customer relationships
   d. Potential of natural resources owned
   e. Market segmentation which is specified
   f. The ease of plant care
2. ***Weakness***
   a. Dependence with wholesalers
   b. Farmers' human resources need to be upgraded, both in cultivation and mastery of information technology
   c. Farm equipment is out dated
   d. the Prices fall during the main harvest
   e. Limited capital
   f. Limited coaching

The following Table 1 is the outcome of the IFAS matrix recapitulation after determining the strengths and weaknesses of guava farmers.

Table 1. Matrix IFAS

Internal Strategy Key Factors	**Value**	**Rating**	**Total**
STRENGHT			
The large production of quantity	0,17	3,5	0,595
Availability of agricultural infrastructure	0,18	3,6	0,648
Well-built customer relationships	0,16	3,4	0,544
Potential of natural resources	0,15	3,4	0,510
The ease of plant care	0,15	3,4	0,510
Specified market segmentation	0,19	3,8	0,722
Sum	**1,00**		**3,529**
WEAKNESSES			
Out dated farmers resources	0,18	2,6	0,468
Dependence with wholesalers	0,21	2,8	0,588
Fall prices during the main harvest	0,12	2,4	0,288
Limited capital	0,14	2,5	0,350
Out dated equipment	0,18	2,4	0,432
Limited coaching	0,17	2,6	0,442
Sum	**1,00**		**2,568**

3.1.2 External factors identification

A. ***Opportunity***

1. The harvest period management
2. Classification of quality red guava
3. Expanding the market
4. Increasing access to capital
5. Guava processed products is small
6. Availability of large vacant land
7. Utilization of information technology

B. ***Threat***

1. The price is played by wholesaler
2. Plant pests and diseases
3. Consumers can turn away
4. Limited of supporting institutions for farming
5. No regeneration of guava farmers
6. Failed to harvest
7. Similar harvest period with other fruit (mango)

Following Table 2 is the outcome of the EFAS matrix recapitulation after determining the strengths and weaknesses of guava farmers.

Table 2. Matrix EFAS

Internal Strategy Key Factors	Value	Rating	Total
OPPORTUNITY			
The harvest period management	0,17	3,8	0,646
Classification of quality red guava	0,13	3,4	0,442
Expanding the market	0,14	3,6	0,504
Increasing access to capital	0,15	3,8	0,570
Guava processed products is small	0,14	3,8	0,532
Availability of large vacant land	0,12	3,6	0,432
Utilization of information technology	0,15	3,8	0,570
Sum	**1,00**		**3,696**
THREAT			
The price is played by wholesaler	0,15	2,8	0,420
Plant pests and diseases	0,12	2,0	0,240
Similar harvest period with other fruits	0,17	2,8	0,476
Harvest failed	0,13	2,2	0,286
No regeneration of guava farmers	0,14	2,4	0,336
Limited of supporting institutions	0,15	2,6	0,390
Consumers can turn away	0,14	2,6	0,364
Sum	**1,00**		**2,512**

Based on the results obtained from the results of the internal and external analysis in the table above, it requires confirmation of the position in the Cartesian diagram, between strengths and weaknesses, as well as opportunities and challenges which are all depicted in positive and negative lines. In terms of coordinate equations, it can be done based on the following calculations:

1. Internal coordinate analysis, the total strength-weakness score divided by 2 then **(3,529-2,568) : 2 = 0,48**
2. External coordinate analysis, the total oportunity score divided by 2 then **(3,696-2,512) : 2 = 0,59**

Based on the results above, it can be determined the coordinate point is located at 0.48 : 0,59

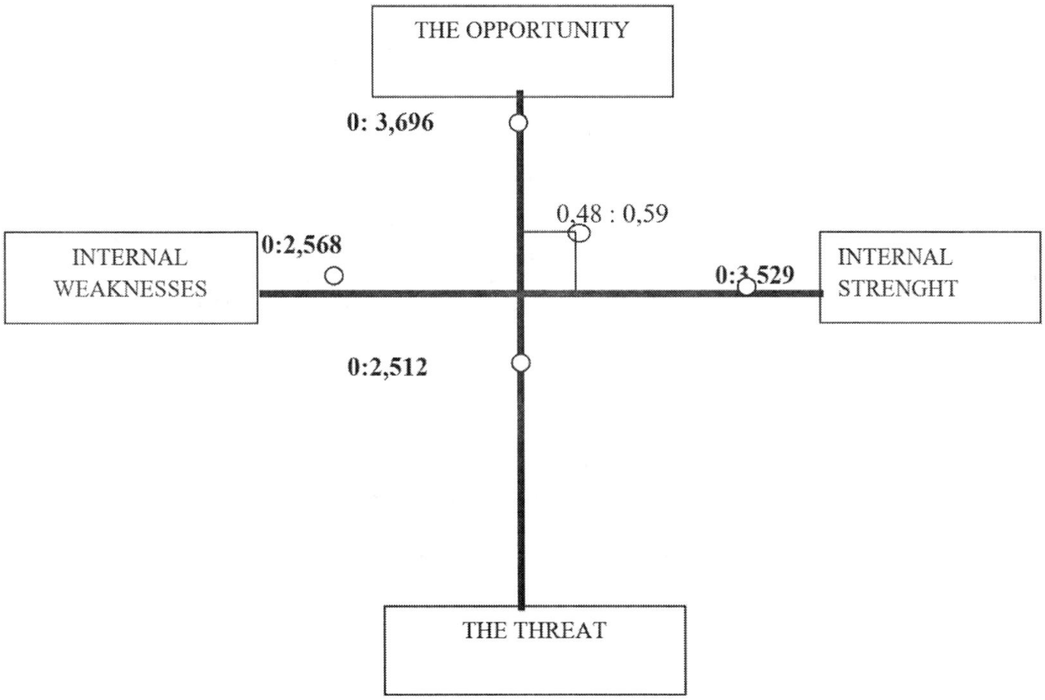

Figure 2. SWOT Diagram of Red Guava farmers

According to the diagram above, guava farmers in four sub-districts are in quadrant I, which indicates that they have a good understanding of growth conditions, as well as good opportunities and strengths to support the production of an increasingly red guava market.

3.2 SWOT Identification on the BMC Block

Furthermore, mapping the strengths, weaknesses, threats, and opportunities of the BMC block was used to determine the current state of the BMC block. The results of SWOT identification on the BMC block are as follows, based on the results of interviews and questionnaires.

IFAS / EFAS	***Strength (S)***	***Weakness (W)***
	1. The availability of agricultural infrastructure(KR) 2. High amount of production (VP) 3. Good relationship with partner and consumers (KP) 4. Natural resources owned (KR) 5. Specified market segmentation (CS) 6. Ease of plant care (KA)	1. Dependence with wholesalers (CS) 2. Farmers' human resources need to be upgraded both in cultivation and mastery of information technology (KR) 3. Farm equipment is still simple (KA) 4. Prices fall during harvest season (CS) 5. Limited capital (CoS) 6. Limited coaching access to farmers (CR)
Oppurtunity (O) 1. Monthly harvest (KA) 2. Classifying red guava quality (VP) 3. Expanding market (CS)	***Strategi SO*** 1. Improving the IT infrastructure (S1,S4,O7)	***Strategi WO*** 1. Creative marketing using social media (W1,O3) 2. Improving human resources of farmers in the cultivation

4. Improving capital access (CoS) 5. Processed red guava product is still limited (VP) 6. Availability of large vacant land (KR) 7. IT utilization (KR)	2. Optimalizing the land and quality in expanding the production (S2,S6,O1,O2,O6) 3. Expanding the new market segment and processed product (S5,O3,O5) 4. Maintaining and improving the capital with good relationship to partners and consumers (S3,O4)	period and IT utilization (W2,O7, O1) 3. Optimizing the main harvest period with processed products (W4,O5) 4. Farmer coaching in optimizing the infrastructures, capital, and guava production. (W3,W5,W6,O2,O4,O6)
Threat (T) 1. Price depended on wholesellers (CR) 2. Plant pest and disease (KA) 3. Consumer can turn away (RS) 4. Limited supporting institution (C) 5. No regeneration of farmers (KA) 6. Failed Harvest (KA) 7. Similar harvest period with other comodity (KA)	***Strategi ST*** 1. improving the infrastructure and maintaining the quality of red guava(S1,S2,S4,S6,T2,T6,T7) 2. Increasing good relationship with partners and consumers (S3,T3) 3. Expanding new market segment besides wholesalers (S5,T1) 4. Establishing farmer groups for coaching and guidance. (S3,T4)	***Strategi WT*** 1. Establishing farmer group and IT Marketing mastery (W1,W2,W4,T1,T3,T5) 2. Giving acces to the farmers for coaching clinic i (W6,T4) 3. Expanding the capital to maintain and increasing the amount of processed product (W3,W5,T2,T6,T7)

Note:
CS=Customer Segment CR=Customer Relationship KA=Key Activities
VP=Value Propositions RS=Revenue Streams KP=Key Partnerships
C=Channel KR=Key Resources CoS= Cost Structure

Figure 3. The Matrix SWOT of Alternative Strategy Formula

3.3 The Alternative Strategy on the Busines Model Canvas of Red Guava Farmers
The next step after the SWOT analysis produces an alternative strategy is the implementation of the resulting strategy in designing a new business model canvas. This stage involves implementing the SO, WO, ST and WT strategies in improving and evaluating the business model canvas as follows:

3.3.1 Customer Segments
Farmers must use the SO-3, ST-3 strategy in this element to overcome the problem of a limited customer segment, namely by evaluating segmentation in order to add new market segments for the growth of this guava sector. Farmers have only sold to collectors and wholesalers so far. Wholesalers, fruit-processed beverage firms, pharmaceutical companies, and modern markets were among the recipients.

Farmers in this element may also use the WO-1, WO-2, and WT-1 strategies to improve consumer segments by optimizing marketing activities with online media and having farmer groups that control product processing and marketing. This strategy is used to address the weaknesses of farmers who are having trouble finding new buyers. This occurs as a result of insufficient human capital and infrastructure to sustain marketing activities. As a consequence, in the future, they will need to update human resources and infrastructure in order to better assist the operational marketing of guava production results.

3.3.2 Value Propositions
The quality of red guava is not classified by farmers at the time of sale or it is commonly called wholesale. Farmers are reluctant to improve quality and will result in low prices. In this condition farmers can apply the SO-2, SO-3, WO-3 strategy, classifying quality into two types, namely super quality or grade A to target wholesale and modern markets as well as standard quality or grade B to be sold to collectors and middlemen as well as made processed red guava products.

3.3.3 Channels
Weaknesses in the channel aspect continue to focus on direct sale and one-to-one communication, as well as less-than-optimal social media marketing. In order to have a successful sales life, farmers should use the SO-4 strategy by maintaining harmonye with partners and customers. They can also use the WO-1 and WO-2 strategies, which include optimizing the use of technology and information in marketing activities in order to see potential markets, especially through online media with supporting human resources and technology. As a result, sales using IT through online media and salespeople managed by marketing staff are alternatives to keeping sales flowing smoothly.

3.3.4 Customer Relationship
This block's weaknesses include reliance on wholesalers and dropping prices during harvest. Implementing the alternative SO-4 strategy, where farmers continue to establish and improve harmony with partners and consumers, both with kinship and others for business continuity is one of the alternative strategies. Furthermore, the ST-4, WO-2, and WT-1, where farmers add human resources, particularly in the field of marketing, to carry out tasks that are more focused on sales, both with kinship and others are also among alternative strategies.

3.3.5 Revenue Streams
In order to maximize benefit, red guava must be graded by the farmer at the time of sale. Farmers can use the SO-2, SO-3, WO-3 strategy, which divides quality into two categories: super quality (grade A) for wholesale and modern markets, and standard quality (grade B) for collectors and middlemen, as well as manufactured goods. guava (red).

3.3.6 Key Resources
Natural resources, human resources, and financial resources are all at a minimum in this block. In terms of natural resources, it has a lot of potential but is still underutilized, so a SO-2 strategy is needed to maximize the land and increase red guava farming production. In terms of human resources, they continue to depend on farmer families and farmer neighbors, especially in terms of red guava production and harvest, resulting in human resources that are inadequate for marketing. A WO-2 strategy is needed to supplement these human resources. Meanwhile, farmers' capital is severely limited due to their reliance on income and bank borrowing. Farmers may use an alternate WT-1 approach in this situation, namely the inclusion of capital assistance from both the government and corporate CSR.

3.3.7 Key Activities
The absence of guava farmer regeneration is a weakness of the key activity factor. In this situation, farmers may use WO-2, ST-4, and WT-2 alternative strategies, such as human resource addition and fulfillment. This strategy is intended to ensure the long-term sustainability of farmers' businesses by providing sufficient human resources via farmer groups as an organization that shelters farmers and collaborates with many schools / universities to meet these needs.

3.3.8 Key Partnership
In this key partnership element, describing how the relationship between farmers and partners is very important to keep the business model running well and smooth. Farmers can apply alternative strategies to maintain the relationship between farmers and partners that have existed so far, so the implementation of an alternative strategy of SO-4, namely maintaining harmonization of partners and consumers, is done to avoid competitors that will hinder farmers' businesses.

Therefore, farmers must employ the ST-4 strategy, which focuses on improving harmonious relationships with partners and customers. The business continuity can run smoothly if there is constant harmonization and strengthening of relationships, and there is also the possibility of finding a new segment for this guava product (SO-3).

3.3.9 Cost Structure

There is no new funding for this aspect, but the demand will grow as farmers increase their operation and management to increase production, pest control, irrigation systems, and regulating the harvest season and processing of red guava.

3.4. The projection of BMC

Alternative strategy mapping implemented on the new Business Model Canvas for Red Guava farmer can be seen in Figure 4.

<table>
<tr><th>Key Partners</th><th>Key Activities</th><th>Value Proporsitions</th><th>Customer Relationships</th><th>Customer Segment</th></tr>
<tr><td rowspan="3">1. Wholesalers
2. Seed plants supplier
3. Fertilizer supplier
4. Workers
5. Land Owners
6. Farmer group
7. The Government
8. Modern Market
9. Capital Supplier</td><td>1. Plantation
2. Fertilization
3. Plant Pest Controlling
4. Covering the guava
5. Harvesting
6. Cutting plant (to manage the harvest period)
7. Classififying the product quality
8. Packaging</td><td rowspan="3">1. Quality
2. Quantity
3. Cheaper price
4. The availability if guava products even in out of harvest period
5. Red guava processed production</td><td>1. Kinship
2. Good price
3. Guarantee
4. Startup capital
5. Good relationship between partners and consument.
6. Good workers harmonization</td><td rowspan="3">1. Personal
2. Collectors
3. Wholesalers, Modern market
4. Company of processed guava/ pharmacy</td></tr>
<tr><th>Key Resources</th><th>Channels</th></tr>
<tr><td>1. Land
2. Good Seed Plants
3. Human Resources
4. Finance
5. Improvement of Agricultural Infratsructure
6. Irigation
7. Land Expansion
8. Marketing Human resources
9. Capital Assistance</td><td>1. Direct selling
2. Partner up
3. Marketing resources
4. IT utilization</td></tr>
<tr><th colspan="3">Cost Struktur</th><th colspan="2">Revenue Streams</th></tr>
<tr><td colspan="3">1. Workers salary
2. Fertilizers cost
3. Covering cost
4. Land rent cost
5. Packaging
6. Red Guava processed product
7. Direct selling transportaion</td><td colspan="2">1. Selling Super Quality /A
2. Selling quality B
3. Selling the processed producs of red guava</td></tr>
</table>

Figure 4. The Alternative Strategy BMC of Red Guava Cultivation

4. Conclusion

The results of the BMC with SWOT indicate that farmer groups should be established, that farmer human resources should be improved in both cultivation and information and processing technology, that market segments should be expanded, that guava quality should be classified, and that capital should be strengthened. Farmers also require guidance and training in guava cultivation, especially in terms of regulating the harvest period.

IOP Conf. Series: Earth and Environmental Science **883** (2021) 012006 doi:10.1088/1755-1315/883/1/012006

References

[1]. Narundana, Vonny T, Aminah, Mimin 2011 Studi kelayakan bisnis tanaman buah jambu kristal pada kelompok tani desa Cikarawang, Kecamatan Dramaga Kabupaten Bogor. Indonesia. IPB University Scientific Repository

[2]. Badan Pusat Statistik Nasional 2018 *Produk Domestik Bruto Indonesia Atas dasar Harga Konstan 2010 Menurut lapangan Usaha (milliard rupiah) Tahun 2014-2017*. www.bps.go.id

[3]. Badan Pusat Statistik Nasional 2018 *Produksi Tanaman Buah-Buahan Berdasrkan Provinsi*. www.bps.go.id

[4]. Badan Pusat Statistik Jawa Tengah 2018 *Banyaknya Pohon Panen dan Produksi Jambu Biji menurut Kabupaten/Kota dan Jenis Buah di Provinsi Jawa Tengah 2015-2018*. www.jateng.bps.go.id

[5]. Osterwalder, Alexander dan Yves Pigneur 2012 *Business Model Generation*. Jakarta : Elex Media Komputindo.

[6]. Sugiyono 2017 *Metodelogi Penelitian Kuantitatif, Kualitatif Dan R&D*. (Bandung: Alfabeta)

[7]. Widyastuti dan Novian 2018 *Pengembangan Usaha Tanaman Hias dengan Pendekatan Business Model Canvas (BMC) (Studi pada Sentra Tanaman Hias di Jakarta Utara)*. Conference on Management and Behavioral Studies Universitas Tarumanagara, Jakarta, 25 Oktober 2018 ISSN NO: 2541-3406 e-ISSN NO: 2541-285X

Isolation and identification of fenobucarb degrading bacteria from Pangalengan farm land

A Akhdiya[1,*], R A Sanjaya[1], and Wartono[1]

[1] Indonesian Center for Agricultural Biotechnology and Genetic Resources Research and Development

* E-mail: alinakhdiya@pertanian.go.id

Abstract. The long term excessive use of pesticides can lead to their residues accumulation in the soils. Soil microbes were considered to convert the residues into harmless compounds, however the indigenous soil microbes having those beneficial properties are limited. Therefore this study aimed to isolate, select and identify the fenobucarb insecticide-degrading bacteria from agricultural soils. The soil samples were collected from the vegetable fields in Pangalengan, West Java, Indonesia. Isolation of the bacteria was conducted using Nitrate Mineral Salt Agar suplemented by 100 ppm of a fenobucarb. The bacteria isolates were selected based on its hypersensitive response, haemolytic activity, and its ability to degrade fenobucarb. The selected isolates was identified base on sequences of 16S rRNA gene. Twenty nine bacteria were isolated from four soil samples and 23 of the isolates were not potentially phytopathogenic and non haemolytic. The best three isolates that could degrade 94.2%, 94.5% and 95.47% fenobucarb residue are B41, B54 and B83 isolates, respectively. The 16S rDNA Sequence analysis showed that B41 and B83 isolates have 100% similarity to *Bacillus thuringiensis* MYBT 18426B54, while B54 isolate has 99% similarity to *Bacillus luciferensis* LMG 18422. These isolates are potential to be developed as a bioremediation agent.

1. Introduction

Farmers commonly use synthetic pesticides as one of the important components to protect their crop. Contrary to benefit of the synthetic agrochemical use, there are so many negative effect on human health and the environment [1]. Uncontrolled use of pesticides will cause adverse changes to ecosystems such as pest resistance, death of beneficial organisms, and pesticides residues in water, soil, food, and organisms [2]. One of the most widely used insecticides in the world is fenobucarb. Fenobucarb is a broad spectrum insecticides which contain BPMC (*butylphenyl methylcarbamate*). The insecticide has high extermination power and rapid knock down effect [3].

Naturally indigenuous microbes reside in a land could degrade pesticides residue. However, the microbial capability will not proportional with the higher increasing of intensity and quantity of agrochemichals input. Efforts are needed to reduce the negative impacts caused by their use such as by implementing integrated pest control and the use of natural pesticides [4]. More over, enrichment of the polluted land by using bioremediation agent is needed. Availability of selected dan fine characterize microbial isolates is important to formulate an efective bioremediation agent. This research was conducted to isolate, screen, and indentify fenobucarb degrading bacteria from farm soil.

2. Methods

2.1.Materials
Soil samples were taken from Paddy, tomato, potato, and cabbage farmland, at Legok Bako village, Margaluyu, Pangalengan, West Java. Bacteria used in this research were isolated from the soil samples, while *Ralstonia solanacearum* is a collection of Biogen Culture Collection, ICABIOGRAD.

2.2.Isolation and Purification of Fenobucarb Degrading Bacteria
Isolation and purification of the bacteria were conducted by following the procedure published by Akhdiya *et. al.* [5] that modified by subtitute the insecticides with fenobucarb. The pure isolates were stored at -20°C in 40% glycerol as stock cultures. While as the working cultures, the isolates were streaked on NA + 100 ppm fenobucarb and stored at 15-18°C.

2.3.Hypersensitive Response (HR) Bioassay
Bacterial isolates were rejuvenated on Nutrient Agar media (Merck, US) for 24-48 hours. A colonies of bacteria were suspended in 1.5 ml of 0.85% NaCl. Half milliter of bacterial suspension was injected into the lower surface of tobacco leaf (*N. tabacum* L.). *Ralstonia solanacearum* was used as positive control. Necrotic tissue formed on the leaf were observed every day for 7 days. The experiment was conducted duplicate

2.4.Hemolytic Activity Testing
The isolates that did not cause necrotic symptoms in the previous bioassay, were inoculated on the surface of blood agar and then incubated at room temperature (25°C) for 24-72 hours. Clear zone formed around the colonies indicated haemolytic activity of the isolates.

2.5.Test to Use Fenobucarb as Exclusive C Source
The test was carried out by following the procedure published by Akhdiya *et. al.* [5] that modified by subtitute the insecticides with fenobucarb. The cultures growth were observed visually for 24-72 hours.

2.6.Fenobucarb Degradation Test
Selected isolates were cultivated using the same procedure for use of exclusive C source. The culture were cultivated for 96 hours before its were centrifugated at 10.000 x g for 10 minutes. Supernatan of the cultures were sended to Agricultural Environmental Research Institute (IAERI) to be determined the residue. Determination of the residues was carried out using Gas Chromatography (Variant Type 450, US) equipped with VF-1701 columns and Electrone Captured Detector. The experiment was conducted in triplicate. Percentage of the degraded fenobucarb, calculated by the following equation

$$Fd = \frac{Fi - Ff}{Fi} \times 100\%$$

Fd : Degraded fenobucarb (%)
Fi : Initial concentration of Fenobucarb
Ff : Final concentration of Fenobucarb

2.7.Identification of the Selected Isolates
DNA of the selected isolates were extracted using the PrestoTM Mini gDNA Bacteria Kit. The DNA were amplified using 27F (5'-AGAGTTTGATCCTGGCTCAG-3') and 1492R (5'-GGTTACCTTGTTACGACTT-3) primers [6]. Steps of the 30 PCR cycles were : activation (94°C, 5 minutes), denaturation (94°C, 5 minutes), annelling (55°C, 45 seconds), elongation (72°C, 1 minute), and final extension (72°C, 1 minute). Two microlitres of the amplification products were runned in

electrophoresis gel (agarosa 1%) using 1X TAE buffer at 80 volt for 40 minutes. Diamond™ Nucleic Acid Dye (Promega, US) was used to visualize the DNA bands under UV transilluminator.

The amplicon DNA were send to sequencing service laboratory. The DNA sequences were aligned with Genbank data using the BLAST-N (Basic Local Aligment Search Tool-Nucleotide) program from the NCBI (National Center for Biotechnology Information) to determine the identity of the isolates. Phylogenetic trees was created using Clustal X [7], neighbor joining method (NJ) with 1000x bootstrap.

3. Results and discussion

Fenobucarb degrading bacteria were isolated from soil samples taken from non-organic agricultural land (paddy, cabbage, tomato, and potato) in Pangalengan area, West Java. Consideration of the sampling locations was that agricultural land is managed intensively with high frequency of pesticide application. The use of NMSA + fenobucarb as the isolation medium was initial screening efforts to suppress growth of satellite colonies and to increase probability to isolate the high fenobucarb degrading capability bacteria. There were 29 fenobucarb degrading bacteria were isolated from 4 soil samples (Table 1).

Table 1. Number of fenobucarb degrading bacteria isolated from the soil

Farm land soil	Number of bacteria isolates
Cabbage	6
Paddy	4
Potato	13
Tomato	6

Hypersensitivity response (HR) bioassay was conducted to eliminate the phytopathogenic potential bacteria. All of the isolates tested did not cause necrotic symptom of tobacco leaves tissues (Fig. 1A and 1B). Contrary to the asymptomatic leaves which injected with the isolates, injection of *R. solanacearum* caused necrotic symptop of the leaf tissues. Hypersensitive response is a rapid defense reaction of plants, when facing pathogens [8]. Plant HR is indicated by formation of leaf browning tissues in the infected area due to the death of local leaf tissues [9]. The bioassay results indicated that all of the isolates are phythopathogenic.

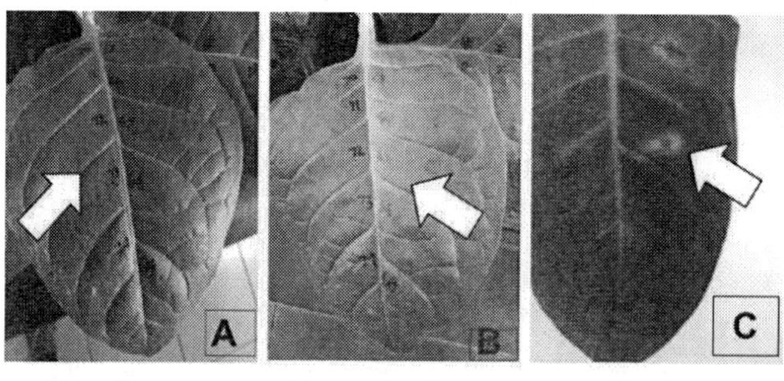

Figure 1. Photographs of the tobacco leaves at 24 hours (A), 48 hours (B) after injected with the bacteria isolates and necrotic leaf after 48 hours after injected with *R. solanacearum* (C).

Next selection base on hemolytic activity of the isolates was conducted to eliminate bacterial isolates potentially pathogenic to mammals and humans. Erythrocytes lysing bacteria are potentially more virulent compared to the non haemolytic bacteria. Ability of bacteria to lysis erythrocytes is determined by an extracellular proteins "hemolisin". Haemolysine producing bacteria will damage erythrocytes in their growth media, forming clear zones and decoloration [10]. Six of the 29 isolates were hemolytic (Figure 2).

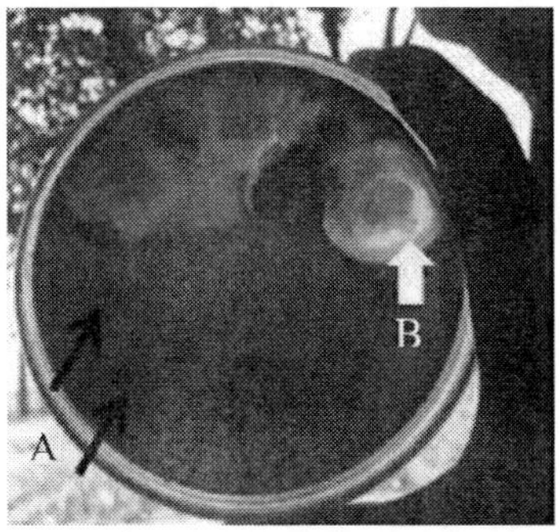

Figure 2. Non-haemolytic bacteria isolates (A) and haemolytic isolates surrounded by clear zone (B)

Among the non-haemolytic isolates, 6 isolates (B41, B54, B83, B95, B96 and B97) showed grow well on liquid NMS media containing fenobucarb as sole of carbon source. Bacteria growth was indicated by increasing culture turbidity after incobation for 48 hours. The turbidity of liquid culture in NMS containing pesticide is caused by the growth of bacteria that utilize pesticides as sole of carbon source for their growth [11]. Fenobucarb resistant isolates but can not use the compound as its C and energy source can not grow in the medium. Carbon is one of the main element of cells macro molecules. There are four organic macro moleculs : proteins, nucleic acids, carbohydrates and fats that have different chemical properties [12]. All organisms needs this element in relatively large quantities for its growth.

Table 2. Fenobucarb residue in liquid culture 6 isolates selected after 96 hours incubation

Isolate code	Degraded pesticides (%)
B41	94.27
B54	94.57
B83	95.29
B95	75.47
B96	69.65
B97	67.19

Analysis of fenobucarb residues in the culture of 6 selected isolates showed its ability to degrade the compound were 67.19% - 95.29% within 96 hours. Isolates B83, B54, and B41 could degrade respectively 95.29%, 94.57% and 94.27% fenobucarb residue (Table 2). Fenobucarb is an aromatic hydrocarbon compound. Ability of microorganisms to degrade aromatic hydrocarbon compounds are varies. The differences can be caused by the type, quantity, and activity of pesticide-degrading enzymes producing by the organisms. Bacterial enzymes that play a role in degradation process of aromatic hydrocarbon compounds include dioxogenases, monooxogenases [13], and hydrolytic enzymes [14, 15]. Monooxogenase and dioxynase enzymes are able to open carbon bonds of the aromatic rings and produce primary alcohol by using oxygen molecules as electron acceptors. The process will produces dicarboxylic and semi-aldehyde acids that are useful for the metabolism of cell intermediates [15]. Decreasing of fenobucarb concentration in the cultivaion media after incubation is the result of bacteria metabolism. Major chemical bonds in the toxic molecules can disrupt by hydrolytic enzymes and results in the reduction of its toxicity. This enzymatic mechanism is effective for the biodegradation of hydrocarbon compounds include oil, organophosphate, and carbamate [15]. Fenobucarb is degraded through a hydrolysis reaction by producing the main product 2-sec-buthylphenol. The hydrolysis product, then metabolized into carbon dioxide and water [16]. Many publications reported that *Bacillus* spp. Have dioxygenase [17], hydroxylase [18], and hydrolases [15, 19, 20]. Phosphatase and esterase are hydrolytic enzymes produced by soil microbes that can break the unstable chain structure such as carbamate [21].

Amplification of 16S rDNA sequences of the three selected isolates using primers 27F and 1492R produced ± 1,500 bp DNA fragments. Molecular identification results based on 16S rDNA sequence analysis showed that B54 isolate has 99% similarity *Bacillus luciferensis* LMG 18422 (99% homology), while B41 and B83 isolates show the highest similariy (99%) with *Bacillus thuringiensis* MYBT 18246. The phylogenetic tree constructed base on the sequences showed the closely related *Bacillus* species with *Bacillus luciferensis* B54 and *Bacillus thuringiensis* B41 and B83 (Figure 3).

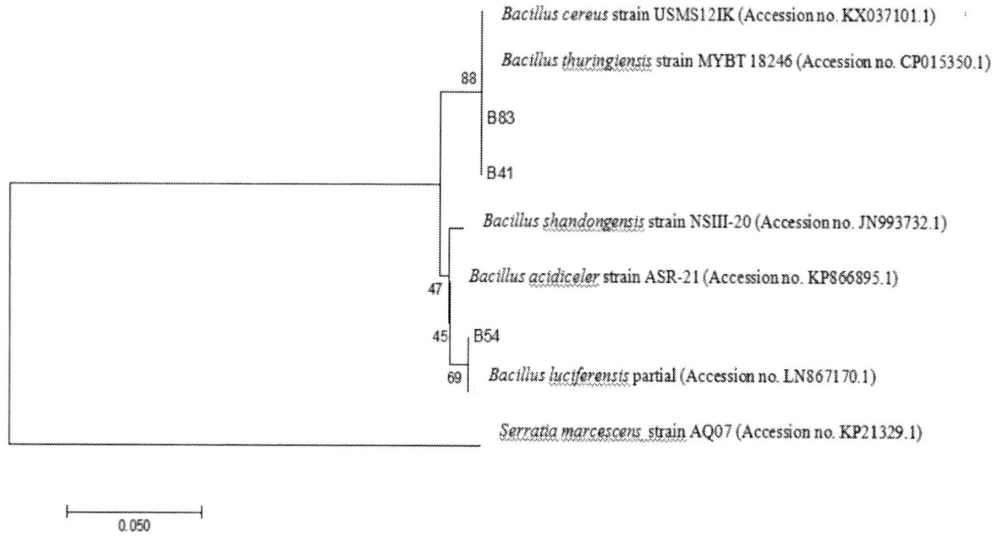

Figure 3. Phylogenetic tree of three phenobucarb-degrading *Bacillus* isolates based on its 16S rRNA sequences.

4. Conclusion

B. thuringiensis B41, *B. thuringiensis* B83, and *Bacillus luciferensis* B54 have high capability to degrade fenobucarb. The *Bacillus* were potential to be develop as bioremediation agents. Further testing is needed to determine its abiliy to degrade other pesticides

Acknowlegments

This study was supported by the DIPA 2016 of ICABIOGRAD-IAARD, Ministry of Agriculture, Republic of Indonesia. Thanks to Ameli Mustika for her assistance in our laboratory.

References

[1] Matthews GA 2015 Pesticides: Health, Safety and The Environment. Ascot, Berkshire (UK): Blackwell Publishing.

[2] Lu Y, Song S, Wang R, Liu Z, Meng J, Sweetman A J, Jenkins A, Ferriere RC, Li H, Lu W. and Wan T 2015 Impacts of soil and water pollution on food safety and health risks in China. *Environment International* **77** 5-15.

[3] Subandi M, Chaidir L, Nurjanah U 2016 Effectiveness of BPMC insecticide and suren leaf extract on brown planthopper (Nilaparvata lugens Stal.) and its natural enemy population on Ciherang rice variety *Jurnal Agrikultura* **27** (3) 160-166.
[4] Peshin R, Zhang WJ 2014 Integrated pest management and pesticide use. Springer: Integrated Pest Management - Pesticide Problems Vol **3**, Chapter 1.
[5] Akhdiya A, Wartono, Sulaeman E, Samudra IM 2018 Characterization of profenofos degrading bacteria. *Jurnal Agro Biogen* **14** (1) 37-46
[6] Weisburg WG, Barns SM, Pelletier DA, Lane DJ 1991 16S ribosomal DNA amplification for phylogenetic study *Journal of Bacteriology* 1991 **173**(2) 697-703.
[7] Thompson JD, Higgins DG, Gibson TJ 1994 Clustal W: improving the sensitivity of progressive multiple sequence aligment through sequence weighing, position gap penalties and weight matrix choice *Nucleic Acids Research* **22** 4673-4680.
[8] Dehgahi R, Subramaniam S, Zakaria L, Joniyas A, Firouzjahi FB, Haghnama K, Razinataj M 2015 Review of research on fungal pathogen attack and plant defense mechanism against pathogen *International Journal of Scientific Research in Agricultural Sciences* **2**(8) 197-208.
[9] Nahar K, Matsumoto I, Taguchi F, Inagaki Y, Yamamoto M, Toyoda K, Shiraishi T, Ichinose Y, Mukaihara T 2014 *Ralstonia solanacearum* type III secretion system effector Rip36 induces a hypersensitive response in the non-host wild eggplant *Solanum torvum Molecular Plant Pathology* **15**(3) 297–303.
[10] Hogg S 2005 Essential Microbiology UK John Wiley and Sons Ltd.
[11] Akhter MA, Laz R 2013 Isolation and molecular characterization of pesticide (fenitrothion) resistant bacteria from agricultural field *Journal of Pharmacy* **3**(5):31-38.
[12] Sinnott M L 2007 Carbohydrate chemistry and biochemistry: Structure and Mechanism. UK RSC Publishing.
[13] Fuentes S, Méndez V, Aguila P, Seeger M 2014 Bioremediation of petroleum hydrocarbons: catabolic genes, microbial communities, and applications *Applied Microbiology and Biotechnology* **98** (11): 4781–94.
[14] Sharma A, Sharma T, Sharma T, Sharma S, Kanwar SS 2019 Role of microbial hydrolases in bioremediation. *Microbes and Enzymes in Soil Health and Bioremediation* **16**: 149-164
[15] Karigar CS, Rao SS 2011 Role of microbial enzymes in the bioremediation of pollutants: a review *Enzyme Research*. 805187 11
[16] Hansen J, Mørk N, Bundgaard H 1992 Phenyl carbamates of amino acids as prodrug forms for protecting phenols against first-pass metabolism *International Journal of Pharmaceutics* **81** 253–61.
[17] Olowomofe TO, Oluyege JO, Aderiye BI, Oluwole OA 2019 Degradation of poly aromatic fractions of crude oil and detection of catabolic genes in hydrocarbon-degrading bacteria isolated from agbabu bitumen sediments in Ondo State *AIMS Microbiology* **5**(4) 308–323
[18] Dokic L, Narancic T, Nikodinivic-Runic J, Bajkic S, Vasiljevic B 2011 Four *Bacillus* spp. soil isolates capable of degrading phenol, toluene, biphenyl, naphtalene, and other aromatic compounds exhibit different aromatic catabolic potentials *Arcieve Biological Science Belgrade* **63**(4) 1057-1067
[19] Arora PJ 2020 *Bacilli*-Mediated Degradation of Xenobiotic Compounds and Heavy Metals *Frontiers Bioengineering and Biotechnology* **8** 570307
[20] Hao J, Liu J, Sun M 2014 Identification of a marine *Bacillus* strain C5 and parathion-methyl degradation characteristics of the extracellular esterase B1 BioMed Research International 863094 7
[21] Rahmansyah M, Sulistinah N 2009 Bacterial Perform in Soil Contaminated with Pesticide *Berita Biologi* **9** (5) 657–64.

IOP Conf. Series: Earth and Environmental Science **883** (2021) 012008 doi:10.1088/1755-1315/883/1/012008

Food consumption patterns among university students in Indonesia during the transition period in new Normal Era of Covid-19 Pandemic

A E Yunianto[1,*], K Kristiandi[2], M Darawati[3], T H Doloksaribu[4], I Anggraeni[5], M Pasambuna[6]

[1] Departement of Nutrition, Siliwangi University, Tasikmalaya, West Java Province, Indonesia
[2] Department of Food Agro-Industry, State Polytechnic of Sambas (Politeknik Negeri Sambas), Sambas, West Kalimantan Province, Indonesia
[3] Department of Nutrition, Health Polytechnic of Mataram (Poltekkes Kemenkes Mataram), Mataram, West Nusa Tenggara Province, Indonesia
[4] Department of Nutrition, Health Polytechnic of Medan (Poltekkes Kemenkes Medan), Medan, North Sumatera Province, Indonesia
[5] Department of Public Health, Mulawarman University, Samarinda, East Kalimantan Province, Indonesia
[6] Department of Nutrition, Health Polytechnic of Manado (Poltekkes Kemenkes Manado), Manado, North Sulawesi Province, Indonesia

* E-mail: andiekayunianto@gmail.com

Abstract. The Covid-19 outbreak that occurred in Indonesia affected all aspects of community life, including students, especially in food consumption patterns. Our study aims to investigate the differences in eating habits of health and non-health students during the transition period to the new normal era. This study used a cross-sectional design with a sample of 5,924 students. The questionnaire was in the form of an e-survey with a google form. The students' diet was obtained using a food frequency questionnaire (FFQ). The Mann Whitney test was used to analyze the differences in dietary patterns between health and non-health majors. There was a significant difference in student consumption patterns including daily eating habits ($p = 0.002$), consumption of animal side dishes ($p = 0.000$), vegetable consumption ($p = 0.000$), water consumption ($p = 0.000$), changes in eating frequency ($p = 0.013$) and changes in the amount of food consumed ($p = 0.008$). There were differences in consumption patterns between students majoring in health and non-health during the transition period. A good diet is needed in increasing immunity in students during the transition period.

1. Introduction

At the end of December 2019 in Hubei, China, several cases of pneumonia were found, and the cause was unknown [1]. The pathogen has been identified as a new RNA Betacoronavirus 2 and named as Severe Acute Respiratory Syndrome Coronavirus 2 (SARS-CoV-2), which bears phylogenetic similarities to SARS-CoV patients [2]. This disease has its main clinical symptoms, such as fever, dry cough, fatigue, myalgia, and pneumonia [3]. The World Health Organization (WHO) recently declared the Coronavirus Disease 2019 (COVID-19) a public health emergency of international concern. This disease spread throughout the world quickly. WHO declared that COVID-19 is a global pandemic with a devastating effect on death. Several cases of death occurred in several countries experienced a significant increase [4].

The COVID-19 pandemic spread rapidly in various countries, in developed and developing countries. According to WHO, the highest case over the world was in May when the United States hit a total of 1,477,516 cases. Indonesia ranks 36 out of 215 countries with 18,496 cases [4]. The COVID-19 epidemic is a public health emergency of international concern and has socio-economic effects in every country, including food consumption [5]. One of its significant effects is about food security that will affect the level of diet consumption [6]. Changes in food production and supply processes result in changes in diet [5]. Also, the implementation of lockdowns in various places has affected the lifestyles, especially their diet and physical activity [7,8].

During the pandemic, the community becomes more active in the house; this also occurred among college students. Previous studies have shown that most students at home have an impact on less physical activity and more sedentary activity [9]. This can be exacerbated by unhealthy eating habits that lead to weight gain and obesity. A study conducted by Yilmaz et al. (2020) shows that most students have significant changes in habits, such as in snack purchasing before and after the pandemic. Student food consumption has increased significantly and is not following the recommended balanced diet [10]. This study aims to investigate the food consumption patterns of college students in Indonesia during the transition in the new normal era of COVID-19 pandemic.

2. Methods

The design in this study was a cross-sectional study and was conducted in August 2020. This study involved 5,924 students from all provinces in Indonesia using the snowball sampling technique. This research has received ethical approval by the Health Research Ethics Committee (KEPK) of the Mataram Health Ministry of Health Poltekkes No. LB.01.03/1.1/2208/2020. The questionnaire was distributed with the google form application via social media, including Whatsapp, Facebook and Instagram. The pattern of student food consumption was obtained by the food frequency questionnaire (FFQ). Data analysis used Microsoft office 2010 and SPSS version 25 for windows. The Mann-Whitney different test was applied to analyze the differences in the food consumption patterns among students at $\alpha=0.05$.

3. Results and discussion

The total number of students involved was 5,924 people who filled out the questionnaire completely. Based on student characteristics, it showed that most of the respondents aged over 19 years, 69.9% were majoring in health, and 30.1% were majoring in non-health. Most respondents were female, and about 72.5% were majoring in health. Most respondents studied at public universities with an undergraduate level. During this transitional period, most students stay at their parents home. Their food expenditure was mostly Rp. 15,000 - 20,000 per day. Besides, non-food expenditure was higher than food expenditure, about Rp. 30,000 per day.

Table 1. Student characteristics

Characteristics	Health		Non – Health		Total	
	N	%	N	%	N	%
Age						
<19 years	1798	64.2	1001	35.8	2799	100
≥19 years	2184	69.9	941	30.1	3125	100
Sex						
Male	300	35.5	545	64.5	845	100
Female	3682	72.5	1397	27.5	5079	100
Institution status						
Public	3481	72.0	1351	28.0	4832	100

Private	501	45.9	591	54.1	1092	100
Level						
D3	2015	92.9	154	7.1	2169	100
D4	1018	83.0	208	17.0	1226	100
Undergraduate	949	37.5	1580	62.5	2529	100
Living with Parents						
Yes	3762	67.8	1786	32.2	5548	100
No	220	58.5	156	41.5	376	100
Food expenditure						
Rp <10.000	193	69.2	86	30.8	279	100
Rp 10.000-15.000	717	64.9	387	35.1	1104	100
Rp 15.000-20.000	874	65.9	452	34.1	1326	100
Rp 20.000-25.000	693	68.6	317	31.4	1010	100
Rp 25.000-30.000	720	68.7	328	31.3	1048	100
Rp >30.000	785	67.8	372	32.2	1157	100
Non-food expenditure						
Rp <10.000	316	60.9	203	39.1	519	100
Rp 10.000-15.000	762	61.2	483	38.8	1245	100
Rp 15.000-20.000	694	65.2	371	34.8	1065	100
Rp 20.000-25.000	588	70.5	246	29.5	834	100
Rp 25.000-30.000	549	72.8	205	27.2	754	100
Rp >30.000	1073	71.2	434	28.8	1507	100

Table 2 shows a significant difference in the dietary patterns of health and non-health majors. Significant differences occurred in student food consumption patterns, specifically daily consumption ($p = 0.002$), animal side dishes ($p = 0.000$), vegetables ($p = 0.000$), water (0.000), changes in eating frequency ($p = 0.013$), and changes in the amount of dietary ($p = 0.008$). However, significant difference in the student's diet of cereal consumption ($p = 0.098$), vegetable side dishes ($p = 0.697$), and fruit ($p = 0.973$) were not found.

Table 2. Differences in food consumption patterns of health and non-health students

Food pattern	Health		Non – Health		Total		P
	N	%	N	%	N	%	
Meals in a day							
1 time	128	53.6	111	46.4	239	100	
2-3 times	3636	67.8	1728	32.2	5364	100	0.002
>3 times	218	67.9	103	32.1	321	100	
Meals frequency changes							
Less often	655	65.6	343	34.4	998	100	
The same	2405	66.6	1208	33.4	3613	100	0.013
More often	922	70.2	391	29.8	1313	100	
Food amount changes							
Less	686	64.7	374	35.3	1060	100	
The same	2473	67.1	1215	32.9	3688	100	0.008
More	823	70.0	353	30.0	1176	100	
Cereals							
1 time	359	62.8	213	37.2	572	100	
2-3 times	3187	67.7	1519	32.3	4706	100	0.098
>3 times	436	67.5	210	32.5	646	100	
Animal-based protein source side dishes							
Never	34	65.4	18	34.6	52	100	0.000
1 time	705	56.1	552	43.9	1257	100	

2-3 times	2749	71.0	1123	29.0	3872	100	
>3 times	494	66.5	249	33.5	743	100	
Plant-based protein source side dishes							
Never	99	52.4	90	47.6	189	100	0.697
1 time	1062	65.3	565	34.7	1627	100	
2-3 times	2395	69.4	1056	30.6	3451	100	
>3 times	426	64.8	231	35.2	657	100	
Vegetables							
Never	84	46.4	97	53.6	181	100	0.000
1 time	709	55.2	576	44.8	1285	100	
2-3 times	2539	71.4	1016	28.6	3555	100	
>3 times	650	72.0	253	28.0	903	100	
Fruit							
Never	413	57.5	305	42.5	718	100	0.973
1 time	1836	65.8	953	34.2	2789	100	
2-3 times	1329	71.4	532	28.6	1861	100	
>3 times	404	72.7	152	27.3	556	100	
Water							
<8 glass	1960	64.6	1074	35.4	3034	100	0.000
≥8 glass	2022	70.0	868	30.0	2890	100	

Diet is a critical key to carrying out a healthy life [11]. During the pandemic, there are differences in the consumption of health and non-health majoring students. However, in general, health majors students have a better level of food consumption patterns than non-health majors students. COVID-19 pandemic affects people's daily lives, including the diet of students [12]. Commonly diet is 2-3 times a day, and it may increase more than three times a day, followed by changes in the eating frequency and the amount of food consumed, which showed significant differences. This is in line with previous research found that during the COVID-19 pandemic, there have been changes in diet, both food frequency and amount of food consumed [8].

There was no significant difference in cereal consumption among the two groups. This is different from previous studies which show that cereal consumption in the community will become higher [13]. Cereal described as a staple food, source of energy, so it is always mandatory and usually stored in the kitchen as a daily stock. Besides, it was followed by the legume consumption, which made some people improve their mood. Usually, some cereal products are used as a snack, i.e., bread, soup, and several desserts [5].

The consumption of animal-based side dishes in students had a significant difference between health and non-health students. Meanwhile, there was no significant difference in the consumption of plant-based protein. This means that health major students care more about their health by consuming higher protein than non-health majors students. Those since health students may have good knowledge, attitudes, and practices in nutritional matters [14,15]. Consumption of plant-based and animal-based side dishes is a source of protein. Protein has an anti-inflammatory effect, and high biological value let it improves immunity [16]. Therefore, it is important during a pandemic that protein consumption is highly recommended according to the needs.

Fruits and vegetables are a source of vitamins and minerals essential for the body's need [17,18]. In general, health major students had a better fruit and vegetable consumption level than non-health majors students. The results showed that vegetable consumption was significantly different from fruit consumption. Vegetables and fruit are natural sources of antioxidants needed to boost the immune system [19]. Vegetables and fruit have a variety of colours, an indicator of antioxidant content [20]. Food components contain very high antioxidants include vitamin C, vitamin E, and phytochemicals such as carotenoid and polyphenol [21]. Fruit functions as an anti-inflammatory and can reduce oxidative stress so it may strengthen immunity [21,22].

Water has an important role in metabolism, transport of substrates across membranes, cellular homeostasis, temperature regulation, and circulatory function. [23]. Water consumption among students has a significant difference between health and non-health majors students. However, in general, it already met the health recommendations, more than 8 cups or 1.5 L per day [24]. Water consumption is an indicator used to see a person's hydration level [25]. Adequacy of water consumed is one of the important during the COVID-19 epidemic. Previous studies have shown that less hydration level is one of the risks to the death of COVID-19 patients, followed by age, sex, race, and congenital chronic diseases. Chronic suboptimal hydration in the weeks before exposure to COVID-19 lead to a greater abundance of angiotensin-converting enzyme 2 (ACE2) receptors in the lungs, which increases the likelihood of COVID-19 infection [26].

4. Conclusion

The COVID-19 pandemic affected the daily food pattern among university students in Indonesia during the transition period in the new normal era. Significant differences occurred in student food consumption patterns include daily eating habits, consumption of animal side dishes, vegetables, and water. It also influenced the changes in eating frequency and the amount of food consumed. There were differences in consumption patterns between health and non-health majors students during the transition period. Also, a good diet is needed in increasing immunity in students during the transition period.

5. References

[1] Huang C, Wang Y, Li X, Ren L, Zhao J, Hu Y, et al. 2020 Clinical features of patients infected with 2019 novel coronavirus in Wuhan, China *Lancet* **395**:497–506. https://doi.org/10.1016/S0140-6736(20)30183-5.

[2] Zhu N, Zhang D, Wang W, Li X, Yang B, Song J, et al. 2019 A novel coronavirus from patients with pneumonia in China, *N Engl J Med* **382** 727–33. https://doi.org/10.1056/NEJMoa2001017.

[3] Zhi ZLXBXZ 2020 [The epidemiological characteristics of an outbreak of 2019 novel coronavirus diseases (COVID-19) in China] *Chin J Epidemiol* **2** 145–51. https://doi.org/10.3760/cma.j.issn.0254-6450.2020.02.003.

[4] World Health Organization. Coronavirus (COVID-19) 2020 Coronavirus (COVID-19). https://covid19.who.int/.

[5] Eftimov T, Popovski G, Petković M, Seljak BK, Kocev D 2020 COVID-19 pandemic changes the food consumption patterns*Trends Food Sci Technol* **104** 268–72. https://doi.org/10.1016/j.tifs.2020.08.017.

[6] Swinnen J, Mcdermott J 2020 Covid19 & Global Food Security.

[7] Ruiz-Roso MB, Knott-Torcal C, Matilla-Escalante DC, Garcimartín A, Sampedro-Nuñez MA, Dávalos A, et al. 2020 Covid-19 lockdown and changes of the dietary pattern and physical activity habits in a cohort of patients with type 2 diabetes mellitus *Nutrients* **12** 1–16. https://doi.org/10.3390/nu12082327.

[8] Di Renzo L, Gualtieri P, Pivari F, Soldati L, Attinà A, Cinelli G, et al. 2020 Eating habits and lifestyle changes during COVID-19 lockdown: An Italian survey *J Transl Med* **18** 1–15. https://doi.org/10.1186/s12967-020-02399-5.

[9] Barkley J E, Lepp A, Glickman E, Farnell G, Beiting J, Wiet R, et al. 2020 The Acute Effects of the COVID-19 Pandemic on Physical Activity and Sedentary Behavior in University Students and Employees *Int J Exerc Sci* **13**1326–39.

[10] Yilmaz HÖ, Aslan R, Unal C 2020 The Effect of the COVID-19 Outbreak on Eating Habits and Food Purchasing Behaviors of University Students *Kesmas Natl Public Heal J* **15** 154–9. https://doi.org/10.21109/kesmas.v15i3.3897.

[11] Rodríguez-Pérez C, Molina-Montes E, Verardo V, Artacho R, García-Villanova B, Guerra-Hernández EJ, et al. 2020 Changes in dietary behaviours during the COVID-19 outbreak confinement in the Spanish COVIDiet study *Nutrients* **12** 1–19. https://doi.org/10.3390/nu12061730.

[12] Husain W, Ashkanani F 2020 Does COVID-19 change dietary habits and lifestyle behaviours in Kuwait: A community-based cross-sectional study *Environ Health Prev Med* **25** 1–13.

https://doi.org/10.1186/s12199-020-00901-5.

[13] Laguna L, Fiszman S, Puerta P, Chaya C, Tárrega A 2020 Since January 2020 Elsevier has created a COVID-19 resource centre with free information in English and Mandarin on the novel coronavirus COVID- 19 . The COVID-19 resource centre is hosted on Elsevier Connect , the company ' s public news and information 2020.

[14] Özdoğan Y, Yardımcı H, Özçelik AÖ 2018 Assessment of Nutrition Knowledge among University Students in Ankara *J Sci Res Reports* **20** 1–8. https://doi.org/10.9734/jsrr/2018/43782.

[15] Hussein S, Naccache H, Sarout S 2018 Nutritional Knowledge, Conception and Behavior of Lebanese Students: a Comparative Study. *Int J Adv Res* **6** 755–63. https://doi.org/10.21474/ijar01/8196.

[16] Fernández-Quintela A, Milton-Laskibar I, Trepiana J, Gómez-Zorita S, Kajarabille N, Léniz A, et al. 2020 Key Aspects in Nutritional Management of COVID-19 Patients *J Clin Med* **9** 2589. https://doi.org/10.3390/jcm9082589.

[17] Slavin JU of M, Lloyd Beate PNGR 2012 Health Benefits Of Cassava-Karrapendalam. WwwYadtekCom › Heal › *Diet Nutr* **3** 506–16. https://doi.org/10.3945/an.112.002154.506.

[18] Marles R J 2017 Mineral nutrient composition of vegetables, fruits and grains: The context of reports of apparent historical declines *J Food Compos Anal* **56** 93–103. https://doi.org/10.1016/j.jfca.2016.11.012.

[19] A. Shetty A 2013 Vegetables as Sources of Antioxidants. *J Food Nutr Disord* **02** 1–5. https://doi.org/10.4172/2324-9323.1000104.

[20] Cömert ED, Mogol BA, Gökmen V 2020 Relationship between color and antioxidant capacity of fruits and vegetables *Curr Res Food Sci* **2** 1–10. https://doi.org/10.1016/j.crfs.2019.11.001.

[21] Iddir M, Brito A, Dingeo G, Del Campo SSF, Samouda H, La Frano MR, et al. 2020 Strengthening the immune system and reducing inflammation and oxidative stress through diet and nutrition: Considerations during the covid-19 crisis *Nutrients* **12** 1–43. https://doi.org/10.3390/nu12061562.

[22] Åsgård R, Rytter E, Basu S, Abramsson-Zetterberg L, Möller L, Vessby B 2007 High intake of fruit and vegetables is related to low oxidative stress and inflammation in a group of patients with type 2 diabetes. *Scand J Food Nutr* **51** 149–58. https://doi.org/10.1080/17482970701737285.

[23] Armstrong L, Johnson E 2018 Water Intake, Water Balance, and the Elusive Daily Water Requirement *Nutrients* **10**1928. https://doi.org/10.3390/nu10121928.

[24] Arnaud M J, Noakes T D 2011 Should humans be encouraged to drink water to excess. *Eur J Clin Nutr* **65** 875–6. https://doi.org/10.1038/ejcn.2011.40.

[25] Popkin B M, D'Anci K E, Rosenberg I H 2010 Water, hydration, and health *Nutr Rev* **68** 439–58. https://doi.org/10.1111/j.1753-4887.2010.00304.x.

[26] Stookey J D, Allu P K R, Chabas D, Pearce D, Lang F 2020 Since January 2020 Elsevier has created a COVID-19 resource centre with free information in English and Mandarin on the novel coronavirus COVID- 19 . The COVID-19 resource centre is hosted on Elsevier Connect , the company ' s public news and information 2020.

IOP Conf. Series: Earth and Environmental Science **883** (2021) 012009 doi:10.1088/1755-1315/883/1/012009

Species of pandan (*Pandanus* sp) in Gorom Island, East Seram Regency

A Sahupala[1,*], T E Siahaya[1], B B Seipala[1], L. Siahaya[1], L. Pelupessy[1], Y.D. Komul[1]

[1]Department of Forestry, Faculty of Agriculture, Pattimura University

* E-mail : ansahupala@gmail.com

Abstract The purpose of this study was to determine the Species of Pandanus (*Pandanus sp*) in Gorom Island, East Seram Regency. The object used in this study was the external body structure of pandanus plants (pandanus sp) such as fruit, flowers, leaves, stems and roots. This research was conducted directly by direct survey in the field and observing the species of pandanus (Pandanus sp), this research includes observations on fruit, flowers, leaves, stems, and roots. Based on the results of the study found 6 species of pandan (Pandanus sp) that is : Pandan Duri, Pandan Mengkuwang, Pandan Kowang, Pandan Sepajam, Pandan Wangi dan Pandan Laut. Of these 6 species we can distinguish from the shape of the leaves, stem shapes, fruit shapes, and root shapes.

1.Introduction

Non-timber forest products (HHBK) have always played an important and large role in the forestry economy in developing countries [1]. No exception in Indonesia. This is inseparable from the many types of NTFPs that can be obtained from forests, both those derived from NTFP plants and from the use of NTFPs in general that meet the needs for food, energy and NTFP products that have become input as well as direct income for meeting the needs of many households and communities around the world [15].

In many countries, the total economic value of NTFPs is estimated to be able to contribute to state revenues as large, maybe even more, than that obtained from logs. in Indonesia alone, the economic value of NTFPs is estimated at 90% of the total economic value that can be generated from forest ecosystems [6]. So far, NTFPs seem to be underestimated and are only considered as forest products. However, in Indonesia, there is quite a lot of plant diversity available that can be used as raw materials for the handicraft industry, including plaiting. To produce woven products from plant materials, knowledge and experience are needed in knowing plants that have long and strong fibers. One of the varieties of plants that meet these two requirements is pandanus, which is a member of the pandan-pandanan (Pandanaceae) tribe, especially from the Pandanus clan.

The types of the Pandanus clan are members of the Pandanaceae with the most extensive distribution and range of habitats they occupy. These plants can be found ranging from sandy beaches to highland forests with an altitude of about 3500 m above sea level and ranging from secondary forests and grasslands with various soil patterns ranging from humid fertile wetlands, limestone, peat swamps to relatively dry and relatively dry sandy soils. poor in nutrients [10]. Several types of pandanus, namely Pandanus: P.tectorius, P.bidur, P.furcatus, the leaves produce plaits, pandanus mats, bags, or food wrapping materials: P.amaryllifolius (fragrant pandanus), as a hair oil fragrance, scent certain dishes (pastries); P.edulis, the fruit is edible.

On Gorom Island, which is a small island, it also has a place to grow pandanus species. These types of pandanus are only found in Mamasa Hamlet and are also used by the community as a woven material, food fragrance. Until now there have not been many researchers on the types of pandanus, based on the description above, the aim of the study was to analyze "Types of Pandanus (Pandanus sp) in Gorom Island, East Seram Regency".

2. Methods

The research was conducted in Gorom Island, East Seram District, Wawasa Village. The research was carried out from July to August 2020. The object used in this study was the structure of the outer body of the pandanus plant (Pandanus sp) such as leaves, fruit, flowers, roots, stems. The types of data used in this study are primary data and secondary data. Primary data is data obtained from the use of pandan leaves by the community for management techniques. Secondary data, namely as supporting data related to the types of pandanus.

The data collection methods used in this research are: Field survey method is carried out directly by observing the species in the growing location of Pandanus (Pandanus sp), including:

A. Leaves consisting of: leaf position, leaf base, leaf surface and leaf bone
B. Trunk
C. Interest
D. Fruit
E. Roots

Observation Plot

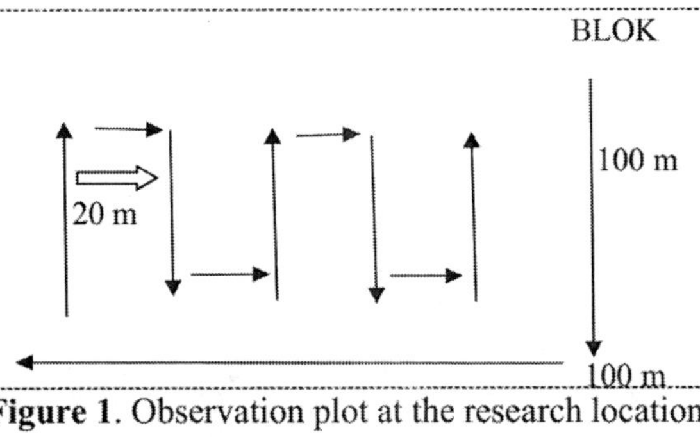

Figure 1. Observation plot at the research location

Description : = ↑ : length of the line of observation
⇨ : Distance Between Lines

3. Results and discussion

3.1. Description of the types of pandanus (Pandanus sp)

Pandanus (Pandanus sp) is a group of monocot plants with the genus Pandanus. Most of the pandanus grows at an altitude of 1500 to 3500 masl on the coasts of the tropics [11]. Usually the most prominent thing about this difference can be seen from the leaves of this plant is characterized by elongated leaves (such as palm fronds or grass), often jagged edges. The roots are large and have supporting roots that support this plant. Pandanus fruit are arranged in round arrangements, like durian fruit, but there are only a few types that have fruit. The size of this plant varies, from 50 cm to 5 meters. The greatest benefit of the equivalent plant (Pandanus sp) is the leaves. There are 6 types of pandan that are known in the community of Wawasa Village, namely: emar or pandan duri (Pandanus tectorius), hodan or pandan mengkuwang (Pandanus artocarpus), iro'ut or pandan kowang (Pandanus furcatus), gogit or sharp pandanus (Pandanus bicornis) , pandan leaves or fragrant pandanus (Pandanus amaryllifolius), and large emar or sea pandanus (Pandanus tectorius), [11].

3.1.1. Types of Pandan Duri (Pandanus tectorius)

This type of pandanus does not have flowers or fruit, it lives widely in open areas in the lowlands and is usually used by the community around the research location to become a raw material for plaiting only when the raw material for pandan kowang starts to run out or is lacking. Pandanus is measuring 4-14 m high and usually grows at an altitude of 15 to 100 meters above sea level, and produces 10-300 leaves per stem per year, [11]. This pandanus has many green leafy branches with a length of 75-90 cm and a width of 4-5 cm, [8].

Figure 2. Types of Pandan Duri *(Pandanus tectorius)*

Based on the results of observations made on Pandan Duri, it has the following characteristics:

A. Leaves

Single leaf type, the position of the leaves in neat circular rows, build leaves or leaves an oval shape to a pointed tip, the base of the leaf is round and the tip of the leaf is tapered slightly rounded. The edges or edges of the leaves are spiny parallel, the surface of the leaves is smooth and uneven, the leaves are parallel and in a straight line, the leaves are parallel and in a straight direction (Figure 2).

Figure 3. Shape of the base and tip of the leaves of pandanus spines *(Pandanus tectorius)*

B. Stem

The shape of the stem is short above the ground (Figure 3)

Figure 4. Pandanus tectorius leaf stem shape

C. Interest does not exist

D. No fruit

E. Roots

The roots are long supporting to the collection of leaves and have spines on the outside of the Root bark (Figure 4).

Figure 5. Root Form of Pandanus Duri (*Pandanus tectorius*)

3.1.2. Types of Pandan Kowang (Pandanus furcatus)

This type of pandanus does not have flowers or fruit, this pandanus grows in the lowlands to the highlands and is usually in rather shady places in the middle of the forest above 100-300 masl. This pandanus has a stem height reaching ± 11 m, erect, dirty white, has a single green leaf and lenset-shaped with pointed leaf tips and spiny leaf edges on the sides of the leaves and behind the leaves and the length of the leaves reaches 90-150 cm and 3- 5 cm, [13].

Figure 6. Pandan Kowang *(Pandanus furcatus)*

Based on the results of observations made on pandan kowang it has the following characteristics:

A. Leaves

- Single leaf type, the position of the leaves or the arrangement of the leaves in a row upward, the shape or shape of the leaves is oval to a sharp tip, the base of the leaf is rounded, the position of the leaves or the arrangement of the leaves is arranged in a straight line (Figure 6), the edges of the leaf edges are spiny and the leaf surface is wavy, bone the leaves are aligned straight and direction (Figure 7).

Figure 7. Shape of the base and tip of the leaves of Pandan Kowang (*Pandanus furcatus*)

Shape

The shape of the stem is not perpendicular but sideways (Figure 7).

Figure 8. Stem Shape in Pandan Kowang (*Pandanus furcatus*)

Interest does not exist
B. Fruit does not exist
C. Roots
The roots are elongated and attached to the stem and are smooth and not too long big.

3.1.3. Jenis Pandan Sepajam (Pandanus bicornis)
In the research location this type of pandanus species has fruit and lives in the lowlands from 15 to 90 meters above sea level, and is usually widely distributed near the sea along with the types of plants that grow on the coast, which are found in high locations of pandanus this sharp reaching 100 -120 cm with leaves 65-90 cm long, 2- 5 cm wide, and usually the leaves are used by the research location community as raw material for spare weaving or as a substitute for pandan mengkuwang but the product is not suitable and the color is dirty green so the people around the research location do not like to use the leaves as a woven material, [13].

Figure 9. Pandan Sepajam *(Pandanus bicornis)*

Based on the results of observations made on pandan mengkuwang it has the following characteristics:
A. Leaves
The leaf edges are arranged in rows, the position of the leaves is arranged according to the branching, the base and tip of the leaves are rounded, the tips of the leaves are tapered, the surface of the leaves is smooth

and wavy, the edges of the leaves are not spiny and smooth, the leaves are smooth and not spiny, the bones of the leaves are straight in the form of fine threads following the leaves (Figure 9).

Figure 10. Shape of the base, tip and surface of pandanus leaves as sharp *(Pandanus bicornis)*

B. Trunk
The shape of the stem is circular and the diameter is large.
C. Interest was not found at the time of the study
D. Fruit
Single upright fruit shape and rounded egg (Figure 10).

Figure 11. Fruit Shape Pandan Sepajam *(Pandanus bicornis)*

D. Roots
The nature of the roots is quite large in length and diameter and also has spines on the outside, the type of roots is branched long and has a large diameter (Figure 11).

Figure 12. The Root Shape of Pandanus Sepajam (*Pandanus bicornis*)

3.1.4. Jenis Pandan Mengkuwang (Pandanus artocarpus)

This type of mengkuwang pandanus is found in the research location, mostly in low to high plains, especially in areas near the sea, does not have flowers or fruit. This mengkuwang pandanus grows up to 15-150 meters above sea level, and grows together with other types on the coast, with leaves reaching 90-150 cm in length, 2-5cm in width, [9]. Why is this pandanus used by the community in the research location because pandan mengkuwang is very strong and its long leaves can produce good quality products and long-lasting kaut, [1].

Figure 13. Pandan Mengkuwang *(Pandanus artocarpus)*

Based on the results of observations made on pandan duri has the following characteristics:

A. Leaves

The leaf type is single, the position of the leaves is tightly arranged and dense, the base of the leaves is wide and the tip of the leaf is tapered, the surface of the leaves is smooth and wavy, the edges of the leaf edges are not spiny and smooth and the leaves are smooth without spines, straight leaf bones and the shape of halur threads follow the leaves

B. Trunk

Figure 14. Stem Shape Pandan Mengkuwang *(Pandanus artocarpus)*

C. Interest does not exist
D. No fruit
E. Roots

The shape of the roots is smooth, not too long and not too big (Figure 14)

Figure 15. Root form Pandan Mengkuwang *(Pandanus artocarpus)*

3.1.5. Types of Fragrant Pandanus (Pandanus amaryllioliu s)

This type of pandanus does not have flowers or fruit, and is not used as a woven material. It grows in the research location at an altitude of 15 to 80 meters above sea level, with a height of no more than 1.5 cm, [8]. The leaf length reaches 80-110 cm and the leaves are 3-6 cm wide. This plant is developed by cutting the leaves, this pandanus is a pandanus with fragrant leaves only used for cooking, [8].

Figure 16. Pandan Wangi *(Pandanus amaryllifolius)*

Based on the results of observations made on fragrant pandanus, it has the following characteristics:
A. Leaves
The leaves are single and neatly arranged, the shape of the leaves is long, the base of the leaf is round and the tip of the leaf is long, the edge of the leaf or leaf type does not have thorns but only at the end of the leaf there are small thorns, the surface of the leaves is wavy and the top is shiny, the leaf bones are long and straight along the length of the leaf.
B. Trunk
The shape of the stem is rounded but not too high just above the ground.
C. Interest does not exist
D. Fruit does not exist
E. Roots
The nature of the roots is supportive and some come out around the base of the trunk and branches, the type of roo finely branched on the stem (Figure 16).

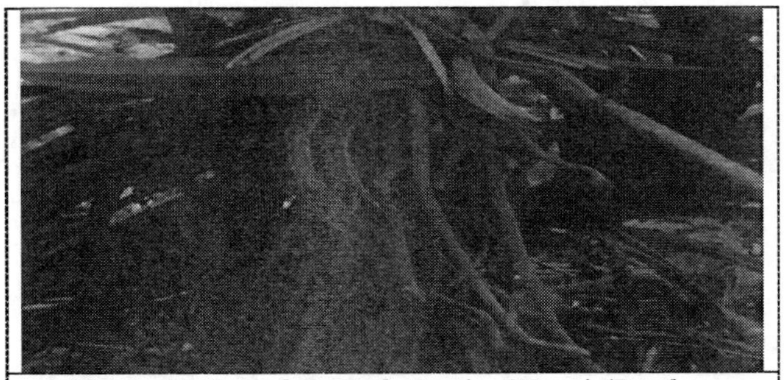

Figure 17. Root form Pada Pandan Wangi *(Pandanus amaryllifolius)*

3.1.6. Type of Sea Pandan (Pandanus odorifer)
In general, what can be found in the research location, this type of sea pandanus grows above an altitude of 15-30 dpml. 125 meters, with a trunk that does not have thorns, grows in the form of a tree or shrub, has a lot of stems, with supporting roots around the base of the trunk and aerial roots from the branches. The length of the leaves is 70-100cm, and the width of the leaves is 3-5cm, with triangular and sharp

edges. It has a fruit where the characteristics of this fruit are a hanging compound fruit in the shape of a ball. The fruit is 2-5 cm in size, the fruit will be yellow orange when ripe, the center of the fruit wall is fibrous with lots of air space, usually lives wild in sandy forests, this pandanus thorn has long been known by coastal communities as a large and large pandanus, 8].

Figure 18. Pandan Laut *(Pandanus odorifer)*

Based on the results of research conducted on sea pandanus, it has the following characteristics:
A. Leaves
Single leaf type, the position of the leaves or the arrangement of the leaves in a row following the branching, build or shape the leaves are oval to sharp edges, the base of the leaves is round and the tip of the leaf tapers slightly, the edges of the leaves are spiny parallel, the leaf surface is smooth and uneven, the leaf bones are parallel to the length of the leaf.
B. Trunk
The shape of the trunk is large and has thorns on the bark
C. Interest
D. Fruit
The shape of the fruit is round
E. Roots
Types of supporting roots (Figure 18).

Figure 19. Root Shape Pandan Laut *(Pandanus odorifer)*

IOP Conf. Series: Earth and Environmental Science **883** (2021) 012009 doi:10.1088/1755-1315/883/1/012009

4. Conclusion

Based on the results of research conducted on Gorom Island, East Seram Regency, it can be concluded that:

1. There are 6 (six) types of pandanus (pandanus sp), namely: emar or pandan duri (*Pandanus tectorius*), iro'ut or pandan kowang (*Pandanus furcatus*), hodan or pandan mengkuwang (*Pandanus artocarpus*), Gogit or Pandan Sepajam (*Pandanus bicornis*), pandan leaves or fragrant pandan wangi (*Pandanus amaryllifolius*), and large emar or pandan laut (*Pandanus odorifer*). Among the 6 (six) types of pandanus (pandanus sp), there are 4 (four) types of pandan that are utilized, namely:

Iro'ut or Pandan Kowang (*Pandanus furcatus*), Hodan or Pandan Mengkuwang (*Pandanus artocarpus*), Pandan Leaf or Pandan Wangi (*Pandanus amaryllifolius*). and Pandan Duri (*Pandanus tectorius*).

5. Acknowledgment

Further research is needed on the utilization and economic value of pandanus (*Pandanus* sp)

References

[1] Arnold J E M 2004 Forest Product Livelihoods and Concervation Case studies of non-timber forest product systems (Kustera, K. And B. Belcher end.) vol 1.

[2] Anonymous 1991 *A Field Guide to Pandanus in New Guinea* The Bismarck Archipelago and the Solomon Islands Publication No 16 of the Christensen Research Institute Madang Papua New Guinea

[3] Anonim 2008c *Pandanus furcatus*. http://id.wikipedia.org/wiki/ Pandanus furcatus. (16 juli 2008) https://id.m.wikipedia. Org > wiki > pandan, diakses pada 16 – 12 – 2018 https://brainly. Teknik pengelolan daun pandan Brainly.co.id > biologi di akses pada 16 – 12 – 2018 https://www.yeastgenome. Org > reference diakses pada16 – 12 – 2018 https://trove. Nla. Gov. Au > Work / Version / 26034525 diakses 16 – 12 - 2018-12-16

[4] Kadri 1982 *UU Kehutanan No 41/1982 Pemerintah RI, Departemen Kehutanan dan perkebunan RI*, Jakarta

[5] Lawrence G H M 1959 *Taxonomy of Vascular Plant* New York The Macmillan Co

[6] Peraturan Menteri Kehutanan RI Nomor P 21/Menhut-II/2009 tentang *Strategi Pengembangan Hutan Bukan Kayu Nasional* tanggal 19 Maret 2009

[7] Prayitno T A 2007 *Peningkatan Nilai Tambah Hasil Hutan Bukan Kayu Melalui Pendekatan Teknologi* Makalah Workshop HHBK Tidak diterbitkan

[8] Purseglove 1972 *Tropical Crops Monocotyledons Vol 1 and 2* (Terjemahan) Jurusan Biologi Universitas Jember Jawa Timur

[9] Purseglove J W 1972 *Tropical Crops Monocotyledons 2* London Longman Group Limited

[10] Santos J 1998 *Pandanus Tectorius* http:/www.geocities.com/The Tropics/Cabana/4705/angio/Pandanus.html. Diakses pada tanggal 20 September 2015

[11] Stone B C 1982 *New Guinea Pandanaceae: First approach to ecology and biogeography* In Biogeography and Ecologgy of New Guinea. Vol. 1. Monographiae Biologice 42

[12] Thomson L A J L. Englberger L. Guarino L Thaman R R Elevitch CR 2006 *Species Profiles for Pacific Island Agroforesty Pandanus tectorius(Pandanus)* http//www.traditionaltee.org 17 Maret 2008

[13] Ulaten E Harpin Tahun 2018 *Identifikasi Jenis Damar (Agathis sp) Di Desa Hunitetu Kecamatan Inamosol Kabupaten Seram Bagian Barat* Skripsi

[14] Widjaja EA *et al* 1989 *Tumbuhan Anyaman Indonesia* Jakarta PT Melton Putra

[15] Wagner W L Herbst D R & Sohmer S H 1990 *Manual of the flowering plants of hawai'i.*

IOP Conf. Series: Earth and Environmental Science **883** (2021) 012010 doi:10.1088/1755-1315/883/1/012010

Estimation rumen degradable protein of local feeds in dairy cattle using *in sacco* method

A Rosmalia[1], I G Permana[1,*], Despal[1] and R Zahera[1]

[1] Department of Nutrition and Feed Technology, Faculty of Animal Science, IPB University, Bogor 16680, Indonesia

[*]E-mail: permana@apps.ipb.ac.id

Abstract. There is a lack of nutrition information on local feeds protein in Indonesia, especially to determine protein fraction of dairy feed. The objective of this study was to determine rumen degradable protein (RDP) of local feeds in dairy cattle using *in sacco* method. The local feeds are copra meal, palm kernel meal, coffee husk, tofu waste, soy-sauce waste, brewer waste, and habbatussauda waste. Two ruminal fistulated male Frisian Holstein were used to determine rumen degradable protein using *in sacco* method. The parameters observed include the estimated kinetic parameters, effective degradability (ED), and RDP. The parameters were evaluated with analysis of variance using SAS University software. The result showed that tofu waste and habbatussauda waste had high potential rapid degradation of CP following by brewer waste, palm kernel waste, soy-sauce waste, copra meal, and coffee husk. Habbatussauda waste had higher RDP followed by brewer waste, tofu waste, copra meal, palm kernel meal, soy-sauce waste, and coffee husk. There was a positive correlation between RDP and crude protein content, and a negative correlation between RDP and crude fiber. It is concluded that local feeds have various characteristics of quality feed protein, which is shown by rumen degradable protein and rumen undegradable protein.

1. Introduction

Protein is one of the most expensive nutrient components. The amount of protein needs to be considered according to livestock requirements so that protein can be used optimally and efficiently. The requirement of protein for ruminant is unique because of the rumen. Feed protein in the rumen will be degraded into ammonia and it will be used to form microbial protein synthesis. The other protein fraction that is not degraded will bypass from rumen (rumen undegradable protein). Rumen degradable protein (RDP) is required for ruminal fermentation to provide peptides, free amino acids, and ammonia for the synthesis of microbial protein. However, rumen undegradable protein (RUP) provides a direct source of amino acid that is digested in the intestine of the animal.

Few methods are developed to estimate RDP and RUP of dairy cattle feed, suc as *in sacco* method, *in vitro* enzymatic method, *in vitro* chemical method, and *in vitro* multi-chemical method [1]. *In sacco* or *in situ* method is commonly used because it can be simulated the ruminal environment [2]. The method consists of incubations of feed samples in nylon bags in the rumen for various times. Variation among feed samples, animals, and laboratories can be observed with this method [3].

Dairy cattle in Indonesia is commonly fed concentrate of local feed as a protein source. There is a lack of nutrition information on local feeds protein in dairy cattle, especially to determine protein fraction of feed. The objective of this study was to determined rumen degradable protein (RDP) of local feeds in dairy cattle using *in sacco* method.

2. Methods

2.1. Location and samples preparation

This study was conducted from September to October 2020 at the Laboratory of Dairy Nutrition, Faculty of Animal Science, IPB University, Indonesia.

Local feed proteins used in this study were copra meal, palm kernel meal, coffee husk, tofu waste, soy-sauce waste, brewer waste, and habbatussauda waste. All local feeds were dried at 60 °C and ground through a 2-mm screen prior to analysis. Nutrient content of local feeds (dry matter, ash, crude protein, crude fat, crude fiber, and nitrogen free extract) were analyzed using NIRs near-infrared spectroscopy [4].

2.2. In sacco procedures

In sacco method was used to determine rumen degradable protein (RDP) and rumen undegradable protein (RUP) [5]. This study was conducted using two ruminal fistulated male Frisian Holstein with average body weight 519±80,61 kg. Dairy cattle fed twice daily at 7.00 a.m and 3.00 p.m. Diets contained 60% Napier grass and 40% concentrate mix on a DM basis. Nylon bags (ANKOM, 5 x 10 cm with ±50 micron porosity) containing 5 g of each local feed were placed in triplicate in the rumen for 0, 3, 6, 9, 12, 15, 24, and 48 h. The nylon bags were placed into the rumen before morning feeding. For 0 h incubation, the nylon bags were only rinsed under tap water. After incubation, the nylon bags were removed from the rumen and washed in the tap water until clear. Then, the nylon bags were dried at 60 °C for 2 days and weighed. The residual samples at each incubation time were taken for crude protein analysis (CP) with the Kjehdal method [6].

Ruminal degradabilities of CP or the kinetic parameters were estimated by fitting the data to the exponential equation [7] (CP versus time) (Eq. 1):

$$y = a + b\,(1 - e^{-ct}) \qquad (1)$$

where y is CP disappearance in the rumen (%) at time t, a is the soluble fraction (%), b is the insoluble fraction but a potentially degradable fraction (%), c is the degradation rate constant of the b fraction (%/h), t is the time of incubation degradation time.

The effective degradation of CP was calculated using the formula [8] (Eq.2) by assuming a rumen outflow rate of 6%/h.

$$ED = a + \frac{bc}{(k+c)} \qquad (2)$$

where (a), (b), and (c) are the same as in Eq.1 and k is the rumen outflow rate. The RDP and RUP values (% CP) were calculated [1] with the following equation (Eq. 3 and 4):

$$RDP = a + b\left[\frac{c}{c+k}\right] \qquad (3)$$

$$RUP = 100 - RDP \qquad (4)$$

where (a), (b), and (c) are the same as in Eq.1, and (k) is the same as in Eq.2. The RDP value was correlated among each nutrient content of local feeds.

2.3. Statistical analysis

One way analysis of variance (ANOVA) was performed using SAS University to evaluate significant differences in the means of all parameters among samples. The animals were treated as blocks. The model was described by [9] and if it was significantly different ($P<0.05$) further tested by Duncan test.

3. Results and discussion

3.1. Nutrient content and protein degradation

The nutrient content of local feeds is present in Table 1. The average dry matter is 93.68%±3.79. The crude protein content of local feeds ranges from 10.08%-27.64%. Habbatussauda waste and brewer waste had the highest crude protein, and coffee husk had the lowest.

IOP Conf. Series: Earth and Environmental Science **883** (2021) 012010 doi:10.1088/1755-1315/883/1/012010

Table 1. The nutrient content of local feeds

Feed	Moisture	DM	Ash	Crude Protein	Crude Lipid	Crude Fiber	NFE
				%			
Copra meal	6.22	93.78	11.11	18.91	7.36	11.09	45.31
Palm kernel meal	1.78	98.22	5.14	12.12	10.49	16.35	54.12
Coffee husk	13.4	86.6	5.97	10.08	0.16	25.08	45.31
Tofu waste	5.49	94.51	6.84	18.35	9.72	14.00	45.60
Soy-sauce waste	4.33	95.67	23.39	21.16	17.87	21.95	11.30
Brewer waste	4.19	95.81	12.30	27.43	12.07	18.92	25.09
Habbatussauda waste	8.85	91.15	8.10	27.64	6.79	3.8	53.07

DM: dry matter; NFE: nitrogen-free extract

The data in Table 2 showed CP degradation of local feed. Ruminal degradabilities of CP local feeds protein were significantly different affected by feed, incubation time, and feed x incubation time ($P<0.01$). Crude protein degradation increased following by incubation time. Degradabilities of CP local feeds are various. Figure 1. showed trends of CP degradation of local feeds protein. The graph shows habbatussauda waste had rapidly degraded, in contrast, coffee husk had slowly degraded. Habbatussauda waste has a lower level of crude fiber, also higher content of nitrogen-free extract. Nitrogen-free extract represents non-nitrogen substances and the soluble carbohydrate included sugars, starches, fructans, galactans, pectins, beta-glucans, and organic acid. Pectin is a rapidly degraded complex carbohydrate and is almost completely degraded in the rumen [10]. Meanwhile, coffee husk has higher crude fiber that is dominated with lignin so difficult to be degraded in the rumen [11]. Lignin has the cross-linking with the other cell wall polysaccharides that can limit CP degradation [12].

Table 2. Degradability of crude protein (CP loss) local feeds

Incubation time (h)	Degradability (%)							P-value		
	Copra meal	Palm kernel meal	Coffee husk	Tofu waste	Soy-sauce waste	Brewer waste	Habbatussauda waste	Feed	Time	Feed x time
0	25.04	23.01	0.00	27.45	10.73	38.43	86.30	<0.0001	<0.0001	<0.0001
3	40.23	41.00	15.09	37.66	24.28	62.94	95.35			
6	44.96	45.20	15.66	45.31	37.68	71.54	96.39			
9	62.08	46.76	20.05	51.04	49.42	77.48	97.43			
12	59.83	64.23	24.84	53.96	66.46	79.70	97.88			
15	60.17	66.36	-6.58	61.96	65.04	82.32	98.31			
24	70.76	65.78	19.64	80.74	80.02	86.23	98.60			
48	78.35	85.61	15.56	94.12	80.49	91.17	98.72			

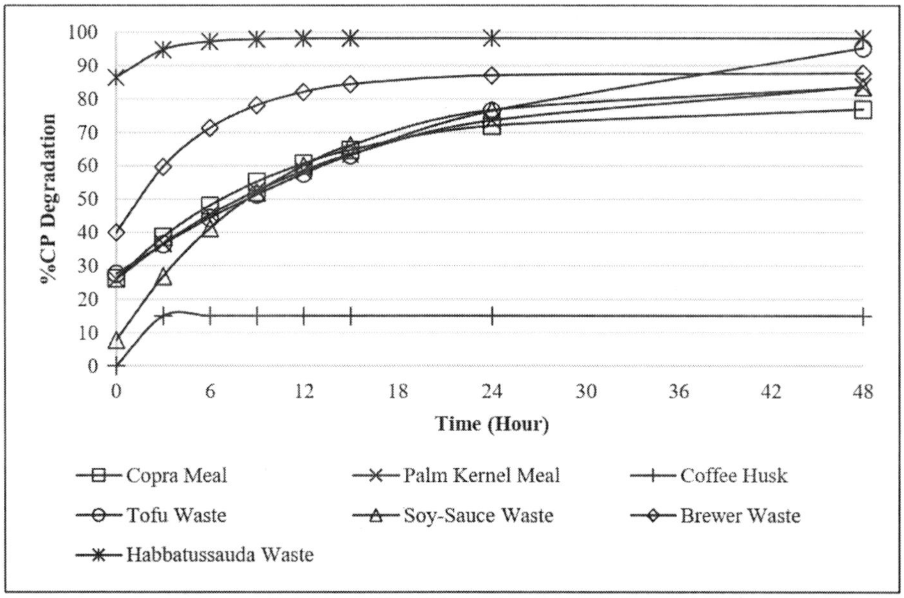

Figure 1. Crude protein degradation of local feeds protein

The result of estimation kinetic parameters and effective degradation (ED) are presented in Table 4. The data showed that soluble fraction (a), potentially degradable fraction (b), potential degradation (a+b) are significantly different affected by feed ($P<0.001$), meanwhile the rate of degradation (c) had no significant ($P>0.05$). The potential degradation is "a+b", which is a summary of soluble material and the potential for material to be degraded. The potential degradation (a+b) CP was higher for tofu waste and habbatussauda, intermediate for brewer waste, palm kernel meal, soy-sauce waste, copra meal, and lower for coffee husk. The calculated ED of local feeds protein shows various ED values with a range from 14.38%-96.70%. Habbatussauda waste the greatest ED value among other samples local feed protein ($P<0.01$).

Table 3. Estimated of kinetic parameters and effective degradation of CP

Parameters	Kinetic parameters (%)						
	Copra meal	Palm kernel meal	Coffee husk	Tofu waste	Soy-sauce waste	Brewer waste	Habbatussauda waste
a (%)	26.23^{c}	26.00^{c}	0.00^{d}	27.53^{c}	7.79^{d}	39.92^{b}	86.40^{a}
b (%)	51.26^{c}	60.44^{b}	15.04^{e}	79.69^{a}	76.53^{a}	47.75^{c}	11.82^{e}
c (%/h)	0.09	0.06	3.66	0.04	0.10	0.18	0.41
a+b (%)	77.49^{b}	86.44^{b}	15.04^{c}	107.22^{a}	84.32^{b}	87.67^{b}	98.22^{a}
ED (%)	57.22^{c}	57.04^{c}	14.38^{d}	59.19^{c}	54.82^{c}	76.62^{b}	96.70^{a}

a: soluble fraction; b: potentially degradable fraction; a+b: summary of a and b values; c: rate of degradation; ED: effective degradation.

3.2. Rumen degradable protein

Data in Table 4 shows the estimated RDP and RUP for local feeds protein. RDP and RUP values are significantly different affected by feed ($P<0.01$). Habbatussauda waste is a local feed protein that had the highest RDP value (96.70%), following by brewer waste, tofu waste, copra meal, palm kernel meal, soy-sauce waste, and coffee husk. RUP value is calculated as 100 minus RDP. In contrast, coffee husk had the highest RUP value (85.62%), following by soy-sauce waste, palm kernel meal, copra meal, tofu waste, brewer waste, and habbatussauda waste.

Table 4. Estimated of rumen degradable protein and rumen undegradable protein

Parameters	Copra meal	Palm kernel meal	Coffee husk	Tofu waste	Soy-sauce waste	Brewer waste	Habbatussauda waste
CP (%)	18.91	12.12	10.08	18.35	21.16	27.43	27.64
				100% CP			
RDP (%)	57.22^{c}	57.04^{c}	14.38^{d}	59.19^{c}	54.82^{c}	76.62^{b}	96.70^{a}
RUP (%)	42.78^{b}	42.96^{b}	85.62^{a}	40.81^{b}	45.18^{b}	24.38^{c}	3.30^{d}
				% of CP			
RDP (%)	10.82	6.91	1.45	13.59	11.60	20.74	26.73
RUP (%)	8.09	5.21	8.63	4.76	9.56	6.69	0.91

CP: crude protein; RDP: rumen degradable protein; RUP: rumen undegradable protein.

The calculated RDP and RUP with the percentage of CP show habbatusauda waste had higher RDP value compare to other local feed. Soy-sauce waste had a higher RUP value (9.56% of CP), followed by coffee husk, copra meal, brewer waste, palm kernel meal, and habbatussauda waste. This result differs from 100%CP because the calculation depends on the CP content of local feeds protein. There was a positive correlation between RDP and crude protein content (r=0.85604). It shows that the high protein content of the feed will increase RDP value. In contrast, there was a negative correlation between RDP and crude fiber (r=-0.78548). High crude fiber content of feed will decrease RDP value and increase RUP value. It can be seen, coffee husk and soy-sauce waste had a high of crude fiber (25.08% and 21.95%), so the RDP values (100%CP) are lower than other local feeds.

4. Conclusion

It is concluded that local feeds have various characteristics of quality feed protein, which showed by rumen degradable protein and rumen undegradable protein.

5. Acknowledgment

The authors would like to thank the Directorate of Higher Education, the Ministry of Education and Culture, Republic of Indonesia, and IPB University for funding this research through the scheme PMDSU Grant.

References

[1] Schwab CG, Tylutki TP, Ordway RS, Sheaffer C, Stern MD 2003 Characterization of proteins in feeds. *J. Dairy Sci.* **86**(1) 86–103

[2] Nocek JE 1988 In situ and other methods to estimate ruminal protein and energy digestibility: A Review *J. Dairy Sci.* **71**(8) 2051–69

[3] Offner A, Bach A, Sauvant D 2003 Quantitative review of in situ starch degradation in the rumen *Anim. Feed Sci. Technol.* **106**(1–4) 81–93

[4] Lengkey LCEC, Budiastra IW, Seminar KB, Purwoko BS 2020 Model pendugaan kandungan air, lemak dan asam lemak bebas pada tiga provenan biji jarak pagar (*Jatropha curcas L.*) menggunakan spektroskopi inframerah dekat dengan metode partial least square (PLS) *J. Penelit. Tanam. Ind.* **19**(4) 203

[5] National Research Council 2001 *Requirements of Dairy Cattle Seventh Revised Edition* (Washington DC: National Academy Press) p 61-62

[6] AOAC 2005 *Official Methods of Analysis Of AOAC International Ed. Ke-18* (Arlington: Association of Official Analytical Chemical)

[7] Orskov ER and Mcdonald I 1979 The estimation of protein degradability in the rumen from incubation measurements weighted according to rate of passage *J. Agric Sci.* **92**(2) 499–503

[8] Agricultural Research Council 1980 *The Nutrient Requirements of Ruminant Livestock* (Farnham Royal: Agricultural Research Council Working Party, Common Wealth Agricultural Bureau)

[9] Falahatizow J, Mesgaran DM, Valiki A, Tahmasbi A, Nazari M 2015 The estimation of ruminal

protein degradation parameters of various feeds using in vitro modified gas production technique *Iran. J. Vet. Res.* **16**(1) 47–52

[10] Zhao XH, Gong JM, Zhou S, Fu CB, Liu CJ, Xu LJ, et al 2015 Effects of degradable protein and non-fibre carbohydrates on microbial growth and fermentation in the rumen simulating fermenter (Rusitec) *Ital J. Anim Sci.* **14**(2) 220–5

[11] Sani F, Nuswantara L, Subrata A 2012 In Sacco degradability of dry matter, organic matter and neutral detergent fiber of waste of agricultural and plantation industry *Anim. Agric. J.* **1**(1) 743–56.

[12] Guadayo GF, Rayos AA, Merca FE, Tandang AG, Loresco MM, Angeles AA 2019 Prediction of in situ ruminal degradability of forages in Buffaloes using the in vitro gas production technique *Trop. Anim. Sci. J.* **42**(2) 128–36

IOP Conf. Series: Earth and Environmental Science **883** (2021) 012011 doi:10.1088/1755-1315/883/1/012011

The effect of light and gibberellic acid concentrations on breaking dormancy of potato micro tuber

A K Karjadi[1,*] and N Waluyo[1]

[1]Indonesia Vegetable Research Institute /IVEGRI, Jl. Tangkuban Parahu No. 517, Lembang – Bandung, West Java. Indonesia

* E-mail: asihkk@yahoo.com

Abstract. Potato (*Solanum tuberosum* L) is one of Indonesia's priority crops considering its nutritive benefit as a carbohydrates source in food diversification. Potato tubers are classified as stem tubers, so the tubers such as micro tuber can appear in the stem of in vitro cuttings. The research aimed to determine the effect of GA3 concentrations (0, 5, 10, 15, 20 ppm) and storage conditions, e.g., Dark (T1) and light (T2), on breaking dormancy of potato micro tubers. This experiment was conducted in the Tissue culture Laboratory of IVEGRI from April until September 2018. The experiment used a randomized block design with 3 replications; each treatment consisted of 40 micro tubers of Granola (var.) with size ranged >3 - <4 g per each. The result showed that storage conditions gave significant differences to the percentage of breaking dormancy for 4-6 weeks, and GA3 concentration did not affect the average number of shoot, average shoot length in concentration 5 until 20 ppm. The average shoot number, shoot lengths were 0.63 – 2.20 per tubers, 7.50 – 20.13 mm, respectively. There was no interaction between treatment. Visual observation concluded that storage in light conditions produced shoots with better quality than ones in dark conditions.

1. Introduction

Potato (Solanum tuberosum L) is an important crop globally, and it is the fourth largest source after rice, wheat, and maize. In Indonesia, potatoes are vegetables in tubers rich in vitamin C and potassium [1].

Tubers of potato are the stem tuber so that it can be produced in the stem of in vitro cutting, i.e., micro tubers [2]. Since the micro tuber is pathogen-free, it can minimize the high failure rate in plantlet acclimatization. Moreover, it has several advantages, such as easier handling of storage materials and planting [3] [4].

According to [5], [6], [[7], a successful in vitro micro tuber production depends on the ability of a variety to produce tubers conventionally. Potato micro tubers develop from the swelling part at the end of the stolons to the node or axillary leaves.

When the potato tubers are harvested in a dormant state, the length of dormancy depends on the variety, the plants' age, and the tubers' quality when harvested. The dormancy period of potato tubers is the beginning of tuber initiation, e.g., slower growth. This kind of 'inactive' period is rather difficult to determine because it is very dependent on the variety, plant growth, and tuber maintenance after harvest [8]. Potato tubers will not sprout even when placed in a supportive environment. The potato dormancy period occurs for 4 - 15 weeks after harvest. The signs of dormancy breaking in potato is when 80% of the tubers have sprouted and produced ≥ 2 mm shoots. The length of the shoots is influenced by the variety, tuber size, and the tubers' growing environment [9]. Meanwhile, according to [10], [11] the dormancy period depends on the plant growth's environmental conditions, the age of the plant, and the quality of the tubers at harvest.

Gibberellic acid (GA3) is a growth hormone that can accelerate the dormancy breaking of potato tubers [12]. It is a growth regulator that can affect plant growth and morphogenesis [13]. To break the dormancy of the micro-tubers, a precise concentration of GA3 is required. GA3 treatment can stimulate shoot growth of potato micro tubers [14]. It is also able to stimulate dormant shoots to grow [15].

This research aimed to determine the effect of GA3 concentration and storage for breaking dormancy of micro tuber cv. Granola. The hypothesis was proposed that the concentration of GA3 and storage would help to break micro-tuber dormancy.

2. Methods

The research was carried out at the Tissue Culture Laboratory of the Indonesia Vegetable Research Institute / IVEGRI from April to September 2018. Micro tubers cv. Granola was initiated by inducing plantlets in MS Media (1962) plus 80 g / l of sucrose. Plant cultures were maintained in an incubation room with a temperature of 21 - 24 °C in a photoperiod of 16 hours light 8 hours dark. The micro tubers were harvested three months after tuber induction.

The treatments included concentrations of GA3 (0, 5 ppm, 10 ppm,15 ppm,20 ppm), and storage treatments, e.g., dark (T1) condition and light (T2) condition, with three replications in a Randomized block design. Each treatment consisted of 40 micro tubers with size > 3 - <4 g tuber. The micro tubers were soaked in GA3 solution for 15 minutes and dried at room temperature, then placed in storage treatment. The results were recorded on 4,6,8,10 WAT (Weeks After-Treatment) e.g., percentage (%) of shoot growth, an average of shoot numbers, and an average of shoot length.

3. Results and discussion

The statistical analysis on the observation of breaking dormancy micro tuber dormancy of cv. Granola, there was no interaction between treatments.

Table 1. Percentage of shoot growth at the age of 4,6,8,10 WAT

Treatment	Percentage of growing shoots (%)			
	4 WAT	6 WAT	8 WAT	10 WAT
T1(dark)	4.75 A	26.50 A	40.74 A	47.00 A
T2 (light)	2.00 B	16.00 B	40.00 A	48 00 A
K1	0.63 c	10.63 b	23.13 b	31.25 b
K2	5.63 a	25.63 a	48.88 a	53.13 a
K3	5.00 a	23.13 a	41.25 a	48.13 a
K4	3.13 b	23.13 a	40.63 a	46.88 a
K5	2.20 b	23.75 a	50.00 a	58.13 a

Note: 1) WAT: Weeks After Treatment . 2) The average value followed by the same letter are no significatly different according to Duncan 's multiple test of 5% . K1-K5 : GA3 (0,5,10,15,20 ppm)

In Table 1, the GA3 treatment was only significantly different at the observation on 4.6 WAT, dark treatment (T1) the growth percentage was higher than the light treatment (T2). For treatment, GA3 concentration was statistically different at the age of 4.6 WAT, then not really different. However, visually, it can be seen that the highest GA3 concentration (20 ppm) resulted in the highest growth percentage at 6.8 10 WAT.

The percentage of sprouting tubers with a length of ≥ 2mm, starting at 4 WAT and 10 WAT, reached 31.25 - 58.13%. The variety and size of tubers strongly influenced the growth percentage of micro tubers. The low percentage of sprouting was probably due to the sub-optimal concentrations of GA_3 or the quality of the cv. Granola micro-tubers. Where the small size of the micro tubers, will lead to loss of weight. This situation is possible because the tubers were harvested late or earlier so it affected dry matter content, especially low content of carbohydrates [16], [14].

The micro tuber can germinate due to a chemical compound in the tuber, especially starch, protein, and shrinkage of micro tuber due to water losses. [17], [18], [19], [20].

The percentage of sprouts from micro tubers is influenced by genotype, and several factors include growing conditions and tuber storage after harvest [21], [22]. At 4-12th weeks after harvest, tubers continue another physiological process called external dormancy. Potato tubers in this condition will sprout if they are placed in a favorable environment for sprouting.

Table 2. Average number of shoots at age 4,6,8,10 WAT

Treatment	Average number of shoots			
	4 WAT	6 WAT	8 WAT	10 WAT
T1 (dark)	0.70 A	1.65 A	1.85 A	1.95 A
T2 (light)	0.45 B	1.60 A	1.90 A	1.95 A
K1	0.00 b	1.25 b	1.50 b	1.75 b
K2	1.00 a	1.38 a	2.00 a	2.10 a
K3	1.50 a	1.63 a	1.88 a	2.00 a
K4	0.75 a	1.88 a	2.00 a	2.20 a
K5	0.63 a	2.00 a	2.00 a	2.20 a

Note: 1) WAT: Weeks After Treatment . 2) The average value followed by the same letter are no significatly different according to Duncan 's multiple test of 5% . K1-K5 : GA3 (0,5,10,15,20 ppm)

The statistical analysis showed no interaction between storage place and GA3 concentration treatment and no significant difference between treatments for GA3. Visually, dark (T1) or light (T2) storage did not affect the number of shoots per tuber. From several studies, it was found that micro tubers, the tubers' size, affected the number of shoots, which depend on the number of eyes on the tubers. The number of eyes is one of the characteristics of the variety. Generally, micro tubers only have one eye [23].

The average number of shoots up to 10 WAT was 1.75 - 2.20 per tuber. There was no effect on GA3 treatment. According to [24], [9], [14] . The number of shoots per tuber was strongly influenced by the genetics of the potato plant and the micro tubers' size. Several factors, e.g., temperature, photoperiod, the concentration of carbohydrate sources, growth regulators used, and nitrogen content in the growth media, gave the effect of micro tuber quality [26.

Table 3. Average shoot length (mm) at age 4,6,8, 10 WAT

Treatment	Average shoot length (mm)			
	4 WAT	6 WAT	8 WAT	10 WAT
T1 (dark)	8.50 A	16.00 A	16.25 A	19.15 A
T2 (light)	5.50B	15.50 A	15.75 A	18.25 A
K1	0.00 b	10.0 b	13.13 b	14.88 b
K2	10.63 a	18.25 a	18.75 a	19.00 a
K3	8.75 a	15.63 a	17.50 a	20.13 a
K4	8.13 a	18.13 a	18.88 a	19.00 a
K5	7.50 a	16.25 a	17.00 a	18.50 a

Note: 1) WAT: Weeks After Treatment . 2) The average value followed by the same letter are no significatly different according to Duncan 's multiple test of 5% . K1-K5 : GA3 (0,5,10,15,20 ppm)

In the observation of 4,6,8,10 WAT, the average shoot length ≥ 2 mm, GA3 treatment (5, 10, 15.20 ppm) statistical analysis results were not significantly different. Although visually there was a difference where at 10 WAT, the average shoot length was 18.50 mm - 20.13 mm, the higher GA3 concentration did not produce the highest average shoot length. The treatment that produced the highest average shoot length was GA3 with a concentration of 10 ppm. According to [25], [23], [27], the average shoot length of potato tubers is influenced by tuber size and storage after treatment.

Treatment of micro tuber storage after GA3 treatment, dark (T1) and light (T2) treatment results of the statistical analysis no different in observations 6 to 10 WAT. Visually, it can be seen that the average length of shoots in a dark place (T1) looks longer, with the quality of shoots not vigorous [12], [10]. It was shown that in the dark places, the micro tubers' shoots were etiolated [20].

Micro tubers are tubers that are produced or obtained from plantlets in vitro. These tubers are easy to handle during the distribution process due to their relatively small size. However, the utilization is not optimal because there are still many constraints on the information of potential yields, solving dormancy. However, there have been many publications that report the use of micro tubers in exchange for germplasm as basic seeds in seed production [28], [29], [30], [31], [32], [33].

4. Concclusions

The results obtained that micro tuber storage treatment after GA3 treatment in breaking dormancy did not interact between treatments. The average length of shoots in the dark (T1) was higher, but the quality was not good; the percentage of shoot growth was treated with GA3 46.88 -58.13 %. The storage place and GA3 concentration were not optimal for breaking the dormancy of cv. Granola micro tubers.

References

[1] Wattimena G A 2000 *Pengembangan propagul kentang bermutu dan kultivar kentang unggul dalam mendukung peningkatan produksi kentang di Indonesia*. Orasi Ilmiah Guru Besar Tetap Ilmu Hortikultusa Fak. Pertania IPB.

[2] Saha S, Ahmed M, Islam M M, Remme R N, Ali M R 2013 Effect of different levels of sucrose on microtuberization and different substrates on mini tuber production resulted from potato meristem culture. *J. Agric. Vet. Sci.* **4** (6) 58 – 62.

[3] Badoni A and Chaukan J S 2010 Importance of potato micro tuber seed material for farmers of Uttarakhand Hills *Int. J. Sustain Agric.* **2** (1) 01 – 09.

[4] Otrosky M, Nazarian F, Struick P C 2009 Effect of temperature fluctuation during in vitro phase on in vitro microtuber productionn in different cultivars of potato (*Solanum tuberosum* L). *Plant Cell. Tiss. Organ Cult.* **98** (2) 213-218 DOI: https.//doi.org/10.1007/s11240-009-9554.6

[5] Wattimena G A, Gunawan L W, Mattjik N A, Syamsudin E, Armini N M, Ernawati A, 1992 *Perbanyakan tanaman dalam bioteknologi tanaman* . (PAU Biotek IPB. Dirjen Dikti. Dept P&K)

[6] Hussey G and Stacey N T 1981 In vitro propagation of potato (Solanum tuberosum L) *Annals of Botany* **48** 787 – 796.

[7] Otrosky M and Struik P C 2006. *Utilization of potato tissue technique in seed potato tuber production scheme*. PhD Thesis (Wageningen Univ. Netherlands)

[8] Aksenova N B, Sergaeva L I, Konstantinova I M, Golya N O, Romannova G A 2013 Regulation of tuber dormancy and sprouting *Russian J. Physiol* **60** 01-312.

[9] Suttle J C. 2004. Physiological regulation of potato tuber dormancy *Amer. J. of potato Res.* **81** 253 – 262.

[10] Ezekiel R and Singh B 2003 Influence of relative humidity on weight loss potato tubers stored at high temperature *Indian J. Plant Physiol* **8** 141–144.

[11] Dhaka M and Naiwal T K 2015 High efficiency micropropagation of potato (Solanum tuberosum L) cv. Kufrijyoti in Kuaun Hills. *J. Plant Breeding and Crop Sci.* **7** (7) 203-2010. DOI. 10.5897/JPBCS 2015.0.507.

[12] Farnshin H, Abbas Z and Enayat R. 2014. Effect of chemicals treatments on dormancy breaking and some sprouting characteristics of two potato cultivars in different tubersize *Europ. J. of Expt. Biol.* **4** (4) 90–102.

[13] Mohammadi M S, Kashani A, Vazan S, Hasani F 2014 Evaluation of potato mini tubers dormancy breaking affected by various chemical, genotype and mini tuber size *Int. J. Bio Sci.* **4** (6) 100–108.
[14] Sasami R, Khazace H R, Nezami A 2009 Effect of gibberelin, benzyl adenin, Zeatin and temperature on mini tuber dormancy breaking *J. of Hort. Sci.* **23** (2); p 61 – 67 .
[15] Ashayi M M , Karrazi A, Sharifi A and Mehrvar 2012 Carnation ring virus elemination through shoot tip culture *J. Biol. Env. Sci* **6** (17) 175–180.
[16] Galun E 2010 Phytohormones and pattering, the role of hormones in plant architecture. *Word Sci. Hormone.*
[17] Colema W K. 1987 Dormancy release in potato tubers. A review *Am. J. Potato Res.* **64** 57–68.
[18] Sonnewald U 2001 Control of potato tuber sprouting *Trends Plant Sci.* **6** 333–335
[19] Bronche F, Sonnewald U, Biemelt S 2007. IN Paul E C , davey M R (eds). *Biotechnology in agriiculture and forestry* (Heidenberg Springer) 297 – 315.
[20] Hassan P D , Yarni M and Khorshidi B 2007 Effect of thiouvea and GA3 on dormncy breaking of Agria potato mini tubers J. *of Agric. Sci.* **4** 81–84.
[21] Turnbull C G N and Hankel D E 1985 The control of bud dormancy in potato tubers *Planta.* **165** 359–365.
[22] Wiltshire J J, and Cobb A H 1996 A review of the physiology of potato tuber dormancy *Ann Appl. Biol.* **129** 553–569.
[23] Classens M M J and Vreugdenhil D 2000 Is dormancy breaking of potato tubers the reverse of tuber initition *Pot. Res.* **43** 347–369.
[24] Rehman F, Lee S K, Kim H S, Jeon J H, Park Ji Y and Young H 2001 Dormancy breaking and effect on tuber yield of potato subject to various chemicals and growth regulators under green house condition *J. of Biol Sci.* **1** (9) 818–820.
[25] Khadige T, Alireza R and Alireza S 2014 Effect of different temperature and hormone treatments on breaking dormancy in potato tubers *J. of Agric. Sci.* **59** (3) 255–264.
[26] Fufa M and Diro M 2014 Micro tuber induction of two potato (*Solanum tuberosum* L) varieties . *Adv. Crop. Sci. Tech.* **2** (2) 122 doi 104172/2329-8863-1000122
[27] Hemberg T 1985 Potato rest. In Li P H (eds) *Potato Physiology* (Acad Press N Y) p. 353 – 388
[28] Zamora A B, Paet C N and Altoveros E C 1994 *Micropropagation and virus elimination procedures in potato for conservation, dissemination and production in the humid tropics*.)Institute of Plant Breeding UPLB Los Banos Phill).
[29] Ranalli P, Bassi F , Ruaro G, Del Re P, Dicandilo M and Mandolino G 1994 Microtuber and mini tuber production and field performance compare to normal tuber *Potato Res.* **37** 383 -391
[30] Donelly D, Coleman W K and Shirlyn E. 2003. Potato microtuber production and performance A review *Am. J. Pot. Res.* **80** (2) 03–115.
[31] Kanwal A, Ali A and Shoaib K 2006 In vitro tuberization of potato (*Solanum tuberosum* L) cultivar Kuroda- a variety in Pakistan *Int. j. Agric and Biol.* **8** (3) 337–340.
[32] Nistor A, Campeam G, Atanasiu N, Chiru N and Karacsonyi 2010 Influence of potato genotypes on production of microtubers- Rommanian *Biotec Lett.* **15** (3) 5317–5324.
[33] Bolandi A R, Hamidi H and Ghavidel R A 2011 The effect of size and micro tuber dormancy on production of potato mini tubers *Am – Eur. J. Agric and Env. Sci.* **10** (2) 69–173.

The evaluation of protease enzyme effectivenes in broiler chicken diet containing jack bean seed (*Canavalia ensiformis*) with different protein level toward internal organ size

B P Mahardhika[1,*] M Ridla[1], R Mutia[1], D N Adli[2]

[1] Department of Nutrition and Feed Technology, IPB University, Faculty of Animal Science, Bogor, Indonesia
[2] Faculty of Animal Science, University of Brawijaya, Jl. Veteran, Malang 65145, Indonesia

* E-mail: brahmamahardhika@gmail.com;
phone:085212345169

Abstract. This study was conducted with the intention of evaluating the use of the protease enzyme in broiler feed containing jack bean seed (*Canavalia ensiformis*) with the different levels of protein towards the internal organs of broiler chicken starter. Bird samples used were 24 broilers of the male lohmann strain taken from 120 chickens. The study design used was a factorial completely randomized design (FCRD) with 2 factors treatment and 3 replications (each replication used 2 samples of bird). The first factor was the level of protein (CP 22% and CP 19.5%) and the second factor was the use of enzymes (non-protease and protease). Data were analyzed by using analysis of variance (ANOVA). The result showed that the use of low protein levels in broiler feed containing jack bean seed significantly ($P < 0.05$) increased the percentage of pancreatic weights but could be improved by the use of protease enzymes. Levels of protein treatment did not significantly affect the percentage of heart, liver and bile weights. The use of protease enzymes significantly ($P < 0.05$) decreased the percentage of pancreatic and bile weights but did not significantly affect the percentage of liver and heart weights.

1. Introduction

The demand for broiler chicken meat in Indonesia is increasing every year. [1]reported that consumption of broiler chicken meat in 2015 to 2016 increased from 4797 kg /capita to 5110 kg /capita. Broilers are in great demand because of their high productivity and for their maintenance in only a short time. The nutritional content of carcasses is also quite good, its protein content reaches 22.15% -23.33% with fat content 1.62% - 2.27% [2]. The good performance especially body weight growth in broilers must be supported with a good feed quality.

Nowadays, Broiler chicken feed in Indonesia uses corn as an energy source and soybean meal as a source of protein. The fulfillment of soybean meal is still difficult to conduct so the fulfillment must be done by import. According [3] in 2016, Indonesia imported soybean meal of 3 million tons in 1994. It is necessary to explore the protein feed ingredients to reduce the use of soybean meal in broiler chicken feed. One of these feed ingredients is the use of jack bean seed (*Canavalia ensiformis*). Jack bean seed has dry matter 86.5% - 96.2%, energy 351.3 kcal / 100 g - 456.5 kcal / 100 g, total starch 24.7-36.9%, crude protein 22.8-35.3%, ether extract 1.6% -12.1%, crude fiber 4.7 - 11.4%, and ash content 2.3-5.8%. According to [4] the use of jack bean seed can increase the metabolic energy of broiler chicken starter.

The use of Jack bean seed as a protein source of feed has a weakness in the form of an anti-nutrient trypsin inhibitor of 7.96 mg / g of material even though it has been processed by heating (the seed by) using an autoclave [4]. The use of 7% of Jack Bean seed in broiler chicken starter feed can reduce body weight but can be improved by the addition of a protease enzyme [4]. The use of the protease enzyme is one of the solutions to improve the quality of the Jack bean seed.

Protease enzyme is one of the enzymes that can be used to improve feed efficiency in broilers. Protease enzymes are able to catalyze the breakdown of peptide bonds in polypeptides and proteins through hydrolysis reactions into simpler molecules like short chain peptides and amino acids [5]. Increased protein digestibility due to the use of protease enzymes in feed can be used to reduce feed costs by reducing the value of protein in the feed formulation. According to [4] the body weight reduction of broiler chickens fed with low protein level (19.5%) can be improve again to match the body weight of broiler chickens fed according to their standard protein level (22%) with supplementation of protease enzymes by 12500 units per kg of feed ration. The use of protease enzymes is very effective to be used in broiler starter feed containing Jack bean seed (*Canavalia ensiformis*) because it can improve the performance of broiler chickens to match the basal feed by using soybean meal [4].

The Evaluation of an enzyme (quality) is not only done by measuring its effectiveness in improving animal performance. Animal health is one *of the ways* that also needs attention. One of the ways that can be done to determine the effectiveness of the protease enzyme as additional feed for broiler chicken is to measure the response of organ size in broilers. Normally, the size of the organ that has no enlargement or swelling indicates that the animal is in healthy condition or their internal organs do not work hard during the metabolic process. This study aimed to evaluate the use of the protease enzyme in broiler feed containing Jack bean seed (*Canavalia ensiformis)* with different levels of protein towards internal organs of broiler chicken starter.

2. Methods

2.1 Material

2.1.1 Animal

Bird samples used in this study were 24 broiler chicken taken from 120 broilers which were previously reared since Day Old Chick (DOC). The broiler chicken used is male broiler chicken strain Lohhman MB 202 PT Japfa Indonesia.

2.1.2 Protease Enzyme

The protease enzyme used in this study from [6] with a row material in the form of carrier ($CaCO_3$, and wheat flour), and silicon dioxide. The brand name for the protease enzyme used was concentrate-P. The protease activity was 12500 hemoglobin units (HUT) / kg diet. The use of this protease enzyme was 0.5 g / kg of diet. This enzyme was extracted from *Bacillus lichenciformis.*

2.1.3 Diet

The diet used during the DOC phase up to age 7 days were commercial feed. Furthermore, broiler chickens were given a treatment of feed containing Jack bean seed (*Canavalia ensiformis*) which was formulated based on recommendation [7]. The composition of feed ingredients and chicken feed for broiler chicken started is presented in Table 1 below.

Table 1. The composition of feed ingredients and feed nutrient for broiler chicken starter

FEED MATERIAL (%)	P1E0	P2E0	P1E1	P2E1
Yellow Corn	48.50	52.85	48.50	52.85
Rice Bran	8.50	9.00	8.50	9.00
Crude Palm Oil	2.20	2.20	2.20	2.20

Corn Gluten Meal	7.00	3.00	7.00	3.00
Soybean Meal	23.70	22.80	23.70	22.80
Jack Bean Seed	7.00	7.00	7.00	7.00
Premix	0.50	0.50	0.50	0.50
Calcium Carbonat	0.40	0.40	0.40	0.40
Dicalcium Phosphate	1.50	1.50	1.50	1.50
L-lysin	0.50	0.50	0.50	0.50
DL- Methionine	0.20	0.25	0.20	0.25
TOTAL	100	100	100	100
Protease Enzyme (g/kg)	0	0	0.5	0.5
Nutrien				
Dry Matter (%)	98.80	99.15	98.80	99.15
Ash Content (%)	5.28	5.29	5.28	5.29
Crude Protein (%)	22.02	19.57	22.02	19.57
Crude Fat (%)	6.53	6.67	6.53	6.67
Crude Fiber (%)	4.58	4.69	4.58	4.69
Metabolizable Energy (Kkal/kg)	3095.07	3091.17	3095.07	3091.17
Calcium (%)	1.04	1.06	1.04	1.06
Phosphor Total (%)	0.57	0.56	0.57	0.56
Phosphor Available (%)	0.41	0.40	0.41	0.40
Lysin (%)	1.33	1.28	1.33	1.28
Methionine (%)	0.56	0.56	0.56	0.56
Protease Enzyme (HUT/KG)	0	0	12500	12500

P1E0 = feed containing jack bean seed, crude protein 22%, without protease treatment, P2E0 = Diet containing jack bean seed, crude protein 19.5%, without protease treatment, P1E1 = Feed containing jack bean seed, crude protein 22%, protease treatment, P2E1 = Feed containing jack bean seed, crude protein 19.5%, protease treatment.

2.2 Procedure

2.2.1 Measurement of Internal Organs

Organs in broilers were measured after slaughtering broilers. The internal organs were separated from the body of broiler chickens after being slaughtered. The internal organs measured were the liver, pancreas, bile and heart. The weight of the internal organ and percentages it with the live weight of broiler chickens before being slaughtered. The formula for calculating the weight percentage of broiler chicken organs is:

$$\textbf{Percentage Organ weight (\%)} = \frac{\textbf{Organ weight (g)}}{\textbf{Body weight (g)}} \times \textbf{100\%}$$

2.2.2 Experiment Design and Data Analysis The experimental design used in this study was Factorial Completely Randomized Design (FCRD). The treatment factors in this study were broiler chicken (feed) ration containing Jack bean seed (*Canavalia ensifromis*) with different crude protein levels (22% and 19.5%) and the use of enzymes (non-enzymes and protease). Data analysis used was Analysis of Variance (ANOVA), The mathematical model used was based on [8] :

$$Yijk = \mu + Ai + Bj + ABij + \varepsilon ijk$$

X_{ij} : Observation values that obtain a combination of i level protein treatment, and j protease enzyme

IOP Conf. Series: Earth and Environmental Science **883** (2021) 012012 doi:10.1088/1755-1315/883/1/012012

μ : means
A_i : The effect of i level protein treatment
B_j : The Effect of i protease enzyme treatment
AB_{ij} : The interaction effect of treatments of i protein level and j protease enzyme
Eijk : Error of this treatment

2.2.3 Treatment

The treatments used in this study were:

P1E0 : Broiler chicken feed ration containing Jack Bean seed, CP 22%, without the protease enzyme.
P2E0 : Broiler chicken ration feed containing Jack Bean seed, CP 19.5%, without the protease enzyme.
P1E1 : Broiler chickens ration feed containing Jack Bean feed, CP 22%, protease enzyme supplementation.
P2E1 : Broiler chickens feed containing Jack bean seed, CP 19.5%, protease enzyme supplementation.

3. Results and discussion

The use of a low protein level in the chicken feed containing Jack bean seed (*Canavalia ensiformis*) significantly ($P < 0.05$) increased the percentage of pancreas weights but could be improved by the use of the enzyme protease. Protein treatment did not significantly affect the percentage of heart, liver and gall bladder weight. The use of protease enzymes significantly ($P < 0.05$) decreased the percentage of pancreatic and gall bladder weights but did not significantly affect the percentage of liver and heart weights. There was no interaction between differences in protein levels and the use of protease enzymes in broiler chicken diet containing Jack bean seed. Internal organ data (of) broiler chickens that were gave feed containing Jack bean seed (*Canavalia ensiformis*) with different levels of protein and the addition of the protease enzyme shown in Table 2.

Tabel 2. Internal organ of broiler chickens with Jack bean seed (*Canavalia ensiformis*) with different protein levels and addition of protease enzymes.

Protein Treatment	Enzyme		Average
	E0	E1	
	Liver (%)		
P1	2.89±0.16	2.79±0.01	2.84 ± 0.07
P2	2.91±0.04	2.87±0.18	2.89 ± 0.03
Average	2.90 ± 0.01	2.83 ± 0.05	
	Gall Bladder (%)		
P1	0.07±0.01	0.05±0.02	0.06 ± 0.02
P2	0.07 ± 0.01	0.05 ± 0.01	0.06 ± 0.01
Average	0.07 ± 0.01b	0.05 ± 0.01a	
	Pancreas (%)		
P1	0.34 ± 0.04	0.30 ± 0.04	0.32±0.04a
P2	0.43 ± 0.02	0.34 ± 0.02	0.39±0.05b
Average	0.38 ± 0.06b	0.32±0.04a	
	Heart (%)		
P1	0.57 ± 0.06	0.44 ± 0.04	0.50±0.09
P2	0.60±0.17	0.53±0.05	0.57±0.07
Average	0.59±0.02	0.49±0.07	

P1 = feed with crude protein 22%, P2 = feed with crude protein 22%, E0 = Non enzyme treatment, E1 = Enzyme protease treatment. Different letters in the same table or column show significantly different values (P <0.05).

3.1 Liver

The treatment of protease enzymes in broiler chicken feed containing Jack bean seed (*Canavalia ensiformis*) with different levels of protein had no effect on liver weight. The protease enzyme did not cause toxicity in broiler chickens. Protease enzymes helped protein digestion. The process of degradation of proteins into simpler compounds was faster with the help of the protease enzyme, so proteins were more easily absorbed in the digestive tract. The protease enzyme used did not contain compounds that endangered the health of livestock so it did not affect the performance of the liver and therefore the size of the liver did not change significantly. The function of the liver is to neutralize toxins found in the body [9]

The size of the liver in this study was 2.79% - 2.91%. The liver was slightly larger than normal size according to [10] which was 1.7% to 2.8%. This is presumably due to the use of jack bean seed in the ration because jack bean seed contain trypsin inhibitors of 7.9 mg / 100 g [4]. Besides trypsin inhibitors jack bean seed contain several other antinutrients, namely choline, hidrozianine acid, troginelin and antichimotripsin, phenols, isoflavones, niacin, saponins [11]

3.2 Gall Bladder

The use of protease enzymes in broiler chicken feed containing jack bean seed (*Canavalia ensiformis*) with different protein levels could reduce the weight of gall bladder. The low weight percentage of gall bladder indicates that bile does not work hard to help the process of digestion and metabolism of nutrients. The weight and volume of bile is influenced by the type of feed, nutritional status, and enterohepatic biliary circulation [12]. Protease enzymes (helped) the process of digestion and absorption of protein so the protein was more easily absorbed by the body. The nutritional status of feed, especially crude protein can be improved by the provision of a protease enzyme. Bile functions to process fat absorption and excretion of toxins from metabolic waste [12]. Gall bladder consists of bile salts, cholesterol, pigments, fats, organic salts and lecithin [13]

The percentage of bile weight in this study ranged from 0.05% - 0.07%. This percentage does not differ greatly from the research of [14] which is 0.029% - 0.097%. Normally the size of this bile indicates that broiler chickens do not experience any disturbance in the metabolic system in the body.

3.3 Pancreas

Reduction of protein level in chicken feed to 19.5% broiler chickens containing Jack bean seed (*Canavalia ensiformis*) have an impact on increasing the size of the pancreas but could be repaired again by the addition of protease enzymes so it produced the same value with the treatment by using 22% PK level. Protein deficiency in poultry would have a negative impact on the body's metabolic system. The deficiency of protein in the feed was thought to stimulate the work of the pancreas to produce digestive enzymes harder so the protein could be used efficiently. Secretion of pancreatic enzyme was stimulated by the hormone cholecystokinin. Cholecystokinin is a hormone secreted by the small intestinal mucosa which functions to stimulate the secretion of the gallbladder [15]. The pancreas secretes pancreatic fluid to digest starch, fat and protein. The pancreas has two functions, namely exocrine and endocrine. Exocrine functions to supply digestive enzymes while endocrine functions to regulate nutrients to be absorbed in the body [9]. Protease releases from the pancreas and it is controlled by the hormone cholecystokinin produced by intestinal mucosal cells [16]

The use of protease enzymes in broiler diets feed containing jack bean seed (*Canavalia ensiformis*) with different levels of protein could reduce the percentage weight of the pancreas. Supplementation of the protease enzyme in the feed would help the digestive process. Protease enzymes are able to catalyze the breakdown of peptide bonds in polypeptides and proteins through hydrolysis reactions into simpler molecules such as short chain peptides and amino acids [5]. It is suspected that the pancreas did not need too much production of protease enzymes in the body if the protease enzyme has been added to the food so it worked lighter. The percentage weight of the organ of the pancreas did not increase if the pancreas did not work too hard. Increased pancreatic weights is a form of adaptation to meet the increased needs of digestive enzymes [17].

The value of the percentage of pancreatic weights in this study was 0.30% - 0.43%. The weight of the pancreas was within the normal range according to [18], i.e. 0.25% -0.40% of body weight. Normally the weight of the pancreas indicated that the body's metabolic system in terms of secretion of digestive enzymes was operated normally.

3.4 Heart
The use of protease enzymes in feed containing Jack bean seed (*Canavalia ensiformis*) did not affect the percentage of heart weight. The heart is an organ that functions as a pump in blood circulation in the body. The heart is very susceptible to poisons and anti-nutrients, heart enlargement can occur if there is accumulation of toxins in the heart muscle [19]. The value of heart weight percentage in this study was 0.44% - 0.60%. This value was included in the value of the percentage of normal heart weight according to [20] which is 0.5% -1.42%. This indicated that the treatment of protease enzymes and different protein levels in broiler chicken feed containing Jack bean seed *(Canavalia ensiformis)* did not interfere with the function of the heart.

4. Conclusion
The use of broiler chicken starter feed containing jack bean (*Canavalia ensiformis*) with a low protein level (19.5%) could increase the percentage of the weight of the pancreas broiler chicken but could be improved by supplementation of protease enzyme. The use of protease enzymes in feed containing Jack bean seed (*Canavalia ensiformis*) could improve the organ size respond in broilers.

References
[1]. DITJENPKH (Direktorat Jendral Peternakan dan Kesehatan Hewan). 2017. Statistik peternakan dan kesehatan hewan 20017. Jakarta (ID) : Kementerian Pertanian Indonesia.

[2]. Azizah, N, A., Mahfudz, L, D., Sunarti, D,. 2017. kadar lemak dan protein karkas ayam broiler akibat penggunaan tepung limbah wortel (*Daucus carota L.*) dalam Ransum. *JSPI* **12** (4) p 389 – 396.

[3]. KEMENDAG (Kementerian Perdagangan). 2017. Perkembangan Impor Menurut HS 6 Digit Periode : 2012-2017. Kementerian Perdagangan RI. Jakarta. Diakses 10 Juli 2020 dari http://www.kemendag.go.id/id/economicprofile/i

[4] Mahardhika, B, P., Ridla, M., Mutia, R., Yuliana, I., Naja, M. 2020. Pemanfaatan kacang koro pedang (*canavalia ensiformis*) dengan perlakuan enzim protease sebagai pakan broiler starter. JIIP.(Procees publish)

[5] Naiola, E , Widhyastuti, N. 2002. Isolasi, seleksi, dan optimasi produksi protease dari beberapa isolat bakteri. *Hayati*. **6** p 467-473.

[6] Canadian Bio System. 2015. Concentrase-P. 31 Juli 2017 available on https://canadianbio.com/Enzyme-Supplements

[7] Leeson, S, Summers J, D. 2005. Commercial Poultry Nutrition. 3rd Ed. Nottingham: Nottingham University Press.

[8] Steel, R,G,D., Torrie, J,H. 1993. Prinsip dan Prosedur Statistika. Ed ke-1. Sumantri B, penerjemah. Translated : Principles and Procedures of Statistics: A Biometrical Approach. Jakarta: Gramedia Pustaka Utama.

[9] Yuwata, T. 2004. Dasar Ternak Unggas. Kanisius. Yogyakarta.

[10] Putnam, P. A. 1991. Handbook of Animal Science. Academic Press. San Diego.

[11] Primawestri, M.,A., Rustanti, N. 2014. Pengaruh pemberian susu koro pedang *(Canavalia ensiformis)* terhadap kadar kolesterol total dan trigliserida serum tikus Sprague dawley hiperkolesterolemia. *J. Nutri College*. **3**: 447-455.

[12] Suprijatna, E., U. Atmomarsono, & R. Kartasudjana. 2005. Ilmu Dasar Ternak Unggas. Jakarta : Penebar Swadaya

[13] Pilliang, W. G. & S. Djojosoebagio. 2002. Fisiologi Nutrisi. Vol 1. Edisi Ke-4. Bogor : IPB Press
[14]. Faria, N., Iriyanti, N., Susanti, E. 2019. Efek penggunaan fermeherbafit enkapsulasidalam pakan terhadap bobot dan persentase organ asesoris ayam broiler effects of fermeherbafit encapsulation in broiler chickens feed on weight and percentage organs acessories. *J Anim Sci Technol*. **1** (3) p 241-251
[15] Hidayat, A,. A. 2010. Metode Penelitian Kesehatan Paradigma Kuantitatif : Jakarta : Heath Books.
[16] Wildman, Medeiros. 2000. Wildman, R.E.C and D. M. Medeiros. 2000. Protein In: Advanced Human Nutrition. New York (UK) : CRC Press . Pp. 123-150
[17] Lisnahan, C, V., Wihandoyo, Zuprizal, Haarimurty, S. 2018. Pengaruh suplementasi dl-metionin dan L-lisin HCl pada pakan standar kafetaria terhadap berat badan, organ dalam dan organ reproduksi ayam kampung fase pullet. Jurnal Ilmiah Peternakan Terpadu. 6 (2) : 128-133.
[18] Sturkie, P.D. 1976. Avian Physiology. 3th. Ed., New York (UK) ; Verlag
[19] Frandson. 1993. Anatomi dan Fisiologi Ternak. Yogykarta (ID) : Gajah Mada University Press.
[20] Nickle R, Schummer A, Seifrle E. 1990. Anatomy of Domestic Bird. Berlin (UK): Verlag Paul Parey.

Effort to reduce ammonia gas in the broiler chicken excreta with the addition of probiotic as substitute for antibiotic growth promoter

B P Mahardhika[1,*], R Mutia[1] and M Ridla[1]

[1] Department of Nutrition and Feed Technology, IPB University, Faculty of Animal Science, Bogor, Indonesia

* E-mail: brahmamahardhika@gmail.com; phone:085212345169

Abstract. This study has been carried out to evaluate the use of drinking water-soluble probiotics as an alternative to *Zinc bacitracin* Antibiotic Growth Promoter (AGP) in an effort to reduce the concentration of ammonia excreta as a result of increased feed digestibility. The probiotic used contains 2.0×10^7 CFU mL^{-1} Lactobacillus sp, 1.6×10^7 CFU mL^{-1} Bacillus sp, and 7.4×10^9 CFU mL^{-1} Streptomyces sp. The excreta sample was obtained from 15 male Lohmann strain broiler chickens aged 35 days from a total population of 300 chickens that had previously been reared since Day Old Chick. The chickens were distributed into three treatments and five replications in a metabolic cage with a size of 50 cm × 30 cm × 56 cm for three days. This study used Completely Randomized Design (CRD) with Analysis of Variance (ANOVA). The addition of probiotic significantly reduced ($P < 0.01$) water content and ammonia concentration of broiler chicken excreta. The addition of drinking water-soluble probiotics significantly increased ($P < 0.01$) the feed intake and feed digestibility of broiler chicken. The use of probiotics was better than *Zinc bacitracin* in reducing excreta ammonia concentration and feed digestibility.

1. Introduction

Ammonia gas is one of the gases that causes air pollution and causes livestock health problems and even human health. Ammonia gas is the result of decomposition of nitrogenous waste material in poultry excreta such as uric acid, excreted proteins and other non-protein nitrogen (NPN) compounds that occur due to excreta microorganism activity [1]. Sources of ammonia emissions are from animal manure, fertilizer, and industrial processes [2]. The most ammonia gas emissions come from the livestock industry, which is 80% to 90%. Ammonia gas concentration greater than 20 ppm will reduce performance and affects broilers health [3]. Ammonia gas concentration that exceeds 25 ppm and has exposure for more than 8 hours will have a negative impact on human health [4].

Ammonia gas is formed due to the conversion of metabolic waste such as uric acid into ammonia gas with the help of the uricase enzyme. The main source of ammonia gas formation in poultry farming environments comes from excreta. Accumulation of nutrient levels that are wasted through excreta causes pathogenic bacteria to grow rapidly because the nutrient intake is fulfilled. The process of converting uric acid to ammonia is triggered by uricase enzyme activity. Efforts should be made to reduce the excretion of excreta and nutrient accumulation in excreta. The method that can be taken is by increasing the feed's digestibility so that the absorption of nutrients is better, and many nutrients are not excreted.

Various methods have been proven to improve the feed digestibility, for example, the addition of Antibiotic Growth Promoter (AGP). In addition to increasing the digestibility of feed, the use of AGP can also improve the health of livestock from pathogenic bacteria in the digestive tract. A long period of AGP addition can cause resistance in livestock and leave residue in livestock organs [5]. Based on this reason, AGP is banned in almost all over the world started by the European Union [6]. This rule worries the farmers because feed efficiency and livestock health will deteriorate. The low feed efficiency will increase excreta production, so it is feared that the cage environment will be unhealthy due to increased in ammonia gas emissions. Another alternative that can be used to replace AGP is by giving probiotics. According to [7], probiotic is a multiplying living microbe in the digestive tract to maintain microflora balance.

Probiotics can increase feed efficiency and digestive tract health as well as benefit the environment. The use of drinking water-soluble probiotics with incubation measurements of 1 hour and 24 hours can reduce ammonia concentration in the faeces and litter of broiler chickens [8]. The use of drinking water-soluble probiotics can reduce ammonia concentration broiler chickens litter from 31.36 ppm to 20 ppm [9]. Decreased ammonia levels due to probiotic addition are due to increased feed digestibility, especially the crude protein. According to [10], the addition of probiotics can increase the growth of broiler chickens, nutrient digestibility and apparent metabolic energy corrected by nitrogen (AMEn) and the composition of microflora in the cecum. According to [3], increased digestibility, and the composition of microflora in the digestive tract due to the addition of the probiotics have an impact on the reduction in ammonia gas derived from the decomposition of nitrogenous wastes in excreta. Addition of additives such as antibiotics, probiotics, prebiotics, and symbiotics can reduce ammonia by reducing pathogenic bacteria's activity in the digestive tract [11].

2. Methods

2.1. Material

2.1.1. Animal and Cages

This study used 15 male Lohmann strain broiler chickens aged 35 days obtained from a total population of 300 chickens that had previously been reared since Day Old Chick. The chickens were distributed into three treatments and five replications in a metabolic cage with 50 cm × 30 cm × 56 cm for three days. During the DOC phase, chickens were kept in a colony cage (2 m × 1 m) with a base of rice husks. One sample of broiler chickens in each test was transferred to the metabolic cage in 35 days old for an excreta collection. Treatment was given from the beginning of maintenance.

2.1.2. Probiotic

This study used probiotics with constituent microorganisms in the form of lactic acid bacteria Lactobacillus sp, and Bacillus sp, and moulds of Streptomyces sp. The probiotic used was tested at the Indonesian Biotechnology and Bioindustry Research Center (2019) which has 2.0×10^7 CFU mL^{-1} Lactobacillus sp, 1.6×10^7 CFU mL^{-1} Bacillus sp, and 7.4×10^9 CFU mL^{-1} Streptomyces sp. The form of probiotic is liquid. The probiotics were dissolved into mineral water in a 1: 1 and fermented for four days–five days before use. The dose was 2 mL probiotics per litre of drinking water.

2.1.3. Diet

The diet used in this study was basal feed for finisher period broilers based on the recommendation by (12). In the antibiotic treatment, AGP supplemented in was the form of 0.04% zinc bacitracin. Data on the composition of feed ingredients and nutrient content of broiler chicken diet are presented in Table 1.

Table 1. The composition of feed ingredients and nutrient content of broiler chicken diet

Material	Basal	AGP
Yellow Corn	59.00	59.00
Rice Brand	6.55	6.55
Soy Bean Meal	16.50	16.50
Meat Bone Meal	6.00	6.00
Corn Gluten Meal	6.50	6.50
Crude Palm Oil	3.00	3.00
CaCO3	0.80	0.80
NaCl	0.20	0.20
Premix	0.50	0.50
DL-Methionine	0.40	0.40
Lysine	0.45	0.45
Tryptophan	0.10	0.10
Total	100	100
Zinc bacitracin (%)	0	0.04
Nutrient		
Dry Matter %	89.94	89.94
Metabolizable Energy (Kkal kg^{-1})	3004.95	3004.95
Crude Protein %	20.17	20.17
Ether Extract %	2.94	2.94
Crude Fiber %	2.98	2.98
Lysine %	1.16	1.16
Methionine %	0.68	0.68
Calcium %	0.95	0.95
Phosphor %	0.49	0.49
Natrium %	0.16	0.16
Chloride %	0.20	0.20

2.2. Procedure

2.2.1. Measurement of Feed Digestibility and Excreta Moisture

Measurement of feed digestibility was carried out using a modified method by [13] the total excreta collected method. Broilers were kept in a metabolic cage with feed adaptation. After that, broilers were fasted 24 hours to empty the digestive tract to avoid the previous feed's influence. Drinking water was given *ad libitum*. Broiler chickens were fed and collected the excreta every day for three days. Every day, feed intake and excreta fresh weight measurement were taken. The collected excreta was put in the freezer for 48 hours. After that, the excreta was dried in an oven in 60°C and 105°C. The dry weight of excreta and excreta moisture was calculated using the [14] method and feed digestibility.

2.2.2. Measurement of Ammonia Excreta Concentration

Ammonia measurement was carried out using ammonia meters. Before use, the ammonia meter was calibrated for 10 minutes. Ammonia gas was measured every day around 11.00 pm–12.00 pm. Ammonia measurement was carried out in each metabolic cage of broiler chickens. The device was equipped with an ammonia gas catching hose close to the excreta storage container in a metabolic cage.

2.3. Experimental Design and Data Analysis

The experimental design used was a completely randomized design (CRD) with three treatments and five replications. This study used Analysis of Variance (ANOVA) with the DMRT (Duncan Multiple Range Test) tests if obtained significantly different data. The mathematical model used is based on (15) as follows:

$$X_{ij} = \mu + \tau_i + \varepsilon_{ij}$$

X_{ij} = The observation value on the treatment and replication
μ = General Mean
τ_i = Effect of treatment
ε_{ij} = Error
i = The number of treatment
j = The number of replication

3. Results and discussion

3.1. Environmental Conditions

This study used a comfortable temperature for poultry in the morning and at night, but the temperature is higher during the day. Humidity in the metabolic cage was relatively high. Data on temperature and humidity on the metabolic cages used in this study presented in Table 2.

Table 2. Temperature and humidity on the metabolic cages

Day	Parameter	Morning	Afternoon	Night
1	Temperature	25.00	32.50	24.00
	RH	87.00	80.00	88.00
2	Temperature	26.50	30.00	23.50
	RH	86.00	80.00	90.00
3	Temperature	24.30	33.50	23.00
	RH	85.00	79.00	90.00
Average	Temperature	25.27	32.00	23.50
	RH	86.00	79.67	89.33

RH= Relative Humidity, Morning (05.00–06.00), Afternoon (11.00–12.00), Night (18.00–19.00)

The temperature in the metabolic cages in this study tended to be a comfortable temperature for the poultry except during the daytime. This study used standard maintenance of broiler chickens based on [16] that the optimal temperature of broiler chicken maintenance in the finisher (22 days–35 days) was not more than 26°C with 60% humidity.

3.2. Feed Digestibility and Ammonia Excreta Concentration

The addition of drinking water-soluble probiotics significantly ($P<0.01$) reduced the moisture and ammonia excreta concentration of broiler chickens. The addition of *Zinc bacitracin* AGP significantly ($P <0.01$) reduced the ammonia excreta of broiler chickens. The addition of drinking water-soluble probiotics significantly ($P <0.01$) increased feed intake and feed digestibility of broiler chicken. Data on ammonia excreta and feed digestibility of broiler chicken are presented in Table 3.

Table 3. Ammonia excreta and feed digestibility of broiler chicken

Parameter	T0	T1	T2
The Ammonia of Excreta (ppm)	37.33 ± 4.51A	31.01 ± 2.31B	25.00 ± 1.74C
The Moisture of Excreta (%)	76.35 ± 5.95A	67.76 ± 1.68B	67.42 ± 0.92B
Dry Mater of Excreta (%)	23.65 ± 5.95A	32.24 ± 1.68B	32.58 ± 0.96B
Feed Intake (g)	106.81 ± 3.78A	112.85 ± 5.52A	138.04 ± 4.57B
Excretion of Excreta (g)	38.08 ± 2.66	35.50 ± 6.46	36.83 ± 1.92
Feed Digestibility (%)	64.34 ± 2.28A	68.49 ± 2.28AB	73.30 ± 1.53B

T0= control (basal diet), T1= T0 +AGP (zink bacitrasin 0.04%), T2 = T0 + drinking water soluble probiotic.

The use of *Zinc bacitracin* antibiotics and drinking water probiotics can reduce ammonia excreta concentration. The decrease in ammonia was due to an increase in feed digestibility. Ammonia excreta concentration was related with the feed digestibility with a coefficient of determination of 0.506. The relationship between feed digestibility and ammonia excreta concentration can be seen in Graph 1.

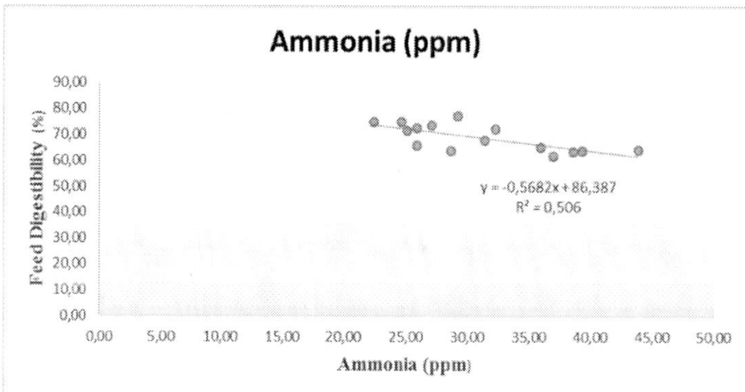

Figure 1. The relationship of feed digestibility and ammonia excreta

An increase in feed digestibility will reduce ammonia excreta concentration. Increase indigestion occurred due to an increase in the balance of microflora in the digestive tract. The addition of probiotics in drinking water can increase the total population of lactic acid bacteria in the digestive tract to improve the ileum morphology by increasing the length of villi and ileum crypt depth (17). The more nutrients ingested shows, the lesser waste of nutrients in the faeces. If more nutrients are left in the faeces, it will increase uric acid conversion to ammonia by the uricase enzyme. According to (18), probiotics can restrain the activity of decomposing proteins in faeces and litter so that a decrease in ammonia gas concentration can occur. Probiotics contain lactic acid bacteria which can produce protease enzymes (19). Proteolytic bacteria such as *Bacillus sp* inhibit uric acid conversion to ammonia by using it as a nutrient (20).

The use of *Zinc bacitracin* antibiotics and drinking water-soluble probiotics can reduce excreta water content. The decrease in the excreta water content will be in line with the decrease in ammonia gas (3). The dry excreta of broiler chickens indicate that the chickens are not in a diarrhoea condition. Wet excreta is caused by the digestive tract in a less favourable condition due to pathogenic bacteria so that the chickens will have diarrhoea or wet excreta. The water content obtained in this study was classified into normal excreta water content based on (12) which is 60%–80%.

Feed digestibility is related to feed intake. The use of drinking water-soluble probiotic on the broiler chicken can increase feed intake during feed digestibility measurements. The more feed consumed and not excreted through faeces will increase in feed digestibility. An increase in feed intake as a result of probiotic supplementation was due to an increase in feed digestibility. An increase in feed digestibility was due to the addition of probiotics that can improve the balance of microflora and improve intestinal morphology so that feed can be digested more quickly.

The use of water-soluble probiotics can decrease ammonia excreta concentration more than the *Zinc bacitracin* antibiotic. This condition happened because *Zinc bacitracin* is a bactericidal antibiotic as the only sensitive to suppression of gram-positive microbial pathogen populations. According to (17), the addition of probiotics can increase the total bacterial population and a total population of lactic acid in ileum without affecting *Salmonella* and *E coli* populations as well as the addition of *Zinc bacitracin* antibiotic. The broiler feed digestibility process with water-soluble probiotics was better than *Zinc bacitracin* in decreasing the ammonia concentration.

4. Conclusion

The addition of drinking water-soluble probiotics can reduce the ammonia concentration in broiler chicken excreta as a result of increased feed digestibility. The addition of probiotics through drinking water was more effective in reducing ammonia excreta concentration compared with the use of *Zinc bacitracin* antibiotic growth promoter.

5. References

[1] Manin F, Hendalia E and Yusrizal. 2012. Potency of *Bacillus* dan *Lactobacillus* bacteria as probiotik to reduce ammonia pollution in poultry house. *JPI.* **14** (2) p 360–367.

[2] Heij G.J and T. Schneider. 1991. Studies in Environmental Science 46. Acidification Research in The Netherlands. Final Report of the Dutch Priority Programme on Acidification. Elsevier Science Publishing Company Inc. 655, Avenue of the Americas. New York, NY 10010, U.S.A

[3] Riza H, Wizna R and Yusrizal Y. 2015. Peran probiotik dalam menurunkan amonia feses unggas. *JPI* **17** (1) p 19–26.

[4] Ritz C W, Fairchild B D and Lacy M P 2004 Implications of ammonias production and emissions from commercial poultry facilities: a review *J. Appl. Poult.* **13** p 684–692.

[5] Marlina, Zubaidah, Sutrisno.2014. Pengaruh pemberian antibiotika saat budidaya terhadap keberadaan residu pada daging dan hati ayam pedaging dari peternakan rakyat. *JIIP*. **25** (2) p 10 –19

[6] Bedford M. 2000. Removal of antibiotic growth promoters from poultry diets: Implications and strategies to minimise subsequent problems. *World's Poultry Science Journal*. **56** p 348–365

[7] Fuller R. 2002. Probiotic what they are and what they do. Sains Mikrobiologi [Internet]. [download at June 5th 2019]. Avaailable on : Htpp://D:/Probiotic. What they are and what do

[8] Hendalia E, Manin, Yusrizal, Nasution. 2012. Aplikasi probiotik untuk meningkatkan efisiensi penggunaan protein dan menurukan emisi gas amonia pada ayam broiler. *Agrinak.* **2** (1) p 29–35

[9] Mahardhika BP, Mutia R, Ridla M.2020. Efforts to reduce ammonia gas in broiler chicken litter with the use of probiotic. IOP Conf. Ser.: Earth Environ. Sci. **399** p 1–6

[10] Mountzouris K.C. P. Tsitrsikos, I. Palamidi, A. Arvaniti, M. Mohnl, G. Schatzmayr and K. Fegeros. 2010. Effects of probiotik inclusion levels in broiler nutrion on growth performance, nutrient digestibility, plasma immunoglobulins, and cecal micrroflora composition. *Poult. Sci.* **89** p 58–67

[11] Yusrizal and TC Chen. 2003 Effect of adding chicory fructans in feed on fecal and intestinal microflora and excreta voatile ammonia. Int. J. of Poult. Sci. 2 (3): 188–194.

[12] Leeson S and J D Summers. 2005. *Commercial Poultry Nutrition. 3rd Edition.* Canada (US) : Notthinghem University pr. P 232

[13] Farrel DJ.1978. Rapid determination of metabolizable energy of food using cockerels. *Br. Poultry Sci.* **19** p 303–308

[14] AOAC 2005 Official Methods of Analysis 18th edition (Gaithersburg, MD: Association Of Analytical Chemists).

[15] Steel RGD., Torrie. 1993. *Prinsip dan Prosedur Statistik (Pendekatan Biometrik). Penerjemah. B. Sumantri*. Jakarta (ID) : Gramedia Pustaka Utama p 205
[16] Pokphand C. 2005. Manual Managemen Broiler CP 707 http://www.charoendpokphand.com
[17] Mahardhika BP, Mutia R, Ridla M. 2020. The Evaluation of Various Methods of Probiotic Addition as an Alternative to *Zinc bacitracin* Antibiotics on the Efficiency of Broiler Production. (Proces Publish)
[18] Sjofjan, O. 2003. Kajian Probiotik (Aspergillus niger dan Bacillus spp) sebagai Imbuhan Ransum dan Implikasi Efeknya terhadap Mikroflora Usus serta Penampilan Produksi Ayam Petelur. Universitas Pajajaran, Bandung
[19] Wang YB, Li Jr, Lin J. 2010. Probiotic cell wall hidropobiciy in bioremediation of aquaculture. *J Aquaculture*. **269** (1): 349–352
[20] Soeharsono. 1999. Prospek Penggunaan Probiotika sebagai Pengganti Antibiotika untuk Ternak.Wacana Ilmu Pengetahuan Teknologi dan Seni Tahun Akademik 1999–2000.

IOP Conf. Series: Earth and Environmental Science **883** (2021) 012014 doi:10.1088/1755-1315/883/1/012014

Salt-induced growth promotion in rice varieties during nursery

B Kurniasih[1,*], N Arini[1], D Alvioliana[1], R I Nisa[1] and R A Wulandari[1]

[1]Faculty of Agriculture, Universitas Gadjah Mada, Yogyakarta, Bulaksumur, Sleman, Yogyakarta 55281

* E-mail: tuti_b@ugm.ac.id

Abstract. Climate change will increase the occurrence of salinity in agricultural land along with the coastal areas. One of the technologies to reduce salinity is NaCl pretreatment. This study aimed to evaluate salinity treatment's effect during nurseries on the growth of lowland rice seedlings. There were three separate experiments, and all the experiments used Randomized Complete Block Design. In the first experiment, local black rice seeds (var. Jelitheng) was used. The nursery was carried out at three salinity levels, i.e. 0.2, 3 and 5 dS/m. The second experiment was conducted using salt-resistant rice seeds (var. Dendang) and salt susceptible rice seeds (var. IR 64). The salinity levels applied were non-saline (0.2 dS/m) and saline (5 dS/m). The third experiment used rice seedling var. IR 64, with the first factor being the salinity level (0.2 and 5 dS/m) and the second factor was a wet nursery and dry nursery. In general, the results from the three experiments showed that giving salinity levels of 3-5 dS/m in several rice varieties improved seedling performance. Although salinity during nursery could increase the concentration of Na^+ and decrease the concentration of K^+ in leaves, salinity during nursery increased the seedlings fresh weight, and dry weight increased the number of seedlings leaves and increased the concentration of leaf chlorophyll. The better seedlings growth variable in the saline nursery will help the plants cope with salinity in the later growth stage in the field.

1. Introduction

The increase in soil salt concentration can be caused by several things, such as poor soil drainage and leaching, either due to inadequate rainfall or irrigation [1]. Salinity can also result from land-use changes from natural forest vegetation to cereal cultivation or changes in cropping patterns [1, 2, 3]. In coastal areas, salinity is generally caused by seawater intrusion through land aquifers [1, 2].

Salinity suppresses plants in two ways, i.e. by inhibiting roots from absorbing water (osmotic stress) and causing toxicity in the plant cells (osmotic stress) [4, 5]. Osmotic stress is generally related to initial response, while ionic stress is the final response [6]. Osmotic stress has the same impact as drought stress, which causes water absorption inhibition, root growth, cell elongation, leaf development, reduced production of new leaves, and damage to plant cells [7, 8]. Therefore, rice plants that are exposed to salinity stress will also be exposed to drought stress.

In general, rice is tolerant to salinity during the germination process [9, 10]. However, it is sensitive at later growth phases. The degree of sensitivity varies during different growths. The tolerance of plants to salt is higher in the tiller formation stage but decreased in the flowering phase [11, 12]. Seedling and flowering phases are the most sensitive to salinity compared to the germination and tillering phase. However, rice plants' response to salinity levels is different for each rice genotype [11, 12]. The damage caused by salinity in rice seedlings includes excess transport of NaCl from the root system to the leaves.

Osmotic stress in rice plants will induce osmoprotectants' production, such as proline [13]. Salinity stress can also be followed by oxidative stress which can damage plants [14, 15]. Therefore, efforts to increase plant productivity need to be done through effective and efficient technological improvements.

Several cultivation technologies have been reported to cope with environmental stress in rice [16]. One promising cultivation technology in saline land is the pre-planting salinity treatment or NaCl pretreatment [17, 18]. This includes seed priming and salt treatment during nursery [19]. Pre-planting salinity treatment is a way to improve the quality of germination in dry or saline conditions by activating the antioxidant defence system and osmotic adjustment [20, 21]. According to [22], low concentrations of NaCl treatment are reported to increase rice tolerance. Seedlings that grow in unfavourable conditions will have different adaptations according to the seeds' tolerance level [23, 24].

The pre-planting salinity treatment on tomatoes resulted in a better plant adaptation to saline conditions. The adaptation ability of tomato plants largely depends on the salt concentration that is applied during the nursery. When the salt concentration used is below the stress threshold, the plants are unlikely to increase the salt tolerance level [25]. NaCl pretreatment increases growth and yield when plants are exposed to salinity [26, 27]. The initial salinity of 15 mM (~ 1.5 dS/m) can increase maize plants' tolerance to 45 mM (~ 4.5 dS/m) at the later growth stage. The same thing happened to *Sorghum bicolor* [28]. Tomato seedlings, exposed to the salinity of 7 dS/m during salt pretreatment, can increase fruit production by 29% in salinity, compared to those without salt pretreatment [25]. Also, mung bean exposed to salt stress of 5 dS/m during nursery can increase growth, chlorophyll content, increase proline accumulation and reduce H_2O_2 concentrations in seedlings [29].

There are two rice nursery methods, i.e. dry bed nursery and wet bed nursery [30]. Wet seeding is carried out on land that contains a lot of water, while dry seeding is done on land with little water content [31]. Instead of longer seeding time, more seeds, and less tolerant to salinity stress, dry seedbed is more water efficient, more comfortable to implement, faster germination, and shorter flowering age [30]. Meanwhile, [32] and [33] showed that dry seedbeds could reduce crop damage when they are inundated. Rice germinated in flooded saline conditions will have higher salt resistance than those germinated in dry conditions [34]. These three experiments were carried out to evaluate rice varieties' response to salt pretreatment, in different nursery conditions.

2. Methods

2.1. *Salt pretreatment on two rice varieties*

The research was carried out in the Baros coastal area, Tirtohargo Village, Kretek District, Bantul Regency, Yogyakarta Province. Rice varieties of IR 64 and Dendang were used in this experiment. Their germination rates were 96% and 95%. A single factor Randomized Complete Block Design was used to evaluate each variety of response to pre-planting salt treatment. The factor consisted of two treatment levels, i.e. the pre-planting salinity treatment and non-saline (without salt pre-planting) treatment. Each treatment was repeated three times. Land preparation includes plowing the soil so that 3 x 3 m plots were obtained. The basic fertilizer application consisted of manure as much as 216 kg is spread evenly over the land. The experiment was terminated at 28 days after sowing when all seedlings were transplanted to the rice field. Observations included plant height, leaf area, seed dry weight, chlorophyll content and nitrate reductase activity in leaf. Observation of leaf chlorophyll content was measured by spectronic 21D [35].

2.2. *Salt pretreatment in wet and dry bed rice nursery*

The research was conducted at the Greenhouse of the Department of Agricultural Cultivation, Faculty of Agriculture, Gadjah Mada University, Yogyakarta. The material used was IR64 cultivar rice seed, which was sown on sand media with Hoagland nutrient solution added with NaCl by the determined treatment. The nursery uses a split-plot design. The main plot was the nursery condition, i.e. wet and dry. Subplots were the salinity treatment, consisted of 0.7 dS/m and seven dS/m. The determination of the salinity concentration of 7 dS/m is based on preliminary research that has been done previously, referring to [36, 37]. Rice is sown in submerged conditions in the wet nursery. While in the dry nursery, the medium was left damp and not inundated. The two factors were combined and repeated three times.

Environmental observations include ambient temperature, light intensity, Relative Humidity and Electrical Conductivity (EC) of seeding media. Soil EC was measured every day for 14 days. The soil EC value was maintained at seven dS/m, with five dS/m and two dS/m were the contributions from NaCl and Hoagland's nutrient solution, respectively. The germinating rate was calculated for 14 days using the following formula:

$$Germination\ Rate\ (\%) = \frac{Number\ of germinated\ seed}{Total\ number\ of\ seed} \times 100\%$$

Observation of leaf Na^+ and K^+ concentrations were carried out at 8, 11, and 14 days after sowing using concentrated HNO_3 extraction method [38].

2.3. Salt pretreatment in black rice

This experiment was conducted at a screenhouse Faculty of Agriculture and was arranged in a Completely Randomized Block Design. Black rice seeds var. Jelitheng was used. It has a fluffier rice texture with an amylose content of 19.6%. This variety also has a high phenolic content and antioxidant content [39, 40]. Black rice seeds are soaked using warm water for 24 hours, before being planted in a nursery tray. The growing medium was a mixture of soil and organic fertilizer. The seedlings were irrigated using saline water of 0.2; 3 and 5 dS/m. The levels of soil Electrical Conductivity (EC) were maintained throughout the experiment. At 21 days after planting, plant height, number of leaves, plumule fresh and dry weight were measured.

2.4. Data analysis

Data were analyzed using analysis of variance with a confidence level of 95% in *SAS Portable* 9.1.3 software

3. Results and Discussion

3.1. Salt pretreatment on two rice varieties

According to [41], the IR64 rice variety is a national superior rice cultivar resistant to brown planthoppers, green leafhoppers, late blight bacteria, and grass dwarf viruses. At the same time, Dendang variety is known as a salt-tolerant variety. These two varieties showed a similar response when exposed to the salt pre-planting treatment. Despite the similar number in leaves and root area between those given the salt pre-planting treatment and the without the treatment, the total dry weight of the seedlings during transplanting showed that seedlings with pre-planting had greater biomass than seedlings without treatment. Meanwhile, IR64 variety, which is relatively sensitive to salinity, showed higher plant and larger leaf area than in salt-tolerant Dendang variety, when both were treated with salt pre-planting treatment (Tables 1 and 2).

Table 1. Plant height, leaf number, total fresh and dry weight, leaf area and root area of IR64 variety under different nursery conditions at 28 days after sowing.

Pre-planting Treatment	Growth variables					
	Plant height (cm)	Leaf number	Total fresh weight (g)	Total dry weight (g)	Leaf area (mm^2)	Root area (mm^2)
Non-priming	29.750 b	4.812 a	0.578 a	0.125 a	17,911 b	2,943 a
priming	41.927 a	4.958 a	1.456 a	0.250 b	40,715 a	3,019 a
CV (%)	2.836	3.930	29.183	6.679	25,287	26,002

The value followed by the same letter in the same column shows no significant difference based on the T-test on α 5%.

Table 2. Plant height, leaf number, total fresh and dry weight, leaf area and root area of Dendang variety under different nursery conditions at 28 days after sowing.

Pre-planting Treatment	Growth variables					
	Plant height (cm)	Leaf number	Total fresh weight (g)	Total dry weight (g)	Leaf area (mm^2)	Root area (mm^2)
Non-priming	29.711 a	4.979 a	0.450 a	0.106 b	21,811 a	2,915 a
priming	31.859 a	4.830 a	1.190 a	0.254 a	33,821 a	2,953 a
CV (%)	4.106	12.017	9.709	10.600	32.524	7.514

The value followed by the same letter in the same column shows no significant difference based on the T-test on α 5%.

Table 3 shows that salt pre-planting treatment did not affect both the chlorophyll a or the total chlorophyll content in IR 64 variety, but increased the chlorophyll content. Whereas in the Dendang variety, salinity increased the total chlorophyll in seedlings with pre-planting treatment [17]. As reported by [42], salinity affects the chlorophyll content of rice. In salt-sensitive rice varieties, salinity reduces the chlorophyll content. In contrast, in salt-resistant rice, mild saltiness increases chlorophyll. To a certain extent, an increase in salinity level will be followed by a decrease in chlorophyll content.

Several biochemical compounds are affected by salinity stress, one of which is nitrate assimilation. Table 3 shows no difference in the Nitrate Reductase Activity (NRA) value between the salt-treated and untreated IR 64 rice plants. This also happened to Dendang variety (Table 4). Nitrate assimilation is catalyzed by nitrate reductase (NR), an enzyme that affects nitrogen absorption and assimilation in the plant body [43]. Several studies reported that nitrate uptake and nitrate reductase activity decreased in plants under salinity [44, 45, 46, 47]. However, other studies showed that salinity stress increased nitrate reductase activity, as one mechanism in rice plants to cope with salinity [48, 49.]

Table 3. Chlorophyll and nitrate reductase activity of IR64 variety under different nursery conditions.

Pre-planting Treatment	Growth variables			
	Chlorophyll a	Chlorophyll b	Total chlorophyll	NRA
Non-priming	27.154 a	30.525 b	57.661 a	1.914 a
priming	28.671 a	31.061 a	59.713 a	3.133 a
CV (%)	2.226	0.153	1.036	20.286

The value followed by the same letter in the same column shows no significant difference based on the T-test on α 5%.

Table 4. Chlorophyll and nitrate reductase activity of Dendang variety under different nursery conditions.

Pre-planting Treatment	Growth variables			
	Chlorophyll a	Chlorophyll b	Total Chlorophyll	NRA
Non-priming	22.467 a	30.609 a	53.058 b	3.878 a
priming	28.481 a	31.132 a	59.593 a	4.415 a
CV (%)	5.505	3.447	0.602	31.643

The value followed by the same letter in the same column shows no significant difference based on the T-test on α 5%.

3.2. Salt pretreatment in wet and dry bed nursery

The concentration of Na^+ and K^+ ions in leaves can be good indicators of salt toxicity experienced by plant cells. Table 5 shows that the salt pre-planting treatment increased the concentration of leaf Na^+ that accompanied by a decrease in leaf K^+ concentration. As reported by [47], K^+ ions play an essential role as a catalyst for various enzymes. Salinity stress causes Na^+ ion toxicity and K^+ ions deficiency.

Large Na^+ concentration inhibits the absorption of K^+. In increasing cell turgor, the deficiency of K^+ can be replaced by Na^+.

Table 5. Concentrations of Na^+ and K^+ (mmol/FW) in wet and dry nurseries under saline and non-saline conditions at 14 days after seeding.

Nursery condition	Na^+ (mmol/FW)	K^+ (mmol/FW)
Wet nursery	3.82 a	1.03 a
Dry nursery	3.78 a	0.73 b
Salinity		
Non-saline	0.90 q	1.06 q
saline	6.72 p	0.72 p
Interaction	(-)	(-)
CV Nursery (%)	20.44	14.57
CV Salinity (%)	19.64	10.06

The value followed by the same letter in the same column shows no significant difference based on Tukey's test at the 95% confidence level. The sign (-) indicates that there is no interaction between the treatment levels tested

Table 6 shows that salinity reduces the germination rate of rice seedlings. Compared to dry nurseries, wet nurseries will increase the germination rate when the seeds are exposed to salinity [18]. It is suspected that cell homeostasis during inundation is more balanced than in moist soil conditions. Inundation conditions are reported to reduce the impact of salinity in rice seedlings. Ion homeostasis in submergence can reduce the Na^+ concentration in rice seedlings in high salinity [48]. High salinity concentrations in moist soil conditions could increase the water absorption rate in rice. If not balanced with the soil's adequate water availability, the increase in water absorption rate will cause cell plasmolysis.

Table 6. The germination rate of seedlings in wet and dry nurseries in saline and non-saline conditions at 14 days after sowing.

Nursery condition	Germination rate (%)	
	Non-saline	Saline
Wet nursery	66.67 a	73.33 a
Dry nursery	53.67 a	30.67 b
Interaction	(+)	
CV Nursery (%)	18,84	
CV Salinity (%)	11,97	

The value followed by the same letter in the same column shows no significant difference based on Tukey's test at the 95% confidence level. The sign (+) indicates that there is an interaction between the treatment levels being tested

3.3. Salt pretreatment in black rice variety

Many studies reported that salinity reduced various growth variables of rice plants. The inhibition of water absorption reduced water availability in plants, thus inhibiting the photosynthesis process and reducing plant biomass accumulation. In some cases, light salinity could stimulate growth, which may be related to Na+ and Cl+ ions' role in contributing to the regulation of osmotic potential in plants.

Figure 1 - 3 shows that giving a salinity of 3 dS/m increased seedling height and plumule biomass. However, increasing the salinity concentration to 5 dS/m lower plant height and plumule dry weight did not differ from those in non-saline treatment. This is following the study by [12] who showed that soil salinity above four dS/m decrease root distribution, plant height [10], and the formation of tillers [19, 49].

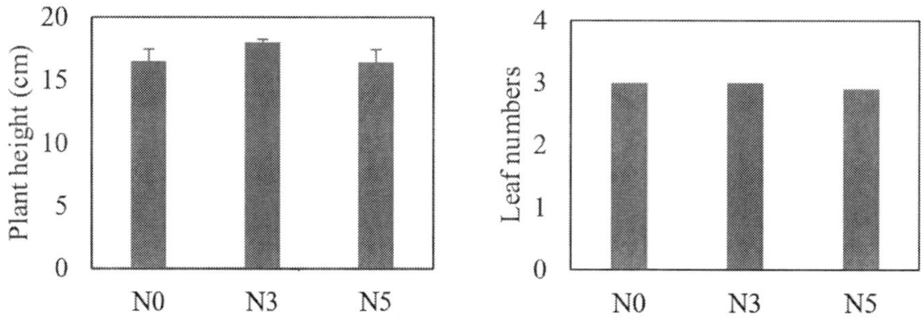

Figure 1. Leaf number (a) and plant height (b) of rice var. Jelitheng at 14 days in the nursery under three different salinity treatments; N0 (Non-saline); N3 (3 dS/m) and N5 (5 dS/m)

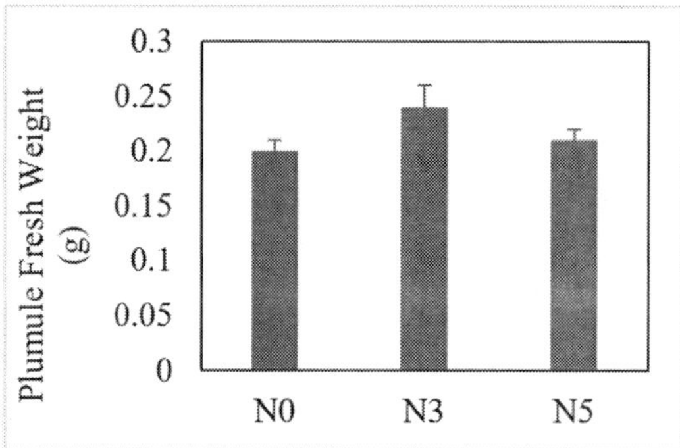

Figure 2. Plumule Fresh Weight of rice var. Jeliteng at 14 days in the nursery under three different salinity treatments; N0 (Non-saline); N3 (3 dS/m) and N5 (5 dS/m)

4. Conclusion

The salinity pre-planting treatment increased the growth of seeds of various rice varieties, both white and black rice. Salt-sensitive rice varieties showed a better response to the salinity pre-planting treatment than salt-resistant varieties. Pre-planting salinity treatment was more effective in the wet nursery than in the dry nursery.

Giving salinity levels of 3-5 dS/m in several rice varieties improved seedling performance. Despite the increase in leaf Na^+ concentrations and decrease the leaf K^+ concentrations, salinity during nursery could increase seedlings fresh weight and dry weight and increase leaf chlorophyll concentration. Good seedling growth during the nursery will support better future plant growth in the field

References

[1] Abrol I P 1988 *Salt Affected Soils: Problems and Prospect in Developing Countries*. Tycooly International Riverton, New Jersey pp 67

[2] Hingston F J and Gailitis 1976 The geographic variation of salt precipitated over Western Australia. *Australian Journal of Soil Research* **14** 319–335

[3] Grattan S R, Diaz F J, Pedrero F and Vivaldi G A 2015 Assessing the suitability of saline wastewaters for irrigation of *Citrus*spp.: Emphasis on boron and specific-ion interactions. *Agricultural Water Management* **157** 48–58

[4] Munns R and Tester M 2008. *Mechanisms of Salinity Tolerance.* CSIRO Plant Industry, Canberra pp 651-672

[5] Horie T, Karahara I, and Katsuhara M. 2012 Salinity tolerance mechanisms in glycophytes: An overview with the central focus on rice plants. *Rice* 5: 11

[6] Muchate N S, Nikalje G C, Rajurkar N S, Suprasanna P and Nikam T D 2016. Plant Salt Stress: Adaptive Responses, Tolerance Mechanism and Bioengineering for Salt Tolerance. *Botanical Review* **82** 371 – 406

[7] Srivastava A K and Suprasanna P 2015 *Redox regulated mechanisms: Implications for enhancing plant stress tolerance and crop yield.* In: Giridhar Kumar Pandey (Eds.), Elucidation of Abiotic Stress Signaling in Plants. Springer, Netherland 191–205

[8] Zhu J K 2002 Salt and drought stress signal transduction in plants. Annual Review of Plant Biology **53** 247–273

[9] Lafitte H R, Ismail A and Bennett J 2004 *Abiotic stress tolerance in rice for Asia: progress and the future.* International Rice Research Institute, Manila pp 13

[10] Rad H E, Aref F and Rezai M 2012 Response of Rice to Different Salinity Levels during Different Growth Stages. *Research Journal of Applied Sciences, Engineering and Technology* **17** 3040-3047

[11] Lutts S, Kinet J M and Bouharmont J 1995 Changes in plant response to NaCI during the development of rice (*Oryza sativa* L.) varieties differing in salinity resistance. *Journal of Experimental Botany* **46** 1843-52

[12] Zhou M, Johnson P, Zhou G, Li C, and Lance R 2011 Quantitative trait loci for waterlogging tolerance in a barley cross of franklin × yuyaoxiangtian erleng and the relationship between waterlogging and salinity tolerance. *Crop Science* **52** 2082-88

[13] Binzel M L, Hess F D, Bressan R A, and Hasegawa P M 1988 Intracellular compartmentation of ions in salt-adapted tobacco cells. *Plant Physiology* **86** 607-14

[14] Lin C C and Kao C H 2000 Effect of NaCl stress on H_2O_2 metabolism in rice leaves. *Plant Growth Regulation* **30** 151-155

[15] Qiang H, Silliman B R, and Cui B 2011 Inconrporating thresholds into understanding salinity tolerance: A study using salt-tolerant plants in salt marshes. *Ecology and Evolution* **16** 6326-33

[16] Suryanto P, Kurniasih B, Faridah E, Nurjanto H H, Rogomulyo R, Handayani S, Kastono D, Muttaqien A S, Alam T. 2020. Influence of furrow with organic material and Chromolaena odorata compost on upland rice productivity in an agroforestry system with Melaleuca cajuputi. *Biodiversitas Journal of Biological Diversity* **21** 780-791.

[17] Arini N 2017 *Pengaruh perlakuan pra-tanam terhadap pertumbuhan dan hasil dua kultivar padi (Oryza sativa L.) tercekam salinitas* Thesis (Yogyakarta, Indonesia: Universitas Gadjah Mada).

[18] Devi Alvioliana 2019 *Pengaruh pesemaian salin basah dan kering terhadap pertumbuhan dan hasil padi (Oryza sativa L.) tercekam salinitas* Thesis (Yogyakarta, Indonesia: Universitas Gadjah Mada).

[19] Budiastuti K, Chalida N, Mohamad N E A P and Tohari. 2020. Seed priming alleviates crop growth inhibition by salinity. *IOP Conf. Series: Earth and Environmental Science* 458 (2020) 012006.

[20] Ehab A B 2016 *Seed priming* to alleviate salinity stress in germinating seeds. Journal of Plant Physiology **192** 38-46.

[21] Liming S, Orecutt D M and Foster J G. 1992. Influence of PEG & aeration method during imbibition on germination and subsequent seedling growth of flatpea (*Lashyrus sylvestris*). Seed Science and Technology **20** 349-357.

[22] Bonilla P, Hirai, Naito T H and Tsuchiya M 1995 Physiological response to salinity in rice plant. Induced salt-tolerance by low NaCl pretreatment. *Japan Journal of Crop Science* **64** 266-72.

[23] Boy R, Indradewa D, Putra E T S, Kurniasih B 2020 Drought-induced production of reactive oxygen species and antioxidants activity of four local upland rice cultivars in Central Sulawesi, Indonesia. *Biodiversitas* 21(6) 2555-65

[24] Villagra P E and Cavagnaro J B 2005 Water stress effects on the seedling growth of Prosopis argentina and Prosopis alpataco. Journal *of Arid Environments* **64** 390–400.

[25] Cayuela E, Estan M T, Parra M, Caro M and Bolarin M C. NaCL pretreatment at the seedling dtage enhances fruit yield of Tomato plants irrigated with salt water. *Plant and Soil* **230** 231-238.

[26] Gebreegziabher B G and Qufa C A. 2017. Plant physiological stimulation by seeds salt priming in maize (*Zea mays*): Prospect forsalt tolerance. *African Journal of Biotechnology*. **17** 154–162.

[27] Suwignyo R, Hayati R, and Mardiyanto. 2009. Pengaruh Salinitas Awal Rendah Terhadap Pertumbuhan dan Toleransi Salinitas Tanaman Jagung. *Agritrop Jurnal Ilmu-Ilmu Pertanian*. **5** 13-19

[28] Amzallag G N, Seligmann H and Lerner H R 1993 A Developmental Window for Salt-Adaptation In Sorghum bicolor. *J.Experimental Botany* **44** 645-652.

[29] Saha P, Chatterjee P, Biswa A K 2010 NaCl pretreatment alleviates salt stress by enhancement of antioxidant defense system and osmolyte accumulation in mungbean (V*igna radiata L. Wilczek*). Indian J Exp Biol. **48** 593-600

[30] Raichur K 2009 Nursery Raising (Dry Nursery Method) in Paddy. <http://agropedia.iitk.ac.in/content/nursery-raising-dry-nursery-method-paddy>. Diakses pada 3 Maret 2018

[31] Patnayak A, Tripathi A K and Munda G C 2008. *Rice seed production manual under IRRI project*. ICAR Research Complex NEH Region. Meghalaya. India.

[32] Liu P, Yin L, Wang S, Zhang M, Deng X, Zhang S, dan Tanaka K. 2015. Enhanced root hydraulic conductance by aquaporin regulation accounts for silicon alleviated salt-induced osmotic stress in *Sorghum bicolor* L. *Environmental and Experimental Botany* **111** 42–51

[33] Sarangi S K, Singh S, Kumar V, Srivastava A K, Sharma P C, and D.E. Johnson D E. 2019. Tillage and crop establishment options for enhancing the productivity, profitability, and resource use efficiency of rice-rabi systems of the saltaffected coastal lowlands of eastern India. *Field Crop Research* **247**

[34] Lennard B E G 2003 The interaction between waterlogging and salinity in higher plants: causes, consequences and implications. *Plant and Soi*l **253** 35–54

[35] Arnon D I 1949 Copper enzymes in isolated chloroplasts, polyphenoxidase in *Beta vulgaris*. Plant physiology **24** 1-15

[36] Yeo A R, Flowers S A, Rao G, Welfare K, Senanayake N, and Flowers T J. 1999. Silicon reduces sodium uptake in rice (*Oryza sativa* L.) in saline conditions and this is accounted for by a reduction in the transpirational bypass flow. *Plant Cell and Environment* **22** 559-65

[37] Kurniasih B, Greenway H, dan Colmer T D 2017 Energetics of acclimation to NaCl by submerged, anoxic rice seedlings. Annals of Botany **119** 129–42

[38] Munns R, Wallace P A, Teakle N L, and Colmer T D. 2010. Measuring Soluble Ion Concentrations (Na^+, K^+, Cl^-) in Salt-Treated Plants. Plant Stress Tolerance. Methods and Protocols, Springer pp 371-382

[39] Anonim, 2019. http://bbpadi.litbang.pertanian.go.id/index.php/info-berita/info-teknologi/padi-hitam-jeliteng-kaya-manfaat. diakses pada 9 Januari 2020.

[40] Pengkumsri N, Chaiyasut C, Saenjum C, Sirilun, Peerajan S M, Suwannalert S P, Sirisattha S and Sivamaruthi B S **2015**. Physicochemical and antioxidative properties of black, brown and red rice varieties of northern Thailand. *J. Food Science and Technology* **35** 331-38.

[41] Anonim 2008. Deskripsi Varietas Tanaman Padi. Balai Besar Penelitian Tanaman Padi. Departemen Pertanian pp. 1

[42] Kibria M G, Hossain M, Murata Y, and Hoque Md A. 2017. Antioxidant defense mechanisms of salinity tolerance in rice genotypes. Rice Science **24** 155-62

[43] Viégas R A, Mello A R B, and Silveria J A G 1999 Nitrate reductaseactivity and proline accumulation in cashew in response to NaCl saltshock. *Revista Brasileira de Fisiologia Vegetal* **11** 21-8

[44] Rao K R and Gnaham A 1990. Inhibition of nitrate and nitratereductase activity by salinity stress

in Sorghum vulgare. Phytochemistry **29** 1047-49

[45] Katiyar S, and Dubey R S **1992** Influence of NaCl salinity on behaviours of nitrate reductase and nitrite reductase in rice seedlings differing in salt tolerance. *Journal of Agronomy and Crop Science* **169** 289-97

[46] Gouia H, Ghorbal M H and Touraine B 1994 Effects of NaCl on flows of N and mineral ions and NO^{3-} reductase rate within whole plants of salt-sensitive bean and salt tolerant cotton. *Plant Physiology* **105** 1407-18

[47] Anuradha S and Rao, S R 2003 Application of brassinosteroids to rice seeds (*Oryza sativa* L.) reduced the impact of salt stress on growth, prevented photosynthetic pigment loss and increased nitrate reductase activity. *Plant Growth Regulation* **40** 29–32.

[48] Kurniasih B, Greenway H and Colmer T D 2013 Physiol Plant 149 222-33.

[49] Aref F 2013 Effect of saline irrigation water on yield and yield components of rice (*Oryza sativa* L.). *African Journal of Biotechnology* **12** 3503-13

Application of Drip Irrigation on Various Planting Media and Cow Urine in the Main Media of Preliminary Oil Palm Nursery

C I Wahyudin[1,*], A S Mahulette[2], V L Tanasale[2], D A Marasabessy[2], N Goo[2], W D Mariati[3], Y Asmi[1], R A Rifqah[1], Michala[4]

[1] Agrotechnology Study Program, Sekolah Tinggi Teknologi Pelalawan, Riau, Indonesia
[2] Agrotechnology Study Program, Faculty of Agriculture, Universitas Pattimura, Ambon, Indonesia
[3] Department of Marine Science, Faculty of Health Agriculture and Marine Science, Universitas Muhammadiyah Palopo, South Sulawesi, Indonesia
[4] Faculty of Islamic Communication and brodadcasting, Institut Agama Islam Bunga Bangsa Cirebon, Jawa Baeat, Indonesia

* E-mail: cecep.i.w@gmail.com

Abstract. The objective of this study was to determine the effect of drip irrigation and biological organic fertilizer treatment on oil palm growth. This study used a randomized complete block design with 3 replications. This research consisted of three experiments, namely drip irrigation, planting media composition, and cow urine dose on the growth of oil palm seedlings. There are three levels of drip irrigation, namely 30 minutes, 60 minutes and 90 minutes. Experiments on the type of planting media composition consisted of two types, namely M1 (ultisol soil + husk charcoal + fern fertilizer) and M2 (ultisol soil + empty bunches + goat manure). The cow urine experiment consisted of three levels, namely 100 ml, 150 ml, 300 ml. The results showed that giving water for 30 minutes gave the best results on stem circumference, plant height increase, plant dry weight increase, plant wet weight increase in oil palm seedlings. M1 media (ultisol soil + husk charcoal + fern fertilizer) had a significant effect on the increase in plant height, wet weight and dry weight of oil palm seedlings. Treatment of 150 ml cow urine liquid fertilizer is sufficient for the needs of oil palm seedlings with a response that is not different from 200 ml of cow urine.

1. Introduction

Oil palm (*Elaeis guineensis* Jacq.) is one of the plantation crops that has an important role in the plantation sub-sector. The development of oil palm provides benefits in increasing the income of farmers and the community, such as by providing raw material for the processing industry that creates added value in the domestic market, Crude Palm Oil (CPO) exports to generate foreign exchange, as well as providing employment opportunities for more than 2 million workers in various fields [1].

Oil palm production is strongly influenced by the quality of the seeds used. Therefore, nursery is the first step in the whole series of oil palm cultivation activities. Through this nursery stage, it is hoped that they can produce good and quality seedlings, so that in the end the seeds will be able to grow well in the field [2].

One of the advances in oil palm nurseries is the improvement of nursery techniques and the development of superior seeds that produce higher and earlier yields, so that farmers do not wait too long for plants to produce, as well as better production and quality. Basically there are two nursery systems: single stage and double stage. The single-stage nursery system is basically the same as double-stage nursery, except it does not go through preliminary nursery first. The increasing demand for oil palm causes production and expansion of oil palm planted areas to increase. With the increasing area of oil palm cultivation, it is necessary to procure large quantities of high-quality seeds. Nurseries are a determining factor for the success of oil palm cultivation. In oil palm nurseries, there is a "double stage" nursery. The preliminary nursery is carried out for three months and requires shade. Preliminary nurseries aim to get plants that grow uniformly when transferred to the main nursery. The main nursery is done to prepare the plants to be strong enough before being transplanting to the field [3].

Ultisol is soil that is commonly found in swampland and has a low nutrient content and pH but can still be used for the cultivation process, namely by adding or mixing other planting media that has high nutrient content such as empty bunches, husk charcoal, cocopeat, fern roots, and manure [4]. The planting medium is the most important component in planting crops because it is the material in which plants grow and develop. The purpose of this study was to determine the effect of drip irrigation, media composition, and cow urine dose on the growth of oil palm seedlings. The use of liquid organic fertilizer from urine metabolized by livestock provides benefits such as helping plant grow, because the high content of N and K contains growth hormones for plants and are easily absorbed by plants [5].

2. Methods

The research used oil palm seedlings in the pre-nursery and kept them for two months in the nursery. This study used a randomized block design and consisted of several experiments. The first experiment used the duration in watering using drip irrigation, divided into three duration levels: 30 minutes, 60 minutes, 90 minutes. The second experiment was to test two types of media for oil palm nurseries, namely M1 (ultisol soil + husk charcoal + fern fertilizer) and M2 (ultisol soil + empty bunches + and goat manure). The study used a dose of cow urine, divided into three levels, namely 100, 150 and 300 ml. Factors observed were the plant height increase, leaf number increase, palm weevil diameter, plant dry weight and plant wet weight. Data analysis used analysis of variance (F test) at α level of 5%. If the F test is significant, a further test is carried out by Duncan's Multiple Range Test (DMRT). Data were analyzed using analysis of variance (F test) at the level of α 5%. If the treatment had a significant effect, it then would proceed with the Duncan Multiple Range Test (DMRT). Data analysis was performed using STAR 2.0.1 software.

3. Results and discussion

3.1 Experiment 1. Effect of drip irrigation duration on oil palm seedling growth.

The duration of drip irrigation was significant for the increase in height, weevil width, dry weight and wet weight of plants. Giving drip irrigation for 30 minutes resulted in a better response than giving drip irrigation for 90 minutes. Giving drip irrigation for 90 minutes inhibited the growth of plants with height gain, weevil width, dry weight and plant wet weight having lower values than that of giving drip irrigation for 30 minutes. The availability of nutrients that can be absorbed by plants by providing sufficient water by plants is one of the factors that affect plant growth, which will increase cell enlargement which affects the diameter of the plants' weevils [6]. Water is used by plants for cell division and enlargement which is manifested in increase in plant height, enlargement of diameter, leaf multiplication and root growth. The condition where the soil water content is above field capacity curbed plant growth due to inhibition of root development caused by a lack of oxygen in the soil [7].

Table 1. Increased observation results from various parameters in drip irrigation

Parameter	Drop Irrigation (minute)		
	30	60	90
PHI	27.67 a	24.10 ab	22.19 b
LNI	7.89 a	7.67 a	7.22 a
WD	19.05 a	18.16 b	16.93 c
CWW	461.97 a	414.43 ab	371.47 b
CDW	160.53 a	140.13 ab	117.30 b

Note: PHI: Plant height increase, LNI: leaves number increase, WD: weevil diameter, CWW: crown wet weight, CDW: crown dry weight, the numbers followed by the same letter in the same column show no significant difference using the test DMRT α 5%.

3.2 Experiment 2. Effect of media composition on oil palm seedling growth

The composition of the planting medium had a significant effect on the parameters of plant height, wet weight and dry weight of oil palm seedlings. Table 2 shows M1 treatment resulted in significant difference from M2 treatment in producing wet weight of oil palm seedlings. This is thought to be related to the physical properties of the plant media in supporting the growth of oil palm seedlings. M1 media with a composition consisting of husk charcoal and fertilizer was thought to increase the porosity and permeability of the planting medium. Husk charcoal media will increase soil porosity and support root growth better than M2 [8]. The role of organic matter is very important in increasing the ability of soil to retain water [9]. By increasing the ability of the soil to hold water, plant roots will more easily absorb nutrients to support the growth of oil palm seedlings.

Table 2. The increase in the results of the observation of various parameters on the provision of planting media.

Parameter	Media	
	M_1	M_2
PHI	29.67 a	23.30 b
LNI	8.00 a	7.89 a
WD	18.27 a	18.26 a
CWW	460.67 a	381.90 b
CDW	152.12 a	125.12 b

Note: PHI: Plant height increase, LNI: leaves number increase, WD: weevil diameter, CWW: crown wet weight, CDW: crown dry weight, the numbers followed by the same letter in the same column show no significant difference using the test DMRT α 5%.

3.3 Experiment 3. Effect of cow urine dose on oil palm seedling growth.

Giving cow urine with a concentration of 150 ml visibly increased plant height compared to giving 100 ml, although it was not significantly different from giving 200 ml. Treatment of 200 ml of cow urine increased the plant's wet weight compared to that of 100 ml of urine, although it was not significantly different from that of giving 150 ml treatment. This shows that the provision of 150 ml of liquid organic fertilizer from cow urine already met the nutrient needs of oil palm seeds. Increasing the concentration of fertilizer to a certain extent will increase plant growth, because plants have a limit of nutrient absorption according to plant growth [10]. Application of fertilizer with a concentration exceeding the optimum concentration will reduce plant growth and production [11]. Plant wet weight is influenced by the absorption process of water and nutrient by the plant. The absorption of water and nutrients by plants depends on the water and nutrient reserves absorbed and the ability to absorb them [12].

IOP Conf. Series: Earth and Environmental Science **883** (2021) 012015 doi:10.1088/1755-1315/883/1/012015

Table 3. Increase in the results of observations of various parameters in cow urine administration.

Parameter	Urine		
	U1 (100 ml)	U2 (150 ml)	U3 (200 ml)
PHI	19.72 b	25.14 a	24.66 a
LNI	7.10 a	7.78 a	7.82 a
WD	16.70 a	17.99 a	17.53 a
CWW	329.88 b	383.67 ab	418.67 a
CDW	101.89 a	134.01 a	134.11 a

Note: PHI: Plant height increase:, LNI: leaves number increase, WD: weevil diameter, CWW: crown wet weight, CDW: crown dry weight, the numbers followed by the same letter in the same column show no significant difference using the test DMRT α 5%.

4. Conclusion

Giving drip irrigation for 30 minutes showed better results than giving drip irrigation for 60 and 90 minutes. Using composition of M2 media (ultisol planting medium + husk charcoal + fern fertilizer) increased plant height, wet weight and dry weight of the plant. The application of 150 ml cow urine liquid fertilizer is sufficient for the needs of oil palm seeds with no significant difference in plant response from the application of 200 ml cow urine.

5. Acknowledgment

The first author thanked Yayasan Amanah Pelalawan (YAP) and Sekolah Tinggi Teknologi Pelalawan (STT Pelalawan) for funding the research.

References

[1] Dradjat B 2008 prospek kebun sawit masih cerah. Lembaga Riset Perkebunan Indonesia. Jakarta.

[2] Sianturi H S D 1993 Budidaya Kelapa Sawit. Fakultas Pertanian Universitas Sumatera Utara, Medan

[3] Mangoensoekarjo S and Semangun H 2008 Manajemen Agribisnis Kelapa Sawit. Universitas Gajah Mada press, Yogyakarta

[4] Prasetyo B H and Suriyadirka D A 2006. Karakteristik Potensi dan Teknologi Pengolahan Tanah Ultisol untuk Pengembangan Pertanian Lahan Kering di Indonesia Bogor. Balai BesarPenelitian dan Pengembangan SumberdayaLahan Pertanian, Bogor.

[5] Budhie D D S 2010 *Aplikasi Urin Kambing Peranakan Etawa dan NASA sebagai Pupuk Organik Cair untuk Pemacu Pertumbuhan dan Produksi Tanaman Pakan Legum Indigofera sp.* IPB, Bogor.

[6] Sarief E S 1985. Kesuburan dan Pemupukan Tanah Pertanian. Pustaka Buana, Jakarta.

[7] Nyakpa M Y, Pulung M A, Amrah G A, Munawar A, Hong G B, and Hakim N Kesuburan Tanah. Universitas Lampung, Lampung.

[8] Lestari G A, Sumarsono, and Fuskhah E 2019 Pengaruh komposisi media tanam dan dosis poc urin kelinci terhadap pertumbuhan dan produksi tanaman melon (*Cucumis melo* L.) Effect. *Pertan. Trop.* **6** 180–189.

[9] Panjaitan C 2010 Pengaruh pemanfaatan kompos solid dalam media tanam dan pemberian pupuk npkmg (15:5:6:4) terhadap pertumbuhan bibit kelapa sawit (*Elaeis guineensis* jacq.) di pre nursery. Fakultas Pertanian Universitas Sumatera Utara, Medan.

[10] Arifianto M, Hartati R M, and Setyorini T 2019 Pengaruh dosis urine sapi dan waktu penyiraman terhadap pertumbuhan bibit kelapa sawit (*Elaeis guineensis* Jacq) di pre-nursery *Agromast* **4** 2–3.

[11] Mappanganro N Sengin E L and Baharuddin 2011 Pertumbuhan dan produksi tanaman strawberry pada berbagai jenis dan konsentrasi pupuk organik cair dan urine sapi dengan sistem hidroponik irigasi tetes *Jurnal Fakultas Pertanian* Universitas Hasanuddin, Makassar.

[12] Moenandir J 1993 Ilmu Gulma Dalam Sistem Pertanian. Raja gafindo persada. Jakarta.

IOP Conf. Series: Earth and Environmental Science **883** (2021) 012016 doi:10.1088/1755-1315/883/1/012016

The Effect of *polyethylene glycol* concentration on some varieties of kenaf (*Hibiscus cannabinus* L.) in enhancing the germination viability

C I Wahyudin[1,*], A S Mahulette[2], J I Nendissa[2], M H Makaruku[2], W D Mariati[3], A Haitami[4], Desrihastuti[1], Febrianti[1], Arby'in Pratiwi[5]

[1] Agrotechnology Study Program, Sekolah Tinggi Teknologi Pelalawan, Riau, Indonesia
[2] Agrotechnology Study Program, Faculty of Agriculture, Universitas Pattimura, Ambon, Indonesia
[3] Department of Marine Science, Faculty of Health Agriculture and Marine Science, Universitas Muhammadiyah Palopo, South Sulawesi, Indonesia
[4] Faculty of Agriculture, Universitas Islam Kuantan Singgingi, Riau
[5] Departemen of Animal Products Technology, Faculty of Animal Sciences, University of Gadjah Mada jl. FaunaNo. 3, Bulaksumur Yogjakarta 5528, Indonesia

* E-mail: cecep.i.w@gmail.com

Abstract. The purpose of this study was to see the interaction between the concentration of Polyethylene Glycol and various varieties of kenaf (*Hibiscus cannabinus* L.) On plant germination. This study used a completely randomized factorial design with 3 replications. The factors used for the study were varieties of kenaf (Karang Proso 6, Karang Proso 9 and Karang Proso 15) and the concentration of Polyethylene Glycol (0, 3, 5, and 7 ppm). Data were analyzed using analysis of variance (F test) using least significant difference (LSD) test of 5%. There was an interaction between the concentration of Polyethylene Glycol and kenaf varieties on germination power. Karang Proso 9 variety was consistent with several concentrations of Polyethylene Glycol which had higher germination compared to other varieties. Polyethylene Glycol concentration did not affect hypocotyl length, seed germination uniformity, dry weight of the varieties of Karang Proso 6, Karang Proso 9 and Karang Proso 15.

1. Introduction

Kenaf (*Hibiscus cannabinus* L.) is an annual plant and belongs to the Malvaceae family. Kenaf tree bark can be used to produce fiber. Kenaf fiber is used as raw material for pulp and paper, bio-composites, drain fiber, particleboard, car dashboards and bio-plastics [1]. In addition, kenaf seeds can also be used as food, because they contain 20% unsaturated fatty acids such as palmitic acid, oleic acid and linoleic acid. The composition of kenaf seed sterol oil is similar to that of soybean and cottonseed sterol [2]. In addition, kenaf leaves contain 24% crude protein, so it is best used as feed for poultry and ruminants.
Program Intensifikasi Serat Karung Rakyat (ISKARA) has been developing kenaf since 1978/1979. Kenaf reached the peak of its plantation in 1986/1987 with an area of 26,000 ha, and the average productivity was around 0.9-1.2 tonnes/ha [3]. Kenaf plantations are spread across Lampung, West Java,

East Java and South Kalimantan. However, the high competition for plastic packaging has led to a drastic decrease in kenaf use. The decline in the use of kenaf has an impact on the decline in the development of the kenaf area. The total area of kenaf in Indonesia is approximately 1,308 ha, and its location is in Lamongan, East Java Province. Fiber production is around 1,254 tons with a productivity of 0.96 tons / ha. This value is far lower than the optimal potential that must be achieved (2-4 tonnes /ha) [4]. Various methods were used to increase kenaf production, such as variety selection, nursery management, fertilization management and testing of kenaf plants on various types of soil (podzolic red-yellow and Incepticol) [5]. Balai Penelitian Tanaman Pemanis dan Serat (Balittas) has produced some of the best varieties including Karang Proso 6, 9, 11, 12, 14, 15, 16 and 19 [6]. However, information on the level of productivity of these varieties is still limited.

On the other hand, a good germination management strategy is an important factor in increasing kenaf production. However, a decrease in seed quality will occur due to errors in germination treatment. The result of these conditions will cause a decrease in seed viability. This problem can be solved by the priming method. The priming method is an accelerating and simultaneous germinating method through water absorption control. Osmoconditioning is a general treatment of the priming methods used for germination. The osmoconditioning process is carried out by immersing the seeds in a solution of Polyethylene Glycol. Polyethylene Glycol solution is a compound to control imbibition and hydration processes in seeds. Although Polyethylene Glycol solution can accelerate germination, the concentration of this solution is still little known. Therefore, this study will examine the concentration of Polyethylene Glycol variety Kenaf (Hibiscus cannabinus L.) in increasing germination viability.

2. Methods

This study used a completely randomized factorial design with 3 replications. The factors used in this study were varieties of kenaf (Karang Ploso 6, Karang Ploso 9 and Karang Ploso 15) and the concentration of Polyethylene Glycol (0, 3, 5, and 7 ppm). Data were analyzed using analysis of variance (F test) at the level of α 5%. If the treatment has a significant effect, it is followed by the least significant difference (LSD) test. Data analysis was examined using STAR 2.0.1 software. After the preparation of kenaf seeds is ready, Polyethylene Glycol 6000 will be made with concentrations of 3, 5, and 7 ppm following the equation N1.V1 = N2.V2. The main solution will be made 100 ppm PEG 6000 = 500 mg which will be dissolved in 500 ml water. PEG 6000 will be diluted into several concentrations. The paper will be dried and damp cotton will be placed on the paper. Parameters were observed 1 week after the seeds were germinated. The seeds will be arranged on wet cotton. In the next stage, the seeds will be covered with wet paper. Observations will be made after seven days. The parameters to be measured were the percentage of germination, hypocotyl length, dry weight and germination time.

3. Results and discussion

3.1 Percentage of Germination

Polyethylene Glycol concentrations of 0, 3, 5 and 7 ppm were not significantly different. There was an interaction between the PEG concentration and the kenaf variety. Varieties of Karang Proso 9 had the highest germination rate compared to Karang Proso 6 and Karang Proso 15 at various concentrations of PEG. It was assumed that Karang Proso seeds had higher viability than the other two varieties. Seed viability was the viability of the seeds as indicated by the ratio of normal sprouts to the total seeds that were tested for germination [7].

Table 1. Percentage of Sprouts (%) in some kenaf varieties and Polyethylene concentrations

Variety	Polyethylene Glycol Concentration (ppm)				Average
	0	3	5	7	
Kalang Proso 6	51,67 d	71,67 bc	56,67 cd	43,33 d	55,83 c
Kalang Proso 9	100,00 a	100,00 a	100,00 a	100,00 a	100,00 a
Kalang Proso 15	80,00 b	71,67 bc	83,33 ab	85,00 ab	80,00 b
Average	77,22 a	81,11 a	80,00 a	76,11 a	

Note: The number for each main treatment and its combination followed by the same lowercase letter, is not significantly different according to the least significant difference 5%.

The high viability of Karang Proso 9 seeds caused the osmoconditioning treatment to be ineffective, as stated in Widajati 1999 study which showed that the osmoconditioning treatment was effective in seeds that had deteriorated and germinated 72-90% [8]. The Karang Proso 6 variety with PEG 3 ppm increased the ability to germinate by 20% compared to the control, then decreased the ability to germinate at a PEG concentration of 7 ppm. PEG concentrations that were too high caused the osmotic potential value around the seeds to become increasingly negative, so that water was difficult to absorb by the seeds. The low osmotic potential of the solution could inhibit the imbibition process [9]. In the Karang Proso 15 variety the germination rate at a concentration of 5 and 7 ppm increased the ability to germinate, this indicated that the optimum concentration to support plant germination was more than 7 ppm for Karang Proso 15 variety.

3.2. Simultaneous growth

The simultaneous growth of Karang Proso 6 variety was lower than Karang Proso 9 and Karang Proso 15. This showed that the seeds of Karang Proso 6 varieties had experienced a decline, supported by table 1 where the sprouts of Karang Proso 6 were lower than those of Karang Proso 9 and 15 varieties. The addition of insignificant PEG concentration increased the simultaneous growth, but there were several interactions between kenaf varieties and the dosage of Polyethylene Glycol. In Karang Proso 9 and 15 varieties, PEG 7 ppm treatment increased the simultaneous growth by 20% compared to the control (0 ppm), while the Karang Proso 15 variety was given 10% compared to the control (Table 2). Nurmauli and Nurmiaty 2010 on soybean plants, giving PEG with a concentration of 20% increased the simultaneous growth compared to PEG treatment of 10% [9]. It was assumed that the increase in PEG concentration would increase water imbibition more quickly, and affect the rate of stimulation of the gibberine hormone in changing food reserves and synchronizing plant growth [10].

Table 2. Simultaneous growth (%)in some kenaf varieties and Polyethylene concentrations

Variety	Polyethylene Glycol Concentration (ppm)				Average
	0	3	5	7	
Kalang Proso 6	31,67 d	28,33 d	46,67 cd	31,67 d	34,58 b
Kalang Proso 9	80,00 ab	78,33 b	65,00 bc	100,00 a	80,83 a
Kalang Proso 15	71,67 b	63,33 bc	73,33 b	81,67 ab	72,50 a
Average	61,11 a	56,67 a	61,67 a	71,11 a	

Note: The number for each main treatment and its combination followed by the same lowercase letter, is not significantly different according to the least significant difference 5%.

3.3. Hypocotyl length and dry weight

Table 3 showed that the hypocotyl length of the three varieties of kenaf and various concentrations of Polyethylene Glycol as well as the combination of the three varieties of kenaf and various concentrations of Polyethylene Glycol showed no significant differences.

Table 3. Hypocotyl length (mm) in some kenaf varieties and Polyethylene Glycol concentration

Variety	Polyethylene Glycol Concentration (ppm)				Average
	0	3	5	7	
Kalang Proso 6	11,13 a	6,60 a	9,10 a	11,10 a	9,48 a
Kalang Proso 9	9,57 a	8,53 a	9,17 a	11,00 a	9,57 a
Kalang Proso 15	9,07 a	10,87 a	9,33 a	9,70 a	9,74 a
Average	9,92 a	8,67 a	9,20 a	10,60 a	

Note: The number for each main treatment and its combination followed by the same lowercase letter, is not significantly different according to the least significant difference 5%.

Table 4 showed that the dry weight of three kenaf varieties and various concentrations of polyethylene glycol as well as the combination of three kenaf varieties and various concentrations of polyethylene glycol were not significantly different. This was because sprouts dry weight was more influenced by the length of the hypocotyl, where there was a very close correlation between the length of the hypocotyl and the dry weight of the sprouts. Increasing hypocotyl length could increase sprouts dry weight [11].

Table 4. Plant dry weight (g) in some kenaf varieties and Polyethylene Glycol concentration

Variety	Polyethylene Glycol Concentration (ppm)				Average
	0	3	5	7	
Kalang Proso 6	73,52 a	43,58 a	60,09 a	73,30 a	62,62 a
Kalang Proso 9	63,17 a	56,35 a	60,53 a	72,64 a	63,17 a
Kalang Proso 15	59,87 a	71,76 a	61,63 a	64,05 a	64,33 a
Average	65,52 a	57,23 a	60,75 a	70,00 a	

Note: The number for each main treatment and its combination followed by the same lowercase letter, was not significantly different according to the least significant difference 5%.

4. Conclusion

Karang Proso 9 varieties were consistent with several concentrations of Polyethylene Glycol which had higher germination rates compared to other varieties. Polyethylene Glycol concentration did not affect hypocotyl length, seed germination uniformity, dry weight of the varieties of Karang Proso 6, Karang Proso 9 and Karang Proso 15.

5. Acknowledgment

The first author would like to thank the Yayasan Amanah Pelalawan (YAP) and Sekolah Tinggi Teknologi Pelalawan (STT Pelalawan) for funding the research.

References

[1] Cheng W Y, Haque Akanda J M and Nyam K L 2016 Kenaf seed oil: A potential new source of edible oil *Trends Food Sci. Technol.* **52** 57–65

[2] Mossello A A, Harun J, Tahir P M, Resalati H, Ibrahim R, Fallah Shamsi S R and Mohmamed A Z 2010 A Review of Literatures Related of Using Kenaf for Pulp Production (Beating, Fractionation, and Recycled Fiber) *Mod. Appl. Sci.* **4**

[3] Natasa A A, Zuhry E and Adiwirman 2016 Pertumbuhan dan kandungan serat beberapa varietas kenaf (Hibiscus cannabinus L.) *JOM Faperta* **3** 99–102

[3] Sudjindro 2004 Budidaya Kenaf (*Hibiscus cannabinus* L.) Seri Edisi Khusus Balai Penelitian Tembakau dan Tanaman Serat. Malang.

[4] Dinas Perkebunan Propinsi Riau. 2014. Riau Kembangkan Potensi Tanaman Kenaf

[6] Balai Penelitian Tanaman Pemanis dan Serat. 2014. Monograf Kenaf

[7] Tefa A 2017 Uji Viabilitas dan Vigor Benih Padi (Oryza sativa L.) selama Penyimpanan pada Tingkat Kadar Air yang Berbeda *Savana Cendana* **2** 48–50

[8] Widajati E 1999 Deteksi vigor biokimiawi dan vigor fisiologi untuk fenomena pemulihan vigor pada tingkat awal deteriorasi dan devigorasi benih kedelai (*Glycine max* (L.) Merr.) Melalui proses invigorasi Institut Pertanian Bogor, Bogor.

[9] Nurmauli and Nurmiaty Y 2010 Kedelai yang telah disimpan selama sembilan bulan *Ilmu Pertan. Indones.* **15** 20–4

[10] Sa'diyah H 2009 Pengaruh invigorasi menggunakan Polietilena Glikol (PEG) 6000 terhadap bernih Rosela (*Hibiscus sabdarifa* var. altisima) Universitas Islam Negeri Maulana Malik Ibrahim, Malang.

[11] Hidayat T R S and Yamin M 2019 Aplikasi Perendaman ZPT terhadap Perkecambahan Benih Kapas Pros. Semin. Nas. 4 295–304

Introduction to a systematic review and meta-analyses in Indonesia nutrition poultry: case study in probiotic

D N Adli[1,3,*], O Sjofjan[1,3], A Jayanegara[2,3], B P Mahardika[2]

[1] Faculty of animal science, University of Brawijaya, 65145, East Java, Indonesia
[2] Department of Nutrition and Feed Technology, Faculty of Animal Science, IPB University, Bogor, Indonesia
[3] Animal Feed and Nutrition Modelling Research Group (AFENUE), Department of Animal Nutrition and Feed Technology, Faculty of Animal Science, IPB University, Bogor, Indonesia

* E-mail: danungnuradli1994@gmail.com

Abstract. The review conducted to Introduction to a systematic review and meta-analysis in Indonesia nutrition poultry: case study in feed additive. A review on the table showed the origin phase research and introduction in Indonesia (representatives for animal sciences) both meta-analyses and probiotic. The result showed first generation origin of probiotic started on Indonesia the oldest one were from *Aspergillus sp* and *Bacillus sp* with meta-analysis started with small ruminant. Meta-analyses origin with author from Indonesia on animal science were started with title Influence of dietary tannin levels on methane production from ruminant livestock: a meta-analysis (Anuraga Jayanegara, 2010). To sum up, the idea help to increase in industry revolution 4.0 were adaptable and applicator.

1. Introduction

The World Health Organization (WHO) announced the coronavirus novel (COVID-19) as a global pandemic outbreak as of March 11, 2020. A total of 4.87 million cases have been confirmed worldwide, with 1.66 million people recovering and 321 thousand people die. The pandemic spread over the world and faced development countries e.g. Indonesia. COVID-19 impact to livestock sector especially poultry production. Intensive poultry production systems demand a supply of high protein- and easy available in development countries. The key current to ad However, during the last decade, since 1997 the banned of antibiotics growth promotes were began at Denmark with *avoparcin* as AGPs. The phase continued, until the last phase when the EU-wide ban on AGPs in animal feed (poultry) took effect since 2006 (EC Regulation No 1831/2003). According to the newest regulation PERMENTAN/14/16/2017 Indonesia has banned the use of antibiotics on poultry both broiler for meat and laying hens for egg product.

The banned were started early periods from 1 January 2018. Probiotics are defined as a living microorganism that is used as feed additives and can benefit their hosts by improving the balance of their digestive microorganism. Probiotics commonly from lactic acid bacteria (LAB), namely; *Lactobacillus* and *Bifidobacterium* that can be found in the intestine. Probiotic feed additives improve growth, feed efficiency, and intestinal health. Earlier studies reported an active role for probiotics in prohibiting or reducing the shedding of *Campylobacter jejuni* load in vivo trials. In recent research that probiotic mixtures also have beneficial effects against a wide range of disorders, although the evidence that mixtures are more effective than their component strains is more limited. Nevertheless, in future a further potential advantage. The feed additive or probiotics era already through by itself from upstream to downstream. A lot of researchers already conduct research on probiotic with data increasing significantly on Scopus data since 2000-2020 but the determination of conclusion from it research were weak. In addition, to answer that the systematic review and meta-analysis were conducted. A meta-analysis approach, which increases the power of the statistical analysis.

2. Methods

2.1. Meta-analysis

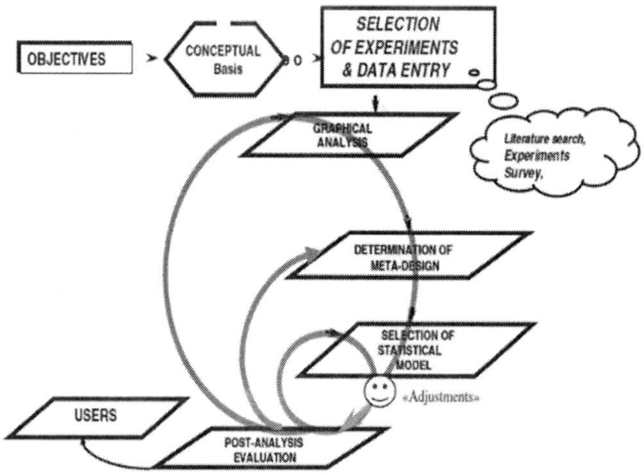

Figure 1. The steps of meta-analyses [23]

The stage for conducting a meta-analyst starts from 1) choosing the objectivity of the study, where the theme selection plays a key role in a study, the hypothesis will be determined. For example the effect of specific nutrients on feed additives. The next step is to start 2) data entry or start entering data, where the selected literature review must be entered into a database, systematically, and each study is given a specific code. The final step is 3) data filtering or sorting out data, which in this stage is further divided into three sub-sections namely a) correlation between studies, b) finding error data on the results of quantitative meta-analyses, c) ensuring that selected publications are mutually related to each other.

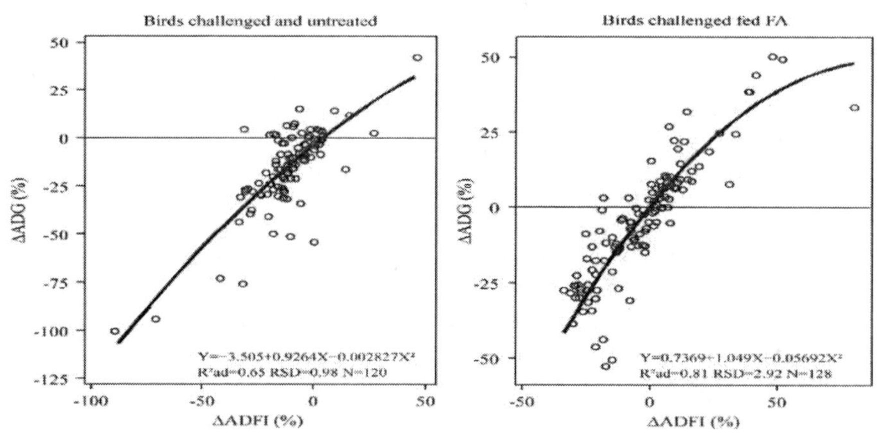

Figure 2. before meta-analyses **Figure 3.** after meta-analyses

2.2. Meta-analyses in animal science from author on Indonesia

Meta-analyses in Indonesia were showed on this table follows sorting from oldest to newest, data adapted from Google scholar, Scopus documents.

Table 1. The practices of meta-analysis in animal science

Animal origin	Purposes	References
Goat and sheep	Summarize and to quantify the tannin effects on CH4 production from ruminants and its associated variables	[1] Jayanegara, *et al.*, (2010)
Sheep, goat, cattle, and Buffaloes	This meta-analysis was performed to evaluate whether voluntary feed intake and digestibility of forage-based diets differ between four domestic ruminant	[2] Riaz, *et al.*, (2014)
Dairy cattle, bulls and heifers, sheep, and goats	To evaluate correlations derived previously by deduction.	[3] Schuba, *et al.*, (2017)
Sheep in Indonesia	The objective of this study was to determine energy and protein requirements, for both maintenance and gain	[4] Jayanegara, *et al.*, (2017)
Dairy	This study aimed to perform a meta-analysis on the effect of 3-nitrooxypropanol (3-NOP) on enteric methane (CH4) emissions from ruminants.	[5] Jayanegara, *et al.*, (2018)
Silages	The present evaluation aimed to analyse the influence of tannins on silage quality	[6] Jayanegara, *et al.*, (2019)
Ruminants	To integrate the published data on different concentrations and types of MCFA such as lauric acid and myristic acid, which investigated ruminal methanogenesis and fermentation in in vitro and in vivo experiments	[7] Yanza, *et al.*, (2020)
Broiler	To determine the effects of zinc on the immune response and production performance of broilers.	[8] Hidayat, *et al.*, (2020)
Laying Hens	To determine the effects of probiotic on the laying hens with it parameters	[26] Sjofjan *et al.*, (2021)

2.3. Probiotic origin from author on Indonesia

Research on probiotic in Indonesia were showed on this table follows sorting from oldest to newest, data adapted from Google scholar, Scopus documents.

Table 2. The probiotics practices started from old to newest in Indonesia

Kind of Probiotic	References	Animal
Aspergillus niger dan bacillus sp.	[9] Sjofjan O 2003	Laying hens
Bacillus sp, Lactobacillus salivarius	[10] Kompiang I P 2009	Poultry industry
Bacillus subtilis	[11] Abbas *et al* 2004	Broiler
Marine yeast (*Saccharomyces sp*)	[12] Sjofjan O 2010	Broiler

Lactic acid bacteria	[13] Afdora *et al* 2010	Quail
Amylolytic bacteria	[14] Sjofjan O and Ardyati T 2011	Quail
Lactobacillus fermentum	[15] Kalsum U Soetanto H, and Sjofjan O 2012	Japanese quail
Saccharomyces sp	[16] Sidauruk H, Ardyati, T, and Sjofjan, O. 2013	Laying hens
Liquid Probiotic Mixed Culture	[17] Pambuka S R, Sjofjan O, and Radiati L E 2014	Laying hens
Lactobacillus sp	[18] Primacitra D Y., Sjofjan O, and Natsir MH 2014	Quail (meat)
Bacillus apiarius	[19] Kompiang P, and Sjofjan O 2014	Laying hens
Bacillus sp	[20] Sjofjan O and Adli D N 2020, April	Broiler

3. Results and Discussion

The references style use APA style. The first generation origin of probiotic started on Indonesia the oldest one were from [9] with A and B concept for probiotics, this author were conducted the research on laying hens and it egg quality. The phase continued when WHO started to banned the antibiotic growth promoters for animal product on 2006. Even though, the Europe countries were already began with *avoparcin* in 1997. The banned spread to development countries e.g. Indonesia. 1999-2000 the researcher association started in on agriculture not animal. However, Sjofjan and Kompiang (2003) started selected for it candidate from crop, intestinal, caecum, and even faeces for it probiotic. *Lactobacillus acidophilus, L. casei, L. fermentum, L. plantarum, L. salivarius, L. reuteri, L. delbrueckti, L. lactis, L. cellobiosus, L. brevis, Aspergillus oryzae, Bifidobacterium longum, B. pseudologum, B. bifidum, B. suis, B. thermophilum, Bacillus subtilis, Enterococcus faecum, Saccharomyces cerevisiae, Streptococcus faecium,* and *S. intermedius*. Not only lactic acid bacteria but also yeast were selected. In addition, probiotic were selected also from marine yeast. Fourth generation from (2003-2020) were still conducted the probiotic until it occur second generation extracted from outer cell wall of *saccharomyces cerevisiae* on arbor acres (Adli and Sjofjan, 2020). In addition, probiotic and selected candidate for probiotics were not only for broiler and laying hens but also on quail, duck, and even native chicken. The regulation PERMENTAN/14/16/2017 Indonesia has banned the use of antibiotics on poultry both broiler for meat and laying hens for egg product [21]. The banned were started early periods from 1 January 2018, the probiotics already implemented in current poultry industries from upstream to downstream [22][24][25]. A huge variation on this research should be in line with the new method to make it strong result with meta-analyses.

Meta-analyses origin with author from Indonesia on animal science were started with title Influence of dietary tannin levels on methane production from ruminant livestock: a meta-analysis (Anuraga Jayanegara, 2010). The author conducted a lot meta-analyses but 80% were for ruminant animal and it first were to determine the effects of zinc on the immune response and production performance of broilers. The author tried wrote the probiotic that already conducted in Indonesia era were already conducted in Indonesia started with this paper for introduction. Hopefully, in industry revolution 4.0 were adaptable and applicated using meta-analyses [26].

4. Conclusion

To sum up, the idea help to increase in industry revolution 4.0 were adaptable and applicated.

References

[1] Jayanegara A, Soliva C R, and Marquardt S 2010 *EAAP* **127** **L** 461
[2] Riaz M Q, Südekum K H, Clauss, M, and Jayanegara A 2014 *Livest. Sci.* **162** L76
[3] Schuba J, Südekum K H, Pfeffer E, and Jayanegara A 2017 *Livest. Sci.* **198** L82
[4] Jayanegara A, Ridla M, Astuti D A, Wiryawan K G, Laconi E B, and Nahrowi N 2017 *Med. Pet.* **40** L118

[5] Jayanegara A, Sarwono K A, Kondo M, Matsui H, Ridla, M, Laconi E B, and Nahrowi 2018 *Ital. J. of Anim. Sci.* **17** L 650
[6] Jayanegara A, Sujarnoko, TU, Ridla M, Kondo M, and Kreuzer, M 2019 *J. Anim. Physiol. Anim. Nutr.* **103** L 456
[7] Yanza Y R, Szumacher - Strabel M, Jayanegara A, Kasenta, AM, Gao M, Huang H, and Cieślak A 2020 *J. of Anim. Phy. and Anim. Nutr.*
[8] Hidayat C, Sumiati A J, and Wina E 2020 *Asian-Austral. J. of Anim. Sci.* **33** L465
[9] Sjofjan O 2003 *JNT* **23**
[10] Kompiang I P 2009 *Pengembangan Inovasi Pertanian* **2** L 177
[11] Abbas M H, Mahata M E, Fauzano R, Rasyaf M, Prasetyo H L, Ketaren P P, and Supiana S 2004 *Int. J. of Poult. Sci.* **16** S58-S61
[12] Sjofjan O 2010 Effect of Addition of Marine Yeast Culture (Saccharomyces Sp) in Diet on Broiler Carcass Quality *JIIP* **18** L2
[13] Afdora P T, Ardiyati, T, Sjofjan, O, and Kalsum, U 2010 *J. of Trop. Life Sci.* **1** L 1
[14] Sjofjan O, and Ardyati, T 2011 *Int. J. of Poult. Sci.* **10** L 411
[15] Kalsum U, Soetanto H, and Sjofjan O 2012 *Int. J. of Poult. Sci.* **11 L** 311
[16] Sidauruk H, Ardyati, T, and Sjofjan O 2013 *Int. J. Biosci* **3** L 36
[17] Pambuka S R, Sjofjan O, and Radiati L E 2014 *J. Worlds Poult. Res* **4** L 5
[18] Primacitra D Y, Sjofjan O, and Natsir, MH 2014 *J. of Trop. Anim. Prod.* **15** L74
[19] Kompiang P, and Sjofjan O 2014 *JITV* **9** L 1
[20] Sjofjan O, and Adli D N 2020 *IOP Conf. Series: Earth and Env. Sci.* **478**
[21] Sjofjan O, Natsir M H, Adli D N, Adelina D D, and Triana L M *IOP Conf. Series: Earth and Env. Sci.* **465**
[22] Adli D N, Sjofjan O, Natsir M H, Nuningtyas Y F, Sholikah N, and Marbun A C 2020 *LRRD* **32** L 120.
[23] Sauvant D, Schmidely P, Daudin J J, St-Pierre NR 2008 *Anim*
[24] Sjofjan O, Adli D N, Natsir M H, Nuningtyas Y F, Bastomi I, Amalia FR 2021 *J. Indonesian Trop. Anim. Agric.* **46** L 1
[25] Sjofjan O and Adli D N 2021 *LRRD* **33** L 17
[26] Sjofjan O, Adli D N, Sholikin M M, Jayanegara A and Irawan A 2021 *J. of Anim. Feed Sci.* 30

Optimization of gamma ray irradiation dose on strawberry plantlets

D Saptadi[1,*], H Arisah[2] and D Agisimanto[2]

[1]Plant Breeding Laboratory, Faculty of Agriculture, Universitas Brawijaya, Jl. Veteran, Malang, Indonesia 65145
[2]Plant Breeding Laboratory, Indonesian Citrus and Subtropical Fruits Research Institute, Jl. Raya Tlekung No.1, Kota Batu, Indonesia 65327

* E-mail: darmawan.fp@ub.ac.id

Abstract. As a commodity that is cultivated vegetatively, strawberries have a relatively low genetic diversity. One effort to increase the diversity of strawberries and genetic quality improvement can be made by using gamma-ray radiation. In vitro explant of strawberry cv Early Bright were radiated with Gamma-ray P3TIR BATAN facility. Starting with a radiosensitivity test to find out Lethal Dose (LD) 50 with dosage level; 5, 10, 20, 30, 40, 60, 80, 100, 125, 150, 175, and 200 Gy. LD50 observations were carried out to get the optimal dose. The treatment of several doses of gamma-ray radiation showed varied results, with the percentage of explants living decrease by increasing the dose of irradiation. Irradiation dose until 10 Gy did not affect the survival rate of explant—the treatment dose of 80 Gy and above causes 100% mortality on the final observation. The dose of 60 Gy had a significant effect on shoot growth, as indicated by curly leaves' growth. The next optimal dose recommended for radiation is set at 20 and 30 Gy to anticipate genetic changes that are not morphologically expressed and to avoid the negative effects of morphological changes.

1. Introduction

Strawberry is not native to Indonesia but can be developed in the highlands of several regions in Indonesia. The economic value of strawberries is very high, both as fresh fruit and raw material for the food processing industry. Nutritional value is also high because the various chemical compounds that are very useful are not indented [1]. Currently, several varieties have been adapted and grown by farmers in Indonesia [2]. Farmers' cultivation constraints are low productivity, fluctuations in production due to environmental fluctuations, less attractive fruit size and color, low storage capacity, and high vulnerability to pests [3].

One of the efforts to overcome cultivation constraints is by providing new superior varieties that have the expected characters. Strawberry is an octoploid plant that reproduces vegetatively. The genetic diversity is relatively low, so conventional breeding is not easy to do [4]. Mutation breeding has proven to be the most feasible approach for creating new genetic diversity in various plant species quickly without changing the overall genetic constitution [5]. Physical mutagens are more advantageous than chemical mutagens because there is no need to remove mutagens after application [6], and leaves no hazardous waste that needs special handling. Mutation induction with physical mutagens has been widely used. 90% of the mutants that exist today are acquired by physical mutations, of which 64% are using gamma rays and 22% by X rays [7][8]. Based on data from the IAEA Mutant Database Variety, 1.665 of the 3.362 mutant varieties currently released are the result of mutations by gamma-ray irradiation.

Selection of mutagen type and the optimal dosage is important to open up plant improvement opportunities through mutation breeding [9]. The mutation dose is more important than the mutagen type. Whatever the mutagens, before application on a larger scale, the dose of the mutation must be determined correctly [10]. High-dose mutations' application opens a higher chance of acquiring mutants but often results in infertility and individual death [11]. In vitro techniques allow the induction of a large number of explant mutations and a relatively short time to separate mutated and non-mutated parts. Also, the mutants obtained are free from the disease [12]. The Indonesian Citrus and Subtropical Fruits Research Institute (ICSFRI) collects strawberry explants ready to be mutated by gamma-ray irradiation. Experiments are needed to determine the LD50 (the dose that causes 50% mortality) to get the correct dose in mutation induction (Predieri, 2001; [14]. This study aims to obtain the right dose of gamma-ray irradiation for strawberry explants as the first step for genetic improvement of strawberries in Indonesia.

2. Methods

The research was carried out at the Plant Breeding Laboratory, Faculty of Agriculture, Universitas Brawijaya and Tissue Culture Laboratory, Indonesian Citrus and Subtropical Fruits Research Institute (ICSFRI), from May to September 2019.

The genetic material used was the Early Bright strawberry explants. The source of the explants is the result of propagation from meristem culture. The main stems obtained were cut to 500 μM and incubated to further select mutation induction materials on MS medium without hormones.

Gamma-ray irradiation activities were carried out at P3TIR National Nuclear Energy Agency of Indonesia (BATAN). Control explants were explants that were not treated with irradiation. A total of 20 explants each were irradiated at a dose of 5, 10, 20, 30 Gy, and 10 explants each were irradiated at a dose of 40, 60, 80, and 100 Gy. The irradiated explants were planted on propagation media (MS + 3% sucrose, 10% myoinositol, 0.05 mgL-1 BA and 0.025 mgL-1 NAA for recovery.

Observation of Lethal Dose (LD) 50 on irradiation was carried out for one month, and further observations were made until the plants showed significant changes in appearance. Observations included the percentage of mortality explants and the qualitative appearance of the planlets. The observations were compared and analyzed descriptively.

3. Results and Discussion

Genetic improvement by mutation is very beneficial if carried out in vegetatively reproduced plants. Once a superior genotype has been obtained, the superiority will be maintained through clonal propagation. Inducing mutations on in vitro explants of strawberries are carried out for several reasons, i.e. (1) the in vitro propagation method is relatively well established [15][16] and (2) chimeras can be isolated and can be developed into whole plants easily [17] so that the selected material will be extensive and the opportunity to get new varieties will be even greater.

The first observation of the survival percentage of explants after irradiation did not show a concomitant trend with radiation dose. Explants treated with 5 Gy irradiation decreased their vitality to about 80%, but at 10 Gy treatment, all explants were able to survive. Contamination contributed to decreased viability at the 5 Gy treatment. Explants irradiated at a dose of 60 Gy, and higher indicated high viability. The treatment that produced a percentage of 50% survives explants was a dose of 20 Gy. At doses 30 and 40, the percentage of live explants was 60% (Figure 1). Because the trend showed an unclear pattern, gamma-ray irradiation was re-treated on the new explants with five new higher doses, which are 125, 150, 175, and 200 Gy. Simultaneously with this second treatment, the treatment was also repeated with doses of 20 and 30 Gy for a higher number of explants.

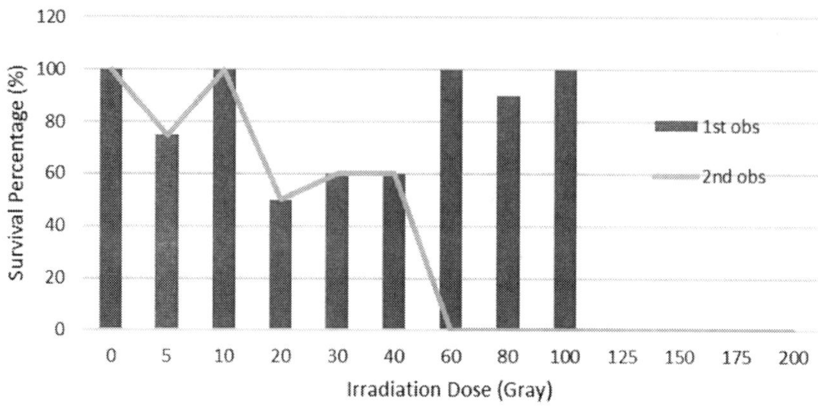

Figure 1. Percentage of survival explant (%) by different gamma ray irradiation dose

In the second observation (about 4 months after irradiation), the propagated plantlets from explants treated above 60 Gy deteriorated and eventually died (Figure 1). At doses of 60 Gy and higher, there were morphological changes in the leaves, in the form of an abnormal leaf shape that became curled (Figure 2).

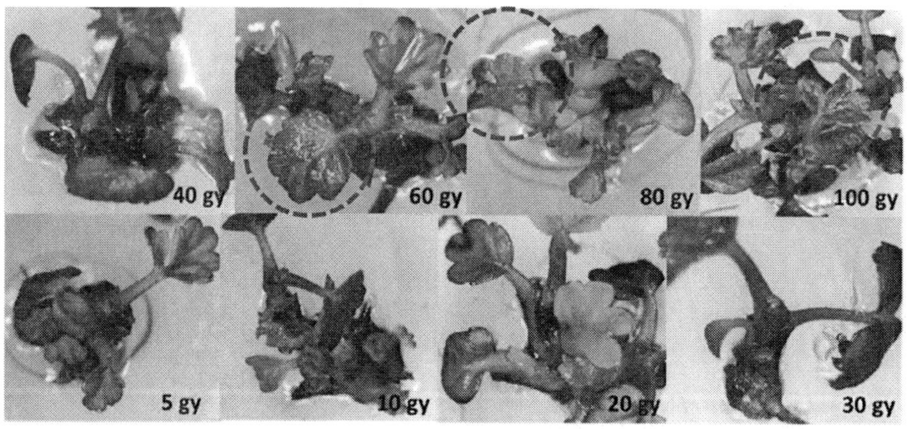

Figure 2. Appearance of planlets irradiated with gamma rays at different doses

In the first gamma-ray irradiation treatment, several plantlets were obtained that could still survive, but there was a difference in their growth performance. Irradiation treatment at a dose of 60-100 Gy caused a decrease in growth performance. This shows that high doses have an impact on the decreasing quality of plantlet growth [18]. Meanwhile, the growth performance at 5-40 Gy dose of radiation treatment can still be categorized as normal (Table 1).

Treatment doses of 125, 150, 175, and 200 Gy caused a low survival rate, namely 0 - 15% and poor growth performance. The explants' growth was stagnant, the shoots were stunted and abnormal, and the number of shoots per cluster was very low. The shoot shape in treatment 175 and 200 Gy was more abnormal when compared to the results of the 150 Gy dose treatment (Figure 3).

Table 1. Growth of plantlets treated with gamma ray irradiation at different doses

No	Treatment (Gy)	Growth performance
1	5	++++
2	10	++++
3	20	++++
4	30	++++
5	40	++++
6	60	+++
7	80	++
8	100	+

Figure 3. Appearance of strawberry plantlets with irradiation treatment of 100 Gy (a) 125 Gy (b) 150 Gy (c) and 175 Gy (d)

Based on the results of observations of the percentage of mortality and the growth performance of plantlets, the dosage between 20 and 30 Gy is thought to be the dosage range that results in 50% mortality, such as the results of previous studies on citrus commodities [19]. This dose will then be recommended to treat mutations with gamma rays on strawberry explants because at this dose. It is expected that genetic changes have occurred even though no morphological changes. The use of low doses is also recommended by [20], which states that low doses of gamma radiation could be used for better survival and growth in strawberries under in vitro conditions. Doses of 60 Gy and above are not recommended because they could cause too much damage. Similar research on strawberry conducted by Thanh [21] shows that doses of 60-100 Gy cause stunted plants, small and abnormal fruit.

The formation of mutated plant populations with optimal radiation doses (20 and 30 Gy) was carried out by maintaining the clusters for further separation, acclimatization, and transfer to the field. Observations of genetic diversity both morphologically and molecularly will then be carried out when the plants are mature and stable in the land.

4. Acknowledgments
We gratefully thank to the Doctor and Professor Research Grant from Universitas Brawijaya for financial supporting to this research

References
[1] Sutopo 2016 Teknologi Budidaya Stroberi di Lahan *Artikel*
[2] Hanif Z and Ashari H 2015 Sebaran Stroberi (Fragaria x ananassa) di Indonesia *Seminar Nasional Pekan Inovasi Teknologi Hortikultura Nasional : Penerapan Inovasi Teknologi Hortikultura dalam Mendukung Pembangunan Hortikultura yang Berdaya Saing dan Berbasis Sumber Daya Genetik Lokal* pp 87–95
[3] Budiman S and Saraswati D 2008 *Berkebun Stroberi Secara Komersial* (Jakarta: Penebar Swadaya)

[4] Nunes C F, Ferreira J L, Generoso A L, Dias M S C, Pasqual M and Cançado G M de A 2013 The genetic diversity of strawberry (*Fragaria ananassa* Duch.) hybrids based on ISSR markers *Acta Sci. - Agron.* **35** 443–52

[5] Khursheed S, Raina A, Parveen K and Khan S 2019 Induced phenotypic diversity in the mutagenized populations of faba bean using physical and chemical mutagenesis *J. Saudi Soc. Agric. Sci.* **18** 113–9

[6] Khan S J, Khan H ullah, Khan R D, Iqbal M M and Zafar Y 2000 Development of Sugarcane Mutants Through in vitro Mutagenesis *Pakistan J. Biol. Sci.* **3** 1123–5

[7] Jain S M 2005 Major mutation-assisted plant breeding programs supported by FAO/IAEA *Plant Cell. Tissue Organ Cult.* **82** 113–23

[8] Jain S M 2010 Mutagenesis in crop improvement under the climate change *Rom. Biotechnol. Lett.* **15** 88–106

[9] Raina A, Laskar R A, Khursheed S, Khan S, Parveen K, Amin R and Khan S 2017 Induce Physical and Chemical Mutagenesis for Improvement of Yield Attributing Traits and their Correlation Analysis in Chickpea *Int. Lett. Nat. Sci.* **61** 14–22

[10] Ulukapi K, Özdemir B and Onus A 2015 Determination of Proper Gamma Radiation Dose in Mutation Breeding in Eggplant (Solanum melongena L .) *Adv. Environ. Agric. Sci.* 149–53

[11] Koornneef M 2002 Classical mutagenesis in higher plants *Molecular plant biology* ed P Gilmartin and B Chris (Oxford University Press.) pp 1–12

[12] Ahloowalia B S 1998 Improvement of Vegetatively Propagated Plants 293–309

[13] Predieri S 2001 Mutation induction and tissue culture in improving fruits *Plant Cell. Tissue Organ Cult.* **64** 185–210

[14] Álvarez-Holguín A, Morales-Nieto, Carlos Raúl Avendaño-Arrazatec C H and Corrales-Lerma, Raúl Villarreal-Guerrero, Federico Santellano-Estrada, Eduardo Gómez-Simuta Y 2019 Mean lethal dose (LD 50) and growth reduction (GR 50) due to gamma radiation in Wilman lovegrass (Eragrostis superba) *Rev Mex Cienc Pecu.* **10** 227–38

[15] Bhandari M and Roy S 2015 Standardization and establishment of an efficient protocol for in vitro multiplication of Strawberry plant and its genetic stability testing *Int. J. Pharm. Sci. Invent.* **4** 7–12

[16] Anuradha, Sehrawat S K, Poonia A K, Kajla S and Bhat S 2016 Production of strawberry plant by in vitro propagation *Res. Crop.* **17** 545–9

[17] Verma A K, Prasad K V., Singh S K and Kumar S 2012 In vitro isolation of red coloured mutant from chimeric ray florets of chrysanthemum induced by gamma-ray *Indian J. Hortic.* **69** 562–7

[18] Sykes P J, Day T K, Swinburne S J, Lane J M, Bhat M, Morley A A and Hooker A M 2006 In vivo mutagenic effect of VERY low dose radiation *Dose-Response* **4** 309–16

[19] Arisah H and Mariana D 2017 Keragaman Buah Jeruk Keprok SoE Mutan Generasi M 1 V 2 Hasil Induksi Mutasi Sinar Gamma (Fruit Diversity of SoE Mandarin Mutant at M 1 V 2 Generation Resulted from Gamma Irradiation) *Bul. Plasma Nutfah.* **23** 69–80

[20] Gupta R, Wali V K, Bakshi P, Singh G, Ahmad Shah R and Rani S 2018 Effects of Gamma Irradiation on Shoot, Root and Survival Percent in Strawberry cv. Chandler under In vitro Conditions *Int. J. Curr. Microbiol. Appl. Sci.* **7** 1173–82

[21] Thanh TC T, Trung H T, Nhi P Van and Trac V T 2016 Induction of materials for mutation breeding of strawberry (FragariaxAnanassa) by gamma irradiation (Phase 2). Annual Report for 2015 (VINATOM-AR--15). Nguyen Thi Phuong Lan (Ed.). Viet Nam

IOP Conf. Series: Earth and Environmental Science **883** (2021) 012019 doi:10.1088/1755-1315/883/1/012019

Physical and mechanical properties of cement-bonded particleboard made from squander of paper

R S Maail[1,*], J J Fransz[1] and J Titarsole

[1] Laboratory of Wood Technology and Wood Working, Department of Forest Products Technology, Faculty of Agriculture - Pattimura University, Jl. Ir. M. Putuhena, Ambon, 97232 Indonesia

* E-mail: rohny_maail@yahoo.com

Abstract. The objectives of this study were to assess and explain the usage of squander paper as a selective crude material in the production of cement-bonded particleboard (CBP) and to decide its physical and mechanical properties. The two treatment factors were used such as squander of paper types (A4 paper sheets, newspaper and cardboard) and $CaCl_2$ as chemical accelerator agents (in concentrations of 2%, 4% and 6% according to the weight of cement) with proportions mixed of cement - squander of paper - water about 2.5:1:1,25. The result indicated that the squander of paper types and the concentration of $CaCl_2$ have affected the physical and mechanical properties of CBP. The board which produced using cardboard with the proportions of materials (2.5:1:1.25) and accelerator $CaCl_2$ about of 6% were satisfied all the requirement standard of JIS A 5417 1992 and have the most physical and mechanical properties values.

1. Introduction

One of the composite board items that are effortlessly made is the cement board which uses cement as covers while the crude material can be either wooden particles or particles of lignocelluloses material. Unfortunately, being developed of cement board items, explicitly cement-bonded particleboard (CBP) faced with the state of the accessibility of wood as crude material have begun restricted, so it turns into an extraordinary research to search for the best answer to resolve it in the forthcoming. The utilization of selective material as crude materials for CBP is one of the promising arrangements. Squander of paper is a material that is conceivably. Squander of paper will be paper that has not been utilized again in its utilization so it is bound to be disposed or reused around the world. Squander is the remainder of the fixings in utilization by people. Paper is dainty material and the subsequent midpoints with pressure of fiber that comes from the mash drying work insight in addition to some extra material that adheres to one another and each other's weaves. Fiber ordinarily utilized common fiber generally containing cellulose and hemicellulose. Though some lignocelluloses material may show some level of contrariness with cement on the manufacture of CBP, there are arrangement the procedures to assess the level of contradiction, which are either thermal characteristics of a cement setting measure or on the physical and mechanical properties of the cement board made in the laboratory, when contrasted with similar boundaries yet got with a slick water-cement combination. On account of an unsatisfactory level of inconsistency, eliminated dissolvable compound by conventional hot or cold water extraction and splashing, respectively,[1-5] long-term putting away of the crude material,[6] numerous substance extraction methods,[7-9] and even treatment by fungi,[10] might be sufficient to redesign the crude material to the ideal level. Likewise, the expansion of modest quantity of cement

IOP Conf. Series: Earth and Environmental Science **883** (2021) 012019 doi:10.1088/1755-1315/883/1/012019

setting quickening agents (accelerators), similar to the chlorides of tin, iron, aluminum, magnesium[11] or calcium[12,13] have the impact on improving the compatibility. They can be applied to make quicker the setting of cement, improving the physical and mechanical properties of the composites, or delivering a given wood types reasonable for making composites with cement. In addition, accelerator empowers the response happens all the more rapidly or permit the response at lower temperatures because of changes that set off against reactant. During the time spent making of CBP, the kind of accelerator utilized was derived from the chloride salts, such as $CaCl_2$.

The objectives of this research were to assess and explain the usage of squander paper as an selective crude material and $CaCl_2$ as an accelerator in the production of CBP and discover the most physical and mechanical properties of CBP.

2. Methods

2.1 Materials

Mixtures with proportionally equal amounts of squander paper particles (HVS paper sheets, newspaper and cardboard) were utilized to manufacture CBP. The average size of the squander paper particles was 3 mm (L), 2 mm (W), and 0.2 mm (T). Ordinary Portland cement of TONASA was utilized as a binder and $CaCl_2$ was used as an accelerator.

2.2 Manufacture and Evaluation of CBP

CBP with a targeted density of 1.2 g/cm^3 was manufactured at a cement/squander paper particle/water weight proportion of 2.5:1.0:1.25. Hand-framed mats of 30 x 30 cm were cold pressed to a targeted thickness of 12 mm and kept in an oven-dry set at 60°C for 4 h. Subsequently, CBP was wrapped with a polyvinylchloride (PVC) film immediately, followed by curing for 24 hours and afterward conditioning at room temperature (1 week) prior to property evaluation. The manufacturing process of CBP as mentioned at Figure 1.

The physical and mechanical properties of the boards were tested in accordance with the Japan Industrial Standard (JIS) A 5908 (1994). The boards were cut into 50 x 50 mm samples for density, thickness swelling (TS), water absorption (WA) tests as well as the internal bond (IB) strength tests, and 50 x 210 mm samples for the static bending test. The static bending test was conducted using a three-point bending test over an effective span of 180 mm (15 times the board thickness) at a loading speed of 10 mm/min. Four test samples were prepared from each treatment group for the foregoing tests, and the average values were calculated.

2.3. Statistical Analysis

For statistical data analysis, this study was conducted using the Multiple Linear Regression model with two independent variables, in particular the squander paper types which consists of three levels, namely A4 paper sheets, newspaper and cardboard; and $CaCl_2$ as chemical accelerator agent comprises of three levels, namely 2%, 4% and 6%. In addition, the values obtained were also compared with the JIS A 5417 1992 standards so that the test boundaries that meet these standards are known.

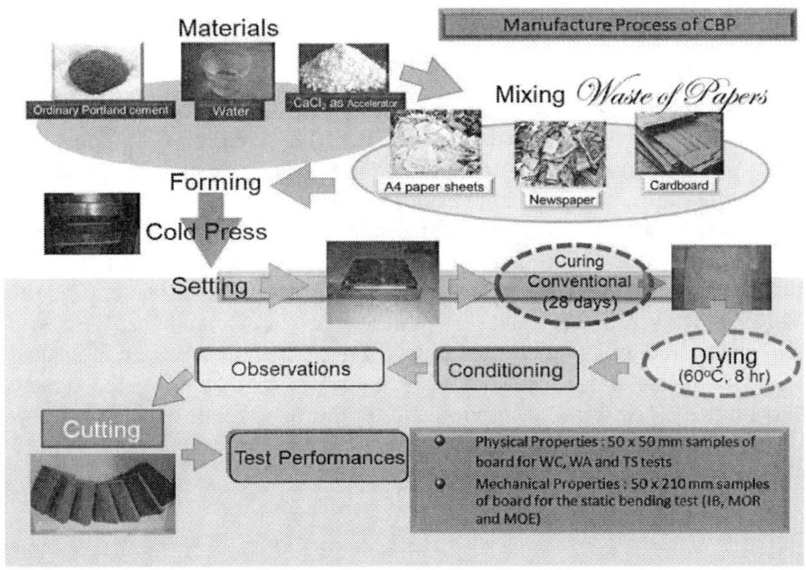

Figure 1. The manufacturing process of CBP

3. Results and discussion

3.1. Physical Properties

The result showed in Figure 2 that the density of the CBP made from squander of paper ranges from 1.18-1.22 gr/cm^3. Such the most density value of 1.22 gr/cm^3 obtained on the CBP - A3B3 with the proportion of 2,5:1: 1,25 were utilized cardboard and accelerator $CaCl_2$ of 6%, while the least density value of 1.18 gr/cm^3 obtained on board A1B1, A1B2 and A2B1 which used HVS paper sheet and accelerator $CaCl_2$ of 2% and 4% as well as newspaper used accelerator $CaCl_2$ of 2% followed by advanced hardening time for 2 weeks.

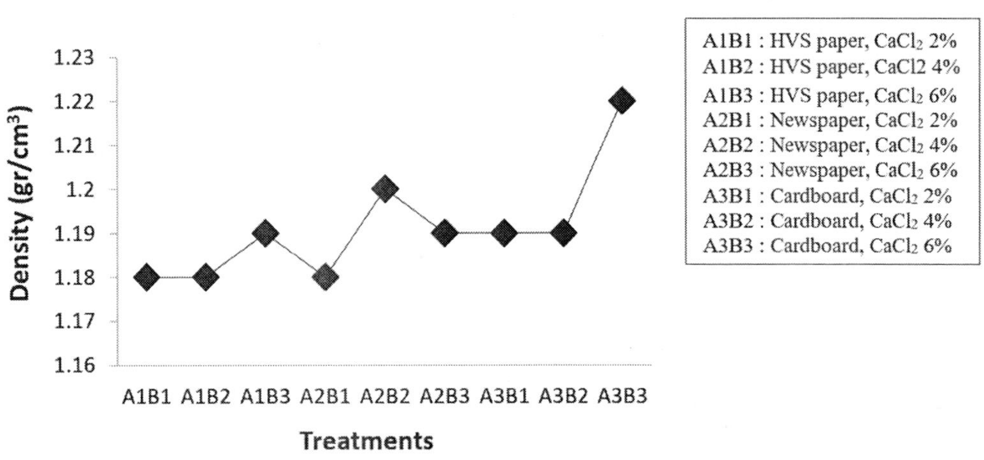

Figure 2. The density of the CBP made from various squander of paper and $CaCl_2$

The results show that the density of CBP produced using different squander of paper is not vastly different, and it is upheld during the time spent making cement board load up where the heaviness of materials are the equivalent. The presence of contrasts of density is just brought about by the distinction of the thickness of the end sheets. Despite the fact that to accomplish target 1,2 cm in thickness of CBP and to accomplish consistency of thick sheets have additionally utilized a steel stick as 12 mm. Furthermore, the accelerator concentration of $CaCl_2$ 6% is the optimum concentration used in the process of making cement boards because with this concentration can accelerate the bond between cement as crude materials with squander paper, particularly for cardboard material and reinforce ties as well as the hardening of the materials.

However, the physical properties change as representative of WA and TS values of CBP after 24 h of water soaking are appeared in Figure 3 and Figure 4. Significant correlations were observed between the WA and TS and treatment conditions of CBP under the board manufacture proportion treatment of 2.5:1:1.25. The lowest average WA and TS values were 15.24% and 0.25%, respectively, at CBP produced using cardboard having accelerator $CaCl_2$ of 6%. This phenomenon indicated that CBP made from cardboard using accelerator $CaCl_2$ of 6% may result in increasing the dimensional stability, apparently through decreased water absorption ability. For this research, the density, water absorption and thickness values of CBP met the JIS A 5417 1992 standard limits.

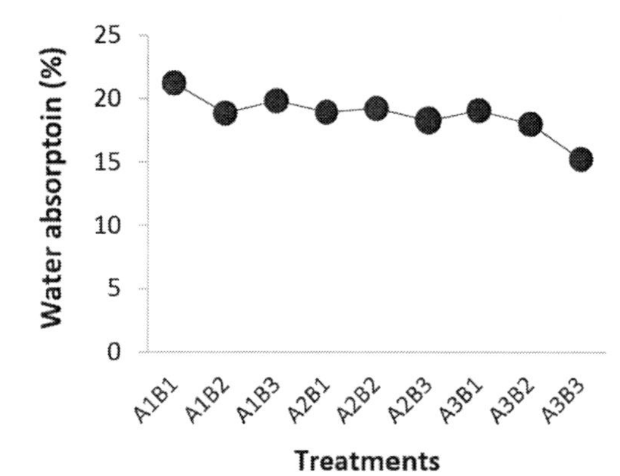

Figure 3. Water absorption test of CBP made. from various squander of paper and $CaCl_2$

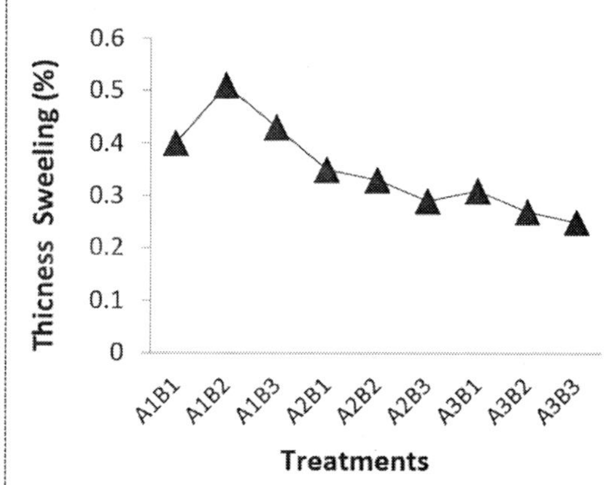

Figure 4. Thickness swelling test of CBP made from various squander of paper and $CaCl_2$

3.2. Mechanical Properties

The average values for the IB strength of CBP cured with conventional curing condition is presented in Figure 5. The linear relationship was found between the concentration of $CaCl_2$ as an accelerator and the IB strength of CBP. With the increase of concentration of $CaCl_2$ within about 6%, the IB values of CBP were significantly increased. The maximum average IB value of CBP made by cardboard was 1.42 MPa after conventional curing treatment. This condition with the proportion of materials such as 2.5:1:1.25 and treated by $CaCl_2$ about of 6% allowed rapid curing of CBP in the range of this experiment and contributed to increasing the IB strength of CBP. Furthermore, the IB value of CBP treated by $CaCl_2$ about of 6% and made by squander of cardboard after conventional curing was improved compared to others board. It should be noticed that CBP treated with $CaCl_2$ about of 6% and continued by conventional curing for 28 days then increased the IB strength. This result relevance with sufficient water for chemical reaction because of w/c ratio should around more than 30% (50%).

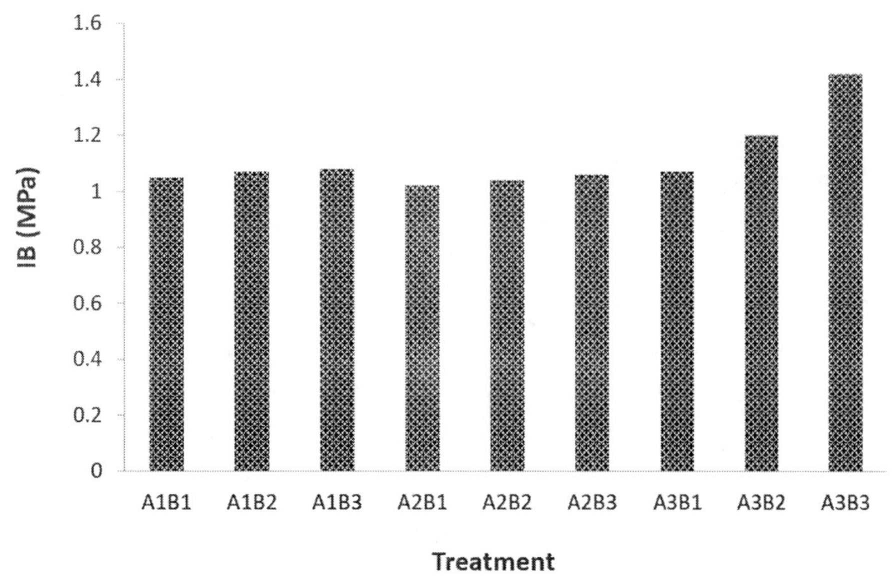

Figure 5. IB test of CBP made from various squander of paper and $CaCl_2$

Figure 6 and Figure 7 show the modulus of rupture (MOR) and modulus of elasticity (MOE) values of CBP treated with some squander of papers and $CaCl_2$. The maximum average MOR and MOE values were 11.2 MPa and 5.1 GPa, respectively, at board made of cardboard having the $CaCl_2$ at around 6%. This finding shows that $CaCl_2$ at around 6% accelerated the curing process and enhanced the MOR and MOE of CBP. The MOR and MOE values of CBP made of cardboard having the $CaCl_2$ at around 6% increased and were remarkably higher than the values of such other boards. Compared to CBP treated with $CaCl_2$ below 6%, in which the $CaCl_2$ from around 2% to 4%, the MOR and MOE values were decreased and they gave a negative effect on board performances. This conditions affected by insufficient of $CaCl_2$ to react with cement which the w/c ratio 0.5, as mentioned earlier, decreasing the internal bending strength of CBP. Wei(14) and Moeslemi(15) also found that the $CaCl_2$ accelerator as an additive can be used to increase the adhesive power of cement against materials containing cellulose such as paper, wood and other materials. $CaCl_2$ catalyst is added to inorganic binders such as cement to improve the quality of the binder bonds to other materials and accelerate the hardening of the binder. In addition, it can reduce the inhibiting effect of materials such as paper or wood used on the hardening of Portland cement. The IB, MOR and MOE test results as described above, meet the minimum standards of JIS A 5417 1992.

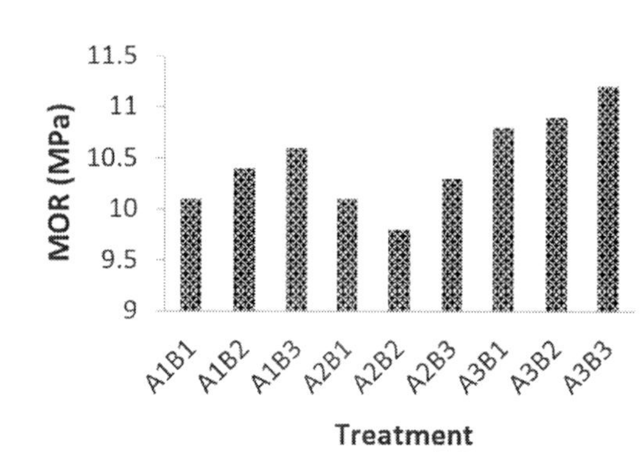

Figure 6. Modulus of Rapture values of CBP

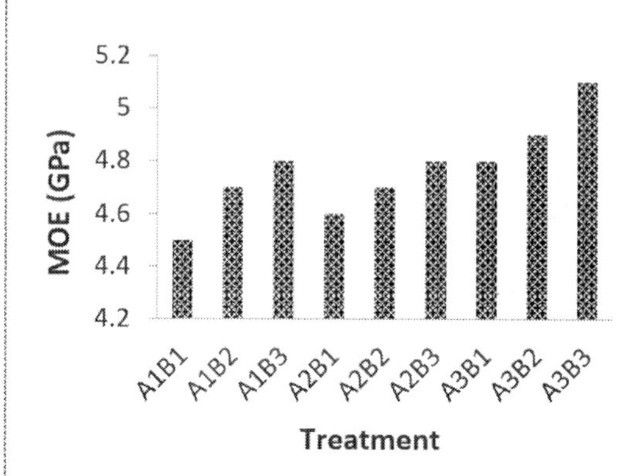

Figure 7. Modulus of Elasticity values of CBP

4. Conclusions

The physical and mechanical properties of CBP are excellent by using squander of papers even after the conventional curing process. These properties, however, have good conditions of boards which made from cardboard. It was considered that CBP made from cardboard using accelerator $CaCl_2$ about of 6% may result in increasing the dimensional stability, through low of water absorption as well as thickness swelling ability such as 15.24% and 0.25%. The maximum average MOR and MOE values were 11.2 MPa and 5.1 GPa, respectively, were found at the same board which is made from cardboard using $CaCl_2$ at around 6%. It was consider that the proportion of materials and the squander of papers as well as the concentration of $CaCl_2$ have affected the physical and mechanical properties of CBP and satisfied all the requirement standard of JIS A 5417 1992.

5. Acknowledgments

The authors thank the Directorate of Higher Education; Ministry of National Education-Indonesia, Pattimura University Ambon for their finance support through the PNBP (Penerimaan Negara Bukan Pajak) in Top Research Areas Grant 2018.

References

[1] Eusebio D A, Soriano F P, Cabangon R J, Evans P D 2000b Manufacture of low-cost-wood-cement composites in the Philippines using plantation-grown Australian species: I. Eucalyptus, "*Wood-Cement Composites in the Asia-Pacific Region, Canberra, Australia*, pp 105-114.
[2] Moslemi A A, Lim Y T 1984 Compatibility of southern hardwoods with Portland cement *Forest Prod J* **34**(7/8) 22-26.
[3] Okino E Y, Souza M R, Santana M A E, Alves M V, Sousa M E and Teixeira D E 2005 Physico-mechanical properties and decay resistance of Cupressus spp. Cement-bonded particleboards *Cement Concr. Compos* **27** 333-338.
[4] Schwarz H G, Simatupang M H 1984 Eignung des Buchenholzes zur Herstellung zementgebundener Holzwerkstoffe *Holz Roh Werkst* **41** 65-69 (in German).
[5] Sutigno P 2000 Effect of aqueous extraction of wood-wool on the properties of wood-wool cement board manufactured from teak (*Tectona grandis*) *Wood-Cement Composites in the Asia-Pacific Region*, Canberra, Australia, 24-28.
[6] Cabangon R J, Cunningham R B, Evans P D 2002 Manual strand orientation as a means of improving the flexural properties of wood-wool cement boards in the Philippines *Forest Prod J* **52**(4) 53-59.

[7] Bertos M F, Simons S J R, Hills C D, Carey P J 2004 A review of accelerated carbonation technology in the treatment of cement-based materials and sequestration of CO2 *J Hazard Mater* **B112** 193-205.
[8] Alberto M M, Mougel E, Zoulalian A 2000 Compatibility of some tropical hardwood species with Portland cement using isothermal calorimetry *Forest Prod J* **50** 83-88.
[9] Kavvouras P K 1987 Suitability of Quercus conferta wood for the manufacture of cement-bonded flakeboards *Holzforschung* **41** 159-163.
[10] Moslemi A A, Garcia J F, and Hofstrand A D 1983 Effect of various treatments and additives on wood-Portland cement-water systems *Wood Fibre Sci* **15**(2) 164-176.
[11] Hermawan D, Subyanto B, Kawai S 2001b Manufacture and properties of oil palm frond cement-bonded board *J Wood Sci* 47:208-213.
[12] Olorunnisula A O, Adefisian O O 2002 Trial production and testing of cement-bonded particleboard from rattan furniture waste *Wood Fibre Sci* **34**(1) 116-124.
[13] Semple K E, Evans P D 2007 Manufacture of wood-cement composites from Acacia mangium. Part II. Use of accelerators in the manufacture of wood-wool cement boards from A.mangium *Wood Fibre Sci* **39** 120-131.
[14] Wei Y, Tomita B 2001 Effects of five additive materials on mechanical and dimensional properties of wood cement-bonded boards *J Wood Sci* **47** 437-444.
[15] Moslemi A A, Plister S C 1987 The influence of cement/wood ratio and cement type on bending strength and dimensional stability of wood-cement composite panels *Wood Fibre Sci* **19** 165-175.

Embryo-gynogenic induction of Indonesia shallot (*Allium cepa* L. Aggregatum group) cultivars using three different protocol unpollinated flower cultures

E S Lestari[1,2], Sulastriningsih[2], D C Prayantini [2], A Purwantoro[1] and E Sulistyaningsih[1,*]

[1] Faculty of Agriculture, Gadjah Mada University, Jl Flora, Bulaksumur, Depok, Sleman, Yogyakarta, Indonesia
[2] PT BISI International, Tbk, Jl. HOS Cokroaminoto 72A, Pare, Kediri, East Java, Indonesia

* E-mail: endangsih@ugm.ac.id

Abstract. Embryo-gynogenic induction was an important step in haploid induction. The research was determined the best method of unpollinated flower culture of Indonesia shallot varieties. Three methods were used including method 1 with two-step cultures using BDS as basic medium supplemented with polyamines 2 mM putrescine on the first step and spermidine 0,1 mM at the second step, method 2 using BDS as basic medium supplemented with 2,4-D 2 mgL^{-1} and BA 2 mgL^{-1} on the first step and NAA 1 mgL^{-1} and 2iP 2 mgL^{-1} on the second step and method 3 using B5 medium supplemented with 2,4-D 2 mgL^{-1} and BA 2 mgL^{-1}. Six shallot cultivars used as flower donor were Katumi, Bima Brebes, Tajuk, Trisula, Superphilip, and Bauji. Flower umbel was collected and pretreated using Yoshida liquid medium then placed on incubator at four °C for overnight. The results showed that all flowers cultured could anthesis properly in the first week of cultured. The embryos could emerge from ovaries from week 7th of cultured in all methods. However, the highest percentage of embryos gynogenic developed to seedlings was achieved using method 3 in Superphilip (6.98%) followed by method 2 in Tajuk (4.44%) .
Keywords: embryo-gynogenic, haploid induction, unpollinated flower

1. Introduction

Shallot (*Allium cepa* L. Aggregatum group) is one of Indonesia's valuable and most widely cultivated crops. Most of them were propagated with bulbs and several of them with seeds. Since shallot is a biennial plant, production inbred lines request at least 6-10 years [1]. Inducing haploid plant followed with chromosome doubling can shorten the time required to create inbred lines. Muren [2] reported induction of haploid onion using ovary culture while haploid culture induction using anther culture has been reported [3]. However, there were no embryos respond using anther culture. The successfully haploid induction using gynogenesis method has been reported by several types of research in several genera of *Allium* such as in *Allium cepa* L [1, 2, 4, 5, 6, 7, 8, 9, 10, 11, 12, 13, 14, 15]. *Allium cepa* L. Aggregatum group [16], *Allium ampelorasum* L. [17], *Allium fistulosum* L. [18] and *Allium chinense* [19]. Several of them used ovule, ovary or flower buds as explants for double haploid induction. Ovary culture and unpollinated flower culture gave similar embryogenic responses, and then unpollinated flower culture was chosen since it was less laborious than ovary culture [1, 20].

Several factors, including genotypes, growing condition of donor plants, and medium composition could be affected the rates of the embryo-gynogenic induction [21]. Gynogenesis induction in shallot has been reported using one-step culture protocols [16] including B5 [22] medium as a primary medium supplemented with 2,4D 2 mgL^{-1} and BA 2 mgL^{-1}. In contrast, research in onion, two-step culture with B5, MS and BDS [23] were used as basic medium supplemented with several plant growth regulators or polyamines [11, 14, 5].

Embryo-gynogenic induction was an important step in haploid induction. Therefore, in this research, the best method of unpollinated Indonesia shallot varieties' flower culture would be determined. Three protocols were used including Protocol 1 [11] which using two-step culture with BDS as essential medium supplemented with polyamines two mM putrescine on the first step and spermidine 0,1 mM at the second step, Protocol 2 [5] which also using BDS as basic medium supplemented with 2,4-D 2 mgL^{-1} and BA 2 mgL^{-1} on the first step and NAA 1 mgL^{-1} and 2iP 2 mgL^{-1} on the second step and Protocol 3 including B5 Gamborg medium supplemented with 2,4-D 2 mgL^{-1} and BA 2 mgL^{-1} [16] (Table 1).

2. Methods

2.1. Growing donor plants

Six shallot cultivars Katumi, Bima Brebes, Tajuk, Trisula, Superphilip, and Bauji, were used as donor plants. Donor plants were planted in the Pujon farm (1100 asl) PT BISI International Tbk, Batu Malang from January to May 2018. Bulbs were used as a source for donor plants. Donor plants were planted in a field with space 15 cm x 15 cm each plant and irrigated with a drip irrigation system.

2.2. Flower bud harvested

Flower buds were harvested when 20-30% flower buds opened or 3-5 days before anthesis. Flowers with stalk (inflorescenses) were placed on beaker consist of Yoshida liquid medium [24] covered with plastic wrapping then incubated in the dark at 4 °C for overnight. Flower buds were selected with the size approximately 3-4 mm in diameter.

2.3. Unpollinated flower buds cultured

The selected flower buds were sterilized using 70% ethanol for 1 minute, followed by 40% commercial bleach (5.25%) for 15 minutes, then rinsed three times with sterile water. Buds were dried on the filter paper then flower stalk was cut off. Buds were placed on three induction medium (Table 1.) based on [11], [4] and [16]. The medium composition of each protocol was shown in Table 1. Protocols based on [16] was one step culture in which there was no transferred to another medium during the culture until embrio emerge and break the ovarium wall. Whereas protocols based on [11] and [4] were two-step culture while flower buds were transferred from fisrt medium (R1) during two week of culture to the second medium (R2) until embrios were emarged. Among three medium were solidified with gelrite 2 gL^{-1}. Flower buds culture were placed on 90 mm x 15 mm Petri dishes consisted of 25 mL of the medium then sealed with plastic wrapping. Culture condition was set at 25 ± 1°C under Philip fluorescent lamps with 16 h photoperiodism. Exceeds gynogenic embryo transferred to the first regeneration medium (R1) consist of B5 supplemented with 1 mgL^{-1} IBA, sucroce 30 gL^{-1} and gelrite 2 gL^{-1}. Regenerants then were transferred to the second regeneration medium consist of B5 medium supplemented with 1 mgL^{-1} IBA, sucroce 30 gL^{-1}, activated charcoal 10 gL^{-1} and gelrite 2 gL^{-1}. All emerged embryos were counted for number of embryos and and calculated the percentage of embryos from all flowers cultured in induction media.

Table 1. Medium composition for gynogenic Induction in *Allium* using three protocols.

Composition	Protocol 1 [11]		Protocol 2 [4]		Protocol 3 [16]
	R I	R II	R I	R II	
Basic media	BDS	BDS	BDS	BDS	B5
Macronutrien (mg/L)					
KNO_3	2530	2530	2530	2530	2500
NH_4NO_3	320	320	320	320	134
$CaCl_2.2H_2O$	150	150	150	150	150
$MgSO_4.7H_2O$	247	247	247	247	250

NH_4H2PO_4	230	230	230	230	
$NaH_2PO_4.H_2O$					150
$NaH_2PO_4.2H_2O$	152	152	152	152	
KH_2PO_4					
$(NH_4)2SO_4$	134	134	134	134	
Mikronutrien (mg/L)					
KI	0.75	0.75	0.75	0.75	0.75
H_3BO_3	3	3	3	3	3
$MnSO_4.H_2O$	10	10	10	10	
$MNSO_4.4H_2O$					10
$ZnSO_4.7H_2O$	2	2	2	2	2
$Na_2MoO_4.2H_2O$	0.25	0.25	0.25	0.25	0.25
$CuSO_4.2H_2O$	0.025	0.025	0.025	0.025	0.025
$CoCl_2.6H_2O$	0.025	0.025	0.025	0.025	0.025
$FeSO_4.7H_2O$	27.8	27.8	27.8	27.8	27.8
Na_2-EDTA	37.3	37.3	37.3	37.3	37.3
Vitamins (mgL^{-1})					
Myo-inositol	100	100	100	100	100
Nicotinic acid	0.5	0.5	0.5	0.5	1
Pyridoxine-HCl	0.5	0.5	0.5	0.5	1
Thiamine HCl	0.1	0.1	0.1	0.1	10
Glycine	2	2	2	2	
ZPT					
2.4 D mgL^{-1}			2		2
BA mgL^{-1}			2		2
NAA				1	
2 iP				2	
Polyamines					
Putrecine (mM)	2				
Spermine (mM)		0.1			
Sucrose gL^{-1}	100		100		75

R1 : Flower buds culture in the first media for two week on Protocol 1 and 2.
R2 : Flower buds culture transferred to the second media after two week in R2 on protocol 1 and 2.

3. Results and discussion

Flowers could anthesis properly in the first week of cultured. Flower petals opened spontaneously within the first week of culture in all protocols. The size of flower buds in protocol 2 and 3 grew 2-3 times bigger than their original size. In contrast, explants in protocol 1 showed a smaller size comparing with protocol 2 and 3.

After seven weeks of culture, ovaries colour in protocols 2 and 3 turned into green, and some of them formed callus at the base of flower buds but there were no embryoids responses observed after seven weeks in culture. While in protocol 1, ovaries showed yellowish colour, no callus response at the base flower and several embryos from Katumi genotype started to emerge from ovaries wall. The embryoids emerged continuously after seven weeks of culture in protocol 1. In protocol 2 and 3 several embryos were emerged from ovarium wall at 80 days of culture. The first embryoid emerged from Philippine using protocol 3. Several embryos emerged directly through septum in the ovarium wall. Most of them coming out with structure like seed coat with black colour. Emerging embryos in the Protocols 3 were shown in Figure 1.

(a) (b) (c)

Figure 1. (a) Flower buds culture after 80 days using Protocol 3. (b),(c) Embryo emerged from the ovarium using Protocol 3.

The number of percentage of embryos responses among three protocols in six genotype of shallot were showed in the Table 2.

Table 2. Gynogenic embryo of six shallot cultivars.

Genotype	The number of flower buds			Number of embryones			% embryos		
	Protocol 1	Protocol 2	Protocol 3	Protocol 1	Protocol 2	Protocol 3	Protocol 1	Protocol 2	Protocol 3
Katumi	4150	367	3745	37	0	49	0.89	0	1.31
Bima brebes	702	297	355	4	0	4	0.57	0	1.13
Tajuk	422	383	68	0	17	0	0	4.44	0
Trisula	514	508	227	0	7	2	0	1.38	0.88
Superphilip	300	337	43	0	3	3	0	0.89	6.98
Bauji	950	623	366	0	2	1	0	0.32	0.27

Protocol 1 employed two different media, first media consisted of BDS supplemented with putrescine 2 mM then were subcultured to the second medium containing BDS medium supplemented with 0.1 mM spermidine. This protocol could induce embryoids in Katumi and Bima Brebes' genotypes with the percentage of 0.89% and 0.57%. Protocol 2 showed that four genotypes produced embryo while both genotypes Bima Brebes and Katumi did not produce embryo formation. The highest percentage of embryo production by this protocol was obtained in genotype Tajuk (4.44%). In protocol 2, flower buds were subcultured to the new media after four weeks of culture in the medium one consist of BDS medium supplemented with NAA 1 mgL^{-1} and 2iP 2 mgL^{-1}. By using protocol 2, four genotypes gave to respond to embryos. By using protocols 3, flower buds were not subcultured to another media until embryo performed. Embryo-gynogenic induction media based on Protocol 3 resulted in five response genotypes in which the highest percentage of embryos production was obtained in genotypes Superphilip (6.98%) followed with Katumi (1.31%), Bima Brebes (1.13%), Trisula (0.88%) and Bauji (0.27%). Meanwhile, the genotype Tajuk did not show embryo production (0%). Protocol 3 resulted that three genotypes produced embryoids higher number than those of in Protocols 1 and 2.

Protocol 3 that employed plant growth hormone 2,4-D 2 mgL^{-1} and BA 2 mgL^{-1} without subculture was effectively increased the number of embryoids in shallot. Moreover protocols 3 was considered since it was more effective and simple to apply. Based on this results, the embryo-gynogenic response was influenced by their genotypes [6, 11, 14, 16] and protocols or culture media. By comparing three embryo-gynogenic induction protocols at the same genotypes, each shallot genotypes showed different responses to embryo-gynogenic in different protocols. In Protocol 1 the addition of polyamines were applied during first 15 days of culture with 2 mM putrescine, followed with 0.1 mM spermine until embrios emerged [11]. By this research, the addition of polyamine in Protocol 1 was only beneficial for few shallot genotypes (Katumi and Bima Brebes). The maximum embryo production was achieved (9.5%) in onion [11]. In contrast, the maximum embryogenic responses in the shallot were achieved

0.89% in the genotype Katumi. However, both genotypes Katumi and Bima Brebes produced embryo-gynogenic lower number than that of conducted by Protocol 3.

According to the protocols 3 [16] reported that unpollinated flowers of three Indonesian shallot cultivars i.e Dili-white, Yogya and Dili-red resulted totaly 89 embryos from 6,812 flower buds culture (1.31%). Comparing the percentage of embryos obtained in this research with previous research [16] it can be concluded that there were two genotypes produced embryos with same or higher than before [16] i.e Superphiplip (6.98%) dan Katumi (1.31%). However four genotypes produced the percentage of obtained embryos lower than previous research results[16]. Furthermore, based on their ploidi level on the research [16] using genotype of Dili-white obtained two plants were haploid indicating that embryos grow from haploid cell or ovule. In other cases, the genotype of Dili-Red obtained the highest number of directly regenerated and callus derived coming out of their somatic tissue. It was needed to distinguish the embryo coming out from ovules or their somatic tissue.

4. Conclusion

Every genotype of shallot has a different response to every three protocols. The embryos could emerge from ovaries from week 7th of cultured in all methods. The highest percentage of embryos gynogenic developed to seedlings was achieved using method 3 in Superphilip (6.98%) followed by method 2 in Tajuk (4.44%).

5. Acknowledgements

The authors would like to thank the PT BISI International, Tbk. who has approved and funded this research.

References

[1] Bohanec B, Jakse M, Ihan A and Javornik B 1995 *Plant Sci.* **104** 215
[2] Muren R C 1989 *Hort. Sci.* **24** 833
[3] Keller E R J and Korzun L 1996 *Curr. Plant Sci. and Biotech. in Agri.* 51
[4] Alan A R, Mutschler M A, Brants A and Earle E D 2003 *Plant Sce* **165** 1201
[5] Alan A R, Brants A, Cobb E, Goldscmied P A, Mutschler M A and Earle E D 2004 *Plant Sci* **167**1055
[6] Campion B and Alloni C 1990 *Plant Cell Tissue and Organ Culture* **20** 1
[7] Campion B, Perri E, Azzimonti M T, Vicini E, Schiavi M 1995 *Plant Breeding* **114** 3
[8] Fayos O, Garces-Claves A, Valles M P and Castillo A 2015 *Frontiers in Plant Science* **384** 1
[9] Geauffriau E, Kahane R, and Rancilac M 1996 *Euphytica* **94** 37
[10] Khar A, Kumar A, Islam S, Kumar A, and Agarwal A *Indian J. Agri. Sci.* **88** 709
[11] Martinez L, Aguero C B, Lopez M E and Galmarini C R 2000 *Plant Sci* **156** 221
[12] Martinez L 2003 *Double Haploid Production in Crop Plants Manual* (London: Kluwer Academic Publisher) pp 275-279.
[13] Mandal J, Acharyya P, Singh W D and Mohanta S 2018 *J. Allium Res.* **1** 28
[14] Michalik B, Adamus A and Nowak E 2000 *J. Pla Phy.* **156** 211
[15] Yarali, F and Yanmaz R 2017 *Afr. J. Biotech.* **16** 1977
[16] Sulistyaningsih E, Aoyagi Y and Tashiro Y 2006 *Plant Cell Tiss. Org. Cul.* **86** 249
[17] Alan A R, F C Toprak and Kaska A 2016 *Plant Tiss. Org. Cul.* **125** 249
[18] Ibrahim A M, Kayat F B, Susanto D, Kashiani P and Arifullah M 2016 *Biotech.* **15** 1
[19] Bah A A, Zhi-Dan W, Xiao-Qing W, Miao-Miao Y, Yeong-Cheol U, Zhen X and De-Ping G 2012 *Sci. Hort.* **147** 105
[20] Bohanec B and M Jakse 1999 *Plant Cell Report* **18** 737
[21] Touraev A, Foster B P and Jain SM 2009 (Slovenia: Springer) pp 35-46
[22] Gamborg O L, Miller R A and Ojima K 1968 *Exp. Cell Res.* **50** 151
[23] Dustan D I and Short K C 1977*Phy. Plantarum* **41** 1
[24] Yoshida S D, Forno A, Cock J K and Gomez K A 1976 *Laboratory Manual for Physiological Studies of Rice* (Manila: International Rice Research Institute) p 61

Mycorrhizal fungi *Glomus* spp. propagation in zeolite enriched with mycorrhiza helper bacteria for controlling nematode in coffee

I N Asyiah[1,*], R Hindersah[2], R Harni[3], B N Fitriatin[2], W Anggraeni[1]

[1] Biological Education Program, Faculty of Teacher Training and Education, University of Jember, Jl. Kalimantan 37 Jember 68121, East Java, Indonesia
[2] Soil Science Department, Faculty of Agriculture, Universitas Padjadjaran, Jatinangor Campus, Jalan Raya Bandung-Sumedang Km 21, Sumedang 45363, West Java, Indonesia
[3] Indonesian Centre for Estate Crop Research and Development, Jl. Parungkuda, Sukabumi West Java 43357, Indonesia

* E-mail: iisnaza.fkip@unej.ac.id

Abstract. Arbuscular mycorrhizal fungi (AMF) play a role in suppressing the nematode *Pratylenchus coffeae*. Mycorrhizal helper bacteria (MHB) can increase the effectiveness of AMF to control the diseases. The experimental purpose was to increase the spore population of AMF *Glomus* spp. in zeolite-based formulation inoculated with liquid consortia of *Pseudomonas diminuta* and *Bacillus subtilis* as MHB. The experimental design was a completely random design with six treatments consisted of 10^6, 10^7, 10^8 and 10^9 CFU/mL MHB liquid inoculants. The control treatments were water and 2% molasses. All treatments were replicated four times. A total of 300 mL/pot Liquid inoculant of MHB have been inoculated a three day before transplanting the maize seedling to the Zeolite inoculated with *Glomus* spp. in the pot. One month after MHB inoculation, *Glomus* formulation in Zeolite with different levels of MHB increased the degree of infection. Three months after MHB inoculation, spore content in Zeolite increased. The density of *P. diminuta* and *B. subtilis* in zeolite-based mycorrhizal inoculant increased at the end of the experiment. Liquid inoculant MHB contained 10^8 CFU/mL enhanced spora number fourth times compared to the control. This experiment suggests that *P. diminuta* and *B. subtilis* were effective to increase the spore density of AMF inoculant.

1. Introduction

Many publications describe mycorrhizae's role in inhibiting parasitic nematodes' penetration and development [1, 2, 3, 4, 5]. Mycorrhizal symbiosis is considered an interaction between plant roots and fungi and must include supporting other organisms. Mycorrhizosphere is a mutual influence that produces what is known as the "mycorrhizosphere" [6, 7, 8]. The mycorrhizosphere is composed of mycorrhizae, external mycelium, and supporting organisms [9]. This mycorrhizosphere effect can lead to increased nutrition, growth, and plant disease resistance [10, 8].

Usually, the AMF and its supporting organisms (bacteria) apply as biofertilizers. Bacteria that can increase mycorrhizal development are defined as Mycorrhizal Helper Bacteria, MHB [11]. Bacteria isolated from mycorrhizal fungi can stimulate mycorrhizal infection, spore production, and plant pathogens' resistance [12, 13].

The MHB is a term used for endophytic bacteria that can help mycorrhizae carry out their role. The bacteria must be in one part of the mycorrhizal body and play a role in developing mycorrhizae. Mycorrhizal symbiosis is not only a relationship between the mycorrhizal-forming fungi and the host plant but involves other supporting organisms such as bacteria.

The bacteria excretes beneficial organic substance that often stimulates the germination of fungal spores. Most bacteria from the AMF spore cell walls were able to increase *Glomus clarum* spores' germination when there was direct contact between spores and bacteria, while some bacterial isolates inhibited spore germination by producing volatile antagonists [14].

Moreover, MHB affects the concentration of antagonistic compounds produced by mycorrhizal fungi [15]. They found that the bacteria were able to detoxify the liquid media from inhibiting fungal metabolites. MHB bacteria may also be able to suppress the production of toxic compounds by soil microbes. Vivas [16] reported that MHB bacteria has positive impact on the spore germination and growth of presymbiotic fungi in a broth contaminated with heavy metals. Bacterial inoculation not only decreased the destruction of *G. mosseae* hyphae but even resulted in increment of root growth by 95% (without Cd) to 254% (with Cd). This effect was as strong as in the Zn treatment, where mycelium growth ranged from 125% (without Zn) to 232% (with Zn solution).

Nunang [17] showed that there were 12 bacteria isolated from AMF spores, seven bacteria from *Gigaspora* sp. and five bacteria from *Glomus* sp. Eight types of bacteria, namely: *P. diminuta, B. licheniformis, B. laterosporus, E. hormaechei, B. brevis, B. subtilis, B. cereus*, and *B. firmus,* can stimulate the development of mycorrhizal hyphae. Seven bacteria have the potential for cellulase and protease activity, namely: B. *subtilis, B. cereus, B. laterosporus, B. pasteurii, P. penneri, B. firmus*, and *B. cereus*. There are four bacteria (*B. subtilis, P. diminuta, P. penneri*, and *E. hormaechei*) which can inhibit the growth of pathogens *Rhizoctonia* sp., *Sclerotium* sp., and *Ganoderma* sp.

The *B. subtilis* and *P. diminuta*) has been reported to increase the degree of mycorrhizal infection in the roots of coffee up to 98.4% [5]. Furthermore, *Pseudomonas diminuta* 10^8 increased the degree of infection by 93.6%; even the density of *P. diminuta* $2x10^8$ increased mycorrhizal infections by up to 98.4 %. Based on the results of this study, we can conclude that mycorrhizal propagation will be more effective with the help of MHB. To develop a more effective mycorrhiza inoculant, the formulation of the inoculant integrated with MHB inoculation is needed. In general, arbuscular mycorrhiza (AM) inoculants are developped by using corn as the fungal host since the mycorrhizal fungal is a host-depend microbes. Our previous experiment demonstrated that molasses-based liquid inoculant of MBH *B. subtilis* and *P. diminuta* supported their cell count up to 10^9 CFU/mL. This liquid inoculant will be utilized to improve the quality of AM inoculant. The objective of this greenhouse experiments was to evaluate the effect of liquid inoculant consortia of *P. diminuta* and *B. subtilis* as MHB on the spores density of AMF *Glomus* spp. in zeolite-based AM inoculant as well as the infection of AMF on the roots of host plants, corn.

2. Methodology

2.1 Materials and Methods

This study used spores of *Glomus* spp; a collection of the Faculty of Agriculture Universitas Gajah Mada. The MHB consortium liquid formula composed of *B. subtilis* and *P. diminuta* with a ratio of 2:3 with the cell count of 10^9 respectively at day three was prepared by Biology education University of Jember Laboratory. The arbuscular mycorrhiza inoculant developed using corn grown in Zeolite with size 1-2 mm. A selective medium of either *B. subtilis* or *P. diminuta* has been used to count bacterial cells in the zeolite-based AMF inoculant.

2.2. Experimental setup

The study was conducted in a greenhouse in a completely randomized design consisting of 6 treatments and four replications. The treatment included an MHB consortium solution with cell densities of 10^6, 10^7, $10^{8,}$ and 10^9 CFU/mL. The control treatment was water and 2% molasses. Each solution is given three days before planting corn seeds as much as 30 ml per pot. The 7-day old corn seeds are planted in polyethyelene pots containing 200 g Zeolite media; each pot consists of two corn seeds. A total of 100 mycorrhizal spores inoculated on maize seedlings after seven days of transplanting.

The corns were watering every day as much as 30 ml/pot. Fertilization of plants using Hyponex (25-5-20) with a 1g/L concentration at a dose of 30 ml/pot a week after planting. Furthermore, fertilization is repeated twice a week with the same amount until the plants are two months old. In the third month, watering is gradually reduced for the stressing process to stimulate spore formation. In the first week, water every other day with a dose of 30 ml of water; on the second week of watering every other day with a dose of 20 ml of water; in the 3rd week of watering every other day with a dose of 10 ml of water, and in the last week of watering is not done watering at all [18].

2.3. Observed Parameters

Observations consisted of 1) the degree of mycorrhizal infection in maize roots, calculated after planting for one month, refers to the Kormanik and Mc Graws method [19]; 2) MHB cell viability by counting the number of bacterial colonies in a selective medium *B. subtilis* and *P. diminuta*, respectively, which observed one month and three months after application refers to Schinner methode [20]; and 3) the number of spores, calculated after planting for three months (harvest) using the extraction method per 10 grams.

2.4. Statistical analysis

All data were subjected to analysis of variance (5% F test) and Duncan Multiple Range Test of 5% by using SPSS program.

3. Results

The addition of the MHB consortium increased the average degree of mycorrhizal infection and bacteria density at one month after treatment, as can be seen in Table 1. The effect of the acquisition of MHB on the propagation medium mycorrhiza on the number of spores and bacterial density (CFU) at harvest time is depitected in Table 2.

Table 1. The percentage of mycorrhizal infection degree and bacterial density in AMF inoculant at one month after treatment

Treatments	Degree of mycorrhizal infection (% ± SD) *	Bacterial density (CFU/g) x10^7	
		P. diminuta	*B. subtilis*
Water	78.00 ± 11.1355 a	-	-
Molase 2%	81.33 ± 11.3725 a	-	-
MHB 10^6	90.67 ± 4.1633 b	58.67	12.33
MHB 10^7	92.67 ± 4.1633 b	46.75	65.17
MHB 10^8	93.33 ± 3.0551 b	15.25	74.67
MHB 10^9	94.67 ± 5.0332 b	26.42	102.33

Note: * the mean number followed by the same letter is not significantly different based on the Duncan test at the 95% confidence level.

Table 2. The number of spores of *Glomus* spp. and bacterial density in AMF inoculant at harvest time

Treatments	Spore number per g inoculant *	Bacterial density (CFU/g) x10^7	
		P. diminuta	*B. subtilis*
Water	33,05 a	-	-
Molase 2%	28,28 a	-	-
MHB 10^6	90,81 b	122,17	257,00
MHB 10^7	115,67 bc	198,75	123,25
MHB 10^8	160,58 c	234,00	107,00

MHB 10^9	144,43 c	276,50	113,33

Note: * the mean number followed by the same letter is not significantly different based on the Duncan test at the 95% confidence level.

4. Discussions

This experiment showed that *B. subtilis* and *P. diminuta* are symbiotic bactera with fungi for carrying out their roles to increse mycorhizal infenction of host plant. The results of this research supported by Nunang [17], which states that several types of MHB were found, including *P. diminuta,* can help develop mycorrhizal hyphae and inhibits several types of pathogens. The *B. subtilis,* assist the development of fungal hyphae and has the potential for cellulase and protease enzymatic activity as well as inhibit the growth of certain pathogens. Garbaye [11] also stated that some bacteria could colonize roots well, for example, various kinds of *Pseudomonas* spp. able to live around the surface of the fungal hyphae. These bacteria live in the mycorrhizal bodies and also around the roots of host plants (rhizosphere).

Bacteria also has a benefit from their symbiosis with mycorrhizal fungi. Linderman [21] state that there is a synergy between bacteria, fungi, and plants. Bacteria get a source of nutrition from root exudates because of their chemical composition due to mycorrhizae's role, so that root exudates contain bacteria suitable. Root exudates are a source of nutrients that are essential for the survival of bacteria. Most of the bacteria in symbiosis with fungi will complete their fungi' life cycle [22]. Bianciotto [23] showed that active bacterial division occurs in the mycelium.

The mechanism of MHB in helping mycorrhizae to infect roots is as follows: 1) increase the rate of roots to mycorrhizal formation, bacteria initiate the formation of IAA hormones to induce short and roots, then bacteria produce enzymes to catalyze the softening of root cell walls prior to AMF-root interaction; 2) Furthermore, MHB plays a role in mediating root biomolecules and also fungi. Roots and fungi can interact based on enzymes or chemical substances produced by fungi or roots, so the part of MHB is to facilitate the introduction of enzymes or chemicals between them by creating certain compounds such as auxins and other enzymes. Mycorrhizal and root interactions can occur because of the presence of *myc* factors released by fungi to be recognized by plants and plants to acknowledge the potential of mycorrhizal symbiosis with strigolactones released by roots along with other root exudates such as sugars, fats, acids amino acids, fatty acids, and hormones growth.

The research showed that mycorrhizal infections in the maize roots were more than 70%. According to [24], the minimum mycorrhizal infections in roots that can increase plant growth and development is 70%. If the degree of infection was less than 70%, then the infection is not optimal. Meanwhile, according to [25], the roots of many host plants with a high degree of root infection by AMF indicate an excellent AMF inoculum source. The exudate produced by MHB will stimulate the germination of fungal spores. Besides, bacteria in symbiosis with fungi can influence plant physiology by increasing the permeability of root cells [10, 11].

P. diminuta and *B. subtilis* have several functions, namely: 1) increasing the effectiveness of mycorrhizal infections against the roots of host plants, 2) as biological control agents, and 3) being able to increase plant growth [5]. These prominent functions are because of the role of two bacteria as Plant Growth Promoting Rhizobacteria (PGPR) as well as a solubilizing phosphate.

5. Conclusions

The conclusions of this study were: 1 Inoculating liquid culture of MHB consortium consisting of *P. diminuta* and *B. subtilis* with a ratio of 2: 3 on mycorrhizal propagation media increased the degree of infection and the number of mycorrhizal spores *Glomus* spp., 2) The MHB consortium liquid culture with a cell density of 10^8 CFU/mL increase the number of spores by 162.067 per gram of Zeolite, 3) The MHB cell viability can survive in Zeolite media for up to / more than three months.

6. Acknowledgments

The authors would like to thank the Agricultural Research and Development Center, the Indonesian Agriculture Department for financing this research through the 2014-2016 KKP3N grant.

References

[1] de la Peña E, Susana RE, Wim H van der Putten, Helena F and Maurice M 2006 *New Phytologist* pp 829-840
[2] Shanthi A and Rajendran G 2006 *Nematol. Medit.* **34**: 69-75
[3] Jumjunidang 2009 *J. Hort.* **19**(2):186-191.
[4] Serfoji,P, Rajeshkumar S and Selvaraj T 2010 *Journal of Agricultural Technology* Vol.6(1): 37-45.
[5] Asyiah IN, Wiryadiputra S, Harni R, Handayani NRH and Narulita E, Adv. *Envr. Biol.* **9**(23), 22-26 (2015).
[6] Foster M 1966 *Australian J Biol Sci* **19**:1027-1038.
[7] Meyer JR and Lindermann RG 1986 *Soil Biol Bioch* **18**:191-196.
[8] Frey-Klet and Garbaye 2005 *New Phytol* 168:48.
[9] Barea JM, Pozo MJ, Azcon R, Azcon-Aguilar C 2005. *J Exp Bot* **56**:1761-1778.
[10] Linderman RG 1988 *Phytopatology* **78**:366-371.
[11] Garbaye J 1994 *New Phytol* **128**:197-210.
[12] Barea JM, Andrade G, Biancioto V, Dowling D, Lohkre S, Bonfante P, O'Gara F, Azcon-Aguilar C 1998 *App Envir Microbiol* **64**:2304-2307.
[13] Budi SW, Van Tuinen D, Negrel J, Martinotti G, Gianinazzi S 1998 in Gianinazzi S, Schueep H, eds. *Arbuscular Mycorrhizas in Sustainable Soil-Plant Systems*. Luxemburg: European Commision. 138-139.
[14] Xavier LJC and Germida JJ 2003 *Soil Biol Biochem* **35**:471–478.
[15] Duponnois R and Garbaye J 1990 *Can J Bot* **68**:2148–2152.
[16] Vivas A, Barea JM, Azcon R 2005 *Microbial Ecol* **49**:416–424.
[17] Nunang, L.M. 2011. Diversitas Bakteri Asal Spora Fungi Mikoriza Arbuskula *Gigaspora* Sp. dan *Glomus* Sp. Serta Potensinya Sebagai *Mycorrhiza Helper Bacteria*. Tesis Sekolah Pascasarjana IP. Bogor. 63p.
[18] Setiawati and Mieke R 2015 Teknik Perbanyakan Kultur Pot FMA. Seminar Fakultas Pertanian. Bandung: Universitas Padjajaran.
[19] Kormanik PP and Mc. Graw AC 1982 Am. Phytopathol. Soc. **46**: 37–45
[20] Schinner F, Ohlinger R, Kandeler E and Margesin R 1995 Methods in Soil Biology. Berlin Heidelberg: Springer-Verlag.
[21] Linderman RG 2000 Effects of Mycorrhizas on Plant Tolerance to Diseases In: Kapulnik Y, Douds DD Jr (eds) Arbuscular mycorrhizas: physiology and function. Kluwer, Dordrecht, pp 345–365.
[22] Bianciotto V, Bandi C, Minerdi D, Sironi M, Tichy HV, Bonfante P 1996 *Appl Environ Microbiol.* **62**:3005–3010.
[23] Bianciotto V, Genre A, Jargeat P, Lumini E, Be´card G, Bonfante P 2004 *Appl Environ Microbiol.* **70**:3600–3608.
[24] Brundret, MC, Melville L and Peterson L 1994 Practical Methods in Mycorrhiza Research. Canada: Mycologue Publications.
[25] Anas I and Tampubolon JLO 2004 *Buletin Agronomi* **32**(1): 26—31.

Effects of physical and chemical treatments on seed germination and soybean seed-borne fungi

E P Ramdan[1,*], A Y Perkasa[1], T K K Azmi[1], Aisyah[1], R Kurniasih[1], P I Kanny[1], Risnawati[1], P Asnur[1]

[1]Agrotechnology Departement, Industrial Technology Faculty, Gunadarma University, Jalan Margona Raya 100, Pondok Cina, Beji, Depok, Indonesia

* E-mail: evan_ramdan@staff.gunadarma.ac.id

Abstract. Soybean is an important food commodity in Indonesia after rice and maize. Plant pathogens still constrain the increase in soybean productivity. One of the plant pathogenic infections can occur during the seed phase. Therefore this study aimed to determine the effect of physical and chemical treatments as control of seed-borne fungi and their impact on soybean seed germination. This study used a completely randomized design consisting of 9 treatments, namely physical therapy by heating the seeds in a microwave at a temperature of 40 °C for 10, 20, 30, and 40 seconds and chemical treatment by soaking the seeds in a fungicide with active ingredient difenoconazole with a concentration of 0.5%, 1%, 2%, and 3%. Seeds without heating and as control are soaking fungicides. Each treatment was repeated three times. The seeds that have been given treatments are then planted using the growing test technique and incubated for seven days. Furthermore, the seed viability and the growth of pathogenic fungi were observed at the end of incubation. The results showed that soybean seed germination was not affected by physical and chemical treatments. The seed viability of 100% with or without treatment. This was confirmed by the findings of seed-borne fungi (*Curvularia, Fusarium, Aspergillus*) with a low infection rate of 0.01-0.19%. Chemical treatment with concentrations of 0.5%, 2%, and 3% had a significant effect on the Fusarium infection level, which was higher than the control, which was 0.18%, 0.17%, and 0.19%. Meanwhile, for *Curvularia* and *Aspergillus*, physical and chemical treatments did not have a significant effect.

1. Introduction

Soybean is one of the essential commodities after rice and corn. Soybeans are also an important and cheap source of vegetable protein. Therefore, the government prioritizes soybeans to increase production. Indonesia is currently a soybean importer because the national need for soybeans has not been met. Moreover, Indonesia's soybean production decreased by 7.16% during the 2011-2017 period [1].

Quality and certified seeds are one of the main components to increase soybean production. One aspect of seed quality is high pathological quality, which means that the seeds are free from infection by seed-borne pathogens, including fungi, bacteria, viruses, and nematodes [2]. The seed-borne fungus can be present on surfaces, in tissues, or freely mixed with seeds. The fungus inoculum source comes from the field, contamination during harvesting, processing, packaging, storage, or distribution of seeds [3]. Soesanto et al. [4] reported that the soybean seed-borne fungi identified were *Aspergillus flavus, Aspergillus niger, Cladosporium oxysporum, Colletotrichum demotion, Curvularia pallescens, Fusarium solani, Melanospora zamiae,* and *Nigrospora sp.* Seed-borne fungi can cause loss of viability and seed weight. In addition, it can also harm mammals with the production of mycotoxins that can cause poisoning. Amza [5] reported some of the mycotoxins are Aflatoxin, Ochratoxin, Trichothecence, Zearalenone, Fumonisense, Ergot alkaloin, and sterigmatysin.

Controls that are carried out to suppress seed-borne pathogens are physical and chemical treatments. Govindaraj et al. [6] have reviewed physical treatment techniques on seeds in the form of magnetic fields, gamma irradiation, electric fields, laser irradiation, sound, healing energy, light, and heat. Meanwhile, seed treatment used chemicals consist of Hexaconacole, Mancozeb, Carbendazim, Thiphanate Methyl, Difenoconazol [7]. Apart from suppressing seed-borne fungi' growth, physical and chemical treatments have also succeeded in increasing seed viability and dormancy [8]. Therefore, this study will examine the physical and chemical treatments of the viability and fungi of soybean seeds.

2. Methods

2.1. Time and Place of Research

The research was carried out from December 2019 to January 2020 at the Middle Agrotechnology Laboratory, Agrotechnology Depertement, Faculty of Industrial Technology, Gunadarma University Campus F7, Ciracas, East Jakarta.

2.2. Soybean-seeds Treatments

The soybean seeds used are commercial seeds that are sold in bulk. Soybean seeds are prepared and separated in advance from damaged seeds and dirt. The seeds in good condition (not damaged and not dirty) as ten seeds for each treatment were randomly selected [9]. This research consisted of physical treatment using a microwave for 10, 20, 30, and 40 seconds with the heating temperature at 40 oC to provide dry heat conditions. In addition, chemical treatment was also given by immersing in a fungicide solution (active ingredient diphenoconazole) with a concentration of 0.5%, 1%, 2%, and 3% for 15 minutes. The seeds without heating in the microwave and soaked in fungicide were used as controls, so there were 9 treatments (Table 1).

2.3. Testing the viability and health of seeds using the Growing on Test Method

The growing on test testing technique follows the method of Ramdan and Kalsum [10] with fan paper media. A total of 3 sheets of frosted paper folded into a fan, then placed on a plastic tray and moistened with water. A total of 10 soybean seeds were placed between the folds of the paper. Then the trays were wrapped in clear plastic so that moisture was maintained. After incubation for 7 days, the seeds were observed for their germination capacity and examined in a stereo microscope to see the presence of pathogens carried by the seeds. Germination using the formula;

$$Seed\ germination = \frac{\sum seed\ germinated}{\sum total\ seeds} \times 100\%$$

while the percentage of infection by seed-borne pathogens was calculated using the formula:

$$Infection\ percentage = \frac{\sum infected\ seed}{\sum total\ seeds} \times 100\%$$

2.4. Fungus Identification from Seed Samples

Each seed observed under a stereo microscope identified a seed-borne fungus. Identification is also carried out under a microscopic compound by placing a glass object containing fungal scrapings on the seeds. Fungi were determined by following the identification key book from [11] and [12].

2.5. Experiment Design and Data Analysis

The experimental design used in this study was a non-factorial, completely randomized design consisting of 10 treatments (Table 1). Each treatment was repeated three times so that there were a total of 30 experimental units. The data obtained were then analyzed using variance fingerprints. If there is a significant difference, then further tested using the Tukey test with a level of 5%.

Table 1. Treatment of soybean seeds

No	Treatment Code	Information
1	K0	Kontrol
2	F1	Physical treatment with dry heating for 10 seconds
3	F2	Physical treatment with dry heating for 20 seconds
4	F3	Physical treatment with dry heating for 30 seconds
5	F4	Physical treatment with dry heating for 40 seconds
6	K1	Chemical treatment by immersing difenoconazole with a concentration of 0.5%
7	K2	Chemical treatment by immersing difenoconazole with a concentration of 1%
8	K3	Chemical treatment by immersing difenoconazole with a concentration of 2%
9	K4	Chemical treatment by immersing difenoconazole with a concentration of 3%

3. Results and Discussion

3.1. Viability of Soybean Seeds in Physical and Chemical Treatment

The results of the analysis of variance showed that physical and chemical treatments did not significantly affect seed viability. This is indicated by the level of soybean seed viability of 100% in each treatment (Table 2). In contrast to the report of Ramdan et al.[13] stated that physical treatment has higher viability than the chemical treatment of corn seeds. Based on these results, it is suspected that the physical and chemical treatments do not interfere with plant physiological processes. In addition, physical treatment with dry heating at 40oC with 10-40 seconds is still safe and does not experience damage. Several studies have reported that dry heat treatment will reduce seed viability due to damage to seed membranes and leakage of electrolyte solutions [14],[15], but Izle and Isik [16] have reported that the optimum temperature in physical treatment that supports the viability of soybean seeds is 45 °C. In this study, the chemical treatment also did not affect seed viability. Nalis et al. [17] reported that chemical treatment could cause symptoms of poisoning in sprouts so that their viability decreases.

Table 2. Seed viability under physical and chemical treatment

Treatments	Seeds Viability (%)
K	100a
F1	100a
F2	100a
F3	100a
F4	100a
K1	100a
K2	100a
K3	100a
K4	100a

The numbers followed by the same letter in the same column show no significant difference based on Tukey's test at the level of $\alpha = 5\%$

Figure 1. Growth of soybean sprouts on a) Physical and b) Chemical treat

3.2. Seed-borne Pathogen Identification

The pathogens carried by jangung seeds that were identified consisted of 3 groups of fungi. Identification was carried out by observing the morphology microscopically. The identification resul ts were microscopic morphological features and fungal colonies on the surface of the seeds (Table 3).

Table 3. Results of identification of soybean seeds fungus

Result	Macroscopic Observation	Microscopic Observation
Curvularia sp.	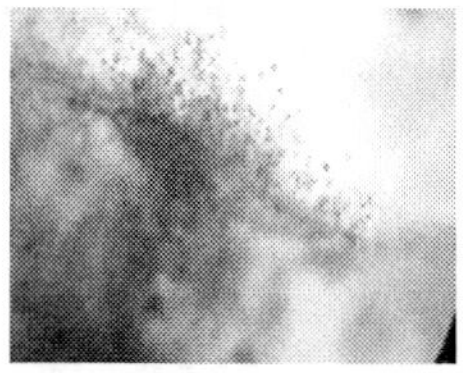	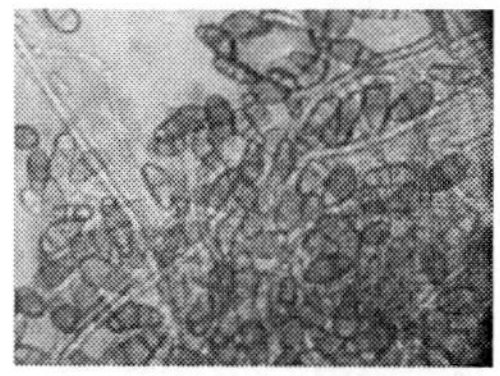
Aspergillus sp.	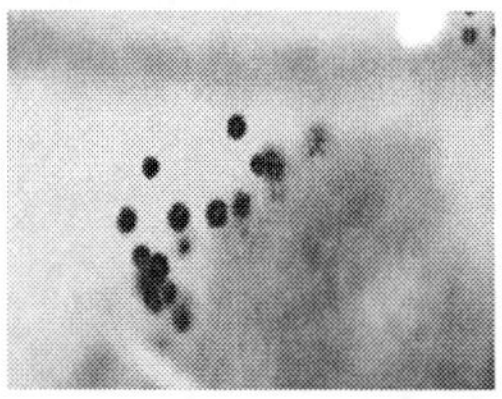	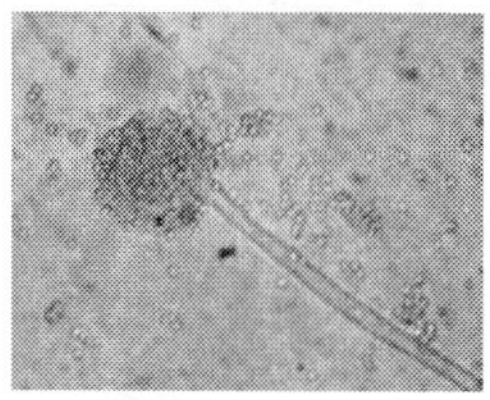
Fusarium p.		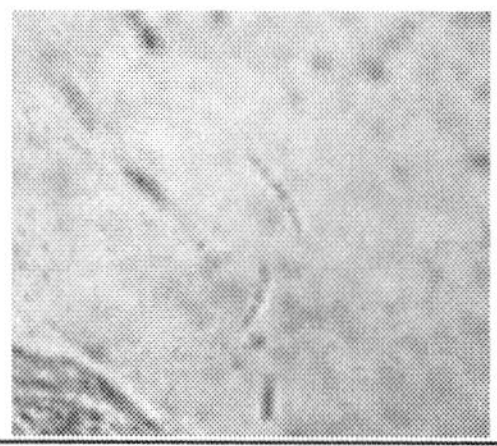

3.2.1. Curvularia sp. Colony characteristics of Curvularia sp. on the seeds' surface, which is characterized by strong conidophore, brownish color. Meanwhile, microscopic observation showed that conidia were light brown to dark brown, mostly straight and slightly curved, with three baffles and four cells.

3.2.2. Aspergillus sp. Colony characteristics of Aspergillus sp. on the surface of the white seed, which will turn green over time. On the surface of the seeds, there are also green powder-like grains. These green grains are the heads of conidia. Meanwhile, on microscopic observations, Aspergillus sp. shows the hyphae structure that has septa, clear (hyaline), with green conidia.

3.2.3. Fusarium sp. Colony characteristics of Fusarium sp. on the surface of the seeds, which is white like cotton. Meanwhile, on the microscopic observations of Fusarium sp. characterized by hyphae with septa and colorless with conidia in the form of macroconidia in the form of a crescent moon.

The three pathogens have been identified as being carried by soybean seeds apart from Alternaria and Rizoctonia [18]. The association of Fusarium and Aspergillus on soybeans has been reported by Ramdan and Kalsum [10] respectively, at 38% and 14%.

3.3. Infection Power of Seed-Carried Pathogens After Physical and Chemical Treatment

The results of the analysis of variance showed that the treatment of K1, K3, and K4 had a significant effect on the infectivity of Fusarium sp. (Figure 2), while the fungus Curvularia sp. and Aspergillus sp. physical and chemical treatment had no significant effect. In general, physical and chemical treatments show a decrease in fungal infection power. This is in line with the high seed viability (100%) in physical and chemical treatments (Table 2). Even though the K1, K3, and K4 treatments were not able to suppress Fusarium sp growth, the infection power is still low 0.17-0.19%. This is consistent with Sonhaji et al. [19], who reported that seed treatment with synthetic fungicides could suppress Peronosclerospora maydis in soybean seeds than other seed treatments. Nurahmi et al. [20] explained that fungicides in seed treatment could work systemically or in contact. Systemic mechanisms work to inhibit pathogens in the seed tissue, while the contact mechanism works to inhibit pathogens on the seeds' surface.

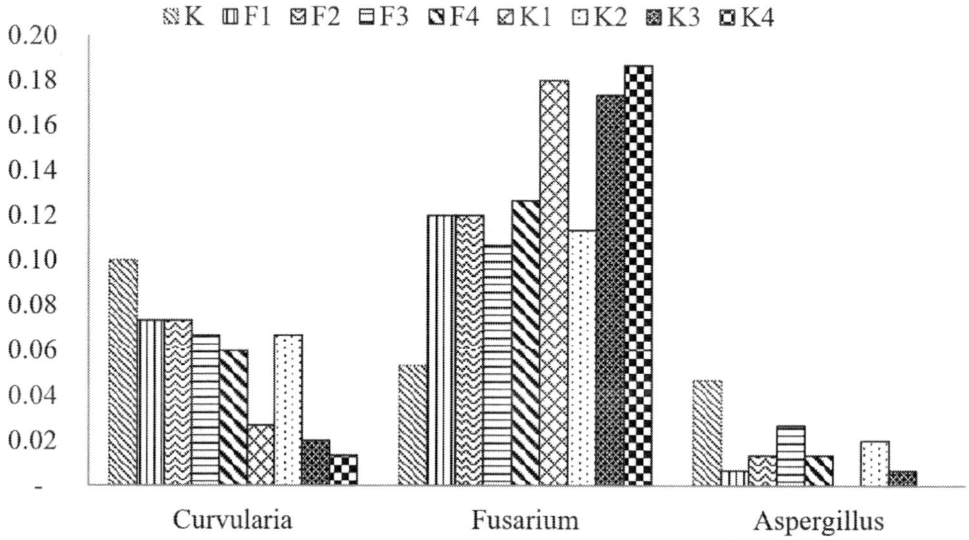

Figure 2. Infection power of seed pathogens in physical and chemical treatments

The use of fungicides as seed treatments has been reported to suppress pathogenic infections, particularly fungi, during 3-4 weeks of storage [21]. The presence of pathogens in seeds causes a decrease in viability, a musty odor, and a change in seed color. Damage to seeds can be exacerbated if stored in a location that supports the development of the pathogen. Factors that support the development of pathogens in seeds include the physical condition and moisture content of the seeds and the storage area's humidity.

4. Conclusion

1. Physical and chemical treatments did not significantly affect seed viability. All treatments showed that the seed viability was up to 100%.
2. The identification results show that the fungi carried by soybean seeds are *Curvularia* sp., *Fusarium* sp., and *Aspergillus* sp.
3. In general, physical and chemical treatments can suppress the growth of fungi. Chemical treatment with a concentration of 0.5%, 2%, and 3% significantly affected the Fusarium infection rate, which was higher than the control, namely 0.18%, 0.17%, and 0.19%, respectively.

References

[1] PS [Badan Pusat Statistik]. 2018. Harvest area, Productivity, and Production of Soybean at East Java.http://jatim.bps.go.id/statictable/2018/10/31/1342/luas-panenproduktivitas-dan-produksi-kedelai- di-jawa-timur-2002-2017
[2] Ilyas S. 2012. Ilmu dan Teknologi Benih (Teori dan Hasil-hasil Penelitian). Bogor (ID): IPB

PressMore references

[3] Agarwal VK, Sinclair JB. 1996. *Principles of Seed Pathology*. Ed ke-2. Florida (US):CRC Press.

[4] Soesanto L, Hartono ARR, Mugiastuti E, Widarta. 2020. Seed-borne pathogenic fungi on some soybean varieties. *Biodoversitas* **21**(9): 4010-4015.

[5] Azma J. 2018. Seed borne fungi; food spoilage; negative impact and their management: A Review. *Food Science and Quality Management* 81: 70-79.

[6] Govindaraj M, Masilamani P, Albert VA, Bhaskaran M. 2017. Effect of physcal seed treatment on yield and quality of crops: A review. *Agriculture Reviews* **38**(1): 1-14.

[7] Gowda B, Kasam R, Doddagoudar SR, Kurnalliker V, Channabassavanna, Ravi ASMV. 2020. Influence of seed treatment with fungicides on seed quality of chickpea cv. GBM 2. *Int. J. Curr. Microbiol. App. Sci* **9**(1): 602-609.

[8] Mutinda YA, Muthomi JW, Kimani JM, Cheminigw'wa GN, Olubayo FM. 2017. Viability and dormancy of rice seeds after storage and pre-treatment with dry heat and chemical agents. *Journal of Agricultural Science* **9**(7): 175-185.

[9] Akonda MMR, Yasmin M, Hossain I. Incidence of seedborne mycoflora and their effects on germination of maize seeds. *International Journal of Agronomy and Agricultural Research* **8**(1): 87-92.. 2015.

[10] Ramdan EP, Kalsum U. 2017. Inventory of seed-borne fungi in rice, soybean, and chilli. *Jurnal Pertanian Presisi* **1**(1): 48-58.

[11] Watanabe T. 2002. Pictorial Atlas of Soil and Seed Fungi Morphologies of Cultured Fungi and Key to Species 2nd Edition. New York (US): CRC Press.

[12] Dugan FM. 2006. *The Identification of Fungi*. Minnesota (US): The American Phytopathological Society.

[13] Ramdan EP, Arti IM, Risnawati. 2020. Evaluation of viability and corn seed-borne pathogens in physical and chemical treatment. *J. Berkala Penelitian Agronomi* **8**(2):16-24.

[14] Lin, R.H., K.Y. Chen, C.L. Chen, J.J. Chen, J.M. Sung.2005. Slow post-hydration drying improves initial quality but reduces longevity of primed bitter gourd seeds. *Sci. Hort.* 106:114-124

[15] Peng, Q., K. Zhiyou, L. Xiaohong, L. Yeju. 2011. Effects ofaccelerated aging on physiological and biochemical characteristics of waxy and non waxy wheat seeds. *J. Northeast Agric. Univ.* **18:**7-12.

[16] Izli N, Isik E. 2010. Determination of economiccost, vigour and rate of germination in batch drying of maize seeds.*Int Agrophys.* **24**:93–96.

[17] Nalis S, Suastika G, Giyanto. 2015. Dry heat and bactericide treatment to supress *Pantoea stewartii* subsp. *stewartii* infection on sweet corn seed. *J. Fitopatol. Indones* **11**(4): 128-136.

[18] Ulemale D, Kalaskar RR, Gore VP, Atre G. 2019. Studies on seed borne mycoflora of soybean (*Glycine max* (L) *Merrill*). *Indian journals* **14**(2): 141-146.

[19] Sonhaji YS, Surahman M, Ilyas S, Giyanto. 2013. Seed treatment improved seed quality, seed production, and controlled downey mildew disease on sweet corn. *J. Agron. Indonesia* **41**(3): 242-248.

[20] Nurahmi E, Sabaruddin, Erlina N. 2010. Effect of fungicide benlate and packaging media in high humidity conditions on cocoa seed vigor and viability after storage. *J. Floratek* **5**: 140-151.

[21] Rahardjo P, Sukamto S. 1987. Maintaining Growth of Cocoa Seeds in Storage with Fungicides. *Pelita Perkebunan* **3**(1): 31-35.

Resistance varieties and pattern of disease progress of rust (*Pucciana horiana* p. henn) in Chrysanthemum

Y A Bety[1,*], R Pangestuti[1]

[1]Assesment Institute for Agriculture Technology in Central Java, Jl. Sukarno-Hatta Km 26, No. 10, Po Box 124, Bergas, Semarang, 50552, Central Java, Indonesia.

E-mail: yayuk.bety@yahoo.com

Abstract. Rust is a major disease in chrysanthemum, which causes degradation of flower quality and results in up to 100% yield loss. Out of all the available strategies, breeding for host-plant resistance and knowledge of the pattern of disease progression in each variety is an effective strategy to control rust disease. The disease progress figured out the level of disease severity from time to time. It would be as guidance to control the disease, especially pesticide application. Six national superior and one introduced varieties were tested for their rust resistance. The varieties were studied to know the pattern of disease progression in each variety from vegetative to generative phase. The study was conducted from August 2014 to January 2015 in the plastic house of chrysanthemum plantation at Garung sub-District, Wonosobo regency, one of the centers chrysanthemum production in Central Java. The disease intensity was measured at 30, 60, and 90 days after planting. The results relayed that based on the rate of disease progress and the average of disease intensity, it could be identified that the superior national varieties of chrysanthemum, Kusumaswasti, Marimar, and Yulimar were more resistant to rust than the four other varieties tested. The rate of disease progress and the level of disease intensity of rust by the plant age determined the resistance of the chrysanthemum varieties to rust. Moreover, the result showed that each variety had a different rate and pattern of disease progression and could be used to control rust effectively.

1. Introduction

One of the major diseases on chrysanthemum is white rust, which is caused by the fungus of *Puccinia horriana P.Henn*. The obvious symptom of the diseases is white to orange pustule at the leaves bottom of and white-yellowish spots or blemishes on top of the leaves. In several chrysanthemum production areas, rust has been reported as one of the destructive diseases that can lower the chrysanthemum's quality and productivity. As a major disease, rust can reduce the yield up to 100% [1], while in Indonesia, it is around 30 to 100% [2][3]. In the other testings carried out at multiple locations and in different climatic conditions, it was reported that rust infection on the susceptible varieties caused the whole part of the plants, including flower, leaves, and stems, to get burn and search, which bring about 100% yield losses [4].

Among the available strategies to *control* rust disease, breeding for host-plant resistance is an effective strategy [5][3]. Resistance varieties are identified as varieties having a slow getting infected and low disease intensity [6][7]. Moreover, resistant varieties can be determined by the rate of infection of the disease on these varieties. It can be explained that the value of rate infection in resistant variety is quite low; even some of them are zero. In contrast, the value of the rate of infection in susceptible variety is high. The study of rust on red bean progress showed that the disease progress of rust in resistant varieties is slower than in susceptible varieties [8].

Furthermore, Vander Plank (1963) stated that the disease progress is a tool to figure out the varieties' resistance, the aggressiveness of pathogen, and the quality of the environment [8]. As an explanation, when r (disease progress) has a value lower than 0,5 units/day in any variety, the variety considered resistant, the pathogen is not aggressive, and the environment does not support the development of the disease, reversally. By using the same control treatment, each variety will give different response to the disease depend on the level of resistance and can be identified based on the magnitude of the disease intensity and the rate of infection. The disease intensity and the rate of infection of rust in Chrysanthemum cv. Giant is lower than of that in Chrysanthemum cv. Cat Eye, and cv. ILK6B1, eventhough it was controlled using the same methods, i.e. roughing, and application of pesticides [9].

Resistance against the disease is an important character of a variety; however, color flower, shape flower, vase life, sometimes are considered more by the farmer, many susceptible varieties are available in the field. Consequently, chemical pesticides are applied intensively to control the disease. Hence to suppress the excessive use of fungicide, information about the peak of disease progression is needed. Freedman and Mackenzie explain that the magnitude of the disease intensity at a certain period of the plant's age is affected by the rate of infection and the disease intensity of the initial attack[10]. The result suggests that the rate of infection may fluctuate with age. The rate of infection of rust in chrysanthemum cv Swarnakencana from age 1 to 2 week, 2 to 3 week, and 3 to 4 week are different. The infection rate is fastest at the age of 1 to 2 weeks. The infection rate is likely to a slowdown in the period 2-3 weeks and flattens out at 3 to 4 weeks. In this research, the fungicides are applied every week, starting at two weeks after planting for up to 2 weeks ahead of the harvest [11].

Moreover, the infection rate can be used as a reference in determining the types of pesticides that are effective in suppressing the disease. On controlling rust in chrysanthemum, bio fungicides Neem-plus can slow down the infection rate of white rust on the plant from age 1-4 weeks for 1 to 3 times compared to other fungicides three times slower than the control [11]. While biopesticide based on kascing (degraded soil by soil worm), isolates of Bacillus subtillis, Pseudomonas florence, Escherichia coli or biopesticides based potato extract plus kascing, molasses, isolates of Bacillus subtillis, Pseudomonas florence, and Escherichia coli can slower infection rate of rust in chrysanthemum plant effectively[12].

2. Methods

2.1 Location and time of study

The research was carried out in the plastic house in Garung District, Wonosobo Regency, Central Java Province, from August 2014 to January 2015. Garung is located at 1000 m above sea level (a.s.l.); the daily temperature ranges from 20-300C, during the cold season from July to August, the night temperature is around 10-15^{0}C. Based on climatological data recorded in 2014 at the closest station WadasLintang, the average humidity in Garung was 81%, average rainfall was 3214 ml/day, and rainy days was 160 days. The soil was classified as Regosol.

2.2 Varieties and methods of testing

Six varieties of chrysanthemum, i.e., Marimar, and Yulimar (originated from mutation breeding), Kusumaswasti, Puspita Pelangi, Puspita Nusantara, and Shakuntala which were released by the Indonesian

Ornamental Crops Research Institute, IAARD, and one variety named Bakardi, an introductory variety, were tested for their rust resistance. The experiments were arranged in a randomized block design with seven treatments and three replications, so 21 experimental plots were used. The disease was inoculated naturally since natural inoculum is available in abundance. Disease intensity of rust in each variety was measured periodically at 30, 60, and 90 days after planting (d.a.p.) by taking five systematic, randomly selected plant samples. Each sample was assessed based on the index of rust disease with criteria as follows [13]:

Table 1. The scale of symptoms of rust disease

Categories	Damage Scale
1	111 211 311
2	122 222 322
3	132 232 332
4	133 233 333

Explanation:
a. The first number indicates the leaf position.
1 = leaves of 1/3 of the plant at the bottom position
2 = leaves of 1/3 of the plant in the middle position
3 = leaves of 1/3 of the plant in the upper position
b. The second number shows the number of pustules on the leaves:
1 = no pustules, 2 = 1-25 pustules, 3 = 26-50 pustules, 4 => 51 pustules
c. The third number indicates the state of the spore
1 = not yet formed spores
2 = the spores have not broken yet
3 = spores broken

Disease intensity in each plot is calculated based on the following formula:

$$IP = \frac{\text{Amount } (n \times v)}{(N \times z)} \times 100\% \quad (1)$$

IP = Disease intensity
n = Number of crops per category
v = Scale value of each category
Z = The highest scale value
N = Number of plants observed

Table 2. Category of resistance based on the disease intensity

Disease Intensity (%)	Category
0.00-10.99	Resistant
11.00-35.99	Moderate resistant
36.00-65.99	Susceptible
70.00-100.00	Very susceptible

The rate of infection is determined based on the Van der Plank formula [8]

$$R = \frac{2{,}30259}{t} \times \log \frac{Xt}{Xo} \qquad (2)$$

R = Infection rate
t = Observation interval
Xt = Disease intensity in time of *t*
Xo= Disease intensity in the initial observation

The flower's productivity and quality are appraised using parameters: ray floret and disk floret diameter, the length and diameter of the flower stalk measured before harvest, morphological character: flower color, flower type, and flower shape.

2.3 Planting chrysanthemum

Land preparation was done by plowing and rotating the soil until loose, and beds were set up for 2 m length, 1 m width, 25-30 cm height, and 40 cm between beds. Let the soil dry and given basal fertilizer in the form of chicken manure (CM) for 100 kg per 100 m^2 (10 $t.ha^{-1}$), compost 105 kg per 100 m^2(10 $t.ha^{-1}$), plant growth-promoting bacteria (PGPR), and red ponska 5 kg 100 m^2 (500 $t.ha^{-1}$). The fertilizer was put on the bed surface and stirred evenly. Planting cutting stems was done one week after land preparation. A day before planting, the soil was reprocessed lightly then covered with the net as plant enforcement. The net was purposed to hold the cutting stems and to determine plant spacing. Then the bed is flooded until it reaches the saturation point. Before planting, the holes were made with a spacing of 12,0 x 12,0 cm (66 plants m^{2-1}). Rooted chrysanthemum cutting stems were then planted on prepared planting holes.

Addition of light: On the day of planting, light TL 18 Watt (yellow ray) was set up with light distance 2.0 x 2.5 m, and height 1.5 m - 2.0 m. The light set in the nite break pattern (20 minutes lights up followed by 10 minutes off for 8 cycles) for 4 hours a day from 10:00 to 02:00 in the morning, arranged using a timer. The addition of light was set for 35-40 d.a.p. or adjust the type of variety or until the plant reaches the height of about 50 cm.

Plant maintenance: Subsequent fertilization was done 2 weeks after planting using NPK fertilizer (16-16-16) as much 50 g m^{2-1} and 5 weeks after planting using NPK fertilizer (18-20-20) as much 50 g m^{2-1}. Leaf and flower fertilizers were given 4 weeks after planting and continued 2 times per week with the appropriate doses indicated on the packaging label. Roguing has done 2-3 weeks after planting, topping (removing the first flower for spray type) was done 3 weeks after planting, and pinching (dumping the next flower for standard type) was done 7 weeks after planting. Pest was controlled using insecticide alternately with 7-day intervals, using ½ recommended doses up to 28 days, and using the recommended doses after 28 d.a.p. Rust disease was controlled with Cabrio with 7-day intervals. Weeding was done every 2 weeks, and watering is done by flooding as needed.

3. Results and Discussion

3.1 Resistance against rust disease

The response of seven varieties of chrysanthemum to rust disease differed depending on the level of resistance of each variety and plant age period (Table 2). On this testing, a variety of Shakuntala, Puspita Nusantara, Puspita Pelangi, and Bacardi always showed a high sensitivity level at 30, 60, and 90 d.a.p. Nevertheless, at 90 d.a.p. only in Shakuntala, the disease increased drastically to 100% and significantly different from six other varieties. Kusumaswasti, Marimar, and Yulimar showed more resistance among varieties tested, with the average disease intensity of 31% to 38%.

At 30 d.a.p, the resistance of seven varieties tested to rust had already been recognized (Tabel 2). At that period, the disease intensity level was quite high (31%); even Bacardi was loaded for 63.3%. Together with Puspita Pelangi, Bacardi is the most susceptible variety. It means that intensive control of rust should be applied earlier on those two varieties. In this period, Marimar was categorized as a resistant variety as the disease intensity only 6.67%. The categorization defined [13] reveals that as resistant variety if the disease intensity on that variety as much 0,00-10,99%. Two other varieties Kusuma Swasti and Yulimar, were also quite durable since those varieties' disease intensity was only 13,33% and 20%, respectively, and did not significantly differ from the disease intensity Marimar. Marimar's obvious rust symptom can be described that 20% of the observed plants had the condition, i.e., the leaves in 1/3 bottom part of the plants had 1-25 immature pustules/leaf to [14], it is categorized as resistant variety. Two other varieties tested, Shakuntala and Puspita Nusantara, also suffer from a massive rust incident. Leaves in 1/3 bottom part of all plants of those two varieties covered partially with unbroken pustules containing and without spores, and 30% of plants covered with pustules containing broken spores.

Table 3. Disease intensity of rust in seven varieties of chrysanthemum at 30, 60, and 90 days after planting. Garung, Wonosobo district, August 2014 to January 2015.

Varieties	Disease intensity of rust (%)			
	30 d.a.p.$^{**)}$	**60 d.a.p.**	**90 d.a.p.**	Average
Puspita Nusantara	33,33$^{ab*)}$	58,33^{a}	75,0 ^{b}	55,55
Shakuntala	40,00^{ab}	43,33^{b}	100,0^{a}	61,11
Puspita Pelangi	46,67 ^{a}	40,00 ^{b}	75,00 ^{b}	53,89
Bakardi	63,33 ^{a}	56,67^{a}	73,00 ^{b}	64,44
Kusuma Swasti	13,33 ^{b}	13,33^{d}	75,00 ^{b}	31,11
Marimar	6,67 ^{b}	28,67^{c}	75,00 ^{b}	36,78
Yulimar	20,00 ^{b}	18,33^{d}	75,00 ^{b}	37,77
Average	31,90	25,33	77,14	

$^{*)}$ the value followed by the same letter not significantly different at $p<0.05$.
$^{**)}$ d.a.p. = days after planting

At 60 d.a.p., the level of resistance to rust of seven varieties was significantly different (Table 2). Consistently, Kusumaswasti, Yulimar, and Marimar endure to rust until 60 d.a.p. at 60 d.a.p., the disease intensity was likely lower, i.e., 13.33, 18.33, and 28.67%, and significantly different from the disease intensity of susceptible varieties Bacardi, Puspita Nusantara, Shakuntala, and Puspita Pelangi. Moreover, in these periods, both susceptible and resistant varieties were less suffering from rust, and the average disease intensity was declined. The treatment of leaves roguing on the bottom part of the plants might successfully reduce the disease. The previous study [9], [15] was found that the treatment of roguing on rust infected leaves was effectively reduced the disease intensity of rust in chrysanthemum. However, the treatment should be appropriately done and intensively; otherwise, it can stimulate infection rate. Unfortunately, in 30 to 60 d.a.p., the disease intensity of the resistant variety of Marimar increased drastically by more than three times as a sufficient amount of healthy leaves.

Moreover, roguing is addressed only for leaves with pustules, while at 30 d.a.p. the pustules in Marimar have not come up yet, so it is a kind of disease escape. High disease intensity of rust in Bacardi, Puspita Nusantara, Shakuntala, and Puspita Pelangi at 60 d.a.p. indicated that these varieties susceptible to rust persistently. Disease control, including roguing and pesticide application, turns out to be less useful to control rust in those varieties. The same result was found from several experiments conducted in different locations and times, indicating Shakuntala as a susceptible variety. Some observations also found that resistance of Puspita Nusantara to rust had broken in recent years, although in 2002, Puspita Nusantara is released as a resistant variety. The ability of Puspita Nusantara and Puspita Pelangi against rust in this research corresponds with the result of the former trials. Adaptation trial conducted in Bandungan in 2012 showed that Puspita Nusantara and Puspita Pelangi had been categorized as a rather susceptible and very susceptible variety to rust [16]. Different results were found in the adaptation trial carried out in Magelang (2008) and Bandungan (2014); in these trials, Puspita Nusantara was considered as moderately resistant to rust [17][18]. It reveals that the resistance of Puspita Nusantara to rust is getting weaker by the year.

The following observation was done ahead of harvesting (90 d.a.p). In this observation, the resistance to rust of seven varieties was significantly different (Table 2). Shakuntala to be the most susceptible with a disease intensity of 100%, and significantly higher than the other six varieties (73 to 75%). The symptom showed that most of the leaves and stems of Shakuntala got burn and scorched, and the broken yellowing pustules came out in widely part of the bottom side of every single leaf. At 90 d.a.p., only Kusumaswasti looked less suffer from rust, and rust did not reduce the flower's aesthetic as pustules did not come out on the top of the flower arrangement.

3.2 The disease progress of rust on chrysanthemum

The pattern of disease progression is governed by initial infection and the rate of infection over time. In this observation, the pattern of disease progress of rust in each chrysanthemum variety is different, so do on the period (Figure 1). At 60 d.a.p., the average disease intensity of rust in seven varieties is slightly lower than at 30 d.a.p. from 32.85 to 31.90% (Table 2), which means that the rate of disease progress becoming slower at the period of 30 to 60 d.a.p. (Figure 1). The disease escalation in susceptible varieties Bacardi, Puspita Pelangi, and Yulimar in this period is going down; in contrast, the disease escalation in resistant variety of Marimar, Puspita Nusantara, and Shakuntala going up as much 4.3, 1.75, 1.08 times (disease intensity 28.67, 58.33, and 43.33%). A similar pattern of the disease progress also arises at 60-90 d.a.p. The disease intensity increases sharply in all varieties and mostly occurs in resistant varieties such as Kusumaswasti, Marimar, and Yulimar, unfortunately (Figure 1). The former sensitive variety of Puspita Nusantara has the least amount of disease progress (only 16,67%) at the periods of 60-90 d.a.p., it might due to limited healthy leaves available. The disease intensity in Kusumaswasti at 30 d.a.p is still low (13,33%) and does not change until 60 d.a.p. It means that the disease progression rate in Kusumaswasti at 30-60 d.a.p is zero ($r_{30\text{-}60}$=0.00 unit/day) (Table 3). The situation is extremely changed to 90 d.a.p.

The dynamic of disease progress of rust in chrysanthemum was influenced by techniqueto control the disease, such as planting resistant varieties and applying natural or chemical pesticides. The application of natural pesticides significantly affected the disease progress of rust in chrysanthemum. According to previous research, the use of varieties and natural pesticides might affect the degree of disease control. A natural pesticide called Neem-plus can reduce disease progression by up to 60% in 7 to 30 days.

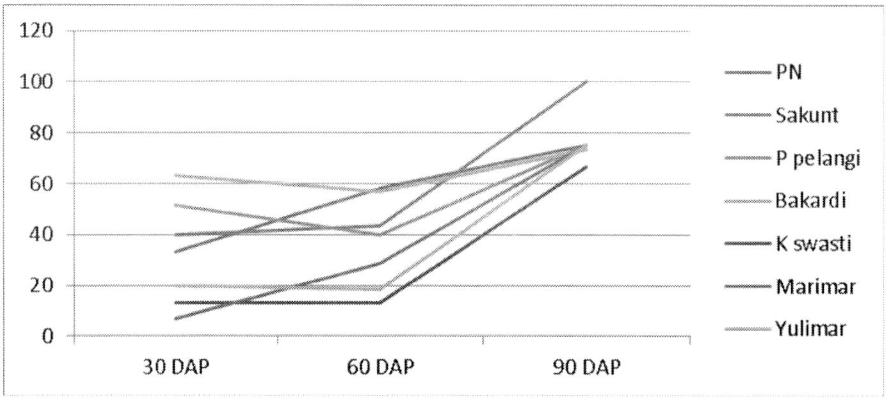

Figure 1. The pattern of rust infection in seven varieties of chrysanthemums. Disease intensity of rust (%) of seven varieties of chrysanthemum at 30, 60, and 90 days after planting (DAP). Garung, Wonosobo district, August 2014 to January 2015.

Disease attack impacts flower size, such as ray floret diameter, which is an important character of the chrysanthemum's standard flower type. Having a standard flower type, Shakuntala's ray floret diameter is usually wide and has a big size. However, in this study, the ray floret diameter was only 5.52 cm, significantly narrower than other standard flower types Marimar and Kusumaswasti (Table 4). Kusumaswasti and Marimar significantly were more resistant to rust than Shakuntala from 60 to 90 d.a.p. or during the generative phase (Table 2). Rust diseases had affected plant growth of chrysanthemum, as shown in the study of the effect of rust in chrysanthemum's vegetative growth. The result showed a significantly negative correlation between disease intensity of rust with plant height with r = -0.837 [20].

Table 4. Flower stalk length, flower stalk diameter; ray floret diameter, disk floret diameter of seven varieties of chrysanthemum, planted at Garung, Wonosobo in August 2014 to January 2015.

Varieties	Flower stalk length (cm)	Flower stalk diameter (cm)	Ray floret diameter (cm)	Disk floret diameter (cm)
Puspita Nusantara	11.76$^{b*)}$	5.83 ^{b}	5.95^{bc}	1.36^{c}
Shakuntala	11.37^{b}	5.71^{b}	5.52^{c}	1.33^{c}
Puspita Pelangi	13.00^{a}	6.13^{ab}	4.65^{d}	1.65^{b}
Bakardi	13.13^{a}	6.42^{a}	5.96 ^{bc}	0.50^{d}
Kusuma Swasti	10.20^{c}	6.21 ^{ab}	6.52 ^{ab}	1.71^{b}
Marimar	10.15^{c}	6.11^{ab}	6.87^{a}	1.97^{a}
Yulimar	13.07 ^{a}	5.84 ^{ab}	6.13 ^{bc}	1.35 ^{c}

$^{*)}$ the value followed by the same letter not significantly different at $p<0.05$.

Table 5. Flower color, flower type, flower shape of seven varieties of chrysanthemum tested at Garung, Wonosobo in August 2014 to January 2015.

Varieties	Flower Colour	Flower type	Flower shape
Puspita Nusantara	Yellow	Spray	Single
Shakuntala	Yellow	Standard	Decorative
Puspita Pelangi	White	Spray	Semi decorative
Bakardi	White	Spray	Single
Kusuma Swasti	Purple	Standard	Semi decorative
Marimar	Yellow	Standard	Decorative
Yulimar	White	Standard	Decorative

4. Conclusion

The rate of disease progress and the level of disease intensity of rust by the plant age determined the resistance of the chrysanthemum varieties to rust. Based on the disease progression rate and the average of the disease intensity, it can be identified that chrysanthemum varieties of Kusumaswasti, Marimar, and Yulimar were more resistant than the four other varieties tested at Garung, Wonosobo. The pattern of disease progress of rust in chrysanthemum can be used as guidance to control that disease.

References

[1] Ellis D 2007 *New Pest Concern in New England Chrysanthemum White Rust Integrated Pest Management* http://www.hort.uconn.edu/ipm/general/biocontrl/chryswhiterust.htm.
[2] Hanudin, Kardin K, Suhardi 2004 Evaluation of the Resistance of Chrysanthemum Clones Against White Rust *J. Hort* **14** 430–5 (in Bahasa Indonesia)
[3] Suhardi 2009 Source of inoculum, response of varieties, and effectiveness of fungicides against white rust disease in chrysanthemums *J. Hort* **19** 207–13 (in Bahasa Indonesia).
[4] Bety YA, Budiarto K, Suhardi 2012 *Performance of Chrysanthemum National Varieties (Dendrathema grandiflora) var. Puspita Nusantara in several test locations* In: Proceedings of the Floriculture National Seminar. Segunung, Cianjur: Puslithorti, Agricultural Research and Development Agency; p. 60–72 (in Bahasa Indonesia).
[5] Hanudin, Marwoto B 2012 White Rust disease and control efforts *J. Agric. Res. Dev*. **3** 51–7.
[6] Rahardjo IB, Suhardi 2008 Disease incidence and disease intensity of white rust on some chrysanthemum clones *J. Hort* **18** 312–8 (in Bahasa Indonesia)
[7] Budiarto K, Rahardjo IB, Suhardi 2008 Selection of resistance of chrysanthemum clones against rust disease *J. Hort* **18** 249–54 (in Bahasa Indonesia).
[8] Manengkey GSJ, Senewe J 2011 Intensity and rate of efficiency of rust disease of Uromyces phaseoli leave on red bean plants Eugenia **2** 218–23.
[9] Silvia YE, Suhardi 2013 Effect of varieties, leaf decomposition, and fungicide spraying on the intensity of rust disease (*Puccinia horriana P.Henn*) on chrysanthemum *plants Agric J*. **25** 19–25.
[10] Freedman, Mackenzie 1991 *Disease progress curves, their mathematical description, and analysis to formulate predictors for loss equations* Teng P, editor. St. Paul Minnesota: APS Press; 37–48 p.
[11] Silvia YE, Nuryani W, Djatnika I, Hanudin, Suhardi, Winarto B 2012 Potential of several

fungicides for vegetables in controlling white rust (*Pucciana horiana P. Henn.*) to improve the quality of chrysanthemums *J. Hort* **22** 385–91 (in Bahasa Indonesia).

[12] Hanudin, Nuryani W, Silvia E 2010 Biopesticide formulations of active ingredients *Bacillus subtilis, Pseudomonas fluorescens, and Corynebacterium sp.* non-patogenic to control rust disease in chrysanthemum *J. Hort* **20** 247–61 (in Bahasa Indonesia).

[13] Norman DJ, Henry RJ, Yuen JMF 1999 Disease resistance of fourteen pot anthurium cultivars to *Xanthomonas campestris pv. Dieffenbachiae Hortic. Sci* **34** 721–2.

[14] Djatnika I, Dwiatmini, Sanjaya L 1994 The resistance of some chrysanthemum cultivars against rust *Ornam. Plant Res. Bull* **2** 19–25.

[15] Mamahit JME, Manueke J 2016 Integrated pest control of ornamental plants in Kakaskasen Village, Tomohon City *J. LPPM Sains Technol.* **3** 81–94.

[16] Bety YA 2016 Adaptation testing and analysis of farming of several superior varieties of chrysanthemums in Central Java *J. Agrivet* **22** 37–49.

[17] Bety YA, Suhardi 2009 Plant performance and consumer response to national chrysanthemum superior varieties in Magelang District *J. Agrosains* **11** 52–7.

[18] Bety YA 2014 *Performance of Chrysanthemum National Varieties (Dendrathema grandiflora) var. Puspita Nusantara in several test locations* In: Proceedings of the National Seminar of the Faculty of Agro-industry. Jogyakarta: UMB; p. 309–18 (in Bahasa Indonesia).

[19] Silvia YE, Nuryani W, Djatnika I 2010 No Effect of carrier material on the effectiveness of Beauveria bassiana in controlling Thrips Parvispinus karny on chrysanthemum plants in plastic houses *J. Hort* **20** 80–5 (in Bahasa Indonesia).

[20] Bety YA, Sarwana T 2007 *The resistance of some varieties of chrysanthemum to rust caused by Pucciana horiana* In: Proceeding seminar on national innovation agricultural technology. Semarang: BB2TP, BPTP Central Java; p. 239–43 (in Bahasa Indonesia).

Improving seed germination and seedling growth of true seed shallot (TSS) using plant growth regulator seed priming

R Pangestuti [1,2], E Sulistyaningsih[1,*], B Kurniasih [1] and R H Murti [1]

[1] Department of Agronomy, Faculty of Agriculture, Universitas Gadjah Mada Jln. Flora no. 1, Bulaksumur, Sleman, Yogyakarta 55281, Indonesia
[2] Assessment Institute for Agricultural Technology of Central Java Jln. Soekarno Hatta KM 26, No 10, Bergas, Kabupaten Semarang 50552, Indonesia

* E-mail: endangsih@ugm.ac.id

Abstract. Seeding is the most important and quite challenging stage in seed/TSS cultivation. Plant growth regulator (PGR) seed priming has the potential to enhance the quality of TSS seedlings. The research had been conducted in the greenhouse of Agriculture Faculty, UGM, from February to April 2020. The factorial treatments of shallot cultivars (Tuk-Tuk, Lokananta and Sanren) and seed treatments (soaking seed with GA3 100 ppm and NAA 50 ppm for 12 hours and untreated seeds as control) were assigned in the RCBD with four replications. Each experimental unit was consisted of 728 seeds per cultivar. Both treated and untreated seeds were sown in soil blocks. The data were analyzed using ANOVA and continued with Tukey HSD procedure at α=0.05. The results showed that PGR priming on seeds significantly increased the germination percentage, plant height, leaf number, leaf area, and hypothetical vigor index of the three cultivars at six weeks after sowing. However, there was no significant difference in the fresh weight and dry weight of seedlings. This research implied that soaking seeds in GA3 100 ppm and NAA 50 ppm for 12 hours could improve seed germination and TSS seedling growth. Additional fertilizers are perhaps needed in soil blocks to increase the effect of hormones on the seedlings.

1. Introduction

Shallot (*Allium cepa* L.Aggregatum group) is categorized as the major condiment in Indonesia that cannot be substituted. It is also a strategic-valued commodity of vegetables that may affect inflation [1,2]. Efforts to increase shallot productivity and production continuity become important to avoid inflation. The use of true seed shallot/TSS potentially enhances the productivity of shallots by >20 tons ha^{-1} from the currently existing productivity (< 10 ton ha^{-1}) by bulb propagation. TSS also has a long shelf life (2 years), easy to store, handling and distribute so that it can become a seed reserve in each growing season [3,4].

Seeding is the most crucial step in TSS cultivation, and it often comes with obstacles. Generally, seeding needs between 30-45 days age before transplanting. The composition of planting media, seeding technique, fertilization, and seed treatment will affect the bulb yield. Good quality of seeding growth may increase the planting success rate in the field and enhance the plant's productivity [5–7]. Adiyoga

IOP Conf. Series: Earth and Environmental Science **883** (2021) 012024 doi:10.1088/1755-1315/883/1/012024

et al. [1] reported that TSS seeding in soil blocks produced the best yield compared to the direct seeding or seeding in soil plots. The use of soil blocks also helps monitored seedling growth until the transplanting stage since the seeds grew harmoniously in uniform seeding holes.

Seed priming is an effective and practical technique to improve seed growth speed and increase the uniformity of seed growth and plant growth [8]. Plant growth regulator (PGR) seed priming may enhance the plant seedling's germination and growth ability. PGR application in seed germination is considered the prior process in plant growth [9]. Gibberellin is the key hormone that plays a significant role in seed germination [9–11]. Gibberellin induces various genes related in producing amylase, including α-amylase, protease, and β-glucanase. These enzymes act by breaking down food reserve compounds into smaller compounds, such as converting starch into sucrose and produce energy for high cellular respiration during germination [9,12].

Treatment by soaking the seeds in gibberellin (GA3) 100 ppm is described to improve growth speed, growth percentage, and seed viability in various plants [11,13–15], including onion [16,17] and shallot [18]. Gibberellin also plays a role in inducing cell division. Wen *et al.* [19] reported that cell division activity supported successful seed germination processes.

The generally regulated genes influence plant growth activity by more than one growth hormone [20]. Auxin indirectly induces signals that activate gibberellin accumulation in the process of germination [21]. In the advanced process after germination, auxin plays a significant role in the growth of young seedlings [22].

Sudaryono [22] reported that soaking TSS seeds of Trisula cultivars in auxin and gibberellin solution of 200 ppm overnight resulted in better plant growth and yield than the soaking treatment of auxin with cytokinin as well as the combination of auxin, gibberellin, and cytokines. The control plant was not used to compare, and the effect of treatment towards TSS seedling growth before transplanting was not reported. Therefore, this current research aimed to obtain the effect of soaking TSS Tuk Tuk, Lokananta, and Sanren cultivars treated by gibberellin and auxin PGR on seed morphology, germination, and seedling growth using soil block as a sowing before transplanting.

2. Methods

The research has been carried out in the Faculty of Agriculture experimental greenhouse, Universitas Gadjah Mada, Banguntapan, Bantul, Yogyakarta-from February to April 2020. This research used a two-factors arranged in a randomized complete block design with four replications. The first factor was TSS of cultivar Tuk Tuk (expired date Mei 2020), Lokananta (expired date July 2020), and Sanren (expired date Augustus 2020) (PT East-West, Indonesia). The second factor was seed priming before sowing by soaking the seeds for 12 hours in gibberellic acid (GA3) (Merck, Germany) 100 ppm + Naphthalene Acetic Acid/NAA (Serva, USA) 50 ppm solution and untreated seeds as a control. Each experimental unit consisted of 728 seedlings. The total population used was 3 x 2 x 4 x 728 = 17,472 seedlings. Environmental conditions were characterized by the average, maximum, minimum, and mean temperature of 34.3; 27.9; 32.1^{0}C, respectively, and the photoperiod of 12 h, the light intensity of 31,119 lux, and relative humidity 75.8%.

2.1. *Seed priming*

Seed priming before sowing was carried out by soaking the three cultivars in the mixture solution of GA3 solution 100 ppm and NAA 50 ppm PGR mixture for 12 hours. The ratio between seed weight and PGR solution volume was 20 % (w/v). After priming, the seeds were strained and dried on tissue paper. The seeds were then treated by a powdered fungicide with 80% mancozeb as an active ingredient (PT Dow AgroSciences, Indonesia). The fungicide was applied to reduce the risk of damping off seedlings [23].

2.2. *Sowing*

The seeds were deep in soil blocks with a size of 2 x 2 x 2.5 cm. The soil blocks were made from cocopeat, compost, manure, guano, and dolomite mixtures in a ratio of 2: 2: 2: 1: 1 (v / v / v / v / v). The soil blocks were arranged in wooden trays, as many as 364 soil blocks per tray, and hollowed out ± 0.8

cm as planting hole. The seeds inserted into the planting hole as much as two seeds per hole. The soil block was then covered with a mixture of cocopeat, manure, and husk charcoal 1: 1: 1 (v/v/v) of ± 1 cm thick. The seedlings were then sprinkled with water using a low-pressure sprayer with fine droplets.

2.3. Seedling Nursery

Watering was every 1-2 days to keep the soil block moist. Fungicide (mancozeb 80%) was applied at 10 and 28 days after sowing (DAS) as much as 2 g L^{-} to prevent disease. Additional fertilization was applied at 28 DAS using NPK 16:16:16 (Saprotan Utama, Indonesia) as much as 5 g L^{-1} per tray by leaking it into the soil block. Weed were controlled manually by hand.

2.4. Seed Morphology

Before and after PGR soaking treatment, the condition of seeds was observed by measuring seed weight and seed morphology (Digital microscope 3D, Kayence VHX-6000, USA). Tetrazolium staining with 1% tetrazolium solution was used in longitudinal sections of mature seeds with viable embryo observation. The observation of seed germination was conducted upon the number of seeds with normal germination on 4, 7, 10, and 14 of DAS. The criteria of normal germination used in the measurement were a seedling with a 2 cm long radicle and 1 cm long cotyledon (cotyledon appeared on the soil block's surface).

2.5. Seedling morphology

Seedling height and the number of leaves were measured in 2, 4, and 6 weeks after sowing (WAS) of the 80 seedling samples per experimental unit. Seedling height was measured from the soil block's surface to the cotyledon's tip or the highest leaf. Seedling fresh weight and dry weight (dry weight was measured after drying process in an oven at 80° C temperature for 72 hours), leaf width, root length (Image Analysis Sistem WinDIAS 3, Delta-T-Devices Ltd, Cambridge UK), chlorophyll a, chlorophyll b and total chlorophyll were measured at 6 WAS (seedling were ready to transplant). Every destructive observation sample unit consists of three seedlings.

Chlorophyll was analyzed using the spectrophotometer method with 80% acetone solvent [24]. A total of 1 g leaf sample was crushed and then mixed with 20 ml acetone 80%. It was then strained using a paper filter. The solvent absorbed was measured using spectrophotometer Uv-Vis (Genesis 10S, China) with a wavelength of 645 nm and 663 nm. Chlorophyll concentration of the leaf was calculated using the formula below:

$$\text{Chlorophyll a} = 0.0127 \times A663 - 0.00269 \times A645 \quad (1)$$

$$\text{Chlorophyll b} = 0.0229 \times A645 - 0.00468 \times A663 \quad (2)$$

$$\text{Total chlorophyll content} = (1) + (2) \quad (3)$$

Seedling quality was analyzed using germination percentage (GP) and hypothetical vigor index (HVI) values. The germination percentage was calculated using the following formulas [11,25] :

$$GP\ (\%) = \frac{G}{S} \times 100\% \quad (4)$$

Where GP is a germination percentage
G is a number of normal germinated seeds
S is a number of total seeds

Hypothetical vigor index (HVI) describes seedling growth for all growth variables components, including seedling height, leaf area, leaf number, fresh weight, and dry weight of seedlings. HVI was calculated using the following equation [26,27] :

(5)

$$HVI = \frac{\log N + \log A + \log H + \log R + \log G}{\log T}$$

where V is a hypothetical relative index of vigor;
A is leaf area per seedling (cm^2);
N is leaf number per seedling;
H is the seedling height (cm);
R is the fresh seedling weight (g);
G is the dry seedling weight (g);
T is seedling age at the time the measurements were taken (weeks after germination).

2.6. Data analysis
The data collected were analyzed statistically using variance analysis performed in the Statistical Analysis Software (SAS version 9.4). Tukey HSD test was used at a significance level of $p<0.05$ when there were any significant differences in anova.

3. Results and discussion

3.1. The effect of plant growth regulator (PGR)seed priming on seed morphology and seed germination
PGR seed priming influenced the germination process in shallot seed. Figures 1.A to 1.D describes the condition of true seed shallot before and after PGR seed priming. The normal mature seed anatomy displayed more information about how the germination process occurs inside the seed.

Figure 1.A shows a shallot seed in the shape of ovate to triangular; one side convex, and the other tends to be flat with a black and granular-textured seed coat that looks like small warts (verruca). Shallot seed anatomy with normal germination was shown in Figure 1.B. The embryo of shallot seed was had a curvy shape with a spiral tip. The embryo was consisted of the radicle (R), meristem tip (SA), that will grow as the first/true leaf, procambium (P), that will grow as the primary vessel connective tissue, placed extending from the radicle tip to the end length of the embryo, cotyledon (C), a seed food storing organ, and at the end of the embryo was haustorium (H), used as the organ that absorbs food reserves from endosperm during the germination process. Endosperm (E) was the main food reserved for the seed with the main constituent of carbohydrates, a small quantity of protein, and fat. The structure of this shallot seed anatomy was identical with what De Mason [28] described as onion seed structure as an interpretation of a study written by Sachs (1863). Shallot coat (SC) was a black-colored and rough coat with a wart-like wavy texture. Under the coat was a layer of ± 0,08 mm thick aleurone (AL). Aleurone was the outermost layer of protein-rich endosperm as a synthesis place of enzymes for the germination process. The layers' structure and characteristics were similar to leek, onions, tomatoes, and pepper [29].

Shallot seed priming with GA3 100 ppm + NAA 50 ppm solution for 12 hours promoted hydrolytic enzyme secretion to weaken the seed coat /testa [30]. The weakened testa accelerated the imbibition process and produced physical changes in the seed, such as swelling and cracking of the seed coat (Figure 1.C). This water uptake caused the three shallot seed cultivars' weight after the soaking treatment to increase (Table 1).

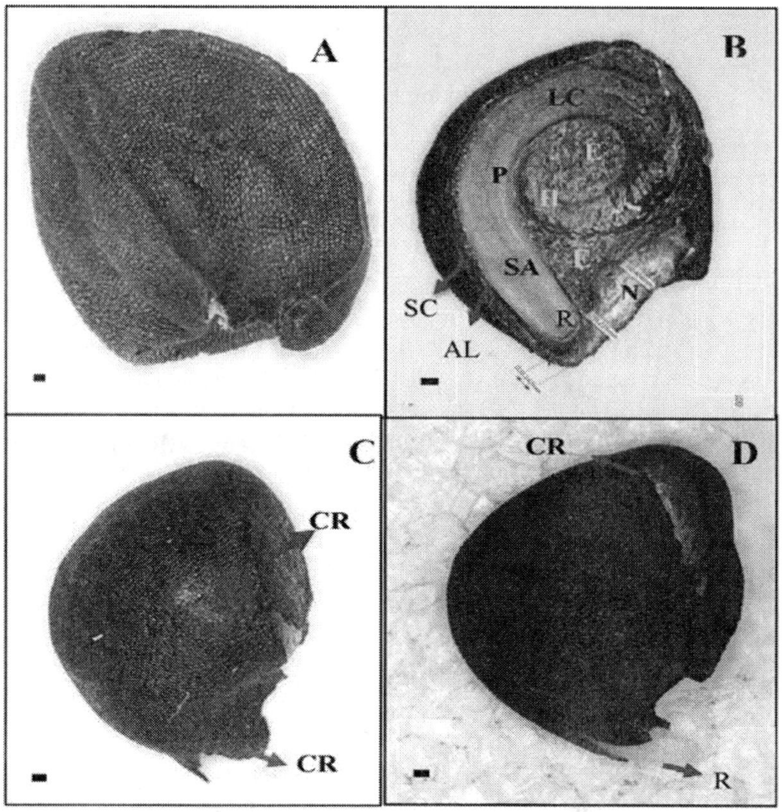

Figure 1. 1.A. Shallot seed/control; **1.B**. Longitudinal section of mature seed with viable embryo containing radicle (R), shoot apex (SA), procambium (P), long cotyledon (LC), haustorium (H), endosperm (E), aleurone layer (AL), nucellus (N) and testa/seed coat (SC) (tetrazolium staining); **1.C**. Shallot seed after PGR seed priming resulting in a swelled seed with cracks on the seed coat (CR); **1.D**. Radicle protrusion (R) and the occurrence of germination. Scale bar = 0.1 mm. Lens Z20X:100

Cracks on the seed coat may activate the gas circulation, especially oxygen, activate metabolism, and synthesize endogenous gibberellin that was transferred into the aleurone layer. Exogenous gibberellin may accelerate the imbibition process and increased the gibberellin's content to be transferred into the aleurone layer [28]. After that, gibberellin will bind to the GID1 protein that interacts with protein DELLA, thus becoming the transcription factors binder that triggers gene transcription. The protein DELLA degradation will occur that caused the transcription factors were released so that gibberellin-responsive genes can be expressed, producing protein and hydrolytic enzyme synthesis, especially α-amylase, proteases, glucanases, and phosphatases. These enzymes may break down the food reserve compounds into smaller compounds, such as turning starch into sucrose and produce substrate for cellular respiration. That mechanism may stimulate faster radicle protrusion from the seed [29] (Figure 1.D). The process in Figure 1.D occurred during the seed germination process in the soil block. This led to the faster germination showed by a higher seedling germination percentage of the three shallot cultivars with PGR seed priming (Table 2).

Table 1. Characters of three TSS cultivars seeds after soaking with GA3 100 ppm and NAA 50 ppm.

Cultivars	Number of seed per 1000 mg	Fresh weight per seed before soaking (mg)	Fresh weight per seed after soaking (mg)	Percentage of seed weight increase by soaking treatment (%)
Tuk Tuk	370.75±0.96	2.70±0.01	4.26± 0.06	57.78±2.27
Lokananta	285.25±1.26	3.51±0.02	5.48±0.06	56.28±1.12
Sanren	264.50±1.30	3.78±0.02	6.05±0.06	60.08±2.20

The three shallot cultivars had different percentages of seed weight gain after PGR seed priming, as shown in Table 1. Sanren cultivar had the highest percentage of weight gain with 60.08%, followed by Tuk Tuk cultivar with 57.78%, and Lokananta with 56.28%. The estimated weight gain for these three seed cultivars was 0.6 times the seed weight. This weight-added value illustrated the ability of live seeds to absorb water imbibition. Yudono [30] stated that the amount of water uptake in the living seeds was influenced by the size of the seeds and their chemical composition, where the available range is 0.5 to 2 times the weight of the dry seeds. The larger the seed size, the more endosperm content was presented, so that the ability to absorb water during imbibition was greater than the small seeds. Sanren cultivar had the most massive seed weight with 3.78 mg, this may cause the largest increased weight added compared to the other cultivars. However, other factors such as differences in seed composition between cultivars or the embryos' size are also suspected in further research

There was no interaction between the TSS cultivars with PGR seed priming in the germination percentage variable four days after sowing (DAS), 7 DAS, 10 DAS, and 14 DAS, as shown in Table 2. PGR seed priming significantly contributed to the increased germination percentage compared to control (untreated seedling). The seed germination percentage given consecutive treatment was 5.75%; 66.85%; 77.77% and 80.22 % at 4 DAS, 7 DAS, 10 DAS and 14 DAS and control seedling had lower germination percentage of 0.03%, 56.78%, 69.51%, and 75.71% respectively. Based on TSS minimum technical requirements, TSS seeds are said to be of good quality if they have a minimum germination percentage of 70% [31], and this value was achieved all treatments at the end of germination observation (14 DAS). However, TSS with PGR seed priming acquired 70% germination faster and produced the highest germination percentage at the final observation. This result was in line with what was found in rice germination [15], guava germination [11], onion germination [16,17] as well as TSS germination [18], where the used of gibberellin 100 ppm may increase the seed germination better than control.

Table 2. Germination percentage (GP) of some TSS cultivars with PGR seed priming using GA3 100 ppm and NAA 50 ppm

Treatment	GP 4 DAS (%)	GP 7 DAS (%)	GP 10 DAS (%)	GP 14 DAS (%)
Cultivars				
Tuk Tuk	2.16 a	53.13 c	69.51 b	74.83 b
Lokananta	2.99 a	61.75 b	74.76 a	79.92 a
Sanren	3.50 a	70.57 a	76.65 a	80.22 a
Seed priming (GA3 100 ppm+NAA 50 ppm, 12 h)				
Without	0.03 y	56.78 y	69.51 y	75.71 y
With	5.75 x	66.85 x	77.77 x	80.93 x
Interaction	(-)	(-)	(-)	(-)
CV (%)	2.45*	9.18	3.01	3.53

Note: Value with the same letters is not significantly different based on HSD Tukey, α=0.05; (-) no interaction between the cultivars and seed priming HVI: hypothetical vigor index; *: transformation data with √x+0.5.

NAA 50 ppm' addition as an exogenous auxin was related to auxin's function in preventing the possibility of a defect in seedling shape or slow growth in seedling production. Furthermore, auxin plays a role in creating lateral roots in the advanced germination stage and initial seedling growth [32], thus increasing seedlings' germination [22]. The accumulation of auxin in cotyledons was the primary source for auxin to proceed to seedling growth [9]. The synergy between auxin and gibberellin could be seen in the advanced growth after the germination process. That was shown by the increased value of the seedling growth variables described in Table 3.

3.2. The effect of plant growth regulatorl seed priming towards seedling growth

The interaction between TSS with PGR seed priming was not found in the plant height, root length, number of leaves, leaf area, fresh weight, dry weight, and hypothetical vigor index after six WAS, as shown in Table 3.

Sanren cultivar had the best seedling growth as seen from the variables of plant height, root length, leaf area, and its hypothetical vigor index value, while the lowest growth rate was found in Tuk Tuk cultivar. PGR seed priming significantly affected the increasing seedling plant height, leaf number, and leaf area of the three TSS cultivars. However, there was no significant difference in root length, fresh weight, and seedling dry weight. The analysis of chlorophyll a, b and total chlorophyll content in Table 4 revealed an interaction between cultivar/genetic factors and seed priming. In Tuk Tuk and Lokananta cultivars, although the chlorophyll content of the seedlings treated with PGR priming was lower than the control, the difference was not significant ($P < 0.05$). In contrast, in the Sanren cultivar, there was a significantly lower difference in the chlorophyll content. The tendency of lower chlorophyll content in leaves treated with PGR priming was the cause of the similar value in fresh weight and dry weight of control seedlings, even though they had higher plant height, leaf number, and leaf area than control seedlings.

Table 3. Character of seedlings in three TSS cultivars with PGR seed priming using GA3 100 ppm and NAA 50 ppm at transplanting stage (6 weeks after sowing)

Treatment	Plant Height (cm)	Root length (cm)	Number of leaves	Leaf area (cm^2)	Fresh weight (g)	Dry weight (g)	HVI
Cultivars							
Tuk Tuk	22.51 b	20.07 ab	3.01 a	13.83 b	0.52 a	0.045 a	1.66 b
Lokananta	23.87 a	16.44 b	3.05 a	16.53 a	0.58 a	0.043 a	1.87 ab
Sanren	24.49 a	21.67 a	3.10 a	18.58 a	0.66 a	0.055 a	2.17 a
Seed Priming (GA3 100 ppm+NAA 50 ppm, 12 h)							
Without	22.44 y	20.50 x	2.97 y	15.23 y	0.56 x	0.043 x	1.75 y
With	24.80 x	18.29 x	3.13 x	16.96 x	0.62 x	0.051 x	2.04 x
Interaction	(-)	(-)	(-)	(-)	(-)	(-)	(-)
CV (%)	5.37	16.05	2.98	11.01	5.87*	1.02*	17.10

Note: Value with the same letters is not significantly different based on HSD Tukey, α=0.05; (-) no interaction between the cultivars and seed priming HVI: hypothetical vigor index; *: transformation data with $\sqrt{x+0.5}$.

Gibberellin could increase seedling height and leaf width by increased cell division, extension, and replication [15]. However, the leaves may become thinner because chlorophyll's formation on the leaves as the main engine of photosynthesis was not following cell division's active process and stretching [33].

Table 4. Chlorophyll content of three TSS cultivars with PGR seed priming (GA3 100 ppm +NAA 50 ppm, 12 h) at transplanting stage (6 weeks after sowing)

Variable	Seed Treatment	Cultivar			Average
		Tuk Tuk	Lokananta	Sanren	
Chlorophyll a (mg. g^{-1})	Without/Control	0.481 a	0.422 ab	0.531 a	0.478
	With	0.414 ab	0.319 bc	0.257 c	0.330
	Average	0.447	0.370	0.394	(+)
	CV = 14.92				
Chlorophyll b (mg. g^{-1})	Without/Control	0.162 xy	0.203 xy	0.259 x	0.208
	With	0.174 xy	0.140 y	0.113 y	0.142
	Average	0.168	0.171	0.186	(+)
	CV = 3.22*				
Total Chlorophyll (mg. g^{-1})	Without/Control	0.643 uv	0.624 uv	0.790 u	0.686
	With	0.587 uv	0.459 vw	0.370 w	0.472
	Average	0.615	0.541	0.580	(+)

Note: Means in each variable with the same letters are not significantly different based on HSD Tukey, α=0.05; (+) an interaction between cultivars and seed priming; *=: transformation data with √x+0.5.

The decreased chlorophyll content in seedlings with PGR seed priming (Table 4) caused no increased in photosynthate in plants. Rogach *et al.* [33] stated that priming with GA3 50 ppm caused decreased in cytokinin content by 24.5% in eggplant. Cytokinin plays a role in the synthesis of chlorophyll in plants [32–35]. Decreased cytokines caused a decrease in chlorophyll synthesis in the leaves. The amount of decrease depends on the cultivar/genetic factor. The chlorophyll content increased can be made using two types of treatment, i.e., cytokinin's addition by spraying seedlings during the nursery or adding fertilizer, especially nitrogen fertilizer. The nitrogen fertilizer addition was strongly recommended since cytokinin may lead to gibberellin function decrease [36,37]. The nitrogen application was reported to have increased photosynthetic pigment content, synthetic enzyme, and chloroplast membrane system [38].

Although there was no significant difference in seedlings' root length, fresh weight, and dry weight, the hypothetical vigor index of the seedlings treated with PGR priming was higher (2.04) and significantly different from the control plant (1.75), as shown in Table 3. Hypothetical vigor index describes seedling quality as the accumulation of all plant growth components, i.e., plant height, root length, number of leaves, leaf area, fresh seedling weight, and dry weight. Vigor is defined as a seed or seedling's ability to grow normally and quickly in suboptimal environmental conditions [25]. Vigor relates to seed production ability, where the seedlings with high vigor content will produce high production as well [27,38]. Bettoni *et al.* [5] stated that onion seedlings with high vigor determined plant growth and bulbs' yield quality. The higher hypothetical index vigor value resulted from seed soaking treatment in GA3 100 ppm and NAA 50 ppm for 12 hours in this study was expected to improve seedlings' viability during the field's transplanting stage and yield of shallot bulbs.

4. Conclusion

Seed priming using a mixture of plant growth regulators GA3 100 ppm and NA 50 ppm could increase seed germination by value of 10.07 and hypothetical vigor index seedling by 2.04. The Sanren cultivars showed the best percentage of germination and seed growth, followed by Lokananta and Tuk Tuk cultivars.

5. Acknowledgment
The authors are very grateful for the support and fund grant of the Indonesian Agency for Agricultural Research and Development (IARRD) scholarship.

References
[1] Adiyoga W, Prathama M and Rosliani R 2020 Partial and farm budged analysis of some sowing techniques in TSS cultivation. *Jhort* **30** 1–10 (in Bahasa Indonesia).
[2] Yudhistira B and Fatmawati A 2020 Diversity of Indonesian soto *J. Ethn. Foods* **7** 1–9
[3] Pangestuti R and Sulistyaningsih E 2011 Potency of True Seed Shallot (TSS) as the material of shallot propagation in Indonesia Pros. Semiloka Nas. "Agro-Innovation Support for Empowerment. Farmers" 258–66 (in Bahasa Indonesia)
[4] Van Den Brink L and Basuki R S 2012 Production of true seed shallots in Indonesia *Acta Hortic.* **958** 115–20
[5] Bettoni M M, Mogor Á F, Pauletti V and Goicoechea N 2014 Growth and metabolism of onion seedlings as affected by the application of humic substances, mycorrhizal inoculation and elevated CO2 *Sci. Hortic. (Amsterdam).* **180** 227–35
[6] Sopha G, Sumarni N, Setiawati W and Suwandi 2015 Sowing technique of true shallot seed to produce seedling and set of shallot) *J. Hortik.* **25** 318–30(in Bahasa Indonesia)
[7] Thoriqussalam A and Damanhuri 2019 Effect of Seedbed Media Composition to the Growth and Production Onion From Seed (True Shallot Seed) *J. Produksi Tanam.* **7** 1314–21(in Bahasa Indonesia)
[8] Hussain S, Khan F, Hussain H A and Nie L 2016 Physiological and biochemical mechanisms of seed priming-induced chilling tolerance in rice cultivars *Front. Plant Sci.* **7** 1–14
[9] Miransari M and Smith D L 2014 Plant hormones and seed germination *Environ. Exp. Bot.* **99** 110–21
[10] Asami T and Nakagawa Y 2018 Preface to the Special Issue: Brief review of plant hormones and their utilization in agriculture *J. Pestic. Sci.* **43** 154–8
[11] Hosseini M S, Zahedi S M, Hoveizeh N F, Li L, Rafiee M and Farooq M 2020 Improving seed germination and seedling growth of guava under heat and osmotic stresses by chemical and hormonal seed treatments *Bragantia*
[12] Lokhande A A, Lokhande A D and Gaikwad D K 2014 Research Article effect of plant growth regulators on germination in two onion varieties *J Adv Sci Res,* **5** 4–7
[13] Sun J, Wang P, Zhou T, Rong J, Jia H and Liu Z 2017 Transcriptome Analysis of the Effects of Shell Removal and Exogenous Gibberellin on Germination of Zanthoxylum Seeds *Sci. Rep.* **7** 1–10
[14] Esan A ., Olaiya C ., Anifowose L ., Lana L ., Omolekan T ., Fagbami O and Adeyemi H R . 2020 Effect of plant growth-promoting rhizobacteria and gibberellic acid on salt stress tolerance in tomato genotypes *African Crop Sci. Journal,* **28** 341–62
[15] Sukifto R, Nulit R, Chee Kong Y, Sidek N, Nuratiqah Mahadi S, Mustafa N and A. Razak R 2020 Enhancing germination and early seedling growth of Malaysian indica rice (Oryza sativa L.) using hormonal priming with gibberellic acid (GA3) *AIMS Agric. Food* **5** 649–65
[16] Somraj B, Reddy K ., Krishna K V . and Srihari D 2012 effect of invigoration treatments on seed germination and seedling vigour in carry-over onion seed (Allium cepa L. *J.Res. ANGRAU* **40** 1–5
[17] Brar N S, Kaushik P and Dudi B S 2020 Effect of seed priming treatment on the physiological quality of naturally aged onion (Allium cepa l.) seeds *Appl. Ecol. Environ. Res.* **18** 849–62
[18] Agung.M.S. A and I.W. Diara 2017 Pre-sowing Treatment Enhanced Germination and Vigour of True Shallot (Allium cepa var. aggregatum) Seeds *Int. J. Environ. Agric. Biotechnol.* **2** 3262–7
[19] Wen D, Xu H, Xie L, He M, Hou H and Zhang C 2017 A loose endosperm structure of wheat seed produced under low nitrogen level promotes early germination by accelerating water

uptake *Sci. Rep.* **7** 1–11

[20] Yang G X, Jan A, Shen S H, Yazaki J, Ishikawa M, Shimatani Z, Kishimoto N, Kikuchi S, Matsumoto H and Komatsu S 2004 Microarray analysis of brassinosteroids- and gibberellin-regulated gene expression in rice seedlings *Mol. Genet. Genomics* **271** 468–78

[21] Sun T ping 2010 Gibberellin-GID1-DELLA: A pivotal regulatory module for plant growth and development *plant Physiol.* **154** 567–70

[22] Bialek K, Michalczuk L and Cohen J D 1992 Auxin biosynthesis during seed germination in Phaseolus vulgaris *Plant Physiol.* **100** 509–17

[23] Arnon D I 1949 Copper Enzymes in Isolated Chloroplasts. Polyphenoloxidase in Beta Vulgaris *Plant Physiol.* **24** 1–15

[24] ISTA 2017 *International rules for seed testing* vol 19–8 (The International Seed Testing Association (ISTA))

[25] Adenikinju S A 1974 Analysis of growth patterns in cocoa seedlings as influenced by bean maturity *Exp. Agric.* **10** 141–7

[26] Purwanti S, Rizky Immawati D and Prajitno D 2018 The Study on The Seed Storability of Black Soybean (Glycine max L. Merrill) Intercropped with Sweet Sorghum (Sorghum bicolor L. Moench) *Planta Trop. J. Agro Sci.* **6** 116–21

[27] De Mason D A 2018 *Onions and allied crops Vol I: Botany, Physiology, and Genetics* ed H D Rabinowitch and J L Brewster (Boca Raton: CRC Pres) pp 27–52

[28] Holdsworth M J, Bentsink L and Soppe W J J 2008 Tansley review Arabidopsis seed maturation , after- *New Phytol.* **179** 33–54

[29] Mena M, Javier Cejudo F, Isabel-Lamoneda I and Carbonero P 2002 A role for the DOF transcription factor BPBF in the regulation of gibberellin-responsive genes in barley aleurone *Plant Physiol.* **130** 111–9

[30] Nelson S K and Steber C M 2016 *Gibberellin Hormone Signal Perception: Down-Regulating DELLA Repressors of Plant Growth and Development* vol 49, ed P Hedden and S G Thomas (John Wiley & Sons, Ltd.)

[31] Yudono P 2015 *Plant Seed: Basic science, technology, and management* (Yogyakarta: Gadjah Mada University Press) p 308 (in Bahasa Indonesia)

[32] Ministry of Agriculture 2015 Kepmentan No 131 the Year 2015 Certification of shallot seeds.pdf 36 (in Bahasa Indonesia)

[33] Liu P P, Montgomery T A, Fahlgren N, Kasschau K D, Nonogaki H and Carrington J C 2007 Repression of auxin response factor10 by microRNA160 is critical for seed germination and post-germination stages *Plant J.* **52** 133–46

[34] Rogach V V., Voytenko L V., Shcherbatiuk M M, Kosakivska I V. and Rogach T I 2020 Morphogenesis, pigment content, phytohormones and productivity of eggplants under the action of gibberellin and tebuconazole *Regul. Mech. Biosyst.* **11** 116–22

[35] Almeida G . and Rodrigues J G . 2016 Development of plants by interference auxins, cytokinins, gibberellins, and ethylene *Appl. Res. Agrotechnology* **9** 111–7

[36] Talla S K, Panigrahy M, Kappara S, Nirosha P, Neelamraju S and Ramanan R 2016 Cytokinin delays dark-induced senescence in rice by maintaining the chlorophyll cycle and photosynthetic complexes *J. Exp. Bot.* **67** 1839–51

[37] Liu Y, Fang Y, Huang M, Jin Y, Sun J, Tao X, Zhang G, He K, Zhao Y and Zhao H 2015 Uniconazole-induced starch accumulation in the bioenergy crop duckweed (Landoltia punctata) I: Transcriptome analysis of the effects of uniconazole on chlorophyll and endogenous hormone biosynthesis *Biotechnol. Biofuels* **8** 1–12

[38] Greenboim-Wainberg Y, Maymon I, Borochov R, Alvarez J, Olszewski N, Ori N, Eshed Y and Weiss D 2005 Cross talk between gibberellin and cytokinin: The Arabidopsis GA response inhibitor SPINDLY plays a positive role in cytokinin signaling *Plant Cell* **17** 92–102

[39] Pramanik K and Bera A K 2013 Effect of seedling age, seeding density, and nitrogen fertilizer on growth and grain yield of rice (Oryza sativa L.) *Int. J. Agron. Plant Prod.* **3** 813489–99

The diversity of bird species based on the altitude of the protected forest area in Sirimau Mountain in Soya Village - Ambon City

C K Pattinasarany[1,*], L Latupapua[1], A Sanduan[1], Y Th Latupapua[1], F F Tetelay[1], F Soselisa[1]

[1] Departement of Forestry, Faculty of Agriculture, Pattimura University

* E-mail: pattinasarany.ceka@gmail.com

Abstract. This study is aimed to discover the diversity of bird species based on the altitude of the area and to discover the impact of vegetation as habitat towards the diversity of bird species in protected forest area in Sirimau Mountain in Soya Village located in Ambon City. The method used to discover the diversity of birds was the IPA-method (Indices Ponctuels d'Abondance) while observing the vegetation used the CSS-method (Continous Strip Sampling) alongside the observation paths. The protected forest consists of three area of different altitudes, namely 300 meters above sea level (masl), 400 masl and 500 masl. The bird species found in the protected forest area in Sirimau Mountain in Soya Village as a whole were 16 species. The forest area at 300 masl was found 14 bird species with diversity index (H') of 2.57, the forest area at 400 masl was found 12 bird species with diversity index (H') of 2.42, while the forest area at 500 masl was found 11 bird species with diversity index (H') of 2.10. There were 8 bird species found in every area of three different altitudes that were dominant such as Perling Ungu (*Aplonis motalica*), Walet Sapi (*Collocalia esculenta*), Perkicit Pelangi (*Trichoglossus haematodus*), Pergam Laut (*Ducula bicolor*), Sesap Madu Sriganti (*Nectarinia jugularis*), Srigunting Lencana (*Dicrucus bracteatus*), Nuri Pipi Merah (*Geoffreyus geoffreyi*) and Tekukur Biasa (*Streptopelia hinensis*). Meanwhile, the dominant types of vegetation found in the area were Damar (*Agathis sp.*), Kayu Merah (*Eugenia jambolana*) and Meranti (*Shorea sp.*)

1. Introduction

Birds have an important role in natural ecosystem control for both the short term such as water control, climate, air, pollination, pest control, etc., or the long term such as food chain, forest formation, etc. An even stage of pollination had made birds as a source of biological wealth that is used in the ecosystem and sensitive to environmental changes [1]. The forest structure and composition can create open spaces in the utilization of available sources (foods, water, cover, and space) that give an effect to the bird's community at some habitat.

Lost and damaged habitat can disturb the life process that was natural before [2]. Habitat change occurs due to management by humans. This is shown by the diversity of local birds, therefore birds can be used as bio-indicator in determining damaged levels in some environments. The bird's diversity is influenced by habitat type diversity. Vegetation structure and availability of feed in the habitat are the main factors that affect the species diversity in some habitats [3]. Therefore, habitats with more diverse vegetation would likely have more bird diversity compared to habitats with fewer vegetation types. The Moluccas area has 348 bird species with 90 species of them are Moluccas endemic species. These species spread in 5 Endemic Birds Area (EBA), with 116 species have limited distribution [4]

Mount Sirimau area was appointed as a protected forest based on the Decree of Forestry Minister number 430/Kpts-II/1996 on August 13th, 1996, which is located in the administrative area of Ambon City, Moluccas Province. The protected forest area in Mount Sirimau plays a significant role like diversity source cover for fauna and flora, also have an economic function for the locals. Soya Village is one of the villages located nearby the Protected Forest Area in mount Sirimau. The bird existence (species diversity) and its habitat as birds diversity support in this Protected Forest have not been known yet.

Based on the comprehension above, this study is aimed to discover the bird's diversity based on the altitude and also to discover the habitat condition (vegetation) that affect bird's diversity existences in the Protected Forest area. The result of this study may benefit as base data in the Protected Forest Area Management, especially for birds management.

2. Methods

2.1. Date and location

This study took place in the Protected Forest in Mount Sirimau, Soya Village, Ambon City, Moluccas Province and lasts for 3 months (December 2019 until February 2020) and the equipment used during this research are GPS, Binocular Spyglass, Camera, Phiband, Haga, Hypsometer, and Birds Identification Book. [5]

2.2. Data collecting

The bird's data collection was conducted by making the observation paths (Transek Lines) in each different altitude type (300, 400, 500 masl). This study was to observe the birds that are around observation paths, to record the species and the amount of them. The observation was done twice which is 06.00-10.00 a.m. and 14.00-18.00 p.m. (based on Eastern Indonesia Timezone). The bird's calculation used [6] IPA's Method (Indices Ponctuels d'Abondance), which in each altitude were made 5 paths and in each path were made 5 observation points. The distance between each path is 200 meters. Meanwhile, the habitat data collection was made alongside the bird's observation paths, record the vegetation type, vegetations amount also habitat condition in each different altitude type.

2.3. Data analysis

2.3.1. Birds data

a) Birds species diversity

The determination of bird's species diversity index based on Species Diversity index (H'), Shannon – Wienner [7]

$$H' = -\sum_{i=1}^{s}(pi)(\ln pi)$$

H' = diversity of species
pi = proportion of individuals of the total sample belongings to the *i*-th species
s = number of species
i = the *i*-th species

Species diversity level category:
H' = 1, Low Diversity; individual amount spreading in each species is low, with low habitat stability
H' = 1-3, Medium Diversity; individual spreading amount in each species is medium, with medium habitat stability
H' = 3, High Diversity; individual spreading amount in each species is high, with high habitat stability.

b) Index of Evenness (E)

Index of evenness (E) is used to discover the evenness of each species (spreading) in each found habitat. The evenness depicts the balance between one community to another [6]

$$E = \frac{H'}{\ln S}$$

E = index of evenness (index between 0 – 10)
H' = diversity of species
In = natural logarithm
s = number of species

c) Similarity Index (IS)

Similarity index is used to compare certain index of bird species in different habitat. Similarity Index – Sorensen [8] is:

$$IS = \frac{2C}{A + B} \times 100\%$$

IS = Sorensen's similarity index
A = number of bird species found in Habitat A
B = number of bird species found in Habitat B
C = number of bird species found in Habitat A and B

Sorensen's Similarity Index criteria

Similarity Index	Information
0 - 20 %	Very Low
21 - 40 %	Low
41 - 60 %	Medium
61 - 80 %	High
81 - 100 %	Very High

2.3.2. Habitat

The vegetation analysis as bird's habitat by inventory the vegetations types using Continous Strip Sampling (CSS) method. The inventory result was analyzed to discover the INP (IVI) index of each habita, [9] that is:

$$\text{Density (D)} = \frac{\text{The amount of one species}}{\text{Sample plot area}}$$

$$\text{Relative Density (RD)} = \frac{\text{One species density}}{\text{All species density}} \times 100\%$$

$$\text{Frequency (F)} = \frac{\text{The amount of plot that found one species}}{\text{The total amount of plot}}$$

$$\text{Relative Frequency (RF)} = \frac{\text{One species frequency}}{\text{All fspecies frequency}} \times 100\%$$

$$\text{Dominance (Di)} = \frac{\text{The total amount of the base area}}{\text{Sample plot area}}$$

$$\text{Relative Dominance (RDi)} = \frac{\text{One species dominance}}{\text{All species dominance}} \times 100\%$$

Important Value Index (IVI) = Relative Density + Relative Frequency + Relative Dominance

IOP Conf. Series: Earth and Environmental Science **883** (2021) 012025 doi:10.1088/1755-1315/883/1/012025

3. Results and discussion

3.1. Bird species diversity and index of evenness

The result showed the amount of bird species that found in the Proected Forest Area are 16 species (Table 1).

Tabel 1. Bird species in protected forest area in Mount Sirimau (Soya Village)

No	Bird Species		Habitat		
	Indonesian's Name	Scientific's Name	A	B	C
1	Kicuit Kerbau	*Motalica flava*	*	*	–
2	Walet Sapi	*Collocalia esculenta*	*	*	*
3	Elang Bondol	*Haliastur indus*	–	–	*
4	Kacamata Ambon	*Zosterops kuehni*	*	*	–
5	Perling Ungu	*Aplonis motalica*	*	*	*
6	Nuri merah	*Eos bornea*	*	–	*
7	Perkici Pelangi	*Trichoglossus haematodus*	*	*	*
8	Pergam Laut	*Ducula bicolor*	*	*	*
9	Pergam Tarut	*Ducula concinna*	*	*	–
10	Sesap Madu Sriganti	*Nectarinia jugalaris*	*	*	*
11	Srigunting Lencana	*Dicrurus bracteatus*	*	*	*
12	Nuri Pipi Merah	*Geoffroyus geoffroyi*	*	*	*
13	Tekukur Biasa	*Streptopelia chinensis*	*	*	*
14	Betet Kelapa Paruh Besar	*Tanygnathus megalorynchas*	*	*	–
15	Nuri Raja	*Alisterus amboinensis*	–	*	*
16	Raja Udang Erasia	*Alcedo atthis*	*	–	–

Notes : A : altitude of 300 masl
B : altitude of 400 masl
C : altitude of 500 masl
* : bird species were found
– : bird species not found

Table 1 above shows that there are 16 bird species were found, their different existence correspond to their habitat condition and where the birds were comfortable in making their habitat, searching for food, playing around, or even to rest. Therefore, some bird species were found only at 300 masl, some were only found at 400 masl, some were only found at 500 masl, and some can be found at every altitude.

Based on the Table 1 above, in protected forest area at 300 masl there were 14 species with the total amount of population were around 185 birds, at 400 masl there were 12 species with the total amount of population were around 110 birds, and at 500 masl there were 11 species with the total amount of population were 64 birds. According to [10]), habitat and altitude are correlated, which in every height addition will affect to species abundance loss. Other factor, like climate and soil fertility also altitude, really determine the bird species diversity on habitat level.

IOP Conf. Series: Earth and Environmental Science **883** (2021) 012025 doi:10.1088/1755-1315/883/1/012025

Table 2. Bird species diversity index (H') and index of evenness (E)

Altitude	H'	E
300 masl	2,57	0,98
400 masl	2,42	0,94
500 masl	2,10	0,88

The result of this study shows the difference between the Species Diversity Index with Index of Evenness at different altitudes (Table 2). Altitude with high Species Diversity Index and high Index of Evenness shows that spreading and habitat control from bird species are better in the lower altitude due to the more complex community composition and structure it has, also have higher diversity than in higher altitude conditions. At 300 masl, the habitat shows the availability of resources that support the life of bird species.

3.2. Bird species similarity

The Altitude at the Mount Sirimau in Protected Forest Area shows the level of species similarity between the 300 masl and the 400 masl of 89%. The level of species similarity between 400 masl and 500 masl is 75%, while the lowest level of species similarity is obtained between 300 masl with 500 m asl of 64%. The similarity in species between the 3 types of altitude shows a value of 64% - 89%, this shows that between 3 habitats have a high level of species similarity, because the species of birds found in the study location are relatively similar.

3.3. Correlation between bird species diversity and altitude (habitat)

The results of the vegetation analysis at 300 masl found 29 tree level vegetation types dominated by Damar (*Agathis damara)* (INP = 51,95 %), Kayu merah *(Eugenia jambolana)* (INP = 28,05 %), Meranti *(Shorea Sp)* (INP = 17, 93 %), Kayu burung *(Elaeoucarpus sphericus)* (INP = 16,125 %), Lenggua *(Petrocarpus Indicus)* (INP = 15,795). At 400 masl found 22 species of tree level vegetation dominated by Damar (*Agathis Damara)* (INP = 31,28 %), Kayu merah *(Eugenia jambolana)* (INP = 25,58 %), Kayu burung *(Elaeoucarpus sphericus)* (INP = 16,52 %). Altitude 500 masl found 18 types of tree level vegetation, which are dominated by Kayu merah *(Eugenia jambolana)* *(*INP = 26,812 %), Lenggua *(Pterocarpus indicus)* (INP = 24,89 %2), Belo hitam *(Diospyros celebica)* (INP = 19,496 %), Kayu burung *(Elaeoucarpus sphericus)* (INP = 20,96 %).

Based on the types of vegetation that are used as bird feed at altitudes of 300 masl, 400 masl, and 500 masl, it is found that the species of birds that are often encountered are fruit / seed-eating birds, such as the type of Pergam laut *(Ducula bicolor),* Perkicit Pelangi (*Trichoglossus haematodus),* Nuri Pipi Merah *(Geoffroyus geoffroyi*), Tekukur Biasa (*Streptopelia chinensis).* Insectivorous birds such as Walet sapi *(Collocalia esculenta),* dan Srigunting lencana *(Dicrurus bracteatus).* Nectar birds likes Sesap madu sriganti *(Nectariana jugalaris)*, while carnivorous birds like Elang bondol *(Haliastur indus)* dan Raja udang erasia *(Alcedo atthis).*

Altitude factors affect the presence of birds because they directly affect the diversity of vegetation which is a source of bird feed. The variety of vegetation types found in a habitat supports the availability of food for birds, so that with a variety of vegetation types, birds will have more choices to choose the type of feed [11]

4. Conclusion

Based on the description of the results and the discussion above, it can be concluded that:

a) In the habitat with the altitude of 300 masl was found 14 species of birds,the altitude of 400 masl was found 13 species of birds, and the altitude of 500 masl was found 11 species of birds. The number of bird species found in the 3 types of altitude was 16 types of birds.

b) The highest bird species diversity index value is the altitude of 300 masl (H' = 2.57) followed by 400 masl (H' = 2.42) and the lowest is 500 masl (H' = 2,10). The altitude of the place affects the level of diversity of vegetation types which also affects the diversity of bird species, due to the complexity of the structure and composition of their habitat.

c) The most frequently used vegetation by bird species at the 3 altitude types is Damar (Agathis damara), Kayu Merah (Eugenia jambolana) and Kayu Burung (Elaeoucarpus sphericus). Vegetation types as a habitat for bird species in the research location massively contributes to the existence of bird species in fulfilling the basic needs of bird species such as food, drink, breeding grounds and bird's cover.

References

[1] Hadinoto, Mulyadi A, Siregar I S 2012 Keanekaragaman jenis burung di Hutan Kota Pekanbaru *Jurnal Ilmu Lingkungan* **6** 25-42.

[2] BalenV S and Prentice C 1997 Bird of the Negara River Basin South Kalimantan, Indonesia.

[3] Rika S D 2007 Kenekaragaman Jenis Burung Berdasarkan Tipe Habitat Taman Nasional Gunung Ciremai *Media Konservasi, Jurnal Ilmiah Bidang konservasi Sumber Daya alam Hayati dan Lingkungan* **12** IPB - Bogor.

[4] Sutjatnika 1995 Melestarikan Keanekaragaman Hayati Indonesia. Pendekatan Daerah Burung Endemik. PHPA/Biordlife Internasional- Indonesia Programme.

[5] Coates B J, Bishop K D 2020 Panduan Lapangan Burung-burung di Kawasan Wallacea (Sulawesi, Maluku dan Nusa Tenggara). Cetakan Kedua. Burung Indonesia & Dove Publication, Pty. Ltd.

[6] Fachrul M F 2007 Metode Sampling Bioekologi. Bumi Aksara, Jakarta

[7] Smith R L 1992 Element of Ecology. Third Edition. Harper Collins Publishers Inc. New York

[8] Maggurran A E 2004 Measuring Biological Diversity. USA: Blackwell Publishing Company.

[9] Soerianegara I, Indrawan A 1988. Ekologi Hutan Indonesia. Fakultas Kehutanan Institut Pertanian Bpgor (IPB). Bogor.

[10] Supriatna J 2008 Melestarikan Alam Indonesia. Jakarta; Yayasan Pustaka Obor Indoneisa

[11} Tews J, Brose U, Grimm V, Tielboerger K, Wichmann M C, Schwager M, and Jeltsch F 2004 Animal Species diversity diven by habitat heterogenety/diversity. The importance of keystone structure *Journal of Biogeography* **31** *79 – 92.*

Characteristics of yellow sweet potato flakes (*Ipomoea batatas* L.) with the addition of moringa leaf flour (*Moringa oleifera*)

G H Augustyn[1,*], V N Lawalata[1], and S G Sipahelut[1]

[1]Agricultural Product Technology, Faculty of Agriculture, Universitas Pattimura, Ambon, Maluku, Indonesia

* E-mail: geloraaugustyn@gmail.com

Abstract. Flakes made from yellow sweet potato flour are one of the products with the addition of moringa leaf flour to increase the added value of the two commodities. The purpose of this study was to characterize the chemical and organoleptic properties of yellow sweet potato flakes with the addition of moringa leaf flour and determine the best treatment. This research was designed using a Completely Randomized Design with the addition of 0%, 2%, 4%, 6%, 8% moringa leaf flour, and repeated twice. The results showed that the addition of 4% moringa leaf flour was the best treatment because it was following SNI, with chemical characteristics of 3.43% moisture content, 1.88% ash content, 7.24% fat content, 2.05% protein content, 80.54% carbohydrate content, 4.81% fiber content and organoleptic characteristics of color (3.93 like), taste (3.76 like), aroma (3.40 like), crispness (3.90 like), and overall (3.90 like).

1. Introduction

Sweet potato (*Ipomea batatas* L.) is the main source of carbohydrates after rice, maize, and cassava, and has an important role in the food supply and industrial raw materials. Among carbohydrates sources; especially rice, cassava, and corn, sweet potatoes are proven to have significant advantages and benefits for Indonesian in terms of high productivity and carbohydrates, their various varieties, relatively cheaper prices and have been known from generation to generation by Indonesian society [1]. Based on their colors, sweet potatoes are classified into 4 groups, namely purple, orange, white, and yellow sweet potatoes. The sweet potatoes that will be used in the study are yellow sweet potatoes.

Yellow sweet potato has a yellow tuber color, which contains quite a high beta-carotene and it also contains high fiber. Besides, it is also a source of vitamins and minerals that good enough to meet the public's nutritional and health needs. Other nutritional contents found in sweet potatoes are protein, fat, and ash. The low utilization of sweet potatoes is due to the lack of post-harvest processing technology being applied. Sweet potatoes processing in Indonesia is relatively simple and the processing is still in the form of fresh sweet potatoes such as roasting, boiling, and frying [2]. Processing sweet potato into flour can produce a different form of processed food product so that it is easily applied in everyday life. The protein content of yellow sweet potato flour is considered low, flour with low protein content is difficult to stir, mix with yeast, and has low absorption. This type of flour is suitable for making flakes.

Flakes are a thin, round, brownish-yellow sheet food product and it is usually consumed with milk or can be consumed directly as a snack [3]. This food is popular with the public because it is delicious, healthy, and practical in the presentation. Flakes can be made from a variety of food ingredients that contain carbohydrates and can be added with other nutritional sources such as protein to meet nutritional needs [4]. Therefore, one of the food ingredients that can be added as a source of protein is moringa leaves.

Moringa leaves are a part of the moringa plant that has been widely studied for its nutritional content and uses. Moringa leaves are very rich in nutrients, including calcium, iron, protein, vitamin A, vitamin B, and vitamin C [5]; [6]; [7]. Besides, Moringa leaves also contain various kinds of amino acids, including amino acids in the form of aspartic acid, glutamic acid, alanine, valine, leucine, isoleucine, histidine, lysine, arginine, phenylalanine, tryptophan, cysteine, and methionine [8]. The moringa leaves processing into moringa flour aims to increase shelf life, facilitate the processing of later products, so that it can provide added value for moringa [9]. According to Dahoklory [10], the protein content in Moringa leaf flour reaches 26.02%.

2. Methods

The ingredients used in making flakes are moringa leaf flour, yellow sweet potato flour, sugar, salt, water. The materials needed for chemical analysis are distilled water, NaOH, a mixed indicator *(Bromcresol green)*, HCl, sulfuric acid solution, ethanol.

This study used a Completely Randomized Design (CRD) with 5 levels of treatment as follows:

T_0 = Without the addition of Moringa leaf flour
T_1 = 2% addition of moringa flour
T_2 = 4% addition of moringa flour
T_3 = 6% addition of moringa flour
T_4 = 8% addition of moringa flour

The percentage moringa flour addition in the treatment was based on the weight of yellow sweet potato flour.

Each treatment was repeated 2 times, so the total experimental unit was 5 x 2 = 10 experimental units with the following mathematical model:

$$Y_{ij} = \mu + \alpha_i + \varepsilon ij \quad (1)$$

The research was conducted in 3 stages; the first stage is the process of making moringa leaf flour, the second stage is the process of making yellow sweet potato flour, and the third stage is the process of making sweet potato flakes. The process of making Moringa leaf flour [11] began with the selection of good raw materials to get a quality product. The leaves used were fresh Moringa leaves. After getting the fresh moringa leaves, then they were sorted and washed. Furthermore, the washed moringa leaves were drained to reduce the moisture in the moringa leaves, then dried using a cabinet dryer for 3 hours at 40°C. The results were then milled and then sieved with an 80 mesh sieve to obtain the moringa leaf flour.

Yellow sweet potato flour [12] is flour made from yellow sweet potatoes, sorted or selected to separate good yellow sweet potatoes from sweet potatoes that are not in good condition. Then they were peeledto separate the skin and the flesh after which the sweet potato meat was washed, and then thinly sliced to speed up the drying process. Then the yellow sweet potatoes were blanched to deactivate the enzymes and to get the fresh color of the product. The blanching lasted for 5 minutes at 82-93° C. After that, they were dried in the sun because it only takes ≤ 1 day. Then the sweet potatoes were milled using a blender, after that they were sifted to get fine grains using an 80 mesh sieve and it became yellow sweet potato flour.

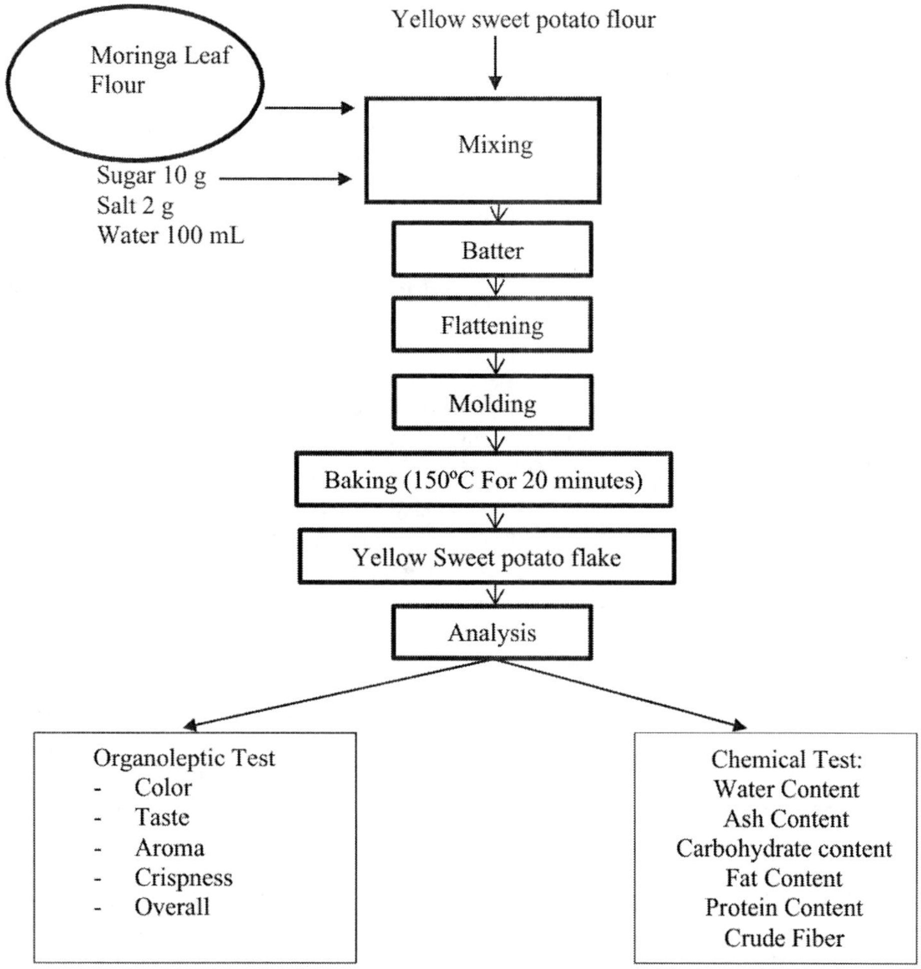

Figure 1. Research Flowchart

3. Result and discussion

2.1. Chemical Test

3.1.1. Water Content. The results of the analysis showed that the addition of Moringa leaf flour had a significant effect on the water content of the flakes. The results showed that the more of the flour being added, the lower the water content of the flakes would be. This result was in line with the research of Putri *et al* [13], namely the more of the Moringa leaf flour being added to the canna cookies, the lower the water content will be. This was due to the increasing number of dry materials so that more water was bound and caused less water content of the flakes produced [14].

3.1.2. Ash Content. The results of the analysis showed that the addition of Moringa leaf flour did not significantly affect the ash content of the flakes. The results of the analysis of the lowest ash flakes content were found in the treatment without the addition of Moringa leaf flour with 1.55%, and the highest value was in the addition of 8% Moringa leaf flour with 2.29%. The ash content of the flakes in this study met the quality requirements of flakes based on SNI 01-4270-1996 with a maximum of 4.0%. Thus, it can be said that the ash content of the flakes with the addition of Moringa leaf flour fulfilled the quality requirements based on SNI.

3.1.3. Fat Content. The results of the analysis showed that the addition of Moringa leaf flour had a significant effect on the fat content of the flakes. The results of the analysis of the lowest fat content were found in the treatment without the addition of Moringa leaf flour with an average score of 6.69%, while the highest score of fat content was in the treatment of the addition of 8% Moringa leaf flour with 8.14%. The results showed that the more of Moringa leaf flour being added, the higher the fat content contained in the flakes product will be. This was due to the high-fat content of Moringa leaves, which is 2.52% [10].

3.1.4. Protein Content. The results of the analysis showed that the addition of Moringa leaf flour had a significant effect on the protein content of the flakes. The results showed that the more Moringa leaf flour was added, the higher the protein content of the flakes will be. This was due to the high protein content in Moringa leaf flour, which is 26.02% [10]. The increase in the protein content of the material was also closely related to the water content of the material. The protein content of the flakes in this study did not meet the requirements of SNI 01-4270-1996 which is a minimum of 5.00%.

3.1.5. Carbohydrate Content. The results of the analysis showed that the addition of Moringa leaf flour had no significant effect on the carbohydrate content of the flakes. The results of the analysis of the lowest carbohydrate content were found in the addition of 8% Moringa leaf flour treatment with 77.24%, while the highest carbohydrate content was in the treatment without adding Moringa leaf flour with 80.99%. The carbohydrate content of the flakes in this study met the requirements of SNI 01-4270-1996.

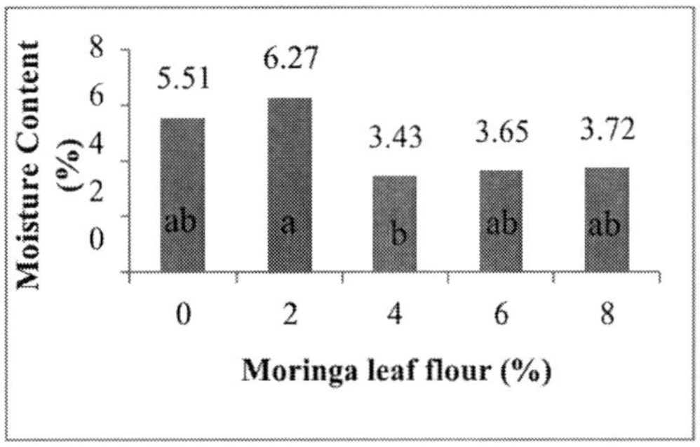

Figure 2. Effect of The Addition of Moringa Leaf Flour Treatment on Flakes' Water Content

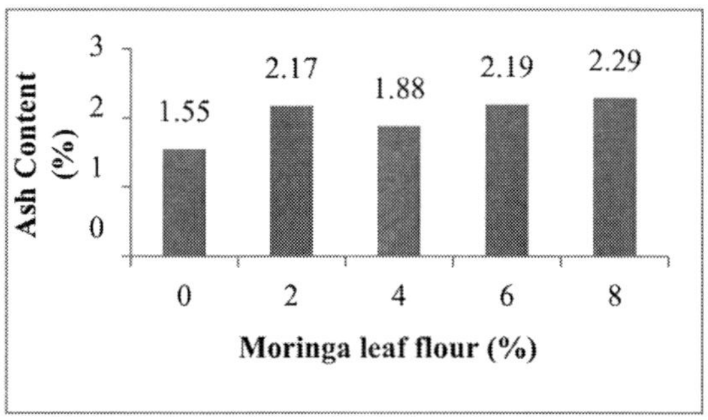

Figure 3. Effect of The Addition of Moringa Leaf Flour Treatment on Flakes'Ash Content

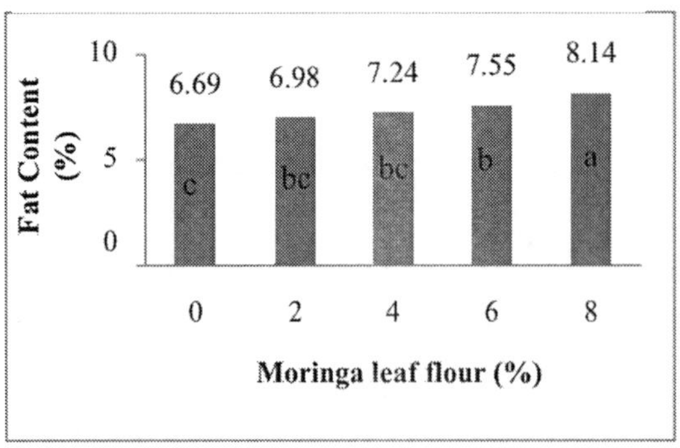

Figure 4. Effect of The Addition of Moringa Leaf Flour Treatment on Flakes' Fat Content

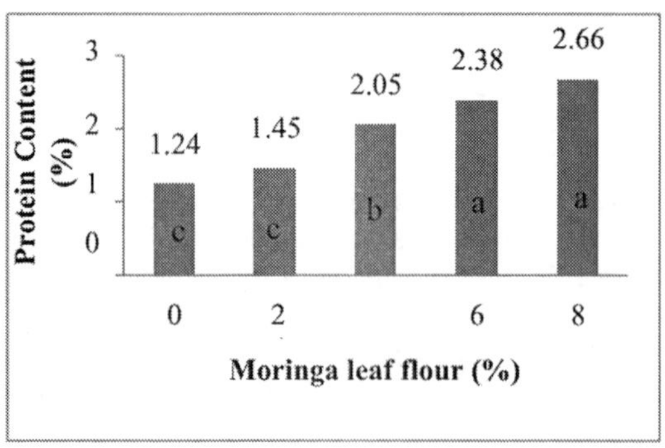

Figure 5. Effect of The Addition of Moringa Leaf Flour Treatment on Flakes' Protein Content

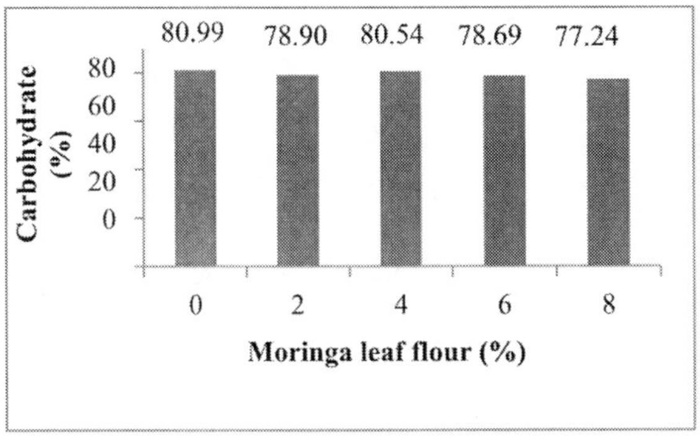

Figure 6. Effect of The Addition of Moringa Leaf Flour Treatment on Flakes' Carbohydrate Content

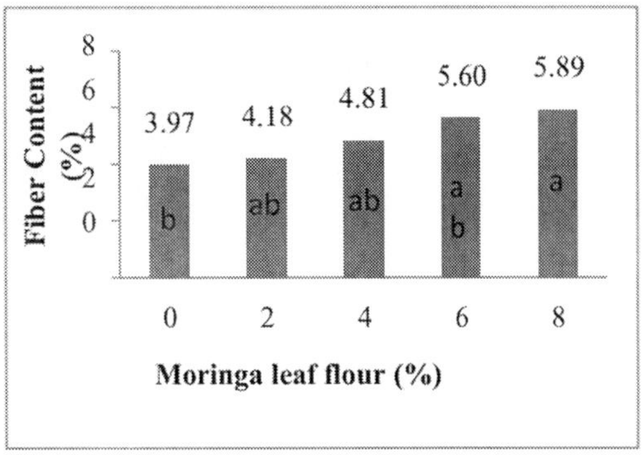

Figure 7. Effect of The Addition of Moringa Leaf Flour Treatment on Flakes' fiber Content

3.1.6. Fiber Content. The results of the analysis showed that the addition of Moringa leaf flour had a significant effect on the fiber content of the flakes. The results showed that the addition of Moringa leaf flour would affect the crude fiber content of the flakes produced [14]. Moringa leaf flour has a high crude fiber content of 4.03% [10]. The fiber content of flakes in this study has met the requirements of SNI 01-4270-1996 which is 7.0%.

3.2. Organoleptik Test

3.2.1. Color. The results of the organoleptic test on the color of the flakes presented in Figure 8 showed that the average like value ranged from 3.33 to 3.93 which was descriptively on the hedonic scale close to *like very much.* From the distribution of these average values, it can be seen that the treatment without adding Moringa leaf flour and 4% addition of Moringa leaf flour treatment had the same average value, namely 3.93. The visual the appearance of the flakes, the higher of the addition of Moringa leaf flour, the color of the flakes tends to be a slightly yellowish brown.

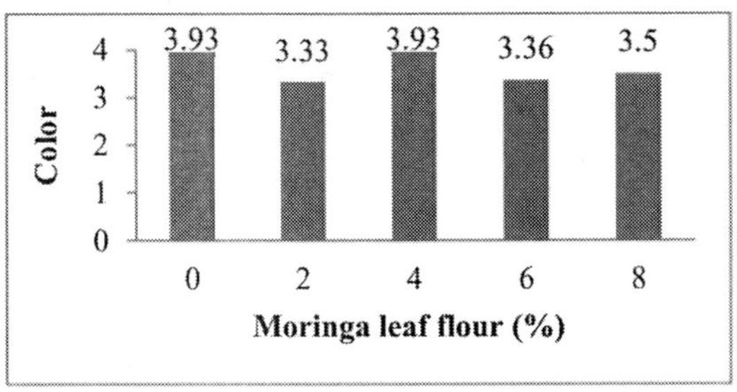

Figure 8. Effect of The Addition of Moringa Leaf Flour Treatment on Flakes' color

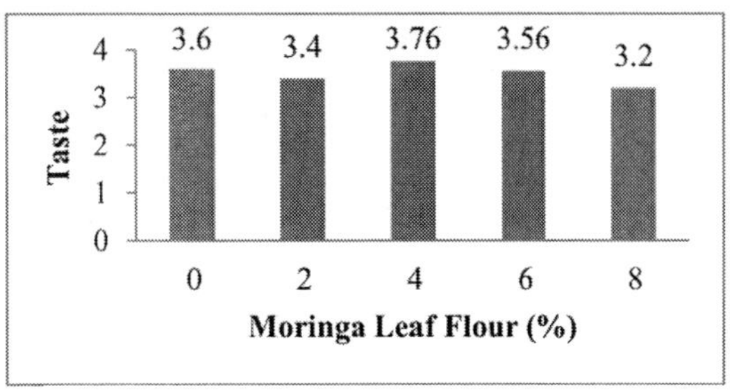

Figure 9. Effect of The Addition of Moringa Leaf Flour Treatment on Flakes' taste

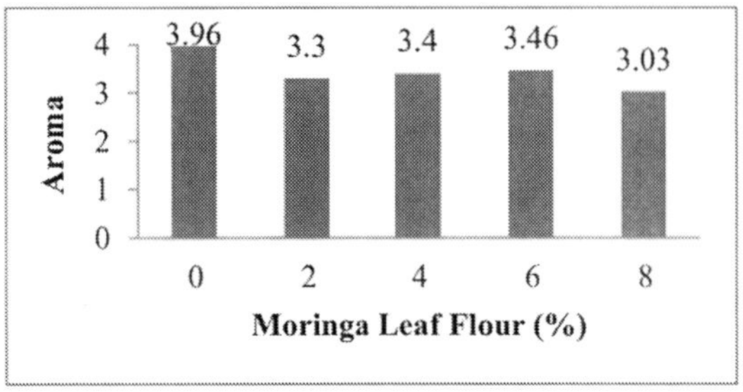

Figure 10. Effect of The Addition of Moringa Leaf Flour Treatment on Flakes' aroma

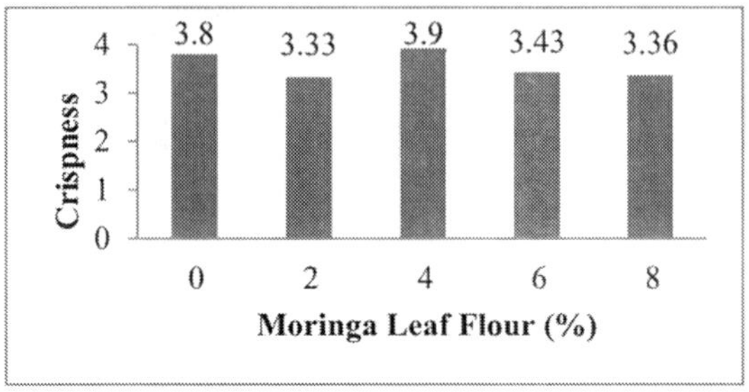

Figure 11. Effect of The Addition of Moringa Leaf Flour Treatment on Flakes' Crispiness

3.2.2. Taste. The results of the organoleptic test presented in Figure 9 showed that the average value of the panelists' liking for the taste of flakes ranged from 3.20 to 3.76 which was descriptively on the hedonic scale close to *like very much*. The average value distribution for most favored by the panelists was in the treatment of adding 4% Moringa leaf flour with an average value of 3.76 (*like*). While the average value that panelists did not like was the addition of 8% Moringa leaf flour with an average value of 3.20. This shows that the more of the Moringa leaf flour being added, the level of preference for the taste decreases. This happened because, with high Moringa leaf flour, it will taste bitter.

3.2.3. Aroma. The organoleptic test results on the aroma of flakes shown in Figure 10 had an average value of 3.03-3.96 which was descriptively on the hedonic scale close to *like very much*. The more of the Moringa leaf flour that was given, the less favor of the flakes will be. The decrease in the level of favor for the aroma of flakes was due to the very dominant aroma of Moringa leaves. Moringa leaf flour has a very strong unpleasant aroma [15]; [16].

3.2.4. Crispiness. The results of the organoleptic test on the crispiness of the flakes with the addition of Moringa leaf flour were presented in Figure 11. It can be seen that the average value obtained ranged from 3.33 to 3.90 which is descriptively on the hedonic scale close to *like very much*. This was because the high water content will decrease the crispiness of the product. Therefore the moisture content of the 4% Moringa leaf flour addition treatment was very low, so the crispiness of the flakes product at 4% treatment increases.

3.2.5. Overall. The results of the overall organoleptic test in Figure 12 showed that the average value of the overall acceptance for flakes ranged from 3.46 to 3.90 which was descriptively on the hedonic scale to be close to *like very much*. The highest average value was found in the treatment of adding 4% Moringa leaf flour with a value of 3.90, this was presumably because the panelists liked flakes with bright color, non-bitter taste, no unpleasant aroma, and crispy. This assessment is in line with the opinion of [17], which states that the overall acceptance of food can be measured in terms of color, taste, aroma, and texture.

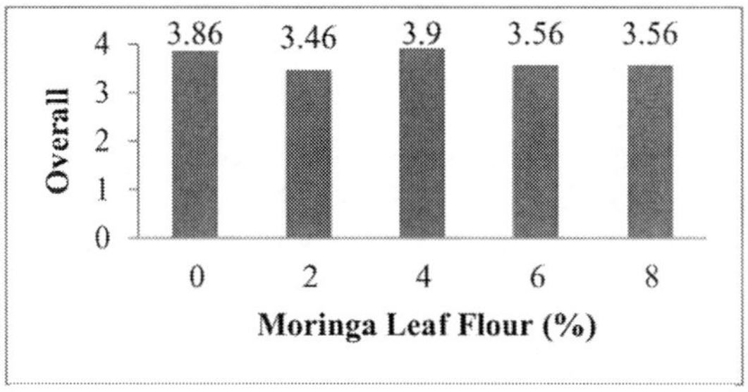

Figure 12. Effect of The Addition of Moringa Leaf Flour Treatment on Flakes' overall

4. Conclusion

Based on the results of the study, it can be concluded that the addition of 4% of Moringa leaf flour treatment was the best treatment because it is following SNI, with chemical characteristics of 3.43% moisture content, 1.88% ash content, 7.24% fat content, 2.05% protein content, 80.54% carbohydrate content, and 4.81% fiber content. Moreover, its' organoleptic characteristics of color (3.93) like, taste (3.76) like, aroma (3.40) like, crunchiness (3, 90) like and overall (3.90) like

References

[1] Wijayanti, S.D., Dewanti T and Widyaningsih D U. 2015. Evaluasi Nilai Cerna *In Vitro* Sereal Flake berbasis Ubi Jalar Oranye Tersuplementasi Kecambah Kacang Tunggak. *Jurnal Teknologi Pertanian.* **16** 1 .

[2] Widyaningtyas, M & Susanto, W. H. 2015. Pengaruh Jenis dan Konsentrasi Hidrokoloid (Carboxy Methyl Cellulose, Xanthan Gum, dan Karagenan) Terhadap Karakteristik Mie Kering Berbasis Pasta Ubi Jalar Varietas Ase Kuning. *Jurnal Pangan dan Agroindustri.* **3** (2) 417-423.

[3] Tamtarini and Yuwanti S. 2005. Pengaruh Penambahan Koro-Koroan terhadapSifat Fisik dan Sensorik Flake Ubi Jalar. *Jurnal Teknologi Pertanian.* **6** 187-192.

[4] Gisca B. 2013. Penambahan gembili pda *Flakes* jewawut ikan gabus sebagai alternative makanan tambahan anak gizi kurang. Artikel Penelitian. Program Studi Ilmu Gizi Fakultas Kedokteran Universitas Diponegoro

[5] Misra S and Misra M K. 2014. Nutritional evaluation of some leafy vegetable used by the tribal and rural people of south Odisha, India. *Journal of Natural Product and Plant Resources.* **4** 23-28.

[6] Oluduro, A.O. 2012. Evaluation of Antimicrobial Properties and Nutritional Potentials of *Moringa oleifer*a Lam. Leaf in South-Western Nigeria. *Malaysian Journal of Microbiology.* **8** 59-67.

[7] Ramachandran, C., Peter, K.V., Gopalakrishnan, P.K. 1980. Drumstick (*Moringa oeifera*): a Multipurpose Indian Vegetable. *Journal Economic Botany* **34** (3) 276-283.

[8] Simbolan J M, Simbolan M and Katharina N. 2007. Cegah Malnutrisi dengan Kelor. Yogyakarta: Kanisius.

[9] Fuglie, Lowell J., ed. 2001. The Miracle Tree: The Multiple Attributes of Moringa. Dakar, Senegal: Church World Service.

[10] Dahoklory M. 2017. Pengaruh Penambahan Tepung Daun Kelor (*Moringa Oleifera)* Terhadap Karakteristik Organoleptik Dan Kimia Biscuit Mocaf (*Modified Cassava Flour*). Skripsi.Ambon.Fakultas Pertanian.Universitas Pattimura.

[11] Suarti, B. A,Evan A.S. M. Masyhura. 2015. Penambahan Tepung Daun Kelor Dan Lama Pemanggangan Terhadap Biscuit Dari Mocaf (Modified Cassava Flour). *Jurnal Agrium.* **19.**

[12] Putri, E.P. 2015. Pembuatan Nastar Komposit Tepung Ubi Jalar Kuning *(ipomoea batatas l)* Varietas Jago. Jurusan Pendidikan Kesejahteraan Keluarga, Universitas Negeri Semarang.

[13] Putri A A E B and Putri A S. 2018. Penambahan tepung daun kelor (*Moringa Oleifera*) Terhadap Fisikokimia Dan Organoleptik Cookies Ganyong, Fakultas Teknologi Pertanian, Universitas Semarang.

[14] Aina Q. 2014. Pengaruh Penambahan Tepung Daun Kelor (*Moringa oleifera*) dan Jenis Lemak Terhadap Hasil Jadi Rich Biskuit. *E-Journal Boga.* **3** (3).

[15] Becker,K. 2003. Moringa oleifera: An Underutilised with Amazing Versatility. Department Aquaculture Systems and Animal Nutritions. Germany: University of Hohenheim.

[16] Kholis, N. & Hadi, F. 2010. Pengujian Bioassay Biskuit Balita yang Dusuplementasi Konsentrat Protein Daun Kelor (*Moringa oleifera*) Pada Model Tikus Malnutrisi. *Jurnal Teknologi Pertanian.* **11** (3): 144-151

[17] Irmawati, F.M, D. Ishartati, & D.R. Affandi. 2014. Pemanfaatan Tepung Umbi Garut (*Maranta arundinacea* L) Sebagai Pengganti Terigu Dalam Pembuatan Biskuit Tinggi Energi Protein Dengan Penambahan Tepung Kacang Merah (*Phaseolus vulgaris* L). *Jurnal Teknosains Pangan.* **3** (1).

IOP Conf. Series: Earth and Environmental Science **883** (2021) 012027 doi:10.1088/1755-1315/883/1/012027

Activity test of *Bacillus Spp* against bacterial wilt (*R. solanacearum*) on tomatoes by in vitro

G N C Tuhumury[1,*], J V Hasinu[1], H Kesaulya[2]

[1] Agrotechnology Department, Faculty of Agriculture, Pattimura University, Ambon, Indonesia
[2] Plant Breeding Department, Faculty of Agriculture, Pattimura University, Ambon, Indonesia

* E-mail: gratianatuhumury@gmail.com

Abstract. *Ralstonia solanocearum* is a pathogenic bacteria that attacks tomatoes and causes wilt disease. Many efforts have been made to control this disease through cultivation, use of chemical pesticides, and development of resistant varieties, but bacterial wilt disease remains a serious problem economically. Nowadays, many biological controls are being developed using microbes. The use of *Bacillus spp* as an unfriendly microbe is very potential to control because it has pathogenic inhibitory activity. This study aims to obtain bacterial isolates of *Bacillus spp* which can suppress the development of bacterial wilt disease in tomato plants. The research was conducted in vitro at the Laboratory of Plant Physiology in the Agriculture Faculty, Unpatti. The results showed that *Bacillus niabensis* strain *PT-32-1* and *Bacillus subtilis* strain *SW116b* could inhibit *Ralstonia solanacearum*, wilt disease in tomato plants in vitro.

1. Introduction

Tomato (*Lycopersicum esculentum* Mill) is a vegetable commodity of fruit that is classified as a seasonal plant (Solanaceae). As a horticultural plant, Tomatoes have a strategic role in daily needs. The content of tomatoes is very useful to meet the body's nutritional needs, in the form of vitamins and minerals. Tomato fruit contains vitamins A, B, and C, while the minerals are contained in the form of iron (Fe), calcium (Ca), and Phosphorus (P) [1]

During its cultivation, tomatoes experienced various obstacles that caused a decrease in quality and production quantity. One of the influential factors is the attack of plant-disturbed organisms. According to [2] harvest failures due to disease attacks are estimated at 14% global. One of the diseases that attack tomato plants is a wilt bacterial disease, that is caused by *Ralstonia solanacearum.* In Indonesia, until now wilt bacterial disease is still one of the common serious diseases in various types of horticultural crops. It was often been reported that this bacterium has many host plants, including tomatoes, potatoes, eggplant, chilies, chickpeas, beans, ginger, and bananas [3]

According to [4] on infested land of *R. solanacearum* naturally, in Germany, this disease caused 75% incidence. Tomato production in Indonesia is constrained by wilt bacterial disease (*R. solanacearum*)where the development of disease increases in the rainy season. This is because tomato production decreased, starting from 30%-60% [5]. This statement is supported by [6] that wilt bacterial disease is a serious disease that attacks tomato crops in Indonesia. Severe attacks can cause cultivated crops to die of crop failure. Wilt bacterial diseases are endemic and quickly develop infected plants to healthier plants in the surrounding area [7] The initial symptoms it causes are wilt on young leaves and shoots, caused plants to die in a short time. The loss due to this pathogen infection can reach 100% [8].

R. solanacearum pathogen interferes with the transport of water and food substances by way of damaging plant cells. Enzymes that play a role in this process are the enzymes cellulase and pectinase. This enzyme destroys the cell walls of plants containing cellulose and pectin. As a result of this attack, there are physiological deviations of plants that are disrupted the process of translocation of water and other nutrients so that plants become wilt and then die [2]

Various efforts have been made in the control of bacterial wilt diseases, including the use of pest and disease-free seeds, planting of non-similar types of crops, non-simultaneous planting on large land, and the use of synthetic pesticides, but this method has not received satisfactory results. Also, the use of pesticides raises concerns about agricultural products that are contaminated due to pesticide residues. Long-term use of synthetic chemical pesticides can kill non-pathogenic microorganisms, poison humans due to improper use of doses, poison animals and pollute the environment, the occurrence of pathogenic resistance, and the emergence of new physiological breeds.

Another control alternative to controlling bacterial wilt diseases is by biological control using microbes. The use of biological agents in the control of plant diseases has several advantages, among others: 1) does not negatively affect the environment, 2) safe for you- such natural certain PESTS, 3) prevent the onset of secondary PEST explosions, 4) produce products that are free of residues synthetic chemical compounds, 5) safe for human health, 6) contained around crops to prevent farmers' exposure to synthetic chemical pesticides, and 7) can lower production costs because APH applications are carried out once or twice in one harvest season [9]

Bacillus sp. is one of the group of gram-positive bacteria that is often used as a biological controller of root diseases. Members of this genus have advantages because bacteria form spores that are easy to store, have a long life expectancy, and are relatively easy to inoculate into the soil. Bacillus spp. has been shown to have the potential as a good biological control agent, for example against pathogenic bacteria such as *R. solanacearum* [10] Some *Bacillus spp.,* species among the B. subtilis are able to prevent the development of anthracnose disease in chili peppers (*Colletotrichum gloeosporioides*) [11]

The purpose of the study was to determine *the ability to isolate Bacillus spp* in suppressing wilt bacteria (*R. solanacearum*) in tomatoes in vitro.

2. Methods

The research was conducted in the Laboratory of Plant Physiology and Plant Diseases Faculty of Agriculture, Pattimura University took place in May - Juli 2018

2.1. Isolate Bacillus sp Bacteria

This study using Bacterial Isolates *Bacillus niabensis* strain *PT-32-1* and *Bacillus subtilis* strain *SWI16b* from rhizosphere potato that is a collection of Plant Physiology Laboratory in Faculty of Agriculture, Pattimura University. Multiplication bacillus is isolated by growing it in agar (NA) media and incubated at a temperature of 28 °C for 24 hours. Isolate rejuvenation in the same medium and is done repeatedly to obtain a pure isolated culture.

2.2. Isolation of Pathogenic Bacteria

Pathogenic bacteria (*R. solanacearum*) used from samples of diseased plants (inoculum) namely tomato plants from farmers' gardens in Waimital, West Seram Regency that shows symptoms of the bacterial wilt disease. Wilt symptomatic plants are taken for the insulation. Pathogenic bacteria are isolated from parts of the plant that show symptoms of wilt in the field. The stems or roots of tomato plants with wilt bacteria on their surface are disinfected with 70% alcohol, then cut into small pieces, and put in a test tube that contains sterile water for 10 to 15 minutes. After sterile water in the test tube is dewy, we use sterile Ose needles. The suspension of the bacteria is scratched on the surface of the NA medium and then incubated. After 48 hours of incubation, a single colony was selected and grown on the NA medium again, then tested its properties, among other Gram test with 3% KOH.

2.3. Bland Zone Testing
This test uses several media to confirm the inhibitory ability of the pathogen through the formed bland zone. The media used in this test are Nutrien Agar (NA) and Yeast Peptone Dextrose Agar (YPDA). Tested using the modified [12-13] methods. Isolate Bacillus spp which is a pure breed cultured in NA media and aged 48 hours taken and then suspended up to a density of 10^9 CFU / ml in sterile aqua dest. Then a sterile piece of filter paper of 5 mm diameter is inserted into the bacterial suspension (Bacillus) for ± 1 minute and dried. Then the piece of filter paper is planted in the middle of the surface of the NA media on a petri dish with 9 cm diameter and incubated for 2 days (48 hours). Once incubated, the petri dish is turned over and on the sterile lid of the petri dish, chloroform 1 ml is given and left for 2 hours. After that, the Petri dish was turned back to its original position. Further, there is a medium surface is poured suspension *R. solanacearum* (0.2 ml in 4 ml for water 0.6%.). Breeds are incubated for 24 hours at a temperature of 300C. The same is done in Yeast Peptone Dextrose agar (YPDA) media. As a control, pieces of filter paper are dipped in sterile aqua dest. Each biocontrol agent was tested three times and the diameter of the resistance zone was measured after incubation at room temperature for 24 hours. The inhibitory zone will be visible around the bad bacteria, are measured and expressed in millimeters

3. Results and discussion
3.1. Isolation of Pathogenic Bacteria
Inoculum pathogen was obtained during observation of symptoms of bacterial wilt disease in tomato plants in Waimital Village, West Seram Regency. Many of the tomato plants show wilt symptoms. When it's taken, this wilt tomato plant, the stem is cut longitude and looks brown. The infected tissue secretes a fine strand of milky white consisting of a mass of bacteria (ooze), which seeps out of the cut-off boundary of an infected plant is pressed. Characteristics of tissues that are brown and if it's pressed there is a mass of this bacteria that distinguishes the wilt symptoms because the wilt bacteria is caused by the other causes/pathogens. This disturbance may be caused by the blockage of cells so that the translocation of water from the bottom (root) to the top of the plant becomes inhibited, as a result of which the plant becomes wilt. The tomato plant is taken and isolated.

The result of isolation from tomato plants obtained bacteria *R. solanacearum* has grown on nutrient media to be as follows: colonies are milky white, rather gloomy, irregularly shaped until round. This is in line with the results of [14] research, that the *R. solanacearum* bacterial colony is milky white, and irregularly shaped when grown on YPGA media. The bacterial properties of the obtained isolates are Gram-negative after being tested with KOH.

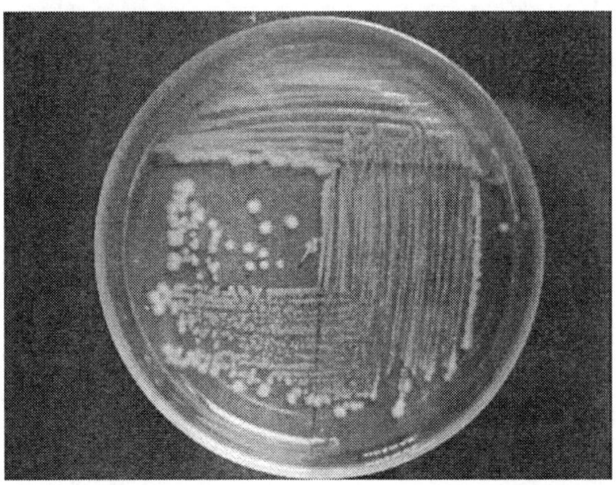

Figure 1. Isolate *Ralstonia solanacearum*

3.2. Reisolation, Rejuvenation and Purification of Bacillus SPP Bacteria

Bacteria that are successfully isolated is done again to multiply, isolate, and later purified. The results of bacterial isolation show that bacteria stored at low temperatures of 4°C after isolation can grow well with colony populations as presented in Figure 2.

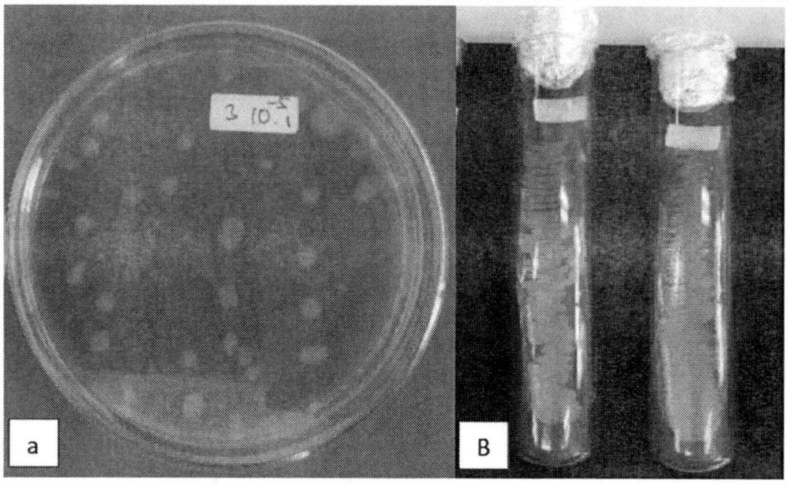

Figure 2. Reisolation, Rejuvenation and Purity of *Bacillus spp Bacteria;* a. Isolate Bacillus *spp* , b. Purification

The morphology of the colony from isolate Bacillus obtained can be seen in Table 1 where the colony is round with intact to jagged edges, has a flat elevation and the colony is white.

Table 1. Morphology Colony of *Bacillus spp* and total population of Reisolation and Rejuvenation Results

Bacteria Active Ingredients	Total Initial Population	Colony Shape	Edge	Elevation	Colony Color
Bacillus niabensis Strain PT-32-1	3.7×10^5	Round	Jagged	Flat	White
Bacillus subtilis Strain SWI16b	3.5×10^5	Round	Intact	Flat	White

To ensure the functional ability of bactery, furthermore carried out bacterial purity test through various tests, such as Gram Reaction, Dissolution of phosphate, Nitrogen Fixation. The results obtained are presented in Table 2.

Table 2. Results Test of Qualitativity Functional Of Bacterial Isolate *Bacillus spp* Qualitative

Bacteria	Gram Reaction	Dissolution of Phosphate	Nitrogen Fixation
Bacillus Niabensis Strain PT-32-1	+	+	+
Bacillus subtilis Strain SWI16b	+	+	+

3.3. Bacillus SPP Antagonist Test Against wilt Tomato Bacteria in vitro
Antagonistic test results showed the presence of inhibition activity by *Bacillus spp* against pathogens wilt diseases in tomato plants. Inhibition ability is indicated by the formation of around clear zone. The size of the inhibition zone in each media can be seen in Table 3.

Table 3. Diameter of Halo Zone On Some Types of Media

Bacillus spp	Diameter of the Halo zone (mm)	
	Na	YPDA
Bacillus niabensis Strain PT-32-1	4.0	6.0
Bacillus subtilis Strain SWI16b	4.0	6.5

Bacillus spp is a bad thing to various plant pathogens including pathogens of wilt diseases of tomato plants. After incubation is seen a clear zone and indicating that the isolate *Bacillus spp* used produces secondary metabolites capable of inhibiting or shutting down pathogens. The formation of halo zones indicates the presence of antibiosis mechanisms and the ability to produce antibiotics in *Bacillus spp* that can be used as bio controller agents. According to [15], *B. subtilis* can produce antibiotics as antibiosis mechanisms such as *Bacillomycin D* and iturin produced by *B. subtilis AU195* and *QST713.*

One of the suppression mechanisms by strains of members of the genus Bacillus is antibiosis indicated by the formation of resistance zones in *Bacillus spp.* cultures that are grown on the medium in layers with pathogenic bacteria. Antibiosis is an antagonistic mechanism by producing secondary metabolites in the form of antibiotics or antibiotic-like compounds such as pelysis enzymes, volatile compounds, siderophores, and other toxic substances [16]

In Table 3. It seems to be a difference in the extent of the halo zones that are formed. It turns out that in the YPDA media inhibitory zone is greater than NA media. Differences in resistance zones in *Bacillus sp.* isolates are thought due to the condition and nutritional content of the media used. The compound and sugar content is contained in the media, that affects the formation of the halo zone. [17] explained that antibiosis compounds are strongly influenced by the composition of the medium, both quantitatively and qualitatively. But in principle, both isolates way show the ability to inhibit the growth of wilt pathogenic bacteria (*R. solanacearum*) in tomatoes.

4. Conclusion
Based on the results, *Bacillus spp* can suppress the development of *R. solanacerum* on tomatoes, so that it can be used as a biological agents.

References
[1] Agromedia 2007 Complete Tomato Cultivation Guide. 2007. Jakarta: PT. AgroMedia Library
[2] Agrios G N 2005 Plant Pathology 5th Edn., Academic Press, New York, US., ISBN-13: 9780120445653,
[3] Eden Green S J & WILT Sastraatmadja 1990 Blood disease present in Java *FAO Plant Protection Bulletin* **38**: 49 – 50.
[4] Rivard C L, S O'Connell, M M Peet, R M Welker and F J Louws 2012 Grafting Tomato to Manage Bacterial Wilt Caused by Ralstonia solanacearum in the Southeastern United States *Plant Disease*. **96**(7): 973-978
[5] Maulida I, E Ambarwati, Nasrullah, and Rudi WILT M 2013 Evaluation of the power of the result of the strain of tomato hope (*Solanum lycopersicum* L.) in the rainy and dry season *Vegetalika* **2**(3): 21- 31
[6] Adriani, A Rahman, Gusnawati S and A Khaeruni 2012 Resistance response of various varieties of tomatoes to bacterial wilt diseases (*Ralstonia solanacearum*) *Agroteknos* **2**(2): 63-68
[7] Hartati, S WILT and N Karyani. 2014. *Ralstonia solanacearum inoculation technique for* patchouli resistance testing against wilt disease *Littro* **25**(2): 127-136.

[8] Purnawati A, Sastrahidayat I R, Abadi A, Hadiastono T 2014 Endophytic bacteria as bio-control agents of tomato bacterial wilt disease *JTLS* **4**(1):33–36 Doi:https://doi.org/10.11594/jtls.04.01.06.

[9] Tombe M 2002. The potential of biological agencies in the control of plant diseases is environmentally sound and its role in the development of the agribusiness sector. pp. 13–34. Proceedings of the National Seminar of PFI Komda Purwokerto.

[10] Soesanto L 2008 Introduction to Plant Disease Biological Control. Supplements to Weeds and Nematodes. 202-222-200

[11] Ashwini N and S Srividya 2014 Potentiality of *Bacillus subtilis* as biocontrol agent for management of anthracnose disease of chilli caused by *Colletotrichum gloeosporioides* OGC1 *Biotechnology* **4**: 127–136.

[12] Kawaguchi A, Inoue K & Ichinose WILT 2008 Biological control of crown gall of grapevine, rose, and tomato by nonpathogenic Agrobacterium vitis strain VAR03-1 *Phytopathology* **98**(11), 1218–1225. https://doi.org/10.1094/PHYTO-98- 11-1218

[13] Arwiyanto T 1997 Biological control of tobacco bacteria wilt diseases:1. isolation of antagonistic bacteria *Indonesian Journal of Plant Protection* **3**: 54-60

[14] Wuryandari Y dan Arwiyanto T 1999 Karakteristik ralstonia solanacearum dari beberapa Inang Solanaceae di Yogyakata. Fakultas Pertanian UGM. Yogyakarta. hal 387-393.

[15] Junaid J M, Dar N A, Bhat T A, Bhat A H, & Bhat M A. 2013. Commercial biocontrol agents and their mechanism of action in the management of plant pathogens *Int. J. Modern Plant & Anim. Sci.* **1**(2): 39-57a.

[16] Haggag W M & Mohamed H A A 2007 Biotechnological aspects of microorganisms used in plant biological control *World J. Agric. Sci.* **3**(6): 771–776.

[17] Al-Zahrani S H M 2007 Studies on the antimicrobial activity of *Streptomyces* sp. isolate from Jazan *JKAU Sci* **19**:127-138.

Characterization of plant growth promoting rhizobacteria of maize

H Kesaulya[1,*], A Talahaturuson [2], A M Kalay[2], E Matatula[1], I J Lawalatta[2], M L Hehanussa[1], S J Nendissa [3]

[1] Departement of Plant Breeding, Faculty of Agriculture, Pattimura University, Ambon, Indonesia
[2] Departement of Agrotechnology, Faculty of Agriculture, Pattimura University, Ambon, Indonesia
[3] Departement of Agricultural Product Technology, Faculty of Agriculture, Pattimura University, Ambon, Indonesia

* E-mail: henry.unpat@gmail.com

Abstract. Isolation and characterization of rhizobacteria are an effort to determine the ability of root colony bacteria to produce various compounds that can be used for various purposes of bio-fertilizer formulations and microbial-based industrial interests. This study aims to characterize biochemically, morphologically and physiologically as well as the ability of root bacteria in maize to produce hormones that can stimulate plant growth. There is a wide variety of isolates morphologically and biochemically, besides that there is the ability of bacterial isolates to physiologically dissolve phosphate, fix nitrogen, produce ACC-deaminase, IAA and GA enzymes.

1. Introduction

The microorganisms in the soil are very diverse including as a bacteria, archaea, yeast, fungi, algae and protozoa. Microorganisms can live in the environment with humans and are also found in extreme conditions such as in hot springs, in oceans for miles away, in rocks and can live in very cold weather and temperatures.

Bacterial populations in the soil play an important role in the nutrient cycle including Nitrosomonas and Nitrobacter (nitrification), Thiobacillus (sulfur and iron oxidation), *Rhizobium* and *Frankia* (N2 fixation), *Bacillus* and *Clostridium* (carbon cycle) and *Caulobacter* and *Pseudomonas* (manganese oxidation). In addition, large numbers of bacteria have been found to produce various types of chemicals that are exploited in the biotechnology industry. Bacteria have been found to be a source of many valuable chemicals including ethanol, acetone, enzymes, perfume, antibiotics. In the last few decades, thousands of antibiotics have been discovered. Rhizosphere bacteria from Pseudomonas and Bacillus on leguminous and non leguminous plants have been identified [1], [2], as well as in Hartapel variety potato plants [3]. These findings make it possible to identify isolates or strains of bacteria from the soil for biochemical, physiological, industrial and pharmacological purposes. The number and species of microbes in the soil depend on environmental conditions such as availability of nutrients, soil texture, presence of moisture in the soil and the type of vegetation cover, and other environmental conditions.

Rhizosbacteria are plant root colonizing bacteria that can provide a plant-stimulating effect through hormone production and facilitate the availability of nutrients for plants and can be antagonistic to soil borne pathogenic bacteria and fungi. There are various types of microbes and their activities are related to roots [4]. The narrow zone of soil around the root system is referred to as the rhizosphere, whereas the term "rhizobacteria" is a group of rhizosphere bacteria that have the ability to colonize roots [5], [6]. These organisms can also benefit plants by stimulating growth. PGPR can affect plant growth by different direct and indirect mechanisms [7]. The rhizosphere is the ecological environment of the soil where plant microbial interactions occur through the colonization of microorganisms where associative, symbiotic, naturalistic, or parasitic interactions occur depending on the nutritional status of plants in the soil, the soil environment, and the types of microorganisms that proliferate in the rhosferosphere zone [8]. Soil bacteria that aggressively colonize the root zone and promote plant growth are commonly referred to as Rhizobacteria. In this study it has been found that some bacterial isolates produce important biochemical and physiological factors.

2. Methods

2.1. Sampling and sources of rhizobacteria

Ten grams of soil from the roots of corn plants were taken at three different points in several planting locations on the islands of Ambon and Seram as test samples. Soil samples were put into sterile bags and immediately transferred to the laboratory for isolation.

2.2. Sample preparation and isolation of bacteria

Every 10 g of soil samples from different maize planting sites were mixed with 90 ml of sterile water in 250 ml different beakers and homogenized in a flask for ten minutes using an orbital shaker at 110 rpm. Subsequently, 1 ml of each sample was transferred aseptically into 9 ml of sterile water and thoroughly mixed using a vortex. Homogenate is serially diluted to 10-5; thereafter, 0.1 ml aliquot of the appropriate dilution is properly spread on a starch petri agar dish. The scattered petri dishes were incubated at 32°C for 48 hours. Bacterial isolates that showed different colony morphology were taken and purified using nutrient agar media.

2.3. Morphological characters of bacterial isolates

The isolated pure culture was cultured on nutrient broth media. The scattered petri dishes were incubated at 32o C for 24-48 hours. The colony morphology of the isolates was observed including: shape, margin, elevation and color.

2.4. *Biochemical character*

2.4.1. Gram reaction test

The gram reaction test uses a 3% KOH solution as much as one drop dripped on the slide. A total of one ose of bacterial isolate is taken which is then transferred onto a slide and mixed or smeared slowly until it forms mucus. A positive test is indicated by no mucus formation and a negative test is indicated by the formation of mucus when mixed with 3% KOH.

2.4.2. Motility test

This test can be used to check the ability of bacteria to migrate away from the inoculation line. To perform this test, the bacterial sample is inoculated into the motility medium using a needle and incubated at 37 ° C for 24 hours and examined for bacterial migration.

2.4.3 Catalase test

This test is carried out to determine the ability of isolated bacteria to degrade hydrogen peroxide by producing enzymes, catalase or peroxidase. A drop of 3% hydrogen peroxide is added to the bacterial colony on a sterile glass slide and mixed well. The production of air bubbles was observed for one minute. The production of air bubbles indicates positive catalase and no bubbles indicates negative catalase.

2.4.4. Oxidase test
The oxidase test is carried out to determine the presence of the oxidase enzyme in bacteria. Oxidase reagents contain reducing agents that change color when oxidized. One loop of bacterial colony was etched on oxidase paper. Observations were made by looking at the reaction caused, if the results of blue streaks on oxidase paper indicate that the tested bacteria have a positive oxidase enzyme and if there is no change in color, it shows a negative oxidase result.

2.4.5. Mac Conkey Agar media Test
This test is to detect gram-negative bacteria. Growing gram-negative bacteria can be distinguished in their ability to ferment lactose. Colonies of bacteria that ferment lactose are brick red. Bacteria that do not ferment lactose are usually pathogenic. This group of bacteria did not show any changes in the media. This means that the colony color is the same as the media color. Colony colors can be seen in a separate colony section. Isolates were grown on Mac Conkey Agar solid media using a scatter plate technique which was incubated at 37oC for 24 hours.

2.5. Ability of phosphate solubility and nitrogen fixation

2.5.1. Phosphate solubitation test
Phosphate dissolution was tested by referring to the Pikovskaya method [9], [10]. The 24-hour-old bacterial isolate suspension was grown on solid Phikovkaya media containing tricalcium phosphate ($Ca3PO4$) by the spread method. The zone of emergence around the bacterial colony shows dissolving activity. The ability of bacterial isolates to report phosphate dissolving by clear zone, phosphate dissolving efficiency and phosphate dissolution index. Measurement of dissolved phosphate concentration using liquid Phikovkaya media. Thirty ml of culture isolate was cultured in Pikovskaya medium for 7 days and in a shaker. The bacterial suspension was filtered (Whatman No.1) then centrifuged for 15 minutes at 10,000 g. Five 5 ml of supernatant pipette and 0.5 ml of concentrated reagent added (Concentrated reagent: 12 g ammonium molybdate, 0.277 g potassium antimole tartrate). Before measuring the absorbance, the level of phosphate dissolution, the pH of the filtrate solution was measured for the acidity. The absorbance was measured on a 693 nm UV-VIS spectrophotometer. The standard PO4 (titrisol) curve was made from dilution with concentration with the regression equation $Y = 0.191 x + 0.048$ where $R2 = 0.957$.

2.5.2. Free nitrogen fixation
The ability of bacterial isolates to fix free nitrogen was tested using free Burk N media [11] and Ashby media [12]. Bacteria that are able to grow on both media are indicated as bacteria capable of fixing nitrogen.

2.6. Production testing of growth enhancing compounds

2.6.1 Screening isolate of producing acc deaminase
Screening of ACC deaminase-producing bacterial isolates was carried out by growing 24-hour culture of rhizoferous bacterial isolates on minimal salt media Dworkin-Foster (DF) [13] enriched with ACC as the only source of nitrogen [7]. Isolates that were able to grow on this medium were indicated as ACC deaminase producing isolates.

2.6.2 Indole Acetic Acid (IAA) production
The production of indole-3-acetic acid (IAA) auxin was tested using by nutrient broth (NB) media and Salkowski's reagent [14]. Bacterial isolates were cultured on NB media supplemented with L-tryptophan (0.1g l-1) at 28 0C. The supernatant was added to 1 ml of Salkowski reagent (12 g l-1 $FeCl3$ in 429 l-1 ml $H2SO4$) [7]. The mixture was stored for 24 hours at 28 ° C in dark conditions and the absorbance was measured at 535 nm on a UV-VIS spectrophotometer. The change in pink indicates IAA production. Auxin concentrations were measured using the IAA standard curve with the regression equation $Y = 0.064 x + 0.09$ where $R2 = 0.995$, made from serial dilutions of IAA stock solutions.

2.6.3 Gibberellic Acid (GA3) production

Gibberelic acid production is based on standard methods [15], which are modified. One ml of bacterial isolate was added to the media and incubated at 37 0C. After incubation the cultures were centrifuged at 8000 g to separate bacterial cells. 15 ml of culture was added to 2 ml of zinc acetate, added to 2 ml of potassium ferrocyanide solution and centrifuged at 8000 g. The supernatant was added to 30 percent hydrochloric acid and incubated at 27 0C for 75 minutes. The absorbance was measured at 254 nm on a UV-VIS spectrophotometer. The concentration of GA3 is measured using the standard GA3 curve with the regression equation Y = 0.888 x + 0.441 where R2 = 0.921, made from serial dilutions of GA3 stock solution.

3. Results and discussion

3.1. Morphological character

Isolation and characterization of bacterial isolates from corn rhizofer obtained as many as 37 bacterial isolates. The morphologically identified isolates showed differences. The bacterial isolates identified colonizing corn roots differed in shape, margin, elevation and color as well as gram reactions. Observation of colony and cell morphology is still needed as an initial stage before further identification such as physiological characterization and DNA identification are carried out. The rhizosphere is the zone between the surface of the roots and soil, and is adjacent to the roots. The bacteria that live in this zone can remain in the soil. Morphologically, the bacterial isloat isolated from the root rhizosphere of maize in various varieties showed diversity in shape, margin, elevation and color. The rhizosbacteria isolates from the rhizosphere of maize were isolated from various varieties, 19 (51.4%) were circular, 16 (43.2%) irregular and 2 (5.4%) filamentous isolates. The results of the observations on the margins and elevation of the rhizosbacteria isolates observed are presented in Table 1. The colony colors observed were dominated by white to dull white. The shape, margins and elevation of observed bacterial isolates refer to Figure 1.

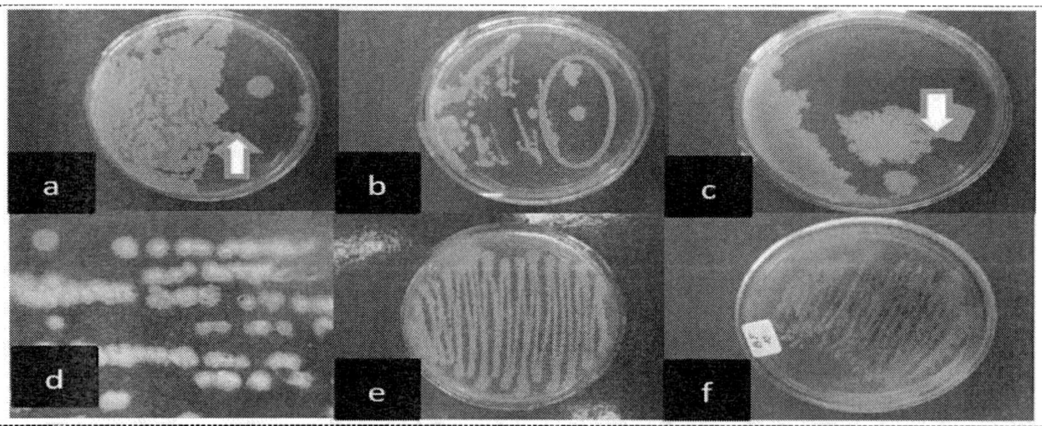

Figure 1. Colony morphology: form: bacteria (a) circular, (b) irregular and (c) filamento and colony color (d) white, (e) dull white and (f) yellowis

Table 1. Morphological characters of bacterial isolates from corn rhizosphere

No.	Isolate Code	Origin of Rhizosphere		Colony of Morphology			
		Location	Variety	Shape	Margins	Elevation	Color
1	PB01	Poka	Bonansa	Iregular	Filiform	Raised	White
2	PU01	Poka	Uric (asal MTB)	Iregular	Undulate	Raised	Yellowish white
3	PU02	Poka	Uric (asal MTB)	Iregular	Undulate	Flat	White
4	PU03	Poka	Uric (asal MTB)	Circular	Entire	Raised	White
5	PU04	Poka	Uric (asal MTB)	Iregular	Filiform	Raised	White

6	PU05	Poka	Uric (asal MTB)	Circular	Undulate	Flat	White
7	PU06	Poka	Uric (asal MTB)	Filamento	Filiform	Umbonate	White
8	PU07	Poka	Uric (asal MTB)	Iregular	Curved	Creteriform	White
9	PU08	Poka	Uric (asal MTB)	Iregular	Undulate	Raised	Putih kusam
10	PU09	Poka	Uric (asal MTB)	Iregular	Filiform	Raised	Dull white
11	PU10	Poka	Uric (asal MTB)	Iregular	Undulate	Raised	Dull white
12	PU11	Poka	Uric (asal MTB)	Iregular	Undulate	Raised	Dull white
13	PU12	Poka	Uric (asal MTB)	Iregular	Undulate	Raised	Dull white
14	PU13	Poka	Uric (asal MTB)	Iregular	Filiform	Raised	Dull white
15	BT01	Telaga kodok	Bonansa	Circular	Entire	Raised	White
16	BT02	Telaga kodok	Bonansa	Circular	Entire	Raised	White
17	BT03	Telaga kodok	Bonansa	Circular	Entire	Raised	White
18	BT04	Telaga kodok	Bonansa	Circular	Entire	Raised	White
19	BT05	Telaga kodok	Bonansa	Circular	Entire	Raised	White
20	BT06	Telaga kodok	Bonansa	Circular	Entire	Raised	White
21	BT07	Telaga kodok	Bonansa	Circular	Entire	Raised	White
22	BT08	Telaga kodok	Bonansa	Circular	Entire	Raised	White
23	BT09	Telaga kodok	Bonansa	Iregular	Undulate	Flat	White
24	BT10	Telaga kodok	Bonansa	Circular	Entire	Raised	Yellowish
25	BT20	Telaga kodok	Bonansa	Iregular	Undulate	Umbonate	Yellowish
26	BT21	Telaga kodok	Bonansa	Iregular	Filiform	Umbonate	Yellowish
27	BT31	Telaga kodok	Bonansa	Circular	Entire	Flat	White
28	BT32	Telaga kodok	Bonansa	Circular	Filiform	Raised	White
29	BW01	Waimital	Bonansa	Circular	Entire	Raised	White
30	BW02	Waimital	Bonansa	Circular	Entire	Flat	White
31	BW03	Waimital	Bonansa	Circular	Entire	Raised	White
32	BW04	Waimital	Bonansa	Circular	Entire	Raised	Yellowish
33	BW06	Waimital	Bonansa	Filamento	Undulate	Flat	White
34	BW07	Waimital	Bonansa	Circular	Entire	Flat	White
35	BW08	Waimital	Bonansa	Circular	Entire	Raised	White
36	PT01	Poka	Maize of Tanimbar	Iregular	Filiform	Umbonate	Yellowish white
37	PT02	Poka	Maize of Tanimbar	Iregular	Filiform	Flat	White

The similarity in shape, margin, elevation and colony color of bacteria does not always show the same gram reaction or similarity in gram reaction does not always show the same morphology of the isolates. Furthermore, it can be assumed that 37 bacterial isolates that have various morphological variations and gram reactions have the potential to also have various different biochemical characters. Table 1 shows that the bacteria colonizing the roots of maize plants varied greatly in terms of mophology, thus indicating the existence of various types of bacteria that cycological plant roots.

3.2. Biochemical character

The gram reaction test and motility test, catalase, oxidase and MagConkey Agar showed differences between isolates. The results of the Gram reaction test using KOH showed that most of the bacteria were Gram positive, reaching 70.3% or as many as 26 isolates. Other test results are presented in Figure 2 and Table 2.

Table 2. Biochemical character of corn plant rhizosphere bacteria isolates

No.	Isolate Code	Gram Reaction Test	Motility Test	Catalase Test	Oxidase Test	Test Mac Conkey Agar Media
1	PB01	G +ve	+	-	+	-
2	PU01	G +ve	+	-	+	-
3	PU02	G +ve	+	-	+	-

4	PU03	G +ve	+	-	+	-
5	PU04	G +ve	+	-	+	-
6	PU05	G +ve	-	+	-	-
7	PU06	G –ve	+	+	+	+
8	PU07	G –ve	+	-	+	+
9	PU08	G –ve	+	-	+	+
10	PU09	G –ve	-	-	+	+
11	PU10	G –ve	+	+	-	+
12	PU11	G +ve	+	-	+	-
13	PU12	G +ve	+	-	+	-
14	PU13	G –ve	+	+	-	+
15	BT01	G +ve	-	-	+	-
16	BT02	G +ve	+	+	-	-
17	BT03	G –ve	-	-	+	+
18	BT04	G +ve	-	-	+	-
19	BT05	G +ve	-	-	+	-
20	BT06	G +ve	-	-	+	-
21	BT07	G +ve	-	+	-	-
22	BT08	G +ve	+	+	-	-
23	BT09	G +ve	-	+	-	-
24	BT10	G –ve	-	+	-	+
25	BT20	G –ve		+	-	+
26	BT21	G –ve	+	+	-	+
27	BT31	G +ve	-	-	+	-
28	BT32	G +ve	-	-	+	-
29	BW01	G +ve	-	-	+	-
30	BW02	G +ve	+	-	+	-
31	BW03	G +ve	-	-	+	-
32	BW04	G +ve	+	-	+	-
33	BW06	G +ve	-	-	+	-
34	BW07	G +ve	+	+	-	-
35	BW08	G +ve	+	+	-	-
36	PT01	G –ve	+	+	-	+
37	PT02	G +ve	-	+	-	-

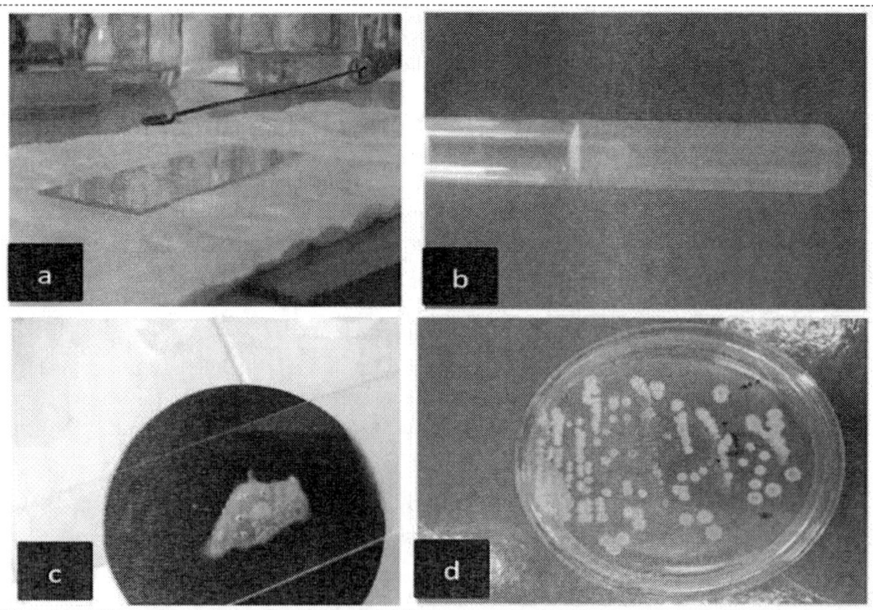

Figure 2. Biochemical character of bacterial isolates (a) gram reaction test, (b) motility test, (c) catalase test and (d) results of Mac Conkey Agar test (formation of mucoid)

3.3. Ability to phosphate solubitation and nitrogen fixation

The ability of phosphate solubitation and producing nitrogen and fixing nitrogen is presented in Table 3.

Table 3. The ability of starch degrading bacteria to phosphate solubitation and nitrogen fixation

No.	Isolate Bacteria	Ability of Phosphate Solubitation			Nitrogen Fixation	
		Hallo Zone (mm)	Hallo Zone Diameter (mm)	Index of Phosphate Solubitation	N Burk Test	Ashby Media
1	PB01	-	-	-	+	+
2	PU01	14.5	7.5	0.48	+	-
3	PU02	-	-	-	-	-
4	PU03	14.5	2.2	0.78	-	-
5	PU04	-	-	-	-	-
6	PU05	-	-	-	-	-
7	PU06	14	8.3	0.41	-	-
8	PU07	12.7	8.7	0.31	+++	+
9	PU08	-	-	-		-
10	PU09	-	-	-	-	-
11	PU10	-	-	-	+++	+
12	PU11	13.5	7.3	0.46	++	+
13	PU12	-	-	-	-	+
14	PU13	-	-	-	++	+
15	BT01	-	-	-	++	+
16	BT02	-	-	-	-	-
17	BT03	15.7	7.7	0.51	++	+
18	BT04	-	-	-	+	+
19	BT05	-	-	-	-	+
20	BT06	-	-	-	++	+
21	BT07	14.3	8.4	0.43	+++	+
22	BT08	14.8	7.8	0.47	+	+
23	BT09	-	-	-	-	+
24	BT10	-	-	-	+	+
25	BT20	-	-	-	-	-
26	BT21	12.7	3.2	0.74	++	+
27	BT31	14.8	9.4	0.36	-	-
28	BT32	-	-	-	-	+
29	BW01	-	-	-	-	-
30	BW02	-	-	-	-	-
31	BW03	12.3	5.1	58.5	+++	+
32	BW04	9.3	5.5	0.41	+++	+
33	BW06	-	-	-	-	-
34	BW07	-	-	-	+	+
35	BW08	-	-	-	-	-
36	PT01	12.6	6.2	0.51	-	-
37	PT02	-	-	-	-	-

Soil is a habitat for a variety of organisms including micro flora and fauna. Soil microorganisms play a very important role in soil fertility not only because of the ability of microbes to carry out biochemical transformations but also because of their importance as a source and reserve of mineral nutrients. This process occurs in general soils due to the exo enzymes released by microbes in the environment and degrade soil components.

The soil in a growing garden is a rich source of microorganisms. Biologically active enzymes can be extracted from living organisms. The enzymatic activity of bacteria correlates with the physiological function of bacteria in promoting growth through hormone production for the benefit of plant growth. It can also be explained that rhizobacteria isolates that are capable of producing hormones such as IAA and GA have the ability to dissolve phosphate and nitrogen fixation.

The phosphate solubitation ability of the tested isolates refers to the formation of a clear zone around the bacteria on Phikovskaya media containing tricalcium phosphate (Figure 3). The formation of a clear zone is a degradation process of tricalcium phosphate as a form of the ability of bacterial isolates to dissolve phosphate. Of the 37 bacterial isolates that were inoculated into the solid Pikovskaya medium, there were only 13 isolates capable of dissolving phosphate with different abilities seen from the wide diameter of the clear zone formed around the bacterial colony, the dissolution index, and the level of solubitated P concentration.

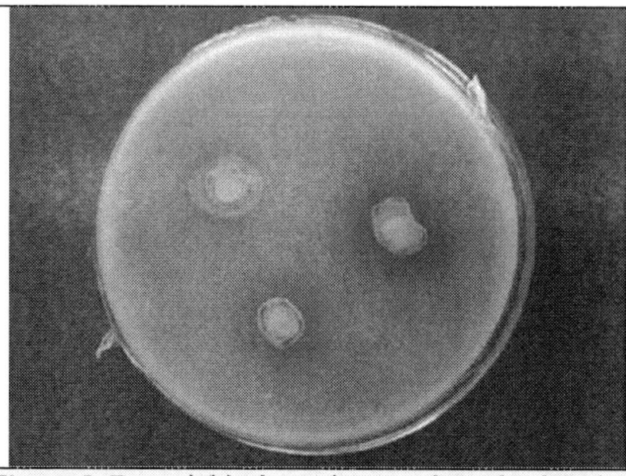

Figure 3. Bacterial isolates that can degrade phosphate

The qualitative size of the diameter of the clear zone indicates the size of the bacteria's ability to dissolve phosphate. The low concentration of dissolved phosphate is thought to be due to the reuse of dissolved phosphate by culture as a source of nutrition for metabolic activity and a decrease in the number of bacterial cell populations which will affect the activity of bacteria in dissolving phosphate. The results presented in Table 3 show that the phosphate dissolution index varies or there is a difference in the phosphate dissolution index. The greater the phosphate dissolution index is thought to have a tendency for the large dissolved phosphate concentration. This shows that the resulting P dissolution index can determine the ability of bacteria to dissolve phosphate.

The test results showed that from 37 bacterial isolates cultured on Burk N-free medium, 16 isolates were able to grow on the media with different fixation levels, while 20 isolates were able to grow on Ashbi media. Based on several research results, bacteria that are able to grow on Burk N-free media are indicated as bacteria capable of fixing nitrogen [16], [17], [18], [19], [20] and can form pellicles [21], [22], can produce ammonium [23], [24] and have nitrogenase activity [22], [25]. The difference in nitrogenase activity is thought to be due to the influence of the nifH gene system as a control for bacterial nitrogenase activity [26], [27], [28], [29], [30], [31], [32], [33].

3.4. Identification of the ability of enzyme and hormone producing bacteria

Bacteria isolated from the roots of corn plants have the ability to produce ACC-deaminase enzymes and hormones. Details of the ability of these isolates are presented in Table 4. The results of testing the IAA production capability of 34 isolates capable of producing IAA with different abilities after being grown on culture media with the addition of L-tryptophan. This difference in ability indicates a difference in the concentration produced. The ability of bacterial isolates to produce IAA was detected by a change in pink color after the addition of Salkowski's reagent in culture (Figure 4), besides that it can be explained that the difference in IAA concentrations produced by various bacterial isolates tested was due to differences in the ability of bacteria to utilize tryptophan or due to different pathways. IAA biosynthetic mechanism.

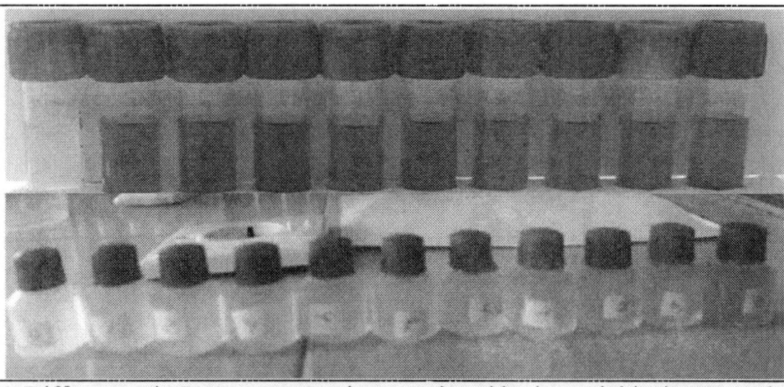

Figure 4. Differences in IAA concentrations produced by bacterial isolates starch degradat

The concentration of IAA produced by bacteria depends on the activity and number of cells, the availability of nutrients and the L-trp substrate in the media. In connection with that, it can be explained that there are three pathways for the formation of IAA, namely the IPyA pathway (Indole-3-Pyruvic Acid), the TAM pathway (Tryptamine) and the IAN pathway (Indole-3-Acetonitril). In bacteria there are only two pathways, namely the TAM and IPyA pathways, which are inducible by tryptophan compounds. Tryptophan is the main precursor in the biosynthesis of IAA. IAA production will increase according to the increase in tryptophan concentration from 1-100 µg / ml (Ahmad et al 2005). Based on the IAA production capability testing method with the addition of L-tryptophan to the culture media as a precursor, it shows the ability of bacterial isolates to produce IAA.

All tested isolates had the ability to produce gibberellic acid (GA3). Gibberellic acid produced by bacterial isolates is presented in Table 4. It is estimated that bacteria can increase GA levels in bacterial isolate cultures due to the production of GAs, deconjugating GAs from root exudates or hydroxylating inactive GA [34].

Table 4. Ability of bacteria to produce ACC-deaminase enzymes and hormones (IAA and GA)

No	Isolate Code	ACC-deaminase	IAA		GA (mgl^{-1})
			Capability of Production IAA	IAA Concentration (mgl^{-1})	
1	PB01	+	+	1.05	6,53
2	PU01	+	+	1.15	5,79
3	PU02	-	-	-	5.25
4	PU03	+	++	2.04	6,89
5	PU04	+	++	2.05	6.14
6	PU05	+	+	1.01	5.23
7	PU06	+	++	2.08	5.87
8	PU07	+	+++	4.05	5.49
9	PU08	+	++	2.51	4.23
10	PU09	+	+	0.16	6.02
11	PU10	+	++	2.02	5.68
12	PU11	+	++	2.62	5.67
13	PU12	+	+	0.17	4.85
14	PU13	+	++	2.59	5.53
15	BT01	+	+	0.03	4.76
16	BT02	+	+++	4.51	6.79
17	BT03	+	+	0.19	4.58
18	BT04	+	++	2.53	5.88
19	BT05	+	+	0.11	4.21
20	BT06	+	+	0.18	4.32
21	BT07	+	+++	5.01	6.07

22	BT08	+	+	1.02	4.56
23	BT09	+	+	1.03	4.86
24	BT10	+	+	1.06	4.35
25	BT20	-	-	-	4.23
26	BT21	+	++	2.05	5.02
27	BT31	+	+	1.06	4.34
28	BT32	+	+	1.14	4.22
29	BW01	+	+	1.52	4.09
30	BW02	+	+	1.42	4.21
31	BW03	+	+	1.33	4.02
32	BW04	+	+	1.23	3.98
33	BW06	+	+	1.87	3.99
34	BW07	+	+	1.97	4.01
35	BW08	-	-	-	4.10
36	PT01	-	-	-	4.03
37	PT02	+	+	1.25	4.01
Note::		*(+)* generate ACC-deaminase	*- Color clear solution* *+ a little pink* *++ bright pink* *+++ very pink*		

4. Conclusion

Thirty-seven bacterial isolates tested had various morphological characteristics ranging from the shape, elevation and colony color and there were various abilities of biochemical reactions through the gram reaction test, the ability of bacterial migration (motility), degradation of hydrogen peroxide, the production of oxidase enzymes, and the ability to sequester phosphate, nitrogen fixation and producing phytohormones (IAA and GA).

5. References

[1] Halverson L J, Handelsman J 1991 *Enhancement of soybean nodulation by Bacillus cereus UW85 in the field and in a growth chamber. Applied and Environmental Microbiology* vol 57 pp 2767-2770

[2] Parmar N Dadarwal KR 1999 *Stimulation of nitrogen fixation and induction of flavonoids like compounds by rhizobacteria J. of Applied Microbiology* **86** 44

[3] Kesaulya H, Baharuddin, B Zakaria, S A Syaiful 2015a *Isolation and Physiological Characterization of PGPR from Potato Plant Rhizosphere in Medium Land of Buru Island Procedia Food Science* vol 3 (Amsterdam: Elsevier) pp 190 – 199

[4] Berendsen RL, Pieterse C, Bakker P 2012 *The rhizosphere microbiome and plant health J.of Trends in Plant Science* **17** 2

[5] Kloepper JW, Scrhoth MN, Miller TD, 1980 *Effects of rhizosphere colonization by plant growth promoting Rhizobacteria on potato plant development and yield J of Phytopathology* **70** 1082

[6] Ahmad M, Kibret M 2014 *Mechanisms and Applications of Plant Growth Promoting rhizobacteria: current perspective J. of King Saud University-Science* **26** 20

[7] Glick BR, 1995 *The enhancement of plant growth by free living Bacteria J. of Canadian Microbiology* **41** 114

[8] Kennedy AC 2005 *Rhizosphere* In: Sylvia DM, Fuhrmann JJ, Hartel PG, Zuberer DA (eds) *Principles and Applications of Soil Microbiology 2nd edn* (Pearson, Prentice Hall, Upper Saddle River, NJ) pp 242-262

[9] Sundara Rao,WCB, Shinha MK 1962 *Phosphate dissolving microorganism in the soil and rizophere* Indian *J. Sci.* **23** 278

[10] SubbaRao NS 1982 *Advances in agricultural microbiology* Oxford and IBH Publications Company, India, pp 229–305

[11] Stella M, Suhaimi M 2010 *Selection of suitable growth medium for free-living diazotrophs isolated from compost J. Trop. Agric. and Fd. Sc.* **38** 219

[12] Subba Rao NS 1984 *Biofertilizers in agriculture* (New Delhi: Oxford & IBH Publishing Co.)
[13] Dworkin, M and J Foster 1958 *Experiments with some microorganisms which utilize ethane and hydrogen J. Bacteriol* **75** 601
[14] Gutierrez CK, Matsui GY, Lincoln DE, Lovel CR 2009 *Production of the phytohormone indole-3acetic acid by the estuarine species of the genus Vibrio Appl. Environ. Microbiol* **75** 2258
[15] Borrow, A M, P W Chester, U E, Curtis, P J Hemming, H G, Jefferey, E C Lloyd, R B, Nixon, I S, Norris, G L F, and Radly, N 1995 *Geberellic acid a metabolic product of fungus Giberella fujikuroi some observation on its production and isolation J. Sci. Food Agric.* **6** 348
[16] Ding Y, J Wang, Y Liu and Chen S 2005 *Isolation and identification of nitrogen-fixing bacilli from plant rhizospheres in Beijing region J. of Applied Microbiology* **99** 1281
[17] Döbereiner J and Pedrosa FO 1987 *The genus Azospirillum. In: Nitrogen-fixing bacteria in non-leguminous crop plants* Madison: Science Tech. Publishers p 155
[18] Döbereiner J, Baldani VLD and Reis VM 1995 *Endophytic occurrence of diazotrophic bacteria in non-leguminous crops. In: Fendrik I, del Gallo M, Vanderleyden J, de Zamaroczy M (eds) Azospirillum VI and related microorganisms* (Berlin, Heidelberg: Springer-Verlag) pp 15-30
[19] Mantilla-Paredes, A J, G I Cardona, C P Peña-Venegas, U Murcia, M Rodríguez and M M Zambrano 2009 *Distribuición de bacterias potencialmente fijadoras de nitrógeno y su relación con parámetros fisicoquímicos con tres coberturas vegetales en el sur de la Amazonia colombiana Rev. Biol. Trop* **57** 927
[20] Sgroy, V, Cassa´n, F, Masciarelli, O, Del Papa, M F, Lagares, A and Luna, V 2009 *Isolation and characterization of endophytic plant growth-promoting (PGPB) or stress homeostasis-regulating (PSHB) bacteria associated to the halophyte Prosopis strombulifera Appl. Microbiol Biotechnol* **85** 381
[21] Lea-Madi, M Kessel, Ester Sadovnik, Y Henis, 1988 *Electron microscopic studies of aggregation and pellicle formation in*Azospirillum *spp. Plant and Soil* **109** 121
[22] Da Silva K, Rafaela Simão Abrahão Nóbrega, Adriana Silva Lima, Alexandre Barberi1, Fatima Maria de Souza Moreira, 2013 *Density and diversity of diazotrophic bacteria isolated from Amazonian soils using N-free semi-solid media Sci. Agric.* Vol 68 (Piracicaba: Braz.) pp 518-525
[23] Shanmugam KT, Valentine RC 1975 *Molecular biology of nitrogen fixation* Science **187** 924
[24] Saribay,G F 2003 *Growth and nitrogen fixation dynamics of Azotobacter chroococcum in nitrogen-free and OMW containing medium* Doctoral Thesis (The Middle East Technical University)
[25] Dilworth, M J 1966 *Acetylene reduction by nitrogen fixing preparations from Clostridium pasteurianum Biochem. Biophys Acta* **127** 294
[26] Ueda, T, Suga, Y, Yahiro, N, and Matsuguchi, T 1995 *Remarkable N2-fixing bacterial diversity detected in rice roots by molecular evolutionary analysis of nifH gene sequences J. Bacteriol.* **177** 1417
[27] Ohkuma, M, Noda, S, Usami, R, Horikoshi, K, and Kudo, T 1996 *Diversity of nitrogen fixation genes in the symbiotic intestinal microflora of the termite Reticulitermes speratus Appl. Environ. Microbiol.* **62** 2752
[28] Widmer, F, Schaffer, B T, Porteous, L A, and Seidler, R J 1999 *Analysis of nifh gene pool complexity in soil and litter at a Douglas fir forest site in Oregon Cascade Mountain Range Appl. Environ. Microbiol.* **65** 380
[29] Kirshtein, J D, Pearl, H W, and Zehr, J P 1991 *Amplification, cloning, and sequencing of a nifh segment from aquatic microorganisms and natural communities Appl. Environ. Microbiol.* **57** 2650
[30] Garbeva, P, van Overbeek, L S, van Vuurde, J W L, van Elsas, J D 2001 *Analysis of endophytic bacterial communities of potato by plating and denaturating gradient gel electrophoresis (DGGE) of 16S rDNA based PCR fragments Microb. Ecol.* **41** 383

[31] Sessitsch, A, Reiter, B, Pfeifer, U, and Wilhelm, E 2005 *Cultivation-independent population analysis of bacterial endophytes in three potato varieties based on eubacterial and Actinomycetesspecific PCR of 16S rRNA genes FEMS Microbiol. Ecol.* **39** 32

[32] Reiter, B, Pfeifer, U, Schwab, H, and Sessitsch, A 2002 *Response of endophytic bacterial communities in potato plants to infection with Erwinia carotovora subsp. atrospetica Appl. Environ. Microbiol.* **68** 2268

[33] Chelius, M K, and Triplett, E W 2001 *The diversity of archaea and bacteria in association with the roots of Zea mays L. Microb. Ecol.* 41 263

[34] Bottini R, Cassan F, Piccoli, P 2004 *Giberellin production by bacteria and its involvement in plant growth promotion and yield increase Appl. Microbiol. Biotechnol.* **65** 503

Edible insects: Alternative protein for sustainable food and nutritional security

H C D Tuhumury[1,*]

[1]Agricultral Product Technology Department, Faculty of Agriculture Pattimura University
Jl Ir. M. Putuhena Poka Ambon Maluku Indonesia 97233

* E-mail: hcdtuhumury@gmail.com

Abstract. Any food systems applied has to ensure that the food is secure for the people. Nutritional needs are considered integral to the food security concept and sustainability is also important. One of the nutritional needs in food is protein. Proteins mostly obtained from animal-based sources such as meat and fish, plant-based proteins, particularly legumes. However, the production of meat, more importantly ruminant meat is more and more argued for some issues including environment, food safety, and animal welfare. In order to ensure sustainable protein production, the dietary change would be to use alternative protein sources such as insects. In general, insects have high protein content and excellent production efficiency compared with other conventional protein food groups. It is necessarily important to develop the use of insects in various forms, including pastes or powders, concentrates or isolates which can be used further as ingredients or fortified agents in new food product formulations. This paper therefore explores the feature of edible insects as sustainable and viable food source that can contribute to food security including, types of edible insects as food and their nutritional compositions; and processing of edible insects. In conclusion, insects can contribute to sustainable food and nutritional security and be a part of the solution to protein shortages. Insect protein concentrates could potentially be an alternative protein source in food formulation.

1. Introduction

World population in 2020 has reached 7.8 billion people (UN data) and has been projected to reach over 9 billion by 2050 [1]. Population would grow geometrically, while food production would grow arithmetically. Therefore, the increase in population requires double the current food production. To make it even worse, the climate change and environmental destruction from industrial development have negatively impact the productivity of food, as more and more land used for food production are gradually decreasing [2]. Food production has a substantial environmental impact and any food system applied need to take this issue into account seriously. Any food systems applied has to ensure that the food is secure for the people. Therefore, food security should be carefully considered in any food production system. Food security is when all people, at all times have access to sufficient, safe, and nutritious food to meet their dietary needs for and active and healthy life. The four important pillars of food security are availability, access, utilization and stability.

Nutritional needs are considered integral to the food security concept and sustainability is also important. One of the nutritional needs in food is protein. Proteins mostly obtained from animal-based sources such as meat and fish, plant-based proteins, particularly legumes. However, the production of meat, more importantly ruminant meat is more and more argued for some issues including environment, food safety, and animal welfare [3]. Deforestation, soil erosion, public health hazard, loss of plant biodiversity and water pollution are mainly the considerable impact of meat production sector. Due to these significant issues, the use of plant-based protein in the diet gained popularity. Yet, plant-based protein have been found to lack certain essential amino acids and to be less digestible than animal-based protein [4][5]. In order to ensure sustainable protein production, the dietary change would be to use alternative protein sources such as insects.

Insects can be considered as human food or as feed. The production of insects seems to be more sustainable than livestock production for some particular reasons: lower greenhouse and ammonia emissions, less land area needed, and potential to be grown on organic [3]. In general, insects have high protein content and excellent production efficiency compared with other conventional protein food groups [6]. Moreover, insects are encouraged as a good source of protein and micro and the production of edible insects in developing countries is supported by FAO of The United Nations. Edible insect protein also meet The WHO essential amino acid requirements [7] and are more digestible than plant-based and only little less digestible than animal-based protein [4]. Nevertheless, eating insects are mostly considered as unfavourable in most countries, especially in western countries. Therefore, it is necessarily important to develop the use of insects in various forms, including pastes or powders, concentrates or isolates which can be used further as ingredients or fortified agents in new food product formulations.

This paper therefore explores the feature of edible insects as sustainable and viable food source that can contribute to food security including, types of edible insects as food and their nutritional compositions; the advantage and disadvantage of edible insect production; and processing of edible insects.

2. Types of Edible Insects and Their Nutritional Composition

Evolution has created a large range of arthropod species adapted to their habitats over the past 400 million years. Of the 1.4 million animal species described on earth, approximately 1 million are insects. Just 5000 out of 1 million insect species mentioned can be considered harmful to crops, livestock or human being [3]. Among the insect species, there are insects which are edible.

Beetles (Coleoptera) are the most common insects eaten worldwide (31 percent). This is not surprising considering that about 40 percent of all recognized insect species are present in the group. The types of edible insects around the world can be seen in Table 1.

Table 1. Types of edible insects

Order	Species	Percentage
Coleoptera	Beetles	31
Lepidoptera	Caterpillars	18
Hemynoptera	Bees, wasps, ants	14
Orthoptera	Grasshoppers, locusts, crickets	13
Hemiptera	Cicadas, leafhoppers, planthoppers, scale insects and true bugs	10
Isoptera	Termites	3
Odonata	Dragonflies	3
Diptera	Flies	2
Other	-	5

Source : Cerritos, 2009 [8]

Caterpillar consumption (Lepidoptera), is estimated at 18%. At 14%, bees, wasps and ants (Hemynoptera) are in third place. Grasshoppers, locusts and crickets (Orthoptera) follow these;

termites (Isoptera) and dragon flies (Odonata) are each 3%[8].Approximately 1,417 species can be eaten. insects are eaten in their adult or larva stage.

Edible beetles are of many types, including aquatic beetles, wood-boring larvae, and dung beetles (larva and adults). [9] described 78 species of edible aquatic beetles, primarily belonging to the families Dytiscidae, Gyrinidae, and Hydrophilidae. Only the larvae of these animals are usually eaten. The palm weevil, Rynchophorus, a significant palm pest throughout Africa, southern Asia and South America, is by Far the most common edible beetle in the tropics. The weevil of the palm R. *Phoenicis* is included in the use of sound in harvesting in tropical and equatorial Africa), R. *Ferrugineus* in Asia (Papua New Guinea, Indonesia, Japan, Malaysia, the Philippines and Thailand. In Indonesia, especially in Papua, and Maluku, there are two species of palm weevil, usually called sago grub/sago worm, and in their adult stage they are distinguished by the color. The R. *billineatus* is black while R. *ferrugineus* is brown[10]. The larvae of this insects are traditional delicacies in both areas.

For Lepidoptera, during their larval periods (i.e. as caterpillars), butterflies and moths are usually consumed. But adult butterflies and moths are often eaten, too. Indigenous Australians have been recorded to eat moths of the cutworm Agrotis infusa (the Bogong moth) and have been observed to eat hawkmoths (Daphnis spp. and Theretra spp.) after cutting the wings and legs in the Lao People's Democratic Republic.

In countries like Thailand and Cambodia, in addition to the arachnid scorpions, people enjoy grasshoppers, crickets, ants, bamboo borers (*Omphisa fuscidentalis*) [11]. These insects are sold in Bangkok in markets that visitors seek as exotic food.Weaver ant larvae, bamboo worms, crickets, and wasps are eaten by many ethnic groups in Laos. In India, some tribes eat red ants, hornets, and termites[12].

The nutritional values of edible insects can very depending on their metamorphic stage, habitat and diet even within the same category of species. They are highly variable due to the large range of species. Also, the preparation and processing methods including drying, boiling or frying used prior to consumption, like most foods, can also affect nutritional composition. Nutrient composition for approximately 200 edible insects were compiled by [13]. Many edible insects supply satisfactory quantities of protein and energy. They also fulfill human needs for amino acids. Not to mention the mono and polyunsaturated fatty acids and they also rich in micronutrients such as iron, magnesium, phosphorous, manganese, selenium, and zinc. Insects also contained significant amount of vitamin including riboflavin, biotin, pantothenic acid, and folic acid.

The protein contents of insects from different orders have been evaluated by [14]. The average protein content was in the range of 13-77 %. The protein contents per insect order can be seen in Table 2.

Table 2. Crude protein content by insect order

Insect order	Stage	Range (% Protein)
Coleoptera	Adults and larvae	23-66
Lepidoptera	Pupae and larvae	14-68
Hemiptera	Adults and larvae	42-74
Homoptera	Adults, larvae, and eggs	45-57
Hymenotepra	Adults, larvae, pupae, and eggs	13-77
Odonata	Adults and naiad	46-65
Ortotepra	Adults and nymph	23-65

Source: XiaoMing, 2010 [4]

The amino acid compositions of edible insects are good in replacing cereal proteins. Amino acids such as lysine and tryptophan,and threonine are well represented in some edible insect species. These amino acids are often lack in cereal protein which are essential staple food around the world. For instance, several caterpillars, palm weevil larvae, and aquatic insects have lysine score higher than 100 mg per 100 g protein[15].

Edible insects are also important source of fat. The quality and structure of oils extracted from several insects were investigated by [16]. Their oils are rich in fatty acids that are polyunsaturated. For example African palm weevil (*Rhynhcophrus phoenicis*) had a fat content of 54% with 38% palmitoleic acid (MUFA) and 45% linoleic acid (PUFA). In landlocked developing countries with lower access to fish, insects could play an important role contributing the polyunsaturated fatty acid such as oleic, linoleic and α-linolenic acid.

The mineral and vitamin contents of edible insects are greatly varied across species and orders. Edible insects are excellent source of iron. Many edible insects have an iron content equal to or greater than that of beef. Some examples, iron content of locusts, and mopane caterpillar have respective iron content of 8-20 mg, and 31-77 mg per 100g compared to beef with 6 mg iron per 100 g [15], [17]. Since they are rich in iron, the inclusion of them in the diet could improve iron status, especially to prevent anaemia in developing countries. Insects are also commonly believed to be good sources of zinc. Palm weevil larvae has 26.5 mg zinc compared to beef with 12.5 g zinc per 100 g of dry weight [15]. Bukkens [15] also showed for a whole range of insects that thiamine (also known as vitamin B1) ranged from 0.1 mg to 4 mg per 100 g of dry matter. Vitamin B12 occurs only in food of animal origin and is well represented in mealworm larvae, *Tenebrio molitor* and house crickets. Many species have very low levels of vitamin B12.

In summary, edible insects are highly nutritious and Highly nutritious and cheapest source of animal protein and fat which is comparable to milk and meat. Insect proteins are highly digestible (between 77% and 98%) Protein has high lysine and threonine and limited levels of either tryptophan or methionine. The fat content ranges from 7% to 77 % , rich in ω-3 and ω - 6 compared to cattle and pigs, a nd has low cholesterol level. They also have high thiamine and riboflavin than whole bread or an egg. Insects can also be a source of fiber due to their high chitin content has high Fe, Zn can alleviate deficiency in women's diet i.e. pregnant women and vegetarian diet. Due to high protein, polyunsaturated fatty acids, and micronutrients of the edible insects, the consumption of these materials are endorsed by FAO and European Commission (EC) as the availability of nutritious food is becoming increasingly difficult [18].

3. Processing of Edible Insects

The FAO has started to encourage edible insects as feasible dietary alternatives for humans, which has ignited a surge of interests in edible insects [19]. By 2023, the global edible insect market is projected to hit USD 522 million [20]. However, negative perceptions of insects limit the expansion of its global market. Novice consumers view insects as a disgust, ignore insects as a daily food in the diet, and fully disregard their nutritional value [3].

Therefore, as the availability of edible insect grows, insect-based ingredients should be developed rather than consuming the products that still retain their original appearance. Incorporating edible insects into already familiar foods maybe more acceptable to an insect-phobic culture, and using insects as food ingredients is beneficial for the development of long-term business models. One example in our context in Maluku, Indonesia is incorporating sago grub as flesh ingredients in poplar snack food, sago crackers [21].

Insects are increasingly being seen as the food of the future to replace conventional animal protein consumption. In the Netherlands, attempts to encourage eating insects include experiments aimed at customizing insects for western tastes, with certain insects such as locusts and mealworm being sold at retail markets. One important effort to eliminate visual associations and improve palatability is to grind insects into powder or meal. Researchers have also looked into the functional properties of insects protein such as gelling ability, foam capacity, emulsion capacity, and solubility in different buffers or solvents [6], [7], [22]. These experiments would make it easier to use edible insects as a food ingredient in conventional foods.

The simplest processing technique for edible insect is the traditional ways of cooking it. Steaming, roasting, smoking, baking, stewing, frying, and curing are some of the typical edible insect cooking methods. Sometimes blanching is used before these procedures to reduce the number of foodborne microorganisms and inactivate enzymes. Obviously, each technologies produces products with distinct sensory and nutritional characteristics [23].

The processing of edible insects begins with insects post-harvesting and ends with the production of food products. Despite the fact that the number of current processes in the edible food industry is enormous and varies depending on the species used and the final product to be produced, the majority of them can be classified into relatively small number of operations based on the same basic principles including blanching and drying.

Blanching is used a pre-treatment for most commercialized edible insects, both on an industrial and artisanal scale, to reduce microbial counts and inactivate degradative enzymes that cause food poisoning and spoilage. It is done by immersing the food in boiling water for a brief period of time, then extracting it and immersing it in ice water or putting it under cold running water to avoid the thermal process [24]. To boost antimicrobial effects with minimal quality loss, blanching treatments should be customized to each specific insect species and be combined with techniques to reduce the presence of bacterial spores However, in order to meet the needs of customers and industrial manufacturing, these procedures must retain the nutritional value of the products as well as other essential quality characteristics including texture and color [25].

Drying methods usually applied in the processing of edible insects including sun-drying, freeze-drying, and oven-drying. These techniques can be used for drying either whole edible insects or insects flours and powders. One of the preferred technologies for increasing human consumption of insects, mostly in Western countries, is drying and grinding whole, perfectly recognizable edible insects into unrecognizable powder[26]. However, all the drying methods caused changes in protein, fat, and fiber content. The most visible quality changes during drying is color. For example, powder made from the larvae of *T. molitor* had decreased lightness when oven-dried compared to freeze-dried [27]. Different drying technologies have an effect on certain consistency parameters in addition to reduction in water content. The drying technology used and the technical conditions applied could affect protein functional properties, lipid oxidation, and color. Therefore, what technology chosen should consider the way insects will be eaten either whole, powdered ingredient, or sole ingredient.

Among all insect products, insect protein is important. However, their functional properties must be carefully tested during transition process in order to integrate them into various foods. Depending on the raw material and its proximate composition, protein-enriched ingredients such as edible insect flour/powder are obtained using various methods and techniques. Nonetheless, a general 5-step food protein processing procedures, including: 1. Pre-treatment; 2. Defatting; 3. Solubilization and recovery of protein; 4. Purification; and 5. Drying [4].

Protein concentrates and isolates have desirable functional properties depending on their origin and processing, and thus may be used in food formulation. Proteins, on the other hand, from complex structures that interact with other elements including proteins, fats, carbohydrates, and minerals. Functional properties of insect proteins also in accordance with other protein functionality such as solubility, emulsification, foam formation, gelation, water-holding capacity, and oil absorption capacity.

Edible insects may be considered a novel protein source in the food industry, as some insect products functioned similarly to traditional food proteins, as mentioned above. However, when these insects are added to food matrices, their functional properties are still largely unknown. Some insect flours, defatted flours and protein concentrates have been incorporated in food matrix including meat products and analogs[28], snack[29], pasta[30], and bread[31].

4. Prospects and Challenges

One of the challenges of consuming edible insects is the food safety. Microorganism, allergic reactions, and toxicity. For microbial safety, insect gut, body surface, and mouthpart is considered as a main habitat for microorganisms which can act as pathogenic vector. Despite the fact that insect-specific pathogenic microorganisms pose no threat to human health, they can only colonize insect cells or tissues. Contamination of human pathogenic microorganisms does, however, occur, and this contamination can be managed by carefully managing breeding conditions[32].

Some edible insects had allergic cross-reactive proteins with arachnids and crustaceans, which are known as arthropods, in allergic reactions. Furthermore, food dye carmine is derived from the bodies of female cochineal insects, and this dye can cause allergic reactions in some patients. We must pay attention to the unknown possible allergens found in edible insects, given that only a few studies on allergic reactions to edible insects have been conducted. More research and investigation is needed to determine the hazards of substances such as allergens and toxic substances, as well as their effects on pathological symptoms in humans[32].

Other important challenge is the consumer acceptance. One of the most significant obstacles to their integration into food products is Westerners' inability to try edible insects.Insects are marketed as a delicacy in some experimental restaurants. Most consumers are repelled by the idea of eating them. Food neophobia and disgust are two major barriers to insect consumption. Therefore, it is necessary to understand consumer psychology and desires, and adjust their preparation strategies accordingly such as innovative development of insect-based food [33].

Despite the challenges mentioned above, a number of scientists believe insects have the potential to be a sustainable food source. The production of insect-based protein powder is expected to be more environmentally friendly than traditional protein-rich foods. In order to maximize the potential of the edible insect proteins, optimizing insect rearing and harvesting is therefore critical. Depending on the insect species, rearing methods, and processing techniques used, data on protein content and functionality varies greatly. Future research should concentrate on determining the best processing conditions for producing insect protein isolates that are ideal for food formulation, have good functional properties, are cost-effective, and are environmentally sustainable.

5. Conclusions

Edible insects with their nutritional composition especially high protein content and all the essential amino acids, are a promising alternative protein source for the world's increasing population. Insects can contribute to sustainable food and nutritional security and be a part of the solution to protein shortages. Since whole insects are still unpopular in Western cultures, insect protein concentrates may be a viable alternative protein source in food formulation.

References

[1] Grafton R Q, Daugbjerg C and Qureshi M E 2015 Towards food security by 2050 *Food Secur.* **7** 179–83

[2] van Huis A and Oonincx D G A B 2017 The environmental sustainability of insects as food and feed. A review *Agron. Sustain. Dev.* **37** 43

[3] van Huis A 2015 Edible insects contributing to food security? *Agric. Food Secur.* **4** 20

[4] Gravel A and Doyen A 2020 The use of edible insect proteins in food: Challenges and issues related to their functional properties *Innov. Food Sci. Emerg. Technol.* **59** 102272

[5] Sarwar G 1997 The Protein Digestibility–Corrected Amino Acid Score Method Overestimates Quality of Proteins Containing Antinutritional Factors and of Poorly Digestible Proteins Supplemented with Limiting Amino Acids in Rats *J. Nutr.* **127** 758–64

[6] Zielińska E, Karaś M and Baraniak B 2018 Comparison of functional properties of edible insects and protein preparations thereof *LWT* **91** 168–74

[7] Bußler S, Rumpold B A, Jander E, Rawel H M and Schlüter O K 2016 Recovery and techno-functionality of flours and proteins from two edible insect species: Meal worm (Tenebrio molitor) and black soldier fly (Hermetia illucens) larvae *Heliyon* **2** e00218

[8] Cerritos R 2009 Insects as food: an ecological, social and economical approach. *CAB Rev.* **v. 4**

[9] Ramos-Elorduy J, Moreno J M P and Camacho V H M 2009 Edible aquatic Coleoptera of the world with an emphasis on Mexico *J. Ethnobiol. Ethnomed.* **5** 11

[10] Köhler R, Irias-Mata A, Ramandey E, Purwestri R and Biesalski H K 2020 Nutrient composition of the Indonesian sago grub (Rhynchophorus bilineatus) *Int. J. Trop. Insect Sci.* **40** 677–86

[11] Hanboonsong Y, Jamjanya T and Durst P B 2013 Six-legged livestock: edible insect farming, collecting and marketing in Thailand *RAP Publ.*

[12] Barennes H, Phimmasane M and Rajaonarivo C 2015 Insect Consumption to Address Undernutrition, a National Survey on the Prevalence of Insect Consumption among Adults and Vendors in Laos *PLoS One* **10** e0136458

[13] Rumpold B A and Schlüter O K 2013 Nutritional composition and safety aspects of edible insects *Mol. Nutr. Food Res.* **57** 802–23

[14] XiaoMing C, Ying F, Hong Z and ZhiYong C 2010 Review of the nutritive value of edible insects. *For. insects as food humans bite back. Proc. a Work. Asia-Pacific Resour. their potential Dev. Chiang Mai, Thailand, 19-21 February, 2008* 85–92

[15] Bukkens S G F 2005 Insects in the human diet: nutritional aspects. *Ecological implications of minilivestock; role of rodents, frogs, snails, and insects for sustainable development* ed M G Paoletti (New Hampshire: Science Publishers) pp 545–77

[16] Womeni, Hilaire Macaire, Linder, Michel, Tiencheu, Bernard, Mbiapo, Félicité Tchouanguep, Villeneuve, Pierre, Fanni, Jacques and Parmentier, Michel 2009 Oils of insects and larvae consumed in Africa: potential sources of polyunsaturated fatty acids *OCL* **16**

[17] Oonincx D G A B, van Itterbeeck J, Heetkamp M J W, van den Brand H, van Loon J J A and van Huis A 2011 An Exploration on Greenhouse Gas and Ammonia Production by Insect Species Suitable for Animal or Human Consumption *PLoS One* **5** e14445

[18] Nowak V, Persijn D, Rittenschober D and Charrondiere U R 2016 Review of food composition data for edible insects *Food Chem.* **193** 39–46

[19] van Huis A, van Itterbeeck J, Klunder H, Mertens E, Halloran A, Muir G and Vantomme P 2013 *Edible insects: Future prospects for food and feed security.* (Rome: Food and Agriculture Organization of the United Nations)

[20] Han R, Shin J T, Kim J, Choi Y S and Kim Y W 2017 An overview of the South Korean edible insect food industry: challenges and future pricing/promotion strategies *Entomol. Res.* **47** 141–51

[21] Tuhumury H, Souripet A and Nendissa S 2020 Effects of Sago Starch Types on Crackers from Edible Larvae of Sago Palm Weevils *Indones. Food Sci. & Technol. J.* **4** 1–5

[22] Kim T-K, Yong H I, Jang H W, Kim Y-B and Choi Y-S 2020 Functional Properties of Extracted Protein from Edible Insect Larvae and Their Interaction with Transglutaminase *Foods* **9** 591

[23] Melgar-Lalanne G, Hernández-Álvarez A-J and Salinas-Castro A 2019 Edible Insects Processing: Traditional and Innovative Technologies *Compr. Rev. Food Sci. Food Saf.* **18** 1166–91

[24] Xiao H-W, Bai J-W, Sun D-W and Gao Z-J 2014 The application of superheated steam impingement blanching (SSIB) in agricultural products processing – A review *J. Food Eng.* **132** 39–47

[25] Caparros Megido R, Desmedt S, Blecker C, Béra F, Haubruge É, Alabi T and Francis F 2017 Microbiological Load of Edible Insects Found in Belgium *Insects* **8** 12

[26] Menozzi D, Sogari G, Veneziani M, Simoni E and Mora C 2017 Eating novel foods: An application of the Theory of Planned Behaviour to predict the consumption of an insect-based product *Food Qual. Prefer.* **59** 27–34

[27] Azzollini D, Derossi A and Severini C 2016 Understanding the drying kinetic and hygroscopic behaviour of larvae of yellow mealworm (Tenebrio molitor) and the effects on their quality

[28] Kim H-W, Setyabrata D, Lee Y J, Jones O G and Kim Y H B 2016 Pre-treated mealworm larvae and silkworm pupae as a novel protein ingredient in emulsion sausages *Innov. Food Sci. Emerg. Technol.* **38** 116–23

[29] Azzollini D, Derossi A, Fogliano V, Lakemond C M M and Severini C 2018 Effects of formulation and process conditions on microstructure, texture and digestibility of extruded insect-riched snacks *Innov. Food Sci. Emerg. Technol.* **45** 344–53

[30] Duda A, Adamczak J, Chełmińska P, Juszkiewicz J and Kowalczewski P 2019 Quality and Nutritional/Textural Properties of Durum Wheat Pasta Enriched with Cricket Powder *Foods* **8**

[31] Osimani A, Milanović V, Cardinali F, Roncolini A, Garofalo C, Clementi F, Pasquini M, Mozzon M, Foligni R, Raffaelli N, Zamporlini F and Aquilanti L 2018 Bread enriched with cricket powder (Acheta domesticus): A technological, microbiological and nutritional evaluation *Innov. Food Sci. Emerg. Technol.* **48** 150–63

[32] Kim T-K, Yong H I, Kim Y-B, Kim H-W and Choi Y-S 2019 Edible Insects as a Protein Source: A Review of Public Perception, Processing Technology, and Research Trends *Food Sci. Anim. Resour.* **39** 521–40

[33] Patel S, Suleria H A R and Rauf A 2019 Edible insects as innovative foods: Nutritional and functional assessments *Trends Food Sci. Technol.* **86** 352–9

Scoring system as an alternative audit method in food safety management system certification body

H D Wahyuni[1,*], S Nurjanah[2 3], W P Rahayu[2 3]

[1] Food Technology Graduate Program, IPB University, Indonesia
[2] Department of Food Science and Technology, IPB University, Indonesia
[3] SEAFAST Center, IPB University, Indonesia

* E-mail: henidwahyuni@gmail.com, sity_nr@apps.ipb.ac.id, wpr@apps.ipb.ac.id

Abstract. Differences between auditor perceptions often occur in the audit process of the management system, even auditors from the Food Safety Management System Certification Body (FSMS CB). The auditor as an assessor is able to influence strongly to the audit report. A scoring system is a quantitative approach in auditing to eliminate subjectivity and impartiality in the auditor's assessment. The purpose of this research was producing an internal audit checklist with a scoring system, obtaining the diversity of the auditors' assessment on the gap analysis in FSMS CB (PT XYZ), and getting information on the sources of the diversity on the gap analysis results. The research scope was ISO/IEC 17021-1:2015 process requirements. The research method consisted of 3 steps, they were: preparation, data analysis, and cause-effect analysis. The auditors' total scores were in the same category, which was the good category ($X \geq 73.2\%$), and not significantly different at the 0.05 significance level. The diversity in the auditors' assessment was found in 4 sub-clauses (9.1, 9.2, 9.5, and 9.9). The sources of diversity in the auditors' assessment were personnel and tool. It could reduce by increasing the auditors' competence and improving the internal audit system from PT XYZ.

1. Introduction

The issuance of the ISO 22000: 2018 standard in replacing ISO 22000: 2005 on 18 June 2018 by the world International Organization for Standardization (ISO) and a transition period for certification changes to its clients for 3 years until 18 June 2021 by Komite Akreditasi Nasional/ National Accreditation Committee (KAN) become a concern for all FSMS CB accredited by KAN to adjust the policies and implementation of the certification process to the new standard. Gap analysis is needed to measure the readiness of FSMS CB for the ISO 22000: 2018 certification process in food industries. By measuring methods and qualified auditors that meet the requirements, the results of the gap analysis are expected to be accurate. The results become the basis of FSMS CB to improve its policies and implementation according to the requirements, and in addition to the accreditation and certification processes in the food industries could be carried out comply to the requirements. So that, an accredited FSMS CB can provide a credible certification process in the assurance of food safety in food industry products.

Differences between auditor perceptions often occur in the audit process of the management system, even auditors from the FSMS CB. The auditor as an assessment is able to influence the audit report. Competence and impartiality are the requirements for personnel to meet the ISO / IEC 17021-1: 2015 that must be implemented by FSMS CB. Inconsistencies of requirements implementation may result in an opportunity for audit activities (including gap analysis) to be conducted not in accordance with the requirements. In addition, the use of the qualitative method in internal audits at FSMS CB produces not measurable audit reports. A qualitative checklist as an audit tool can provide opportunities for subjectivity to the auditor's assessment.

Gap analysis can be done both qualitatively and quantitatively. The qualitative method is carried out using a checklist method which consists of a series of requirements in the form of statements with "appropriate" or "inappropriate" answers related to compliance with the requirements. The quantitative method is carried out with the scoring system as a measure of compliance with the requirements. The scoring system is designed to produce measurable audit reports, which can eliminate subjectivity and impartiality to the assessment of compliance with requirements.

Quantitative approaches are generally less debatable than qualitative. This is because the quantitative approach is more measurable and involves the use of data such as numerical data, so as to reduce subjectivity or bias [1]. The scoring system has the ability to provide information in the form of numbers so that it can provide an evaluation of the feasibility of the tested subjects in the form of an appropriate value [2], [3]. The scoring system is carried out by establishing quantitative criteria (generally with a rating/score for each answer in the list of questions to measure the significance of the audit findings and the overall conclusion of the audit results [4].

Different scoring systems are used in the gap analysis of compliance with management system requirements. In previous research, scoring of 1 to 5 is most widely used [5], [6], [7], [8], [9]. [5] describes the method of scoring 1 to 5 by decision diagram to determine the level of implementation of requirements.

The purpose of this research was: 1) producing an internal audit checklist form equipped with a scoring system, 2) obtaining the diversity of the gap analysis results conducted by auditors which were related to compliance with the FSMS CB process requirements according to ISO / IEC 17021-1: 2015 and 3) obtaining information on the sources of diversity on the gap analysis results in FSMS CB (PT XYZ). The research scope was ISO/IEC 17021-1:2015 process requirements. Process requirements were chosen in this study because they were used in carrying out the process in the product certification of the FSMS CB to meet ISO 22000: 2018 requirement so that the certification process in food industries could be carried out credibly.

2. Methods

2.1. Research Location

The study was conducted at the PT XYZ located in Bogor, West Java, Indonesia. PT XYZ is a FSMS CB accredited by KAN and will upgrade its accreditation for the needs of ISO 22000: 2018 certification for clients.

2.2. Material and Tool

The materials used in this study were: 1) primary data obtained from the gap analysis; 2) secondary data was ISO / IEC 17021-1:2015 and the documents and records of PT XYZ. The tools used in this study were: 1) gap analysis checklist, 2) MS Office Excel 2010 statistical tool for data analysis and 3) cause-effect diagram tool for cause-effect analysis.

2.3. Research Steps

This study consisted of 3 steps. The first step was the preparation, consisting of a literature study of process requirements in FSMS CB, auditor verification, and the development of a checklist. The second step was data analysis of the gap analysis results in PT XYZ. The third step was the cause-effect analysis of the diversity in the gap analysis results. The explanation was as follows:

IOP Conf. Series: Earth and Environmental Science **883** (2021) 012030 doi:10.1088/1755-1315/883/1/012030

2.3.1 Step 1

The first step was the preparation. It began with the literature study of the FSMS CB process requirements by accessing the KAN and IAF websites and other literature studies to ensure that these requirements were updated and valid [7]. Then, the auditor verification. It was carried out to ensure that the auditor met the competence according to the requirements of PT XYZ internal auditors [10]. Verification was carried out by matching the auditors' track record with the requirements set by PT XYZ. The sample of the auditors who complied were 6 auditors consisting of 5 internal auditors and 1 external auditor. This number was based on the consideration of a certain criterion, which was a sample of auditors who met the requirements of PT XYZ internal auditors. In addition, the total number of internal auditors who met the ISO / IEC 17021-1:2015 standard at PT XYZ was 5 auditors. The external auditor was used as a comparison to the internal auditor in the assessment. This consideration was supported by the statement that purposive sampling could be carried out with a choice of emphasis on a sufficient number or tight limitations on the sample [11]. The sampling was a suitable technique for a common sample with certain criteria and also sampling for experts who were needed to support the study at common objectives [12]. The final step was to study and develop the ISO / IEC 17021-1: 2015 internal audit checklist from PT XYZ by completing the checklist with scoring 1-5 as a measure of assessment [5]. This scoring was chosen because it was easy to understand and there was also an explanation of scoring through a decision diagram. The description of the score was according to Table 1.

Table 1. Description of the scores used in the checklist

Score	Description
1	no available documents, not understanding
2	no available documents, understanding, not implementing
3	available documents, not implementing
4	available documents, implementing, but not consistent no available documents, understanding, implementing
5	available documents, implementing, consistent

2.3.2 Step 2

The second step was the implementation of gap analysis at PT XYZ [5], and data analysis of the gap analysis results. The gap analysis began with an audit plan which has been approved by PT XYZ. The implementation of the gap analysis was arranged by the researcher in accordance with the agreement with PT XYZ. In this gap analysis implementation, the method used in collecting information was in accordance with the ISO/IEC 17021-1:2015 process requirements (clause 9.4.4.2), which were interviewing, observing activity processes of the certification body, and reviewing sources and records [10]. The tool was a checklist that had been prepared in the first step. The interviews were conducted with the personnel in charge of process requirements, such as document controller, certification manager, and technical staff. Observation of the processes and activities were carried out during the gap analysis at PT XYZ. Documents and records were reviewed on all quality management system documents and records of the implementation of ISO 22000 certification at PT XYZ. After carrying out the gap analysis, the auditor submitted the results in a checklist that had been filled in a hard copy to the researcher. Furthermore, the researcher calculated the results of the assessment in the audit report from each auditor and made a soft copy of the audit record for data analysis. The data analysis consisted of 1) determining the categorization of the auditors' assessment score, and 2) the auditors' score test. The determination of the auditor's assessment score categorization was carried out according to the previous study with the calculations in Table 2 [2]. The auditors' score test was carried out with a one-way analysis of variance (one-way ANOVA) at a 0.05 significance level to determine whether the auditor's score was significantly different or not [13].

Table 2. Determination of the auditor's assessment score categorization

No	Items	Formula
1	Determination of the gap analysis score categorization	
a	Determined descriptive statistical data which were consisting of the number of questions (n), the number of category score (low, medium, good), minimum range (X_{min}) (1), maximum range (X_{max}) (2), distribution distance (3), deviation standard (σ) (4), and mean theoretical (μ) (5)	(1) X_{min} = n x minimum score (2) X_{max} = n x maximum score (3) Distribution distance = $X_{max} - X_{min}$ (4) $\sigma = \frac{\text{distribution distance}}{6}$ (5) μ = n x number of categories
b	Calculated p-value using normal distribution table with first determine the value of Z_{min} (6) and Z_{max} (7). The p-value is seen in the normal distribution table with a confidence level of 95%. If $p_{max} > p_{min}$ then the next calculation will be used p_{max}, meanwhile if $p_{max} < p_{min}$ then the next calculation will be used p_{min}	(6) $Z_{min} = \frac{(X_{min} - \mu)}{\sigma}$ (7) $Z_{max} = \frac{(X_{max} - \mu)}{\sigma}$
c	Determined the categorization of the gap analysis score. Three categories consisted of low (8), medium (9), and good (10)	(8) Low, $X < (\mu - (p \times \sigma))$ (9) Medium, $(\mu - (p \times \sigma)) \leq X < (\mu + (p \times \sigma))$ (10) Good, $X \geq ((\mu + (p \times \sigma))$
2	Categorization of the auditor's assessment accordance with 1c	

2.3.3 Step 3
The third stage was carried out to identify the sources of the diversity in the auditor's assessment and made recommendations for corrective actions that should be taken by PT XYZ. The analysis was carried out using the cause-effect diagram tool with the consideration that it was easy to understand visually, systematically and could identify all sources of causes completely [14].

3. Results and discussion

3.1 Information on FSMS CB process requirements, PT XYZ's internal auditor and gap analysis checklist with scoring system

Based on information in the KAN and IAF websites and other conducted literature studies, information was obtained that ISO / IEC 17021-1: 2015 is still valid and is used as a requirement for FSMS CB accreditation. The ISO / IEC 17021-1: 2015 process requirements consisted of 9 sub-clauses covering: 9.1 pre-certification activities, 9.2 audit planning, 9.3 initial certification, 9.4 audit implementation, 9.5 certification decisions, 9.6 certification maintenance, 9.7 appeals, 9.8 complaints, and 9.9 client records. Verification of the auditors' track record stated that all auditors have met the requirements of PT XYZ's internal auditors.

Verification of the auditor's track record stated that all auditors met the requirements of PT XYZ's internal auditors. All auditors had different educational backgrounds and periods of working experiences. The auditor of working experiences period was: > 3 years for 2 auditors, > 5 years for 1 auditor, and > 10 years for 3 auditors (including external auditors). The level of internal auditors consisted of 2 lead auditors and 3 auditors. Some auditors had not FSMS CB auditing experience according to ISO / IEC 17021-1: 2015, but all of the auditors have attended sufficient training for the competency of FSMS CB internal auditors.

The result of developing PT XYZ's internal audit checklist was a checklist that has been completed with a score of 1-5. The total of all questions was 53. The comparison of the development results checklist with the existing checklist in PT XYZ was in accordance with Table 3.

Table 3. The comparison of the checklist development with the existing checklist from PT XYZ

Item	The existing checklist	The development checklist
Language	English and Indonesian	Indonesian
Question	The original text and translation of the original text of the requirements in ISO / IEC 17021-1: 2015	Translation of the original text of the requirements in ISO / IEC 17021-1: 2015
Question number	All clauses were according to ISO / IEC 17021-1: 2015	Clauses and sub-clauses were according to ISO / IEC 17021-1: 2015, questions from the sub-clauses was using numbers, for example, question in clause 9.1.1 was numbering with 1 / No 1, clause 9.1.2 was numbering with 2/ No 2
Checklist section	The checklist consisted of the clause number, the contents of ISO / IEC 17021-1: 2015, the reference in the quality document, the text in the quality document, and the results of the auditor's evaluation	The checklist consisted of sub-clause number/question number, the contents of ISO / IEC 17021-1: 2015 process requirements, auditor's assessment, description of non-conformity, and recommendation
Assessment	Qualitative with "appropriate" and "inappropriate" answers and a description of non-conformity if there was a discrepancy. Observation were written in the non-conformity column	Quantitative used a score of 1-5 and a description of nonconformity if there was a non-conformity. Observation were written in a separate column

3.2 Diversity of auditors' assessments of compliance with ISO / IEC 17021-1: 2015 process requirements at PT XYZ

3.2.1 Categorization and differences in auditor scores

According to the methods in [2], the three groups of score categories (X) were: low score category if $X < 46.8\%$, medium score category if $46.8\% \leq X < 73.2\%$ and good score category if $X \geq 73.2\%$. Based on these categories, the total score of all auditors included in the good category (Table 4). The auditors' score test using the one-way ANOVA also gave results that were not significantly different at the 0.05 significance level (Table 5). The difference in the scores of internal auditors compared to external auditors was relatively not significant, which was 1.51% (b), 1.88% (c), and 2.26% (d, e) so that it did not provide a statistically significant difference (Table 4). The results of categorization of scores and differences in the auditors' score indicated the consistency of the auditor's assessment using a checklist form with a scoring system as a tool for gap analysis at PT XYZ.

Table 4. Categorization of the auditors score

Auditor	Total score	Total score (%)	Category
a	259	97.74	good
b	263	99.25	good
c	264	99.62	good
d	265	100.00	good
e	265	100.00	good
f[1]	259	97.74	good

[1] external auditor (comparator)

Table 5. One-way ANOVA test result

Source of variation	*SS*	*df*	*MS*	F	F crit
Between groups	4.38889	5	0.87778	0.00386	2.40851
Within groups	10915.11111	48	227.39815		
Total	10919.50000	53			

3.2.2 The diversity of the auditors' assessment

There was the diversity in the auditors' assessment from the gap analysis results at PT XYZ in 4 sub-clauses, which were 9.1, 9.2, 9.5, and 9.9, while in the other 5 sub-clauses were obtained the same results, which were 9.3, 9.4, 9.6, 9.7 and 9.8 (Table 6). Based on Table 6, the total score of auditor a was 259, auditor b was 263, auditor c was 264, auditor d and e were 265, the average score of PT XYZ's internal auditors was 263.2 and the total score of external auditors was 259. The assessment came from the score given by the auditor on each sub-clause of the process requirements. The diversity of scores in the 4 sub-clauses was supported by the difference between the average of internal auditors' score and external auditor's score in that sub-clause which were: in sub-clause 9.1 was 1.6, in sub-clause 9.2 was -0.2, in sub-clause 9.5 was 1.6, and in sub-clause 9.9 was 1.2. When compared with external auditors, there was 1 auditor with the same total score, it was 259, while the other 4 auditors were different. Four auditors produced higher total scores of 263, 264, and 265. This indicated that fewer gaps were generated and/ or did not get gaps on requirements. The results also showed a low standard deviation value both on the score for each sub-clause and in the total score (Table 6). The standard deviation value in sub-clause 9.1 was 1.0, in sub-clause 9.2 was 0.4, in sub-clause 9.5 was 1.0, and in sub-clause 9.9 was 0.9 while the total score was 2.8. These scores indicated a slight variation in scores and total scores among auditors. The deviation standard was a number that indicated the degree of variance of a data set. The greater of the deviation standard, the numbers in the quantitative data were more varied or heterogeneous [15]. The smaller of the deviation standard, the data was better [16]. The total score of the auditor was the value of PT XYZ's compliance with the ISO / IEC 17021-1: 2015 process requirements so that the accurate total score could affect the increase in PT XYZ s policies and implementation level.

Table 6. Score of the results auditors gap analysis

Sub-clauses of process require-ments	Internal auditor score					The Average score a-e	f[1] score	The difference between the average score a-e and f	Deviation standard
	a	b	c	d	e				
9.1	48.0	50.0	50.0	50.0	50.0	49.6	48.0	1.6	1.0
9.2	39.0	40.0	40.0	40.0	40.0	39.8	40.0	-0.2	0.4
9.3	20.0	20.0	20.0	20.0	20.0	20.0	20.0	0.0	0.0
9.4	25.0	25.0	25.0	25.0	25.0	25.0	25.0	0.0	0.0
9.5	23.0	25.0	25.0	25.0	25.0	24.6	23.0	1.6	1.0
9.6	55.0	55.0	55.0	55.0	55.0	55.0	55.0	0.0	0.0
9.7	15.0	15.0	15.0	15.0	15.0	15.0	15.0	0.0	0.0
9.8	15.0	15.0	15.0	15.0	15.0	15.0	15.0	0.0	0.0
9.9	19.0	18.0	19.0	20.0	20.0	19.2	18.0	1.2	0.9
Total score	259.0	263.0	264.0	265.0	265.0	263.2	259.0	4.2	2.8

[1] external auditor (comparator)

The diversity of the auditor's scores on the 4 sub-clauses of the process requirements was derived from the scores given by the auditors on each question in the gap analysis checklist. The diversity in scores was in sub-clause 9.1 question no.7, sub-clause 9.2 question no. 12, sub-clause 9.5 question no. 32, and sub-clause 9.9 questions no. 51 and 53. The difference between the average of internal auditors' score and external auditor's score was: in sub-clause 9.1 question no. 7 was 1.6, in sub-clause 9.2 question no. 12 was -0.2, in sub-clause 9.5 question no. 32 was 1.6, in sub-clause 9.9 question no. 51 was 0.6 and in sub-clause 9.9 question no. 53 was 0.8. The question in checklists no. 7, 12, 32, and 53 consisted of one question, while question no. 51 was divided into several points. In this study, the score was given by the auditor on each question and was not given on each point in the question. This method of assessment could lead to bias in the auditors' assessment so that there was a diversity in the question score no. 51. The questions in checklist no. 7, 12, 32, 51, and 53 contained process requirements according to ISO / IEC 17021-1: 2015, and there were no modifications from the researcher. The diversity of scores on these questions indicated that there were differences in the auditors' understanding of the process requirements' implementation at PT XYZ. It was supported by the explanation that the accuracy of the score given by the auditor depended on the selection of questions or critical matters examined at the time of the audit [4].

The used of a checklist form with a scoring system had both advantages and disadvantages. The advantages of a checklist with a scoring system were: 1) the audit result reported the score of the compliance with the requirements and description of non-conformity, 2) data analysis was easier and could be processed statistically 3) could reduce the subjectivity of auditors because there were assessment standards and 4) it was easy to use. The disadvantages of a checklist with a scoring system were: 1) accuracy of the auditors' score depended on the auditors' good understanding of the requirements and 2) there was not a scoring system that could use as a reference for management system audits.

3.3 The source of the diversity in the PT XYZ auditors' assessment and recommendation of corrective actions

The sources that might be the cause of the dversity in the FSMS CB auditors' assessment were personnel, methods, tools, materials, environment, maintenance, and management. Personnel, in this case, the auditor could become a source of diversity if the competencies were not appropriate and were not free from bias. The method, in this case, the procedure of the FSMS CB could become a cause of the diversity in the auditor's assessment if it was not consistent in its application. In addition, a tool in the form of an audit checklist could become a cause of the diversity if it was difficult to understand and was used as a tool for assessing compliance with requirements. Material, in this case, documents and records of certification clients, could be the cause of diversity in the auditors' assessment if there were inconsistencies in the documents and records from the auditee. The environment, in this case, the location or place where the audit was conducted could become a cause of diversity in the auditors' assessment if the auditee provided different places between auditors for audit activities. The maintenance could become a cause of diversity in the auditors' assessment if the maintenance of the tools used in the audit process was not carried out properly. Meanwhile, management could become a cause of diversity in the auditors' assessment if there was no commitment from management for the internal audit activity. In this study, the sources of diversity in the auditor's assessment at PT XYZ were personnel and tools (Figure 1).

The first source was personnel, in this case, the internal auditors of PT XYZ. Based on the study, the gap of PT XYZ's process requirements implementation was low so it required good competence from PT XYZ's internal auditors to set scores and non-conformity in the implementation of requirements. Inconsistencies in the auditors' understanding of the process requirements implementation in the question were showed in data of the diversity in total auditor score (Table 4), diversity in auditors' assessment at 4 sub-clauses of process requirements (9.1, 9.2, 9.5, and 9.9) (Table 6), and the difference between the average of internal auditors' score and external auditor's score in the question no. 7, 12, 32, 51 and 53. The biggest differences between the average of internal auditors' score and external auditor's score were found in questions no. 7, 32, and 53 because there was the greatest diversity in the auditor assessments. Competency was a requirement that must be met by FSMS CB personnel according to ISO / IEC 17021-1: 2015 requirement 7 (resources requirements). Auditor competence and experience were important things to consider in selecting auditors [17]. Auditors who meet the specified competencies could perform outstanding standard evaluations [18]. According to the verification results of the auditors' track record in step 1, it showed that all auditors have met the competence of the PT XYZ's of ISO / IEC 17021-1: 2015 internal auditor. Based on the verification results, it was found that all auditors had different educational backgrounds, periods of working experience, and periods of auditing experience. Based on PT XYZ's ISO / IEC 17021-1: 2015 internal auditor competency policy, the specified educational background was D3 or S1 all majors, and a minimum of 1-year work experience in the food field industry or the relevant one. Some auditors had not audit experience on the process requirements meanwhile in this study, all of the internal auditors of PT XYZ were occupied. The difference in educational background and the periods of experience of the auditors could become the cause of the contrary in the auditors' assessment. It was supported by the results of the previous study which stated that one of the problems in audit activities was the limited number of qualified personnel according to the scope of the audit carried out [19]. Auditor competence and experience could affect audit results [17], [18]. The periods of the auditor's experience, influence the quality of audit result was better [20].

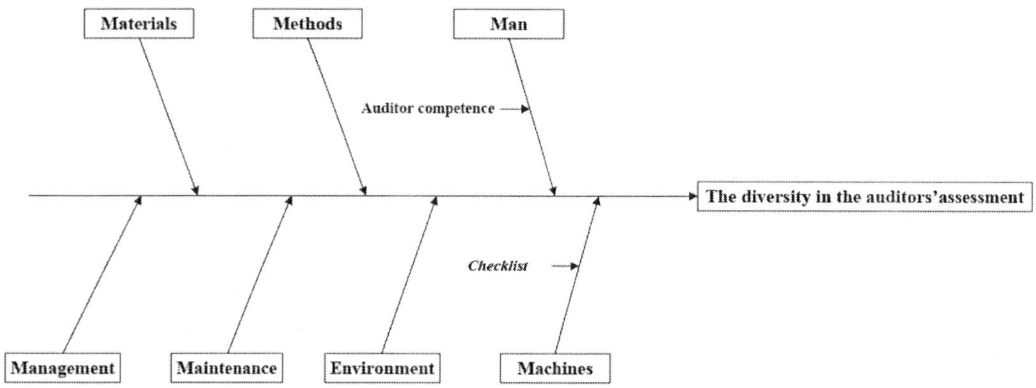

Figure 1. Cause effect diagram of the diversity in the auditors' assessment

The second sources was the method used in the implementation of the gap analysis at PT XYZ. The method, in this case was the development checklist from step 1. A quantitative checklist with a scoring system could help to reduce subjectivity in the score and non-conformity of auditors. According to the results of the study, there was variation in the scores for questions that consist of many points. The score variation was found in the score of question no. 51 with the difference in the average score of internal auditors and external auditors was 0.6. The questions of a similar score (many points) on the developed checklist were found at no. 1, 13, 15, 16, 24, 25, 35, 40, 48 but there was no diversity in the auditor's assessment. Although the variability only occurred in 1 of 10 similar questions and the difference in the average score of internal auditors and external auditors was relatively small, this data could indicate bias in the auditor's assessment because all points had the same score. The variety of the auditors' scores on question no. 51 related to the diversity of non-conformity reported by the auditor in sub-clause 9.9. The variety scores were in points a, b, and j of question no. 51. The variety in the auditors' score was 3 auditors gave a score of 4 while the other 2 auditors gave a score of 5. The scoring system used for questions with many points could provide opportunities for auditor subjectivity so that auditors might give the same score on different implementations at the point of the requirement.

Based on the source of the diversity in the auditors' assessment, the recommendation for corrective actions that should be taken by PT XYZ was to increase the competence of internal auditors and improve the internal audit system (Table 7). This corrective action aimed to close the causes for the diversity in the auditors' assessment and avoid to reoccur in other audit activities at PT XYZ.

Table 7. Recommendations for corrective actions on the diversity of the auditor's assessment in the gap analysis process at PT XYZ

Cause category	Main cause	Recommendation of corrective action
Personnel	Auditor competence	Providing additional audit experience field on the process requirements or other requirements to internal auditors
Tools	Scoring system on the multiple point questions in checklist	Assessment on each point of the multiple-point questions in checklist

4. Conclusion

The auditors' total scores were in the same category, it was in the good category ($X \geq 73.2\%$), and were not significantly different at the 0.05 level. The diversity in the auditors' assessment was found in 4 sub-clauses (9.1, 9.2, 9.5, and 9.9). The sources of diversity in the auditors' assessment were personnel and tool. Personnel was related to the auditors' competence that came from difference educational background and periods of auditing experience as an auditor. The tool was related to multi-point questions in the checklist that cause bias in the auditors' assessment. PT XYZ could reduce the diversity of auditor assessments by increasing the competence of auditors and improving the internal audit system.

References

[1] Suaedi 2013 *Quantitative Methods for Policy Analysis* Bogor (ID): IPB Pr
[2] Darwis D 2019 Comparison of scoring system methods and profile matching to measure employee performance at PT Wahana Raharja *Journal of Computing* **7**(2): 43-54 Jakarta (ID)
[3] Setiawan R 2011 The decision system to determine the suitability of the police in holding firearms uses the scoring system *National Seminar on Applied Information and Communication Technology* [time and place of meeting unknown] Malang (ID): Universitas Ma Chung
[4] Arum 2013 Internal audit of the quality management system uses a "scoring system" *Business Lounge Journal* [internet] [downloaded on Apr 20th, 2020] Available at http://www.blj.co.id/2013/08/05/audit-internal-sistem-manajemen-mutu-menggunakan-scoring-system
[5] Rachmawati R, Supartono W dan Suyantohadi A 2017 The evaluation to the fulfillment of ISO 22000 on frozen fish fillet product (case study PT XYZ) Inside: Isnansetyo A dan Nuringtyas TR, editor *Proceeding of the 1st International Conference on Tropical Agriculture* [time and place of meeting unknown] Yogyakarta (ID): Springer International Publishing hlm 587-598 doi: 10.1007/978-3-319-60363-6_60
[6] Ahmudi, Purwanggono B, Handayani NU 2018 Effectiveness analysis of ISO 9001:2015 implementation at manufacturing industry *SHS Web of Conferences* **49**, 01008 (2018)
[7] Crismanto Y, Noya S2018 Gap analysis on the implementation of the quality management system ISO 9001: 2015 at CV Tirta Mangkok Merah *Journal of Information and Industry Technology* **1**(2): 73-81 Malang (ID): Universitas Ma Chung
[8] Handayani NU, Mustikasari A, Hutagalung S, Monica B 2018 Analysis of implementation of ISO 9001:2015 in SME Aldin Natural Home *SHS Web of Conference* **49**, 01007 (2018)

[9] Putri ZT, Fahma F, Sutopo W, Zakaria R 2019 A framework to measure readiness level of laboratory for implementing ISO/IEC 17021: a case study *IOP Conference Series: Materials Science and Engineering* **495**(2019)012011 doi: 10.1088/1757-899x/495/1/012011

[10] ISO/IEC 17021-1:2015 Conformity Assessment-Requirements for Bodies Providing Audit and Certification of Management Systems

[11] Hidayat A Purposive sampling [internet] [downloaded on Apr 26th, 2020] Available at http://www.statistikian.com/2017/06/penjelasan-teknik-purposive-sampling.html

[12] Etikan I, Musa SA, Alkassim RS 2016 Comparison of convenience sampling and purposive sampling *American Journal of Theoritical and Applied Statistics* **5**(1): 1-4

[13] Quirk TJ 2016 Excel 2016 for business statistics, excel for statistics Springer International Publishing doi:10.1007/978-3-319-38959-2_8

[14] Liliana L 2016 A new model of Ishikawa diagram for quality assessment *IOP Conference Series: Material Science and Engineering* **161** 012099

[15] Ananda R, Fadhli, M 2018 Educational Statistics (Theory and Practice in Education) Saleh S, editor Medan (ID): Widya Puspita

[16] Aden, Waryanto H, Setiawan TH, Ilmadi 2019 Quality Control Statistics Tangerang (ID): UNPAM Pr

[17] Dominigues JP, Reis AM, Fonseca LM, Avila P dan Putnik G 2019 The added value of the ISO 9001:2015 International Standard from an auditor's perspective a CB-SEM based evaluation *International Journal for Quality Research* **13**(4): 967-986

[18] Akhmedjanov KB 2016 Internal audit and its development prospect. *International Journal of Economics, Commerce and Management* **4**(8): 555-560

[19] Rivera DE, Villar AS dan Marrero JS 2017 Auditor selection and audit team formation in integrated audits *Quality –Access to Success Jurnal* **18**(157): 65-71

[20] Zahmatkesh S, Rezazadeh J 2017 The effect of auditor features on audit quality Tekhne-*Review of Applied Management Studies* http://dx.doi.org/10.1016/j.tekhne. 2017.09.003

Indigenous maize seed storage using buffalo dung ash practiced by smallholder farmers in Kisar Island

H Jesajas[1,*], E Kembauw[2], M J Matatula[1], A D Tagueha[1], I J Liur[1]

[1] Department of Animal Husbandry, Faculty of Agriculture. Campus Poka, Jl. M. Putuhena. Ambon. Indonesia.

[2] Department of Agriculture Social-Economics, Faculty of Agriculture. Campus Poka, Jl. M. Putuhena. Ambon. Indonesia.

* E-mail: hery10jesajas@gmail.com

Abstract. Kisar island is a small island located on the tip of Timor island. Corn is the main agricultural crop, and livestock husbandry also plays an important role in the livelihood of the farmers as a source of income, provide mainly manure for maintaining soil fertility. and in the local culture as a `bride price' and for the exchange of goods. Considering periodical seasonal hazards, during eight months dry season people suffer from a serious shortage of water. Failure of the harvest is not an unusual phenomenon on the small, dry, eroded coral island. Seasonal pest invasion on maize cultivation threaten their seed security and food security. The objective of this review is to describe the validity of farmers' indigenous knowledge (IK) of using buffalo dung as a preservative agent in maize storage and its relevancy in the contemporary living of farmers in the island. A phenomenological study was held in June to August 2018. The results showed that applying buffalo dung ash makes the maize dry faster, cleans and kills maize weevil and preserves the grain, It was concluded that the knowledge and practice is scientifically valid and benefited community in the past but has less relevancy to the contemporary living of the farmers in the island.

1. Introduction

The small island of Kisar is located in the most south-western part of the Province of Maluku, Indonesia just 15 miles north of the eastern tip of Timor island. Upon arriving at Kisar via a sea going vessel, the island looks like a fort. The outer perimeter is a coral rock cliff with a hill separating the inner island (where everyone lives) and this outer cliff. Moving around the island, the occasional corridor cuts through the outer hill to the interior. The only town, Wonreli, is centrally located and facilitates government officials.

Kisaresses depend mainly on activities that utilize their natural resources to generate income for their living. Approximately 80 percent of population of Kisar were farmers. Corn is the main agricultural crop, and is the staple diet supplemented by beans, green vegetables and root vegetable (cassava) for Kisaresse. Data from the [1] shows that the total area used for cultivation for food crops was approximately 3,510 ha, and it accounted for 30 % of the total area of Kisar, while the area for cash crops was approximately 720 ha or 6 % of the total area. Corn is traditionally planted in the homestead twice a year, just before the westerly and easterly monsoons. Other crops such as coconut and oranges are also planted for cash.

Furthermore, livestock husbandry also plays an important role in the livelihood of the Kisaresse. Besides being a source of income, livestock provide mainly manure for maintaining soil fertility. Livestock also have a ritual meaning in the culture of the Kisaresse as a `bride price' and for the exchange of goods. Domestic pigs are also kept in the village, whereas goats are kept in fields outside the village. Goats dominate the population of livestock on Kisar Island. and it accounted for around 52 % of the total livestock in Kisar. Besides goats and sheep, pigs, buffaloes and chickens are also raised by farmers, accounting for the remaining 48 % of livestock.

Considering its geographical aspect and periodical seasonal hazards, during eight months dry season people suffer from a serious shortage of water. Failure of the harvest is not an unusual phenomenon on the small, dry, eroded coral island. In the past, hunger was an almost annual occurrence. It is exacerbated by other conditions such as seasonal pest invasion on maize cultivation, which potentially affect their food security. Seed security is key to the attainment of household food security among the farmers in Kisar island. In the tough conditions arising from shocks caused by drought, pest, and other vulnerable situations which attack their seed security, the farmers as conscious beings have the capability to overcome the situation. Through their praxis capability, expressed as critical reflection on the crises and assets they have, farmers build their indigenous knowledge of seed storage as a part of adaptation mechanisms in the island. The objective of this review is to describe the validity of farmers' indigenous knowledge (IK) of using buffalo dung as a preservative agent in maize storage and its relevancy in the contemporary living of farmers in Kisar island.

2. Methods

The notion of the research was a phenomenological study by which the farmer's IK of using buffalo dung ash in maize storage as a focus phenomenon to be explored. Here, every knowledge holder was considered possess unique experience and knowledge to be explored The visit into the island was held in June to August 2018. The research that conducted the field activities was made up of nine people, five university academics, two third year undergraduate students and two farmers. The academics had a background in agronomy, agriculture economics, rural sociology and animal husbandry, whereas the farmers had an intimate knowledge of their situations. In addition, a number of individuals contributed their knowledge through some casual discussions. The researchers worked with individuals and small group of farmers, locally named *roson,* as a `pilot' group to avoid raising local expectations about what the research might yield. Accidental sampling was applied which rely heavily on snowball technique to find the knowledge holders. In an effort to explore the reality, the 'Apa itu ?' principle, that is asking `what is it ?' question, was an important technique. This question proved effective in encouraging the farmers to express their experiences and knowledge about particular practices, and as an entry question for the `how' and the `why' questions. The qualitative data collected were analysed by manual content analysis and descriptive statistics, such as frequencies. Content analysis is a systematic, replicable technique for compressing text into content categories based on explicit rules of coding and identifying themes or patterns.

3. Results and discussion

3.1. Maize Seed Storage

There were four ways of maize storage practiced by farmers in Kisar island, namely using "bising" (a traditional storage container made of dried sugar palm leaf), using aluminum drum, hanging over fireplace and store in the upper-room, The most traditional storage treatment was applying buffalo dung ash.

Indigenous practice of using buffalo dung ash as preservative agent for indigenous corn storage, locally called "bising" practice. Dried buffaloes' dung collected and burned to produce dusts. The dusts then applied by using a `sandwich' method by which the dust was placed a layer by a layer. The theory behind using the dust produced by buffaloes' dung rather than others, as the farmers believe, because it was softer than others, and easier to penetrate into the spaces between the grains but also more `strong' to kill the pest. Using buffalo dung ash is considered as the oldest fashion of maize seed storage has been practiced by the farmers in Kisar island for centuries. The detail process as showed in the Figure 1 below.

1. Collect buffalo dung from common pasture area
2. Combust buffalo dung
3. Collect ash
4. Centrifuge (riddle) ash
5. Spread ash to the surface of maize layer by layer while pound gently to compact it
6. Store the maize seed in warm and dry place (barn or storage room)

Figure 1. Steps in maize seed storage treatment using buffalo dung ash

According to the farmers, applying buffalo dung ash has several advantages. Buffalo dung ash makes the maize dry faster, cleans and kills the bug [maize weevil] and preserves the grain, Furthermore, it has no toxic because it has burnt completely, so the maize is safe to consume, no effect on the odor, no effect on the color of maize (compared to use goat dung ash, maize color changed reddish). A key respondent farmer explained in an informal conversation that: "the key of success in applying buffalo dung ash as a preservative ingredient is that we have to spread the ash in balance to the volume of the maize, approximately a half of maize volume would be better, so it has enough capacity to adsorb the water of maize and make it dry faster and our maize seed will safe until next season...beat it properly so maize and ash become tightly cemented...then at the end, the bising [container] should be tightly closed and placed in warm room to avoid water or rain...if we placed into wrong room it would be useless because when it wet the mold will grow soon and destroy the maize then we have no seed to plant next season, meaning ready to starving next season or buy rice if we have money"

The farmer commented further that in practice preserving maize seed by using buffalo dung ash treatment produced higher germination capacity than by just hanging the maize seed over the fire-stove because the heat and smoke from fire-stove is detrimental to the seed

Scientific findings have been proved the validity of farmers' IK of applying buffalo dung ash to maize seed preservation. However, there are some debates among the scientists on the effects of using buffalo or cow dung in seed preservation, particularly its effects on moisture content, insect damage and germination capacity. In the practice of maize storage treatment using buffalo dung ash, ash is used both as an inert filler and for its other negative effects on insects. As an inert filler, ash works by filling up the space around the seed and impeding the movement of insects as well as, in sealed containers, reducing the volume of air available to the insects for respiration [2]. Animal (buffalo) dung ash contain silica which behave as adsorbent to absorb water content [3,4] and become unavailable to microbes to live and damage the cuticle of insects causing them to dehydrate and also has detrimental effect on egg development [5-7]. In addition, it prevents the seed from insect attack and maintain seed rigour [8].

Futher [2] found that the effectiveness of ash to damage insect and viability of the seed possibly influenced by the kind and amount of ash used and method and time of application. He suggested that the thickness of ash/grain layer 3 cm may maintain viability of seed for long period of time. Meanwile, [9] and [5] suggested that the quantity of ash used should be around 25-50% of maize volume in order to effectively damaging insect.

3.2. The Shrinkage of Farmer IK and Practice

However, utilization of buffalo dung ash has been shrunk. The practice was famously applied when buffaloes were raised by majority farmer households in the island. One of the key respondents responded that it lost its famous when huge degradation of buffaloes happened in the island because of the value of buffalo in Kisar culture as bride price has been replaced mostly by pigs and other material things. This situation reflected that the farmers' knowledge is not static but dynamic. New meaning of farming practices continually emerges during farmers' interaction with other farmers and with their environment. It may result in two-folds: expands or the shrinkages of IK and practice.

Currently, there are only few farmers who are still practicing using buffalo dung to preserve maize seed. The practice of using buffaloes dung for grain preservation has been changed by using drum. Since buffaloes have been almost extinct, farmers were no longer relied on buffaloes dung anymore for grains preservation. The farmers are using drums combined with sun drying to store their grains including corn, beans and rice. It was seen more effective in terms of time required for preparations. However, the disadvantage of using drums was due to the obligations for sun drying the grains as dry as possible in order to ensure the longevity of the storage grains. In other word, as long as it was evaluated satisfied their need and relevant enough for storing grains including maize and beans, the technology in term of technique, knowledge, and product management, will be continuously framed into daily practice of the farmers. In this case, the inherited knowledge may shrink or expand depending on the need and relevance of the knowledge in its realm of everyday life of a given generation [10]. "shrinkage ensues when one sort of knowledge becomes superfluous to the everyday requirements of a successor generation. Expansion may be due to either of two factors. On the one hand, new social conjunctures, new social demands generate new social experience takes root in the form of everyday knowledge. This new intake is then transmitted to the successor generation as part of the ancestral estate. In the second place, knowledge may infiltrate or may be consciously induced into the realm of the everyday from the sphere of species-essential objectivation *for itself....* which arise from the intention of satisfying needs felt by members of a given society " [10].

4. Conclusion

Farmer indigenous knowledge of using buffalo dung ash as inert filling in maize seed storage processing has been benefited the community in Kisar island for century in terms attaining seed and food security. Such practice has been proved scientifically valid to reduce moisture content, reduce insect damage and maintain seed viability. However, it has less relevance to the current living situation in the island and has been replaced by the combination of using drum and sun dried which considered easy to apply now and satisfied the need of the farmers to secure their maize seed for seasons of planting.

5. References

[1] District Statistical Bureau 2019 Maluku Barat Daya Regency in Figure

[2] Wambugu P, Mathenge P W, Auma E O and van Rheenen H A 2009 Efficacy Of Traditional Maize (*Zea mays* L.) Seed Storage Methods In Western Kenya *African Journal of Food Agriculture, Nutrition and Development* **9** (4)

[3] Garba J, Samsuri W A, Othman R and Hamdani M S A 2019 Evaluation of Adsorptive Characteristics of Cow Dung and Rice Husk Ash for Removal of Aqueous Glyphosate and Aminomethelphoshonic Acid. Scientific Reports. *Nature Research* **9** 17698.

[4] Szymadja, Laska A G and Majewski M 2020 Characteristics of Ashes from the Combustion of Cow Dung Biomass. Proceeding MDPI, 51,14.

[5] Almekinders C J M and Louwaars N P 1999 Farmers' seed production. New approaches and practices. IT publications, London.

[6] Sivekumar G and Amutha K 2012 Studies on Silica Obtained from Cow Dung Ash *Advaced Material Research* **584**(1) 470-473.

[7] Manadhar, Malindi A P and Shah A 2018 An Overview of the Post-Harvest Grain Storage Practice of Smallholder Farmers in Developing Countries.

[8] Lougraimzi H, Bouaiichi A, Sou S M, Kotba I and Fadli M 2019 Potential Effect of Natural Products in the Management of Seed Infestation of Cereals During Storage *Journal of Microbiology, Biotechnology and Food Science* **8** (5) 1151-1164.

[9] Wright M, Donaldson T, Cromwell E and New J 1994 The retention and care of seeds by small-scale farmers *NRI Report* R2103.
[10] Heller A 1984 Everyday Life. *Translated from the Hungarian* by G.L. Campbell. Routledge and Kegal Paul Publication, London.

Sustainable development strategy for agroforestry

I S Ruhimat[1,*] and A Widiyanto[1]

[1] Agroforestry Technology Research and Development Institute
Jl. Raya Ciamis-Banjar Km 4 Ciamis 46201, Indonesia

* E-mail: idintea@yahoo.co.id

Abstract. This research aims to determine the status of sustainability, key factors that influence, and the strategy of developing sustainable agroforestry farming. The study was conducted from January to December 2016 in the Sodonghilir subdistrict, Tasikmalaya Regency, West Java Province. Data is analyzed using sustainability analysis, prospective analysis, and hierarchy process analysis. The results showed that some of the ecological, economic, and social problems would reduce agroforestry farming's sustainability. The sustainability of agroforestry farming in the Sodonghilir sub-district is less sustainable, both in general and in each sustainability dimension (ecological, social, and economic). Two conclusions were obtained from the study. First, six key factors determine sustainable agroforestry development's success. There is farmers' knowledge in soil and water conservation, farmer group performance, availability of agroforestry technology packages, farmers' knowledge in farming activities, level of pest and plant diseases attacks, farming capital sources vital. Second, increasing farmer capacity and strengthening farmer group institutions are priority strategies that the government implement, private, and community in conducting sustainable agroforestry farming development programs in the Sodonghilir subdistrict.

1. Introduction

Agroforestry is a form of multidirectional land use that combines woody, seasonal crops, and/or livestock activities on one stretch of land simultaneously which is resulting in economic and ecological interactions between tree crops and other agroforestry constituents [1,2,3]. Land use with an agroforestry system provides several financial, ecologically, and socially [4]. Economically, agroforestry systems can increase land productivity, increase farmers' income, develop local economies, and meet community needs for clothing, food, and shelter [3,5,6]. Ecologically, the systems can efficiently use natural resources, improve soil fertility, and improve land hydrology functions [2,7]. Socially, agroforestry can create jobs in rural areas, reduce population pressure on natural forests, and reduce the rate of migration of people from villages to cities [2].

The benefits obtained from this system have encouraged various regions in Indonesia to develop agroforestry farming systems to optimize land use [8]. One area that has implemented the agroforestry system is Sodonghilir District, Tasikmalaya Regency, West Java Province.

In general, the dominant and developing agroforestry system in Sodonghilir Subdistrict is involved in a mixed garden form. Complex agroforestry is an agroforestry system consisting of many elements of trees, seasonal crops, and/or grass in one area of land with an unlimited amount and time [3]. Therefore, their physical appearance and dynamics are similar to those of primary and secondary natural forest ecosystems.

The problems faced in implementing the agroforestry system in Sodonghilir Subdistrict, including (1) the high attack of pests on sengon plants as the main crop such as bagworm pests and gall rust disease, (2) fluctuations in the price of seasonal produce, (3) common knowledge of farmers concerning agroforestry, (4) limited farming capital, (5) weak farming institutional system, both government, private and community institutions, and (6) the conversion of land to seasonal crops and/or settlements [9]. These problems will have the potential to reduce the sustainability of agroforestry farming.

Agroforestry farming management using a sustainable agricultural system approach is expected to be one solution to overcome the various problems farmers face in developing the old farming system in Sodonghilir District. Sustainable agriculture is defined as the ability of an agricultural business that is economically viable, safe for the environment, and socially acceptable to the community [10]. In general, the concept of sustainable agriculture is based on a triangular framework of sustainable development, namely development oriented towards three dimensions of sustainability that are mutually supportive and interconnected. The three dimensions are economic, social, and ecological [11].

Based on the background of the problems that have been raised, this research aims to determine the status of sustainability, the key factors that influence, and strategies for developing sustainable agroforestry farming. The research results are expected to be used as a material consideration for policymakers in developing sustainable agroforestry farming policies.

2. Methods

2.1 Time and Location of Research

The research was conducted from January to December 2016 in Sodonghilir District, Tasikmalaya Regency, West Java Province. The selection of research locations was carried out purposively, considering that Sodonghilir District is one of the areas targeted for developing sustainable agroforestry farming in the Tasikmalaya Regency [9].

2.2 Data Collection and Analysis Techniques

The data collected in this study consisted of secondary data and primary data. Secondary data were obtained from the literature studies and documentation studies on various related data such as research results, data from the Central Statistics Agency, village monographs, sub-district monographs, extension programs, and associated agency programs.

Primary data were obtained from interviews, research questionnaires, and discussions. Purposive sampling was used to identify respondents and they were categorized as an involved men, master, and people with knowledge of agroforestry farming n Sodonghlir District. Thirty respondents were assigned from the Sodonghilir District Agricultural Extension Center, village government, community leaders, sub-district government, Tasikmalaya Regional Forestry Service Branch VI, and agricultural / forestry traders products, farmer groups, and farmer group associations.

There are three data analysis tools used in this research, namely sustainability analysis consisting of multidimensional scaling (MDS) and leverage analysis, prospective analysis, and analytical hierarchy process (AHP) analysis. The analysis stages are shown in Figure 2.

The sustainability analysis was carried out using the Rapid Appraisal for Fisheries approach developed by the University of British Columbia. This sustainability analysis consists of Multidimensional Scaling (MDS) analysis and leverage analysis. MDS analysis serves to determine the index and status of the sustainability of agroforestry farming. The study identifies various factors that influence agroforestry farming in social, economic, and ecological dimensions. There are three final values generated from the MDS analysis: the sustainability index value, the coefficient of determination (R^2), and the s-stress value (S). The value of the agroforestry farming sustainability index in the data analysis is grouped into four levels of sustainability status, namely 0-25.00 (unsustainable), 25.01-50.00 (less sustainable), 50.01-75.00 (moderately tolerable), and 75.01-100.00 (ongoing) [13]. The model obtained from MDS analysis is said to be good enough (goodness of fit) if it has a value of $S < 0.25$ and $R^2 > 80\%$ or close to 100% [12,14].

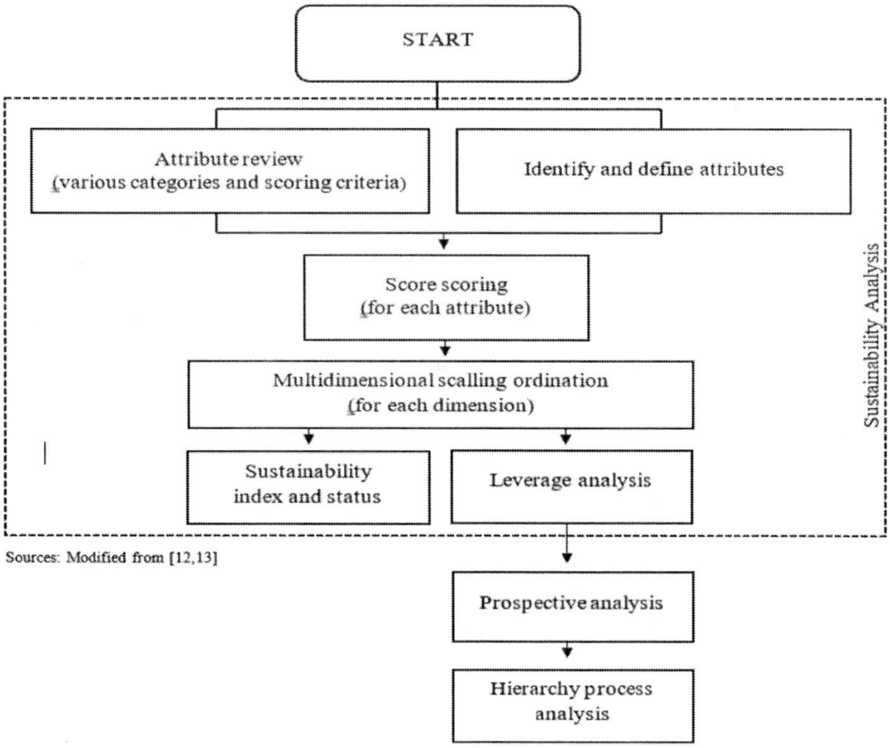

Figure 1. Data analysis stages

The sensitivity analysis (leverage analysis) is carried out to determine the research attributes that function as a leverage factor in agroforestry farming development. The Root Mean Square (RMS) change is the value obtained from the leverage analysis's final results. The more significant the difference in RMS value, the more sensitive this attribute's role is in increasing agroforestry farming's sustainability status [13]. The leverage factor is determined from an RMS value higher than half the sustainability dimension's highest weight [12].

These sensitivity analysis results are used to conduct a prospective analysis in determining the critical factors for agroforestry farming's sustainability by mapping the level of influence and the level of dependence between factors. The results of the prospective analysis are mapped in a diagram divided into four quadrants, namely [12]: a) the first quadrant contains the driving variables that have a strong influence with low inter-factor dependence, b) the second quadrant contains the leverage variables with the power strong and high inter-factor dependence, c) the third quadrant contains the dependent factor (output variables) which has a small effect with high inter-factor dependence, and d) the fourth quadrant contains marginal variables which have a little impact with support between factors low. Factors that enter into quadrants I and II are the critical factors in systems with strong influence characteristics with quiet/strong dependence on systems [12].

Analytical Hierarchy Process Analysis (AHP) analysis was used to determine priority strategies in agroforestry farming development. The priority strategy is determined based on the strategy's level of ability, described in priority weights in managing and meeting the critical factors generated in the prospective analysis. AHP analysis simplifies the various complex, strategic, dynamic, and multicriteria problems into a hierarchy consisting of three levels: goals, criteria, and alternative strategies [15,16,17]. The goal is a strategy to be achieved in conducting AHP analysis. The requirements are filled with all the key factors resulting from the leverage analysis and alternative designs are determined based on the results of discussions with various parties.

3. Results and discussion

3.1 Agroforestry Farming Sustainability

Based on the results of the sustainability analysis using MDS (Table 1 and Figure 4), it can be seen: 1) the level of agroforestry farming sustainability in Sodonghilir District is in a less sustainable status, both in general and each of the dimensions of sustainability (ecological, social, economic). Three critical factors cause the result. First, the index value on all dimensions of sustainability is in the range of 25.01-50.00. Second, the model generated from the MDS analysis has a value of S <0.25. It shows the effect of the error on assessing an attribute is minimal to be ignored. Lastly, the coefficient of determination (R^2) generated by each sustainability dimension has a value between 0.9320-0.9487. The resulting value shows that all the attributes (social, economic, and ecological) used in measuring agroforestry farming's sustainability status have explained 93% - 95% of the current system. The sustainability index for each sustainability dimension is presented in the kites diagram, as shown in Figure 4b.

Table 1. Results of analysis of the sustainability of agroforestry farming

Dimensions	Sustainability index	The coefficient of determination (R^2)	s-stress value
Multidimensional	42.80	0.9572	0.1305
Economic	40.11	0.9544	0.1343
Ecology	35.03	0.9453	0.1338
Social	43.25	0.9516	0.1308

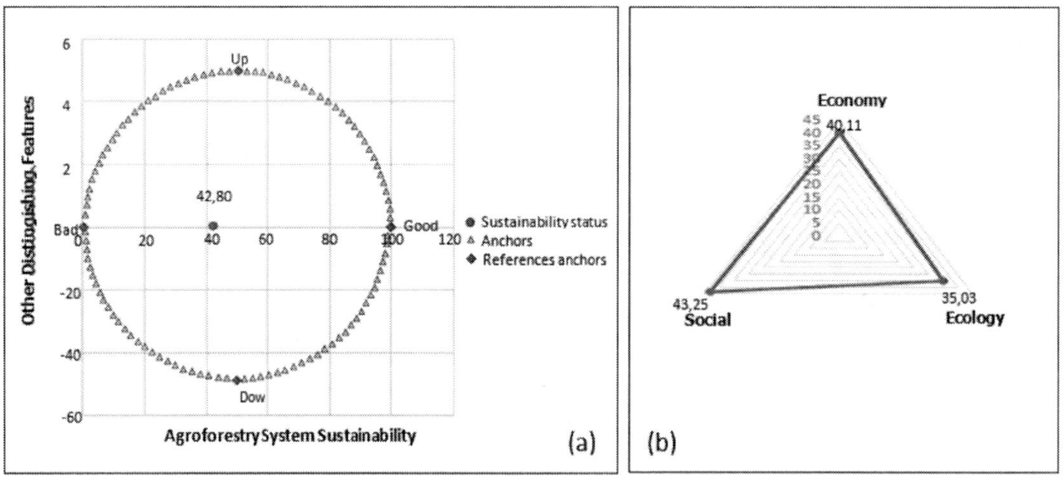

Figure 2. Sustainability index of agroforestry farming in multidimensional (a) and kite diagrams of sustainability index in each dimension (b).

The sustainability analysis for the ecological dimension considers 12 research attributes that affect agroforestry farming's sustainability, as presented in Figure 5. Based on the sustainability analysis results using MDS analysis, the ecological dimension has a sustainability index of 35.03, so that its status is categorized as less sustainable. The resulting S and R2 values are 0.1338 and 0.9453, which means that the MDS analysis model on the ecological dimension is good enough (goodness of fit). The model that uses these attributes can explain 94.53 percent of the real condition.

The leverage analysis results show that there are two attributes in the ecological dimension that function as leveraging factors for agroforestry's sustainability status, namely pest and plant disease attacks and organic material in cultivation activities agroforestry system. Management of these two leverage factors is expected to improve agroforestry farming's sustainability status in the future. The results of the leverage analysis for the ecological dimension are presented in Figure 5.

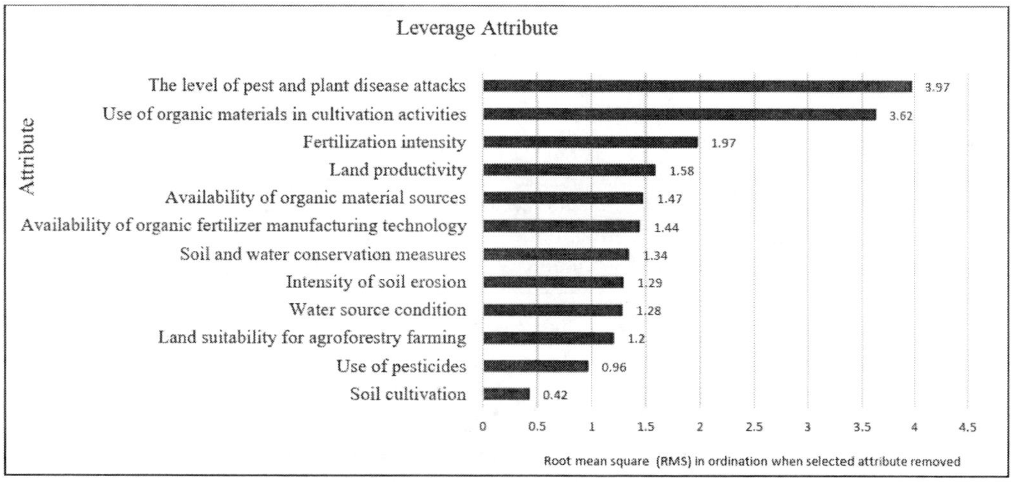

Figure 3. Leverage attributes affecting the sustainability of ecological dimension.

The ordination analysis results for the economic dimension involving 14 research attributes revealed a sustainability index value of 40.11 with an s-stress value of 0.1343 and a determination coefficient of 95.44. There were two pieces of information based on the research criteria. First, the sustainability index value was 25.01 - 50.00 and predicted it as a less sustainable category. Second, the s stress value and the coefficient of determination implied the resulting model has a good enough value (goodness of fit) to predict the level of sustainability in the current research condition.

Leverage analysis of the economic dimension attributes produces three leveraging factors for agroforestry farming's sustainability: farming's ability to generate tiered income for farmers, a farming capital source, and the contribution to farmers' total revenue. The leveraging attributes that affect the sustainability of the economic dimension are presented in Figure 6.

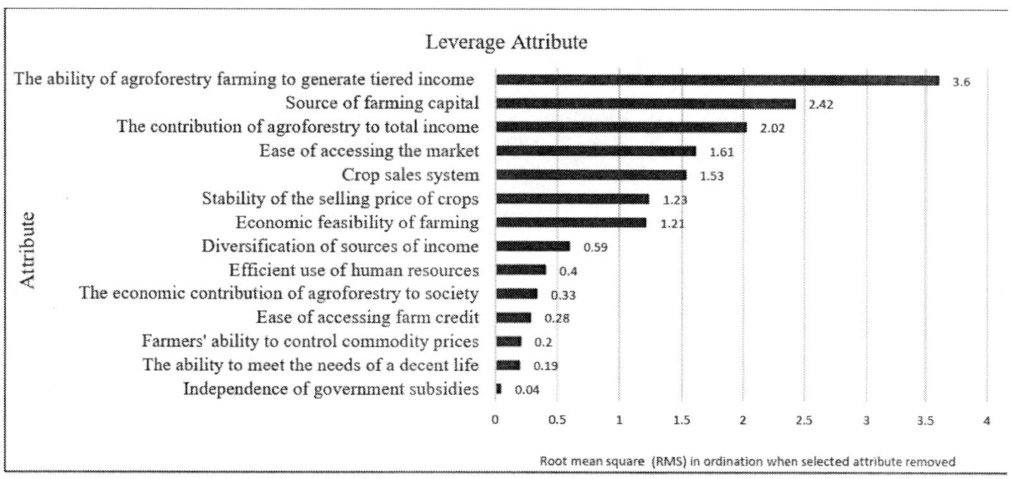

Figure 4. Leverage attributes affecting the sustainability of economic dimension.

Kite diagram in Figure 2b. shows that agroforestry farming in the social dimension, the result of the analysis of sustainability with MDS, is less sustainable because it has a sustainability index of 43.25 with an s-stress value of 0.1308 and a coefficient of determination of 95.16 percent. The s-stress value and the coefficient of determination show that the model with 18 attributes that make up the social dimensions analyzed can explain 95.16 percent of agroforestry farming's sustainability conditions in the research location.

The results of the leverage analysis on the social dimension (Figure 7) show that of the 18 attributes analyzed, seven features can be used as leveraging factors for the sustainability status of agroforestry farming, namely farmer group performance, availability of agroforestry technology packages, farmer's knowledge in agroforestry farming, farmer's knowledge in soil and water conservation, frequency of agroforestry extension, farmer's experience in controlling pests and plant diseases, and participation of farmers in farmers groups. The seven attributes were chosen as leverage factors because they have a Root Mean Square (RMS) value higher than half the maximum RMS value. Therefore, the management of these three leverage factors will affect the improvement or improvement of agroforestry farming's sustainability status.

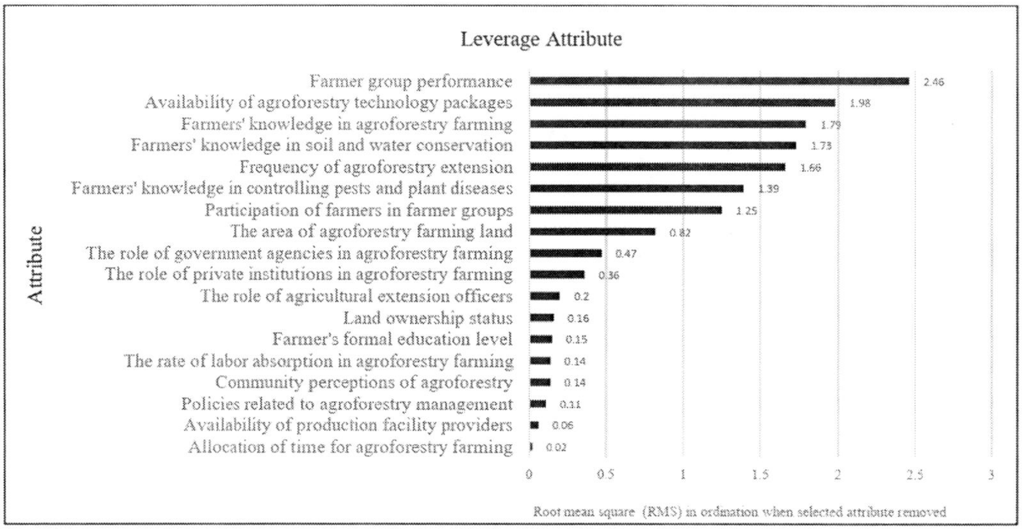

Figure 5. Leverage attributes affecting the sustainability of social dimension.

3.2 Key Factors in Agroforestry Farming Development

A prospective analysis revealed each leverage factor could be distributed in four quadrants: the determinants in quadrant I, connecting factors in quadrant II, dependent factors in quadrant III, and free factors in quadrant IV. Based on the results of a prospective analysis (Figure 8.), six key factors have a dominant influence on the development of agroforestry farming, namely: 1) five key factors that have a substantial effect between factors with low dependence, namely farmers' knowledge in soil and water conservation, farmer groups performance, availability of agroforestry technology packages, farmer's knowledge in agroforestry farming, level of pest and plant disease attacks and 2) one key factor that has a strong influence among factors with high dependence is the source of farming capital. The management of these six key factors will strongly influence agroforestry farming's sustainability in Sodonghilir District.

Farmers' knowledge in soil and water conservation is the first leverage factor that affects sustainable agroforestry development. Farmers need knowledge of soil and water conservation in conserving the carrying capacity, quality, function, and ability of soil, water and biodiversity resources in a harmonious and balanced manner [18]. Farmers who know soil and water conservation will always consider aspects of ecological sustainability and their farming's economic factors.

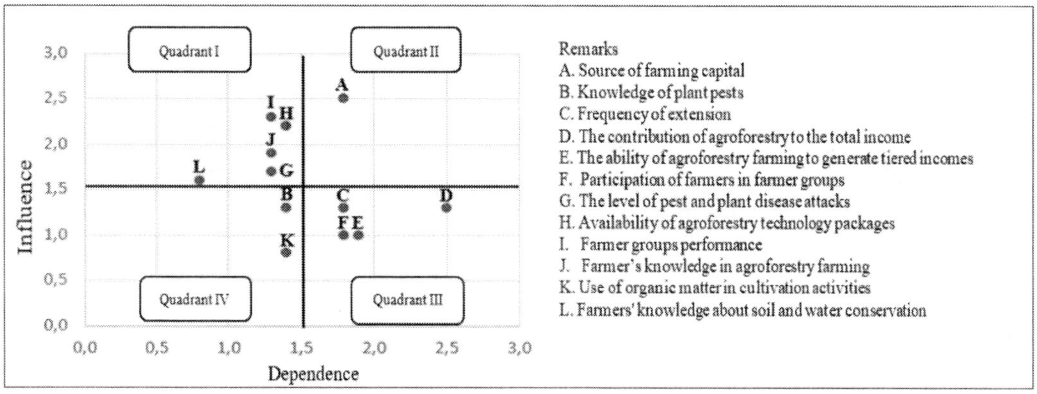

Figure 6. The critical factor that affects the sustainability of agroforestry farming

Based on the discussions with relevant stakeholders and field observations, it can be seen that farmers' knowledge about soil and water conservation is still lacking. One indication of the lack of knowledge about soil and water conservation is the presence of several sloping lands owned by farmers that have been converted into settlements and/or seasonal crops. Some land is still owned by farmers who have not been perfect in applying soil and water conservation principles, both vegetative, chemical, and mechanical conservation methods [9].

The performance of farmer groups is the second leveraging factor that affects the sustainability status of agroforestry farming. Farmer group performance is defined as the level of farmer group ability to plan, manage, and evaluate group activities [19]. Farmer groups that have high performance will be able to achieve the group goals that have been formulated jointly by all members of the farmer group.

In general, farmer groups in Indonesia have low performance and are passive, so that it becomes a significant problem in the success of farming development [20]. This problem also occurs in several farmer groups in the Sodonghilir District area.

Availability of technology packages and farmers' knowledge about agroforestry farming are vital factors affecting agroforestry farming's sustainability. Agroforestry technology packages are a collection of innovations produced by research and development institutions. Those collaborations aim to answer various stakeholders' various problems, especially farmers, in agroforestry management [13]. The Sodonghilir Subdistrict is still not available, so an agroforestry technology package is needed, which contains various technical, social, economic, and ecological innovations.

One of the other determining factors for agroforestry farming's sustainability is the level of pest and plant disease attacks. The attack of pests of agroforestry composting plant diseases, bagworms, and tumor rust on sengon plants is a serious problem that threatens agroforestry farming's sustainability in Sodonghilir District [9]. The inability of farmers to cope with these pests will threaten the sustainability of agroforestry farming. Due to damage caused by pests and diseases, which are already epidemic and have not been resolved, the losses are quite significant [21].

Farming capital is one of the factors of production that determines the success of agroforestry farming. The inability of farmers to provide farming capital will risk failure or low income earned by farmers to affect the sustainability of agroforestry farming [22]. Farmers can use two capital sources for the farming process, namely, external and internal farmers.

In general, farmers in Sodonghilir Subdistrict are religious people, so they must accept capital assistance from other parties, especially in interest loans (usury) either from banks or loan sharks. Therefore, it takes efforts from farmers and other stakeholders' support to help farmers capitalize from sources that are following their values. One of these efforts is to implement a business partnership program with a profit-sharing system.

3.3 Sustainable Agroforestry Farming Development Strategy

The agroforestry farming development strategy is carried out to manage the key factors resulting from the prospective analysis using analytic hierarchy process analysis (AHP). Based on the results of AHP analysis, it is obtained a sequence of strategies that can be carried out in the development of agroforestry farming in Sodonghilir District, namely (1) increasing the capacity of farmers with a priority weight of 0.461, (2) strengthening farmer group institutions with a priority weight of 0.331, (3) conducting farming partnerships with priority weights 0.134, and (4) providing farm capital incentives with a priority weight of 0.073. Therefore, increasing farmers' capacity and strengthening farmer group institutions has the first and second priority weights. It is used as the primary strategy for the development of sustainable agroforestry in Sodonghilir District. The results of the AHP analysis are presented in Figure 9.

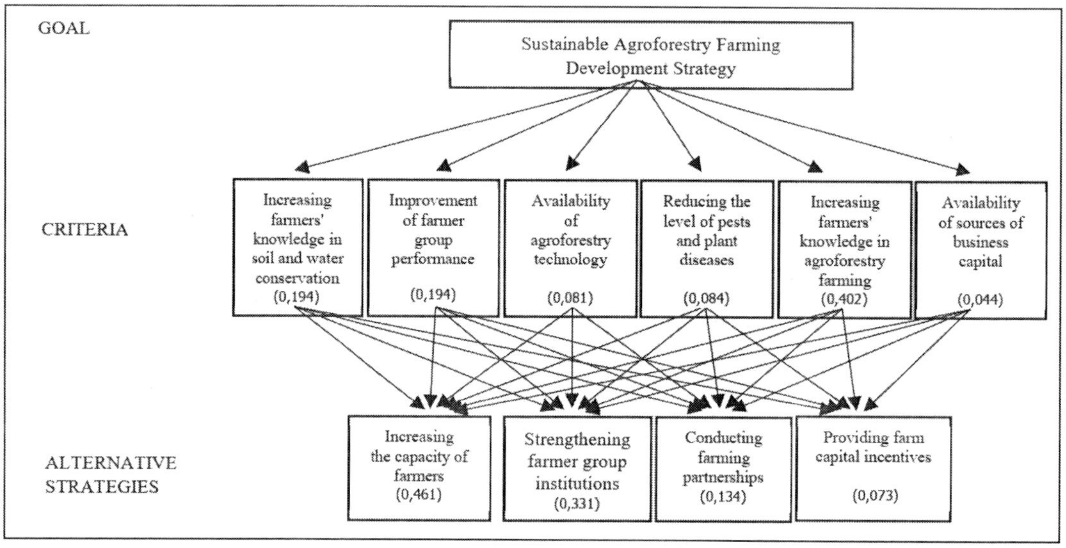

Figure 7. Hierarchy of sustainable agroforestry farming development strategies

Increasing the farmers' capacity in agroforestry farming management (technical, social, managerial capacity) is the priority strategy. There are three capacities that farmers need to improve in agroforestry development, namely technical ability, managerial capacity, and social capacity.

Technical capacity is the ability of farmers to recognize, know, understand, and implement various technical principles, knowledge, and plant cultivation skills [23,24]. Managerial capacity is farmers' ability to plan, manage, implement, and evaluate agroforestry farming. In the decision-making process at each stage of agroforestry farming, managerial skills are needed by farmers, starting from providing business capital, technical crop cultivation, packaging, and marketing of products to manage its resources [23,24]. Social capacity can be defined as farmers' ability to interact with various parties outside of themselves, such as fellow farmers, government, private parties, traders, and others. Farmers' social abilities function to foster cooperation with other parties in overcoming various problems faced by farmers in the agroforestry farming process [23,24].

Strategies for increasing farmers' capacity in agroforestry development can be carried out by optimizing education, training, and counseling programs for farmers in a systematic, planned, targeted, and sustainable manner. The optimization of the education and training program can be done in various ways, including [24]: (1) increasing the performance of extension through expanding the capacity of extension workers and the availability of innovation in training methods and materials, (2) optimizing coordination and intensive communication among various stakeholders related to the program training, such as between extension agents and government and private research and development agencies in updating agroforestry technology innovations and (3) providing policy support in the form of a legal umbrella, programs and budgets for the implementation of an effective, efficient and sustainable education and training program.

Strengthening farmer group institutions is the second priority strategy that can be carried out in agroforestry farming. Strong farmer group institutions are one of the determining factors for the success of agroforestry farming. Strengthening of these farmer groups can be done by optimizing the role of farmer groups [25].

There are three primary roles of farmer groups in agroforestry farming: farmer groups as a vehicle for learning, production units, and cooperation [25]. As a learning vehicle, the farmer group functions as a teaching-learning platform for all members to increase their farming knowledge, skills, attitudes, and independence. The farmer group as a production unit serves as a farming unit in achieving more profitable economies of scale, both in terms of quantity, quality, and continuity. Farmer groups as a vehicle for cooperation make farmer groups strengthen relationships between members of one group or different groups in dealing with all farming problems faced by technical, managerial, and social issues [20, 26, 27].

4. Conclusion

Agroforestry farming in Sodonghilir District is still less sustainable due to the inadequate public attention to social, economic, and ecological aspects. Six key factors determine sustainable agroforestry farming's success: farmers' knowledge in soil and water conservation, farmer group performance, availability of agroforestry technology packages, and farmer knowledge in agroforestry farming pest and plant diseases attack, and sources of farming capital. The management of these six key factors will have a strong influence on sustainable agroforestry farming's success. Increasing farmer capacity and strengthening farmer group institutions are priority strategies that can be implemented by all stakeholders (government, private, and community) in managing the six critical factors in agroforestry farming development in Sodonghilir District.

5. Acknowledgment

The author would like to thank those who have helped the smooth running of this research, especially members of farmer groups involved in this study, Head of the Agroforestry Technology Research Institute, the Center for Agricultural Extension and Forestry, Sodonghilir District, Darsono Priono, Dede Rahman, and Kurnia. Those people have provided facilities, support, attention, and assistance in the research data collection process.

References

[1] Murthy IK, Dutta S, Varghese V, Joshi PP, and Kumar P 2016 Global *Journal of Science Frontier Research* **17**(5) 1-15

[2] Bijarpas MM, Shahraji TR, and Linaei SM 2015 Socioeconomic evaluation of agroforestry system *Journal of Forest Science* **61**(11) 478-84

[3] Alao JS and Shuaibu RB 2013 Agroforestry practice and concept in a suitable land-use system in Nigeria *Journal of Horticulture and Forestry* **5**(10) 156-9

[4] Santoro A, Venture M, Bertani R, and Agnoletti M 2020 A review of the role of forest and agroforestry system in the FAO Globally Important Agricultural Heritage System Program *Journal Forest* **11**(860) 1-21

[5] Pasaribu SW, Kaskoyo H, and Safeii R 2019 The gender role in agroforestry management in Sungai Langka Village *Journal of Sylva Indonesiana* **2**(2) 62-9

[6] Sobolo OO, Amdi DC, and Jamala GY 2015 The Role of agriculture in environmental sustainability *Journal of Agriculture and Veterinary Science* **8**(5) 20-5

[7] Saha S, Sharmin A, Biswas R and Ashaduzzam M 2018 Farmers perception and adoption of agroforestry practice in Faridpur District of Bangladesh *International Journal of Environment Agriculture and Biotechnology* **3**(6) 1987-94

[8] Rohadi D, Herawati T, Firdaus N, Maryani R, Permadi P 2013 *Strategi nasional penelitian agroforestry*, ed Rusli Y and Pradhan U (Jakarta: Forda Press) pp 1-59

[9] Hani A, Achmad B, Ruhimat I, Handayani W, Suhaendah E, Winara A, Rahmawan B, dan Priono D 2016 Penerapan model agroforestry kayu pertukangan jenis sengon dan manglid *laporan hasil penelitian* pp 1-100

[10] Roberton GP and Harwood RR 2013 *Sustainable Agricultural* vol 1, ed Levin SA (Massachusset: Elsevier) pp 111-8
[11] Munasinghe M 2001 Sustainable development and climate change: applying the sustainomic transdisciplinary meta-framework *International J. Global Environment* **1**(1) 13-55
[12] Suwarno, J. (2011). Pengembangan kebijakan pengelolaan berkelanjutan DAS Ciliwung Hulu, Kabupaten Bogor. (Disertasi). Program Studi Pengelolaan Sumberdaya Alam dan Lingkungan, Sekolah Pasca Sarjana, Institut Pertanian Bogor, Bogor.
[13] Ruhimat, I.S. (2015). Status keberlanjutan usahatani agroforestry di lahan masyarakat: studi kasus di Kecamatan Rancah Kabupaten Ciamis Provinsi Jawa Barat. Jurnal Sosial Ekonomi Kehutanan, 12 (2) : 97-108
[14] Saida, Sabiham, S., Sutjahjo, S.H., & Widiatmaka. (2012). Analisis keberlanjutan usahatani hortikultura buah-buahan pada lahan berlereng di hulu DAS Jeneberang. Jurnal Ilmiah Bertani, 6 (2): 162-177.
[15] Taherdoost H 2017 Decision making the analytic hierarchy process (AHP): Step by step approach *International Journal of economics and management system* **2**(1) 244-6
[16] Tecle A 2018 Analytic hierarchy process application for multiple purpose forest resources management budget allocation in Durango Mexico *International of Journal Analytic Hierarchy Process* **10**(1) 36-63
[17] Al-Rahbi AKH 2020 Application of the analytic hierarchy process for management of soil erosion in oman *International of Journal Analytic Hierarchy Process* **12**(1) 104-16
[18] Yanti R dan Ibrahim H 2018. Kajian sosiologi perilaku konservasi dengan wanatani wilayah semi arid khatulistiwa *Journal of Applied Agricultural Science and Technology* **2**(2) 55-70.
[19] Sukanata IK, Dukat, Yuniati A 2015 Hubungan karakteristik dan motivasi petani dengan kinerja kelompok *Jurnal Agrijati* **28**(1) 17-34
[20] Hermanto, Swastika KS 201 Penguatan kelompok tani: langkah awal peningkatan kesejahteraan petani *Jurnal Analisis Kebijakan Pertanian* **9**(4), 371-390
[21] Anggraeni I dan Lelana NE 2011 *Penyakit karat tumor pada sengon*, ed Darma T et al (Jakarta: Forda Press) pp 1-57
[22] Mandry SV, Salmiah, dan Sihombing L 2016 Analisis kemampuan permodalan usahatani palawija di dan hortikultura di pedesaaan *Jurnal on Social Economic of Agriculture and Agribusiness* **5**(11) 1-10
[23] Suprayitno A, Sumardjo, Darwis S dan Basita G 2011 Model peningkatan partisipasi petani sekitar hutan dalam mengelola hutan kemiri rakyat *Jurnal Penelitian Sosial Ekonomi Kehutanan 8*(3), 176-195
[24] Ruhimat IS 2014 Faktor-faktor untuk peningkatan kemandirian petani dalam pengelolaan hutan rakyat *Jurnal Sosial Ekonomi Kehutanan* **11**(3) 239-52
[25] Rahmanadih, Bulkis S. Arsyad M, Amrullah A, Viantika NM 2018 *IOP Conf. Series. Earth and Environmental Science* Vol **157** pp 1-8
[26] Rahaman AA 2018 Do farmer groups impact on farmers yield and efficiency of smallholder farmers? Evidence from rice farmers in Northern Gana *Food Policy* **81**(1) *95-105.*
[27] Rustinsyah 2019 The significance of social relations in rural destination development: a case study of a beef cattle farmer groups in Indonesia *Journal of Cooperative and Management* **7**(2)1-7

Soybean varieties tolerance to intercropping with maize

I Permanasari [1], E Sulistyaningsih [2,*], B Kurniasih [2], D Indradewa [2]

[1] Postgraduate Student of Graduate School of Agriculture, Gadjah Mada University, Jl. Flora, Bulaksumur, Karang Malang, Caturtunggal, Depok, Sleman, Daerah Istimewa Yogyakarta 55281, Indonesia.

[2] Faculty of Agriculture, Gadjah Mada University, Jl. Flora, Bulaksumur, Karang Malang, Caturtunggal, Depok, Sleman, Daerah Istimewa Yogyakarta 55281, Indonesia.

* E-mail: endangsih@ugm.ac.id

Abstract. There is no available information in determining soybean varieties that shows tolerant when intercropped with maize. The objective of this study was to determine soybean varieties suitable for intercropping, based on changes in plant growth and yield when intercropped. The field experiment was conducted from October 2019 to February 2020 based on RCBD with 2 factors. The first factor was soybean with 16 national varieties. The second was cropping systems i.e., monoculture and intercropping. Morphology and yield data were observed and subsequently analysed based on performance difference (D) of each parameter between intercropping and monoculture. Tolerance level to cropping systems with maize was grouped based on the yield-based and selection index calculations. The results showed that different cropping systems affected morphological and yield parameters on 16 tested soybean varieties. Seven varieties i.e., Derap 1, Devon 1, Devon 2, Demas 1, Dena 1, Dena 2, and Wilis were grouped as tolerant to intercropping with maize. Among these, Derap 1 was the most tolerant variety.

1. Introduction

Soybean (*Glycine max* L.) is the third most important crop in Indonesia [1]. In the last decades, soybean consumption in Indonesia has increased significantly, and currently becomes one of the national strategic commodities. Soybean is also known as source of protein with high nutrition contents such as isoflavone, saponin, fatty acid and fiber [2]. On average, our national consumption of soybean reached 2.3 million tons per year. However, we are still lacking soybean production, in which in the last 5 years, the national production was only below one million tons per year, or only 43% of the soybean demand [1].

Soybean production, its productivity and the planted area were fluctuated from 2014 to 2018. While there was an 82.39% increase of the planted area in 2017-2018, its productivity on the other hand, showed a drop of 4.62% [1]. Intercropping is believed beneficial in minimizing the competition on land use for production of other crops. Soybean-maize intercropping is projected to form an integrated agriculture system consisting of four pillars i.e., production, stability, food security and sustainability [3].

In intercropping system, above and below ground competitions affect plant growth. One of the above ground factors which important is the light availability. In intercropping system of soybean-maize, the maize shading reduced the light intensity received by soybean, as its morphological stature is lower. Therefore, a response or adaptation of soybean to maize shading are of important index for those who selecting soybean varieties suitable for maize-soybean intercropping system [4]. Study on the variability of intercropping-tolerance among intercropped crops is therefore one of great benefit for selecting high-yield soybean varieties under the intercropping system [5]. The objective of this study was to determine soybean varieties suitable for intercropping, based on the changes in plant growth and yield characteristics when intercropped.

2. Methods

2.1. *Site description*

The field experiment was conducted at Rural Extension Center (Balai Penyuluhan Pertanian) of Seyegan, Sleman, Yogyakarta, Indonesia, (latitude: -07 deg 43 min 49.7621 sec, longitude: 110 deg 18 min 16.4783 sec). The soil was composed of silt loam with pH 6.5 with Nitrogen total (0.19%), Nitrogen availability (0.02%), Kalium availability (180 ppm), and P_2O_5 (131.50 ppm).

2.2. *Experimental design*

The research was arranged in RCBD with 2 factors and experimented from October 2019 to February 2020. First factor was sixteen soybean varieties, while the second factor was cropping systems i.e., monoculture and intercropping. In total, there were 32 treatment combinations (16 varieties × 2 cropping systems) with 3 replications (96 plots) applied in this research. We utilized a 5 m ×1 m plot size. Soil tillage was done and manured using chicken manure (10 ton ha^{-1}). Prior planting, soybean seeds were mixed with leguminosae inoculum (Legin®) (30 g 10 kg^{-1} seed). The planting arrangement for monoculture followed a 20 cm ×20 cm with 2 seeds per planting hole and 20 cm ×100 cm spacing for soybean and maize, respectively. For intercropping, the soybean was planted between row with maize by 40 cm spacing and ratio of 1:2 (maize: soybean). For each monoculture and intercropping plot contained 220 (2 seeds×110 planting holes) and 100 (2 seeds×50 planting holes) plants, respectively. Soybean fertilizations were employed using Urea (50 kg ha^{-1}), TSP and KCl (each 100 kg ha^{-1}) [6], completed at planting time by spreading it around the planting hole. As for maize, the fertilizations were conducted at 1 week-after-planting (WAP) by using Urea (100 kg ha^{-1}), SP36 (50 kg ha^{-1}), and KCl (75 kg ha^{-1}). Additional Urea (150 kg ha^{-1}) and KCl (50 kg ha^{-1}) were applied at 4 WAP [7]. Watering was scheduled on 15-21, 25-35, and 56-63 day-after-plantings (DAPs). Harvesting was done when it reached physiological maturations, i.e., brown and falling leaf, pod and stem turned brown, and seed had fully contained.

2.3. *Datal analysis*

Morphological data of stem diameter, plant height, and number of productive branches were obtained at certain schedules. Stem diameter was recorded at 7 WAP, whereas plant height and number of productive branches were measured and counted at 2, 4, 6 and 8 WAPs and harvesting time, respectively. Yield variables included the number of pod/plant, number of seed/plot, number of seed/plant, 100-seed weight and seed weight/plant were obtained at harvesting time. Sigmoid curve for growth based on plant height parameter was then constructed. The performance difference (D) of each parameter between intercropping and monoculture in all varieties, was calculated in a formula [14]: D = ((Intercropping-Monoculture)/Monoculture) × 100%.

To screen soybean varieties tolerant to intercropping with maize, the sixteen soybean varieties were clustered into groups of tolerant and intolerant based on two measurements. The first was yield-based calculation. The variety was defined as tolerant if the yield on intercropping system (Ys) was higher than those of the average from sixteen varieties tested in both cropping systems. Secondly, the grouping was based on screening index. Six different screening indexes including stress tolerance index (STI) i.e. higher STI value was more tolerant [8]; stress susceptibility index (SSI) i.e. SSI value <1 was more tolerant [8]; tolerance (TOL) i.e. lower TOL value was more stable [9]; mean productivity (MP) i.e. higher MP value was more desirable [9]; geometric mean productivity (GMP) i.e. higher GMP value was more desirable [10], yield stability index (YSI) i.e. higher YSI value can be regarded as more stable [11], and yield index (YI) i.e. high value of YI will be suitable for stress condition [12], were evaluated. Only those with significant coefficient correlation at $P<0.05$ in both cropping systems were further employed to categorize the tolerance level based on similar criteria as used for yield-based calculation. These analyses were conducted by using Microsoft Excel for Mac 2016 (Microsoft, USA).

3. Results and discussion

3.1. Morphological characters

Based on performance difference calculation (14), the 16 varieties were divided into 3 groups: a). low plant height-high stem diameter, i.e., Pangrango, Dena 1, Grobogan, Dega 1, Demas 1, Devon 1, Derap 1, and Gepak Kuning; b) low plant height-low stem diameter, i.e., Dena 2, Devon 2, Detap 1 and Mahameru; and c). high plant height-low stem diameter, i.e., Anjasmoro, Wilis, Deja 1 and Slamet (Table 1). Characters of varieties in Group a and b were contradictive compared to other studies [13], [14], [15], in which intercropping had increased plant height and decreased stem diameter. Varieties in Group c showed characters fairly in line with [13, 17] in which intercropping increased internode length (78.50 - 82.30%) at the expense of reducing stem diameter (-40.00 to -50.00%), even though its difference level for the same variable in this current research was high (0.68-2.10% and 0.83-15.75%, respectively). Cassal [16] reported that the increasing plant height in this group was due to the reduced light intensity as part of the shading effect, and thus reducing the photosynthetic-active-radiation (PAR) by 50.0 - 54.0%. This microclimate change was considered as the main driver for the difference of morphological parameters between cropping systems. We considered that the shading stress had occurred at low level. Therefore, the varieties did not respond it by etiolation and elongating the stem as an avoidance strategy (16). Yan et al. (2010) [4] further confirmed that soybean grown under maize shading in intercropping system has resulted thin stems and easy to lodging. Plant height, first internode length, stem diameter, and dry weight were reported to directly correlate with lodging resistance [18]. However, no lodging incidences were found on these tested varieties in the field. Liu et al. (2017) [17] stated that light quality (than its quantity) had great impact on morphology, as evidenced by the increase in specific leaf area and plant height (internode length) and less branching. Lower light intensity led to less growth, causing smaller final leaf size and stem diameter of soybean in the two intercropping treatments, which resulted in lower yield. Besides shading avoidance, plants usually optimize light captured and utilization under shading, which needed for their high efficiency of carbon gain [13]. Muneer et al. (2004) [19] reported that the light is the main competition factor in which the plant height was negatively affected in intercropping.

Table 1. Performance difference (D) of 16 soybean varieties for morphological and yield components (plant height, stem diameter, no. pod/plant, no of seed/pod, no. seed/plant, 100-seed weight and seed weight/plant) between intercropping and monoculture systems

Varieties	Plant height (%)	Stem diameter (%)	No. pod/plant (%)	No. seed/pod (%)	No. seed/plant (%)	100-seed weight (%)	Seed weight/plant (%)
Pangrango	-1.38	5.32	5.41	-17.44	-19.29	3.93	-25.28
Anjasmoro	1.98	-0.83	-1.38	-4.60	-11.53	-0.58	-30.24
Dena 1	-1.40	2.53	14.08	-18.14	6.35	-4.66	-5.06
Dena 2	-9.51	-5.90	21.84	-29.26	-5.79	-0.95	-18.34
Wilis	1.33	-14.83	-27.14	-3.31	-28.88	1.13	-31.52
Grobogan	-10.16	1.84	38.05	-50.07	3.78	-9.21	52.80
Dega 1	-13.43	7.59	1.93	-1.42	-0.45	-9.32	-18.82
Demas 1	-9.36	0.84	2.24	-15.49	-22.38	-7.42	-29.91
Deja 1	0.68	-15.75	4.10	-8.53	-6.88	4.36	-43.34
Devon 1	-1.77	9.80	21.63	-10.53	6.60	-0.65	1.91
Devon 2	-9.09	-9.14	13.20	11.18	24.14	-8.95	31.17

Detap 1	-2.75	-5.69	33.66	-34.19	-24.98	1.17	-40.22
Derap 1	-14.25	9.30	59.07	0.19	18.09	-0.54	13.25
Mahameru	-1.53	-4.90	-1.54	17.97	14.40	-4.21	-0.72
Gepak Kuning	-2.86	14.55	4.84	18.12	28.71	-0.17	46.12
Slamet	2.10	-13.37	-23.89	2.03	-36.00	-3.90	-42.06

3.2. Yield Parameters

When two cropping systems were compared, it observed that there were range of increasing and decreasing yield parameters in each soybean varieties. The decreasing seed weight (-0.72 to -42.06%) was lower than those of [14] whose reported a higher decrement for up to -23.00 to -42.42%. Across the calculated performance difference on yield parameters, Slamet revealed the most decreasing performance (Table 1) (number of pod per plant: -23.89%, number of seed/plant: -36.00%, and seed weight/plant: -42.06%) from monoculture compared to intercropping. Especially Derap 1 showed highest increased on the number of pod/plant (59.07%) and number of seed/plant (18.09%) in intercropping. Intercropping increased seed weight/plant in some soybean varieties tested i.e. Grobogan (52.80%), Devon 1 (1.91%), Devon 2 (31.17%), Derap 1 (13.25%) dan Gepak Kuning (46.12%). Seed weight per plant in Grobogan dan Gepak Kuning were higher in intercropping compared to that of in monoculture. These varieties showed low yield compared to the average of all varieties tested (Figure 1). Dena 1 and Devon 2 gave better seed weight per plant in intercropping. On the other hand, Derap 1 gave the highest seed weight per plant in both cropping system. It argued that this high yield in Derap 1 might be due to the capability of the variety on optimizing the photosynthetic yields to be distributed mainly to the seed rather than to other vegetative organs, such as pods development, stem growth, plant height, and productive branches. As the results, Slamet was assigned as group varieties with low capacities to compete with maize on microclimates and other growth factors. In addition, Grobogan, Devon 1, Devon 2, Derap 1 and Gepak Kuning were assigned as group varieties with high capacities to compete with maize on microclimates and other growth factors. This in turn, showed that there was a different level of tolerance to compete with maize for light, nutrition, and water [21] as plant response on competition for light and water can be an additive, synergic and antagonist [22]. Liu *et al.* (2019)[23] stated that the maize when intercropped with faba bean produced exudate which lead to root formation, increase root nodules and assist the Nitrogen fixation on faba bean, With the better supply of nitrogen, it showed that both crops can utilize it efficiently. On the contrary, Pangrango, Deja 1, Detap 1 and Slamet suffered a declining seed weight per plant on intercropping. This was in line with the report of Balitkabi (2015) [24], mentioned that Pangrango, Kipas Putih and Slamet had lower yield potential compared to Wilis. This result implied that these 4 varieties were unable to survive and complete their life cycle when intercropped with maize. The lack of sunlight, nutrients, and competition for water were the reasons of this low performance. Other varieties, however, showed a stable but low performance. We also confirmed that Dena 1 was categorized as shading tolerant which was in line with the previous report [24], which reported the similar results.

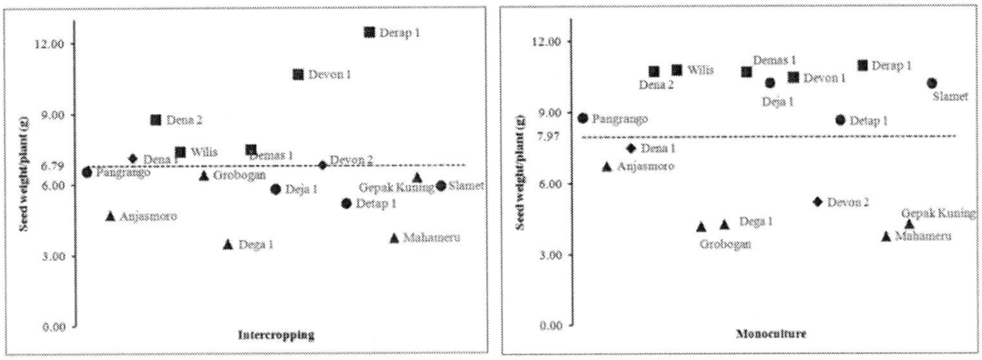

Figure 1. Grouping of 16 soybean varieties based on seed weight/plant (g). Thresholds of averages among all varieties were set as borderline to categorize the varieties into tolerant (above threshold) and intolerant (below threshold) in intercropping and monoculture systems, respectively.

In each of intercropping and monoculture, the correlation of seed weight per plant and number of pod per plant was significantly and positively correlated (r= 0.56; *P*=0.024 and r= 0.56; *P*= 0.022, respectively). The correlation of this parameter to number of seed per plant also showed same trend (r= 0.67 *P*= 0.004 and r= 0.79; *P*= 0.000). Number of pod per plant and seed per plant therefore gave a significant impact on seed weight/plant. This result showed that the greater number of pod and seed/plant would increase seed weight per plant in both cropping system. As for its correlation to the other parameters, even though positive, the level was not significant except for the number of branches on monoculture (number of seed per pods: r= 0.47; *P*= 0.069 and r= 0.33; *P* =0.217; stem diameter: r= 0.29 *P*= 0.269; and r= 0. 0.46; *P*= 0.070); number of branches: r= 0.42; *P*= 0.105 and r= 0.57; *P*= 0.022). Seed weight per plant was negatively (but not significant) correlated with 100-seeds weight (r= -0.05; *P*= 0.866 and r= -0.44; *P*= 0.089) in intercropping and monoculture, respectively. Number of pods produced in intercropping system was positively correlated with all parameters, except for 100 seeds weight. This implied that the increasing 100-seeds weight had suppressed other parameters. These results were similar to [20] which reported the yield of shaded soybean was positively correlated to seed weight per plant, number of branches (r= 0.34), number of internodes (r= 0.27), number of contained seed (0.55) and 100 seed weight (r= 0.42). Plant height between two cropping systems had correlated differently to seed weight per plant. The decrement of plant height on intercropping was also followed by decrement on seed weight per plant (r = -0.20; *P*= 0.46). Whereas, the plant height increased on monoculture was also followed by the increament of seed weight per plant (r= 0.28; *P*= 0.29). These two cropping systems however, showed a negative correlation with 100 seeds weight (r = -0.83).

3.3. Tolerance Selection

To determine the soybean tolerance to intercropping, the performance difference calculation based on seed weight per plant and selection index were employed. The result of selection index based on yield (Ys) in intercropping showed that 7 soybean varieties i.e., Derap 1, Devon 1, Devon 2, Demas 1, Dena 1, Dena 2, and Wilis were grouped as tolerant to intercropping with maize, as their yield were above the average of those 16 varieties in both of monoculture and intercropping (Figure 1).

The 7 varieties belongs to tolerance group were in accordance with its selection indexes results (Table 2). Among 7 selection indexes evaluated, there were 4 selection indexes (STI, MP, GMP and YI), which showed significant correlation coefficient to both of Ys and Yp at 1% and 5%. Among these varieties Derap 1 were the most tolerant variety as it has the highest yield (12.42 g) in intercropping and selection index values for STI (2.15), MP (11.69), GMP (11.67), and YI (1.83), respectively. On the other hand, Dega 1 and Mahameru were categorized as intolerant as its Yp and Ys and 4 selection indexes showed the lowest. The low seed weight per plant on these two varieties showed that the varieties were low yield varieties in both cropping systems. The remaining was grouped as intermediate tolerant in intercropping.

The tolerant varieties indicated that they had less interspecific competition between two species for basic growth resources, which reflected positively on yield attributes of soybean compared with those of its monoculture [5]. The yield increment was supported by other parameters such as number of pods/plant and number of seed produced. The more pod available, more seed weight was expected.

Table 2. Selection indexes of 16 soybean varieties. Pearson correlation coefficient between selection indexes and yields in monoculture and intercropping was depicted on the lower table

Varieties	Yp	Ys	STI	SSI	TOL	MP	GMP	YSI	YI
Pangrango	8.77	6.55	0.91	1.72	2.22	*7.66*	*7.58*	0.75	0.96
Anjasmoro	6.75	4.71	0.50	2.05	2.04	5.73	5.64	0.70	0.69
Dena1	7.50	*7.12*	0.84	0.34	0.38	7.31	*7.31*	0.95	*1.05*
Dena2	10.74	*8.77*	*1.48*	1.24	1.97	*9.76*	*9.71*	0.82	*1.29*
Wilis	10.79	*7.39*	*1.25*	2.14	3.40	*9.09*	*8.93*	0.68	*1.09*
Grobogan	4.19	6.41	0.42	-3.58	-2.21	5.30	5.18	1.53	0.94
Dega1	4.29	3.48	0.23	1.28	0.81	3.88	3.86	0.81	0.51
Demas1	10.69	*7.49*	*1.26*	2.03	3.20	*9.09*	*8.95*	0.70	*1.10*
Deja1	10.24	5.80	*0.94*	2.94	4.44	*8.02*	*7.71*	0.57	0.85
Devon1	10.46	*10.66*	*1.75*	-0.13	-0.20	*10.56*	*10.56*	1.02	*1.57*
Devon2	5.20	*6.82*	0.56	-2.12	-1.62	6.01	5.95	1.31	1.00
Detap1	8.66	5.18	0.71	2.73	3.48	6.92	6.70	0.60	0.76
Derap1	10.97	*12.42*	*2.15*	-0.90	-1.45	*11.69*	*11.67*	1.13	*1.83*
Mahameru	3.76	3.73	0.22	0.05	0.03	3.74	3.74	0.99	0.55
Gepak Kuning	4.29	6.27	0.42	-3.13	-1.98	5.28	5.19	1.46	0.92
Slamet	10.22	5.92	*0.95*	2.85	4.30	*8.07*	*7.78*	0.58	0.87
Average	7.97	6.79	0.91	0.59	1.17	7.38	7.28	0.91	1.00
Correlation Coefficient									
Yp	1	0.62*	0.86**	0.59*	0.60*	0.92**	0.91**	-0.59*	0.62*
Ys		1	0.92**	-0.22	-0.26	0.88**	0.89**	0.22	1.00**

Yp, mean yield in monoculture; Ys, mean yield in intercropping; STI, stress tolerance index; SSI, stress susceptibility index; TOL, tolerance; MP, mean productivity; GMP, geometric mean productivity, YSI, yield stability index, YI, yield index. Italic indicated values above its respective average. * and **, significant at $P= 0.05$ and $P= 0.01$, respectively.

4. Conclusions

Seven varieties i.e., Derap 1, Devon 1, Devon 2, Demas 1, Dena 1, Dena 2, and Wilis were grouped as tolerant to intercropping with maize. Derap 1 was the most tolerant varieties when intercropped with maize.

References

[1] Indonesia Ministry of Agriculture 2018 The recent five years data of soybean production and planting area http://www.pertanian.go.id/home/?show=page&act= view&id=61.
[2] Wan Y, Yan Y, Xiang D, Ye M, Yang W and Liu J 2015 Isoflavoid accumulation pattern as affected by shading from maize in soybean (*Glycine max* (L.) Merr.) in relay strip intercropping system Plant Production Science, **18** (3) : 302-313.
[3] Du J, Han T, Gai J, Gong T, Sun X, Wang X, Yang F, Liu J, Shu K and Yang W 2018 Maize-

soybean strip intercropping: Achieved a balance between high productivity and sustainability Journal of Integrative Agriculture, **17** (4): 747-754.

[4] Yan Y, Gong W, Yang W, Wan Y, Chen X, Chen Z and Wang L 2010 Seed treatment with unicobazole powder improves soybean seedling growth under shading by corn in relay strip intercropping system Plant Production Science, **13**: 367-374.

[5] Metwally AA, Sa na SA, Sherief MN and Abo-Hegazy RE 2019 Productivity and some competitive relationships of intercropping soybean with three corn cultivars under alternating and mixed systems J J Agriculture **5** (1): 016.

[6] Adie MM, Nugraheni N, Sundari T, Marwoto, Kariyasa IK, Widiarta IN and Harnowo D 2013 *General guidance of production and distribution of breeder seed.* Indonesia Ministry of Agriculture ISBN 978-602-1280-04-1.

[7] Subekti NA, Syafruddin, Effendi R dan Sunarti S 2008 *Morphology and growth phase of maize.* Cereal Plant Research Centre. Maros South Sulawesi.

[8] Fischer RA and Maurer R 1978 Drought resistance in spring wheat cultivars I Grain yield response Aust J Agric Res, **29**:897–907.

[9] Hossain ABS, Sears AG, Cox TS, and Paulsen GM 1990 Desiccation tolerance and its relationship to assimilate partitioning in winter wheat Crop Science, **30**:622-627.

[10] Fernandez GJ 1992 Effective selection criteria for assessing stress tolerance. Proceedings of the International Symposium on Adaptation of Vegetables and Other Food Crops in Temperature and Water Stress Tolerance. Asian Vegetable Research and Development Centre, Taiwan, 257-270 p.

[11] Bouslama M, and Schapauch WT 1984 Stress tolerance in soybean I Evaluation of three screening techniques for heat and drought tolerance Crop Science, **24**:933–937.

[12] Gavuzzi P, Rizza F, Palumbo M, Campanile RG, Ricciardi GL, and Borghi B 1997 Evaluation of eld and laboratory predictors of drought and heat tolerance in winter cereals. Can J Plant Sci, **77**:523-531.

[13] Gong WZ, Jiang CD, Wu YS, Che HH, Liu WY and Yang WY 2015 Tolerance vs avoidance: two strategies of soybean (*Glycine max*) seedlings in response to shade in intercropping Photosynthetica, **5** (2): 259–268.

[14] Su BY, Song YX, Song C, Cui L, Yong TW and Yang WY 2014 Growth and photosynthetic responses of soybean seedlings to maize shading in relay intercropping sistem in Southwest China Photosynthetica, **52** (3): 332–340 https://doi.org/10.1007/s11099-014-0036-7.

[15] Sundari and Mutmaidah 2018 Suitability of Soybean Promising Lines for Maize + Soybean Intercropping Jurnal Agronomi Indonesia, **46** (1):40-47.

[16] Casal JJ 2012 Shade avoidance The Arabidopsis Book **10**: e0157.

[17] Liu X, Rahman T, Song C, Su B, Yang F, Yong T, Wu Y, Zhang C and Yang W 2017 Changes in light environment, morphology, growth and yield of soybean in maize-soybean intercropping systems Field Crops Research, 38–46.

[18] Huang ZW, Zhao TJ, Yu DY, Chen SY and Gai JY 2008 Lodging resistance indices and related QTLs in soybean Acta Agronomica Sinica, **34**: 605-611.

[19] Muneer A P, Fida H M, Mumtaz A K and Muhammad I S 2004 Performance of maize in intercropping system with soybean under different planting patterns and nitrogen levels Journal of Applied Sciences, **4**: 201–204 https://doi.org/10.3923/jas.2004.201.204.

[20] Sundari, T and Artari R 2018 Response of soybean lines to shading condition Buletin Palawija, **16** (1): 27-35).

[21] Lv Y, Francis C, Wu P, Chen X, and Zhao X 2014 Maize-soybean intercropping interactions above and below ground Crop Science, **54**:914–922.

[22] Zhang J, Smith D L, Liu W G, Chen X F, and Yang W Y 2011 Effects of shade and drought stress on soybean hormones and yield of main-stem and branch African Journal of Biotechnology, **10**:14392-14398.

[23] Liu Y, Yin X, Xiao J, Tang L and Zheng Y 2019 Interactive influences of intercropping by nitrogen on flavonoid exudation and nodulation in faba bean Science report **9**:4818.www.nature.com/sciencereports.

[24] BALITKABI 2015 Dena 1, A New improved variety tolerant to shading is suitable for tempe preparation Info Teknologi http://balitkabi.litbang.pertanian.go.id/infotek/dena-1-a-new-improved-variety-tolerant-to-shading-is-suitable-for-tempe-preparation/ Accessed; 17th Desember 2015.

Farmers empowerment level analysis in farming during the Covid-19 pandemic and its impact on farm income

I P N Damanik[1,*], M E Tahitu[1], M Turukay[1], F P Adam[1]

[1]Agriculture Social Economics Department, Faculty of Agriculture, Pattimura University

* E-mail: intadamanik@ymail.com

Abstract. This study aims to analyze the level of farmer empowerment in carrying out farming activities as the main livelihood and its impact on farm income during the Covid-19 pandemic. The research was conducted online by distributing questionnaires in the form of google form in various affordable areas. There were 113 responses obtained by returning the google form on time. The three regions that sent the most responses were farmers in Ambon City, West Seram Regency, and Central Maluku Regency. After analyzing using qualitative descriptive analysis, the results showed that the level of farmer empowerment during the Covid-19 pandemic decreased, especially in terms of marketing of crop products for fruit and vegetable commodities, while for food commodities such as rice tended to be stable. This has an impact on farm productivity and income. The main factor causing a decrease in the level of farmer empowerment is the decreasing purchasing power of the community during the Covid pandemic19; Accessibility of farmers to markets is difficult, there are even times when farmers cannot go to the market due to large-scale social restrictions, the frequency of presence of collectors is reduced, access to plant production facilities is hampered unlike before the Covid pandemic19. This condition requires serious attention from the government so that farmers do not experience difficulties in marketing agricultural products and can maintain the stability of farm income. The solution at the farmer level is to form farmer groups so that the marketing of farm products and the fulfillment of plant production facilities can be done through farmer groups.

1. Introduction

The Covid-19 pandemic has brought changes to almost the entire world community, especially social and economic life, including in Indonesia. Almost all development sectors in Indonesia cannot realize the planned programs, including the agricultural sector. Even so, the agricultural sector is one of the sectors that is able to continue to grow during the Covid-19 pandemic. Data from the Ministry of Agriculture of the Republic of Indonesia shows that the growth of the agricultural sector's Gross Domestic Product (GDP) increased by 16.4% every quarter and 19% per year in the second quarter of 2020[1]. This condition illustrates that there is hope for agricultural development at a time when the number of farm households continues to decline from year to year. This means that the agricultural sector is a sector that can accommodate workers who are forced to lose their jobs due to the COVID-19 pandemic.

The agricultural sector cannot be separated from the farmer's existence. Farmers as the main actors are one of the determining factors for the success of agricultural development. Prior to the Covid-19 pandemic, most farmers in Indonesia had indeed faced various problems, including a lack of capital, limited knowledge and skills, inability to innovate, and difficulties in marketing their agricultural products. These various problems are caused by the level of farmer empowerment is still relatively low so that they have not been able to solve the existing problems. The empowerment of farmers is closely related to their capacity. Farmers with high capacity will be more empowered in managing the farm and achieve good performance in farming. This is in line with the opinion [2] that the capacity of a person or organization will affect the performance of the person or organization. In addition, this problem also causes farm management to be unable to maximize its impact on farm productivity and income.

Similar to farmers in Indonesia in general, farmers in Maluku also face the same problems. If before the Covid-19 pandemic Maluku farmers had to face sharing farming problems, what about during the Covid-19 pandemic? This is interesting to analyze considering that during the Covid-19 pandemic, all aspects of people's lives were affected

Based on this, this study aims to analyze the empowerment of farmers in managing their farms during the Covid-19 pandemic and their impact on farm income. In this case, farmer empowerment is intended as the ability of farmers to manage their farms by fulfilling their input needs including labor and marketing of farm products. The results of the analysis are complemented by solutions to increase farmer empowerment in facing the Covid-19 pandemic that has not ended and is in the future.

2. Methods

The research was conducted during the Covid-19 pandemic. The existence of large-scale social restrictions and government policies requiring work from home have made online surveys chosen in an effort to collect data.

The survey was conducted using google form in July 2020 for one full month. Thus the location determination was not carried out because the expected respondents were farmers who could access the google form. Information about the distribution of questionnaires via google form is provided to parties who can be contacted via telephone and social media, namely farmer group leaders, agricultural extension agents, students who at that time returned to their home areas due to the policy of learning from home, and acquaintances/team. Furthermore, this information is conveyed to farmers so that they can participate in the survey they are conducting.

It is done every day to check the forms returned by the farmers while tabulating the data in the questionnaire. The form checking ends after a full month and the survey is closed.

There are obstacles to going to the research location and only relying on data that is distributed online, causing data analysis to only be carried out using qualitative descriptive analysis with simple tabulation support.

3. Results and discussion

3.1. Number and Distibution of Responden Locations

The number of forms (questionnaires) returned during the data collection period was 113 forms, meaning that there were 113 farmers as respondents in this study. All respondents come from the Maluku region with the distribution of numbers and regions shown in Table 1.

Table 1. Number and Distribution of Respomden Locations

No.	Responden location	Number	
		People	%
1.	Ambon city	37	33
2.	West Seram District	47	42
3.	Center Maluku District	18	16
4.	Tual city	7	6
5.	Buru District	4	3
	Total	113	100

The distribution of respondents' areas is influenced by the availability of internet networks and the ability of farmers to use information technology, in this case the use of cell phones. There are three regions that mostly send back questionnaires online, even though in each district there are farming centers.

Table 1 indicates the need for improvements in the field of information and communication technology, both the availability of a relatively stable internet network throughout the Maluku region and the ability of farmers to use information and communication technology. Mastery of information and communication technology will be able to broaden farmers' insights through available extension messages and can be accessed by farmers. Especially during the industrial revolution 4.0 and even more so when the Covid-19 pandemic could not be predicted when it would end, making information and communication technology one of the right choices to help farmers survive or develop their farming. In addition, the high cost of communication via the internet also makes farmers limited in using internet-based information technology.

3.2. Respondents Distibution by Commodity

There are three main types of commodities cultivated by farmers, namely fruits, vegetables, and rice, with the distribution of respondents as shown in Figure 1. Farmers with the main commodity of rice (lowland rice) are in West Seram Regency, precisely in Waimital Village and Buru Island dominated questionnaire returns.

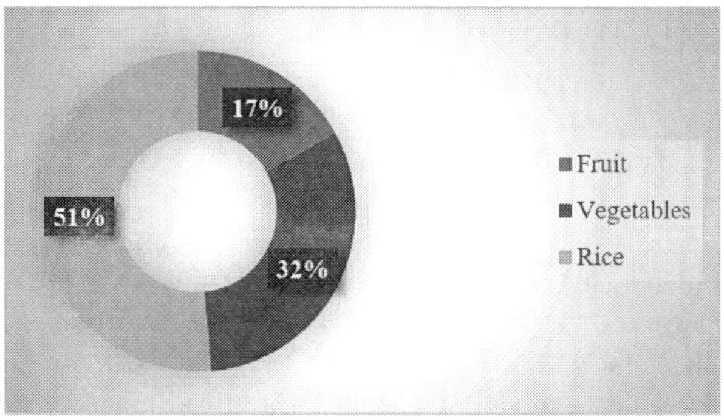

Figure 1. Respondents Distibution by Commodity

Waimital village is one of the lowland rice production centers in Maluku with farmers who are familiar with farming technology and information and communication technology, in this case communication via the internet. Vegetables and fruits are dominated by farmers in Ambon City, Central Maluku Regency and Tual City. Vegetables that are predominantly cultivated by farmers are kale, mustard greens, spinach, long beans, green beans, tomatoes, chilies, and eggplant. Fruits are dominated by papaya, banana, mango, pineapple, avocado, coconut, durian, langsat, and duku.

3.3. Empowerment of Farmers Before and During the Covid-19 Pandemic

Farmers' empowerment is the farmer's ability to carry out farm management starting from cultivating land, planting, to marketing farm products. Farmers with a high level of empowerment have the ability to solve various farming problems, which in this case are grouped into three problems, namely: (1) fulfillment of production inputs, (2) fulfillment of labor, and (3) marketing of farm products. For these three problems, the empowerment of farmers before and during the Covid-19 pandemic is presented in Figure 2.

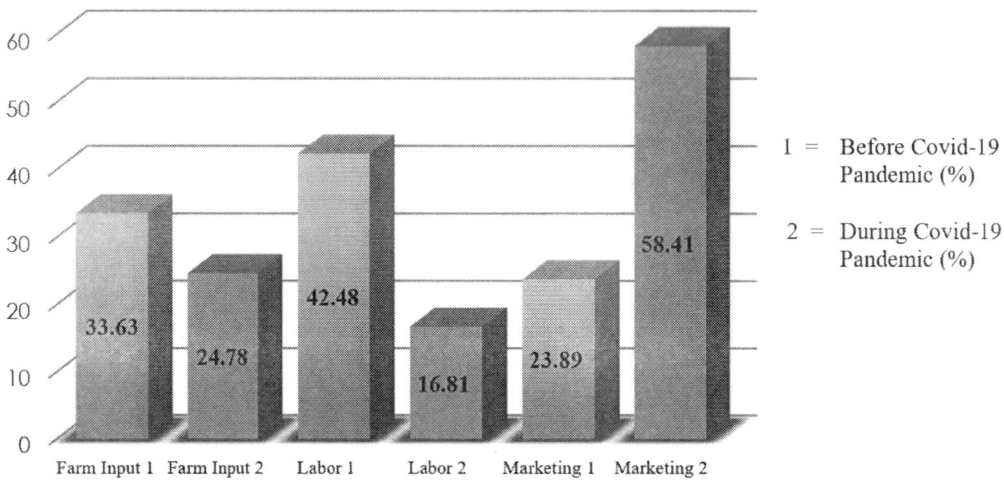

Figure 2. Farmer's Main Problems Before and During Pandemic Covid-19

Figure 2 shows that the main problems faced by farmers before the Covid-19 pandemic were dominated by meeting labor needs. This condition can be explained that before the Covid-19 pandemic, most of the young workforce preferred to work outside the agricultural sector, such as migrating to the city to become shopkeepers, working in restaurants, cafes, and other service sectors. This causes the availability of labor in the village to decrease. This is especially true for rice farmers during the growing season and fruit farmers at harvest time. On the other hand, during the Covid-19 pandemic, the availability of labor in villages increased due to the implementation of several policies that had an effect on reducing economic activity in cities. The reduction in economic activity has resulted in many business sectors laying off employees for an unknown period of time (until activities return to normal as before the Covid-19 pandemic). Likewise, the application of learning from home causes many students and students to return to the village and have the opportunity to become laborers in the agricultural sector. This causes the fulfillment of labor needs not to be the dominant problem for farmers during the Covid-19 pandemic, but rather the marketing of agricultural products which is a problem for most farmers (Figures 3a and 3b).

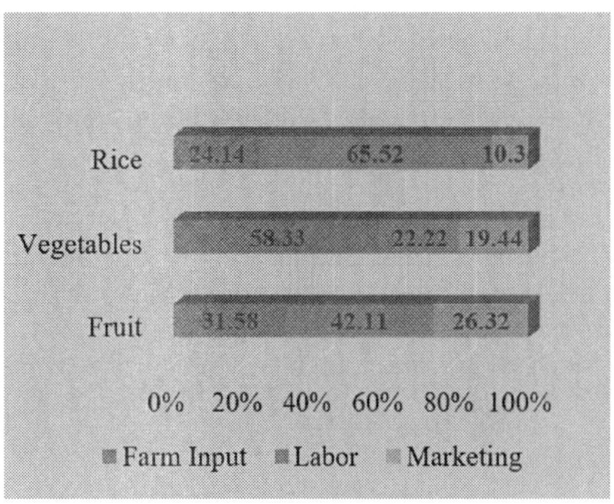

Figure 3a. The Main Problem of Farmers by Commodity Before Covid-19 Pandemic

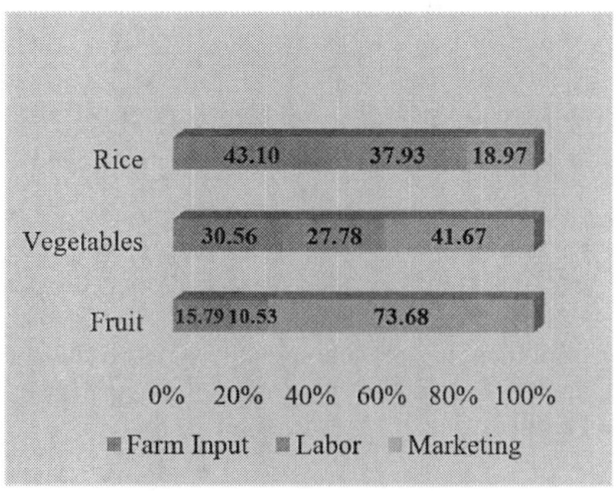

Figure 3b. The Main Problem of Farmers by Commodity During Covid-19 Pandemic

As is well known, another policy taken by the government was large-scale social restrictions that resulted in prohibiting public transportation activities between cities within districts and within provinces. Meanwhile Ambon City is the main marketing destination for agricultural products not only for farmers in Ambon City, but also outside Ambon City, such as those from West Seram District, Central Maluku, Buru Island, and Southeast Maluku including Tual City. This will affect the supply chain and decrease demand [3]. In other words, the Covid-19 pandemic has an impact on the increasing number of farmers who face obstacles in marketing their crops, both for rice, vegetable and fruit farmers. The worst conditions are felt by farmers of fruit and vegetables, which are perishable commodities [4], [5]. Papaya farmers on the outskirts of Ambon City bear huge losses because they cannot sell their crops as usual. Likewise, vegetables and fruit that are usually imported from outside the city of Ambon, namely from the West Seram District, Central Maluku, Buru Island, and Tual City cannot enter the market in Ambon City, while the markets in the village and in the capital of the local sub-district / regency cannot afford accommodate all agricultural products from the local area. As a result, farmers lose their income because their crops are wasted [5]. This condition is also experienced in various countries such as Pakistan [6].

Fulfilling production facilities problem before the Covid-19 pandemic was higher than during the Covid-19 pandemic (Figure 2), especially for vegetable and fruit farmers (Figure 3a and Figure 3b). This is because the intensity of planting and planting area continues to increase but is not followed by an increase in business capital so that the fulfillment of production facilities, especially fertilizers and pests and disease eradication drugs, becomes an obstacle for farmers. During the Covid-19 pandemic, the problem of fulfilling production facilities decreased due to reduced planting intensity and planting area as a result of constraints in marketing crop yields. In other words, vegetable and fruit farmers (except for long-lived fruit crops) reduce their farming activities for fear of not being able to sell their crops. Constraints to the fulfillment of production facilities during the Covid-19 pandemic are not only caused by limited capital capacity of farmers, but also because farmers cannot access production facilities due to restrictions on transportation activities and the closure of shops/stalls selling plant production facilities during large-scale social restrictions such as experienced by rice farmers. Based on the conditions faced by farmers according to Figure 2, Figure 3a, and Figure 3b, it can be said that the Covid-19 pandemic has an impact on the level of farmer empowerment, especially in the marketing aspect of agricultural products.

3.4. The Covid-19 Pandemic Impact on Farm Productivity and Income

In systems theory, disruption of one or more subsystems will affect the system as a whole. Likewise in the farming system, disturbances in the marketing system of farm products will interfere with the whole farm management, especially those managed by farmers with limited ability (empowerment). This disruption has an impact on decreasing farm productivity and income during the Covid-19 pandemic. The impact of the decline in productivity and farm income due to the COVID-19 pandemic was experienced by fruit and vegetable farmers (Table 2).

Table 2. The Covid-19 Pandemic Impact on Farm Productivity and Income

Commodities	Productivity decreased (/harvest) (%)	Income decreased (/harvest) (%)
Vegetables	15 – 25	30 – 40
Fruits	10 – 20	50 – 60
Rice	5 – 10	10 – 15

3.5. The Main Causes of Decreasing Farmers' Empowerment Levels during the Covid-19 Pandemic

There are four factors that cause the decline in the ability of farmers to overcome farming problems, namely: (1) farmers find it difficult to reach the market, even during large-scale social restrictions, access to markets is closed for a certain period of time. (2) The purchasing power of the local market for farmers' crops is limited. When access to local markets is opened, the ability of local markets to accommodate farmers' crops is limited so that not all farmers can sell their crops or not all of their crops

IOP Conf. Series: Earth and Environmental Science **883** (2021) 012034 doi:10.1088/1755-1315/883/1/012034

can be sold; (3) traders rarely even sometimes don't come, like before the Covid 19 pandemic, to buy farmers' crops. This is also due to large-scale social restrictions that close all community activities outside the home including economic activities; (4) shops/kiosks for plant production facilities are closed and if any are still open but the availability of goods needed by farmers is limited.

The decline in the ability of farmers to manage their farms during the Covid-19 pandemic which directly impacts on farm productivity and income (Table 2) if it continues in the long term will have an impact on decreasing the quality of life of farmer family members and can increase the poverty rate as happened in China. 7]. In addition, it can also affect food security [8], [9], [4], both at the household and regional and national levels. In this regard, a strategy is needed to increase farmer empowerment, not only in facing the Covid 19 pandemic, but also in a sustainable manner to create resilient farmers in every situation and condition of life.

A strategy that can be implemented is to make farmers, especially vegetable and fruit farmers aware of the importance of farmer groups, so that farmers can form farmer groups with their own desires. Farmer groups are tools that can help farmers achieve farmer goals through their function as a vehicle for learning and collaboration. Through farmer groups, farmers can collaborate with cooperatives or other parties that can help farmers manage their farms. An example is a rice field farmer group that has been proven to be able to help farmer group members manage their farming. The strategy from the government's point of view is seriousness in fostering farmers, including strengthening the capacity of farmers by improving the quality of human resources (farmers) by increasing the quantity and quality of agricultural extension agents. The quantity and quality of agricultural extension workers in Maluku have not been able to meet the needs of farmers [10], even though the presence of agricultural extension agents must be felt by farmers and beneficial to farmers. It is time to improve the capacity of extension workers in conducting internet-based extension which must also be balanced with the ability of farmers to access information via the internet. The covid-19 pandemic places information and communication technology as one of the right choices for farmers [11], [4] and agricultural extension agents. More than 60% of farmers in Ambon City have low knowledge and skills to use cell phones [12] and need to be improved, which of course needs to be followed by the provision of a supportive internet network in every agricultural production center. In 2020, Government attention is also needed in overcoming difficulties for farmers to market their crops by providing buyers who will buy farmers' crops through existing farmer groups or farmer cooperatives. Apart from that, the purchase of plant production facilities through farmer groups is also facilitated. Thus, there is continuity between the formation of farmer groups and the implementation of the functions of farmer groups with the benefits that can be obtained by farmers by becoming members of farmer groups [13]. This is important because many members of farmer groups do not feel the benefits of being members of farmer groups.

4. Conclusion

The Covid-19 pandemic has an impact on decreasing the ability of farmers, especially in carrying out marketing functions. Before the Covid-19 pandemic, farmers had various choices in marketing their farm products, including selling to collectors who came during the harvest period, selling to traders in the market, or selling directly to consumers in the market; however during the Covid-19 pandemic there was no single option available at the time large-scale social restrictions were imposed. This condition is very detrimental to farmers of vegetables and fruits, which are perishable products. Apart from the marketing aspect, the decline in farmer empowerment is also in the fulfillment of plant production facilities due to the absence of public transportation and restrictions on community economic activities so that plant production facilities shops / kiosks also do not serve buyers.

Reflecting on this, the government and related parties need attention so that they can help farmers in situations such as the Covid-19 pandemic. Farmers play a very important role in maintaining food security, therefore farmers should be supported in order to keep farming optimally in all situations where it is possible to do business by increasing farmer empowerment. Increasing farmer empowerment can be done by encouraging farmers to form farmer groups to facilitate farmer development, in this case agricultural extension agents play an important role.

It is time to equip farmers with mastery of information and communication technology so that farmers can broaden their horizons about modern digital-based agriculture to reduce the range of control over the absence and absence of agricultural instructors.

References

[1] Makki S 2020 *Mentan klaim sector pertanian tumbuh di era corona* CNN Indonesia Jakarta.
[2] Liou J. 2004. Community capacity building to strengthen socio-economic development with spatial asset mapping. *3rd FIG Regional Conference*; 2004 Oct 3-7. Jakarta Indonesia.
[3] Yusuf A A, Suganda T, Hermanto, Mansur F and Hadisoemarto P 2020 *Strategi Ekonomi Sektor Pertanian di Tengah Pandemi Covid-19* Center for Sustainable Development Goals Studies Padjadjaran University Bandung Indonesia
[4] Food and Agriculture Organization (FAO) 2020. *The impact of COVID-19 on food and agriculture in Asia and the Pacific and FAO's response* Asia and the Pacific Regional Conference Secretariat
[5] International Labour Organization (ILO) 2020. *Risalah Sektoral ILO: COVID-19 dan dampaknya terhadap pertanian dan ketahanan pangan*
[6] Asian Development Bank (ADB) 2020 *COVID-19 Impact on farm households in Punjab, Pakistan: Analysis of data from a cross-sectional survey* ADB Briefs No.149. Manila Philippines
[7] Ji-kun H 2020. Impacts of COVID-19 on agriculture and rural poverty in China *Journal of Integrative Agriculture* **19(12)** pp 2849–2853
[8] Ayanlade A, Radeny M, 2020. COVID - 19 and food security in Sub-Saharan Africa: implications of lockdown during agricultural planting seasons. *npj Science of Food* **4:13** pp 1-6
[9] Poudel P D, Poudel M R, Gautam A, Phuyal S, Tiwari C K, Bashyal N, and Bashyal S 2020 COVID-19 and its Global Impact on Food and Agriculture *Journal of Biology and Today's World* **9(5)** pp 221-224
[10] Damanik I P N, Tahitu M E. 2019. Disparity between the availability and agricultural counselor need to improve the product of prime commodity in Maluku, Indonesia *Bioscience Research* **16(1)** pp 251-261
[11] Timilsina B, Adhikari N, Kafle S, Paudel S, Poudel S and Gautam D 2020 Addressing impact of COVID-19 post pandemic on farming and agricultural deeds *Asian Journal of Advanced Research and Reports* **11(4)** pp 28-35
[12] Damanik I P N, Tahitu M E 2020 Perilaku komunikasi petani dan strategi penguatan kapasitas mengakses informasi pada era revolusi industri 4.0 di Kota Ambon *Jurnal Penyuluhan* **16(01)** pp 92-104
[13] Adong A, Mwaura F and Okoboi G 2013. What are the predominant members of Uganda? Evidence from the Uganda census of agriculture 2008/9 *Journal of Sustainable Development* **6(4)** pp 37-55.

IOP Conf. Series: Earth and Environmental Science **883** (2021) 012035 doi:10.1088/1755-1315/883/1/012035

Contribution of home-garden farming to household income and its sustainability in Yogyakarta City, Indonesia

Irham[1,*], D S Gusfarina[2], A W Widada[1] and A Nurhayati[1]

[1] Department of Agricultural Socioeconomics, Universitas Gadjah Mada, Jl. Flora, Bulaksumur, Yogyakarta 55281, Indonesia
[2] Agricultural Technology Research Agency, Jambi Province, Jl. Samarinda Paal V Kotabaru 36128 Jambi, Indonesia

* E-mail: irham@ugm.ac.id

Abstract. Farming is nowadays intensively developed in urban areas. Home-garden has excellent potential in supporting household food sufficiency and providing aesthetic value for the household economy. This study's objectives are: (1) to estimate home-garden farming's contribution to household income and (2) to determine the Sustainability of home-garden farming in Yogyakarta City. The research was carried out in the city center and fringe area covering seven districts of Yogyakarta City. The sample was 113 households that utilized their home-garden for farming as the members of farmer groups. Data were analyzed using a simple statistical calculation to determine the contribution of home-garden farming income. Farming sustainability data were categorized into ecological, economic, and social dimensions using a Likert scale. The results of the study showed that the contribution of home-garden farming to household income was 0.81%. This value is relatively small because the land is very narrow. Also, the primary purpose is not for sale. Home-garden farming, both in the city centre and in the suburbs, has a high level of Sustainability, which is indicated by the economic dimension (79.11%), the ecological dimension (78.71%), and the social dimension (84.12%).

1. Introduction

The need for food in an urban area tends to rise, in line with the increase in population number. Simultaneously, however, urban sprawl around the city of Yogyakarta has reduced the area of fertile agricultural land that reduced substantially [1][2]. Therefore, preserving agricultural land in urban areas, including homegarden, is vital to provide a food supply for urban communities by themselves [3]. To cope with such a phenomenon, the government is currently encouraging urban agriculture development [4][5]. Agriculture in urban areas can be carried out in cities and suburban areas, with residential, housing complex, or open areas [6]. Management also varies, such as gardens on public land assigned to individuals or households, gardens on abandoned vacant lands as community gardens, or individual gardens on homegardens, balconies, or even on roofs of buildings [7]. It is expected in the future; each household is encouraged to optimize its resources, including homegarden, in providing food for households [8][9]. Homegarden has a great potential in supporting ecosystem services [10], household food production [11], providing aesthetic value, and sustainable household economy [12]. Although the size of homegarden is generally small, both in urban and rural areas [13], the utilization of that small size land has great benefit and value for households, including fulfilling household nutrition improvement, ensuring quality, and hygienic foodstuffs, reducing expenditure and increasing household income. The problem is that the household's attention to the use of their homegarden is still limited. The study's objectives are: 1) to estimate home-garden farming's contribution to household income; and 2) to determine the Sustainability of home-garden farming in Yogyakarta City.

2. Literature review

2.1. Farming contributions

The concept of household refers to the economic meaning of the household unit. This activity includes how the household manages its economic activities, division of labour and functions, how much income is earned or spent, and the types of products and services produced [14]. The contribution of income from a kind of activity to the household depends on the productivity of the production factors. Household income stability tends to be influenced by primary income sources. The types of income generated from outside the agricultural sector are generally not dependent on the season and can be done at any time of the year [15]. Farming in the homegarden contributes to income in the form of commodities that are sold and commodities consumed by themselves.

2.2. Farming sustainability

Research has described the homegarden as traditional farming that is ecologically, socially, and economically sustainable [16]. The primary ecological processes that occur are a competition between plants, plants' consumption by humans, livestock, and pets, and predation of pests by their natural enemies. However, the current ecological process is governed by the farming process in fertilizers, water control, pests and diseases, and harvest regulation. In turn, economic and social decisions manage agricultural operations. General indicators regarding the socio-economic Sustainability of homegarden by looking at the fact that homegarden farming contributes to nutritional security and energy needs contributes to household income even under high population density conditions. Besides, from a socio-economic perspective, adapting to significant transformations from rural to urban development is required.

2.3 . Urban homegarden

In the city, there is currently a conversion of land to a settlement. Housing development will be provided with a specific area for homegarden. The homegarden as the land around the house with clear boundaries and ownership is an important landscape that supports the beauty of housing, supplies healthy food for households, and supports the household economy. Plants in the homegarden consist of a mixture of cultivated plants such as vegetables, fruit, plantations, spices, and medicinal plants [17]. The functions of horticultural crops as productive urban landscapes can be grouped into four parts [18] namely:

1. Providing food is related to vitamins, minerals, fibre, and other compounds in fulfilling nutrition.
2. Economic function, horticultural commodities have a high enough economic value to be a source of income for farmers, traders, industrial circles, etc.
3. Health functions, especially the benefits of biopharma commodities to prevent and the threat of various diseases.
4. Socio-cultural functions, the role of horticultural commodities as an element of the beauty or environmental comfort, and their role in various ceremonies, tourism, etc.

3. Methods

3.1. Data collection

The sampling technique for the research location was carried out using *a purposive sampling technique*. The research took place in 7 districts of the 11 districts in the city of Yogyakarta. This district is divided into 2 locations, namely the city centre and the suburbs. To make it easier to find respondents who use the homegarden for vegetable and/or fruit farming, each sub-district with a farmer group that operates the homegarden for agriculture is part of several farmer groups. The seven sub-districts were sampled because they have many farmer groups working on homegardens for horticultural agriculture. In contrast, for selecting the village's location by looking at the diversity of superior horticultural commodities, it is hoped that the sample will meet diversity.

The sampling method of respondents was done by using *purposive random sampling*. Respondents in this study were households that had used their homegarden for vegetable and/or fruit farming. In this study, Respondents were 113 respondents (77 respondents were taken from the city centre and 36 respondents from the suburbs).

3.2. Measures

3.2.1 Contribution of home farming to household income. This study's data analysis is a quantitative analysis by calculating the average revenue, income, and contribution. To find out the amount of income obtained by subtracting the total revenue from the total cost, with the equation [15]:

$$I_f = TR - TC \quad (1)$$

Where: I_f = *income from farming*; TR = *Total Revenue*; TC = *Total Cost*

The total household income by using homegardens for farming in urban areas is calculated by summing up the income of all household members, namely:

$$I = I_f + I_{nf} \quad (2)$$

Where: I = Total Income of Household; I_f= Income from Farming; I_{nf}= Income from Non-Farming (salaries, wages, etc.)

The contribution of income from the homegardens to total household income is calculated as follow:

$$P = \frac{I_f}{I} \times 100\% \quad (3)$$

Where : P = contribution of home garden income to total household income (%); I_f= income from using the homegarden for agriculture; I = Total household income

The amount of contribution of income from each economic activity to total household income is used by the following criteria: a) If the revenue contribution is <25%, then the contribution is small; b) If the contribution of income is (25-49)%, then the contribution is moderate; c) If the revenue contribution is (50-75)%, then the contribution will be large; and d) If the income contribution is >75%, then the contribution is immense.

3.2.2. Sustainability of farming in urban homegarden. The Sustainability of farming in the homegarden was measured by employing a qualitative method, namely by looking at the sustainability indicators of farming using a Likert scale. The dimensions on the Likert scale are translated into measurable indicators. This indicator is then used as a starting point for making instruments in the form of statements that respondents need to answer by selecting the answers provided. The answers provided are SL = Always, SR = Often, KK = Sometimes, JR = Rarely, and TP = Never. The score is categorized 1-5 for each statement. This answer is presented with scale intervals to be a high, medium, and low reference category. The indicators used include ecological dimensions, economic dimensions, and social dimensions. Details on each dimension can be seen in Table 1.

Table 1. Dimensions, indicators, and sub-indicators of sustainability

Dimensions	Indicator	Sub indicator
Ecology	Use of input	Use of organic fertilizers
		Long time using organic fertilizers
		Use of inorganic fertilizers
		Long time using inorganic fertilizers
		Use of biological pesticides
		Duration of using biological pesticides
		Use of chemical pesticides
		Duration of use of chemical pesticide

Dimensions	Indicator	Sub indicator
		The number of commodities cultivated
	Biodiversity (biodiversity)	Number of commodity types cultivated
Economy	Household income	Increase in income
		Reducing expenses
		Reducing transportation costs
		Reducing cooling costs
		Increase space use value
	Food security	Employment Opportunity
		Increase household food security
		Food diversification
Social	Togetherness	Increase interaction between communities
		A sense of community among residents
		Extension intensity
		Access to information
	Comfort	Increase physical activity
		Improve health
		Improving aesthetics

To determine the sustainability index, the following formula is used.

$$\text{Sustainability index} = (\text{total score}/\text{maximum score}) \times 100\% \quad (4)$$

By using the above formula, the level of farming sustainability of the homegarden is categorized as follow [19]: (a) Index value 00.00-25.00 (poor/unsustainable); (b) Index value 25.01-50.00 (less/less sustainable); (c) Index value 50.01-75.00 (sufficient/sufficiently sustainable); and (d) Index value 75.01-100.00 (good/highly sustainable).

4. Results and discussion

4.1. Respondent characteristics

Most of the respondents in this study were women both in the city centre and the suburban. The condition means that the use of homegarden is indeed an area of work for mothers, especially housewives at home. The percentage of respondents in the city centre location for male respondents is lower than the suburban respondents, which means that men in the suburbs are more active in utilizing a homegarden than men in the city centre. This could be because men in suburban locations have more flexibility in using their homegardens to do business.

The lowest age of respondents is 25 years, and the highest is 77 years divided into three scales so that the interval is 17. Respondents aged 25-42 years in the city centre are very few compared to respondents in the suburbs; however, those aged 61-78 years are more in the city centre than in the suburbs. Respondents aged 25-42 years were less in the city centre because people at this age did not use their homegardens to grow vegetables and/or fruit much. There were many other side jobs available in the city centre for people of this age than in the suburbs. The age of the respondents in the city centre are relatively older than in the suburbs. Some elderly respondents in the city centre are retirees, civil servants, and private employees who are used to working productive hours. Farming in the homegarden is one of the productive jobs that can be done for the elderly; besides filling their spare time, it also provides psychological and physical benefits. For the elderly, this city centre is also a place to gather and socialize, and self-actualize. In the suburbs, there may be informal jobs that do not have age and education requirements. In general, more than half of Yogyakarta City respondents are between 43-60 years old. At that age, most of the respondents already have independent children or retired, so that they used the free time to take care of the plants in the home garden.

Respondents in Yogyakarta City have relatively high education. Only 7.79% of respondents have an elementary education in the city centre only, and half of the respondents have a high school education/equivalent. Likewise, with suburban locations, the smallest percentage is an elementary school, and the highest is SMA/equivalent. The level of education will affect motivation with the general consideration that education level reflects the level of understanding of the power of reason and knowledge of the society in assessing the right and threatening environment. This high level of education also reflects a reasonably high income. Some respondents are civil servants / retired civil servants and private employees who receive pension funds every month; other jobs are traders, laundry workers, social workers, homemakers, and sell services.

Respondents' experience in farming vegetables and/or fruit in Yogyakarta's homegarden tends to be low, with an average age of 53 years and an average experience of 11 years. It means that the average respondent starts working in the homegarden at the age of 43 years, at which time it is estimated that they have begun to have a lot of free time, for homemakers they have children who are no longer toddlers. Experience in general will affect motivation; the more experience, the more benefits we get from farming in the homegarden, the higher the motivation.

Suburban locations grow more vegetables alone than in suburban areas. Respondents in the suburbs tend to choose vegetables because they can be enjoyed quickly; for example, leaf vegetables, spinach, and kale can be harvested within a month, or fruit vegetables such as chilies can be harvested every day. Respondents who cultivated fruit trees in the city centre were more than those of the suburbs, which mostly cultivated here are potted fruit plants. City centre respondents thought potted fruit plants were more comfortable to maintain than growing vegetables. More than half of the respondents in the city centre and suburban locations cultivate vegetable and fruit crops in their homegardens; this reflects the respondents' desire to give different colors to their homegardens.

A house in the suburbs is wider than a house in the city centre, with a wider homegarden area. The homegarden's utilization in the city centre is 53.86% of the area of the homegarden, while in the suburbs, it is 48.49%. This explains that the urban community uses the homegarden more than the suburban community; this can be because the city center community values land more than the suburban community. After all, the economic value of land in the city center is much greater than that of the suburbs. The average house in Yogyakarta City is very large [8], namely 152.37 m 2 and the area of the homegarden in the city of Yogyakarta (53.58 m^2). About half have been used to cultivate vegetables and/or fruit of the homegardens' total area.

4.2. Home-garden farming's contribution to household income

Farming income in this study results from farming in the homegarden, whether it is consumed or sold. In general, in Yogyakarta, the homegarden's produce is not sold but rather for household consumption. For this reason, the analysis of farming in this study is to calculate the total production for one year of farming in the homegarden multiplied by the price prevailing in the market. Farming analysis can be seen in Table 2. Farming in the homegarden is a positive activity for urban communities; besides getting health benefits, the environment also gets economic benefits. The yield from the homegarden done optimally can meet household consumption needs, reducing the burden of spending on daily vegetable shopping. The total average income per respondent from farming in the homegarden is Rp. 400,000.00/year. The contribution is the amount of contribution given from an activity or job to household income. Table 3 showed that homegarden farming's contribution to household income in urban areas, especially the city of Yogyakarta is very small, which is only about 0.81 percent.

Table-2. Average income, costs, and income from farming in the homegarden per household

Location	Revenue (IDR / year)	Cost (IDR / year)	Income (IDR / year)
City center	582,200.00	211,100.00	371,200.00
Suburbs	683,900.00	222,200.00	461,700.00

Location	Revenue (IDR / year)	Cost (IDR / year)	Income (IDR / year)
Yogyakarta City	614,600.00	214,600.00	400,000.00

The average total income per respondent in the suburbs is greater than that of the city center respondents; this is because homegardens on the city's outskirts are indeed wider than homegardens in the city center. The commodity is grown in the city center and the suburbs are not much different namely vegetables. Vegetables widely grown in Yogyakarta are mustard greens, lettuce, spinach and kale, eggplant vegetables, cayenne pepper, curly chilies, and tomatoes. Some leafy vegetables such as lettuce and kale are also grown using a hydroponic system.

Fruit plants seen from the planting medium in Yogyakarta City are planted in pots such as; Crystal guava, lime, sapodilla, and star fruit. The plants planted directly include mango, rambutan, banana, guava, papaya, and durian. The income of farming in this study is also obtained from fruit trees planted directly, although more is distributed to neighbours than from selling.

Table 3. Contribution of homegarden farming to household income per year

Location	Farm income (IDR)	Household Income (IDR)	Contribution (%)
City center	371,200.00	51,334,545.00	0.72
Suburbs	461,700.00	45,301,388.00	1.02
Yogyakarta City	400,000.00	49,412,389.00	0.81

The contribution of homegarden farming to household income in urban areas, especially Yogyakarta City, is very small, which is only 0.81 percent. In another study by Putri [19] in rural DIY, home farming contributes slightly more than urban areas, due to the large income of urban people compared to rural communities. In town, homegarden farming is a side business, only for entertainment and filling spare time. Several other studies conducted in rural areas contributed around 3.45% to 20.37%. The fact is caused by the location of homegardens in rural areas, which tend to be wider than in urban areas.

We cannot deny the potential of urban agriculture for low-income urban communities. It's just that overcoming food insecurity should not be overemphasized because the contribution of income from the utilization of homegardens is minimal (tiny) [20].

4.3. Sustainability of homegarden farming in Yogyakarta City

4.3.1. Sustainability of homegarden farming based on ecological dimension. The index of Sustainability for homegarden agriculture (Table 4) based on the ecology dimension in the city centre is higher than that of in the suburbs in preventing pests and diseases by making biological pesticides but both are remained in the same category of sustainable enough.

Table 4. The average score of homegarden farming sustainability from the ecological dimension

Ecological dimension	City Center		Suburbs		Yogyakarta City	
	Score	Index	Score	index	Score	Index
Make your own organic fertilizer	3.61	72.2	3.69	73.8	3.64	72.8
Using biological pesticides for pest control	3.70	74.0	3.67	73.4	3.69	73.8
Do not use chemical pesticides	3.83	76.6	3.94	78.8	3.87	77.4
Make use of every possible location for growing vegetables or fruit	4.48	89.6	4.36	87.2	4.44	88.8
Can take the time to clean the homegarden	4.40	88.0	4.22	84.4	4.35	87.0

Ecological dimension	City Center		Suburbs		Yogyakarta City	
	Score	Index	Score	index	Score	Index
Always look for ways to maximize the growth of vegetable and/or fruit crops	4.29	85.8	4.03	80.6	4.20	84.4
Planning what commodities to plant	3.75	75.0	3.36	67.2	3.36	67.2
The average number of scores	*28.06*		*27.27*		*27.55*	
The highest total score	35.00		35.00		35.00	
Ecological dimension sustainability index (%)		80.17		77.91		78.71

Agricultural Sustainability in Yogyakarta City must be given an understanding of the ecosystem because people's behavior must be ecologically oriented. The use of homegardens is beneficial for humans and a habitat for fauna such as butterflies, birds, worms, and even ants, so you must be careful in using pesticides, which will also disturb the survival of the existing fauna. Besides, the preservation of fruit trees planted directly in Yogyakarta's homegarden is preserved by providing counseling for the urban community on the importance of these trees as a producer of fruit and the habitat for other living things. In the field, it was found that cutting down fruit trees was not the house owner's wish but was caused by complaints from neighbours who were disturbed by the tree. This is in line with Soemarwoto and Conway's research in 1992 [21], where the homegarden's role as a genetic resource has not been widely recognized until now.

4.3.2. The Sustainability of homegarden farming based on the economic dimension. Based on the homegarden farming sustainability index (Table 5), in the future, suburban respondents are more willing to sell their homegarden products than city-centre respondents. There is enough market for homegarden farming in the suburbs to the city centre. This is still in a fairly sustainable category.

Table 5. The average score of farming sustainability from the economic dimension

Economic dimension	City Center		Suburbs		Yogyakarta City	
	Score	index	Score	Index	Score	Index
Selling the homegarden	2.79	55.8	3.36	67.2	2.97	59.4
Can save money to get vegetables/fruit from the homegarden	4.27	85.4	4.14	82.8	4.23	84.6
The frequency of buying from the shop/ supermarket will be reduced	4.14	82.8	4.14	82.8	4.14	82.8
The cost for farming in the homegarden is not proportional to the benefits obtained	3.66	73.2	3.50	70.0	3.61	72.2
It is very easy to get input for farming	4.10	82.0	4.00	80.0	4.07	81.4
Free time will be more useful for maintaining plants in the homegarden	4.40	88.0	4.33	86.6	4.38	87.6
Will always harvest vegetables for the household from the homegarden	4.44	88.8	4.31	86.2	4.40	88.0
Households will eat various vegetables from the homegarden	4.21	84.2	4.11	82.2	4.18	83.6
The market for home crops is always available	3.53	70.6	3.81	76.2	3.62	72.4
The average number of scores	35.54		35.70		35.60	
The highest total score	45.00		45.00		45.00	
Economic dimension sustainability rate (%)		78.98		79.33		79.11

City centre respondents hope to save money on vegetables and/or fruit, reducing the frequency of buying at stalls/supermarkets and suburban respondents, in the highly sustainable category. City centre respondents also quickly get input for farming in their homegardens through an online shop. Respondents in the city centre also felt that free time would be more beneficial; in the future, households could harvest and enjoy different vegetables from the homegarden. These items in the city center fall into a highly sustainable category and in the suburbs.

City centre and suburban respondents feel that the cost for farming in the homegarden is not proportional to their benefits. They get a lot of intangible benefits; this item falls into a highly sustainable category. Overall, the Sustainability of the suburbs' economic dimension is greater than that of the city center. In Yogyakarta City, Sustainability from an economic dimension is good and highly sustainable.

All items from the economic dimension fall into the very continuous category except selling home produce both in the city centre and in the suburbs. However, this does not affect the Sustainability of the economic dimension because most Yogyakarta City respondents are not yet interested in selling their homegarden products. Their motivation is only to save money or reduce expenses for buying vegetables and/or fruit.

4.3.3. Sustainability of homegarden farming based on the social dimension. The sustainability index of homegarden farming for the social dimension is higher in the city centre than in the suburbs, as in Table 6. The sustainability index of homegarden farming from suburban respondents' social dimensions hopes that everyone will be entertained to see the homegarden. In contrast, the city centre respondents want to share the harvest. Respondents in the city centre were higher in participating in community service, caring for group gardens, exchanging experiences with others in farming, and participating in farming counselling. Suburban respondents were more active in providing input and suggestions on what crops to plant on the group's land. All are in a highly sustainable category.

Table 6. The mean score of homegarden farming sustainability from the social dimension.

Social Dimension	City Center		Suburbs		Yogyakarta City	
	Score	Index	Score	Index	Score	Index
Want anyone to be entertained to see the homegarden	4.43	88.6	4.53	90.6	4.46	89.2
Will be happy to share the crop with neighbours	4.49	89.8	4.17	83.4	4.39	87.8
Will always participate in community service to care for group gardens	4.32	86.4	4.19	83.8	4.28	85.6
Will exchange experiences with others about farming	4.38	87.6	4.19	83.8	4.32	86.4
Will always attend farmer group meetings	4.19	83.8	4.06	81.2	4.15	83.0
Will often follow counselling about agriculture	4.09	81.8	3.89	77.8	4.03	80.6
Will provide input and suggestions on what crops to plant on the group's land	3.91	78.2	4.08	81.6	3.96	79.2
It will be very easy to get information about farming in the homegarden	4.21	84.2	4.19	83.8	4.20	84.4
It is important to educate the environment about the importance of consuming vegetables	4.27	85.4	4.00	80	4.19	83.8
Will have a good relationship with extension workers and production facility providers	4.12	82.4	4.00	80	4.08	81.6
The average number of scores	42.41		41.3		42.06	
The highest total score	50.00		50.00		50.00	
Social dimension sustainability rate (%)		84.82		82.60		84.12

City center respondents can get information about farming easily. Besides, there is also a higher level of importance for the homegarden in the city center to educate the environment about consuming vegetables and maintain good relations with extension workers and production facilities providers. Each of these items is in a highly sustainable category.

The Sustainability of urban agriculture has three interrelated dimensions as described in this study, ecological, economic, and social. Compared to other dimensions, the low index of ecological Sustainability can illustrate the existence of an ecological crisis. For this reason, policymakers and urban planning, especially concerning the homegarden, can take advantage of urban agricultural development opportunities in a more just and ecologically sustainable manner [22] and socially balanced [18].

Table 7. Farming sustainability index in Yogyakarta City.

Dimensions	City center	Suburbs	Yogyakarta City
Ecology / Environment	80.17	77.91	78.71
Economy	78.98	79.33	79.11
Social	84.82	82.60	84.12

Overall, the city center's social dimension is higher than that of the suburbs (84.82% and 82.60%). The Sustainability of homegarden farming in Yogyakarta City from the social dimension is profitable and highly sustainable (84.12%). The future benefits expected from using the homegarden are strengthening the community by providing a place to meet and enjoy quality time with households and friends in a clean and calm natural environment [23]. The order of farming sustainability indexes in the city centre is, respectively, the social dimension (84.82%), the ecological dimension (80.17%), and the economy (78.98%). Meanwhile, the sequence of the index in the suburbs is the social dimension (82.60%), the economic dimension (79.33%), and the ecological dimension (77.91%). The agricultural sustainability index in Yogyakarta City as a whole for the environmental, economic, and social dimensions is useful and highly sustainable can be seen from Table 7. The highest is the social dimension (84.12%), followed by the economic dimension (79.11%) and finally the ecological dimension (78.71%). This result is highly different compared with Jakarta's city case studied by Cahya (2016), where the index of ecological dimension for the Sustainability of agricultural development was less sustainable (43.67%) [24].

5. Conclusion and suggestions for further research

The study results showed that the contribution of homegarden farming in Yogyakarta City was relatively very small but still beneficial for the urban people from the ecological, economic, and social perspectives. Seen from ecological, economic, and social dimensions, the level of urban farming sustainability index in Yogyakarta City was categorized as highly sustainable. Researchers cannot dismiss the potential of urban agriculture for low-income people to overcome urban food sufficiency. Therefore, other studies are needed to take samples of low-income people in urban areas. Besides, it is also necessary to develop other potential commodities in urban areas such as ornamental flowers and spices, and medicinal plants to contribute to household income.

6. Acknowledgements

Thank you to the Agricultural Research and Development Agency, Ministry of Agriculture, Republic of Indonesia, for the research financial support.

References

[1] Irham 2012 *Urban sprawl, food security, and Sustainability of Yogyakarta City, Indonesia*, Urban Environment) p 71-81

[2] Sudirman S, Irham, Slamet H dan Azwar M 2010 *Analisis Faktor Penyebab Dan Dampak Perubahanpenggunaan Lahan Pertanian Pinggiran Kota Yogyakarta*. J-SEP **4** 1

[3] Rusida R 2016 *Potensi pengembangan pertanian perkotaan untuk mewujudkan kawasan perkotaan Belopa yang berkelanjutan* J. Perencanaan Wilayah dan Kota **5** p 125-135

[4] Puriandi F 2013 *Proses Perencanaan Kegiatan Pertanian Kota Yang Dilakukan Oleh Komunitas Berkebun di Kota Bandung Sebagai Masukan Pengembangan Pertanian Kota Di Kawasan Perkotaan* J. Perencanaan Wilayah dan Kota **24** p 227-240

[5] Cahya D L 2014 *Kajian peran pertanian perkotaan dalam pembangunan perkotaan berkelanjutan (Studi kasus: pertanian tanaman obat rumah tangga di Kelurahan Slipi, Jakarta Barat)*. In Forum Ilmiah 11

[6] Mougeot L J 2000 *Urban agriculture: Definition, presence, potentials and risks, and policy challenges*

[7] Orsini F, Gasperi D, Marchetti L, Piovene C, Draghetti S, Ramazzotti S and Gianquinto G 2014 *Exploring the production capacity of rooftop gardens (RTGs) in urban agriculture: the potential impact on food and nutrition security, biodiversity and other ecosystem services in the city of Bologna* Food Security p 781-792

[8] Badan Litbang Pertanian 2012 *Pedoman Umum Pengembangan Model Kawasan Rumah Pangan Lestari (M-KRPL)* (Badan Litbang Pertanian: Jakarta)

[9] Gusfarina D S dan Irham 2019 *Mengukur Tingkat Motivasi Masyarakat Terhadap Pemanfaatan Pekarangan Untuk Pertanian Perkotaan di Kota Yogyakarta*. Jurnal Kawistara **9** 22 p 2018-2019

[10] Mohri H, Lahoti S, Saito O, Mahalingam A, Gunatilleke N, Irham, Hoang V T, Hitinayake, G, Takeuchi K, Herath S, 2013. *Assessment of ecosystem services in homegarden systems in Indonesia, Sri Lanka, and Vietnam* (Ecosystem Services) p 124-136

[11] Yusiana E, Slamet H, Irham. 2017 *Home Garden Eficiency in Kulon Progo District, Yogyakarta. Ilmu Pertanian* J. Agricultural Science **1** 3 p: 111-114

[12] Irham dan S Sudirman 2017 *Farm Coversion and The Sustainable City: The Case of Yogyakarta, Indonesia*. Sustainable Lamdscape Planning in Selected Urban Regions.

[13] Irwan dan Sarwadi 2015 *Lanskap Pekarangan Produktif di Permukiman Perkotaan dalam Mewujudkan Lingkungan Binaan Berkelanjutan* (Prosiding Seminar Sains dan Teknologi; Universitas Muhammdiyah Jakarta)

[14] Yulida R 2012 *Kontribusi Usahatani Lahan Pekarangan terhadap Ekonomi Rumah Tangga Petani di Kecamatan Kerinci Kabupaten Pelalawan Indonesian* J.of Agricultural Economics

[15] Suratiyah K 2009 *Ilmu Usaha tani* (Penebar Swadaya: Jakarta)

[16] Wiersum K F 2006 *Diversity and change in homegarden cultivation in Indonesia*. (In: Kumar B.M., Nair P.K.R. (eds) Tropical Homegardens. Advances in Agroforestry, **3**. Springer, Dordrecht) p 13-24

[17] Galhena D H., Freed R and Maredia K M 2013 *Home gardens: a promising approach to enhance household food security and wellbeing* J.Agriculture & Food Security p 8.

[18] Putri A S 2017 *Analisis pemanfaatan lahan pekarangan di Desa Taman Tirto, Kabupaten Bantul, Yogyakarta* (Fakultas Ekonomi dan Manjemen: IPB).

[19] Kavanagh P and Pitcher T J 2004. *Implementing microsoft excel Software for rapfish: A technique for the rapid appraisal of fisheries status*. Fisheries Centre University of British Columbia Research Reports **12** p 75.

[20] Zezza A and Tasciotti, L 2010 *Urban agriculture, poverty, and food security: Empirical evidence from a sample of developing countries* (Food policy) p 265-273

[21] Soemarwoto O dan Conway G R 1992 *The Javanese Homegarden* J. for Farming Systems (Research-Extension) p 95-118

[22] McClintock N 2010. *Why farm the city? Theorizing urban agriculture through a lens of metabolic rift* Cambridge journal of region, economy, and society)\ p 191-207.

[23] Alkon A 2008 *Paradise or pavement: the social constructions of the environment in two urban farmers markets and their implications for environmental justice and Sustainability* (Local Environment) p 271-289

[24] Holmer M 2010 *Environmental issues of fish farming in offshore waters: perspectives, concerns and research needs* AEI (1) 1

[25] Cahya D L 2016 *Analysis of urban agriculture sustainability in Metropolitan Jakarta (case study: urban agriculture in Duri Kosambi)* CITIES 2015 International Conference, Intelligent Planning Towards Smart Cities, CITIES 2015, Procedia - Social and Behavioral Sciences p 95-100

Level of escherichia coli contamination of broiler chicken meat in Ambon City Market

I J Liur [1,*], M Veerman[1]

[1] Animal Science and Technology Laboratory, Livestock products Technology Animal Science Department, Faculty of Agriculture, Pattimura University Jl. Ir. M. Putuhena, Kampus Poka – Ambon 97233

* E-mail: isye.jean@gmail.com

Abstract. Chicken meat is a source of animal protein that is very popular with all people, especially in Indonesia. The meat also contains nutrients that are good for the body. However, broiler meat also has a weakness because it is a food that is easily damaged; most of the damage is caused by poor handling, thus providing opportunities for growth microbes. Broiler meat is easily contaminated by various microorganisms that are in the environment. This study aimed to determine the level of microbial contamination in broiler chicken meat in the Ambon city market. The study used a random sampling technique and used 12 samples from the market in Ambon city. Parameters of microbiologic status observed total plate count and E. coli. Data were described descriptively. The result showed that total plate count (TPC) and E-coli was not more than the standard. The microbial contamination of broiler chicken meat markets in Ambon city was not in excess with the National Standardization Agency (NSA, 2009). Thus, broiler meat is safe for consumption.

1. Introduction

The data on the consumption of purebred chicken is according to the 2017 Central Bureau of Statistics (CBS) Staple Consumption Study of 12.13kg/capita/year. Based on statistical data, the consumption of purebred chickens in 2017 has increased. Based on statistical information, the consumption of purebred chickens in 2017 has increased by 11.2% compared to the previous year [1]. In reality, the increase in demand for chicken meat has not been in line with the rise in meat quality, especially in terms of food safety related to nutritional and microbiological content of meat-related to microbial contamination and from a health perspective.

Broiler chickens are a source of chicken meat in Indonesia. The high market demand for chicken meat is directly proportional to chicken meat production. People tend to consume chicken meat because it does not require a long processing time, relatively affordable prices, and low-fat content [2]. The increase in demand for chicken meat impacts the spread of diseases originating from animal origin to humans or foodborne disease.

Escherichia Coli (E. coli) is a pathogenic bacterium that attacks poultry meat [3] worldwide. During animal slaughter, contamination can occur through animal intestines and spread to humans, causing disease [4]. In normal circumstances, E. coli can grow in the digestive tract [5] [6] but can be pathogenic and able to attack animals and humans in certain conditions such as digestion disorders. Diseases in humans, including urinary tract infections, neonatal meningitis, and sepsis [7][8][9][10].

One of the factors in the spread of disease is through chicken meat due to trade between regions. To obtain quality chicken meat or processed products is to meet the quality requirements of poultry products set by the National Standardization Agency for Indonesia. One of the requirements for poultry products' quality is that they are free of pathogenic microbes, including E. coli, which cannot exceed the maximum microbial contamination limit for chicken meat [11]. Pathogenic microbial contamination in chicken meat and its processed products can cause various diseases for the human who consume them. Chicken meat can be contaminated with pathogenic microbes due to water from poor sanitation for the cutting and processing of chicken meat [12].

Poultry meat is suitable as a medium for microbial development because poultry tends to be dirty. Poultry meat contamination in Indonesia can also be caused by the low level of knowledge of breeders, cleanliness of cages, water, and feed sanitation. Poor cage microbes, such as E. coli in chicken meat, can cause public concern about the dangers of consuming chicken meat. Thus it becomes essential to ensure the health of consumers. Based on chicken meat's potential as a medium susceptible to E. coli bacteria contamination, the researchers wanted to know E. coli contamination in broiler chicken meat found in Ambon City markets.

2. Methods

This study used an experimental method, namely the analysis of microbial contamination in chicken meat samples scattered in several markets in Ambon City. The research was started by taking chicken breast and thigh samples from 5 market locations in Ambon City. The sample used was chicken meat that had been stored for three days. E. coli then carries out the contamination test at the Laboratory of Industrial Research and Standardization Center Ambon.

The sample is weighed as much as 25g of solid and semi-solid samples, measured the liquid sample as much as 25 ml aseptically, and then put in a sterile container. Add 225 ml of 0.1% BPW solution into the clean bag containing the sample, homogenize it with a stomacher for 1 minute to 2 minutes (except for liquid milk samples). This is a solution with a 10-1 dilution.

Testing using a series of 3 tubes, isolation-identification test, and biochemistry test. Transfer 1 ml of the 10-1 dilution solution with a sterile pipette into a 9ml 0.1% BPW solution to obtain a 10-2 dilution. In the same way as above, a 10-3 dilution is made. Pipette 1 ml each from each dilution into a series of 3 LSTB tubes containing Durham tubes. Incubate at 35^0C for 24 hours to 48 hours. Note the presence of gas-forming in the Durham tube. The test result is positive if gas is formed. Testing should always be accompanied by using positive controls. Transfer the positive culture from 4.3.5.1.1 d) using an inoculation needle from each LSTB tube into the ECB tube containing the Durham tube. Incubate the ECB at 45.5^0C for 24 hours ±2 hours; if the result is negative, incubate for 48 hours ± 2 hours. Note the presence of gas-forming in the Durham tube. The test result is positive if gas is formed. Use the Most Probable Number (MPN) table to determine the MPN value based on the number of positive ECB tubes containing gas in the Durham tube as the number of E.coli per milliliter or per gram.

The number of coliforms in the test sample is interpreted by matching the number of tubes that show a positive result, based on the MPN value table (Appendix A). The combination is taken, starting from the highest dilution, which still produces all positive tubes, while in the next dilution, there are negative tubes. The combination has taken consists of three dilutions. The sample MPN value is calculated as follows :

MPN samples (MPN/ml atau MPN/g $= \dfrac{\textit{nilai MPN table}}{100}$ x dilution factor in the middle

The data obtained were described as descriptively.

3. Results and Discussion
The research data in Table 1 show that E. coli contamination in broiler chicken meat in the Ambon City market ranges from < 2 - < 3 cfu/g. The presence of E. coli in meat shows a sign that the treatment process carried out is not hygienic because E. coli can be transferred directly from hand to mouth or passively through food and tools, and materials. Equipment contaminated with E.coli can be transmitted into the meat to became a means for E.coli bacteria to enter the blood and passively spread throughout the livestock body. E.coli that contaminated chicken meat generally comes from the room, equipment, and table where the chicken is slaughtered and the water used during chicken meat slaughter.

Table 1. E. coli content of chicken meat in Ambon City Market

Samples	E. coli Contamination	
	Chest	Thigh
A	< 2	< 2
B	< 2	< 2
C	< 3	< 2
D	< 2	< 3
E	< 3	< 3

In addition, the increase in the number of E. coli is also influenced by the intrinsic factor of the food product. E.coli contamination in chicken meat is possible due to the use of water that has been contaminated with E. coli. The samples; background also affects the total number of bacterial contaminants [12]. Research conducted by Romanda [13] states a significant relationship between personal and E. coli in food. The factors that can affect the number of E. coli contaminants in chicken meat samples include the hygienic location and processing. In addition, the water used can be a direct cause of contamination in meat as well as indirect contaminants from the place of slaughter and sanitation from the origin of the meat during the slaughtering process that has not implemented proper hygiene and sanitation as well as personal hygiene that acts in slaughter and processing and treatment meat after slaughter.

Furthermore, the environment where chicken meat is sold is also very influential on the number of E. coli bacteria contaminants. Hence, it is necessary to pay attention to the place/market the meat comes from because the slaughterhouse does not experience contamination. It can still provide an opportunity for E. coli bacterial contamination when it is in the sales place or market jerking on the meat.

In the market as a place for selling meat, it is not enough to implement hygiene and sanitation for meat containers. The temperature in the market, which is a temperature above the temperature for meat storage, can impact the growth of E. coli bacteria [14]. Temperature, humidity, and sanitation will significantly determine whether E. coli bacteria will contaminate the place or meat or not. If the environmental conditions are optimal and suitable, it will be possible for E. coli to grow well.

Research conducted by [15] stated that the percentage of positive samples for E. coli in chicken slaughterhouses was 11.5%, while the percentage of negative samples for E. coli was 88.5%. Based on the research results by Selfiana et al. [16], the number of E. coli contamination in broiler chicken meat sold in Banda Aceh Market was carried out on four traders, all of which exceed the threshold set by NSA. In addition, research [17] stated that broiler meat in supermarkets was contaminated with E. coli with an amount exceeding the NSA limit. From the study's conclusion, E. coli probably came from contamination with the environment (mostly water) during treatment.

Poor sanitation from the handling and processing management processes results in contamination by E. coli bacteria, which are environmental contamination bacteria [18]. E. coli contamination's height is the lack of hygiene and sanitation from slaughter, treatment, and selling meat.

E. coli bacteria's increased contamination will significantly reduce the meat's quality not to be consumed and cause digestive disorders for humans such as diarrhea. One of the pathways for E. coli bacteria to enter the human body is through the digestive tract. Meat that has been contaminated with E. coli bacteria if the processing is not carried out adequately, such as cooking until cooked, the bacteria can enter and develop in the human body who consume it. Consumption of food that E. coli bacteria have contaminated will undoubtedly be very detrimental because diarrhea is a deadly disease. The diarrhea disease can occur to E. coli bacteria's presence, resulting in fatal consequences such as death due to the toxins produced.

Acute diarrhea caused by E. coli bacteria accompanied by dehydration, causing nausea and vomiting, and can even cause bleeding. Some of the symptoms that often appear in sufferers of diarrhea include vomiting, heat, weakness, no appetite. Even in the feces, there are mucus and blood.

Acute diarrhea that can occur due to the E. coli bacteria, over the limit, is the second cause after rotavirus. This diarrhea will be accompanied by dehydration, which will result in morbidity and mortality, especially in children.

The sample is said to be high microbial if the number of bacterial contaminants exceeds the predetermined limit, and vice versa if the sample is said to be low microbial if the number of contaminants follows the NSA's limits 7388:2009, namely $1\text{x } 10^1$ colony/g.

Hygiene requirements are essential in the chicken slaughtering business, including cleanliness of the buildings, equipment, and tools used not to become a source of chicken meat contamination. Sanitation management includes cleaning using detergents and disinfectants to clean the room and the equipment used. In producing chicken meat, producers and consumers are expected to apply Good Manufacture Practices (GMP) and apply a food safety system or HACCP. The resulting chicken meat is safe and healthy for consumption [19].

Thus, the data obtained shows that E. coli contamination does not exceed the NSA's (2009) contaminant limit, which means that broiler chicken meat is fit for consumption to reach consumers safely. The low E. coli contamination can be caused by adequate handling and sanitation. Personal hygiene and environmental sanitation are essential factors in contaminating E. coli bacteria, including the hygiene of the equipment used in meat processing, even locations/places for meat management such as sales or markets.

4. Conclusions

Escherichia coli bacteria contaminants in 12 samples of chicken meat tested were at the NSA's threshold, making it safe for public consumption.

5. References

[1] Directorate General of Animal Husbandry and Animal Health 2018 *Animal Husbandry and Animal Health Statistics* Jakarta: Indonesian Ministry of Agriculture

[2] Alvarez-Astorga, M Capita, R Alonso-Calleja, C Moreno, B Garcia-Fernandez C 2002 Microbiological quality of retail chicken by-products in Spain *Meat Sci* **62**(1), 45-50.

[3] Bhunia A 2008 *Foodborne Microbial Pathogens* Springer, New York.

[4] Panisello P J, Rooney R, Quantick P C, Stanwell-Smith R 2000 Application of foodborne disease outbreak data in the development and maintenance of HACCP systems *Int J Food Microbial* **59**(3):221-34.

[5] Miranda J M, Vazquez B I, Fente C A, Barros-Velazquez J, Cepeda A, Franco C M 2008 Evolution of resistance in poultry intestinal Escherichia coli during three commonly used antimicrobial therapeutic treatments in poultry *Poult Sci* **87**(8):1643-8.

[6] J L Kornacki and J L Johnson 2001 *Enterobacteriaceae, coliforms, and Escherichia coli as quality and safety indicators* Compendium of methods for the microbiological examination of foods (4th ed); 69-82.

[7] Avasthi T S, Kumar N, Baddam R, Hussain A, Nandanwar N, Jadhay S 2001 The genome of multidrug-resistant uropathogenic Escherichia coli strain NA114 from India *J Bacteriol* **193** 4272-4273. Doi:10.1128/JB.05413-11.

[8] Hussain A, Ranjan A, Nandanwar N, Babbar A, Jadhav S, and Ahmed N 2014 Genotypic and phenotypic profiles of Escherichia Coli lates belonging to clinical sequence type 131 (ST131), clinical non-ST131, and fecal non-ST131 lineages from India *Antimicrob Agents Chemother* **58** 7240-7249. Doi: 10.1128/AAC.03320-14.

[9] Nandanwar N, Hussain A, Ranjan A, Jadhav S, and Ahmed N 2016 Population structure and molecular epidemiology of human clinical multi-drug resistant (MDR) Escherichia coli strain from Pane, India *Int. J. Infect. Dis.* **45** 343-344. Doi: 10.1016/j.ijid.2016.02.741.

[10] Ranjan A, Shaik S, Nandanwar N, Hussain A, Tiwari S K, Semmler T 2017 *Comparative genomics of Escherichia coli isolated from skin and soft tissue and other extraintestinal confections* Mbio 8:e01070-17. Doi: 10.1128/mBio.01070-17.

[11] National Standard Agency 2009 *Quality of Carcass and Chicken Meat* NSA. 3924:2009. Jakarta.

[12] Nugroho W S 2005 *Veterinary Public Health Aspects of Staphyloccocus Bad Bacteria that are Often Overlooked* http//:weesnugroho.staff.ugm.ac.id [8 December 2020].

[13] Romanda Fitka 2016 *Personal Hygiene Relationship With The Existence of Escherichia coli in Food at the Buffer Area of Adi Soemarmo Airport, Surakarta* Publication Text: Faculty of Medicine, Muhammadiyah University of Surakarta.

[14] Suardana I W dan I B Swacita 2009 *Food Hygiene* Udayana University Press, Bali.

[15] Kartikasari A M, Iwan S Hamid, Muhammad T E Purnama, Ratna Damayanti, Faisal Fikri, Ratih Novita Praja 2019 Isolation and identification of escherichia coli as bacterial contamination in broiler chicken meat in poultry slaughterhouse Lamongan district *Medik Veteriner Jurnal* **2** (1):66-71. DOI:10.20473/jmv.vol2.iss1.2019.66-71.

[16] Selfiana D R, Rastina, Ismail, Cut N T, Darniati, Muttaqien 2017 The level of Escherichia coli contamination in chicken meat sold in Rukoh Traditional Market, Banda Aceh *JIMVET* **01**(2):148-154. ISSN: 2540-9492 148.

[17] Ngitung R 2008 *Level of Microbiological Contamination of Broiler Meat at Supermarkets in Makassar City* Makassar Public University.

[18] Mundi N 2018 *Characterization of Antibiotic Resistance Profile in Escherichia coli Isolated From Chicken Meat Sold in Several Markets in Surabaya* [Thesis]. Faculty of Veterinary Medicine. Airlangga University.

[19] Abubakar 2003 Quality of Traditional Slaughtered Chicken Carcasses and Applying a Hazard Analysis Critical Control Point system *Journal of Agricultural Research and Development* **22**(1):33-39.

Livelihood strategy of coastal households during covid-19 pandemic: case study in Wermaktian District, Tanimbar Islands Regency

J F Sopamena[1,*], L O Kakisina[2], A E Pattiselanno[3]

[1] Agriculture Extension Program, Faculty of Agriculture, Pattimura University, Ambon
[2] Agribusiness Program, Faculty of Agriculture, Pattimura University, Ambon
[3] Magister Agribusiness, Pattimura University, Ambon

* E-mail: junianitasopamena@gmail.com

Abstract. This research aims to analyze coastal households' livelihood strategy in coping with the covid-19 pandemic, especially for the households who live by coastal forest in Wermaktian District, Tanimbar Islands Regency. Research is conducted at two villages, namely Marantutul Village and Batuputih Village. These villages are selected because the households in that village undergo both farming and fishery activities. The sample is determined purposively, involving 60 households with 30 households taken from each village. The research shows that households who live by coastal forest have used several strategies for fulfilling household necessities during the covid-19 pandemic. Those strategies are classified into three categories, respectively on-farm (intensifying farming works), non-farm (working at lumber industry), and off-farm (as a trader). Working in the lumber industry has the biggest contribution to household income, which is 53 percents. The remaining percentage is contributed by copra commodity, crops, fishery, and household services.

1. Introduction

1.1. Background

Previous research generally viewed livelihood strategy as a very important measure that enables every household to maintain life necessities. Those researches took its context from either large islands or small islands and found that the households on those islands have many options to fulfil their life necessities. Coastal households have several options for their livelihood, but mostly, their livelihood option always involves farming and fishery that require them to access land and sea. Therefore, farming and fishery are dominant activities among these households. Indeed, their livelihood activities are the product of their adjustment to the natural and environmental conditions, and this is done through the so-called adaptation process.

[1] Distinguished the livelihood strategy into three categories: farming intensification, non-farming extensification, and livelihood strategy diversification. These categories add the diversity of livelihood strategies for the households. [2] obtained several findings from their research in Playen District, Gunung Kidul Regency, Jogyakarta. It was found that most farmer households used a mixed livelihood strategy,

which comprises of three elements such as younger age of husband, higher expense per capita, and more substantial capital of household physicality. All those elements can affect household welfare. Meanwhile, the coping strategy and location-based strategy did not affect household welfare. [3] Explained that to fulfil their daily necessities and to maintain their livelihood during the unfortunate condition, sugarcane large holder (big scale farmer) households used both their natural and financial capitals, while sugarcane smallholder (small scale farmer) households spent their financial capital more dominantly. All the researches above were carried out in the context of large islands.

Other researches were conducted in the context of small islands. In general, the findings showed that farmers who live by the coast could also act as fishers if the climate makes them do so. Role switching from farmer to fisher often occurred in small islands, signifying that farming and fishery are hardly separated. [4] explained that farmers in small islands tend to use a collaborative livelihood strategy that involves role switching between farmer and fisher with climate as the main factor influencing this role adjustment. Among the livelihood strategies used in small islands are the lending-borrowing initiative (63%), the marketing of farming and fishery products (59.5%), and the use of farming outputs as collateral (42%). Kinship bonding is the most influencing base in the economic activity of households in small islands.

Central Bureau of Statistics [5] reported that the Wermaktian District population is as much as 13,396 persons consisting of 6,765 men and 6,631 women. Farming activity in Wermaktian District is dominated by crop commodities, such as dry field rice, edible tubers, beans, and corn. Those commodities are subsistence in nature or precisely used to fulfil household food necessities. Coconut is the leading commodity in the District, with a planting area width of 3,120 hectares, and the main output of this commodity is copra. The production of captured-fishery increased to 1,500 tons in 2019.

There is a diversity of livelihood strategies that the households in the Wermaktian District can use. But, only two strategies are dominant, respectively farming and fishery. Those strategies are widely used to manage livelihood sources for the fulfilment of household necessities.

1.2. Problem Formulation

Several research types on the livelihood strategy of the households in small islands have been conducted in the last two years. When the climate changes, farmers sometimes must switch their role in the household into fishers. When such role switching happens, it might be difficult to determine which activity is the main job of coastal households in small islands. Therefore, farmers-fisher households become the designation given to households to implement the economic activity in two sources of livelihood, precisely land and sea.

The researches intended by the paragraph above are [6-12]. All those researches have the same findings, more or less. Households in small islands rely on their activities on land and sea to fulfil their life necessities. Therefore, households in small islands have two natural resources for their livelihood, respectively land and sea. The utilization of those resources becomes the cultural legacy across generations. It has been recognized as local wisdom by the people of small islands to be used to fulfil their livelihood. The latest research by [12) showed that the households in Tomra Village, Leti Island, Northwest Maluku Regency, anticipated rainfall anomaly through three livelihood strategies, respectively on-farm activity by breeding livestock, off-farm activity by working at livestock ranch, and non-farm activity by working at service sector.

During the covid-19 pandemic, coastal households must choose the proper livelihood strategies to ensure that their life necessities are fulfilled. The covid-19 pandemic has impeded many livelihood activities. The Government restrains households from keeping them staying at home for a while, but this restraint disturbs financial stability. Apart from this condition, there are still a few research types on livelihood strategy during the covid-19 pandemic. The reason is possibly because this pandemic is new and becoming a serious outbreak since March 2020. Therefore, the research problem is: "What is the livelihood strategy used by coastal households to cope with covid-19 pandemic".

1.3. Research Objective
This research aims to analyze the livelihood strategy of coastal households in coping with the covid-19 pandemic and examine the contribution of livelihood source on the income of coastal households in Wermaktian District, Tanimbar Islands Regency.

2. Methods
2.1. Time, Location, and Sampling Method
The research was conducted in Wermaktian District, Tanimbar Islands Regency. Two villages were used as sample village, namely Marantutul Village and Batuputih Village. These villages are selected because coastal households in the village are implementing both farming and fishery activities. The research was carried out for two months, from May to June of 2020. The sample was determined purposively with a criterion that the households had conducted farmer and fisher activities in a year. Thirty households were taken from each village, and therefore, the total household sample was 60 households. The key informant was selected from each village to explore the answer given by the respondent. Researchers believe that the key informant has information regarding the households in the sample village.

2.2. Data Collection and Data Analysis
Primary data are usually collected by giving questionnaires to respondents [13] or conducting depth interviews with key informants [14]. Secondary data are usually obtained with participative observation [15], [13]. Researchers must participate in daily life and then listen to what the people (research subject) say and differentiate what the people and not do. The current research is processing both types of data. Data are analyzed using Simple Tabulation to describe the condition and characteristic of the research location. Data are then displayed in the table and diagram to facilitate readers to understand the research location.

3. Result and Discussion
3.1. Source of Livelihood
There are three dominant livelihood sources for the households in Wermaktian District, respectively forest clearance, fishery, and farming (planting crops). In the sample village, respectively Marantutul Village and Batuputih Village, forest clearance has the highest household income contribution. Forest land in those villages is still vast and providing timber species with high economic value. Forty individuals are working as the *chain saw operator* in each sample village. Being a *chain saw operator* had given economic value to the village people since Trans-Yamdena Road's opening. Forest clearance has become rampant in the last ten years due to high market demand every year. The greatest demand for timber-based commodity comes from traders in Wermaktian District's capital and the capital of Tanimbar Islands Regency, who are mostly Chinese descendants. The trade on timber outputs (plank and bar) is made to supply market demand not only in the locale but also in other regions. Timber species with high economic value, such as linggua timber, iron timber and torem timber, are highly demanded. Besides supply market demand, household demand for wood products (plank and bar) is also high because people in sample villages usually use them for house construction. It is supported by the fact that almost all inhabitants in the sample village cannot afford cement for constructing house wall, and they replace cement with wood plank made from linggua timber.

However, a lucrative opportunity in forest clearance works simply the people to neglect other livelihood works such as farming, fishery, service and others. If it is examined thoroughly, the latter sectors are not less important than the former in contributing to household income. The current research has few findings regarding this comparison. Among the findings is that the second-highest contributor to household income, after forest clearance as the first, is coconut commodity that is processed into copra. The next contributor is occupied by edible tubers and fishes, but these commodities are mostly produced

IOP Conf. Series: Earth and Environmental Science **883** (2021) 012037 doi:10.1088/1755-1315/883/1/012037

for household consumption or precisely not for sale. The following table indicates the detail of the contribution of various sectors to household income.

Table 1. Distribution of Sectoral Contribution to Household Income

No	Sector	Income Level (IDR/ Month)	Contribution (%)
1	**Farming**		
a	*Coconut*	600,000	6.5
b	*Crop*	200,000	2.2
	Sub Total	800,000	8.7
2	**Fishery**		
a	*Fish*	500,000	5.4
b	*Cuttle*	300,000	3.3
c	*Shrimp and Sea Cucumber*	500,000	5.4
d	*Sea Weed*	300,000	3.3
	Sub Total	1,600,000	17.4
3	**Service**		
a	*Trader of Forest and Fishery Products*	500,000	5.4
b	*Small Store*	300,000	3.3
	Sub Total	800,000	8.7
4	**Forest Clearance**		
a	*Chain Saw Operator*	5,000,000	54.3
b	*Wood Carrier*	1,000,000	10.9
	Sub Total	6,000,000	65.2
	TOTAL	9,200,000	100.00

Source: Data of Research, 2020

3.2. Contribution of Livelihood Source to Household Income

Based on the contents of Table 1, the highest contributor to household income is coming from a job called *chain saw operator*, either in the way of monthly or annually (54.3 percents). If every operator produces two cubics of timber per month, the operator will get an income of IDR 5,000,000.00 by a condition that the average sale price of timber per cubic is IDR 2,500,000.00. The commercial timber type in the market is first-class species, such as linggua timber, iron timber and torem timber. These species are usually traded in the forms of plank and bar. Plank products are often made from linggua timber, whereas bar products are produced from iron timber and torem timber. The sale price for wood products is the result of shared commitment between the *chain saw operator* and trader/buyer. If trader/buyer agrees that timber is collected at the place of their choosing, then the average sale price is IDR 3,000,000.00. If the location of timber is determined by the *chain saw operator*, for example, the timber is collected from *sawmill*, then the sale price can only be as much as IDR 2,500,000.00 per cubic. Besides first-class timber, *chain saw operator* also processes second-class timber such as white timber, yellow timber and others. The sale price per cubic of second-class timber ranges from IDR 1,200,000 to 1,500,000. Builders usually buy this second-class timber to be used as construction material, particularly as storey supporter during concrete pouring. Therefore, when the households plan to build or renovate the house, second-class timber becomes a popular option because it is cheap and reliable. The high demand for timber among the locals is reasonable because almost all inhabitants' houses in the research location use wood plank wall. If the house owner must hire the work of *chain saw operator*, then the owner must pay the costs of fuel (gasoline and lubricant oil) and meal (food/beverage and cigarette).

Trans-Yamdena Road's opening that connects many villages in Wermaktian District indirectly increases the frequency of lumber logging. The road helps the households to access other villages for selling their products, but the road is built after clearing several forest lands, which signifies that the number of critical lands is increasing and the households must take the long walk to find woods for their house or to buy first-class timber for commodity. For instance, they must take a journey for 10-15 kilometres or spend 1-2 hours of journey to go to the logging site.

Moreover, the fishery sector also contributes to household income in Wermaktian District, and its contribution level is 17.4 per cent. Shrimp and sea cucumber are commodities harvested once a year. The regular customer for those commodities is traders who usually act as a reseller in the regency capital. Fisher's breed shrimp nearby the downstream or along the bay of sample village. The harvest is done in October – December using net and *Bubu* (fish trap). In the village market, shrimp is traded with a sale price per bundle of IDR 20,000 – 30,000. Sea cucumber is traded usually in dry form, with a sale price of IDR 400,000 – 500,000/kg for the superclass. Regular sea cucumber is sold at the price of IDR 50,000 – 75,000/kg.

Two sectors give the same contribution to household income, respectively the service sector and farming sector. The contribution level of both sectors is 8.7 percents. Two players dominate the service sector, respectively traders of forest and fishery products and small stores (that sell nine staples). In each sample village, usually, there are 8 – 10 households that open staple small store. Not many villagers are dared to be traders of forest and fishery products because this profession is expensive. Being a trader requires them to take a journey from the trade centre, usually the capital of regency, to the potential buyer, signifying that it involves huge early capital.

The outputs of the farming sector are not for trade but household consumption (subsistence). The only farming output for trade is copra, but the designation for sale is not given to edible tubers such as cassava, sweet potato, yam (*gembili*), and others. Fruits, such as mango and orange, are sometimes traded, but the trade is often done once a year and only relies on collector trader in the village/district to buy the fruits and then sell them in the regency market.

4. Conclusion

Coastal households use several strategies to fulfil their livelihood during the Covid-19 pandemic. Those strategies are classified into three categories, respectively on-farm (intensifying farming works, especially for crop commodity and forest products), non-farm (working at lumber industry), and off-farm (as a trader). Working in the lumber industry, which allows utilizing timber products, has biggest contribution to household income, which precisely is 65.2 percents. The remaining contribution is given by other sectors, respectively fishery sector through shrimp and sea cucumber, service sector through traders of forest and fishery products, and farming sector through copra and subsistence crop.

The utilization of timber products must be considered because improper logging management may impair the forestland in Wermaktian District. Soil degradation in the corrupted forestland may badly impact environmental sustainability. The local Government must regulate the lumbering activity and utilize timber products, at least in favour of the next generation.

References

[1] Fridayanti, Novia dan Dharmawan A H 2013 Analisis Struktur dan Strategi Nafkah Rumahtangga Petani Sekitar Kawasan Hutan Konservasi Di Desa Cipeuteuy, Kabupaten Sukabumi *Sodality Jurnal Sosiologi Pedesaan* **1**(1) 26 – 26.
[2] Sugiharto, Adam, Hartoyo, Muflikhati I 2016 Strategi Nafkah dan Kesejahteraan Keluarga pada Keluarga Petani Tadah Hujan *Jurnal Ilmu Keluarga dan Konsumen* **9** (2) 33-42.
[3] Budiyanti, Indah dan Dharmawan A H 2018 Strategi Nafkah Dan Relasi Sosial Rumahtangga Petani Tebu (Studi Kasus: Desa Jenar, Kecamatan Jenar, Sragen) *Jurnal Sains Komunikasi dan Pengembangan Masyarakat [JSKPM]* **2** (1) 105-122; DOI: https://doi.org/10.29244/jskpm.2.1.105-122; Copyright ã 2018 Departemen SKPM – IPB; http://ejournal.skpm.ipb.ac.id/index.php/jskpm ISSN: 2338-8021; E-ISSN: 2338-8269
[4] Pattiselanno, August E, Tuhumury M T F, Wenno N F, Sopamena J F 2017 Collaborative Livelihood Strategy: A Reflection of Social Network in Economic Activity (Case Study in Small Islands, Maluku Province, Indonesian) *International Journal of Environment, Agriculture and Biotechnology (IJEAB)* **2** (5), *http://dx.doi.org/10.22161/ijeab/2.5.25 ISSN: 2456-1878*
[5] BPS 2019. Kecamatan Wermaktian dalam Angka Tahun 2019. Saumlaki : BPS Kabupaten Kepulauan Tanimbar
[6] Sopamena J F, and Pattiselanno A E 2018 Tnyafar: Women, Livelihoods Strategy in Selaru Island, West Southeast Maluku District *International Journal of Environment, Agriculture and Biotechnology (IJEAB)* **3** (5) 1685-1690.
[7] Sopamena J F 2019a Peran Gender Dalam Rumahtangga Masyarakat Pulau Kecil (Studi Kasus Kecamatan Teluk Ambon Baguala Kota Ambon) *Jurnal Agribisnis Terpadu* **12** (1) 72 – 86.
[8] Sopamena J F 2019b Resiliensi Nafkah Rumahtangga Masyarakat Pulau Kecil (Studi Kasus Pulau Wetar Kabupaten MBD) *Jurnal Habitat* **11** (2) 79-87.
[9] Sopamena J F 2019c Kontribusi Perempuan Terhadap Penerimaan Rumahtangga Masyarakat Pulau Kecil (Studi Kasus Kecamatan Teluk Ambon Kota Ambon) *JEPA* **3** (4) 720-729.
[10] Sopamena J F 2020a Women and Onion Farming in Lakor Island, Southwest Maluku Regency *SOCA (Jurnal Sosial Ekonomi)* **14** (2) 265-274.
[11] Sopamena, Junianita F., 2020b. Domestication Of Women Role In Small Island (Case Study Woman At Tnyafar In Selaru Island, Tanimbar Island Regency) *International Journal of Innovation, Creativity, and Change (IJICC)*1021-1036
[12] Sopamena J F, Pattiselanno A E, and Semuel Leimehariwa 2020 Strategi Nafkah Dan Pola Adaptasi Terhadap Anomali Curah Hujan Di Desa Tomra Kecamatan Leti Kabupaten Maluku Barat Daya *Jurnal SEPA UNS* **17** (1) 24-32
[13] Babbie, Earl., 2004. The practice of social research. Publisher : Belmont, CA : Thomson / Wadsworth.
[14] Debus, Mary and Novelli, Porter. 1996. Methodological Review: A Handbook for Excellence in Focus Group Research. Washington D.C: Academy for Educational Development.
(15) Denzin dan Lincoln, 1994. Handbook of Qualitative Research. Publisher: Thousand Oaks : Sage Publications.

Analysis of macro nutrient content in biofouling waste organic fertilizer pearl oyster (*Pinctada maxima* L.)

J I Nendissa[1,*], M H Makaruku[1], V L Tanasale[1], A K Kilkoda[1], J Taribuka[1]

[1] Agrotechnology Study Program, Faculty of Agriculture, Universitas Pattimura, Ambon, Indonesia

* E-mail: jinendissa@gmail.com

Abstract. Biofouling development is a major problem in bilvave aquaculture around the world. The purpose of this study was to obtain macro nutrient analysis data for organic fertilizer from pearl oyster (*Pinctada maxima* L) biofouling waste. Where is the manufacture of solid organic fertilizer products that have quality equivalent to other commercial organic fertilizers by utilizing pearl oyster (*Pinctada maxima* L.) biofouling waste so that high quality solid organic fertilizers are produced which are in accordance with the Quality Standards for solid organic fertilizers and the Indonesian National Standard. Organic fertilizer from pearl oyster biofouling waste is obtained by using pearl oyster biofouling waste which is taken from the results of cleaning microorganisms attached to the pearl oyster shell area. Another major concern regarding biofouling is the potential for food competition caused by the attachment of the filter-feeder. Besides, the waste of biofouling pearl oysters is extracted and soaked in fresh water for one week at a time. Afterward, the pearl oyster biofouling waste is dried in the sun to dry, odorless, blackish brown in color. Analysis of macro nutrient content using Atomic Absorption Spectrophotometer and UV-Vis Spectrophotometer. The results of the analysis of macro nutrient content were Water Content (5.76%), C (14.48%), Organic Matter (24%), N-total (0.92%), P-total (0, 25%), K-total (1.02%), C/N ratio (15.72).

1. Introduction

One of the Indonesian exports to Marine farming and its high selling value is pearl oysters (*Pinctada maxima* L.) that produce pearls. Pearl oyster (pinctada maxima) is so highly prized by society, that so many newly interested ones from pearl oysters (pinctada maxima) and have the prospect of future development.

The high market demand for pearls both from within and outside the country has prompted cultivators to increase production. In addition, pearl oysters are one of the mainstay marine biota for marine cultivation that has developed in several locations in Indonesian waters and their production is increasing but their supply in nature is not proportional to the rapid market demand, so that the pearl oyster population is decreasing and the price continues to increase.

Biofouling development is a key issue in bivalve aquaculture worldwide [1]. Among the problems faced in raising pearl oysters are attacks of parasitic organisms on shells that can interfere with such vital functions as opening and closing the shell or reducing the economic value of pearl oysters and may prevent streams in and out of the crowding as well. Many synthesizing organisms can compete with such cultured organisms as pearl oysters for food. The negative effect of this downdraft can be greater in the tropics. Generally, harmful traits are attributed to stamping organisms and a way to minimize their effect especially on pearl oyster farming.

Besides another barrier to the pearl oyster culture is the biofouling or the organisms (phytoplankton or zooplankton) attached to shells or nets and thus having an effect on the quality of the pearl or the oyster productivity [2]. Physical damage would result if fouling organisms such as mushrooms or porifera disturbed nutiara oysters, thus reducing pearl production. Barnacles also interfere with mechanical functions such as opening and closing valves of pearl oysters in the mouth or hinges so as to have a negative impact by being a proxy (phytoplankton) [3].

The waste from pearl oyster shell (*Pinctada maxima* L) is left over from the pearl clam treatment industry. If it is not cared for and run properly, it is potentially a waste that accumulates and results in environmental harm. Based on the observations I have made of coastal areas especially on the western shore, the village of stone trees has become a pile of untreated, untreated, pinctada maxima. If it is not cared for and run properly, it is potentially a waste product and adverse to the environment.

Waste can be used as organic fertilizer, compost, and others. The production of organic fertilizer from the effluent of pearl oysters has never been done, but results found to contain high edible nutrients that can be used to promote plant growth and productivity. Growing crop production [4] is recommended.

2. Methods

Basicly, before organic fertilizer was made, the solid waste of the pearl oyster was analyzed. The test for macro nutrient content of pearl oyster biofouling waste was carried out at the Integrated Research and Testing Laboratory of Gadjah Mada University. The removal of a sample of the pearl oyster biofouling waste was taken from the remnants of the pest and pearl oyster that attacked the shell. The cleaning was done using both manual and spray. After that the sewage is taken and soaked in fresh water for a week. Soaking water is replaced every day. The goal of submersion is to remove seawater. Then the waste of biofouling pearl oysters was left to dry in the sun for one week. Observations are made of color, smell and water content. It should be dark brown, dry and odorless (SNI 19-7030-2004). Before organic fertilizer was made, the solid waste of the pearl oyster was analyzed. The gynecological exam of macrowaste biofouling oysters was done in the research lab and integrated test of Gadjah Mada University. After that the sewage is taken and soaked in fresh water for a week. Soaking water is replaced every day. The goal of submersion is to remove seawater. Then the waste of biofouling pearl oysters was left to dry in the sun for one week. Observations are made of color, smell and water content. It should be dark brown, dry and odorless (SNI 19-7030-2004).

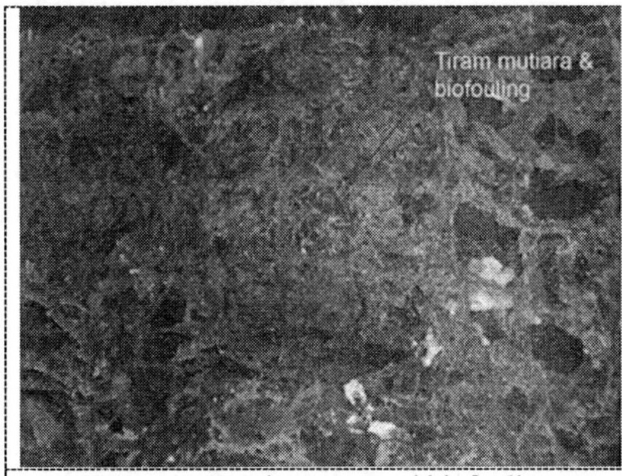

Figure 1. Pearl oysters and biofouling

Figure 2. Manual shell cleaning

Figure 3. Cleaning by spraying

Figure 4. Biofouling drying

3. Results and discussion

Table 1 shows that the nutrient content contained in pearl oyster biofouling waste can be used as material for the manufacture of organic fertilizers.

Table 1. Data from analysis of macro nutrient content of pearl oyster biofouling waste

No.	Parameter	Result	SNI 19 – 7030- 2004
1.	Water Content	5,76 %	0 – 50 %
2.	Organic Matter	14,48%	15 – 25 %
3.	Carbon-organic	24,96 %	9,8 – 32 %
4.	N-Total	0,92 %	Minimum 4 %
5.	P-Total	0,25 %	Minimum 1 %
6.	K-Total	1,02 %%	Minimum 2 %
7.	C/N ratio	15,72	10 - 25

Source: Integrated Research and Testing Laboratory of Gadjah Mada University and SNI-2030-2004 in Wellang (2015)

Based on Indonesian National Standardization SNI-19-7030-2004, analysis of the waste biofouling pearl oyster (LBTM) was obtained that the macro-contained nutrients contained within them qualify them for use as dense organic fertilizer. Organic fertilizer is the result of decomposition of organic materials both dry (humus) and sewage from disstored livestock manure by microbes to the point that it provides the nutrients that plants need for their growth and development. It means so much as a buffer of the physical, chemical, and biological properties of the soil that it can increase the efficiency of soil fertilizer and productivity [5]. The pearl oyster biofouling, deeper than it is, contains a considerable number of macro nutrients and a potential in organic fertilizer. It is thought to contain multiple microorganisms attached to the shell. According to [6] 36 types of biofouling from the two great kingdoms of the animal kingdom of 21 types came from six phyla (arthropoda, bryozoa, annelida, cnidaria, porifera and mollusca), and a 15-type plant world of Divisio Spermatophyta and Thallophyta. Found 14 varieties of Thallophyta and one of Spermatophyta. Of the Arthropoda are found 11 types, three types of the Bryozoa, four of Annelida, and one of Phyla Cnidaria, Porifera and Mollusca. [7] stated that plants will thrive when nutrients needed by plants are available in a balanced proportion to particularly macro nutrients such as N, P and K. This element has a specific role to play in plant growth. In the growth of the nitrogen (N) elements, plants play a role in the production of chlorophyll in photosynthesis, where photosynthesis is good, more carbohydrates will be produced [8]. The function of phosphorus (P) as one of the elements of protein reduction, needed for the formation of flowers, fruit and seeds, stimulates the growth of roots to lengthen and grow strong so that plants will survive drought. A deficiency of phosphorus would cause the plant to dwarf, impeding and stunted seed growth. The potassium element (K) has a role in characteristics such as photosynthesis and respiration that are essential to plant growth. C-organic produced from the waste of biofouling pearl oysters at table 1 is 24,96%. When compared with Indonesia's National Standard of 15-25%, it can meet the standard of organic fertilizer. C-organic has a function that enhances soil fertility and provides micro-nutrients and other growth factors. The level of water contained in the biofouling waste of 5.76% suggests that the level of maturity in organic fertilizer is higher. The ripe organic fertilizer smells like soil, as the material it contains resembles dark, brown soil, which is formed by the influence of stable organic material. The organic material formed from the waste of biofouling pearl oysters is 14.48%.

The C / N ratio of pearl oyster biofouling waste is in the range of the C / N ratio indicated by SNI 19-7030-2004, namely 15.72. This is due to the pearl oyster biofouling waste consisting of constituent materials containing algae, worms, plants and other microorganisms that can increase soil fertility. This is due to the pearl oyster biofouling waste consisting of constituent materials containing algae, worms, plants and other microorganisms that can increase soil fertility. This is due to the pearl oyster biofouling waste consisting of constituent materials containing algae, worms, plants and other microorganisms that can increase soil fertility. This is due to the pearl oyster biofouling waste consisting of constituent materials containing algae, worms, plants and other microorganisms that can increase soil fertility. The latter disintegrates because it is destroyed by the natural breakdown of the microorganisms that live inside it. This conforms to SNI Standards 19-7030-2004. The price of LPG in 12-kg and 50-kg containers in the second quarter of 2007 stood at us $58.9 million, he said. This is because the effluent of pearl oysters consists of shrink material containing algae, worms, plants and other microorganisms that can increase soil fertility.

4. Conclusion

The quality of organic fertilizer produced from the pearl oyster wastes are those: the stuff at macro, P and K are 0.92%, 0.25%, and 1,02%. Nisbah C/N 15.72, C-organic 24.96%, 14.48% Organic Matter and Water Levels of 5.76% have met the established standard of organic fertilizer.

References

[1] Lacoste E, Le Gilles G, Moullac, Levy P, Gueguen Y, and Gaertner-Mazouni N 2014, Biofouling development and its effect on growth and reproduction of the farmed pearl oyster Pinctada margaritifera *Aquac. J.* **434** 18–26.

[2] Pit J H and Southgate P 2003 Should slow growing pearl oyster (Pinctada margaritifera) spat (Runts) be discarded.

[3] Southgate and Lucas 2008 Seasonal fouling stress on the farmed pearl oyster, Pinctada fucata, from shoutheastern Arabian sea *J. World Aquac. Soc.* **43** 514–525.

[4] Juliutomo D, Mirawati B, and Imran A 2018 Media tanam campuran limbah cangkang kerang mutiara (*Pinctada maxima*) untuk pertumbuhan tanaman jagung (Zea mays) *J. Ilm. IKIP Mataram* **5** ISSN:2355-6358.

[5] Wijana N Y and Adnyana G M 2012 Aplikasi jenis pupuk organik pada tanaman padi sistem pertanian organik *E-jurnal Agroekoteknologi Trop.*

[6] Jefri E, Yasir I, and Syafiuddin 2017 The composition of biofouling types in pearl oysters (*Pinctada maxima* L.) on the cultivation land of Pt. Autore Pearl Culture Lombok *Spermonde J.* **2** 9–16.

[7] Priangga and Suwarno 2013 Pengaruh Level Pupuk Organik Cair Terhadap Produksi Bahan Kering dan Imbangan Daun-Batang Rumput Gajah Defoliasi Keempat. Fakultas Peternakan Universitas Jenderal Soedirman Purwokerto.

[8] Kurniawati H Y, Karyanto A, and Rugayah 2015 Pengaruh pemberian pupuk organik cair dan dosis pupuk NPK (15:15:15) terhadap pertumbuhan dan produksi tanaman mentimun (*Cucumis sativus* L.) *J. Agrotek Trop.***3** 30–35.

Potential of *Bacillus spp* as a biocontrol agent against Ralstonia bacterial wilt in bananas

J V Hasinu[1,*] , G N C Tuhumury[1], Henry Kesaulya[2]

[1] Department of Agrotechnology, Faculty of Agriculture, Pattimura University, Ambon, Indonesia
[2] Department of Plant Breeding, Faculty of Agriculture, Pattimura University, Ambon, Indonesia

* E-mail: jeffijhasinu@gmail.com

Abstract. Bacterial wilt disease in banana plants caused by *Ralstonia solanacearum* is an important disease that can reduce banana production. Until now there is no effective method to control this disease. The use of biocontrol agents such as *Bacillus spp* is an alternative method of controlling *R. solanacearum* in bananas. The aim of this study was to determine the activity of *Bacillus spp* against banana wilt disease caused by *R. solanacearum*. In vitro testing was carried out in the Plant Physiology Laboratory, Faculty of Agriculture, Pattimura University. The study used *Bacillus subtilis* strain SW116b and *Bacillus subtilis* strain HPC2-1 isolates. The results showed that the SW116b stain *Bacillus subtilis* has the highest activity against *R. solanocearum*, which is 10.5mm, so it has the potential as a biological control agent in suppressing the development of ralstonia wilt disease in bananas.

1. Introduction

Banana is a plant that is widely cultivated in regions that have hot and humid climates such as Southeast Asia, the Pacific Islands, Papua, South America and Africa. The nutritional content of Bananas for every 100 grams is 89 calories, 22.8g carbohydrates, 75% protein, 2.6g fiber and 8.7mg vitamin c. Based on data from the [1, Indonesia produced 7 million tons of bananas, in 2017 it was 7.16 million tons and in 2018 it increased to 7.26 million tons. Banana production in Maluku Province in the last three years has decreased as follows: in 2017 amounted to 35,196 tons, in 2018 decreased to 33,319 tons and in 2019 amounted to 5.33 tons. One of the factors of decreased production in this area is the presence of disease.

In its growth and development, banana plants often experience disease disorders such as fusarium wilt and bacterial wilt. This disease is caused by two very dangerous types of soil borne pathogens, namely fusarium wilt caused by *Fusarium oxysporum* f. sp. cubense (Foc) [2], [3] and bacterial wilt caused by the bacterium *Ralstonia solanacearum* [4]. The two pathogens can attack one banana stem at the same time [5].

Blood disease is still a major obstacle in banana cultivation in Indonesia. This disease is often found in Kepok bananas but other varieties can also be infected. The symptoms caused are very similar to those of Moko's disease in Latin America which is caused by *Ralstonia solanacearum* race 2 [6]. In adult plants the base of these leaves will break so that the leaves will hang around the stem. The pups show wilting, although the infection is not always systemic. In banana plants that have already produced fruit, the heart of the banana looks dry and shriveled and blackened. To see other symptoms, if the stem is cut, it appears in the vascular vessels there is necrosis that is brownish red. In addition, from the cut part, white to reddish brown or blackish bacterial mucus will appear.

When viewed from the outside, bananas that are stricken with blood diseases often look healthy. However, if the banana is cut, it will rot and contain bacterial mucus that is reddish yellow or blackish red. Blood disease is spread through several intermediaries, namely from one place to another by humans through plant material or fruit obtained from infected plants. Another method of spread is through pollinating insects stopping at sick flowers and then stopping at healthy flowers. This disease can also occur because through sick seeds, through agricultural means, watercourses, and means of transportation. The way to control blood diseases can be done through preventive methods, controlling the infectious insects / disease vectors and curative methods, namely by eradicating plants.

Bacterial wilt control still uses fungicides and technical cultures, but has not been able to reduce the incidence of disease in the field. To overcome this, it is necessary to develop biocontrol methods that are safer and more environmentally friendly. Biological control is directed as research on antagonist agents develops. *Bacillus spp* is a genus of bacteria that is reported to be able to increase plant resistance [7].[8] Morphological and physiological characters as well as the inhibition mechanism of *Bacillus spp* need to be studied to determine their effect on banana bacterial wilt infection. Recent studies on microbial siderophores in the rhizosphere are associated with their biocontrol activity due to their competitive effects with plant pathogens. Bacteria produce a wide variety of siderophores under iron-limiting conditions: hydroxamates, phenol-catecholates, and carboxylates. In nature, different types of siderophores such as hydroxymate, catecholets and carboxylates, are produced by different bacteria [8]. In addition to siderophore activity, siderophore bacteria have an important role in inducing resistance in plants to pathogenic infections [9]. The purpose of this study was to determine the effectiveness of thestrain culture *Bacillus spp* bestfor its development in suppressing Ralstonia wilt disease in banana plants.

2. Methods

The research was conducted at the Laboratory of Physiology and Disease Pathology, Faculty of Agriculture, Pattimura University.

2.1. Sources Bacteria Ralstonia solanocearum

Samples of banana plants suspected of being infected with *Ralstonia solanocearum were* obtained from banana plantations owned by farmers in Taeno village, Ambon city. Planting the sample was carried out by pour plate on the media (NA). identification is done macroscopically and microscopically.

2.2. Sources bacteria Bacillus spp

Bacteria *Bacillus subtilis and Bacillus subtilis Strain Strain SWI16b* HPC2-1 origin rhizosphere potato used in this study were obtained from the collection of the Laboratory of Plant PhysiologyFaculty of Agriculture, Pattimura University. Bacterial isolates were propagated using nutrient agar medium (NA) and incubated at 28°C for 24 hours. The purity of the isolates was then rejuvenated using the quadrant method. Test the effectiveness of the culture of *Bacillus sp*. against *R. solanocarum.*

2.3. The Implementation of The Test

Media used in this test uses two media to confirm the inhibitory ability of the pathogen through the suppression activity that is formed. The media used in this test were: NA (meat extract, 10 g; peptone, 10 g; NaCl, 1.5 g; Agar, 15 g; Aquadest, 1000 ml); King`s B Agar 10% (Proteose peptone No.3, 2 g; glycerol, 1 g; K2HPO4, 0.15 g; MgSO4.7H2O, 0.15 g; Agar, 15 g; Aquadest, 1000 ml). In vitro antagonism test against *R. solanacearum was* carried out using a modified technique developed by [10,11]. Pure cultures of isolates *B. subtilis* aged 48 hours were used. Sterile filter paper pieces with a diameter of 5mm were put in the bacterial suspension for ± 1 minute and drained. After incubation, the plates were turned over and added 1 mL of chloroform, then left for 2 hours at room temperature. After all the chloroform vapor had evaporated, the petri dishes were reversed to their original position, then 200 μL ofsuspension was poured *R. solanacearum* in 4 mL to 0.6% water. The cultures were incubated for 24 hours at 29°C to observe the resistance zone. After the incubation period, isolates that produced inhibitory compounds were recorded as indicated by the presence of an inhibition zone around the antagonistic bacterial colony.

2.4. Observation Parameters

Parameters observed were colony morphology on NA and King's B media. Functional test of bacteria *Bacillus spp* and their inhibitory activity against *isolates R. solanocearum* in both media. Qualitative analysis was used on the growth of *Bacillus subtilis* stain HPC2-1 and *Bacillus subtilis* stain SW116b on different growth media.

3. Results and Discussion

From the results of colony morphological observations, the isolates of *Bacillus subtilis* strain SW116b and *Bacillus subtilis* HPC2-1 showed the same morphological characters in shape, color, edges and colony elevation. Colonies of Bacillus isolates that grew on nutrient media were generally round in shape, smooth edges, raised elevations and colony colors were white to brownish. (Figure 1). Bacillus colonies can have different shapes and structures because colony growth is influenced by environmental conditions and the character of each species and specific strains [12]. The results of the population count of the two Bacillus strains after incubation for 1 day can be seen in Table 1.

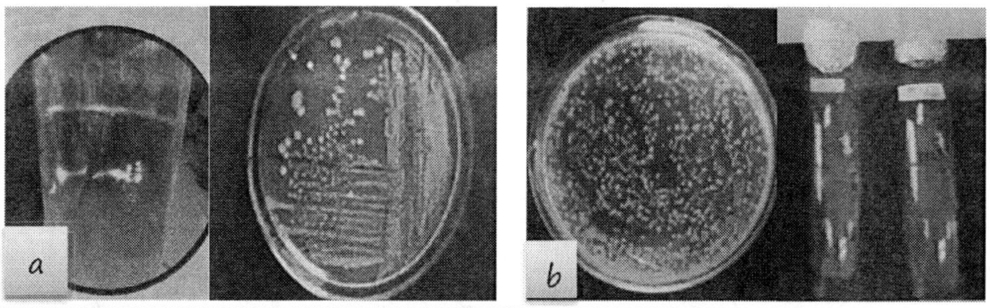

Figure 1. a.isolate *R. solanocearum* from banana; b.isolates *B. subtilis*

Table 1. Colony Morphology Bacillus spp Results Reisolasi and Rejuvenation

Bacteria isolates	Total population	form colonies	Bank	elevation	Color colony
Bacillus subtilis strain SW116b	3.5×10^5	Round	Whole	Average	White
Bacillus subtillis HPC2-1	3.8×10^5	Round-Up	Whole	average	White

In the table above is known to the average population in the total population of each Bacillus strain is a strain of *Bacillus subtilis* SW116b total population of 3.5×10^5 lower than the *Bacillus subtilis* HPC2-1, 3.8×10^5.

Gram staining of detected Bacillus isolates was carried out to ensure that all of these species were Bacillus. Morphological observations were made to ensure that all the isolates were Bacillus.

3.1. In Vitro Activity Test Against Banana Bacterial Wilt

The antagonistic test showed that *Bacillus spp.* had inhibitory activity against wilt disease pathogens in banana plants. Inhibition ability is indicated by the clear zone. The differences in inhibition by eachstrain are *Bacillus spp.* shown in Table 1. The *Bacillus spp.* showed same order of biocontrol determinants production [13] including the wilt disease pathogen in banana plants.

Table 2. Inhibition Zone Formation by Biocontrol Agents on Media Type

Bacillus spp	Inhibition zone diameter (mm)	
	Nutrient Agar	King, s B
Bacillus subtilis Strain SWI16b	7.5	10.5
Bacillus subtilisstrain HPC21	7.5	9.5

The formation of the inhibition zone byisolates was *Bacillus spp* different from one another. Based on the results of the research, the *isolates of Bacillus subtils strai SW116b* and *Bacullus subtilis strain HPC 2-1* had the tapping activity of inhibition zone in King, B media better than. The formation of the inhibition zone indicates the presence of an antibiotic mechanism and the ability to produce antibiotics in *Bacillus spp* which can be used as a biocontrol agent. The compounds and sugars contained in the media have an effect on the formation of the inhibition zone.

The difference in the zone of inhibition in the Bacillus strain tested is thought to be due to the conditions and nutrient content of the media used. [14] explained that antibiosis compounds are strongly influenced by the composition of the medium, both quantitatively and qualitatively. The results of the in vitro antagonism showed that the two Bacillus isolates were able to inhibit pathogen growth. The highest percentage of inhibition was produced by *Bacillus subtilis Strain SWI16b* in King's B while the lowest was produced by *Bacillus subtilis Strains* HPC2-1 in NA. This shows that Bacillus isolates tested in vitro have the potential to inhibit the growth of pathogens in culture media. Bacillus spp. has been known as a bacterium capable of producing various antibiosis compounds. According to [15] that *Bacillus spp.* can produce polypeptide-subtilin antibiotics, gramicidine, bacitracin, polymyxin, phytoactin and bulbiformin. This is clearly seen from the results of the test for *Bacillus subtilis* in-vitro. According to [16] *Bacillus subtilis* on NA and King, s, B medium can suppress the growth of *Ralstonia solanacearum*. Bacillus isolates can produce antibiotic compounds capable of suppressing colony growth which are toxic to other microbes, one of which is bacitracin which can inhibit the formation of pathogenic cell walls [17, 18]. Furthermore, the *Bacillus subtilis* strains SW116b and *Bacillus subtilis* strains HPC2-1 in addition to producing antibiosis also have a high ability to colonize, so that these strains are able to compete in space and nutrition with pathogenic bacteria, including soil borne pathogens such as *R. solanacearum*. This is because of *Bacillus subtilis*. able to use various substrates as a source of nutrition, and have a much faster growth than pathogenic bacteria, so that they can maintain the population optimally [19]. Space competition between *Bacillus subtilis* strain SW116b and *Bacillus subtilis* strain HPC 2-1 with pathogenic bacteria occurs through restriction of secondary development and spread of pathogenic bacteria by *Bacillus subtilis is* thus widely distributed [20]. In addition, nutritional competition also occurs as a result of a high population increase of *Bacillus subtilis,* especially in using carbon, nitrogen, and $Fe3 +$ sources for growth and activity which can result in limited nutrient sources available for pathogen needs [21]. In this case, the effectiveness of *Bacillus subtilis* in suppressing pathogens is determined by its ability to produce antibiosis, induce resistance, competition, and colonize the root system over a long period of time and environmental factors and the spread of bacteria in the soil [21]. This shows that the Bacillus isolates tested in vitro have the potential to inhibit the growth of pathogens in culture media.

3.2. Bacillus spp. Functional Test

Table 3. Results of the functional test for Bacillus spp.

Bacteria Active Ingredients	Gram reaction	Solvent phosphate
Bacillus subtilis strain SW116b	+	+
Bacillus subtillis HPC2-1	+	+

The ability of Bacillus isolate to dissolve phosphate qualitatively was seen from the ability of Ca3 (PO4) 2 as a source of phosphate bound to the media. The clear zone that forms around the colony indicates that the bacteria are able to dissolve phosphate. The two Bacillus isolates had different clear zone diameters, namely on King medium, s B 9.5 to 10 mm, while on NA medium only 7.5 mm. This is thought to be influenced by the organic acids produced by these bacteria. These organic acids released by bacteria then react with Ca3 (PO4) 2 and form chelates (stable complexes) with P-binding cations (Ca4 +) accompanied by the release of HPO4 [23].

4. Conclusion

Based on the research results, it can be concluded that the isolate *bacillus subtilis* strains SW116b and HPC2-1 can inhibit the growth of *R solanocearum*. Isolate *bacillus subtilis* strain SW116b in King; s B medium had the highest inhibitory activity, which was 10.5mm. Both isolates have the potential as biological agents to suppress the development of disease *Ralstonia solanocearum* in bananas.

References

[1] Badan Pusat Statistik (BPS) dan Direktorat Jenderal Hortikultura pada tahun 2016.
[2] Moore N Y, Pegg K G, Allen R N and Irwin J A G 1996 Characterisation of strains of Fusarium oxysporum f.sp. cubense by production volatiles *Australian J. Botany* **39** 161-166.
[3] Ploetz R C and Pegg K G1997 Fusarium wilt of banana and Wallace's line: Was the disease originally restricted to his Indo-Malayan region *Australian Plant Pathol* **24** 38-43.
[4] Tjahyono H B dan Sugiharso 1993 Uji resistensi varietas pisang terhadap bakteri layu *J. Hort* **3**(1) 37-42.
[5] Nasir and Jumjunidang 2003 Karakterisasi ras Fusarium oxysporum f. sp. cubense dengan metoda Vegetative Compatibility Group Test dan identifikasi kultivar pisang yang diserangnya *J. Hort* **13**(4) 276-284.
[6] Sequeira L 1997 Bacterial Wilt: The Missing Element in International Banana Improvement Programs, p. 6–14. In P. Prior, C. Allen, & J. Elphinstone (eds.) Bacterial Wilt Disease: Molecular and Ecological Aspect, Reports of the Second International Bacterial Wilt Symposium. France.
[7] Rebib H, Abdeljabbar H, Marc R, Abdellatif B, Ferid L, Najla S N 2012 Biological control of fusarium foot rot of wheat using fengycin-producing Bacillus subtilis isolated from salty soil *Af J Biotech* **11**(34) 8464-8475.
[8] Bholay A D, Jadhav Priyanka U, Borkhataria B V, Mayuri V, Dhalkari 2012 *Fluorescent Pseudomonads* as Plant Growth Promoting Rhizobacteria and Their Siderophoregenesis *IOSR J. Pharmacy and Biol. Sci., (IOSRJPBS)* **3** 27-32.

[9] Mercado-Blanco J, Koen M G M, Van Der Drift, Per E. Olsson, Jane E. Thomas-Oates, Leendert C. Van Loon, and Peter A H M Bakker 2011 Analysis of the *pmsCEAB* Gene Cluster Involved in Biosynthesis of Salicylic Acid and the Siderophore Pseudomonine in the Biocontrol Strain *Pseudomonas fluorescens* WCS374 *Journal Of Bacteriology* **183** (6) 1909–1920.

[10] Arwiyanto T 1997 Biolocal Control of Tobacco Bacterial Wilt In South East Asia. Isolation of Antagonistic Bacteria *Journal of Indonesian Plant Protection* **3** 54–60

[11] Kawaguchi A., Inoue K, and Ichinose Y 2008 Biological control of crown gall of grapevine, rose, and tomato by nonpathogenic Agrobacterium vitis strain VAR03-1 *Phytopathology* **98** (11) 1218–1225. https://doi.org/10.1094/PHYTO-98- 11-1218.

[12] Schindler J 1993 Dynamics of Bacillus colony growth. Elsevier Science 1:333-338.

[13] Rais A, Jabeen Z, Shair F, Hafeez F Y, Hassan M N 2017 Bacillus spp. a bio-control agent enhances the activity of antioxidant defense enzymes in rice against Pyricularia oryzae. Department of Biosciences, COMSATS Institute of Information Technology, Park Road, Islamabad, Pakistan)

[14] Al-Zahrani S H M 2007 Studies on the antimicrobial activity of Streptomyces sp. isolat from Jazan *JKAU Sci* **19** 127-138.

[15] Sastrosuwignyo S 1988 Dasar-Dasar Perlindungan Tanaman. Bagian Ilmu Penyakit Tanaman. Jurusan Hama dan Penyakit Tumbuhan . Fakultas Pertanian. Institut Pertanian Bogor. 153p.

[16] Chrisnawati, Arwiyanto T, dan Nasrun 2006 Pengendalian Hayati Penyakit Layu Bakteri Nilam Menggunakan Kombinasi Bakteri Pseudomonad fluoresen dan Bacillus spp. Laporan Penelitian Hibah Bersaing Tahap Pertama Dikti (Tidak Publikasi).

[17] Mochizuki M, Nishijima T, Toyota K 2005 Predominant culturable Bacillus species in Japanese arable soils and their potential as biocontrol agents *J Microb Environ* **20** (1) 61-68.

[18] Soesanto L, Rokhlani, Prihatiningsih 2008 Penekanan beberapa mikroorganisme antagonis terhadap penyakit layu gladiol *J Agrivita* **30** (1) 75- 83

[19] CampbelL R 1989 Biological Control of Microbial Plant Pathogens. Cambridge University Press, Cambridge. p.218.

[20] Weller D M and Cook R J 1983 Suppression of Take-all of Wheat by Seed treatments with Fluorescent pseudomonads *Phytopathology* **73** 463-469

[21] Bull C T, Weller D M, and Thomashow L S 1991 Relation between root colonization and suppression of Gaeumannomyces graminis var. tritici by Pseudomonas fluorescens strain 2-79 *Phytophatology* **81** 954-959.

[22] Janisiewicz W J, Tworkoski T J, and Sharer C 2000 Characterizing the mechanism of biological control of postharvest disease on fruits with a simple method to study competition for nutrients *Phytopathology* **90** 1196-1200

[23] Kumar J, Saraf M 2015 Plant growth promoting rhizobacteria (PGPR) *Journal of Agricultural Research and Development* **5** 108-119.

Digital extension and the development of agricultural performance in Indonesia

M A Nasir[1*], Ismiasih[2], and Jamhari[3]

[1] Postgraduate student, Department of Agricultural Socio-economics, Faculty of Agriculture, Universitas Gadjah Mada, Jl. Flora, Bulaksumur, Special Region of Yogyakarta, 55281, Indonesia
[2] Lecturer, Department of Agribusiness, Faculty of Agriculture, Institut Pertanian Stiper (INSTIPER) Yogyakarta, Jl. Nangka II, Krodan, Maguwoharjo, Depok, Sleman, Special Region of Yogyakarta, 55281, Indonesia
[3] Lecturer, Department of Agricultural Socio-economics, Faculty of Agriculture, Universitas Gadjah Mada, Jl. Flora, Bulaksumur, Special Region of Yogyakarta, 55281, Indonesia

*E-mail: jamhari@ugm.ac.id

Abstract. The agricultural sector plays an essential role in providing foodstuffs for the population of every country. However, the lack of innovation in agricultural technology and the limited distribution channels for crop yields have caused farmers' welfare levels to be relatively low in various regions. Meanwhile, Smart Agriculture and 4.0 Technologies play a considerable role in increasing the agricultural sector's efficiency. Digitalization has had a significant influence on transforming traditional agricultural systems into data management-based approaches. The Desa Apps is an open access agricultural extension platform developed in Indonesia that facilitates farmers to interact with other farmers to share ideas, innovations, and solutions to face various farm and off-farm obstacles. This paper aims to describe the development platform and identify opportunities related to developing the Desa Apps platform. The results showed that Java and Sumatra islands have good potential for developing the Desa Apps platform. Besides, the registered members with the most significant reporting level are farmers, extension agents, and traders.

1. Introduction

Digitalization is one of the innovations that is widely used in the agricultural sector. In the agricultural sector, digitization can be applied in various forms, both in the production system, supply chain, and food system [1]. Currently, the use of digitalization or digital machines has been widely applied in the agricultural sector, as evidenced by the increase in review articles in the fields of smart farming, extensive data analysis, artificial intelligence, and IoT [2,3]. Digitalization has had a massive influence in transforming traditional agricultural systems into data management-based approaches [4]. Besides, digitization can change the gender structure in the agricultural sector [5].

Digitalization in the agricultural sector is currently considered more comfortable to do in line with mobile phones. The development of smartphones is currently able to support various applications based on virtual tools, which are considered capable of increasing productivity in the agricultural sector [6]. Besides, the application of digitization in the agricultural sector can also overcome sustainability challenges [7]. Increased productivity and sustainability in the agricultural sector can be caused by optimizing various resources through various innovations such as sensor technology, digital image processing, positioning systems, etc.

In this paper, we choose to focus on digitization to promote agricultural extension approaches in Indonesia. The digitization of the extension system has advantages in reaching several simultaneous actors [8]. The extension approach can connect various agricultural sectors with different background knowledge and know-how (farmers, credit and input providers, buyers, researchers, and development organizations). This approach can become a medium for various agricultural sector needs, such as crop varieties, animal breeds, planting seasons, quality seeds, disease, pest control mechanisms, the best market prices, and favorable credit providers [9]. On the other hand, traditional extension approaches have a lower capacity to connect various agricultural sectors to solve problems and increase farmers' access to information to markets and credit [10,11].

This paper examines Desa Apps (Digital Extension Society for Agriculture) platforms in Indonesia as an integrated agricultural extension media platform. The paper analyzes the characteristics of the user and actual platform usage. This paper aims to describe platform development and identify opportunities related to platform development.

The rest of the paper is structured as follows. Section 2 provides results and a discussion and overview of the Desa Apps as an agricultural extension application. Section 3 will present the conclusions.

2. The Desa Apps as an Agricultural Extension Application

Desa Apps is a Mobile Application developed to help farmers from planting, maintaining, harvesting to selling. Farmers can interact directly with experts for question and answer and consultation and share information with other farmers [12]. Desa Apps applies a machine learning algorithm where farmers can be registered into the platform to ask questions regarding various on-farm and off-farm activities, which will then get feedback in the form of an online chat service. Desa Apps is an open-access platform where farmers can interact with other farmers to share ideas, innovations, and solutions to face various obstacles on-farm and off-farm.

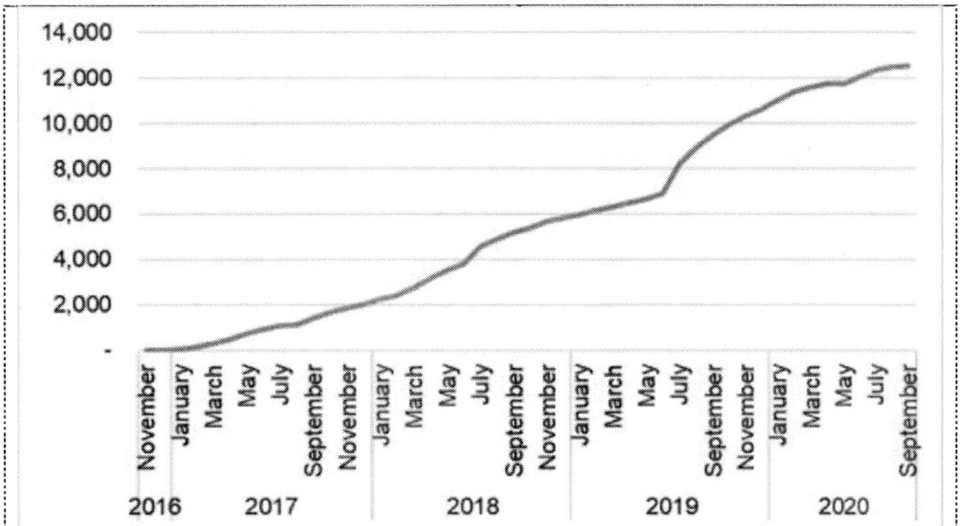

Figure 1. Growth in the Number of Registered Members of Desa Apps.

Based on Figure 1, it is known that the number of registered members of Desa Apps has increased from November 2016 to September 2020. Since its inception, Desa Apps has several registered members of 12,509 people nationally. The most significant number of registered members of Desa Apps occurred in July 2019, which reached around 1,303 people. On the other hand, the lowest increase in the number of registered members occurred in December 2016 and May 2020. The increasing number of registered members implies that this extension application is quite attractive to the public. Figure 2 shows that there are around 5,454 discussions that have been elicited. The most significant number of discussions per month occurred in August 2018, which reached 810 cases. The dominant theme is the problem of rice plants' pests and diseases, with a complete discussion of 77 cases. Meanwhile, the dominant discussion time took place in August 2018, which reached around 173 cases.

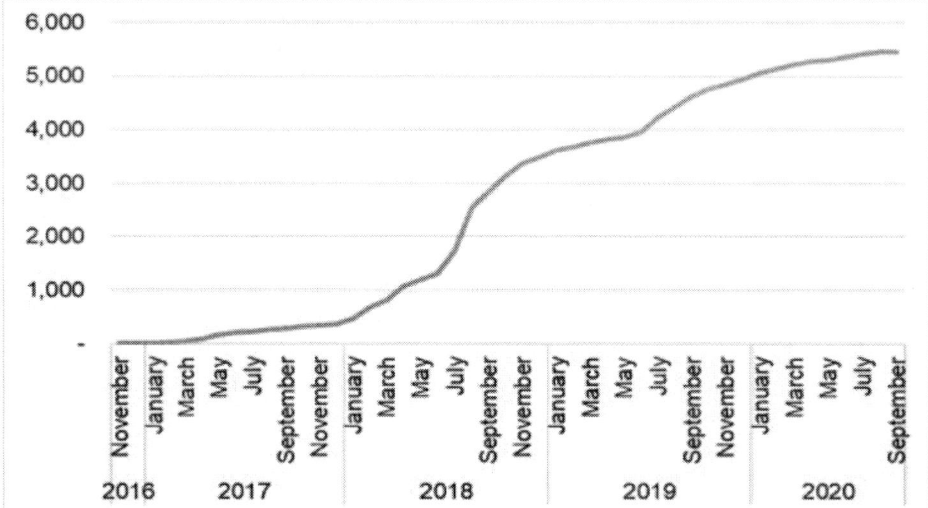

Figure 2. Number of Discussions at Desa Apps.

Based on Figure 3, Figure 4, and Figure 5, It is known that Java Island is the area with the most considerable discussion level in Desa Apps, with the main actors being farmers, extension agents, and traders. This condition means that Java Island has the potential to develop adequate digitalization of agricultural extension. Besides, Java Island can be used as one of the areas with the potential for platform development. Java Island is the region with the most extensive reporting rate because it has a substantial proportion of businesses in the agricultural sector, especially food crops in rice. The island of Sumatra is the region with the second-largest reporting rate after Java, which implies that this area can be developed into one of the areas with the potential for developing Desa Apps, which is quite potent.

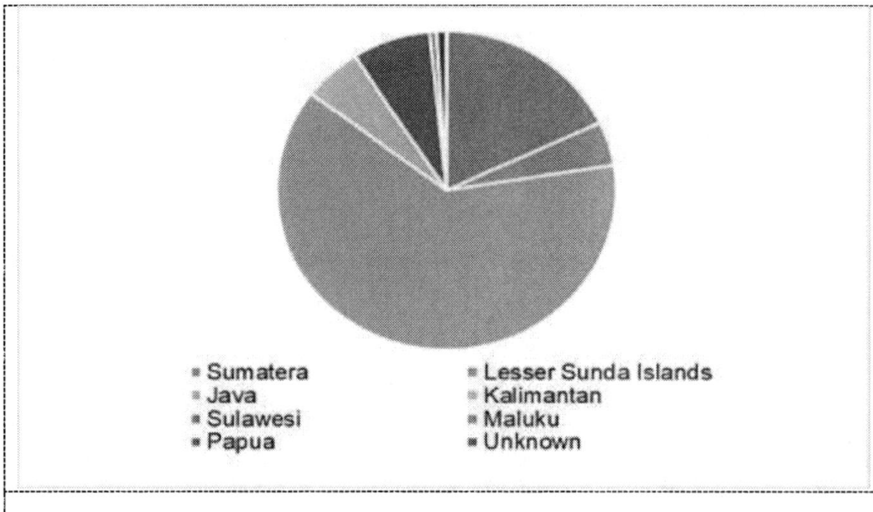

Figure 3. Desa Apps Member Report (Farmer).

IOP Conf. Series: Earth and Environmental Science **883** (2021) 012040 doi:10.1088/1755-1315/883/1/012040

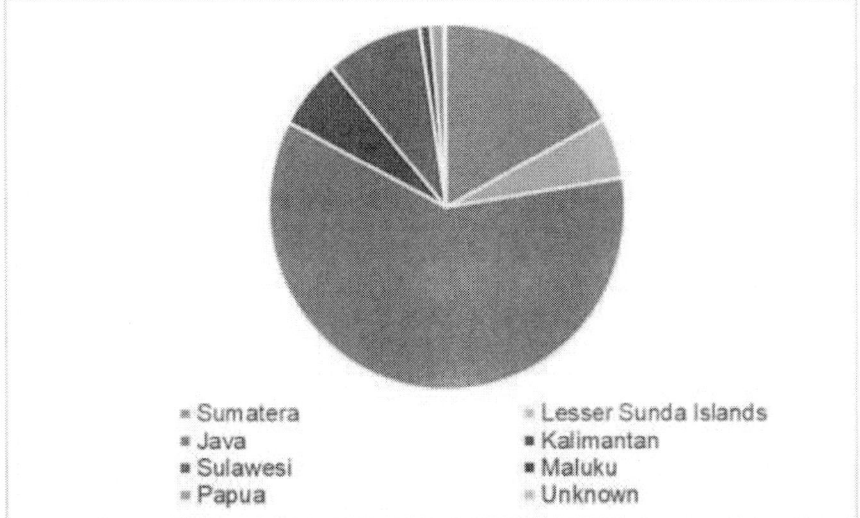

Figure 4. Desa Apps Member Report (Extension agent).

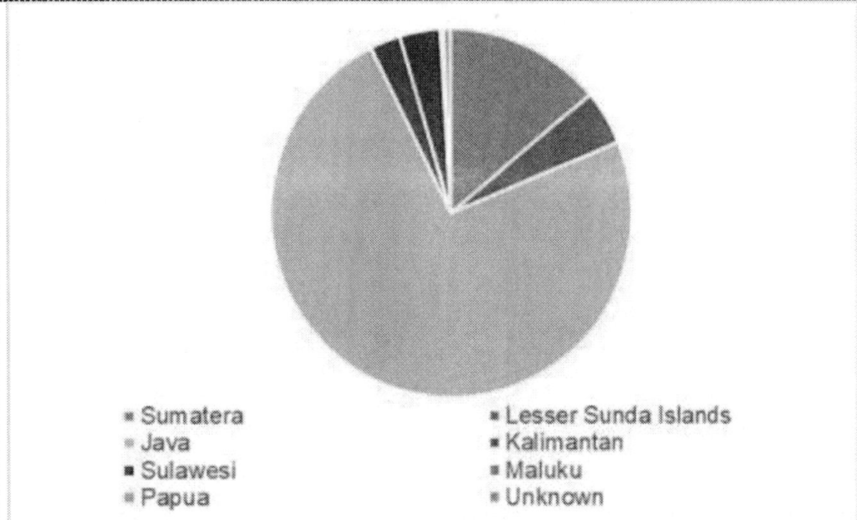

Figure 5. Desa Apps Member Report (Trader).

3. Conclusion

This paper aims to describe platform development and identify opportunities related to platform development. Java Island and Sumatra Island are areas with good potential for the development of Desa Apps. The registered members with the most extensive reporting level are farmers, extension agents, and traders.

4. Acknowledgment

This research was supported by the Ministry of Research and Technology / National Agency for Research and Innovation (Kementerian Riset dan Teknologi/ Badan Riset dan Inovasi Nasional) through funding the Penelitian Dasar Unggulan Perguruan Tinggi (PDUPT) funding plan for the 2020 budget year.

References

[1] Klerkx L, Jakku E and Labarthe P 2019 A review of social science on digital agriculture, smart farming and agriculture 4.0: New contributions and a future research agenda *NJAS - Wageningen Journal of Life Sciences* 90-91

[2] Annosi M C, Brunetta F, Monti A and Nat F 2019 Is the trend your friend? An analysis of technology 4.0 investment decisions in agricultural SMEs *Computers in Industry* **109**: 59-71

[3] Balafoutis A, Beck B, Fountas S, Vangeyte J, Van Der Wal T, Soto I, Gómez-Barbero M., Barnes A and Eory V 2017 Precision agriculture technologies positively contributing to ghg emissions mitigation, farm productivity and economics *Sustainability* **9** 1339

[4] Butler D and Holloway L 2016 Technology and Restructuring the Social Field of Dairy Farming: Hybrid Capitals, 'Stockmanship' and Automatic Milking Systems *Sociologia Ruralis* **56** 513-530

[5] Bear C and Holloway L 2015 Country Life: Agricultural Technologies and the Emergence of New Rural Subjectivities *Geogr Compass* **9** 303–315

[6] Daum T, Buchwald H, Gerlicher A and Birner R 2018 Smartphone apps as a new method to collect data on smallholder farming systems in the digital age: A case study from Zambia *Computers and Electronics in Agriculture* **153** 144-150

[7] Wolfert S, Ge L, Verdouw C, and Bogaardt M J 2017 Big Data in Smart Farming – A review *Agricultural Systems* **153** 69-80

[8] Omulo G, and Kumeh E M 2020 Farmer-to-farmer digital network as a strategy to strengthen agricultural performance in Kenya: A research note on 'Wefarm' platform *Technological Forecasting and Social Change* **158** 120120

[9] El Bilali H and Allahyari M S 2018 Transition towards sustainability in agriculture and food systems: Role of information and communication technologies *Information Processing in Agriculture* **5** 456-464

[10] Kilelu C W, Klerkx L, Leeuwis C and Hall A 2011 Beyond knowledge brokering: an exploratory study on innovation intermediaries in an evolving smallholder agricultural system in Kenya *Knowledge Management for Development Journal* **7** 84-108

[11] Klerkx L, Schut M, Leeuwis C and Kilelu C 2012 Advances in knowledge brokering in the agricultural sector: Towards innovation system facilitation *IDS Bulletin* **4** 53-60

[12] Desa Apps 2020 Desa Apps Retrieved September 9, 2020, from https://desa-apps.ugm.ac.id/

IOP Conf. Series: Earth and Environmental Science **883** (2021) 012041 doi:10.1088/1755-1315/883/1/012041

Response of soybean (*Glycine max* L.) that was applied by various liquid organic fertilizer in climate change at acid soil

Jamilah[1,*], Rapialdi[1], M Ernita[1]

[1] Magister Agrotechnology, Faculty of Agriculture, University of Tamansiswa Padang, Jl. Tamansiswa No. 9 Padang, 25138

* E-mail: jamilahfatika@gmail.com

Abstract. Acidic soils are formed from rapid climate change. Acid soils have problems in it, among others; high levels of Al, Fe, and Mn and the unavailability of elements P and N for staple crops. Indonesia imports almost 70% of its domestic requirement. The purpose of this study was to determine the response of soybean plants that were given various types of liquid organic fertilizer (LOF) to climate change in acid soils. Experiments have been carried out in acid soil in the Kuranji area of Padang City. Materials needed include; Mutiara-1 seed. The lime and LOF trials were arranged in a completely randomized design (CRD), with 9 treatments namely; without calcification and without LOF (A); give lime equivalent to 1 x Exchangeable Al and without LOF (Control) (B); Crocober (C); Crocoberma (D); Crocoberderma (E); Titocroco (F); Titocrocoma (G); (H); Trichoderma harzianum (I). The observational data were analyzed statistically by using an F level of 5% significance level if the treatment had a significant effect, it was followed by a LSD test of 5% significance level. Observation parameters include; agronomic, pH soil. Experimental results established that given of dolomite lime can increase soil pH, Titocrocoma LOF accompanied by liming gives the best effect on soybean varieties Mutiara-1 at vegetative stadium, produces 3.45 Mg ha^{-1} forage dry weight equivalent to107,64 kg N ha^{-1} or 618,93 kg ha^{-1} crude protein.

1. Introduction

Climate change has the potential to affect changes in soil chemical properties and the dynamics of food crop growth. The extreme climate will make crop yields decline. moreover to disruption to water availability, the development of pests and diseases. In extreme wet rainy conditions and even very high puddles in a certain planting season, the plants fail to grow, and the same thing happens if the weather is very dry [19] tated that the rainy season in November-March has also shifted. Global climate changed has resulted in projecting to reduce food crop yields [1]. According to [28] continuous climate changed quickly will easily leach all nutrients, especially on dry land. The acid reaction is one of the characteristics of soil that has undergone intensive washing and is classified as sub-optimal land [21]. [4] reported that the rainfall throughout June is included in the Normal category which reaches 240 mm, [10] explained that it is classified in the Oldeman category it is classified as wet because of > 200 mm $month^{-1}$

The acid soil had become the largest agricultural land in Indonesia until now it is used as a development area for agricultural land. Acid soil has various other problems, among others; high levels of Al, Fe, and Mn and unavailability of P and N elements for staple plants. Some of them are classified as sub-optimal land, among others; Ultisol, Oxisol, part Inceptisol, Entisol, and Histosol. The soybean plant is a very popular product for the Indonesian people. [23] explained that of all Indonesia's soybean requirements, 70% was imported. Furthermore, he explained that farmers' soybean products were low and could still be improved if there were technological innovations, such as using superior seeds and needed liming and proper fertilization in that cultivation.

[2] has released a new superior variety Mutiara-1, with a potential yield of 4.1 Mg ha^{-1}. Varieties with high yield potential should be tried in West Sumatra on sub-optimal soil or Ultisols. The expectation is that soybean yields can achieve the target and even exceed the target. Therefore, this must be complemented by appropriate technological engineering than the farmer's method.

Currently, how many farmers have complained about the cost of farming costs that are already greater than the results they will reach. Therefore it is necessary to assist farmers by providing adequate understanding and technological innovation so that yields can increase. Lime application is important for soybean cultivation. Besides the application of lime, fertilizer application is also needed for optimal results. The use of *C.odorata* Liquid Organic Fertilizer (LOF) has been evidenced to save up to 50% of artificial fertilizers on various rice, corn, chili, onion, melon, cayenne, watermelon, and strawberry crops [11], [13], [14], [15], [16], but the soybean crops have not been evidenced.

However, LOF production engineering quality must continue to be improved. The enrichment of constituent raw materials is considered to improve these qualities, including the addition of *Tithonia diversifolia (T.diversifolia*), and *Trichoderma sp.* According to [22] the benefits of *Trichoderma sp*; as a bio fungicide are used to inhibit the growth and development of toxic fungi that can raise disease in plants. Therefore, it is necessary to study the effectiveness of these materials mixed together to become a quality LOF so as to increase the resistance of soybean plants in acid soils. The research objective was to determine the responsiveness of soybean plants given various types of LOF to climate change in acid soils.

2. Methods

The experiment was carried out in Belimbing, Lalang Village, Kuranji District, Padang City, in April-July 2020. The materials required include; soybean seed variety Mutiara-1 , *C.odorata, T. diversifolia, Trichoderma sp.* All materials were obtained from the experimental location environment, while Trichoderma sp was obtained from the Laboratory of the Faculty of Agriculture, Tamansiswa University (Unitas) Padang. chemical analysis of N, P, and K nutrients of the plant will be carried out in the Unitas Padang laboratory and in the soil laboratory of the Faculty of Agriculture, Andalas University. The experiment was arranged in a completely randomized design (CRD), consisting of 9 combined treatments of lime and liquid fertilizer. Data generated were subjected to analysis of variance (ANOVA) at $P \leq 0.05$ using SPSS 20 for windows. When significant differences were observed treatment means were separated using the Least Significant Difference (LSD) according to [24]. Various types of liquid organic fertilizers with various ingredients, complete with the composition:

Table 1. The treatment of giving Liquid organic fertilizers (LOF) along with the composition of the ingredients

Number	Type of LOF and ingredients
0	- liming and - LOF
1	+ liming equivalent 1 x Exchangeable Al and - LOF
2	**Crocober** (30% *C.odorata* +10% cattle urine + 20% coconut fiber + 20% local micro organism from coconut water +20% manure)
3	**Crocoberma** (30% *C.odorata* +10% cattle urine + 20% coconut fiber + 20% micro organism from coconut water +20% manure + *Trichoderma*)
4	**Crocoberderma** (30% *C.odorata* +10% cattle urine + 20% coconut fiber + 20% micro organism from coconut water +20% manure + (2x *Trichoderma*)

5	**Titocroco** (30% *C.odorata* + 30% Tithonia + 10% coconut fiber + 10% local micro organism from coconut water +20% manure)
6	**Titocrocoma** (30% *C.odorata* + 30% Tithonia + 10% coconut fiber + 10% local micro organism from coconut water +20% manure + *Trichoderma*)
7	**Titocrocoderma** (30% *C.odorata* + 30% Tithonia + 10% coconut fiber + 10% local micro organism from coconut water +20% manure + (2x *Trichoderma*)
8	***Trichoderma*** dissolved in water (2 kg/20 L liquid) took 50ml/L applied

Implementation; Liquid fertilizer is made with the composition of raw materials that have been prepared based on treatment. The coconut fiber had been added to enrich the potassium nutrient in it [9]. All organic substance was finely chopped and then composted for 1 month, turning them every week so that oxygen can be penetrated. The compost was black and odorless and the biomass has crumbled. Then the material as much as 20 kg each was put into a container then coconut water was added. The addition of water was carried out as much as the solid material (1:1) specified in the container. Each dose of Trichoderma according to the treatment was put into the liquid fertilizer container then 2.5 kg of brown sugar was added, then fermented for 1 month. In order for fermentation to run properly, an aerator was used and equipped with a gas exhaust hose to avoid too high gas pressure.

Soybeans were planted at a spacing of 20 x 15 cm with 2 seeds per hole, in each plot the size of 200 cm x 200 cm. Soil reacted with acid and was given lime equal to 1 x Exchangeable Al , 2 weeks before planting. Liquid organic fertilizer was given starting 2 weeks after planting (MST). Subsequent applications were carried out by spraying evenly every 2 weeks with the same dose, namely 50 ml L^{-1} of water. Spraying was stopped after the plants entered the flower primordia phase. At this time, LOF was applied only 3 times. Urea, SP36, and KCl are still given but the dose was 50% of the recommended by BPTP West Sumatera, given the whole initial planting.

Observation parameters, among others; determination of pH of liquid organic fertilizers, plant height, number of branches and nodules, fresh and dry weight of stover of crops, length of primary roots, number of secondary branches of roots, and fresh and dry weight of roots, initial soil pH and 43 days after planting (dap) in the rhizosphere.

3. Results and discussion

3.1 Soil chemical analysis

The results of soil chemical analysis before the experiment was carried out are presented in Table 2. The chemical reaction of the soil before the treatment (initial soil pH) was considered acidic, therefore if it was to be planted with soybeans it needs to lime. The value of the cation exchange capacity was also classified as moderate, the exchangable-Mg level was also low. The content of other elements was classified as moderate, but the phosphorus content in the soil was very high. This was because the soil used in this experiment was former rice fields. This also proved that the high levels of phosphorus in the soil were thought to be due to the intensive use of phosphate fertilizers and that there were a lot of P deposits in the soil.

Table 2. Initial chemical soil analysis at Ultisol Kuranji, Padang City, West Sumatera

Soil Characteristic	Unit	Nutrient content in the soil	
		Value	Classification
pH (KCl)		4,18	-
pH (H2O)		4,66	acid
Nitrogen	%	0,278	moderate
Fospor (P)	ppm	21,175	Very high
KTK	Cmol kg^{-1}	19,55	moderate
Na	Cmol kg^{-1}	0,376	low

Ca-dd	Cmol kg^{-1}	0,327	Very low
Mg-dd	Cmol kg^{-1}	0,619	low
K-dd	Cmol kg^{-1}	0,502	moderate
Exchangeable Al	Cmol kg^{-1}	1,591	-
H-dd	Cmol kg^{-1}	0,53	-
C-organic	%	2,682	moderate
C/N		9,647	low
Organic Matter	%	4,624	

Note: Criteria for soil chemical properties based on [25]

Producing healthy soybeans with high production really requires proper treatment, including maintaining adequate nutrition and disturbances from pests and plant diseases and expanding the planting area [17], [29] explained that acid mineral soils had sufficient potential for the development of soybean crops in the future. The problem of various acid mineral soils such as Ultisols is that the content of Al and Fe is quite high and the soil pH is low. Some nutrient content is also low, especially base saturation. If this soil to be used for the cultivation of food crops, especially soybeans, it needs special treatment, by liming. Based on the chemical reaction conditions of the less fertile soil in general, liming was carried out. This experiment carried out liming equal to 1 x Exchangeable Al with dolomite ($CaMg(CO_3)_2$. Producing healthy soybeans with high production really requires proper treatment, including maintaining adequate nutrition and disturbances from pests and plant diseases and expanding the planting area [22] explained that acid mineral soils have sufficient potential for the development of soybean crops in the future. The problem of various acid mineral soils such as Ultisols is that the content of Al and Fe is quite high and the soil pH is low. Some nutrient content is also low, especially base saturation.

3.2 Liquid organic fertilizer pH

The chemical reactions (pH) of liquid organic fertilizer (LOF) of various ingredients are presented in Table 3. pH of Liquid organic fertilizer was determined after the LOF was harvested and will be applied to soybean crops. From the results of the analysis, it can be explained that in general, the LOF reaction was slightly alkaline to alkaline, while the *Trichoderma* solution reacted slightly acidic. In the types LOF of Crocober, Crocoberderma to Titocrocoderma, they contained more diverse microorganisms than the LOF which only come from *Trichoderma sp* fermentation. It was proven that liquid fertilizers containing various types of microorganisms can increase higher the pH than only Trichoderma liquid fertilizer. However, this study did not identify the types of microorganisms contained in the LOF. It will be hoped that future research will be carried out to analyze the types of microorganisms contained in it. [27] indicated that *Trichoderma sp* requires a growth medium with an optimal pH of 3-7, between very acidic to neutral.

Table 3. Reaction (pH) of various LOF with ingredients at 2 months after fermentation

Number	Type of fertilizers	Average of pH	Criteria*)
0	- Liming and - LOF	-	-
1	+ liming equivalent 1 x Exchangeable Al and – LOF	-	-
2	**Crocober**	7,635	rather alkaline
3	**Crocoberma**	8,120	rather alkaline
4	**Crocoberderma**	8,610	alkaline
5	**Titocroco**	8,685	alkaline
6	**Titocrocoma**	8,630	alkaline
7	**Titocrocoderma**	8,665	alkaline
8	***Trichoderma***	6,495	rather acid

*) Reference [25]

The Minister of Agriculture Regulation No. 70/ PERMENTAN/SR.140/10/2011 determined that regarding organic fertilizers, biological fertilizers, and soil repairers, that the pH of LOF range from 4 - 9. In terms of pH, LOF was very suitable as fertilizer that was sprayed on plants as a provider of nutrients for plants. The content of the LOF raw material comes from various shrubs of *C.odorata*. *T. diversifolia* which are added with manure, fermented coconut water, and coconut fiber. All of these materials contain nutrients that are very useful for plants. [12] had succeeded in proving the positive effect of providing liquid organic fertilizer from *C.odorata* on the production of cayenne pepper plants; black rice and Cisokan rice. The provision of liquid fertilizer from *Tithonia diversifolia* had also been proven by [18], which can be higher the yield of upland rice on Ultisol soil by 13.33% than to those not given LOF. Liquid organic fertilizers did not have a negative effect on the environment and food crops. Therefore, the use of LOF must be developed in the management as well as taking care of plant nutrition. The selection of ingredients for LOF must be careful because the wrong composition and the incorporation of useless materials will have a low effect and will not increase the yield of food crops. [20] had proven that the composition of *C.odorata* + coconut fiber + manure + urine + local microorganism) was able to exchange the use of artificial fertilizers up to 50% in lowland rice plants.

The usage of coconut fiber was indeed appropriate in making LOF because its potassium content is quite high. Apart from agriculture, coconut fiber was also used as a storage for K_2CO_3. This was agreed by [9] from the results of studies on the use of catalysts using coconut fiber ash, palm bunch ash, and K_2CO_3 to convert castor oil into biodiesel with methanol as solvent. Biodiesel was made by converting unrefined castor oil, using a catalyst, through a transesterification reaction. The catalysts were annealed at temperatures of 500, 600, 800, and 900° C for 10 hours. The reaction was carried out in a three-neck flask at a constant temperature of 60° C for 3 hours. The results showed the use of a coco ash catalyst without annealing and with annealing at 800°C gave the highest biodiesel yield (87.05 and 87.97%) with low soap content (0.23-0.26%). The characteristics of the biodiesel produced from the use of these catalysts complied with the quality requirements set by Indonesian and International Standards so that these catalysts were suitable for use as commercial K_2CO_3 catalysts.

[22] explained that the benefits of *Trichoderma sp*; as a bio fungicide which is used to inhibit the growth and development of toxic fungi that can cause disease in plants. Examples of these pathogenic fungi include the fungus *Rhizoctonia solani, Sclerotium rolfsii, Rigdiforus lignosus, Fusarium moniliform*, and also *Fusarium oxysporum. Trichoderma sp.* also can be used as a decomposer in the manufacture of organic fertilizers. Apart from being a decomposing organism, Trichoderma sp can also function as biological agents and growth stimulators in plants. This fungus also had the ability to increase the production of healthy roots and increase the depth of roots below the soil surface so that it can increase the rate of growth and development of plants.

3.3 Soil pH and development of agronomy

Soybean plants reaching the age of 43 days after planting (dap) were presented in Table 4 and Figure 1. There was a lot of information that can be conveyed regarding this treatment. Plant growth on soils that were not given lime nor given LOF looked stunted. If the soil was given lime equivalent to 1 x Exchangeable Al, the growth was better than it did not apply lime, and there was a significant difference. It was proven that plants need lime for better growth. There was a higher of 47.06% in height plant, 45.17% in the number of branches, and 85.41% by weight of fresh stover of the Mutiara-1 soybean plant which was given lime than to those that were not given lime.

Table 4. Effect of lime and LOF on changes in soil pH and the shoot of the Mutiara-1 soybean plants at 43 dap

Treatments	Soil pH	Plant height	number of plant branches	Fresh weight of shoot
		cm	piece	g/clump
Inisal Soil chemical	4,66			
- Liming and - LOF	4,63 c	44,00 c	3,83 c	43,33 d
+ liming equivalent (1 x Exchangeable Al)	***5,18***	***64,71***	***5,56***	***80,34***
(+ liming and -LOF)	5,00 abc	58,83 b	5,16 a	68,40 cd
Crocober	5,23 ab	63,16 ab	5,50 ab	74,40 bc
Crocoberma	**5,33 a**	**68,50 a**	**6,20 a**	75,56 bc
Crocoberderma	5,25 ab	62,16 ab	5,33 ab	90,26 ab
Titocroco	4,95 bc	63,50 ab	5,66 ab	63,93 cd
Titocrocoma	5,18 ab	68,33 a	4,50 b	**90,80 a**
Titocrocoderma	5,23 ab	**68,50 a**	**6,36 a**	**102,70 a**
Trichoderma	5,28 ab	67,16 a	5,73 ab	76,63 bc
CV (%)	3,07	5,19	11,04	14,42
LSD at 5%	0,37	5,58	1,02	18,86
P-value	0,018	$14{,}4 \times 10^{-5}$	0,00045	0,000038

Note: The number followed by the same supersript in each column is significantly different according to the LSD at 5% significance level.

The treatment of soil treated with lime equivalent to 1 x Exchangeable Al, giving LOF also had various impacts on soil pH. The lowest soil chemical reaction (pH) was in the experimental plot that was not given lime or the LOF application. Soil given dolomite lime equivalent to 1 x Exchangeable Al increased by 0.52 from the initial pH of only 4.66 to 5.18 during the generative phase of plant growth. Soils that were calcified with the application of various types of LOF on soybean plants, it was found that the soil pH of the rhizosphere also experienced various changes. Giving Crocoberma LOF gave the highest soil pH impact of 5.33. The soil was treated with lime, without LOF application resulted higher in the soil pH reached 5, than soils was unlimed and not applied LOF was only 4.63.

There were also indications that liming which raises soil pH increases the general growth of soybean plants. *Trichoderma sp* added to any type of LOF can increase the pH of the soybean plant rhizosphere. From table 4, it was also proven that the addition of tithonia to the raw material of *C.odorata* LOF can reduce the pH of the soybean rhizosphere. The highest plants were those given LOF Crocoberma and Titocrocoderma. The same pattern also occurred in the formation of the number of plant branches. The plant's fresh weight was higher than the Titocrocoderma LOF treatment. There had been a higher of 16.44% plant height, 23.25% number of branches, and 50.15% weight of stover in plants given LOF than to those not given LOF.

Figure 1. Soybean plants applied with LOF at 43 dap

The Titocrocoderma LOF was a mixture of *C.odorata* which is added with Tithonia and coconut husk and *Trichoderma sp.* All complete ingredients were contained in it, thus producing nutrients that meet the growing needs of soybean plants. The positive impact started on height, the number of branches, and fresh shoot weight per clump. However, several other types of fertilizers also have the same positive impact as this type. Likewise, it was unless great if pure *Trichoderma* which was fermented and then applied as LOF was still insignificant with other ingredients. This showed that *Trichoderma sp* was able to provide useful nutrients for soybean plants.

The application of lime and LOF affected the length of the primary roots, the number of nodules, and the fresh weight of the roots (Table 5) and (Figure 1). The highest fresh root weight was influenced by Titocrocoderma LOF although other types of fertilizers were able to significantly increase the primary root length, the number of root branches, and root nodules. The highest root weight resulted in the highest shoot weight (Table 4). It turns out that the LOF effect was not only able to affect the top of the plant but also the root of the soybean plant. Adequate nutrition from LOF Titocrocoderma because there was a mixture of *C.odorata* and *T. diversifolia* raw materials, while the other ingredients contain *Trichoderma sp.*

Plants treated with lime showed a significant increase in primary root length, a number of nodules, and root fresh more weight than to those not given lime. Crops that were without liming will grow unhealthy and morphologically the leaves will also appear more yellow, than those given lime even though they are not given LOF. The impact of liming on crops had been explained by many experts, among others [3], [5], [8], [26], stated that lime application could reduce the excessive availability of Al, Fe and other microelements, and increase the availability of P, Ca, Mg and K elements. They also reported liming on acidic soils can increase nutrient uptake, especially phosphorus nutrients.

Table 5. The effect of lime and LOF application on the root development of soybean plant at 43 dap

Number	Treatments	Primary root length (cm)	Number of root branches / plant (numbers)	Number of Root nodules (grain)	fresh Root weight (g/clump)
0	- Liming and - LOF	9,16 ^{c}	28,67	9,33 ^{c}	6,56 ^{c}
	+ liming equivalent (1 x Exchangeable Al)	***12,58***	***26,29***	***25,83***	***8,38***
1	(+ liming and -LOF)	12,66 ^{abc}	23,00	17,66 ^{bc}	7,46 ^{bc}
2	**Crocober**	**15,66** ^{a}	31,00	14,33 ^{bc}	8,40 ^{ab}
3	**Crocoberma**	11,00 ^{bc}	27,00	19,00 ^{bc}	8,06 ^{abc}
4	**Crocoberderma**	**13,33** ^{a}	29,00	**52,00** ^{a}	7,73 ^{bc}
5	**Titocroco**	10,16 ^{bc}	23,00	16,00 ^{bc}	8,46 ^{ab}
6	**Titocrocoma**	13,00 ^{ab}	26,66	**46,33** ^{a}	8,66 ^{ab}
7	**Titocrocoderma**	11,33 ^{bc}	28,66	20,66 ^{b}	**9,50** ^{a}
8	***Trichoderma***	13,50 ^{ab}	22,00	20,66 ^{b}	8,76 ^{ab}
	CV (%)	17,22	16,37	19,19	10,78
	LSD 5%	3,61		7,90	1,51
	P-value	0028	0,171	23x 10^{-12}	0,023

Note: The number followed by the same superscript in each column is significantly different according to the DMRT at 5% significance level.

The growth of longer plant roots was able to produce more nodules. Longer roots would make the range of roots in getting nutrients much wider too. This indicated that the nodule will develop more like the root length increases. Living root nodules depended on the plant nutrients it received from the roots. Therefore, for better root growth, it was certain that the nutrients that are circulated were much more rooted than the short roots. Giving Crocoberderma LOF was able to produce nodules of up to 52 pods, outperforming plants not given LOF which were only 17.66, or an increase of up to 3 times. The formation of root nodules increasingly affected the growth of soybean crops. According to [7], [8] stated that root nodules are the site of N fixation in plants, so the availability of N was largely determined by the presence of root nodules. If nitrogen was available and absorbed by plants, soybean plant growth will increase.

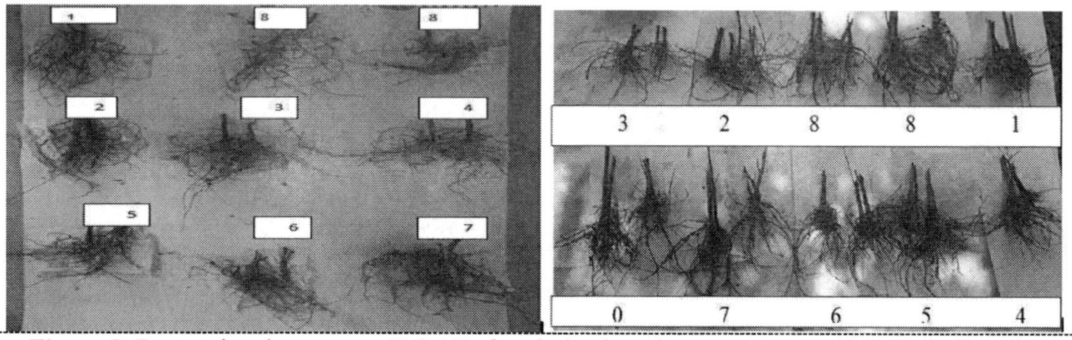

Figure 2. Roots of soybean crops 43 DAP after drying based on treatment at 43 DAP (left) and at harvest time (right) (numbers according to treatment can be seen in Table 5).

Figure 2, shows that the roots of soybean plants that were not given lime had fewer and shorter roots than those treated with lime. However, in the treatment given lime, the LOF application did not have a significant difference in root development [5], [26] reported that applying lime can increase nutrient availability in acid soil and reduce the detrimental effects of trace elements.

If calculated based on the dry matter weight of soybean plants, according to [6] with N content, while according to [27] conversion of N to crude protein in soybean plants by multiplying 5.75. So the nitrogen and protein content in the forage for soybeans was presented in Table 6. Soybean plants maintained up to 43 days after planting (dap), besides being able to grow, will produce pods that were useful for harvesting seeds. However, if you are not going to harvest the seeds, the soy forage will become a source of protein for ruminants.

3.4. The crude protein yield of the soybean plant

The highest crude protein content in soybean forage was found in the Titocrocoma LOF treatment which produced 618,93 kg ha^{-1} of crude protein. The higher the dry matter content in soybean crops, the higher the crude protein content. The crude protein content in soybean crops that were not given lime and not given LOF was only 192,53 kg ha-1, while those given lime but not given LOF increased by 52,07%. If the crude protein content was due to the effect of the same treatment with a lime even those added with the Titocrocoma LOF application, there was a higher 77,78% than those not given LOF.

Table 6. Effect of lime and LOF on transport of nutrients N and crude protein for soybean plants at 43 dap.

Treatments	Plant shoot per hectare			Crude protein (kg)
	Dry matter (mg ha^{-1})	N - plant (%)	N-uptake (kg)	
- Liming and - LOF	1,646 ^{c}	2,21 ^{b}	34,56 ^{c}	192,53
+ liming equivalent (1 x Exchangeable Al)				
(+ liming and -LOF)	2,598 ^{bc}	1,96 ^{c}	50,92 ^{b}	292,79
Crocober	2,828 ^{b}	3,33 ^{a}	93,324 ^{a}	541,49
Crocoberma	2,872 ^{b}	3,07 ^{ab}	88,17 ^{ab}	506,98

Crocoberderma	3,430 [ab]	2,91 [b]	99,81 [a]	573,93
Titocroco	2,430 [bc]	2,98 [ab]	72,41 [ab]	416,38
Titocrocoma	3,450 [ab]	3,12 [ab]	107,64 [a]	618,93
Titocrocoderma	3,902 [a]	2,32 [bc]	90,53 [ab]	520,53
Trichoderma	2,912 [b]	2,84 [b]	82,70 [ab]	475,53
CV (%)	14,42	7,46	18,59	-
LSD 5%	4,91	0,36	184,19	-
P -value	0,00004	0,00000067	0,00496	-

Note: The number followed by the same superscript in each column is significantly different according to the LSD at 5% significance level.

In this experiment, it was proven that LOF containing *C.odorata* raw materials combined with *Thitonia* and *Trichoderma* as a Titocrocoma LOF were able to produce nutrients that were better absorbed by the top of the crop and promote the formation of crude protein which is useful for producing pods if maintained until the physiologically mature phase or as a source of animal feed. for ruminants if the forage is harvested. [20] indicated that quality animal feed is if the crude fiber content is low and the crude protein was high. However, the most important thing that must be considered was that the source of forage to be used as animal feed must be economical. It still needs further study regarding the efficiency of using a ration by livestock which is known from the difference between the amount of crude protein from soybean forage was consumed and excreted by the body either through feces or urine. If you look at the high rainfall in West Sumatra, it had the potential to increase the development of pests and diseases in soybean plants, especially pod borer or fungal attack which will harm pod production, so forage for soybean had the potential to be used as a source of animal feed.

4. Conclusion

Experimental results established that given of dolomite lime can increase soil pH, Titocrocoma LOF accompanied by liming gave the best effect on soybean varieties Mutiara-1 at the vegetative stadium, produced 3.45 Mg ha^{-1} plant shoot or forage dry weight equivalent to107,64 kg N ha^{-1} or 618,93 kg ha^{-1} crude protein.

5. Acknowledgment

Gratitude is conveyed to the Ministry of Research, Technology and Higher Education who has funded this research activity as a magister grant scheme No. B/87/E3/RA.00/2020 and 092/LL10/PG/2020, Maret 11th 2020. Gratitude is also conveyed to the chancellor of the University of Tamansiswa Padang who has agreed to facilitate all these activities.

References

[1] Atmaja T, Adi R F, and Pangan K 2018 Adaptasi perubahan iklim dan ketahanan pangan: telaah inisiatif dan kebijakan *Jurnal Hukum Lingkungan Indonesia* **5**(1) 60–68.

[2] BATAN 2018 *Kedelai Varietas Baru, Hasil pemulilaan Mutasi Radiasi*. Jakarta: Media Informasi Ilmu Penegtahan Teknologi Nuklir. Retrieved from http://drive.batan.go.id/kip/documents/leaflet2018/Kedelai Varietas Baru.pdf

[3] Bolan N S, Adriano D C, and Curtin D 2003 Soil A Cidification And Liming Interactions With Nutrient And Heavy Metal T Ransformation *Advance in Agronomy* **78**(2) 215–272.

[4] Damayanti R H, and Denata 2020 *Analisis Hujan Juni 2020 dan Prakiraan Hujan Agustus, September dan Oktober 2020*. Jakar. Retrieved from https://cdn.bmkg.go.id/web/Buletin_Hujan_Bulanan_Bmkg_Edisi_JULI_2020.pdf#viewer.action=download

[5] Fageria N K, and Baligar V C 2008 Ameliorating Soil Acidity of Tropical Oxisols by Liming For Sustainable Crop Production *Advance in Agronomy* **99**(8) 345–399. https://doi.org/10.1016/S0065-2113(08)00407-0

[6] Faozi K, Yudono P, Indradewa D, and Ma A 2019 Serapan Hara N , P , K dan Hasil Biji Kedelai (*Glycine max* L. Merrill) pada Pemberian Bokashi Pelepah Pisang pada Tanah Pasir Pantai Nutrient uptake of N , P , K and Soybean Seed Yield by Application of Bokashi made from Banana Stem in the Coastal Sand *Vegetalika* **8**(3) 177–191.

[7] Fatima Z, Zia M, and Chaudhary M F 2006 Effect of Rhizobium strains and phosphorus on growth of soybean (*Glycine max*) and survival of Rhizobium and P solubizing bacteria *Pak. J. Bot* **38**(2) 459–464.

[8] Heide B. Van Der 2004 Pengelolaan Kesuburan Lahan Kering Masam Untuk Tanaman Kedelai *Bulletin Palawija* **50**(7) 39–50.

[9] Husin H, Mahidin, and Marwan 2011 Studi Penggunaan Katalis Abu Sabut Kelapa , Abu Jarak Menjadi Biodiesel *Reaktor* **13**(4) 254–261.

[10] Ilmu Geografi.com. 2020. 5 Kategori Klasifikasi Iklim Oldeman (Paling Lengkap). https://ilmugeografi.com/ilmu-bumi/meteorologi/klasifikasi-iklim-oldeman, akses Agustus 2020.

[11] Jamilah 2017 Pengaruh Pupuk Organik Cair Asal *C.odorata* Terhadap Serapan Hara Kalium Dan Hasil Padi Ladang *Jurnal Bibiet* **1**(1) 17–26.

[12] Jamilah, Ben Kurniawan Z 2018 Pengaruh Pupuk Organik Cair Unitas Super Asal *Chromolaena odorata* Terhadap Pertumbuhan Dan Hasil Padi Hitam (*Oryza sativa* L.) *Jurnal Agroteknologi* **8**(2) 15–20.

[13] Jamilah, Hamdani Nusri, Zahanis M E 2018 Penetapan Konsentrasi Pupuk Organik Cair Unitas Super Yang Tepat Pada Tanaman Cabai Rawit Lokal (*Capsicum frutescens* L.) *EnviroScienteae* **14**(1) 33–37.

[14] Jamilah, Ramdan Soleh W H 2017 Respon Tanaman Padi (*Oryza sativa* L.) Kabir 07 Terhadap Pupuk Organik Cair Crocober Plus Khusus Kota Padang dengan Iklim Af *Jurnal Solum* **14**(1) 18–26.

[15] Jamilah A D 2016 Potensi tanaman jagung yang dipangkas dan dipupuk kompos *C. odorata* sebagai model integrasi tanaman pangan dan peternakan. In *Seminar Nasional Dampak Perubahan Iklim Terhadap[Biodiversitas Pertanian Indonesia* (pp. 27–34). Payakumbuh, Sumatera Barat: Politani Pertanian Negeri Payakumbuh.

[16] Jamilah, Fadhila R, and Mulyani S 2017 Farm Analysis Of Rice Crop Trimmed Periodically. In *International Conference on Social, Humanities and Government Science (ICSHGS) 2017* (pp. 202–211). Palembang: Universitas Tamansiswa Palembang.

[17] Jamilah and Herman W 2018 *Karakteristik sifat fisika dan kimia tanah yang mendukung ketahanan pangan di Sumatera Barat*. Padang.

[18] Jamilah, Maradona C, Zahanis, and Ernita M 2014 Penetapan konsentrasi dan intreval pemberian pupuk organik cair asal sabut kelapa dan tithonia untuk meningkatkan hasil padi ladang (*Oryza sativa* L.). In P. P. N. Payakumbuh (Ed.), *Seminar Nasional Pebangunan Bio-Industri untuk Mewujudkan Kedaulatan Pangan Indonesia* (pp. 53–59). Payamkubuh: Politani Pertanian Negeri Payakumbuh.

[19] Khudori 2011 Sistem Pertanian Pangan Adaptif Perubahan Iklim *Pangan* **20**(2) 105–119.

[20] Munir J, Juniarti, and Srimulyani 2017 Mowing Rice Crop as Ratoon and Applying *Chromolaena odorata* Compost to Support Food Security *International Journal of Appled Engineering Reserach* **12**(21) 11738–11748.

[21] Nazir M, Syakur, and Muyassir 2017 Pemetaan Kemasaman Tanah dan Analisis Kebutuhan Kapur di Kecamatan Keumala Kabupaten Pidie *Jurnal Ilmiah Mahsiswa Pertanian Unsyiah* **21** 21–30.

[22] Rizal S, Susanti T D 2018 Peranan Jamur Trichoderma sp yang Diberikan terhadap Pertumbuhan Tanaman Kedelai (*Glycine max* L.) *Sainmatika* **15**(1) 23–29. https://doi.org/10.31851/sainmatika.v15i1.1759

[23] Satria 2015 *Produksi Kedelai Nasional Masih Rendah*. Jogyakarta.

[24] Steel R G D, Torrie J H 1980 Principles and Procedures of Statistics: A Biometrical

Approach *Biometrics* **37**(4) 859. https://doi.org/10.2307/2530180

[25] Tanah B P 2009 *Analisis Kimia tanah, Tanaman, air dan pupuk* (1st ed.). Bogor: Balai Penelitian Tanah, Bogor.

[26] Taufiq A, Kuntyastuti H, Prahoro C, and Wardani T 2007 Pemberian Kapur dan Pupuk Kandang pada Kedelai di Lahan Kering Masam *Penelitian Pertanian Tanaman Pangan* **26**(2) 78–85.

[27] Taufiq A, Manshuri A G 2005 Pemupukan dan Pengapuran pada Varietas Kedelai Toleran Lahan Masam di Lampung *Penelitian Pertanian Tanaman Pangan* **24**(3) 147–158.

[28] Notohadiprawiro T and Soeprapto S, Endang S 2006 Pengelolaan kesuburan tanah dan peningkatan efisiensi pemupukan 1. In *Ilmu Tanah* (pp. 1–19). Jogyakarta: Universitas Gadjah Mada.

[29] Thorik, Aqwamit 2018 Masalah Impor dan Hama Jadi Penyebab Banyaknya Petani di Jatim yang Enggan Menanam Kedelai. Tribun.com Jatim https://jatim.tribunnews.com /2018/11/05/masalah-impor-dan-hama-jadi-penyebab-banyaknya-petani-di-jatim-yang-enggan-menanam-kedelai.

Accelerated adoption of sugar palm farming technology to supports sustainable resource utilization in North Sulawesi

J B M Rawung[1,*], J G Kindangen[1], R Indrasti[2] and A Gaffar[3]

[1] The Assessment Institute for Agricultural Technology of North Sulawesi, Manado, Indonesia.
[2] Centre of Assessment and Technology Agricultural Development, Bogor, West Java. Indonesia
[3] The Assessment Institute for Agricultural Technology of Maluku, Ambon, Indonesia.

* E-mail: jbmarkusrawung2000@yahoo.com

Abstract. The purpose of this study was to examine the characteristics and opportunities of accelerating the adoption of palm sugar farming technology in sustainable resource use in North Sulawesi Province on July to November 2016 . The data used are secondary data and primary data derived from respondents as many as 120 palm-based farming households in the central district of palm plantations in Tareran, Tomohon, and Motoling, Analysis used cross tabulation for farmer characteristics and binary logistic regression approach for accelerated adoption opportunities. The results showed that the farmers cultivated sugar palm plants with an intercropping farming system with a variety of plantation and forestry food crops. Good knowledge of palm plant cultivation system (64.84%) in cultivating sugar palm with a good understanding of the sustainability of palm plant-based farming. Opportunities for accelerating the appreciation and adoption of farmers to sugar palm-based farming technology can be realized in the form of participatory technology assistance on all technology components, both basic and optional technology, by expanding the business scale, increasing various processed products, bringing the location of farming closer to residential and information sources technology. To increase the production of aren and their derivative products and to maintain the continuity and preservation of the environment, it is necessary to cultivate and expand the palm area in harmony with regional spatial planning for conservation and sustainable management.

1. Introduction

Arenga pinnata MERR (Arenga pinnata MERR) is one of the multipurpose plantation crops because almost all parts of the plant have benefits and high economic value. Agronomically, this plant is able to live in the lowlands to the plains with an altitude of 1500 meters above sea level and grows optimally at an altitude of 500-800 meters above sea level with a minimum evenly distributed rainfall of 1200 mm a year. The spread and growth of sugar palm (*Aren*) generally takes place naturally and can grow well in mountainous areas, valleys, near rivers, springs and is commonly found in forests [1,2]. This plant is one of the most potent plantation crops for use in tackling land degradation and reforestation, because it can grow well in various ecosystems, tolerates mixed cropping patterns, grows relatively fast, has dense roots and canopy, and does not require intensive maintenance making it suitable for use on marginal land, and the result can be a source of additional income for local farmers who generally have limited income.

In general, sugar palm still grows wild in several areas including North Sulawesi and even though it has been cultivated by farmers, it is not optimal because most of it is still part-time and still depends on the dominant commodity. In fact, this sugar palm plant still grows wild on the slopes or riverbanks. The cultivation of palm trees in Indonesia has actually been going on for a long time, because Indonesia is one of the regions of origin of sugar palm, but its development into an agribusiness commodity is still slow. This is because most of the sugar palm population has not been cultivated. Palm cultivation has only been carried out in a small number of areas in North Sulawesi, Maluku, East Kalimantan, West Java and Banten [3, 4]. Sugar palm plantations in central areas generally spread sporadically. Sugar palm has begun to get serious attention from various parties in the last ten years to be developed into an agribusiness commodity. Aren turns out to be able to produce about 60 types of products with economic value and some have the potential for export. One of the mainstay products of palm plants, namely palm sugar, either in the form of printed sugar (brown sugar), sugar or crystal sugar. Sugar palm also acts as an energy supplier (bioenergy) and a component of environmental conservation [5]. The intensity of sugar palm planting is increasing and utilizing forest areas and mountainous slopes. Data from 2008 in the Forestry and Plantation Office of Tomohon City stated that the area planted with palm sugar in Tomohon City was around 981.27 ha and in South Minahasa District it was 1,613.2 ha (2006). So whether consciously or not, the behaviour of farmers with initial awareness to lead to the sustainability of their farming is already owned. As a conservation plant, planting sugar palm certainly makes an important contribution to forest conservation and soil conservation. Traditional processing technology produces limited consumer products such as brown sugar, *kolang kaling* and vinesugar or alcohol. As a result, the role of this sugar plant in the national economy is not prominent so that the attention on the development of this commodity is inadequate. Motivation of farmers in optimizing palm-based farming has not been clearly identified, including the factors that influence farmers in adopting technology. The purpose of this study was to examine the characteristics and identify the factors that influence farmers to adopt sugar palm farming technology. By knowing the characteristics of sugar palm farmers and knowing the factors that influence farmers in adopting cultivated sugar palm farming technology, it is hoped that the development of palm plants in the future will be more widespread and be calculated as a commodity that brings benefits to farmers in sustainable management.

2. Methods

The research was conducted from July to November 2016 in the central district of palm sugar farming in Tareran, Tomohon, and Motoling, North Sulawesi Province. The purpose of this study was to examine the characteristics and opportunities of accelerating the adoption of palm sugar farming technology in sustainable resource use in North Sulawesi. The data used primary data derived from respondents as many as 120 palm-based farming households. Data analysis used cross tabulation for farmer characteristics and binary logistic regression approach for accelerated adoption opportunities, where the adoption gap is the dependent variable and 15 other variables are independent.

To determine the factors that influence the adoption of sugar palm technology by farmers, logit regression analysis is used. According to [6], logistic regression can facilitate understanding of research so that the discussion will start with the simplest logistic regression model, namely the model without predictors or zero regression, which will form the basis of the next model. [7] states that logistic regression analysis is a probability modeling research technique. In detail, the use of this logit function has been described and discussed by [8' 9, 10, 11]. The logit function model formula can be formulated as [12,13]:

Regression model formulation $$Pi = \frac{1}{1 + Exp(-Z)} + e_i \qquad \text{..............(1)}$$

In logarithmic form, the equation can be written as follows:

$$\ln \frac{P_i}{(1 - P_i)} = \alpha + \sum_{j=1}^{n} \beta_j X_{ji} + \sum_{k=1}^{m} \gamma_k D_{ki} + \varepsilon \quad \text{..........................(2)}$$

P_i = Opportunity for farmers to apply technological innovations faster.
P_i = 1 if the farmer adopts the innovation for less than one year,
P_i = 0 if farmers adopt the innovation for more than one year
$1 - P_i$ = Opportunity for farmers to adopt the innovation for more than one year
α = Constant

βnxjn = independent variable in certain units is arranged as follows:
Xj = independent variable vector (j = 1, 2, 3,... n)
Dk = dummy variable vector
(k = 1, 2, 3,... n)
α, βj, and γk = estimated parameters of the logistic function
ei = random error
The model includes adoption acceleration as the dependent variable. There are 15 independent variables in the model, including two dummy variables in it. The free variables are as follows:
X1 = Age of farmer (years), X2 = Basis of formal education (years), X3 = Farming experience (years), X4 = dependents of the family (soul), X5 = scale of business (ha), the area of farm land owned, X6 = Distance from farm to settlement (km), X7 = Distance of settlement to main technology source /AIAT (km), X8 = Distance of settlement to the nearest information source (*Regional Agriculture Service Office*) (km), X9 = Distance from settlement to source of capital (km), X10 = Distance from settlement to market (km), D1 = infotech flow (category 1 = short, less than three knots); category 0 = other), D2 = land tenure status (category 1 = owner, category 0 = other (not owner).

The completion of the logit function estimation is done by using the Maximum Likelihood method using SPSS version 17 for windows. To determine the effect of independent variables used in the model together on dependent variables, the test is used at the 95% confidence level. The chances of farmers applying technology are faster to predict based on the Odds Ratio coefficient value derived from the formula:

$$\hat{Y} = \frac{ODDS}{1+ODDS} x100\% \qquad \text{Odds} = exp^{a+bx}$$

a = constant; b = the coefficient of the x-variable. In the output of the SPSS version 17 program, the Odds Ratio coefficient is identical to the Exp (B) coefficient which can be interpreted as an opportunity in percentage units. It is based on the obtained coefficients calculated using the formula:
$\hat{Y}$ = prediction of chances of accelerated adoption
In testing the hypothesis, the following criteria are used:
(a) If the P-value <α, then Ho is rejected. This means that the opportunities for farmers to adopt innovation more quickly (<1 year) are influenced by the independent variables formulated in the model.
(b) If the P-value> α, then Ho is accepted. This means that the assumed independent variable has no effect on the opportunities for farmers to adopt innovation more quickly.

3. Results and discussion

3.1. Characteristics of Respondents and Adoption of Palm Plant Farming Technology

Characteristics of individual respondents that are suspected of being associated with technology adoption were identified consisting of age, education, family dependents, and land tenure status. From the results of the interview, it was revealed that the age of the adopter varies from the youngest 21 years to the oldest 48 years. Judging from that age range, overall adopters are in productive age. This means that the opportunity to increase farm productivity is quite large. Even if we look at the grouping, the majority (63.3%) of the total adopters are in the very productive age range, namely 21-40 years. In addition to support for productive age, all adopters have a formal education base of 5-12 years, 60% of whom have a junior high school level education and above (9-12 years). With the basis of formal education like this will strengthen the capability of farmers in running farming.

Tabel 1. Characteristics of sugar palm farmers

Rrespondent characteristics	Minimum	Maximum	Average
Age (years)	21	48	33,3
Education (years)	5	12	9
School-age dependents (persons)	0	5	1,2
Dependent family of working age (person)	0	6	2,2
Farming experience (years)	1	28	17,3

In relation to the number of family dependents, adopter identification can be divided into school age (<15 years) and working age (> 15 years). This distinction relates to support for family labor (for working age) and workload (for non-productive age). The working age group is a potential workforce on and off farm. On the other hand, the pure school age group is a burden on the family. The description of the characteristics of the respondents can be seen in Table 1.

Farming experience ranged from 1-28 years with an average of 17.3 years. Farmers think that farming experience for more than 10 years already has a fairly good basis for cultivation. They are able to solve and find solutions to farming problems based on their empirical experience. With this experience capital, farmers have experienced a dynamic mindset in adopting technological innovations. The acceleration of the adoption of this technology is being carried out by farmers who want to increase the production and added value of palm-based farming.

Table 2 shows that almost all respondents have good knowledge in terms of cultivating sugar palm. In some places, especially those with a habit of making sugar or consuming alcoholic drinks, sugar palm has often been planted deliberately, although generally as a peripheral or intercropping plant between existing tree crops. The types of plants found in palm oil plantations in the field are plantation crops (*cloves, coconut, coffee, cocoa, nutmeg, cinnamon and vanilla*), fruit trees (*avocado, durian, langsat, duke, rambutan, longan, jackfruit, mango, banana*), forestry plants *(campana, nandu, gemmal, senton, gmelina, and mahogany*) and food crops (corn and peanuts). Of the various types of plants, there are 36 palm-based farming patterns developed by the respondent farmers. All identified farming patterns can be grouped into three types of farming patterns, namely (1) sugar palm is integrated with food crops, (2) sugar palm is integrated with various non-timber plants and (3) sugar palm is integrated with woody plants. The component that is still weak is the selection of seeds from superior palm types. The yield of sap from superior palm types can reach 15-20 liters / day or equivalent to 3-4 kg of sugar / day / tree.

Table 2. Respondents Knowledge component of cultivating

No.	Knowledge Components	High		Middle		Low	
		Respondent	%	Respondent	%	Respondent	%
1	Requirements growth	96	80,00	14	11,67	10	8,33

2	Selection of parent trees and provision of seeds	20	16,67	71	59,17	29	24,17
3	Nursery and Nurseries	98	81,67	10	8,33	12	10,00
4	Land preparation and making planting holes	88	73,33	26	21,67	6	5,00
5	Planting	70	58,33	46	38,33	4	3,33
6	Farm diversification	119	99,17	1	0,83	0	-
7	Weeding	116	96,67	2	1,67	2	1,67
8	Fertilization	31	25,83	60	50,00	29	24,17
9	Tree sanity	96	80,00	11	9,17	13	10,83
10	Pest and Diseases Control	97	80,83	12	10,00	11	9,17
11	Harvest and Post Harvest	13	10,83	100	83,33	7	5,83

4. Conclusions

The adoption gap from when farmers know about new technology to implement it on farms as a proxy for accelerated adoption ranges from less than 1 year to more than 4 years. There are five independent variables that partially affect the acceleration of palm farming technology adoption (Wald value> 0), namely business scale, distance to settlement, distance to BPTP, distance to the nearest information source, and land tenure status, two of which have a negative effect, namely distance to BPTP and to the nearest information (BPP). Each additional planting area of 1 hectare, there is an opportunity for farmers to accelerate technology one fold. Likewise, the distance to the settlement, the reduction in the distance from the farm to the settlement of 1 km has the opportunity to accelerate adoption six times. The increasing distance to BPTP and to BPP as the closest source of technology information has the opportunity to slow down adoption by 1.8 times and 0.2 times, respectively. With regard to land status, owner farmers will adopt faster technology than farmers with non-land owner status.

The implication of the results of this study recommends that the acceleration of farmers' appreciation of palm sugar farming technology be realized in the form of participatory technology assistance for all basic and optional technology components. The gap in technology adoption is focused on effective extension methods and seeking assistance in the form of price guarantees and a pro-farmer market. Opportunities for accelerating the adoption of technology can be done by expanding the scale of the business, closer to the location of the farm and settlements and bringing farmers closer to sources of technological information.

References

[1] Polakitan A L and Akuba R H 1993 *Buletin of* Indonesian Coconut Research Institute **20** 73-80

IOP Conf. Series: Earth and Environmental Science **883** (2021) 012042 doi:10.1088/1755-1315/883/1/012042

[2] Allorerung D 2007 Arenga : Multipurpose Plant *Workshop* on Cultivation and utilization of sugar palm for food and energi
[3] Akuba R H 2004 *Pros National Seminar on Aren* Indonesian Coconut and Other Palm Crops Research Institute Manado
[4] Rawung J B M 2016 *Bul Agrosaintek* Manado
[5] Smith W 2004 Development of Arenga in Indonesia. (*unpublished*)
[6] Reed P and Wu Y 2013. Logistic regression for risk factor modelling in stuttering research. *Journal of Fluency Disorders* **38** 88-101
[7] Loh 2006 Logistic Regression Tree Analysis in H Pham *ed. Handbook of Engineering Statistic* London
[8] Simatupang P 1991 Qualitative dependent variable regression: The theory of predictive procedures. *Workshop Agro Economic Research Methods* Bogor
[9] Pakpahan A dan S Nizwar 1991 *J. Agro Economi* **10** (1; 2)
[10] Hendayana R 2004 *Pros National Seminar on Agriculture Technologi* BPTP Papua Jayapura 5-6 Oktober 2004 Centre for Agricultural Socio-economic Research and Development Bogor
[11] Astuti U P, Sugandi D dan Hamdan 2015 *Pros Seminar Nasional* Bogor 10 November 2015 (Jakarta: IAARD Press) 2016.
[12] Pindyck R S and Rubinfeld D I 1981 *Econometric Models and Economic Forcast* (Singapore: Mc Graw-Hill International Editions) *3rd* Edition
[13] Gujarati 2002 *Basic Econometrics*. (Jakarta: Erlangga Publisher) 418 hlm.

Breed availability analysis of local beef cattle in Seram Utara Timur Seti District Maluku Tengah Regency

J F Salamena[1], Rajab[1]*, B J Papilaya[1] and R Sarfan[1]

[1] Department of Animal Husbandry, Agriculture Faculty, University of Pattimura Jln. Ir. M. Putuhena Kampus Unpatti-Poka, Ambon - Indonesia.

* E-mail: rajab.amir@gmail.com

Abstract. This study aim were to analyze the potential availability of breeds for beef cattle development in Seram Utara Timur Seti District. The research method used was a desk study and survey method. Desk studies were conducted by utilizing relevant secondary data from various related agencies, as well as relevant research results. Meanwhile, surveys and direct observations in the field were carried out in the Animal Husbandry Production Center (KSP) areas, namely Wailoping, Kobisonta and Waitila Village. The variables observed included population structure and potential availability of beef cattle. The results showed that the proportion of the beef cattle population for the categories of calves, youth and adults was 23.18; 22.18 and 54.64% respectively. The sex ratio of male to female in all age categories is still below the normal ratio of 1 male to 5. Total breed growth (net increase) was 70 tails per year from the total population of beef cattle in Seram Utara Timur Seti District (13,032 tails). The total growth of beef cattle is still less than 1% (around 130 heads) of the total population, so it is necessary to pay attention for the development of the source area for beef cattle breeding in Seram Utara Timur Seti District.

1. Introduction

The livestock sub-sector in Indonesia has a strategic role in developing the structure of the national economy. Apart from being a producer of animal protein products and the formation of Gross Domestic Product (GDP), the livestock sub-sector also absorbs labour, foreign exchange reserves and people's income [1]. In the context of the livestock industry policy, development is expected to lead to the formation of a livestock industry while still paying attention to aspects of environmental, ecological sustainability and community welfare. The components involved in the livestock development process consist of three main components, namely the farmer (community) as the subject, livestock as an object and land as a medium for business development [2]. The integration of these components is critical to the success of the development of the livestock industry.

The facts show livestock development is still faced with several of problems. One of them is that the community's need for livestock products increases in line with the rapid increase in population improved welfare and changes in people's lifestyles. Animal husbandry products, especially beef from beef cattle, are unable to meet the community's needs, so it is predicted that the supply of imported meat will increase and reach 70% by 2020 [3][4]. Therefore, more intensive efforts are needed to increase productivity and livestock population, especially for beef cattle in Indonesia [5]. Given this, the local beef cattle breeding program is considered very strategic [6]. The productivity of local cattle still has the potential to be improved, as is the case with Bali cattle [7].

The problems in the beef cattle business sector include the underdevelopment of the cattle breeding business, which is currently still being carried out by farmers with a small scale of ownership and is part-time in nature. So far, many cattle breeding have been carried out by the community, while the private sector has been reluctant to run a breeding business because of the relatively long turnover of business capital. The existence of calves and feeders is very necessary. If this business does not develop, feeder cattle availability for fattening will be hampered [8]. The lack of supply of superior local cattle for breeders results in the quality of existing cows to decrease from the predetermined standard which is due to the lack of availability of superior bulls in the field because generally good cows are sold by breeders to get a higher selling price [5].

Maluku Province is geographically was an archipelago which has several quite large islands including Seram Island. In terms of the availability of natural resources, Seram Island is very potential for the development of agricultural commodities, including the livestock sub-sector, especially large ruminants (cattle). Souhoka *et al.* (2020) stated that in October 2018 alone, a total of 52.2% of the cattle slaughtered in the Ambon City RPH came from Seram Island, this is because the island is a beef cattle development area in Maluku province due to land potential and availability of animal feed in the form of grass and agricultural waste for adequate animal feed [9].

One of the potential development areas is Seram Utara Seti District, which is known as one of the centres of agricultural and livestock production in Maluku, considering that this area has a rice planting area of 3,151 ha (in 2018), with the population of Bali cattle as local beef cattle in 2018 was 11,433 heads [10]. This research was conducted to study and analyze the potential availability of seeds to develop beef cattle in Seram Utara Timur Seti District.

2. Methods

This study was conducted in the Animal Husbandry Production Center (KSP) area in the Seram Utara Timur Seti District, Central Maluku Regency, Maluku Province. The study has been conducted for four months.

The research method used is a desk study and survey methods. Desk studies are conducted by utilizing relevant secondary data from various related agencies, as well as relevant research results. While the survey was conducted using a list of questions and field observations in several villages in the livestock production centre areas as the focus and locus for development, namely Wailoping Village, Kobisonta Village and Waitila Village.

The research variables observed included (1) population structure in livestock, including male and female broodstock, male and female calves, and (2) potential availability of beef cattle. Descriptive analysis of the data needed to answer the objectives to be achieved in this study, there was to analyze the potential availability of seeds.

3. Results and Discussion

3.1 Beef Cattle Population Structure

The potential availability of beef cattle is very much determined by the population structure in an area at a certain time period. The population structure in livestock includes male and female broodstock, as well as male and female calves [11]. The structure of the beef cattle population in Seram Utara Timur Seti District, as presented in Table 1.

Table 1. Population Structure of Beef Cattle In Seram Utara Timur Seti District

Category of Beef Cattle	The Rate	Composition (%)	Sex Ratio (Male:Female)
Calves (< 1 year of age):			
- Male	1,424	47.14	1 : 1.12
- Female	1,597	52.86	
- Total	3,021	23.18	
Veal (1 - 2 year of age):			
- Male	1,208	41.80	1 : 1.39
- Female	1,682	58.20	
- Total	2,890	22.18	

Adult Cows (> 2 years of age):			
- Male	4,025	56.52	1 : 0.77
- Female	3,096	43.48	
- Total	7,121	54.64	
Overall:			
- Male	4,791	36.76	1 : 1.72
- Female	8,241	63.24	
- Total	13,032	100.00	

Based on the distribution of the number of beef cattle by age category (Table 1), there was a difference in the number between the category of children (23.18%) with the category of young (22.18%) and the category of adults (54.64%). The categories of children and young people tend to be the same, but the percentage is lower (about 50%) than the adult category. It is suspected that this is due to the relatively low birth rate and the relatively high mortality rate. The birth rate depends on the number of females who can mate, get pregnant and give birth normally. If the mating season occurs during the dry season, the tendency for marriage rates to be low, and an impact on the birth rate is also low [12].

Based on the calculation of male and female sex ratios, it can be seen that in all age categories the sex ratio is still below the normal ratio of 1 male to 5 females for natural mating in extensive maintenance. Based on the minister of agriculture's regulation on guidelines for good beef cattle breeding, for a breeding business with natural mating, the ratio of male/female in the population is 1 male to 8-10 female. Male/female sex ratio is influenced by (1) the ratio of male / female at the beginning of birth that occurs in the population, (2) the number of slaughter, sale and death of livestock at young and adult ages [13], and (3) livestock entry and expenditure in various categories of age in livestock population [14]. The sex ratio of male and female beef cattle in Seram Utara Timur Set District is more determined by the start of birth in the population. For the development of the cattle breeding area in the future, it is necessary to control the sex ratio at the beginning of birth through controlling marriage both naturally and by artificial insemination.

3.2. Availability of Beef Cattle Seeds

The potential availability of beef cattle in Seram Utara Timur Seti District, Central Maluku Regency, was presented in Table 2 below.

Table 2. Potential Availability of Beef Cattle Seeds in Seram Utara Timur Seti District

No.	Parameter Item	Parameter Value	Percentage (%)
A.	Beef cattle population	13,032 tails	-
B.	cows (from a population)	8,241 tails	63.24
C.	Productive cows (from cows)	5,666 tails	43.48
D.	Productive cows slaughtered	1,033 tails	18.23
E.	Productive cows insertion	-	-
F.	Total of productive cows	4,633 tails	81.77
G.	AI mating (from the total of productive cows)	-	-
H.	AI Natality	-	-
I.	Natural Mating (from the total of productive cows)	4,633 tails	100.00
J.	Natural Mating Natality	4,633 tails	100.00
K.	The total of Natality	4,633 tails	35.55
L.	Mortality (from a population)	652 tails	5.00
M.	natural increase	3,981 tails	30.55
N.	Cows population increased	1,592 tails	40.00
O.	Bull population increased	2,387 tails	60.00

P.	Cow seeds increased	1,274 tails	80.00
Q.	Bull seeds increased	477 tails	20.00
R.	Cows seed replacement	718 tails	-
S.	Net increase in cows	54 tails	4.27
T.	The National target of cow seeds growth (%)	43 tails	80.00
U.	1. the gap of cow seeds growth	-	20.37
	2. the gap of cow seeds (tails)	11 tails	-
V.	Bull seeds replacement (from a productive cow)	35 tails	-
W.	Net increase of bull seeds (%).	16 tails	3.39
X.	The Local target of bull seeds growth (%).	3 tails	20.00
Y.	1. The gap of bull seeds growth	-	81.25
	2. The gap of bull seeds (tail)	13 tails	-
Z.	Total net increase of beef cattle seeds	70 tails	-

From the results of the research, it was found that the percentage of mortality was 5% so that the natural increase was 30.55 % which was obtained from the percentage of total births (35.55%) minus the percentage of deaths. Of these, there is a 40 % increase in the female population and 60 % increase in the male population, in contrast to the increase in female seedlings an increase of 80 % and an increase in male seedlings by 20 %. This indicates that female beef cattle are retained for seed while males are mostly slaughtered (sold, consumed, donated and customary dowry). For female seed growth (net increase) is 4.27 % obtained from the ratio of female seed increase and female seed replacement divided by population. The growth of male seedlings (net increase) was 3.39%. These results illustrate that the growth of female seedlings is 0.88% greater than the growth of male seedlings which are expected to support the availability or sustainability of beef cattle breeds in Seram Utara Timur Seti District. From the existing beef cattle population, it is estimated that it is 12.5% and 3% to replace old and advanced cattle, respectively. The final result shows that the total seed growth (net increase) of 70 heads per year from the total population of beef cattle in Seram Utara Timur Seti District (13,032 heads). The total growth of beef cattle is still less than 1% (around 130 heads) of the total population, so it is necessary to pay attention to the development of the source area for beef cattle in Seram Utara Timur Seti District.

4. Conclusion

Based on the results of the study, it can be concluded that the proportion of beef cattle population for the categories of children, young and adults is 23.18; 22.18% and 54.64%. The sex ratio of male to female in all age categories is still below the normal ratio of 1 male to 5. Total seed growth (net increase) of 70 head per year from the total population of beef cattle in Seram Utara Timur Seti District (13,032 heads). The total growth of beef cattle is still less than 1% (around 130 heads) of the total population, so it is necessary to pay attention to the development of the source area for beef cattle in Seram Utara Timur Seti District.

References

[1] Setyowati N 2011 Strategi Pengembangan Subsektor Peternakan dalam Rangka Memperkuat Sektor Pertanian di Kabupaten Boyolali *Sains Peternakan* **9(1)** 32-40.

[2] Santoso U 2015 *Pengembangan Peternakan Terpadu Berwawasan Agribisnis dalam Rangka Mewujudkan Ketahanan Pangan.* https://sivitasakademika.wordpress.com/2015/03/30/. [08/10/202].

[3] Soeharsono, Rustijarno S, and Triwidyastuti K 2008 Pembibitan Ternak Sapi Potong dalam Sistem Integrasi Tanaman Ternak di Kawasan Pantai Selatan Kabupaten Bantul *Sains Peternakan* **6(1)** 49-55.

[4] Diwyanto K, Sitompul D M, Manti I, Manthius I W, and Soentoro 2003 *Pengkajian Pengembangan Sistem Integrasi Kelapa Sawit Sapi* (Prosiding Lokakarya Nasional Sistem

Integrasi Kelapa Sawit – Sapi Bengkulu, 9 - 10 September 2003 Pusat Penelitian dan Pengembangan Peternakan Bogor) pp 11-16.

[5] Romjali E 2018 Program Pembibitan Sapi Potong Lokal Indonesia *WARTAZOA* **28(4)** 190-210.

[6] Sayaka B 2012 Pengembangan Perbenihan Sapi Potong dan Perannya Dalam Pencapaian Swasembada Daging *Forum Penelitian Agro Ekonomi* **30(1)** 59-71.

[7] Ishak A B L, Nurhayu A, Ella A, Sariubang M, and Rahmawati T 2014 *Seleksi performans induk sapi bali sebagai upaya pembentukan populasi dasar pada program pembibitan dan pemurnian sapi Bali di Kabupaten Barru Provinsi Sulawesi Selatan* (Prosiding Seminar Nasional Optimalisasi Sumberdaya Lokal pada Peternakan Lokal Berbasis Teknologi. Makassar, 9-10 Oktober 2014) pp 59-65.

[8] Ilham N, Indraningsih K C, and Elizabeth R 2017 Kinerja Berbagai Pola Usaha Pembibitan Sapi Lokal di Beberapa Daerah Pengembangan Sapi Potong *Analisis Kebijakan Pertanian* **15(1)** 67-82.

[9] Souhoka D F, Tagueha A D, and Rajab Tingkat Insidensi Pemotongan Sapi Betina Bunting dan Karakteristik Fetus Di Rumah Potong Hewan Kota Ambon *Agrinimal Jurnal Ilmu Ternak dan Tanaman* **8(1)** 44-50.

[10] Badan Pusat Statistik Kabupaten Maluku Tengah 2018 *Kabupaten Maluku Tengah Dalam Angka Tahun 2018* (Masohi: BPS Kabupaten Maluku Tengah).

[11] Riwukore J R., Yani A, Priyanto R, Purwanto B P, Abdullah L, Fuah A M, and Habaora F 2020 Analysis of Development of Cattle Population in Nusa Tenggara Timur Province of Indonesia (Case Study in Timor Island Forest Agroecosystem) *Veterinary Research* **13(3)** 33-39.

[12] Muthalib R A 2015 Pengaruh Musim Kawin Terhadap Produktifitas Induk Kambing PE Pada Kondisi Pedesaan *Jurnal Ilmu-Ilmu Peternakan* **18(1)** 36-42.

[13] Agus A, and Widi T S M 2018 Current Situation and Future Prospects for Beef Cattle Production In Indonesia – A review *Asian-Australas J. Anim. Sci.* **31** 976-983.

[14] Kusuma S B, Ngadiyono N, and Sumadi 2017 Estimasi Dinamika Populasi Dan Penampilan Reproduksi Sapi Peranakan Ongole Di Kabupaten Kebumen Provinsi Jawa Tengah *Buletin Peternakan* **41(3)** 230-242.

Study of dryland cultivation farming technology to support sustainable food independence on small islands

J Riry[1,*], A S Mahulette[1], A M Tapotubun[2], W A Riry[3]

[1]Agrotechnology Study Program, Faculty of Agriculture, Universitas Pattimura, Ambon, Indonesia
[2]Fisheries Product Technology Study Program, Faculty of Fisheries, Universitas Pattimura, Ambon, Indonesia
[3]Law Study Program, Faculty of Law, Universitas Pattimura, Ambon, Indonesia

* E-mail: riryjohan@gmail.com

Abstract. The agricultural sector's strategic roles for economic growth, including, among others: providing food for the Indonesian population, earning the country's foreign exchange through exports, providing industrial raw materials, increasing employment and business opportunities, increasing regional revenues, and alleviating poverty. The Food and Agriculture Organization (FAO) stated that food self-sufficiency occurs when everyone in every moment can access food both physically and economically to meet their daily needs. Drylands have characteristics such as water shortage, erosion sensitivity, low land productivity, high variability in soil fertility, limited plant species, low adoption of advanced technology, minimal capital availability, and inadequate infrastructure. Small islands are islands ecologically separated from the mainland island, have clear physical boundaries, and are isolated from the parent island's habitats, so they are insular. Small islands usually have relatively small catchment areas, are vulnerable to global warming, sensitive to natural disasters, isolated and far from the main market, open to small-scale economic systems, have high population growth rates, have limited infrastructure and education, and limited skills of its inhabitants. Small islands have a high rate of land degradation that threatens the long-term sustainability of agriculture. Technologies that can be applied to dryland on small islands are conservation farming, LIESA system, Biointensive Gardening, Dusung, Agropasture, Sloping Agricultural Land Technology (SALT), Rounders type, No-tillage, and small island weed management. This paper is an ideal contribution to overcome the food problem in small islands that generally have drylands. It was presented at the National Seminar of the Indonesian Agronomy Association (PERAGI) in Bogor

1. Introduction

Based on Law No. 18/2012 of the Republic of Indonesia, food is defined as anything that comes from biological sources of products from agriculture, plantations, forestry, fisheries, livestock, aquatics, and water, both processed and unprocessed which are designated as food or drink for human consumption, including food additives, food raw materials and other materials used in the process of preparing, processing and/or making food or beverages. Food sovereignty is the right of the states and nations, which independently determines the food policy that guarantees the right to food for the people and gives the community the right to determine a food system following the potentials of local resources.

The food conditions of a country significantly affect its stability. The food problem is a crucial problem because it is directly related to various groups of people. Food determines the people's life and death; that was the statement of the first President of the Republic of Indonesia (Sukarno) when he laid the groundwork for the IPB campus in 1953. We have passed 64 years, and the food problem is still a serious problem to be discussed and developed. The indication from the world's food organization (FAO) concerning the possibility of a world food crisis in 2025 has prompted agrarian countries to struggle to utilize food resources to meet their country's food needs. With a population growth of 1.6% per year, this includes Indonesia, a severe burden for its food supply. The government policy about rice commodity as a food source has resulted in the staple food pattern of the previously diverse people (rice, sweet potato, corn, sago, banana, etc.) according to local potential and culture, now experiencing changes that tend towards a single staple food pattern—namely rice [1].

Adequacy of food is the right of every human being, which means that everyone has the right not to be hungry and to have access to adequate, nutritious, and safe food for their health. Food security can be defined as access to adequate food for everyone at any time to obtain a healthy body and an active life. Food security includes food production and availability, distribution and affordability by all people, individual consumption to meet nutritional needs, and monitoring food insecurity. Weak food security can be caused by several factors, such as low production of agricultural food products that cause low supply; this factor can be caused by nature or climate that is not supportive and mismanagement of agricultural food production and natural and human disasters. Also, dependence on one type of food, changes in earth temperature, and new food pathogens can cause weak food security [2]. Food and family economic needs encourage the community to continue to manage land with shallow solum, rocky lands, narrow and terraced fields, and steep slopes with hilly and mountainous terrain. Not only that, they use dry areas with wet months of only 3-5 months, which is enough to plant corn or other crops. If there is no better alternative, any land will be utilized; that is the community's choice to maintain and fulfill their daily needs [3]. The government's policy to boost food self-sufficiency that promoted rice has resulted in increasingly narrow food diversity. Lowland paddy requires much water, a good permanent irrigation network, and generally requires a high initial cost. The policies that lead to one type of food have caused other types of food to experience degradation, especially local food, to decline so that areas experiencing rice shortages will be called food insecure areas, even if local food sources are quite abundant.

One of the causes of low food production in Indonesia is decreasing productive land availability due to land conversion, use of productive lands as locations to build factories, housing, sports facilities, and other public facilities. In Indonesia, it is estimated that there is a loss of productive lands of approximately 23-25 thousand hectares every year. In addition to land conversion, many lands are classified as marginal due to limited nutrients, rocky structure, and many of them are acidic soils [4]. With the decreased agricultural land in Indonesia, it is not easy to expect the farmers to produce optimally [5]. Apart from the problem of agricultural land availability, climate change has also affected low agricultural productivity. Global climate change has become an important issue and attracts attention.

Table 1. Drylands that are suitable for agriculture in Indonesia

Province	Lowlands (ha)			Highlands (ha)			Total
	Annual crops	Perennial Crops	Sum.	Annual crops	Perennial Crops	Sum.	
Sumatera	4.899.476	15.848.203	20.747.679	1.103.176	992.055	2.095.231	22.842.910
Jawa	925.412	3.982.008	4.907.420	200.687	484.960	685.647	5.593.067
Bali and Nusa Tenggara	1.091.878	1.335.469	2.427.347	58.826	201.761	260.587	2.687.934
Kalimantan	10.180.151	14.340.956	24.521.107	592.129	389.521	981.650	25.502.757
Sulawesi	1.801.877	3.664.040	5.465.917	70.780	1.134.320	1.205.100	6.671.017
Maluku and Papua	4.360.318	8.282.809	12.643.127	43.094	233.981	277.075	12.920.202
Indonesia	23.259.112	47.453.485	70.712.597	2.068.692	3.436.598	5.505.290	76.217.887

Source: Center for Land and Agro-climate Research and Development, 2001

Based on the islands' physiography, Indonesia is an archipelago consisting of large islands (Java, Sumatra and Kalimantan, Sulawesi, and Papua), medium islands (Halmahera, Bali, Timor and

Yamdena, Seram, Buru, and Tanimbar), and numerous small islands. Management of large island agriculture is different from those of small islands. In general, small islands have weaknesses in crop production. These weaknesses include the fertile soil layer is very thin, having extreme weather, and limited water resources. Also, the islands formed in a cluster of islands have different characteristics from one island to another. Some islands are located far away from each other, but they have the same characteristics (distant similarities). However, on the other hand, some islands are close to each other, but they have different characteristics (close differences).

Policymakers often ignore the utilization of dry land for agriculture since they are more interested in increasing rice production in paddy fields. This may be because there is an assumption that increasing rice production is easier and more promising than crops such as upland rice, which have a higher risk of failure. Despite the available drylands, this is quite extensive and can produce upland rice of more than 5 t/ha. Potential drylands can produce sufficient and varied staple foodstuffs, upland rice, and other crops if managed using effective technology and appropriate development strategies. The main food source is rice and corn, sorghum, soybeans, green beans, cassava, sweet potatoes, and the like, all of which can be cultivated on dry land [9]; [10].

Low production of agricultural food products causes low supply, which factor can be caused by nature or climate that does not support and mismanagement of agricultural food production as well as natural and human disasters. In conducting agricultural business on dry land, one of the important questions that arise is how to get water for crop irrigation. The irrigation system suitable for dry land is a micro-irrigation system, which uses emitters. Micro-irrigation has the main objective of making efficient use of water by plants and reducing evapotranspiration. The next upland irrigation technology is the 'jug fertigation system'.

This later system has the advantage of up to 25% fertilizer efficiency and reduces mechanical damage to crops. There are various challenges in conducting agricultural cultivation on dry land, including limited water sources, requiring plant varieties resistant to water stress, and relatively low soil fertility. This writing aims to describe the potentials for dry land technologies in Indonesia to develop food crops towards food self-sufficiency. According to (tetap harus ADA nama) [11], to be able to produce optimally and benefit farmers, planting upland rice using the SRI (System of Rice Intensification) method in dry land should be done in the rainy season because farmers do not need to pay for water supply, including rental pump machines and labor. Ultisols, oxisols, and inceptisols generally dominate dry land's soils. These soil types are generally yellow to red in color and have undergone further weathering with the content of the sand fractions, which are dominated by quartz and opaque. In contrast, the clay fractions are dominated by kaolinite, goethite, and hematite, have low clay activity, are acidic, have low nutrient contents, and a high Al saturation [12]. Organic fertilizers and agricultural wastes, such as biochar, can increase productivity on drylands [13]

2. Methods

This paper is based on a desk study and library research. The information collected relates to the various technologies applicable to dry land. This paper is also based not only on a literature review but also on the author's experience as an observer in agriculture, arid land agriculture.

3. Results and discussion

Based on identifying various agricultural problems that occur in the dry land, the discussion is directed at finding a solution for increasing crop production aimed at achieving food self-sufficiency.

The discussion concerns technology offers that can be applied to solve dry land problems generally found on small islands based on the topic. Apart from the technology offered, there are also other solutions to deal with low agricultural production in Indonesia.

Some simple agricultural technologies that have been implemented in Indonesia and developed in the Maluku Islands are discussed as follows:

3.1. LIESA (Low Input External Sustainable Agriculture)

LIESA system is a rural agricultural technology applied in Indonesia, especially in Maluku, since centuries ago. People only use natural resources in situ for their agricultural needs. The use of agricultural production supplies in fertilizers and pesticides from outside (ex-situ) is only slightly or even rarely done. LIESA in Maluku is classified as simple organic farming. The term simple organic agriculture is deliberately conveyed by the author to understand that between organic and inorganic agriculture, there is what is called simple organic or semi-organic farming.

3.2. Bio-intensification (use of biological fertilizers).

Biointensification or the so-called intensification of biological agriculture, i.e., by utilizing living organisms both above the soil surface, such as pollinating insects, natural enemies, and other organisms, as well as below the soil surface, such as earthworms that can fertilize the soil, mycorrhiza which can fix nitrogen from the air which in turn helps growth and crop production. Biointensification technology is a form of intensive agriculture that uses biological materials and environmental management to revive productive and sustainable natural agricultural ecosystems t. Several technologies that use biological materials to restore natural agricultural ecosystems include planting green manure crops, alley cropping with legume crops, and crop-livestock integration (agropastoral). All these technologies provide an opportunity for the agricultural ecosystem to restore its condition to become productive again [14].

3.3. Bio-Intensive Gardenig (BIG) such as Alley Cropping

BIG is a gardening technique widely applied in the Maluku region, especially West Southeast Maluku, by using appropriate spacing technology between staple crops and other plants that produce nutrients for the crops. Alley cropping is a planting technique that regulates plants with a certain distance between staple crops and side plants so that sunlight can be used as much as possible.

3.4. Dusung System Agriculture (Agroforestry)

Dusung is a food crop conservation agricultural model on dry land applied in several regions in Indonesia. Dusung is a sedentary agricultural system that optimally utilizes space, a system that has been implemented by the people of Maluku from generation to generation.

This system is a complex agricultural one because it grows various crops and other plants in an area (high intercropping system). In various high-stratum plants, such as durian and canary nut, other trees to produce housing materials, such as salawaku, titi, and teak, are grown. Those are deliberately planted because they have multiple benefits, namely economic and ecological benefits. The ecological benefit is that the trees have a large stature that can withstand strong winds (win breaks) to protect cloves and nutmeg. Besides, in the same area, cloves, nutmeg, coconut, and medium-sized fruit trees, such as langsat, duku, cempedak, jackfruit, and others, are also grown. In addition to the types of plants above, food crops, such as cassava, taro, gembili (lesser yam), and other types of tubers, as well as several types of nuts and beans, are also grown as lower-stratum (bottom) plants [15].

3.5. Agropasturale

The term agropastoral or mix farming that combines agriculture and livestock in the same area. The relationship between livestock and crops/plants is very close because the plants need livestock, and the livestock also needs the plants. Residues of the harvested crops can be used as animal feed. Livestock can take advantage of harvested crop residues, such as corn stalks and leaves, legumes, cassava leaves, and others. Plants need nutrients from animals; feces and urine released by animals will become good fertilizers used by plants..

3.6. Sloping Agricultural Land Technology (SALT)

SALT is a farming system that is suitable for use in areas with sloping land conditions. In this system, plants are planted in a contour following the direction of the slope, the plants are arranged in such a way as to hold water when it rains, thereby reducing the impact of erosion that can make the soil critical due to soil degradation.

3.7. *Rounders Type (with Crop Rotation)*
Rounders type is a system developed from the agropastoral system. The agropastoral has a small land area, and the livestock can move around. On the other hand, the rounders type has a larger area, the number of plants is more, and the pen system is fixed in the middle. Plants are grown in rotation alternately between forage crops (fodder) and food crops (human food) [16].

3.8. *No Tillage or Minimum Tilage (TOT or OTM) by using a shail*
Cultivation technology without tillage (no-tillage) or minimum Soil Cultivation (OTM) is a conservation tillage technology developed in the 1980s. However, this technology is not new in Maluku because, since time immemorial, our ancestors have used simple tools such as suang. (Maluku Tengah) and Shail (MTB) as a substitute for a hoe for limited tillage. This technology is very suitable for Maluku because the soil's topography tends to be sloping, so it is susceptible to erosion. In perfect soil cultivation (OTS), the soil becomes looser and lighter so that it is not resistant to beating raindrops. On the contrary, with TOT or OTM, the soil is heavier and resistant to beating raindrops.

3.9. *Alley Cultivation Technology*
Alley cultivation technology is a practical approach in conserving soil fertility. Perennials are planted in rows at a distance, which creates space between the rows for planting short-lived perennials and biennials to allow space between perennial plants' rows. It is advisable to plant the perennial rows in an east-west direction so that the row plants' shadow does not cover the intercrops. The selection of perennial plants and perennials in alley cultivation should pay attention to mutually beneficial interactions (symbiotic mutualism) or at least one that benefits the other without experiencing a loss at all (commensalism). For example, plants such as gamal, lamtoro, calliandra, or sengon planted in rows can fix nitrogen and serve as a green manure source for food crops. Likewise, on the other hand, for example, annual plants and biennial green manure are planted between rows of perennial plants instead of green manure.

3.10. *Wise weed control with critical period calculations*
If we look at the time a farmer takes in gardening, we can conclude that most of the time is spent pulling weeds or clearing grass from their plants. Until now, the habit of our farmers is to control weeds or weeds (grass) from their crops at any time even though they have not yet crossed the economic threshold [17]. They do not want any grass on their crops, especially food crops. This is what takes up their time, even though weed control should pay attention to the crop's critical period. The critical crop period is the period in which the presence of weeds must be eliminated, so weed control only occasionally, not all the time [18]. Control at any time can cause losses in the form of time required a lot, and if the grass is often cleaned, the soil is more easily eroded. One of the benefits of weeds is to resist erosion [19]; [20].

3.11. *Jug Irrigation*
The jug irrigation system has actually been practiced for a long time in several places in Indonesia, estimated since 1996. The jug irrigation system is more efficient because the plants' water is all absorbed by the plants, and none is wasted. This system is implemented simply by immersing a jug filled with water in the ground. The jugs used are made of clay and other materials so that the water in the jugs can seep into the plant roots. The planted jug is filled with water through a hose connected to the water barrel. The water in it will seep into the surrounding soil through its permeable walls. The ability of the jug walls to infiltrate water is designed in such a way as to compensate for the evapotranspiration needs of the plant at any time and with due regard to the hydraulic properties of the soil. Filling water into the jug can be done in two ways, namely manually and automatically. The manual method can be done if the number of jugs is limited or there is sufficient labor. Meanwhile, the automatic method is to apply the principle of the connected vessel. Namely, on the one hand, it is a jug

and on the other is the water source tank. The jug irrigation system can be used to cultivate seasonal and annual crops and young perennial crops. This method is one of the techniques in water-saving irrigation on dry land [21].

3.12. *Drip Irrigation*
Drip irrigation provides water with low discharge and high frequency in a sustainable manner to crops either through the soil surface or directly to the root zone using either a single emitter or in the form of a drip line (perforated hose). Water flow in drip irrigation utilizes capillary and gravitational forces that move vertically and horizontally in the soil profile. Drip irrigation has an efficiency value of 80-95 percent compared to bulk irrigation and surface irrigation. Provision of water in small and sustainable volumes through drip irrigation aims to maintain soil moisture and avoid water loss through percolation and runoff processes so that water availability for crops is still fulfilled. Drip irrigation technology can manage water provision in the root zone of crops in a sustainable manner to increase land productivity, and cultivation activities can last all the time. The application of a drip irrigation system based on an automatic control system can increase the efficiency of water use in crops because it can work based on the actual conditions of agricultural land through soil moisture levels. This is very suitable for dryland agriculture, which is generally water-limited [22].

3.13. *Implementing Sustainable Agriculture*
The vision of agriculture in the future is sustainable agriculture, which aims to increase farmers' productivity and welfare while preserving environmental resources. Sustainable agricultural systems rely on the empowerment of local agricultural resources, both physical and human resources, supported by site-specific sustainable technology managed by agribusiness [23]. The concept of sustainable agriculture itself is derived from the basic concept of sustainable development. Namely, the needs of human life today can be met without sacrificing the ability to meet the needs of the next generation. The boundaries of sustainable agriculture themselves are very diverse. However, the common thread that becomes the basis is that (1) agriculture must be more productive and efficient, (2) biological processes in situ must play a more important role, and (3) internal nutrient recycling is prioritized [24]. The definition of sustainability contains the meaning of income sustainability (agribusiness perspective) and natural resource sustainability (environmentally sound). In simple terms, these limits can be formulated as follows:

$$\text{Sustainability} = \text{Production (Income)} + \text{Resource}$$

Sustainable agriculture is the successful management of agricultural enterprises' resources to change human needs while maintaining or improving environmental quality and conserving natural resources. [25]. By defining and assessing more broadly, agriculture can be said to be sustainable agriculture if it includes the following characteristics:

a) Ecologically stable, which means that the quality of natural resources is maintained and the agroecosystem's overall capacity from humans, plants, and animals to soil organisms is enhanced. Local resources are used so that the loss of nutrients, biomass, and energy can be reduced as low as possible and can prevent pollution. The emphasis is on using renewable resources.
b) Can be economically sustainable, which means that farmers can produce enough to meet their own needs and/or income and get sufficient income to return the energy and costs incurred. This economic sustainability can be measured in terms of ongoing agricultural products and other functions such as conserving natural resources and minimizing risks.
c) Fair, which means that resources and power are distributed so that all community members' basic needs are met, and their rights in land use, adequate capital, technical assistance, and marketing opportunities are guaranteed. All people who have the opportunity to participate in decision making, both on the ground and in the community.

d) Humane, which means that all life forms (plants, animals, and humans) are respected. The basic dignity of all living things is respected, and relationships and institutions incorporate fundamental human values, such as trust in honesty, dignity, cooperation, and compassion.
e) Flexible, which means that rural communities can adapt to ongoing changes in farming conditions, such as population growth, government policies, market demand, and so on. This includes not only the development of new and appropriate technology but also innovation in a social and cultural sense.

In agricultural development, increasing production is often given the main attention. However, there is a maximum limit to ecosystem productivity. If this limit is exceeded, the ecosystem will be degraded and likely to collapse so that few people can survive with the remaining resources. Production and consumption must be balanced at an ecologically sustainable level. Although sustainability is viewed from a dynamic concept that allows the changing needs of an ever-increasing global population, basic ecological principles oblige us to realize that agricultural productivity has limited capabilities. According to FAO, environmental problems in developing countries are largely due to over-exploitation of lands, expansion of planting, and deforestation [26]. Several large irrigated areas have been damaged by sanitation. In particular, the degradation of soil fertility and the scarcity of wood fuel show concern about this problem.

3.14. *Use of Drought Resistant Varieties*
One of the government's efforts through the Food Crops Research and Development Center (Puslitbangtan) in overcoming water shortages in the dry season or the impact of extreme climate change is by researching to obtain drought-resistant crop varieties. Many varieties that have been successfully developed include: (a) lowland rice, such as Inpari 18, Inpari 19, and Inpari 20; (b). apart from lowland rice, there is also upland rice, such as Inpago 4, Inpago 5, Inpago 6, and Inpago 8; (c) droughts tolerant maize Bima 4 for hybrid types and Lamuru for composite types. Apart from lowland rice, upland maize, and maize, various varieties of mungbean, peanut, and soybean have also been found to grow on dry land.

3.15. Determination of the Right Planting Time
Climate change has various impacts on life on earth, including farming techniques. In agriculture, climate change, among others, causes a shift in the planting season so that various anticipatory measures need to be made. The Center for Research and Development of Agricultural Land Resources (Balai Besar Penelitian dan Pengembangan Sumberdaya Lahan Pertanian) has produced a planting calendar map useful in planning cropping patterns according to climatic conditions in each area. Determination of planting time is closely related to groundwater availability for agriculture, which is a cultivation technique by utilizing water stored at the soil surface for plant needs. This technique is beneficial for areas that have low rainfall levels or have climate types D to F (Q value 0.6 to 3) based on the Schmidt-Ferguson climate classification [27].

3.16. Utilization of Embung (water reservoir)
In principle, agriculture is synonymous with water and sunlight harvesting; these two elements are essential in the photosynthesis process as a sophisticated factory producing various agricultural products. Limited water can reduce yield to a total crop failure (puso). Embung is a pond or reservoir for collecting and storing rainwater during the rainy season, and the collected water can be used as irrigation during the dry season. Thus, rainwater can be harvested during the rainy season to store it temporarily and use it in the dry season [28]. Such is the importance of reservoirs for people with minimal water that the government, through the Ministry of Agriculture and the Ministry of Health, annually builds reservoirs in villages in small islands that have limited groundwater availability.

3.17. *Utilization of Mulches*
Mulches are ground cover materials, both organic and inorganic (plastic). The use of mulch is useful for maintaining soil moisture, reducing evapotranspiration, suppressing weed development, reducing

fertilizer loss due to evaporation, suppressing the development of plant pest organisms, increasing sunlight intensity, avoiding fertilizer loss, saving pest control costs, reducing surface runoff. Moreover, it finally increased yield [29].

In a study with shallot, the application of cow manure at ca. 30 tons per hectare significantly affected the growth and yield of bulbs per hectare, which increased both with or without mulching. The rice straw mulch effect was more dominant than the effect of cow manure on tuber dry weight per hectare [30]. Straw mulch and silver, black plastic mulch resulted in increased leaf area, plant dry weight, tuber number, and tuber weight per crop. However, there was no difference between the mulch treatments in a study with potato [31].

3.18. *Government Policies that Support Dry Land*

Most of Indonesia's agricultural lands have a hilly topography and have increasingly smaller flat lands available for agriculture. The consequence of utilizing a hilly land is that it is difficult to irrigate since it tends to dry out. Most of the farmers who are outside Java Island cultivate their land with various crops in intercropping patterns. It is hoped that the central government policy will not leave its support dry land because dry land is identical to the local food supply. Abandoning dry land is synonymous with abandoning local food supply. It is necessary to remember that a weak local food supply will lead to weak national food security. Therefore, it is suggested not only to build rice field irrigation but also build dry land irrigation.

4. Conclusion

The conclusion obtained from this paper's discussion is that dryland can make a major contribution to the achievement of Indonesia's food independence through the application of dryland technology. Communities that inhabit small islands can survive by sustainably managing local natural resources by applying upland technology wisely that has proven beneficial for people on small islands. It is suggested that the central and regional governments need to continue to pay attention to the food needs of small island communities with a policy that supports good utilization of drylands so that there is no impression of food shortages on an island.

5. Acknowledgment

Prof. Dr. Ir. S. H. T. Raharjo is hereby acknowleged for his support in translating this paper. Thank you.

References

[1] Utami P and Sulistyani Budiningsih S 2005 Potential and Availability of Local Food Materials Sources of Non-Rice Carbohydrates in Banyumas Regency *Journal of Economic & Business Dynamics* **12** 150-158.

[2] Mulyani A, Nursyamsi D, Irsal L 2014 Accelerating Development of Dry Land Agriculture in Dry Climate in Nusa Tenggara *Journal of Agricultural Innovation Development* **7** 187-198.

[3] Mulyani A and Mamat H S 2019 Management of Dry Climate Dry Land for Corn Development in Nusa Tenggara. *Journal of Land Resources* **13** 41-52.

[4] Riry J 2016 *Weed Management to Support Sustainable Food Self-Sufficiency in Small Islands*. Permanent Professor Inauguration Speech in the Field of Agricultural Science (Weed Science), Faculty of Agriculture, Unpatti. Ambon. Not published

[5] Prabowo R 2010 Government Policy in Achieving Food Security in Indonesia *Mediagro Journal (Agricultural Sciences)* **6** 62-73.

[6] Damasius S 2016 Changes in the Working Culture of Dryland Agriculture in Atoni Pah Meto in North Central Timor Regency. *Paradigm:Journal of Cultural Studies* **6** 145-164.

[7] Amiruddin S 2003 Dryland Management System in the Upper Watershed *Journal of Agricultural Research* **22** 162-171.

[8] Center for Research and Development of Soil and Agro-climate 2001 *Atlas of Directions for Indonesian Agricultural Spatial Planning, Scale 1: 1,000,000.* Center for Research and Development of Soil and Agro-climate, Bogor. 37 pages.
[9] Abdurachman A, Dariah A, and Mulyani A 2008 Dry Land Management Strategy and Technology Supports National Food Procurement *Journal of Agricultural Research***27**43-48.
[10] Mulyani A, and Hidayat A 2009 Increasing the Capacity of Food Crop Production on Dry Land *Journal of Land Resources***3**73-84.
[11] Christanto H, and Agung I G AMS 2014 Number of seeds per hole and planting distance have an effect on the yield of upland rice (*Oryza Sativa* L.) with the system of rice intensity (SRI) in dry land *Bumi Lestari Journal***14**1- 8.
[12] Antonius K 2019 Soil Improvement to Increase the Effectiveness and Efficiency of Balanced Fertilization and Productivity of Acid Dry Soils*Journal of Land Resources***13**27-40.
[13] Urida and Neneng L 2016 The Potential of Biochar Utilization for Dry Land Rehabilitation in Indonesia*Journal of Land Resources***10** 57-68.
[14] Jambormias E 2017*Plant Breeding for the Sustainability of the Archipelago Island Agricultural System in Maluku*. Scientific Oration at Unpatti's 54th Anniversary. Ambon. Not published.
[15] Matinahoru J M 2011*Development Strategy of Dusun Model as Special Agroforestry System for Small Islands in Maluku [Development Strategy of Dusun Model as Agroforestry System Especially in Small Islands of Maluku*. Paper presented. Ambon. Not published
[16] Jambormias E 2017 *Strengthening Arin System with Biointensification and Plant Breeding for Productivity and Agricultural Sustainability in West Southeast Maluku Regency*. Department of Cultivation, Faculty of Agriculture, Unpatti. Ambon. Not published.
[17] Riry J 2014 *Determining the Correct Time of Weeding for Soybean Production*. Research Report. Not published.
[18] Nyarko K A, De Datta S K 1991 *Handbook for Weed Controle in Rice*. IRRI. Pp. 113
[19] Hejda M, Pysek P and Jarosik V 2009 Impact of invasive plants on the species richness, diversity and composition on invaded communities*Journal of Ecology***97**393–403.
[20] Neito H J, Brondo M A and Gonzales J T 1998 Critical Periode of The Crop Growth Cycle for Competition From Weeds*Pest Articles and News Sumaries*159-166.
[21] Sutrisno N, and Nani H 2019 Development of Water Saving Irrigation to Increase Agricultural Production in Dry Climates *Journal of Land Resources***13** 7-26.
[22] Adhiguna RT, Amin R 2018 Drip Irrigation Technology in Optimizing Water Use Efficiency in Agricultural Land in*Proceedings of the National Seminar on World Water Day*. e-ISSN: 2621-7449, Pp. 107-119
[23] Hadianto W, Ariska N 2019 Cultivation System of Corn (*Zea Mays* L.) Growth and Yield. *Sustainable Agrotech Journal***5**39-47.
[24] Daniel J T, Templeton G E, Smith R J, Jr and Fox W T 1993. Biological control of northerm jointveth in rice with an endemic fungal disease *Weed sci***21**303-307.
[25] Reijntjes C, Haverkort B, and Waters-Bayer A 1999 *Agriculture of the future. introduction to sustainable agriculture with low external inputs*, Canisius. Yogyakarta,pp 304 Pages.
[26] Rao V S 2000 *Principles Of Weed Science*. Science publisher, Inc. Enfield.USA. pp.viii + 555 pp.
[27] Schmidt F H, and Ferguson J H A 1951 *Rainfall Types Based on Wet and Dry Period Ratios To Indonesia,* Bogor. Pp 77.
[28] Kurnia U 2004 Prospect of Dry Season Crops Irrigation Agriculture*Journal of Agricultural Research***23** 130-138.
[29] Wisudawati D, Muhammad A, Iskandar L 2016 The Effect of Type of Mulch on the Growth and Yield of Shallots (*Allium ascalonicum* var. Lembah Palu) Given a Cover *Agrotechbis Journal* **4** 126-133.
[30] Mayun I A2007 The Effects of Rice Straw Mulch and Cow Manure on the Growth and Yield of Shallots in Coastal Areas *Agritrop Journal***26**33 - 40.

IOP Conf. Series: Earth and Environmental Science **883** (2021) 012044 doi:10.1088/1755-1315/883/1/012044

[31] Jajang S H 2009 Effect of Mulch Types on Growth and Yield of Three Cultivars of Potato (*Solanum tuberosum* L.) Grown on Medium Plains*Agron Journal Indonesia***37** 14-20.

Food security among pandemic covid 19

J F Sahusilawane[1], A M Sahusilawane[2,*]

[1] Department of Forestry, Faculty of Agriculture, Pattimura University, Ambon, 97232
[2] Department of Social Economic of Agriculture, Faculty of Agriculture, Pattimura University, Ambon, 97232

* E-mail: inakalapori@gmail.com; johnsahusilawane01@gmail.com

Abstract. Covid 19 which occured around the world has brought great sacrifice to humanlife, including changes in the natural order of the human life style. While, new normal or new life has been harderd following application doesn't mean this deadly virus is able to blocked at terminated. It's impact even goes every where bringing is one risk to another one of them was about food. People getting worried about having enough to eat. It is possible to continue in the habbit it consuming imported foodstuff does the diet still important from the island continue mainly for people in small islands such the Moluccas that have grown accustomard to eat. The decision can give the answer that 1) local food become very urgent for to survive life in pandemic 2) Woman as food supplier for the families still keep a lot of local wisdom of local food as for production as for keeping food during pandemic

1. Introduction

The pandemic of covid 19 has become a major problem for the world, if it is not resolved quickly and appropriately, then the danger of extinction awaits. The cause is still unknown but what we know for sure is that the virus has become very viral, and it always attacks the body's resistance. One of the patented antidote strategies is to increase the endurance of the body. That means that the food and drink is very important for the immunity system. Therefore, it has to be available continuously, and also the habit of keep things sanitized and clean is a part of life insurance.

This paper aims to reveal the importance of food security amid the Covid 19 pandemic. The Covid 19 pandemic is not only threatening human lives but also, can lead to the danger of hunger due to limited activities to work outside the home, increased unemployment, reduced supply of outside food which has been the main food source. There is no other way but to go back to local food and preserve it, which means being independent on one's own feet. The development of local food as a source of food as well as a source of household livelihood is a strategy that prevents the danger of hunger or reducing the economic crisis due to unemployment. Women as housewives have a big role in this. Empowering and providing opportunities for them to work more, so that the production and vertical diversification of local food will increase. The source of livelihood is increasingly open. A number of traditional culture possessed by women are part of their working capital as well as their social capital. The people of Maluku Province spread out and live on small islands. In the midst of the pandemic, food security is an important thing that needs to be addressed and anticipated. Brainstorming about this issue is carried out using a qualitative descriptive approach.

The impact of the pandemic of covid 19 seems to have contributed to burdening women's work in preparing household food. Usually she works together with her friends, but now it is limited (keep the distance). The garden where women work while strengthening family social relations is rarely or no longer used. Also, the habit of eating local non-food foods (rice, noodles, flour etc.) at increasingly expensive prices. Not only that, another question that arises, can these materials continue to be supplied? If the pandemic of Covid 19 continues, what will happen eventually? Starvation threatens! Even though Maluku has so much wealth of plants and animals in the sea, on land, and in the mountains. There are various types of endemic food plants such as sago, tubers (purple sweet potato, white sweet potato) hotong, banana, corn and so on. There is a saying that says "where there is sago, there is no one ever going hungry". It is undeniable that the Covid 19 currently happening is actually an opportunity to strengthen local food security. The most appropriate and easy strategy is to continue to maintain and produce it. Producing also means preserving nature with all its biological resources as well as being a form of respect and glorification to God the Life-Giver. It is possible that women in the city can invite their families to use the yard to become small gardens by planting various vegetables, herbs and others.

Women still have a number of local cuture that can be used as capital for environmental preservation, so they need to be involved in strategic programs concerning the realities of environmental change. Eventually, local food remains available, natural food barns will still guarantee the availability of food for the household. Covid 19 will not bring extreme danger of hunger to people's lives in Maluku.

2. Methodology

This paper use methology by Analysis Discription with Data collection techniques are carried out through in-depth interviews, observation, participation and documentation ***

3. Results and discussion

3.1 The Urgency of Local Food Security in the Middle of the Pandemic of Covid 19

People in Indonesia, including Maluku, have felt the impact of the pandemic for almost 1 year. One of them can cause a local food security crisis. Talking about local food, it is a food produced by the local community which is the result of adaptation to the environment as well as learning outcomes from one generation to the next. Food such as okuwaku, sukun santan, papeda, dried sago, tutupola sago, arwan sirsir and so on are local foods which has been consumed long ago, as well as corn, sago,tubers and nuts. Women process it into food every day and strong work ethic. Women realize that without working, household members will not eat. In any culture, she always serves her household [1].

Currently, the traditional food consumption pattern is experiencing degradation. Boiled sweet potatoes, okuwaku, plate sago, hotong, papeda and others were displaced by the growing interest of the people in consuming rice, bread, instant noodles, pizza and so on, (felt in rural areas). How to solve it? As a housewife who is responsible to fulfill the food and nutritional needs, she is in a difficult position. Consuming local food is cheaper than rice, french fries, fried chicken, and others. for that she then participated in various vertical diversification activities [2] carried out by the government, community organizations, or by trying various recipes through printed media, online media etc. In this way, she can make various types of food and cakes from local ingredients which taste just as good as non-local food. Local food continues to be sovereign at its own home.

Covid 19, which is currently hitting Maluku, where we have to stay at home is an opportunity to build a sense of pride, love for local food, as well as report the identity of the people of Maluku, for example by supplying local food, listening to regional songs (Pohon Sagu, Papeda Dingin) or local food-based traditional expressions (e.g. Sagu salempeng Dipatah Dua, "sago tree philosophy") and folk tales. Through religious education at home, children are given the understanding that nature with all the richness and biodiversity that exists in Maluku is a gift from God to us. Therefore, protecting and preserving nature with all its biological resources is a form of respect and glory to God.

3.2 Local Culture in Maintaining Food Security

Women in Kisar Island realize that their area has a short rainy season (4 months) and a relatively long summer (8 months) with dry, rocky soil conditions, so the ideal crops to cultivate are corn and beans which are grown using itawase/hapa technique, [3] All seeds (corn, red beans, green beans, wood beans) are inserted into one planting hole with the number of seeds between 3 to 5 seeds, where the corn seeds are deliberately exaggerated with the assumption that if there are corn seeds that die / are damaged because the ants still eat it, there's another corn seed. Corn as the main food must still be available in the barn so it has to become a priority. It is certain that the plants will continue to grow simultaneously and in time will be harvested together.

If corn and beans are grown using the itawase / hapa cropping system, women will apply the relay planting system for tubers by preparing live food barns. Root crops such as taro, petatas, and kumbili which are planted at harvest time (summer), although they will die, during the rainy season they will grow back. It can be said that in the cultivation process, almost no seeds fall to the earth without the touch of a woman's fingers [4]. Relay planting is generally implemented by female farmers in Maluku [5]. These planting methods do not cause damage to the natural environment, otherwise its adaptation and improvisation make the work become more easier and efficient [6]. In line with that, the itawase / hapa planting system can be an example for people on other small islands that have the same natural conditions as on the island of Kisar.

It is realized that the activity of caring for and maintaining various kinds of food plants using traditional methods is far more successful, therefore, mostly the use of pesticides is currently avoided. Some female farmers on the island of Kisar feel sad because nowadays, the Kisar lemons which are one of the endemic plants no longer have a sweet taste and even the pulp of the lemons has dried up due to the emergence of diplodia due to using pesticides, whereas according to them, by moving up a special group of ants to trees that get a diplodia can be ascertained that within one or two weeks the diplodate on the lemon tree has disappeared because it has been eaten by these special ants. This actions are very profitable, effective and efficient and not have a detrimental impact on the environment.

Regarding obedience and respect for supernatural powers, rural women / female farmers tend to believe in this. They always try to protect and maintain places that are considered sacred and have magical powers such as big trees, springs, small rivers. even if there were dry leaves or twigs that fell around the place, women often didn't dare touch them. The agricultural pattern in small islands is without burning and without chemical fertilizer input, but recycling leaves and manure [7]. The habit of carrying out special ceremonies before carrying out activities to open the garden or planting, so that their work is successful and while working, all family members are protected from danger. It makes them work carefully, quietly, seriously even calmly and not doing other activities that can damage natural environment because they believe there are supernatural powers around the place. Without realizing it, the behavior of female farmers has supported and improved the environment that stores a lot of biological resources in the area such as water and other environments.

4. Conclusion

This is the time for the government to make revitalization of local food maps according to the availability (Nahosi) of each island cluster. Therefore, if it is developed appropriately, the richness of local food diversity will be relied upon as a potential for food security towards food self-sufficiency during the pandemic of covid 19. The hope is it can open the way for improving community welfare.

The work ethic and local culture owned by women will make it easier for them to work as agents of reform in supporting government programs to achieve food security during the pandemic of covid 19.

References

[1] Tong Rosemarie Putnam, 1998, *Femenist Thought The Most Comprehensive Introduction to the Mainstream of Feminist Thought,* 4th Indonesian edition, 2008, Chapter Eight page 359 - 366. The Translator, Aquarini Priyatna Prabasmoro, Jalasutra, ISBN 979 - 3684 - 23 - 2, Publish by (2) Nurhayati Aniek, 2011. *The Dynamics of Rural Communities*, IAIN Sunan Ampel Press, Surabaya JALASUTRA, Yogyakarta

[2] Nurhayati Aniek, 2011. *The Dynamics of Rural Communities*, IAIN Sunan Ampel Press, Surabaya

[3] Sahusilawane A.M.,Uluputty Muhammad Riadh, Kembauw Esther, Djoko Sri Wahyuni, 2015 *Hapa, a local wisdom of the Meher Tribe in Kisar Island in Maintaining Food Security*, Journal of Social Welfare Research, Center for Research and Development of Social Welfare Services (B2P3KS) Accredited No 607 / AU2 / P2MI-LIPI / 03/2015 **14**(3) 305-316 Yogyakarta ISSN1412-6451

[4] Wijaya Hesti 1995 *International Trade in the Rural Economy and Women in the Era of Globalization.* (Women in Political Discourse of the New Order. Selected Articles of Prisma 2004). Page 102 - 124 Editor by Lisa Hadiz Jakarta Pustaka LP3ES Indonesia ISBN 979 - 3330 - 19 - 8

[5] Ayawaila J W 2005 *Cultural Dynamics of Maluku People*, in *Maluku Welcoming the Future*, page 168 -177. Maluku Cultural Institute. 2005 Ambon.

[6] Sahusilawane A M 2019 *Women and Sheep Farming in Purpura Village*, https:/ojs. Unpatti.a.c.Id/index php/agrilan/article/download/944/454 Pdf file publish 2019.

[7] Sahusilawane A M 2012 *Portrait of Oirata Women in Kisar Island Keeping the Food, Yogyakarta* Titah Surga ISBN:978-602-19023-3-2, Page Cetakan Ke dua.

Nutmeg Cultivation Intensity (Myristica fragrans Houtt) in Banda District

A Y Wattimena[1,*], M H Makaruku[1], E Kembauw[2], A S Mahulette[1]

[1]Agrotechnology Study Program, Faculty of Agriculture, Universitas Pattimura, Ambon, Indonesia
[2] Agribusiness Study Program, Faculty of Agriculture, Universitas Pattimura, Ambon, Indonesia

* E-mail: annawattimena@yahoo.com

Abstract. This study aims to determine and analyze the components of nutmeg cultivation at the farm level. The method used in the study was a survey method of plant cultivation. Sampling was done by using the multistage random sampling method. The area of reserach selected uses the pivate area startification. Each village was randomly sampled as many as 10 farmers, the total number of which was 20 sampled farmers. The results showed that farmers in Lonthor Village and Selamong Village, Banda Subdistrict, had not fully carried out intensive nutmeg cultivation that included land cultivation, seeding, fertilization, maintenance and post-harvest processing. This is because the existing nutmeg plants generally inherited from their parents so that it affects the decreased nutmeg production.

1. Introduction

Nutmeg is a specific commodity that is superior in Maluku region apart from cloves, which expected to provide an economic contribution to the region, especially for farming communities in order to increase their income and welfare. Apart from being a regional superior commodity, nutmeg is also an export commodity for Indonesia. Until now, Indonesia is the largest nutmeg exporting country in the world market, 75% and 25% of Grenada [7]. Although the nutmeg plant has long been known and cultivated, a common problem of nutmeg in several regions in Indonesia is the problem of crop cultivation. Nutmeg cultivation problems in North Halmahera Regency: Garden sanitation, poor nutmeg cultivation practices and harvesting [4]. The productivity of nutmeg in Maluku is classified as low, which is less than 1500 - 3000 seeds / tree / year or equivalent to 0.70 t / ha. The low productivity is due to the seeds used in the form of random seeds and the farmers do not fertilize or control plant pests [8].

The results of Legoh's research showed that generally the seeds used for seedlings did not come from the mother tree but came from sweeping seeds. The spacing applied by farmers was still very varied, farmers did not fertilize due to farmers' lack of knowledge regarding fertilization. Although there are farmers who receive fertilizer assistance, fertilizers only used to fertilize other crops [4]. Parliansyah stated that the average nutmeg production in HKm Rangai Sejahtera is up to 60 kg of wet nutmeg / tree / year so it can categorized as high nutmeg production [6]. The Central Bureau of Statistics of Central Maluku Regency stated that the average production of wet nutmeg throughout

Indonesia was only 48.4 kg / tree / year [1]. Meanwhile, for the Banda Islands, which are the original habitat for nutmeg plants, the yield of wet nutmeg only reaches 54 kg / tree / year [1]. Hamka stated that the production of nutmeg in Central Halmahera was only 3 kg / tree / year [2]. This study aims to determine the intensity of the nutmeg cultivation technique, especially in Central Maluku Regency, whether this plant is cultivated in accordance with good cultivation techniques or whether this plant is allowed to grow and farmers only take the results.

2. Methods

The method used in the study was a survey method of plant cultivation. The research was conducted in Banda Subdistrict, Lonthor Village. The selected area is determined with the consideration that the area is a nutmeg production center, has a large population of nutmeg plants and has a different cropping style and type. The material used was a sample of the nutmeg plant population. The tools used were clinometers, voice recorders, questionnaires, stationery and cameras.

Sampling has taken by using the multistage random sampling method. The area selected for the study area uses the land ownership area startification. 10 farmers were randomly sampled in each village so that the total number of farmers was 20 farmers. Information will collected from sample locations, namely primary data by means of direct observation on how to cultivate nutmeg and interviews, while secondary data comes from related agencies. The research data were analyzed using qualitative and quantitative methods. Quantitative data using tabulation system to analyzed descriptively. Descriptive analysis used to determine the general description of the object of research, the characteristics of farmers including age, gender, education level, farming experience and farming characteristics.

3. Results and discussion

3.1. Regional Characteristics

Central Maluku Regency is part of Maluku Province which has ten regencies and one city. Banda sub-district is an island cluster area because there are 11 inhabited islands and 5 uninhabited islands. the sub-districts area of Central Maluku district of 172.0 Km^2 and has a coastline length of 90,377 Km. The topography of Banda sub-district is a sloping sea level ranging from 25 to 75 m from the lowest ebb and the others are steep waters. The sea was 38 - 3,125 m depth while the highest land surface is in the volcano island with an altitude of $\pm$ 600 m and is an active volcano [1].

Nutmeg in its growth requires a hot climate with bulk high rainfall and evenly with ambient air temperature around 20-30 ° C. Nutmeg is also a type of plant that is susceptible to dry spells for several months, loose and fertile soil is needed for nutmeg and is suitable and productive in well-drained volcanic soils [5]. The climate in Maluku Tengah Regency is a tropical marine climate and a monsoon climate. This climate occurs because the Central Maluku islands surrounded by a wide sea, so the climate in this area is very much influenced by the sea that goes in tune with the existing seasons. Apart from local climatic conditions, the development of nutmeg also related to soil conditions. In the Maluku area which partly volcanic soil, it is very beneficial for the nutmeg plant. Volcanoes on the island of Banda that for a certain period give off volcanic ash and other materials are very helpful in replacing mineral nutrients that have long used by nutmeg plants for their growth.

Table 1. Soil Analysis and Soil Textu Results in Lonthor Village and Selamong Village, Banda District

Sub district Maluku Center	pH	C-Org	N Tot	P Tot	K Tot	Ca Tot	Tekstur class
Lonthor Village	5,8	6,24	0,32	18,12	0,02	0,23	loam sandy
Selamong Village	6,1	2,89	0,35	160,8	0,04	0,52	loam sandy

Source : UGM Soil Science Laboratory, Yogyakarta

Plant growth directly or indirectly determined by soil pH. Soil pH directly affects plant growth through the direct influence of H^+ ions and indirectly, namely soil pH can cause nutrient unavailability. Most plants are tolerant of extreme low or high pH, but in the soil there are sufficient nutrients that are greatly influenced by pH, where some nutrients are not available at extreme pH and others at toxic levels [3].

Soil conditions/soil fertility in Lonthor Village, Selamong Village characterized by physical and chemical characteristics of soil pH ranging from 5.8 to 6.1 so that the soil classified as acidic. For C-organic also, Lonthor Village classified as very high and Selamong Village classified as medium. The N-total for both villages is moderate. P-total classified as low, very high. Soil in Banda Subdistrict, Lonthor Village and Selamong Village, the soil type is podsolic. It was should explained that most of the nutmeg plantations in Banda Subdistrict are located on mountain slopes that form nutmeg forest stands.

3.2. Nutmeg Cultivation Level in Banda District

The results of the percentage of farmers who answered the questionnaire are presented in Figure 1 and Table 2. The results of the questionnaire showed that, on average, most of them used nutmeg from the seeds and its fulsion. Selamong Village Maintenance is limited to cleaning weeds around the plant area as well as eradicating pests and diseases without using pesticides this is because farmers do not have enough capital to buy pesticides

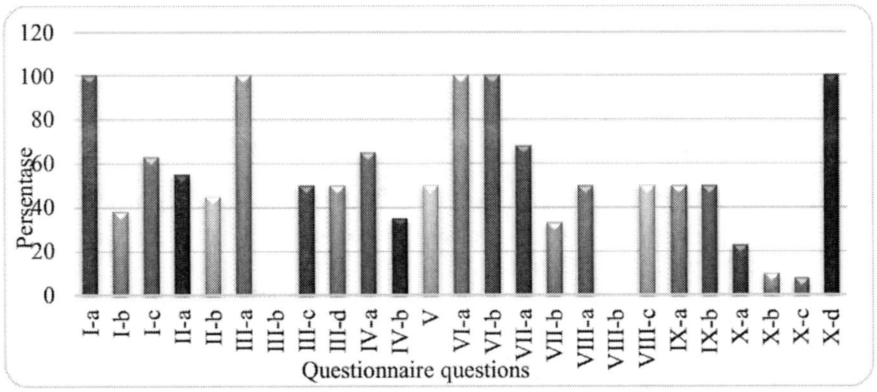

Figure 1. Percentage of Farmers Who Carry Out Nutmeg Cultivation in Banda District

Table 2. Picture1 description List

No.	Uraian
I-a	Nutmeg Cultivation
I-b	Follow the cultivation method recommended by the Dinas
I-c	Unfollow the cultivation method recommended by Dinas
II-a	Mixed cropping patterns
II-b	Monoculture cropping patterns
III-a	do your own nursery
III-b	swep seeds
III-c	seeds selected
III-d	using spacing
IV-a	using spacing
IV-b	Not using spacing
V	Do not vertilize
VI-a	Maintenance through weed weeds

VI-b	Eradicating pests and diseases without pesticides
VII-a	Harvest time 6 months after flowering
VII-b	Harvest time 7 months after flowering
VIII-a	The process of drying the seeds through drying
VIII-b	The process of drying the seeds through fumigation
VIII-c	The process of drying the seeds through drying and smoking
IX-a	The process of drying the mace through drying
IX-b	The process of drying the mace through fumigation
X-a	The pulp is used in the form of sweets
X-b	The pulp is used in the form of syrup
X-c	The pulp is used in the form of jam
X-d	Seeds and mace are not processed into essential oils

The percentage of farmers who answered the questionnaire showed that 100% of farmers cultivated plants (question no I-a). Banda Subdistrict for Lonthor Village and Selamong Village are farmers who follow the cultivation method recommended by the Department for 38% and do not follow the cultivation method recommended by the Office 63% this is because the nutmeg plants currently in existence are inherited from their parents so that they have not been able to carry out proper cultivation. Well done by farmers. The cropping patterns used by farmers in Lonthor and Selamong Villages are mixed cropping patterns and monocultures (questions no II-a and II-b) this is because most people in Banda District use the dusung system so that annual and annual crops are both plants. Plantations or fruit trees or other crops of economic value planted in one area to form a dusung.

Most of the Lonthor and Selamong Villages carry out their own nurseries 100% (question no III-a) but the seeds used as seedlings are quite diverse. There are farmers who use selected seeds, sweeping seeds and some are using selected seeds and sweeping seeds. 50% used selected seeds and 50% used sweeping seeds (questions no III-b, -c, -d) and this greatly affected the yield of nutmeg. The seeds used, if we use the sweep seeds as seeds, where the sweep seeds are the seeds taken for sowing without knowing the origin of the mother tree so that the production of the nutmeg plants is also greatly influenced, determine the yield of the nutmeg plant. In addition to the seeds used, the spacing also greatly affects the production of the nutmeg plant. If the spacing is too close, the resulting fruit will be small. 65% of Lonthor Villages use spacing and 35% of Selamong Villages do not use spacing (Question no IV) The cropping pattern used in Lonthor Village is a monoculture cropping pattern because it is a nutmeg plantation so that the spacing can be adjusted. The cropping pattern used in Selamong Village is a mixed cropping pattern in which the nutmeg plant is in one area with other annual plants.

The growth and development of the nutmeg plants in both villages both cultivate nutmeg plants but in processing, the nutmeg plant and the results of the nutmeg plant are still better. Selamong (question no V). For pest and disease eradication, 100% do not use pesticides, either chemical pesticides or bio-aggressive pesticides, while for weeding 100% of farmers do it (questions no VI-a and VI-b). The harvest time carried out by Lonthor Village farmers and Selamong Village Farmers is six months after flowering and some seven months after flowering, marked by the breaking of the nutmeg, many of these things are done so that the quality of the seeds and mace of nutmeg is good (question no VII). The process of drying the seeds and mace and utilizing the nutmeg pulp (questions no. IX and X) 50% of the farmers do fumigation and 50% of the farmers dry and fumigate them. In utilizing the nutmeg pulp, farmers in Lonthor Village and Selamong Village farmers process it into snacks in the form of sweets, jams and syrups.

4. Conclusion

Farmers in Banda Subdistrict have not fully carried out intensive nutmeg cultivation including land cultivation, fertilization, pest and disease control. This results in decreased nutmeg production.

5. References

[1] Badan Pusat Statistik Kabupaten Maluku Tengah 2014 Kabupaten Maluku Tengah Dalam Angka Tahun 2014

[2] Hamka 2015 Analisis perbandingan pendapatan petani pala basah dan kering di desa paniti halmahera tengah *Jurnal Ilmiah Agribisnis dan Perikanan* **8** (1) 36-41.

[3] Hakim N, Nyakpa M Y, Lubis A M, Nugroho SG, Diha M A, Hong G B, Bailey H H 1986 Dasar-Dasar Ilmu Tanah. Universitas Lampung. 488 hal.

[4] Legoh W L, Kojoh D dan Runtanuwu S 2017 Kajian Budidaya Tanaman Pala (*Myristica fragrans* Houtt) Di Kabupaten Kepulauan Sangihe *J Cocos*

[5] Lawalata M, Thenu S F W dan Tamaela M 2017 Kajian Pengembangan Potensi Perkebunan Pala Banda Di Kecamatan Banda Neira Kabupaten Maluku Tengah *Jurnal Agribisnis Kepulauan* **5** (2) 132-150.

[6] Parliansyah R, Riniarti M, dan Duryat 2019 Kajian Produksi Tanaman Pala Di Hkm Rangai Sejahtera Kph Rajabasa Study Of Nutmeg Plant Production In Hkm Rangai Sejahtera Kph Rajabasa Gorontalo *Journal Of Forestry Research* **2**.

[7] Wattimena A Y 2017 Analisis Korelasi Variabel Jumlah Buah Yang Mempengaruhi Produktivitas Tanaman Pala (Myristica sp.) Pada Dua Kabupaten Di Provinsi Maluku *J Agroekoteknologi dan Agribisnis* **1**.

[8] Wattimena A Y 2018 Pengaruh Pemberian Pupuk Npk Terhadap Pertumbuhan Bibit Tanaman Pala (*Myristica fragran* Houtt) *J. Agriment* **2.**

IOP Conf. Series: Earth and Environmental Science **883** (2021) 012047 doi:10.1088/1755-1315/883/1/012047

Implementation study of Good Agricultural Practices (GAP) of Red Fruit (*Pandanus conoideus* Lamk.)

M H Makaruku[1,*], A Y Wattimena[1], A S Mahulette[1], E Kembauw[2]

[1]Agrotechnology Study Program, Faculty of Agriculture, Universitas Pattimura, Ambon, Indonesia
[2] Agribusiness Study Program, Faculty of Agriculture, Universitas Pattimura, Ambon, Indonesia

* E-mail: litaerlin@gmail.com

Abstract. This study aims to determine and analyze the GAP components that have implemented by red fruit farmers. The method used in the study was a survey method of plant cultivation. The multistage random sampling method used to taking sampling sampling. The area selected for the study area uses the land ownership startification. Each village was randomly sampled as many as 10 farmers, the total number of which was 30 sampled farmers. The results showed that the farmers in the Taniwel District had not fully implemented the GAP guidelines in the cultivation of red fruit plants which included seeding methods, maintenance and post-harvest processing. This is due to the absence of outreach or counseling from the relevant Dinas regarding the GAP guidelines for red fruit plants.

1. Introduction

Red fruit is an endemic plant that grows wild in Maluku and Papua [2] [6]. Red fruit is the fruit of a type of pandanus plant. Red fruit has become one of the medicinal plants that attracted attention and has been widely studied since the end of 2004 [3] [7]. In Maluku and Papua, red fruit plants found in coastal areas to the highlands. Red fruit plants are found growing in areas with an altitude between 0-2300 meters above sea level (asl). Red fruit can also found growing in the northern part of Maluku that spreads from coastal areas to mountainous areas [4] [7]. The results of Makaruku's research (2016) show that the distribution area of red fruit plants in Taniwel District, West Seram Regency grows at an altitude of 1 - 700 m asl and has the largest population of red fruit plants at an altitude of 300 - 400 m asl. Fruit plants grow from areas with topography / slopes ranging from gentle / wavy areas 3 - 8%, wavy areas 8 - 15%, slightly steep areas 15 - 30% to steep areas 30-45%. The soil in Taniwel District is acidic to slightly acidic with a pH ranging from 5 to 6. The soil in Taniwel District has a gelatinous sand texture with cambisol, rendzina and podzolic soil types. This soil condition is supportive of the growth and development of red fruit plants. The distribution of red fruit plants only found in certain areas in West Seram Regency [7]. GAP is a guide on how to cultivate fruit and vegetable crops appropriately, well, correctly, environmentally and produces products that are safe for consumption. The application of GAP reflects the three pillars of sustainability, namely economically feasible, environmentally friendly and acceptable to the community [8]. Mahyuda et al (2018), stated that the application of GAP for Gayo arabica coffee cultivation has an effect on increasing coffee production and quality [5] [8].

In West Seram District, red fruit plants can grow and produce well, but the cultivation system is still traditional [7]. The facts in the field show that of the entire GAP component, it is possible that only farmers have adopted a few parts, but the majority of farmers have not implemented GAP optimally because it requires high costs, has a complicated procedure and a high level of order. In addition, the socio-economic condition is still low and the lack of socialization about GAP from the government causes farmers to have no further interest and understanding about GAP [8]. This study aims to determine and analyze the GAP components that have implemented by red fruit farmers.

2. Methods

The method used in the study was a survey method of plant cultivation. The research was conducted in Riring Village, Rumahsoal Village, Lohia Sapalewa Village, Taniwel District. The selected area is determined by considering that the area is a red fruit production center, has a large population of red fruit plants and has a different pattern and type of planting. The material used was a sample of the red fruit plant population. While the tools used are clinometers, voice recorders, questionnaires, stationery and cameras.

The multistage random sampling method used to sampling system. The area selected for the study area uses the land ownership startification. Ten farmers randomly sampled in each village so that the total number of farmers was 30 sampled farmers. Information will be collected from sample locations, namely primary data by means of direct observation on how to cultivate red fruit plants and interviews, while secondary data comes from related agencies. The research data were analyzed using qualitative and quantitative methods. Quantitative data input tabulation system. The qualitative analysis just has analyzed descriptively. Descriptive analysis used to determine the general description of the object of research, the characteristics of farmers including age, gender, education level, farming experience and farming characteristics.

3. Results and discussion

3.1. General conditions of Respondens

The population of West Seram Regency based on the Population and Civil Registration Service in 2017 was 208,009 people, consisting of 106,410 male residents and 100,599 female residents. Compared to the total population in 2016, the population of West Seram experienced a growth of 0.85%. Meanwhile, the 2017 male to female population ratio was 105. The population density in West Seram Regency in 2017 was 30 people / km2. Population density in 11 sub-districts is quite diverse with the highest population density located in Kairatu Barat Subdistrict with a density of 106 people / km2 and the lowest in Elpaputih District at 4 people / km2. From 2015 to 2017, the sex ratio of the population of West Seram was always above 100. This means that the number of male population is more than the number of female population [1].

Table 1. Percentage based on age and education of farmers in Taniwel District

Description	Village (%)		
	Rumah Soal	Riring	Lohia Sapalewa
Farmers age			
a. 20-30 years old	0	0	20
b. 31-40 years old	30	20	30
c. 41-50 years old	60	50	40
d. 51-60 years old	0	10	0
e. 61-70 years old	10	10	10
f. > 70 years old	0	10	0
Last education			
a. Been to school	10	20	10
b.Elementary school	20	0	20
c. Junior high school	30	30	30
d. High School	40	50	40

The age of the farmers in the three villages in Taniwel District evenly distributed from the age of 20 to over 70 years. The largest number of farmers is aged 41-50 years, spread across three villages, namely the Rumah Questions Village at 60%, Riring 50%, Lohia Sapalewa 40%.

Farmers' education is closely related to the farmer's age, this is because on average younger farmers go to high school level while farmers who are over 70 years old have only been educated for only 3-4 years.

3.2. Implementation of the Red Fruit GAP Component in Taniwel District

The results of the percentage of farmers who answered the questionnaire are presented in Figure 1 and Table 2.Red fruit plants cultivated in Taniwel District are still in different planting locations, namely in hills, riverbanks and roadsides resulting in unreached plants often not being managed properly. In Figure 1 and Table 2, it shows that generally farmers who cultivate 100% red fruit have not followed the cultivation recommendation from the Dinas (question no. I). This is because extension officers do not conduct counseling and socialization activities regarding good red fruit cultivation techniques to farmers so that generally the cultivation techniques used by farmers are only based on hereditary experiences that have been carried out by their parents.

In Taniwel District, the ownership of plants was 26.7%, and more were farmers who continued to cultivate crops belonging to their parents or inherited by 73.3% (questions no. II-a and II-b). This is because for them the marketing process for red fruit is still difficult, so the harvested red fruit generally used for their own purposes. For Kairatu Subdistrict, the ownership of plants itself is more, namely 80% and only 20% is still cultivating plants owned by parents (questions no. II-a and II-b).

The cropping pattern used by farmers in Maluku in cultivating red fruit plants in Taniwel District is still a mixed cropping pattern (question no. III) This is because most Maluku people use the dusung system so that seasonal and annual crops, both plantation crops and crops Fruits or other plants of economic value planted in one area to form a dusung.

Most of the Taniwel sub-districts carry out their own nurseries 100% (question no. IV) But the seeds used are quite diverse. Farmers do not use seeds from seeds because of their relatively low growth capacity and based on their experience who have tried to use seeds as seeds, it turns out that most of the seeds sown do not grow. In Taniwel District 16.6% used seedlings from cuttings and 83.3% used seedlings from tillers (questions no. V-a and V-b) and this greatly affected the population and production of red fruit plants. Generally, farmers in Maluku cultivate five types of red fruit plants, namely saune Ellah, saune Ole, saune Porol, saune Oti and saune Batu. In Taniwel District, the farmers who cultivate Ellah saune are 60% and Oti saune as much as 40% (questions no. VI-a, VI-b, VI-c, VI-d and VI-e). Farmers in Taniwel District do not use spacing 100% at the time of planting (questions no. VII-a and VII-b). Based on the picture above, it can be seen that in Taniwel District, farmers do not fertilize with manure (question no. VIII-a). Farmers do not carry out the red fruit plant maintenance by regular watering. Watering done only at the beginning of planting until the plants can survive in the land and then the plants only depend on groundwater sources, rainwater and humidity that can help the growth of red fruit plants. Only 16.7% of the farmers in Taniwel District do watering (question no. VIII-b). In the Taniwel District, the farmers who did weed were 83.3% (question no. VIII-c). Farmers do weeding only by cleaning the weeds around the plants and cutting off some of the leaves that will used as plaiting materials at the same time. Red fruit plants shading done at the beginning of planting until the plants are able to adapt to the land.

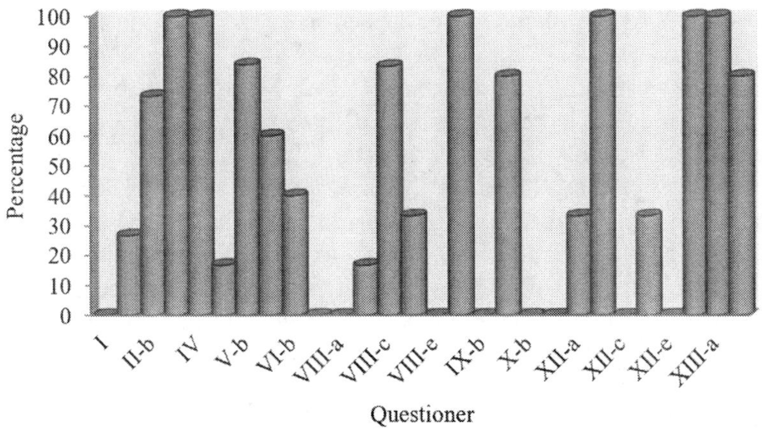

Figure 1. Percentage of farmers who carry out the cultivation of red fruit plants for Taniwel District

Tabel 2. Picture 1 description List :

No.	Description
I	Not following the cultivation method from the Dinas
II-a	Owned plants inherited from parents
II-b	Not a self-owned plant
III	Mixed cropping patterns
IV	Do your own nursery
V-a	Seedlings of cuttings
V-b	Seedlings
VI-a	Types planted (Saune Ellah)
VI-b	Types planted (Saune Oti)
VII	Using spacing
VIII-a	Fertilize with manure
VIII-b	Doing watering
VIII-c	Doing weed weeding
VIII-d	Use shade
VIII-e	Eradication of pests and diseases with pesticides
IX-a	Harvest 1 time a year
IX-b	Harvest 2 times a year
X-a	Sold in the form of fresh fruit
X-b	Sold in the form of oil
XI	Simple oil processing
XII-a	The fruit is processed as a medicinal ingredient
XII-b	Fruit is processed as cooking oil
XII-c	The fruit is processed as essential oil
XII-d	Fruit is processed as a sauce
XII-e	Fruit is processed as lunkhead
XII-f	The fruit is processed as a natural dye
XIII-a	The use of leaves as plait
XIII-b	Use of trunks as buildings and fences

In Taniwel District, 33.3% of the farmers used shade (question no. VIII-d). Eradication of pests and diseases is very rarely carried out because red fruit plants are rarely attacked by pests and diseases, most of the plants are collapsed as result of being hit by strong winds. Farmers in Kairatu and Taniwel Districts do not use 100% of pesticides, either chemical pesticides or biological pesticides (question no. VIII-e). Meanwhile, for the control of pests and diseases in fruits attacked by caterpillars and birds, farmers only use mechanical pest eradication with bird traps, look for, and kill caterpillars that attack red fruit. Farmers in Taniwel Subdistrict have planting locations that are very far away and can only reached by foot, which is about 66.7% of farmers who have red fruit planting locations located in the hills, causing farmers not to take intensive action to take care of their plants. Farmers in Taniwel District 100% (question no. IX-a) harvest red fruit only once a year, namely in October, so that it becomes an obstacle in increasing the yield of red fruit plants.

The motive for selling red fruit crops for sale seen that 100% of farmers did it in the form of fresh fruit (questions no. X-a and no. X-b). The 100% processing of red fruit oil by farmers in Taniwel District (question no. XI) still done simply or traditionally using bamboo. The use of red fruit oil produced by farmers in Taniwel District still uses red fruit oil for their daily needs in place of coconut oil. Very limited knowledge about good red fruit oil processing techniques and inadequate technology is a huge obstacle in the use of red fruit. This can seen in Figure 1, namely the utilization of red fruit in Taniwel District 100% of farmers do not use and process red fruit as essential oil and as a food ingredient such as dodol (questions no. XII-c and XII-e). Farmers in Taniwel District (question no. XII-f) use red fruit as a 100% natural dye because this knowledge has obtained from their parents from generation to generation and can be done using traditional and simple tools. Farmers in Taniwel District 100% use red fruit as cooking oil instead of coconut cooking oil, 33.3% use red fruit as a sauce and 33.3% use red fruit as a medicinal ingredient (questions no. XII-a, XII-b and XII-d). In addition, 100% of farmers in Taniwel District use the leaves of the red fruit plants as plait for sale and daily needs (question no. XIII-a). The use of red fruit stems as building material and fence material by farmers in Taniwel District is around 80% (question no. XIII-b) (question no. XIII-b).

4. Conclusion

Farmers in Taniwel Subdistrict have not fully implemented GAP guidelines in the cultivation of red fruit plants which include seeding methods, maintenance and post-harvest processing. This is due to the absence of outreach or counseling from the relevant Dinas regarding the GAP guidelines for red fruit plants.

References

[1] Badan Pusat Statistik Kabupaten Seran Bagian Barat 2018 Kabupaten Seram Bagian Barat Dalam Angka Tahun 2018.

[2] Bourke R M 2005. Indigenous Fruit in Papua New Guinea. Departement of Human Geography The Australian University. Canbera.

[3] Duryatmo S 2005 Buah merah penakluk penyakit maut. Trubus XXXVI (422) : 11-15.

[4] Heyne K 1987 Tumbuhan Berguna Indonesia. Jilid I. Badan Litbang Kehutanan. Jakarta.

[5] Mahyuda et al 2015 Tingkat Adopsi Good Agricultural Practice Budidaya Kopi Arabika Gayooleh Petani di Kabupaten Aceh Tengah *Jurnal Penyuluhan* **14** (2) 321.

[6] Makaruku M H 2011Hubungan parameter pertumbuhan tanaman buah merah (*Pandanus conoideus* Lamk.) (Studi kasus di Kabupaten Seram Bagian Barat Propinsi Maluku) *J. Agroforestri.*

[7] Makaruku M H 2016 Kajian sebaran tanaman buah merah (*Pandanus conoideus* Lamk.) di Kecamatan Taniwel Kabupaten Seram Bagian Barat *J. Agroforestri*.

[8] Nahraeni W, Masithoh S, Rahayu A A, Awaliah L 2020 Penerapan Good Agricultural Practices (GAP) Jeruk Pamelo *(Citrus maxima* (Burm.) Merr.) *J. Agribisains* 50-59.

Factors affecting the capacity of farming rice farmers in managing rice field in the District Tangerang

Khaerul Saleh[1],*, Suherman[1]

[1] Department of Agribusiness, Faculty of Agriculture, Sultan Ageng Tirtayasa University, Jl Raya Jakarta Km 4 Pakupatan Serang Banten Kode post 42124

* E-mail: khaerulsaleh63@yahoo.co.id

Abstract. Farmers' capacity to manage rice paddy farming is closely related to individuals' behaviour in managing the farm. Farmer capacity building program has been widely and often conducted relevant government innovation lowland rice farm and delivered to farmers. Still, the farmers cannot give a proper response to lowland rice farm management innovation, especially in terms of strengthening farmers' capacity. This study aims to: (1) determine the level of farmer capacity in managing rice paddy farmers and (2) to analyze the factors related to the capacity level of rice paddy farmers. The research was conducted in the northern part of Tangerang Regency (Pantura), the centre for rice production in the Tangerang Regency. The research was conducted from July to November 2020. Data was collected through FGD interviews and observations of 150 farmers. The data analysis used descriptive test and multiple regression. The results showed that the farmer capacity in managing lowland rice farming in Tangerang district was classified as moderate, both in resource utilization, problem identification, farming planning, and adaptability. The main factor determining the level of farmer capacity in managing rice paddy farmers is formal education and nonformal education.

1. Introductions

Farmers' capacity to manage rice paddy farming is closely related to the individual farmer's behaviour to innovation. Lowland rice farming activities they describe activities that hereditary and conventional as the previous generation, most farmers commercialize farming is their parents' legacy. Farm management paddy field farmers such as the use of quality seeds, integrated pest management, crop rotation and balanced fertilization through the farming pattern of System of Rice Intensification (SRI), Low External Input Sustainable Agriculture (LEISA/LISA) and the Agricultural Input Organic [1]. However, most paddy farmers have not been able to respond to the rice field of innovation management. The ability or capacity of farmers is not conducive to produce products of profitable farming.

One of the internal factors is individual characteristics that are thought to be related to farmer capacity. The characteristics of individual farmers are characteristic for each farmer with various aspects of his life, often used as a distinguishing element in shaping behaviour between individual farmers. They can support their survival in managing their farming. The results of research Subagio [2] state that every individual naturally has a capacity inherent in himself. Furthermore, it is stated that personal characteristics affect farmers' capacity. Rogers [3] states that these characteristics underlie one's behaviour in work situations and other situations. [4,5] stated that the factors that affect the capacity of farmers in managing lowland rice farming are the characteristics of farmers (age and farming experience), social, environmental characteristics, level of empowerment, cosmopolitan level, land area and performance of agricultural extension agents.

Veronice [6] stated that the low capacity of lowland rice farmers of paddy is due to the weak level of technical knowledge of farmers in managing rice farming. While Sudana and Subagyono [7], states that the factors that significantly affect the chances of accelerating the adoption of innovation are the age and level of education and the number of existing extension agents. Further Sudana and Subagyono [7], farmers' factor characteristics positively correlated with the level of fulfilment of farmers' needs. Farmers' capacity to manage the farm includes age, formal education, nonformal education, and farmers' experience in agribusiness [8].

Fatchiya [9] states that fish cultivators' characteristics that significantly affect nonformal education, experience, income, and business scale. Meanwhile, according [10] revealed the real effect of individual characteristics, including age, formal education, nonformal education, long effort, motivation, and scale farming. Capacity also influences the pace of innovation adoption, among others: the characteristics of the farmers such as age, education, farming experience, contact extension, annual income, economic motivation, and engagement counselling agencies [11, 12].

Farmers' capacity to manage rice paddy farming is farmers' ability in business planning, troubleshooting, and use resources. [13] argued that farmers know wetland rice PTT component 76-91%. Still, in practice, not all components recommended by the known technology can be applied very well, so paddy farming's success rate is still very low. This condition is seen from the level of productivity of farm produces.

Referring to the importance of the success of applying the principles of lowland rice farming, the purpose of this study were (1) to analyze the factors that determine the capacity of farmers to manage to the farm of paddy; and (2) analyze the pattern of the relationship of each element in supporting the farmers' capacity levels.

2. Methods

The research was conducted in Tangerang Regency. The research location was determined purposively, namely in three BPPs, eight BPPs were selected in Tangerang Regency (Table 1), namely 1) BPP Kronjo; 2) BPP Tegal Kunir; and 3) BPP Sukatani. Data collection was carried out from July to October 2020.

In this study, Respondents were selected purposively with the criteria of farm families who do rice farming rice fields. The study population as many as 850 people lowland rice farmers, selected respondents as many as 150 people—mechanical determination of sample using techniques random proportional sampling based on the number of populations. Types of data collected included primary data and secondary data. Primary data was collected through interviews using questionnaires, interviews, observation and documentation study.

The collected data were analyzed using descriptive statistics and inferential statistical analysis. The parameters used in this study of the characteristics of the farmers are limited to formal education, nonformal education, experience farming, tenure farming, and the capacity of farmers, the ability to utilize the resources, ability to plan to farm, ability to identify farming opportunities, ability to overcome the problem, ability to adapt with the environment. In this study, selected activities are generally carried out paddy farmers, among others: 1) perform rotation or rotation crop varieties; 2) perform intermittent irrigation (intermittent); 3) a balanced fertilizer that is inorganic and use of organic fertilizers and biofertilizers; 4) the application of PHT with emphasis on biological control with botanical pesticide and conserve natural enemies (parasitoids, predators and pathogens).

Table 1. Several members and farmer groups based on group class in Tangerang Regency in 2019.

No.	WKBPP	Total members	Class group P	L	M	U	Total group
1	Kp Melayu	1.857	22	39			61
2	Sepatan	8.701	48	101	1		150
3	Tg Kunir	6.261	75	30			105

4	Sukatani	5.913	52	59		111
5	Kaliasin	12.260	59	97		156
6	Kronjo	11.213	98	56	2	156
7	Carenang	9.093	8	114	4	126
8	Caringin	6.064	56	105		161
Total in district		61362	418	601	7	1026

Source: The Tangerang Regency Agriculture and Food Security Service 2019

3. Analysis of Data

Indicators and parameters were analyzed using a scoring system assessment. Scoring follows the Likert scale with a scale of 1 to 3. Scores obtained from respondents further specified interval scale ranges or categories of high, medium, low. To determine the level of farmer capacity and characteristics, using the formula [14], namely:

$$\text{Hose (Range)} = \frac{\text{(maximum value-minimum value)}}{\text{(number of answer categories)}}$$

The following formula is used to determine the influence factors of each variable on the level of capacity of rice farmers,

$Y = a + bX_1 + bX_2 + bX_3 + bX_4 + bX_5 + bX_6 + e$

Where: Y = Farmer of Capacity; a = Constant; X_1: Age of farmer; X_2: The farmer's formal education level; X_3: Farmers' nonformal education; X_4: Farming experience; X_5: Area of land tenure; X_6: extension activities; b: Regression Coefficient; e = Error.

4. Results and discussion

4.1. Characteristics of lowland rice Farmers

Analysis of the characteristics of lowland rice farmers in Tangerang Regency shown in Table 2 shows that the level of formal education, nonformal education, land ownership area and farming experience is relatively the same, namely in the medium category. Age (age) of farmers is in the Adult Age group with an average age of 50.58 years. The young age group (20-40) years old still colour rice farming activities. This condition shows that agricultural activities are still in great demand by young age groups. The high interest of young farmers in running lowland rice farming indicates that agriculture is not just about sustaining life but able to be a source of community income.

Judging from the formal education level, 87.33 percent were in the moderate category between 6-12 years of age while those with low education (less than 6) were 12.66 percent. Similarly, the nonformal education amounting to 50.67 percent of farmers carry out training activities between 4- 10 times a year. The level of education both formal and nonformal farmer indicates the power of innovation adoption are run by farmers.

The level of education experienced by respondents has an average of 7.86 years, meaning that the level of formal education that has been attended by farmers has graduated from Elementary School (SD) to Junior High School (SMP) so that farmers are quite literate. However, the research results by [7] state that in general, formal education has not been able to digest the technology conveyed through printed media independently. Meanwhile, the level of nonformal education obtained through training or extension activities is 2.78 times a year, meaning that farmers generally have a reasonably good technical ability to manage lowland rice farming. This was also revealed by [5] near related to the high level of nonformal education.

Table 2. Characteristics of lowland rice farmers.

Variable	Category	Total (Percent)
1) Age Average 50,58 year.	1) Young (20– 40)	20,33
	2) Adult (41 – 55)	64.00
	3) Old (> 55)	15.67
	Total	100,00
2) Formal educations Average, 86 year.	1) low (<6)	12,66
	2) Moderate (6- 12)	87,33
	3) Hight (> 12)	1,00
	Total	100,00
3) Nonformal educations Avarage 2,78 times/year.	1) never	32,67
	2) Rarely (4- 10)	50,67
	3) often (> 10)	16,67
	Total	100,00
4) Ownership of Farm LandAvarage 0.23 Ha	1) Low (<1)	54,67
	2) Moderate (1- 2)	32,00
	3) Hight (> 2)	13.33
	Total	100,00
5) Farming experience Average 19,56 year	1) Low (<10)	28,66
	2) Moderate (10-20)	36.00
	3) Hight (>20	35,33
	Total	100,0

The results showed that almost half (32.67%) of the respondents answered that they rarely participate in extension activities, even if there are farmers who have never attended nonformal education in the form of extension activities or training specifically in agriculture. The low level of nonformal education is caused by the lack of intensity of extension activities. Both carried out by the extension agents themselves and other parties carrying out technology promotion. This condition is also the same as the results [15] research that farmers in participating in organic rice counselling (nonformal education) are only 0-2 times. In line with the results of research by [9, 10] shows that low nonformal education is correlated with a low level of extension services even tends to be the same training participants. Such as group leaders or members who are successful but are not followed or followed up technically so that it does not benefit participants/farmers.

Rice fields are used as farming land farmers generally are in the form of sakap with the result. Still, some farmers are farming landholding of own property, farming landholding 54.67 percent of respondents are currently on a narrow category tenure of less than 1 ha. The average area of land cultivated to paddy farmers is 0.23 hectares, including a narrow land category (Table 2). Farmers Tangerang district does have a narrow field. The condition is caused due to high land conversion to non-agricultural (residential, industrial and Tourism) acquisition of farmland by farmers in some areas experiencing the transfer of ownership and is only used as farmland for the land has not been used by its owner.

The low level of land ownership also results in the insufficient application of farming technology. The application of farming technology (technological innovation) is mostly driven by high land tenure as a result of research [5] states that land ownership is positively correlated with the level of an innovation application, so that farm productivity is higher. Ownership of land controlled by someone shows the level of one's social structure in the community. According to [16.17], states that ownership of land (resources) shows the source of one's power in rural communities.

The success of farmers in managing lowland rice farming cannot be separated from farmers' experience in cultivating this commodity. It is known that the experience of respondent farmers in the study area varies, the farming experience of respondents is 36.00 percent in the medium category (10-

20 years) with an average experience of 19.56 years (Table 2). This shows that farmers' average experience in the research location is good enough in conducting lowland rice cultivation activities.

The farmers' experience can be a measure of farmers' knowledge or lessons in running lowland rice farming, especially in making decisions about implementing innovations such as controlling pests and plant pests, applying superior seeds, or using balanced fertilizers, and can be used as a guide in achieving their farming success.

4.2. Extension support in increasing the capacity of lowland rice farmers

Extension role is the main tasks and functions of counseling in analyzing, providing stimulation, facilitating and motivating farmers. Extension activities in increasing the capacity of lowland rice farmers are included in the medium category (60.17%). Extension activities in lowland rice farmer capacity building include the method, the material intensity and extension capabilities, which contribute to the capacity building of farmers. In general, extension capability in performing their duties well enough, but due to lack of agricultural extension personnel field (PPL) resulting in a lack of extension services, farmers will need innovation to address both technical and social farming.

The goal of the outreach activities, as expressed in the Law System of Agriculture Fisheries and Forestry Extension No. 6 in 2006 is to increase knowledge and skills of entrepreneurs and protagonists, so has the readiness to enter the wider area. Figure 1 explained that the method, the material extension and extension capabilities respectively by 64.00 percent, and 76.66 percent were in the moderate category. Simultaneously, the intensity of the new extension reaches 50.67 percent of medium in the category, the need to increase both the capacity and the amount of extension of existing personnel.

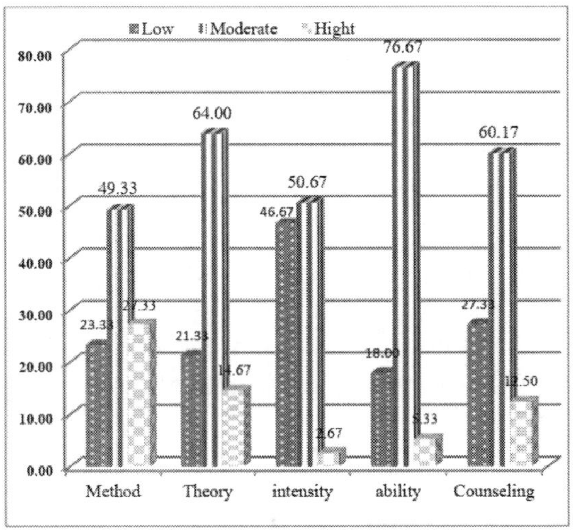

Figure 1. The education function in supporting the capacity of lowland rice farmer

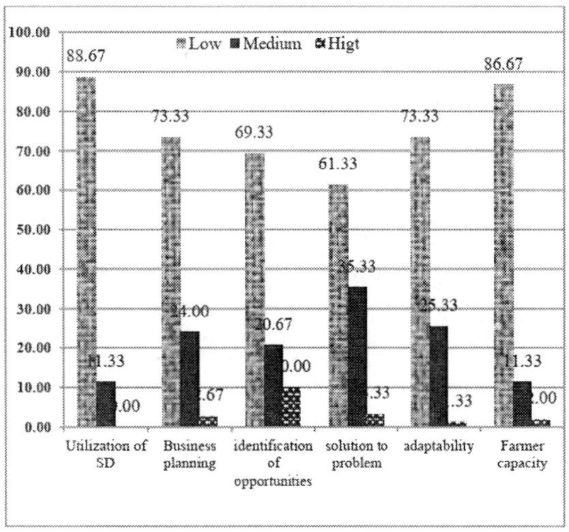

Figure 2. Level capacity lowland rice Farmers

4.3. Paddy Farmers' Capacity Level

The power capacity of farmers is a picture or personal capacity owned by farmers in supporting farmers' activities. The ability needs to be continuously improved by the time, and it is by the philosophy of the education continuum that starts from the idea, willing and able. The capacity of farmers studied includes; 1) The Ability to utilize resources; 2) the Ability to Busines plan farming; 3) the ability to identify farming opportunities; 4) the ability to address the issue of farming, and 5) the ability of environmental adaptation (Figure 2).

Five indicators of a lowland rice farmer's capacity in Tangerang district entirely within the category of being, as Figure 2, Indicators capacity have a higher category of indicators coping skills and the ability to identify business opportunities other indicators are in the moderate category. In general, the capacity of a lowland rice farmer in Tangerang Regency that is in the category of being the 81.3 percent (Figure 2).

4.4. Factors Affecting Farmers capacity of paddy in running the farm
The estimation model of factors that affect farmers' lowland rice in running the farm using Regression Model Ordinary Least Squares (OLS). Regression analysis obtained the following predictive model:
$Y = 2.452 + 0.047X_1 + 0.160X_2 - 0.227X_3 - 0.050X_4 + 0.037\,X_5 + 0.031\,X_6$

Before the models used testing, data used in statistical and econometric criteria. Econometric testing includes the assumption of normality, multicollinearity and heteroskedastic. While statistical testing includes statistical tests include T-test and F-test:

4.4.1. Assumption testing
4.4.1.1. Normality assumption test. The normality test is carried out to detect whether the data used is normally distributed or not. The normality test results show that the data can be said to meet the requirements for multiple linear regression analysis.

4.4.1.2. Multicollinearity assumption test. Collinearity test is conducted to determine the collinearity among the independent variables. From the calculation, it is known that there is no multicollinearity in the multiple regression line model.

4.4.1.3. Heteroscedacity assumption test. The analysis results showed that the data were homoscedasticity or heteroscedasticity did not occur and met the regression analysis requirements.

4.4.2. Results of estimated factors affecting lowland rice farmers in managing rice farming. The results of the F test show that simultaneously the variables of age (X_1), level of formal education (X_2), nonformal education (X_3), farming experience (X_4), the area of land tenure (X_5) and, participation in extension activities (X_6), have a significant effect. On the level of farmer capacity in running rice farming (Y).

The value of $r = 0.290$ indicates a very strong relationship. Meanwhile, the value of the coefficient of determination was 60.3%, which means the capacity of farmers in carrying lowland rice farming of 60.3% can be explained by the performance variables of age (X_1), the level of formal education (X_2), Nonformal education (X_3), the experience of farming (X_4), land tenure area (X_5), and participation in extension activities (X_6). In comparison, the remaining 39.7% was influenced by other factors not examined. These results indicate that the model is quite good.

While the partial test results showed variable levels of education followed by farmers, both formal and nonformal factors influence farmers' capacity to carry out paddy farming (Table 3).

Table 3. Regression test results factors that are the capacity of farmers in managing lowland rice farming

Variabel	Coefficient	Standard Error	Significant
Constant	2.452	0,641	0,049
Age	0,047	0,066	0,476
Farmers Education Level	0,160	0,095	0,094*
Non formal education	-0.227	0,075	0,003**
Farming Experience	-0,050	0,066	0,448
Control of farm land	0,037	0,044	0,399
Extension Activities	0,031	0,189	0,868

**) Sign at $\alpha < 0{,}05$, *) Sign at $\alpha < 0{,}1$

The regression analysis in Table 3 above shows that farmers' age, farmers 'level of formal education, tenure and extension activities positively affect the level of farmers' capacity in managing the lowland rice farm. This shows that the better farmers' education level, the wider the mastery of farming and extension activities, the farmers tended to have a high capacity. While nonformal education and experience of farming can negatively affect the level of farmers' ability.

The level of education both formal and nonformal significant effect on the level of capacity in managing the Farmers' lowland rice farming, [9, 5]; said that farmers with higher education are relatively faster in adopting an innovation. Most of the research location lowland rice farmers have a relatively good education level, namely junior high school and several people educated to high school level and even Bachelor.

The level of education-related knowledge and the ability of farmers in analyzing the information received before they apply it into paddy fields. Some research also suggests that the education level influence the farmers' decision to adopt an innovation [18; 19; 20; 21]

The experience of farmers in carrying out farming of paddy correlated negatively on the capacity level farmers. This is possible because, it experienced a change in both the techniques and ways of cultcultivating pady fields, the development of farming technology lowland rice can only be applied if the farmer has good knowledge and qualified as seed selection fertilization, and in an organic and pest control (OPT).

Most farmers in the location has not yet fully using certified seeds, and low use of inorganic fertilizers for reasons other than price is relatively higher too rare in the market, as well as pest control generally more farmers look at the effect produced when the pesticide was applied and tended to ignore the environmental sustainability.

Level tenure positive effect on the capacity but not significantly, the majority of wetland that is used as a farm instead of property rights, but through a system of revenue sharing, in which when the owner of wetland uses the land for other activities then the farmers will easily relinquish control of the land.

Higher than function land occurred in Tangerang Regency [22] resulting in increasingly smaller mastery paddy fields owned by farmers of paddy, for it needs the intervention of the government towards governance paddy fields either through a policy of zoning (Industrial zone, residential and agricultural) and given sanctions if the fields are transformed to enable land to non-agricultural, so the green area as a place of farming of paddy farmers can sustainably [23].

Extension activities have a positive effect on the level of farmer capacity. Extension activities as part of nonformal education activities can increase farmers' knowledge, change attitudes, and skills in running rice farming. Several studies also show that extension activities have a significant effect on farmer behaviour in adopting new technology [24; 17; 5]

The instructor's role is to implement the primary duties and functions of the extension agent in analyzing, stimulating, facilitating and motivating farmers. The role of extension in enhancing the capacity of respondents to manage rice farming rice fields was included in the classification. This condition shows that extension activities in increasing the capacity of lowland rice farmers in Tangerang Regency are still not optimal, and it would be excellent.

5. Conclusion

The farmer capacity level needs serious attention, especially in resource utilization capabilities and the ability to plan business considering the limited amount of land available. Farmers growing lowland rice is not marginalized by changes in the environment are continually changing. Simultaneously lowland rice farmer capacity is determined by age, formal education, tenure and extension activities. Simultaneously, partially lowland rice farmer capacity level is determined by farmer education, both formal and nonformal. Counselling activities primarily related to the extension materials need to be tailored to the needs of more applicable farmers and extension capability, which is still in the moderate category that received special attention from stakeholders through technical training as well as an increase in service education if they were supported by routine extension intensity accompanied by the increased capacity of extension workers and equipped with the availability of infrastructure and supporting facilities for the implementation of the extension. Because, to increase the capacity of lowland rice farmers, it is necessary to support extension activities.

6. Recommendation

For local governments to establish land use is mainly agricultural land that sustainable rice paddy farming through spatial planning policy. For the extension, mostly freelance Power status (THL) will be improved gradually and supported with education and training programs for field extension.

Acknowledgment

Our thanks to the PIU of food security (Food security) University of Sultan Ageng Tirtayasa which has provided funding to carry out this research through Internal Research scheme.

References

[1] Istiantoro, Bambang A N, Soeprobowowati T R 2013 Analysis of socio-economic factors that affect the control of pests and diseases of lowland rice in a sustainable agricultural system *Journal of Ecoscience* **5** (2) 16-20.
[2] Subagio H 2008 The role of farmer capacity in realizing the success of farming: The case of Vegetable and Rice Farmers in Malang and Pasuruan Districts, East Java Province. [Dissertation]. Bogor (ID): Bogor Agricultural University.
[3] Rogers E M 2003 Diffusion of Innovations (5th ed). New York (US): The Free Press.
[4] Yunita, Sugihen B G, Asngari P S, Susanto D, Amanah S 2012 Strategies for Increasing Household Capacity of Lebak Lowland rice Farmers Towards Household Food Security (Case in Ogan Ilir and Ogan Komering Ilir Districts, South Sumatra) *Extension Journal* **8** (2) 10-16.
[5] Herawati, Aida Vitayala Hubeis, Siti Amanah and Anna Fatchiya 2017 Capacity of Technical Irrigation Paddy Paddy Farmers in Applying the Principles of Environmentally Friendly Agriculture in Central Sulawesi *Journal of Agricultural Technology Assessment and Development* **20** (2) 155-170
[6] Veronice, Helmi, Henmaidi, Ernita Arif 2018 Capacity Development and Institutionalization of Small Farmers in Agricultural Areas Through Knowledge Management Approaches *Journal Of Applied Agricultural Science And Technology* **2** (2) 1-10
[7] Sudana W and Subagyono K 2012 Study of Determinants of the Adoption of Integrated Rice Management Innovations through the Integrated Crop Management Field School *Journal of the Assessment and Development of Agricultural Technology* **15** (2) 94-106.
[8] Marliati 2008 Empowerment of Farmers to Fulfill the Need for Capacity Building and Independence of Agribusiness Farmers (Case in Kampar District, Riau Province). [Dissertation]. Bogor (ID): Bogor Agricultural University
[9] Fatchiya A 2010 Capacity Level of Fish Cultivators in Managing Sustainable Aquaculture Businesses *Extension Journal* **6** (1) 11-18.
[10] Tahitu M E 2015 Capacity Building for Sago Managers in Increasing Sago Utilization in Central Maluku, Maluku Province. [Dissertation]. Bogor (ID): Bogor Agricultural University.
[11] Singha A K, Baruah M J, Bordoloi R, Dutta P, Saikia U S 2012 Analysis on nfluencing Factors of Technology Adoption of Different Land Based Enterprises of Farmers under Diversified Farming System *Journal of Agricultural Science* **4** (2)123-139
[12] Howley P, Donoghue C O, Heanue K 2012 Factors Affecting Farmers' Adoption of Agricultural Innovations: A Panel Data Analysis of the Use of Artificial Insemination among Dairy Farmers in Ireland *Journal of Agricultural Science* **4** (6) 171.
[13] Suharyono, Nurfathiyah P, Wahyudi E 2015 Factors affecting the acceleration of the adoption of lowland rice PTT technology innovation in Sarolangun Jambi District. Jambi University *Research Journal: Science Series* **17** (2) 09-17.
[14] Saleh Khaerul 2014 Strengthening Social Capital in empowering Women Home Industry Actors (case of Cassava Tape craftsmen) *Jurnal Agribisnisn Terpadu* **7** (1) 27-41
[15] Wahyuni, S.; Sumardjo; Lubis, D.P.; Sadono, D. 2017. Factors Affecting Group Dynamics and Farmers' Capacity in Organic Rice Agribusiness. Journal of Socioh umaniora, 19 (1): 21-28.
[16] Sayogyo 1982 Interest in Village Economy. Yogyakarta: Gadjah Mada University Press.

[17] Saleh Khaerul. Sumardjo, Aida Vitayala S Hubeis dan Herien P 2017 The Effect of Sosial Capital in Improving The Capacity of Rural Women As "Emping Melinjo" Cottage Industry Actors in Banten Province, Republic of Indonesia *IJSBAR* **33** (1) 300 – 312.
[18] Susanti, Lisiana Widi, Sugihardjo and Suwarto 2008 Factors Affecting Farmers' Decision Making in the Application of Orgaik Rice Farming in Sukorejo Village, Sambirejo District, Sragen Regency. Agritext Journal No 24: 1-14 December 2008.
[19] Farid A and Novita Dewi Kristanti 2009 Factors Affecting Farmers' Capacity (Case of Vegetable Farmers in Malang and Pasuruan Districts) *Agriextensia* **8** (1) 36-48
[20] Putri, Gijayana A K, Sulistyaningsih 2014 Analysis of Factors Affecting Farmers' Decisions in Implementing Organic Rice Farming. [Internet] can be downloaded at Jurnal.unars.ac.id/artikel/2014]
[21] Listiana I 2017 The Capacity of Farmers in the Application of Integrated Pest Management Technology (Pht) for Rice Paddy in Situgede Village, Bogor City *Agrica Extensia* **11** (1) 46-52
[22] Dyah May Karini 2013 Impact of the Change in the Function of Paddy Fields on Rice Production in the Context of Food Security (Case Study in Tangerang District) *National Defense Journal* **XIX** (1) 12-18
[23] Suandi 2012 Social Capital and Sustainable Food Security Development *AGRISEP* **11** (2) 270 - 281
[24] Ali, Akhter and Rahut, Dili Bahadur 2013 Impact of Agricultural Extension Services on Technology Adoption and Crops Yield: Empirical Evidence from Pakistan *Asian Journal of Agriculture and Rural Development* **3** (11) 801-812.

Impact of salicylic acid and biosilica application on plant growth of shallot under water deficit

L D Indarwati[1,*], E Sulistyaningsih[1] and B Kurniasih[1]

[1] Department of Agronomy, Faculty of Agriculture, Universitas Gadjah Mada. Jl. Agro No.1, Bulaksumur, Sleman 55281, Yogyakarta, Indonesia.

* E-mail: endangsih@yahoo.com

Abstract. Shallot is a horticulture crop with a fibrous root system, which is susceptible to water deficit, particularly in the bulb formation stage. This study was carried out to examine the effects of salicylic acid (SA) and biosilica (Si) exogenous induction on plant growth of shallot grown under water deficit in plastic baskets. A factorial 4×2 experiment was laid out in a randomized complete block design with four blocks. The first factor of treatments was a four-level exogenous induction, i.e., 0.5mM SA, 6mM Si, combination (0.5mM SA and 6mM Si), and control (0 without SA and Si). The second factor was a level of interval irrigations i.e, one-day interval and three-day interval. The results showed that the treatment of combination 0.5mM SA and 6mM Si can maintained a plant height and the number of leaves was better than the control under water deficit. Avoidance was one of the mechanisms of shallot in dealing with water deficit, namely by reducing the stomatal density. Decreased stomatal density was negatively correlated with water use efficiency.

1. Introduction

Shallot is a crop with nutritional value that must be cultivated economically. The largest shallot production is in Asia at 54.7%, followed by Africa 28.3%, Europe 6.8%, America 5.5% and Oceania 4.7% [1]. Shallot production in Southeast Asia in 2017 is 137,949 tons and decreased to 131,381 tons in 2018 [1]. Climate change causes an increased temperature, and the average temperature has increased by 0.9 °C since the 19th century. The increase in temperature caused irregular rain patterns, increased droughts, floods, heatwaves and other extreme events affecting crop production [2]. Shallot will produce optimal bulbs if planted in suitable environmental conditions. The requirements for growing shallot are a minimum of 70% irradiation, 25-32 °C air temperature, 50-70% relative humidity, planted on crumb-structured soil with medium clay texture, and adequate drainage and aeration, with enough organic matter as well as a neutral soil pH (5.6-6.5). Shallot can be cultivated from lowlands to highlands (0-1000 mpdl) [3]. Shallot belongs to the Allium genus. More sensitive drought stress than onion and garlic, where they have a better mechanism in increasing moisture content in leaves and osmotic regulation than shallot [4]. The condition of water deficit is one of the constraints that affect the yields quality and quantity. Four mechanisms commonly occur in drought-stricken plants, i.,e drought avoidance, escape, tolerance and recovery mechanisms [5]. The avoidance mechanism is by increasing water absorption by increasing root elongation and depth of root

IOP Conf. Series: Earth and Environmental Science **883** (2021) 012049 doi:10.1088/1755-1315/883/1/012049

penetration and decreasing water loss by controlling the stomata's opening and closing and reducing leaf surface expansion, reducing the transpiration rate. The escape mechanism is by shortening the life cycle, while the tolerant mechanism is by a better osmoprotectant, antioxidant capacity and drying tolerance by accumulating biochemical compounds in osmotic adjustment [6]. The response of plants to drought is by changing the distribution of assimilates and to regulate the opening and closing of stomata [7]. Drought is oxidative stress that can increase the accumulation of reactive oxygen species (ROS). When the plants in water deficit condition, some of the stomata closed, which causes CO_2 to be inhibited from entering, reducing the photosynthetic activity [8] [9]. The plant resistance mechanism by administering salicylic acid (SA) induced by abiotic stress is few to understand at the molecular level. However, recent research shows that the plant resistance mechanism caused by abiotic stress has similarities with the resistance mechanism to biotic stress, namely by accumulating Pathogenesis Related (PR) protein [10][11][12]. Exogenous salicylic acid is a synthetic organic chemical exogenous induction compound. The application of exogenous salicylic acid can reduce water pressure by maintaining leaves moisture content, modulating canopy temperature and more efficient water use to increase the yields [13]. One of the mechanisms for drought tolerance in SA-induced plants is the interaction with the other molecules such as ABA, JA, ethylene, Ca^{2+}, NO, H_2O_2 etc. [14]. Salicylic acid can increase the closed stomata, and drought tolerance is also caused by ABA accumulation, so the possibility of SA and ABA in guard cells can synergistically regulating plant tolerance to drought stress by regulating the opening and closing of the stomata which is possible through Ca^{2+}[15]. Silica (Si) is a beneficial element for plants considered to act as an essential nutrient for several plants. However, not all plants need it. Silica has various functions for different plant species [16]. Plants absorb Si in the form of mono silicic acid or orthosilicic acid (H_4SiO_4). Si accumulates in shoots varied by plant species, ranging from 0.1% -10% Si in dry weight [17]. The absorbed silica will accumulate in epidermis cell walls as polymerized monosilicate acid or silica amorph ($SiO.nH_2O$) [18]. Si is absorbed by plants and enters the xylem then transported together with water and translocated to the leaves. The water evaporated and the silica stored in the epidermis as monosilicate acid [19]. Si application produced a silica-cuticle double layer on the leaves epidermal tissue responsible for the higher leaves water potential during water deficit [20].

2. Methods

The research was conducted in Tri Dharma experimental station, Agriculture Faculty, Universitas Gadjah Mada, Yogyakarta, from November 2019 to April 2020. The experiment was arranged in factorial randomized complete block design. The first factor was application of salicylic acid, i.e., 0.5 mM salicylic acid, 6 mM biosilica, combination (0.5 mM salicylic acid and 6mM biosilica) and control (without salicylic acid and biosilica). The second factor was watering, i.e., one day watering with field capacity (moisture content 18.41%) and three days watering (moisture content 10.46%).The cultivar used Bima local Brebes was planted in a plastic basket with a dimension of 45 × 32 × 16 cm. There are 12 bulbs each basket. Shallot fertilizer consists of primary and supplementary fertilizer. The primary fertilizer was applied three days before planting by NPK fertilizer (16:16:16) at 250 kg/ha, KCL at 30 kg/ha, and SP-36 at 50 kg/ha. The first supplementary fertilizer was applied at 15 days after sowing (DAS) using ZA at 400 kg/ha. The second supplementary fertilizer was applied at 30 DAS using urea at 180 kg/ha [3]. Exogenous induction as salicylic acid, biosilica, or combination of salicylic acid and biosilica was applied at 14, 17, 20 and 23 DAS. Salicylic acid was applied in concentration 0.5 mM foliar at ± 1ml per plant. Biosilica was applied in a concentration of 6 mM on the soil at ± 2ml per plant and a combination treatment with salicylic acid 0.5 mM at 1ml per plant foliarly 6 mM biosilica 2ml per plant applied on the soil. Irrigation was carried out until the plants are 14 DAS, watering was carried out based on treatment in one-day and three-day. The data were statistically analyzed using the appropriate analysis of variance with the honestly significant difference value, calculated as Tukey's statistic at the 5% level. Observations was started at 49 DAS, including the anatomical observation in epidermal cells of leaves, stomatal density, stomata width, relative water content (RWC) (%) = $\frac{[Fresh\ Weight-Dry\ Weight]}{[Turgid\ Weight-Dry\ Weight]}$ x 100%, water use efficiency (WUE) (g/plastic basket) = $\frac{Dry\ weight\ per\ plastic\ basket}{Total\ water\ applied\ per\ basket}$, while the plant growth was observed at 2 weeks after planting (WAP) till 7WAP. As described in the Bima local Brebes cultivar description, harvesting was done on the 60th DAS based on the criteria of 60% leaves are withered.

3. Results and discussion

Water is an important molecule that supports different physiology and metabolism processes in plant cells and tissues. Water deficit in plants can be reduced the water potential of leaves, loss of cell turgor and stomata closure, thereby reducing transpiration and photosynthesis can inhibit plant growth. In conditions of severe water deficit, it is possible plants withered and deceased [21]. The study conducted a significant interaction at $P<0.05$ between the treatment of salicylic acid and the irrigations interval factors in stomatal density (Table 2). However, there was no significant interaction in stomata width (Table 1).

Table 1. Anatomical and physiological data

Treatment	Stomata width (μm)	Relative water content (%)
Salisylic acid		
0,5 mM salicylic acid	5,55 a	54,89 a
6 Mm biosilica	6,03 a	50,78 a
Combination	6,05 a	51,96 a
Control	4,98 a	52,57 a
Watering		
One day interval	5,67 p	57,26 p
Three day interval	5,63 p	47,84 q
CV (%)	18,24	15,16
Interaction	(-)	(-)

Noted: a number followed by the same letter in the same column shows no significant difference based on Tukey's adjustment at the 5% level.

The highest average value of stomatal density was control treatment (without salicylic acid and biosilica) with one-day watering at 61.50 cells/mm^2, while the lowest average value of stomatal density was the combination treatment (0.5mM salicylic acid and 6mM biosilica) with a three-day watering at 48.50 cells/mm2 (Table 2). Stomata have a role as a controller for exchanging gases and CO_2 that enter and leave the leaves [22]. Transpiration can be through stomata, cuticles and lenticels. However, the highest transpiration occurred in stomata [23]. Plants that have a water deficit would decrease the stomatal density, thereby reduced transpiration. Meanwhile, plants with adequate irrigation have tighter stomata (Table 2). Naturally, plants would reduce stomatal density and activated the ABA (abscisid acid) pathway to closed stomata in water deficit conditions [24]. On the other hand, the closure of stomata could also reduced CO_2 fixation which caused disruption of the photosynthesis process. However, based on the analysis shows that stomata width was significantly no difference, so the CO_2 concentration that entered the stomata might have been the same (Table 2). Shallot with combination application (0.5 mM salicylic acid and 6 mM biosilica) with a three-day watering could reduce transpiration by reducing stomatal density. So that plants can avoid water deficit conditions and can still perform better growth and development than the control.

Table 2. Interaction of salicylic acid treatment and watering at variable stomatal density (cells/mm^2) 7 MST

Watering	Salicylic acid				Mean
	0,5 mM Salicylic acid	6 mM Biosilica	Combination	Control	
One day	56,25 ab	55,00 ab	52,50 b	61,50 a	56,31
Three day	60,00 a	49,50 b	48,50 b	55,75 ab	53,44
Mean	58,13	52,25	50,50	58,63	(+)
CV (%)	5,91				

Noted: a number followed by the same letter in the same column shows no significant difference based on Tukey's adjustment at the 5% level.

There is a significant interaction between salicylic acid treatment and watering on the water use efficiency variable (Table 3). On one day watering, salicylic acid aplication did not have a significant effect on WUE. Meanwhile on three days watering, salicylic acid aplication gave a significant difference on WUE. Plantas with combination of SA 0,5 mM and Si 6 mM had a better water use efficiency than on three days watering control. The highest mean WUE value in combination treatment of SA 0,5 mM Si 6 mM with the three days watering was 112,38 g/basket. While the lowest mean WUE value was 50,34g/basket on three days watering control treatment (Tabel 3). Increased plants tolerance through the increased water use efficiency possibly caused by manipulating sugar metabolism in guard cells. The sucrose was functioned as osmoticum guard cells and acted as a substrate on guard cells. The decreased sucrolitic activity was accociated with the decreased stomata conductance which it could increased WUE, and conversely with the increased sucrolitic activity, it could decreased WUE. This is a metabolic manipulation on guard cells that improves stomata turgor regulation to increase the plants tolerance of water deficit [25].

Table 3. Interaction of salicylic acid treatment and watering at variable water use efficiency (g/plastic basket)

Irrigation interval	Salicylic acid				Mean
	0,5 mM Salicylic acid	6 mM Biosilica	Combination	Control	
One day	58,41 a	70,04 a	75,97 a	55,24 a	64,92
Three day	88,47 ab	70,00 b	112,38 a	50,34 b	80,30
Mean	73,44	70,02	94,18	52,79	(+)
CV (%)	26,77				

Noted: a number followed by the same letter in the same column shows no significant difference based on Tukey's adjustment at the 5% level.

The stomatal density negatively correlates to water use efficiency with a Pearson correlation at -0.37 (weak correlation). So, the higher the stomatal density value, the lower water use efficiency by plants. Conversely, the lower the stomatal density value, the higher water use efficiency by plants. Water defisit could inhibit growth and significantly lead to decreased crop yields [26]. There was no significant interaction between the salicylic acid treatment and irrigations interval on relative water content variable, but there was a significant on main effect of irrigation interval. The highest relative water content in the one-day interval and the lowest relative water content in the three-day interval (Table 1). This is supported by the previous study, that plants grown under water deficit conditions showed lower relative water content values than plants grown under normal conditions (no water deficit) [27].

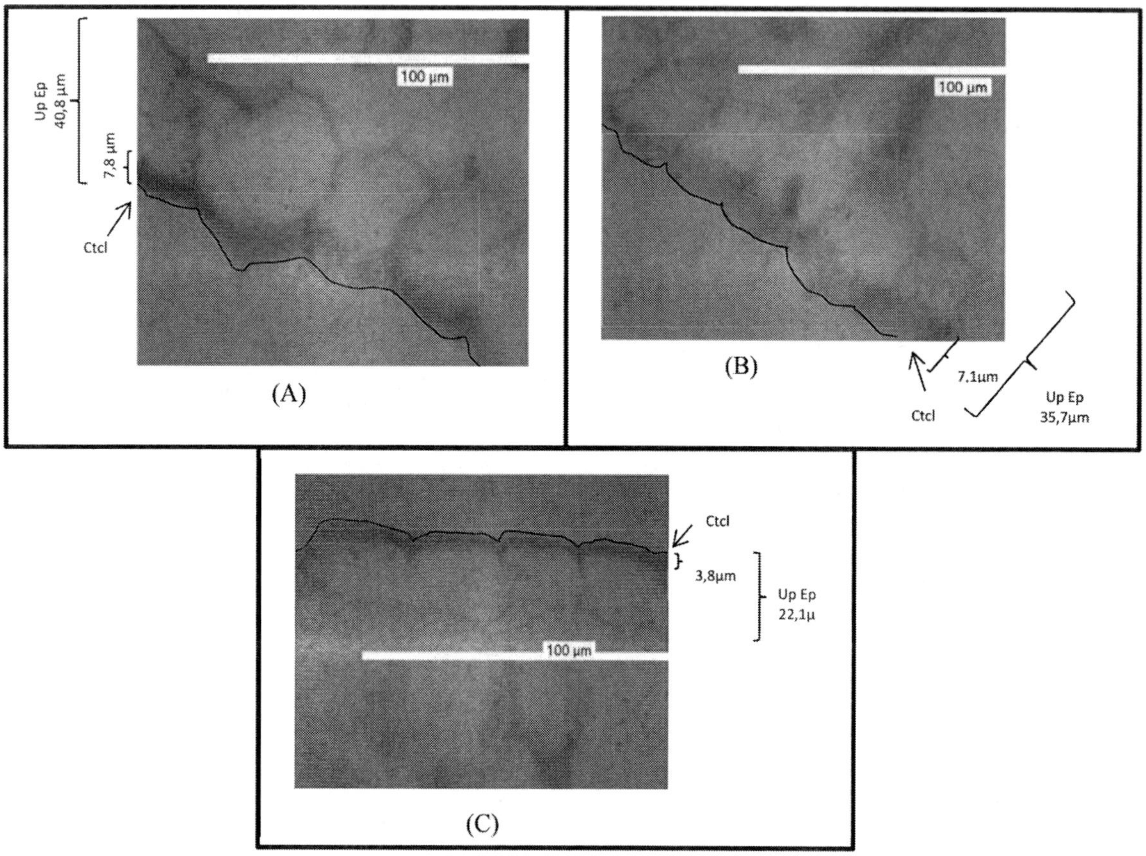

Figure 1. A-C: Longitudinal cross section of shallot leaves 100 × (µm) magnification. (A) Biosilica 6mM with three-day interval, (B) Combination (Salicylic acid 0,5mM and biosilica 6mM) with three days watering, (C) Control (without salicylic acid and biosilica application) with three days watering. Ctcl (Cuticle); Up Ep (Upper Epidermis).

Anatomical observation showed that the shallot leaves given the 6mM biosilica with three days watering of the thickest cuticle layer at 7.8 µm. This is followed by a combination (0.5mM salicylic acid and 6mM biosilica) with a three-day watering with a cuticle thickness at 7.1 µm. The control treatment (without salicylic acid and biosilica) with three days watering is the lowest cuticle layer thickness at 3.8µm (Figure 1. A-C) and (Table 4). The combination (0.5Mm salicylic acid and 6mM biosilica) with three days watering had a smaller cuticle thickness than the 6mM biosilica with three days watering. However, the combination treatment (0.5Mm salicylic acid and 6mM biosilica) with three days watering has a higher water use efficiency value. This occured because the main plant transpiration happen in stomata, where is the combination treatment (0.5 mM salicylic acid and 6 mM biosilica) with three days watering has a low stomatal density, so it is better at preserving the water content in the leaves and water use efficiency compared with 6 mM biosilica under water defisit (three days watering). Silica does not have an essential role for plants grown under controlled conditions. However, under stress condition, silica is known to affect

IOP Conf. Series: Earth and Environmental Science **883** (2021) 012049 doi:10.1088/1755-1315/883/1/012049

plant health by strengthening the cell walls through sedimented silica which is translocated from the roots as silica acid from xylem until it settles under the cuticle and intercellular space [28]. Application of Na_2SiO_3 in wheat under water deficit can increase the relative water content of leaves, the leaves water potential, make a wider leaf area which may be useful for maintaining high assimilation ability, as well as leaves thickening which is useful to reduce loss transpiration compared with plants without Si application [29][30].

Table 4. Anatomy data of shallot leaves.

Exogenous induction	Watering	Cuticle width (μm)
0,5Mm salicylic acid	One day	4,4
6mM biosilica	One day	6,1
Combination (0,5mM salicylic acid + 6mM biosilica)	One day	6,1
Control (without salicylic acid and biosilica)	One day	4,0
Salicylic acid 0,5mM	Three days	5,7
Biosilica 6mM	Three days	**7,8**
Combination (0,5mM salicylic acid + 6mM biosilica)	Three days	7,1
Control (without salicylic acid and biosilica)	Three days	**3,8**

Note: Numbers in bold indicate the highest and lowest value.

The water deficit condition in shallot plants could reduced plant height (Figure 2. A-B) and number of leaves (Figure 3. A-B). Shallot plants were experienced a rapid increase in height, namely at 2 to 3 week after planting (WAP), while at 4 to 7 WAP the height of shallot plants is relatively constant. Generally, the plants experience three phases in their life, namely the lag phase, the log phase and the stationary/fixed phase. The lag phase of plants was experienced slow growth, on log phase the plant growth was fast while the stationary phase the plants was no longer doing vegetative growth [31] [32]. The height pattern of shallot plants was formed a sigmoid curve, where the application of 6 mM Si could maintain the plant height with a higher trend compared to other treatments under one day watering conditions at the last observation (7 WAP) with a height of 32.81 cm (Figure 2.A). When the watering application is carried out every three days, it had be seen that the combination treatment of 0.5 mM salicylic acid and 6 mM biosilica has a plant height with a higher trend of 32.47 cm, while the control treatment has a plant height with the lowest trend of 29.04 cm at the last observation (7 WAP) (Figure 2.B). This shows that the aplication of salicylic acid from a combination of 0.5 mM salicylic acid and 6 mM biosilica can reduce the negative effects of drought so it can maintain plant height better than control under three-days watering conditions. The graph on the leaves number observation parameter shown that in one day watering, the combination treatment of 0.5mM salicylic acid and 6 mM biosilica had the lowest number of leaves with an average of 8.95 leaves, while the highest number of leaves was the control treatment with the avarage of leaves was 10.22 leaves (Figure 3.A). The number of leaves in each treatment were formed a sigmoid curve with a fast rapid increased number of leaves, namely at 2 to 3WAP while the average number of leaves from 4 to 7WAP had a relatively fixed number of leaves (Figure 3. A-B). The number of shallots leaves with the lowest trend was in water deficit conditions (three days of watering), namely the 6 mM biosilica treatment with an average number of leaves of 8.16 leaves. While the number of shallots leaves with the highest trend was in the combination of 0.5 mM salicylic acid and 6 mM biosilica with control treatment which had relatively the same number of leaves since 4WAP to 7WAP (Figure 3.B).

IOP Conf. Series: Earth and Environmental Science **883** (2021) 012049 doi:10.1088/1755-1315/883/1/012049

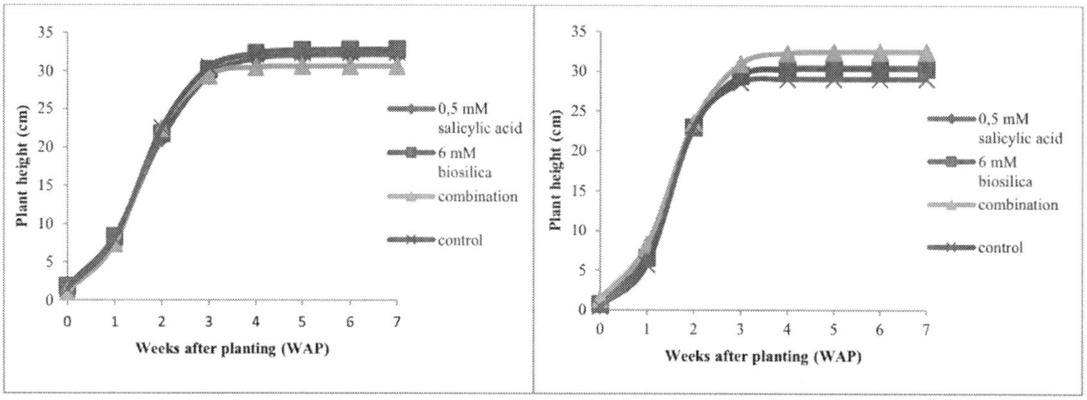

A. B.

Figure 2. A-B: (A.) Shallot plants height growth graph 0-7 WAP with salicylic acid on one day watering; (B.) Shallot plants height growth graph 0-7 WAP with salicylic acid on three days watering.

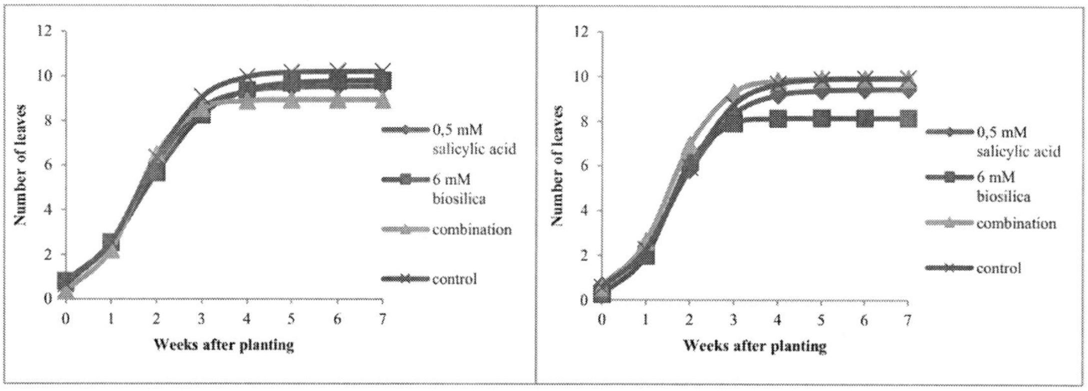

A. B.

Figure 3. A-B: (A.) Shallot plants leaves number graph 0-7 WAP with salicylic acid on one day watering; (B.) Shallot plants leaves number graph 0-7 WAP with salicylic acid on three days watering.

In general, the plant growth was decreased when there is a water deficit (three days of watering), compared to normal condition (one day watering). The decreased growth includes a decrease in plant height and number of leaves. However, the shallot plants with a combination treatment of 0.5 mM salicylic acid and 6 mM biosilica were able to maintain higher growth than shallot plants with other treatments under water deficit conditions (three days of watering). The results showed that the application of a combination of 0.5 mM salicylic acid and 6 mM biosilica was able to maintain the growth of shallot plants with a higher trend compared to the control treatment under water deficit conditions. This was supported by the previous studies, that the effect of simultaneous application of salicylic acid and biosilica was greater than that of salicylic acid or biosilica separately under drought stress conditions in wheat [33].

4. Conclusions

The growth and development of shallot plants under water deficit (three days watering) had a higher trend with the combination of 0.5 mM salicylic acid and 6 mM biosilica compared to the control. The plant resistance to water deficit can be done through an avoidance mechanism, namely by reducing the stomatal density, so the plants can increase water use efficiency. Stomatal density has a weak negative correlation (-0.37) on water use efficiency of shallot under water deficit (three days watering).

5. Acknowledgement

Thank for the final-thesis funding program from Universitas Gadjah Mada.

References

[1] FAO (Food and Agriculture Organization of the United Nations) 2020 [Online] Available at http:// www. fao.org /faostat/ en/#data/ QC/ visualize Verified 17 November 2020
[2] Arora N K 2019 Impact of Climate Change on Agriculture Production and itsh Sustainable Solutions *Environ. Sustain.* **2** 95-96
[3] PUSLITBANGHORTI (Pusat Penelitian dan Pengembangan Hortikultura). 2015. Budidaya Tanaman Bawang Merah. [Online]. Available at http:// hortikultura. litbang. pertanian. go. id/ teknologi-detail-42.html. Verified 17 November 2020.
[4] Csiszar J, Lantos E, Tari I, Mados E, Wodala B, Vashegyi A, Horvath F, Pecsvaradi A, Szabo M, Bartha B, Galle A, Lazar A, Coradini G, Staicu M, Postelnicu S, Mihacea S, Nedelea G and Erdei L 2007 Antioxidant Enzyme Activities in Allium species and Their Cultivars Under Water Stress *Plant Soil Environ.* 53 **12** 517-523
[5] Fang Y and Xiong L 2015 General Mechanism of Drought Response and Their Application in Drought Resistance Improvement in Plants *Cell. & Mol. Life Sci.* 72 **4** 673-689
[6] Nio S A, Cawthray G R, Wade L J and Colmer T D 2011 Pattern of Solutes Accumulated During Leaf Osmotic Adjusment as Related to Duration of Water Deficit for Wheat at The Reproductive Stage *Plant Physio. & Biochem.* 49 **10** 1126-1137
[7] Ai N S dan Banyo Y 2011 Konsentrasi Klorofil Daun sebagai Indikator Kekurangan Air pada Tanaman *Jurnal Ilmiah Sains* 11 **2** 166-173
[8] Amin B, Mahleghah G, Mahmood H M R and Hossein M 2009 Evaluation of Interaction Effect of Drought Stress with Ascorbate and Salicylid Acid on Some of Physiological and Biochemical Parameters in Okra (*Hibiscus esculentus* L.) *Res. J. Biol. Sci.* **4** 380-387
[9] Lestari E G 2006 Hubungan Antara Kerapatan Stomata dengan Ketahanan Kekeringan pada Somaklon Padi Gajahmungkur, Towuti, dan IR 64. *Biodiversitas,* 7 **1** 44-48
[10] Ali S, Ganai B A, Kamili A N, Bhat A A, Mir Z A, Bhat J A, Tyagi A, Islam S T, Mushtaq M, Yadav P, Rawat S and Grover A. 2018. Pathogenesis-related proteins and peptides as promising tools for engineering plants with multiple stress tolerance *Microbiol. Res.* **212-213** 29-37
[11] Singh N K, Kumar K R R, Kumar D, Shukla P and Kirti P B 2013 Characterization of a Pathogen Induced Thaumatin-Like Protein Gene AdTLP from Arachis diogoi, a Wild Peanut *PLOS ONE* 8 **12** 1-18
[12] Hong J K and Hwang B K 2005 Induction of enhanced disease resistance and oxidative stress Tolerance by over expression of pepper basic PR-1 gene in Arabidopsis *Physiol. Plant* **124** 267–277
[13] Wakchaure G C, Minhas P S, Meena K and Singh N P 2018 Growth, Bulb Yield, Water Productivity and Quality of Onion (*Allium cepa* L.) as Affected by Deficit Irrigation Regimes and Exogenous Application of Plant Bio-regulators *Agric.Water Manage.* **199** 1-10
[14] Hasanuzzaman, M, Nahar K, Bhuiyan T F, Anee T I, Inafuku M, Oku H and Fujita M. 2017 Salicylic Acid : An All-Rounder in Regulating Abiotic Stress Responses in Plants Chapter 3 *INTECH* pp 32-75

[15] Prodhan M Y, Munemasa S, Nahar M N E N, Nakamura Y and Murata Y 2018 Guard Cell Salicylic Acid Signaling is Integrated into Abscisic Acid Signaling via the Ca2+/CPK-Dependent Pathway. *Plant Physiol.* **178** 441-50

[16] Pilon-Smits E A H, Quinn C F, Tapken W, Malagoli M and Schiavon M 2009 Physiological Functions of Beneficial Elements *Curr. Opinion in Plant Biol.* 12 **3** 267-274

[17] Ma J F and Takashi E 2002 Soil, Fertilizer, and Plant Silicon Research in Japan. Amsterdam : Elsevier Science.

[18] Malhotra C H, Kapoor R T and Ganjewala D 2016 Alleviation of Abiotic and Biotic Stresses in Plants by Silicon Supplementatio *Sci. Agric.* 13 **2** 59-73

[19] Ma J F and Yamaji N 2006 Silicon Uptake and Accumulation in Higher Plants *Trends in Plant Sci.* 11 **8** 392-397

[20] Matoh, T, Murata S and Takahashi E 1991 Effect of Silicate Application on Photosynthesis of Rice Plants (in Japanese) *Japan. J. of Soil Sci. & Plant Nutrition.* **62** 248-251

[21] Araya Y N 2007 Ecology of Water Relations in Plants. In: ed. *Encyclopedia of life sci.* Wiley. **26** 1-16

[22] Kim T H, Böhmer M, Hu H, Nishimura N and Schroeder J I 2010 Guard cell signal transduction network: advances in understanding abscisic acid, CO2, and Ca2+ signaling. *Annual Review of Plant Biology* **61** 561–591

[23] Lendzian K J 2006 Survival Strategies of Plants During Secondary Growth: Barrier Properties of Phellems and Lenticels Towards Water, Oxygen and Carbon Dioxide *J. of Exp. Bot.* **57** 11 2535-2546

[24] Anosheh H P, Moucheshi A S, Pakniyat H and Pessarakli M 2016 Stomatal responses to drought stress. *Water Stress and Crop Plants: A Sustainable Approach* 1 **1** 24-40

[25] Antunes W C, Provart N J, Williams T C R and Loureiro M E 2012 Changes in Stomatal Function and Water Use Efficiency in Potato Plants with Altered Sucrolytic Activity *Plant, Cell & Environ.* **35** 747-59

[26] Nemeskeri E, Molnar K, Vigh R, Nagy J and Dobos A 2015 Relationships between stomatal behavior, spectral traits and water use and productivity of green peas *(Pisum sativum* L.) in dry season *Acta Physiol Plant.* 37 **2** 34

[27] Chowdhury J A, Karim M A, Khaliq Q A, Ahmed A U and Mondol A T M A I 2017 Effect of Drought Stress on Water Relation Traits of Four Soybean Genotypes *SAARC J. Agri.* 15 **2** 163-175.

[28] Heckman, J 2013 Silicon: A Beneficial Substance *Better Crops* 97 **4** 14-16

[29] Gong H, Chen K, Chen G, Wang S and Zhang C 2003 Effects of Silicon on Growth of Wheat Under Drought *J. of Plant Nutrition.* 26 **5** 1055-1063

[30] Sacala E 2009 Role of Silicon in Plant Resistance to Water Stress *J. of Element.* 14 **3** 619-630

[31] Ramulifho E, Goche T, Van As J, Tsilo T J, Chivasa S and Ngara R 2019 Establishment and Characterization of Callus and Cell Suspension Cultures of Selected *Sorghum bicolor* (L.) Moench Varieties: A Resource for Gene Discovery in Plant Stress Biology. *Agron.* **9** 1-18

[32] George E F, Hall M A and Klerk G J D 2008 Stock Plant Physiological Factors Affecting Growth and Morphogenesis Plant Propagation by Tissue Culture 3rd Edition. pp 403-422.

[33] Maghsoudi K, Emam Y, Niazi A, Pessarakli M and Arvin M J 2018 P5CS Expression Level and Proline Accumulation in The Sensitive and Tolerant Wheat Cultivars under Control and Drought Stress Conditions in The Presence/Absence of Silicon and Salicylid Acid *J. of Plant Interaction.* 13 **1** 461-471.

The characteristic anda feasibility of Banda'S nutmeg agro-industry in Banda Island of Maluku Province

M Lawalata[1,*], N R Timisela[1], M Turukay[1], E D Leatemia[1] and J M Luhukay[1]

[1] Agribusiness Study Program, Faculty of Agriculture, Universitas Pattimura, Ambon, Indonesia

* E-mail: marfinlawata@gmail.com

Abstract. This research aims to identify the characteristics of the agro-industrial business of Banda's manufactured nutmeg pulp products and to analyze the income and feasibility of the agro-industrial business of Banda's manufactured nutmeg pulp products. The results show that agroindustry is feasible to develop because the average value of ratio B-C is 1,8. The feasibility of investment analysis shows that the NVP value is Rp.2.569.750, the IRR value is 36,24%, the value of Ratio Net B/C is 1,8, and the payback period is 3,44. Based on the analysis of the factors that influence nutmeg agroindustry producer's income rates, the factors include raw material prices, supporting material prices, production, labor wages, and working capital. The determination coefficient is 55,7, which means that the income rates are influenced by the variable of ages, the education levels, the number of family members, the business experiences, and the business capitals counted as 55,7 %. The rest 44, 3 %, is influenced by the other factors out of the model.

1. Introduction

Agricultural commodities, in general, are commodities in the form of raw materials and quickly expired, so they must be consumed directly. To overcome this problem, various manufacturing processes have emerged that can assign added value to the agricultural commodities. This process is a part of the Agro-industry sub-system. Agro-industry is an activity that processes raw materials from agricultural commodities into manufactured products that are more attractive and have added value. Agroindustry emphasizes the food processing management in a manufactured product company which the main raw materials are agricultural products [1].

States that manufacturing in agro-industry is a stage of the transformation and preservation process through physical or chemical transformation, storing, packaging, and distribution of the generated products. In other words, manufacturing is an operation or series of operations which is performed on a raw material to change its shape and/or composition [2].

Manufacturing of agricultural commodities consists of simple manufacturing, such as cleaning, grading, packaging, as well as more sophisticated manufacturing such as milling, powdering, extracting, roasting, spinning, canning, and other manufacturing processes. Recently, one of the up-and-coming Agro-industries in Maluku is the well-known nutmeg pulp manufacturing business in the Banda Islands. Nutmeg, as a spice plant and source of essential oil, is a type of plants that has an important role in the life of the people of Banda. The part of the nutmeg plant that has economic value is the fruit [3]. The status of the nutmeg plant as an important material for industry and as a trading commodity had caused European nations to fight over the areas of nutmeg producer in Indonesia, one of which is the cluster of Banda Neira Islands [4].

Indonesia is the world's largest producer of nutmeg (70-75%), of which around 98% is generated by smallholder-owned plantations. The area of nutmeg plantations in Indonesia in 1996 had reached 60,735 acres, which then decreased to 43,873 acres in 2000 [5].

As a potential export commodity for the Maluku local government, Banda Nutmeg (Myristica fragrans Houtt) is a source of economic and regional income growth.

Nutmeg is a spice plant native to the Maluku islands [6] which has been cultivated from generation to generation in the form of smallholder plantations in most of the Maluku's region. Indonesia's nutmeg commodities are among the leading commodities in the world market because the distinctive aroma and high level of oil yield. Nutmeg plantations are dominated by smallholder-owned plantations which operated by farmers in small scale (area less than 0.5 acres). In 2017, the plantation area for smallholder-owned plantations (PR) was 169,103 acres and state-owned plantations (PBN) was 484 acres, with a total area of 169,587 acres (Directorate General of Plantation, 2015-2017).

The added value of the nutmeg can be increased through the diversification of the nutmeg pulp manufacturing. Banda is a center of household-scale business which is engaged in the manufacturing of nutmeg pulp into candied, syrup, juice and jam. In addition, the manufacturing results can provide additional income for the people. Based on data from Maluku's Central Bureau of Statistic (BPS Maluku) in 2016, the nutmeg production reached 464 tons and the number of households involved this business was 1,211 The purpose of this research is to identify the characteristics of the agro-industrial business of Banda's manufactured nutmeg pulp products and to analyze the income and feasibility of the agro-industrial business of Banda's manufactured nutmeg pulp products.

2. Methods

This research is conducted from October to December 2018, in the Sub-district Banda witthe focus on four villages, namely Nusantara Village, Dwiwarna Village, Merdeka Village and Kampung Baru. The sample selection is performed using the method of purposive sampling, because the majority of the household industries in those four villages produce nutmeg candied, nutmeg syrup, nutmeg juice, and nutmeg jam. The sampling process is carried out by applying census techniques to all nutmeg pulp producers in Banda Island. The number of samples in each village is as follows: Kampung Baru (1 respondent), Dwiwarna Village (1 respondent), Merdeka Village (10 respondents), and Nusantara Village (18 respondents). The total respondents in this research are 30 people.

Qualitative and quantitative analysis methods are applied to analyze the data which have been obtained through sampling. Qualitative analysis is used to identify data on the characteristics of the producers of manufactured nutmeg pulp products, while quantitative analysis is used to analyze the feasibility and income of the producers of manufactured nutmeg pulp products.

3. Results and discussion

3.1. Characteristics of Banda's Nutmeg Agro-Industry Producer

Age can affect a person's performance in work both physically and non-physically. The collected data in the research state that most of the producers are in the category of productive age: the oldest is 68 years old and the youngest is 27 years old. It can be observed in Figure 1 below. Due to this condition, innovations are more easily accepted by the producers in the research location.

The level of producers' formal education in the research locations is classified as low. A total of 46.6% producers only went to elementary school. Figure 1 shows that a low level of education means that respondents have limited options for jobs. This is the main reason for most respondents to be involved in the nutmeg agro-industry business. However, despite being low educated, producers are willing to embrace innovation to ensure the continuity of their business.

The number of family members affects the welfare of producers. The more family members a producer has, the greater his/her needs are. In order for all these needs to be fulfilled (whether it is clothes, foods, or educations), a producer must be supported by adequate income in each production process. Figure 1 shows that the percentage of family members of the medium category (4 - 6 people) is 46%. Experience in managing business is also an important determinant of a producer's income. Adequate experience enables producers to calculate and anticipate risks in managing the business. Figure 1 shows that the highest percentage of the period of managing the business among the respondents is in the category of 0-10 year, namely 64.3 percent.

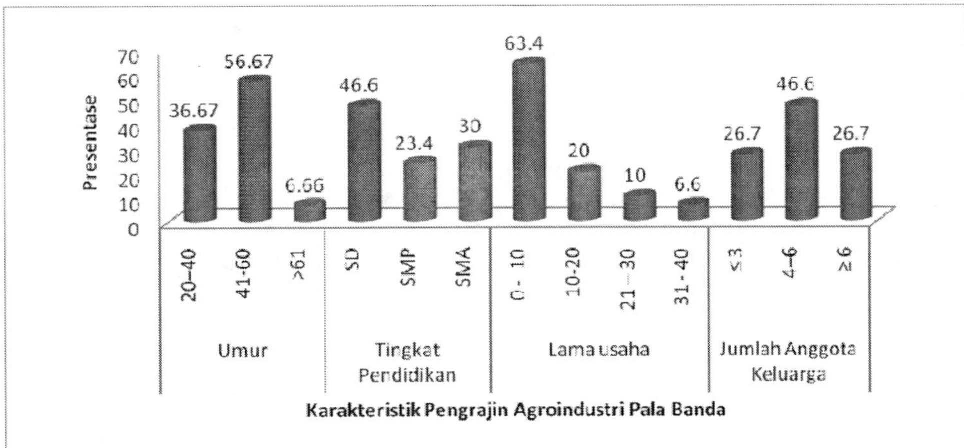

Figure 1. Chart of Characteristics of Banda's Nutmeg Agro-Industry Producers

3.2. Characteristics of Banda Nutmeg Agro-Industry Business

The distinctive fragrant aroma and the considerable amount of essential oil content within the nutmeg and mace make this plant an export plantation commodity which is largely targeted by international market [7]. These two commodities can be manufactured essential oil which is used as raw materials for the beverage, medicine and cosmetic industries. Besides that, it can also be used as a mixture in soft drinks. Meanwhile, the nutmeg pulp can be manufactured into snacks such as candied, syrup, juice, and jams.

Raw Materials: The main commodities of the nutmeg plant, which have high economic value, are the nutmegs and maces, while the pulp is considered a by-commodity or even waste. However, along with the development of science and technology, the nutmeg pulp is no longer considered a waste. In Banda, almost all households have nutmeg trees in their yards, so most households in this area are involved in agro-industrial business of manufactured nutmeg pulp products such as wet nutmeg candied, dried nutmeg candied, nutmeg syrup, nutmeg juice and nutmeg jam. Currently, there are thirty home industries engaged in the manufacturing the nutmeg pulp. The monthly requirement for this business is 300 kg or 6000 pieces which can be obtained from the producer's own harvest or through purchases from other nutmeg tree owners. Although nutmeg trees are very common in Banda, the availability of raw materials for this industry is highly dependent on the harvest season. It is abundant in the harvest season but rare in the other periods. The price of the nutmeg itself is relatively stable at around Rp. 5,000.00 / kg.

Supporting Materials: Supporting materials are other materials which are manufactured in unison with raw materials in a production process. The Supporting material used in the manufacturing of nutmeg pulp is sugar which functions as a preservative materials. The use of sugar for each type of product varies in terms of quantity and composition. Consequently, the required amount of sugar for each product is different, namely around 1 kg to 1.5 kg. Another Supporting material is Sodium Benzoate for the manufacture of nutmeg juice products. Sodium Benzoate functions as a preservative material, so that the products is durable. The required amount of sodium benzoate ranges from 0.5 g to 0.8 g for each nutmeg juice products.

Sources of Capital and Equipment: The decision on choosing the source of capital is strongly influenced by the considerations of each producer, with the main consideration being the most profitable source of capital. Since the start of their business until now, most producers prefer to use their private funds as the source of capital. There are 90 percent of respondents who use their private funds as the source of capital. Concerns over the risk of loan repayment and an unwillingness to bear an extended debt burden are the main reasons the respondents choose this source of capital. Meanwhile, 10 percent of respondents obtain their capital through business capital loans from the National Community Empowerment Program (PNPM-Mandiri).

With the existing business scale, capital is not an obstacle to running the business, because the amount of capital required is not too sizable, both for the procurement of equipment (fixed assets) and current assets. The unstable availability of raw materials also limits production capacity so that the producers may still rely on their private funds to support the continuity of the production. Although the desire to increase the amount of capital from external sources, provided the capital does not have interest, producers prefer to borrow money from family or neighbors if necesarry.

Labor: Usually, the labors in this business are family-based labors. The non-family-based labors are merely employed when the demand for products is increasing at certain times such as during the holiday season at which the number of visiting tourists is increased.

These workers can be categorized as permanent workers and temporary workers. In average, the existing business employs two permanent workers and two temporary workers. Their wages are IDR 30,000 / day and paid in cash. They are usually employed in one production process, for example in the phase of nutmeg peeling as a part of the manufacturing of wet and dried nutmeg candied.

Technology: The technology applied in the nutmeg agro-industry business is classified as simple, both from the fruit harvesting phase to the labeling and packaging phases. This is related to the amount of available capital and educational level of the workers. In the production process, most producers still work manually or use equipment with very simple technology (mostly household tools), so that the quality of the generated products cannot be controlled. In addition, the small scale of the business causes the producers to be unable to apply technology optimally in the production process.

For the manufacturing process, producers are fostered by government institutions in Maluku Province such as the Food Security Agency, the Office of Agriculture, the Office of Health, and the Office of Industry and Trade.

Production: Production activities can take place on condition that raw materials and Supporting materials are available. The number of products generated depends on the number of materials available, while the ability of the equipment to support the production process is relatively equal. The products produced are of good quality and attractive to consumers. The type of product and the amount of production in the nutmeg agro-industry can be seen in Tables 1 and 2.

Table 1. Types of Products and Amount of Nutmeg Agroindustry Production (Wet Nutmeg Candied and Dried Nutmeg Candied).

Products	Production Quantity (Packs/Months)	Percentage (%)
Wet Nutmeg Candied	805	64,81
Dried Nutmeg Candied	437	35,18
Total	1.242	100

Source : Primary Data, 2019 (processed).

Table 1 shows that the most generated product is wet nutmeg candied. It is due to the fact that its production process is simple and does not require particular knowledge. In addition, because of its simplistic nature, the production process does not consume too much time. Therefore, producers of wet nutmeg candied can easily and quickly master the manufacturing process of this product.

Meanwhile, the number of dried nutmeg candied products generated is less than the wet nutmeg candied, because its production process consumes more time and requires larger quantities of sugar.

Table 2. Types of Products and Amount of Nutmeg Agroindustry Production (Nutmeg Syrup, Nutmeg Juice, and Nutmeg Jam).

Products	Production Quantity (Bottles/Months)	Percentage (%)
Nutmeg Syrup	78	29,21
Nutmeg Juice	150	56,17
Nutmeg Jam	39	14,62
Total	267	100

Source : Primary Data, 2019 (processed).

The number of syrup, juice and nutmeg jam products is relatively small because it is influenced by several factors, such as: lengthy production time, minimum knowledge of the manufacturing process of the products, a small number of producers, low consumer's interest on the products, high production costs and selling prices, small marketing opportunities, and difficult business licensing process. The producer self-determines the selling price of the nutmeg manufactured product based on the calculation of the production costs. Consumers of manufactured nutmeg pulp products are retailers, ship passengers and tourists. The retailers will then resell the products in Ambon and other cities around Banda. Table 3 shows the selling price of the nutmeg manufactured product.

Table 3. Prices of Manufactured Nutmeg Pulp Products at the Production Site.

Products	Price (IDR)
Wet Nutmeg Candied	5,000/Packs
Dried Nutmeg Candied	5,000/Packs
Nutmeg Syrup	30,000/625 ml
Nutmeg Juice	8,000/110 ml
Nutmeg Jam	20,000/250 gram

Source : Primary Data, 2019 (processed).

Table 3 shows that the product that has the highest selling price is nutmeg syrup. This is due to several factors, namely: more syrup volume, the composition of nutmeg's pulp essence which is only mixed with granulated sugar and does not use preservatives, and long production times.

Marketing: In contrast to marketing on the primary commodities, such as seeds and mace, nutmeg manufactured products are more likely to attract consumers directly, thus it is typically traded single-handedly by the producers and the scope of marketing locations for the products varies widely.

Producers usually set the price independently by each product. The price of each type of the product depends on the price of raw materials and supporting materials. If the price of raw materials and supporting materials rises, the producer will practically increase the selling price of the product in order to keep on gaining benefit.

The price of each type of product depends on the price of raw materials and Supporting materials. If the price of raw materials and auxiliary materials rises, the producer will practically increase the selling price of the product in order to continue to benefit. Pricing for each product is usually carried out independently by producers and traded individually. Most of the sales of manufactured nutmeg pulp products are carried out at Banda's harbor during "ship days" (the schedule of PT Pelni's ship to dock at Banda's harbor). Producers usually sell their manufactured nutmeg pulp products at previously constructed stalls / tents that are aligned along the harbor, or directly on the ship that is docked. Generally, those interested in nutmeg manufactured products are those from outside of Banda Island, such as retailers, ship passengers and tourists.

The problem that occurred is depending on the ship schedule of PT. Pelni, the product marketing opportunities are very minimal, because sales activities will end when the ship departs. This has even led to most producers would rather start production when they get orders, because the certainty of benefits is more guarenteed, than starting production and waiting for the ship's schedule to dock which is only twice a week and lasts 2-3 hours per ship.

Meanwhile, a small portion of them are marketed in leading supermarkets in Ambon City such as Sinar Makro, Citra, Planet 5000, Indojaya, and Toko Hidayat as well as in other cities around Banda such as Masohi, Dobo and Tual. Promotional activities can affect the sustainability of the Banda nutmeg agorindustry in the future, thus promotion is one of the efforts that need to be performed by the current Banda nutmeg producers [8]. The promotion of Banda nutmeg products usually occurs at certain activities in Banda. Therefore, in order to deal with market uncertainty, which impacts on product negligence and losses for producers, producers must actively promote products and establish cooperative relationships with supermarkets in Ambon city and other cities around Banda.

Products ordered from the cities of Ambon, Masohi, Tual and Dobo are usually sent by the businessman via ships that stop at these delivery areas. Commonly, the producer will entrust it to an acquaintance within the ship's crews, who will then deliver it to the customer upon arrival at the destination. Shipping costs are borne by the manufacturer as a token of remuneration because the product has been purchased by the customers.

Payment transactions in this marketing channel are carried out after a decided period of time between producers and supermarkets, based on the number of products sold. Proceeds from sales of products on this channel are paid to the producer by transfer. The large number of supermarkets selling the same product results in the risk of unsold products, which lead to losses, becoming a significant risk that must be faced by producers as a result. Thus, the strategy of developing a distinctive taste and hygienic packaging of the products are important things that must be considered by producers Product marketing channels can be described as follows:

i. Producer ⟶ Retailer ⟶ Consumer
ii. Producer ⟶ Consumer
iii. Producer ⟶ Supermarket ⟶ Consumer

Figure 2. Marketing Channels of Banda's Nutmeg Agro-industry Business

The scope of the marketing area for processed nutmeg products is very limited, only in cities within Maluku province. Some of the factors that become obstacles for producers to market their products outside Maluku are: difficult business licensing, unsecured product quality, limited working capital and production capabilities, lack of or no cooperation, and risk of loss due to the payments which is based on the number of sold products. According to the research, there are far more producers who choose to directly sell the products to the consumers than the ones who use other marketing channels. The percentage of respondents in each marketing channel for manufactured products can be seen in the following table:

Table 4. Number of Producers Involved in Each Marketing Channel in the Banda Neira Sub-district.

Marketing Channel	Producers (people)	Persentage (%)
i. Producer, Retailer, Consumer	4	13,34
ii. Producer, Consumer	18	60
iii. Producer, Supermarket, Consumer	8	26,67
Total	30	100,00

Source : Primary Data Analysis, 2019 (processed)

Table 4 shows that the number of marketing channels most often used by producers is the second marketing channel (producers → consumers), which is 18 respondents (60 percent). This is due to the tendency of the producers to sell their products directly to consumers at the harbor of Banda. Meanwhile, the second largest marketing channel is the marketing channel through supermarkets in Ambon city and cities around Banda. There were 8 respondents who chose to use this channel (26.67 percent). Meanwhile, the marketing channel chosen by the fewest respondents was the marketing channel through retailers with 4 respondents (13.34 percent). Meanwhile, the marketing channel chosen by the fewest respondents was the marketing channel through retailers with 4 respondents (13.34 percent). The resulting products can be seen in Figure 3.

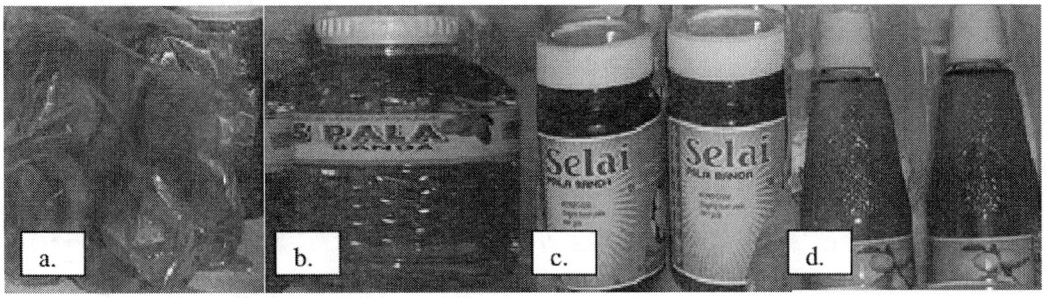

Figure 3. Banda's Manufactured Nutmeg Pulp Products ;(a) Wet and Dried Nutmeg Candied, (b) Nutmeg Juice (c) Nutmeg Jam (d) Nutmeg Syrup

3.3. Feasibility Analysis of Nutmeg Agro-Industry Business

To determine the feasibility of the nutmeg agro-industry in terms of income, the R / C ratio and B / C ratio measuring instruments are used. The results of data analysis in Table 4 show that the R / C ratio and B / C ratio of each product are different. The product with the highest R / C ratio and B / C ratio was the wet nutmeg candied product, while the smallest was the nutmeg jam product. The RCR value in the table means that if the production cost of production is IDR 1, it will generate an income of IDR 3.19 for wet nutmeg candied, IDR 3.12 for dried nutmeg candied, IDR 2.32 for the nutmeg syrup, IDR 2.67 for nutmeg juice and IDR 2.64 for nutmeg jam. The RCR value in the table means that if the production cost of production is IDR 1, it will generate a profit of IDR 2.19 for wet nutmeg candied, IDR. 2.12 for dried nutmeg candied, IDR 1.32 for the nutmeg syrup, IDR 1.67 for nutmeg juice, and IDR 1.64 for nutmeg jam.

Tabel 5 . Profitability of Nutmeg Agro-Industry Business Per Month

Products	Revenue (IDR)	Total cost (IDR)	Benefit (Rp)	R/C	B/C
Wet Nutmeg Candied	3830000	1200000	2630000	3,19	2,19
Dried Nutmeg Candied	3900000	1250000	2650000	3,12	2,12
Nutmeg Syrup	4750000	2050000	2700000	2,32	1,32
Nutmeg Juice	2540000	950000	1590000	2,67	1,67
Nutmeg Jam	2115000	800000	1315000	2,64	1,64

Sumber: Data Primer, 2019

Table 5 shows that of the five agro-industrial products of nutmeg, three of them, namely sweetened wet nutmeg, dried nutmeg candied and nutmeg syrup are considered feasible. Meanwhile, nutmeg juice and nutmeg jam are considered not feasible, because it is influenced by several factors, namely:

the absence of continuity of production (only producing when there is an order), the lack of expertise in the use of technology so that the generated products are not durable, and the low purchasing power of consumers.

3.4. Feasibility Analysis of Nutmeg Agro-Industry Investment

To determine the financial feasibility of nutmeg agro-industry, a business analysis was carried out using investment criteria measuring instruments including: NPV, Net B / C Ratio, IRR, and Payback Period. Each investment criteria is based on the assumption that individuals or companies would rather have benefits that can be gained instantly than benefits gained in a long-term. Therefore, the value of NPV, Net B / C ratio, IRR, and Payback Period shows future results but is calculated using its present value by multiplying it with the discount factor. In the business feasibility analysis, the data used are primary data for 12 months with a discount factor equal to the current interest rate in the government-owned banks (16%). NPV shows the level of profit that will be gained by producers in the nutmeg agro-industry for 12 months which is calculated at present value The results obtained from the financial analysis of the nutmeg agro-industry are as follows:

Tabel 6. Calculation Result NPV, Net B/C Ratio, IRR, and Payback Period

No.	Kriteria Investasi	Hasil
1	NPV (16%) (Rp)	2.569.750
2	Net B/C Ratio (%)	1,80
3	IRR (%)	36,24
4	Payback Period	3,5 bulan

Sumber: Data Primer,2019 (dianalisa).

From the results of data analysis, it can be seen that the difference between the benefit (revenue) and the cost (expenditure) which has been multiplied by the discount factor is 16% or represented by 2,569,750 (> 0 or positive). This means that at the current interest rate, which is 16%, this business will generate a profit of IDR 2,569,750, - so this business is feasible to be maintained and developed. With the NPV of IDR 2,569,750 in a period of 1 year, the business was able to generate substantial profits, considering that the initial capital issued was only Rp. 1,560,000.

Based on the result of the analysis, it is known that the comparison between the presentvalued net profit from month 0 to month 12, with the presentvalue net cost in the same year (Net B / C ratio) is 1.8. That means, every IDR 1 spent for investment in the nutmeg agro-industry for 1 year will generate benefits of Rp. 1.8 and because the Net B / C Ratio is> 1, the agro-industry investment is considered feasible.

In IRR calculation, a discount factor of 16% and 34%, with a difference between the discounts of 18% is used to obtain the correct interpolation of the numbers. IRR can also be used to compare the percentage of profits if deposited in the Bank. From the analysis it is known that the IRR is 36.24%, which means that the nutmeg agro-industry provides a return on capital of 36.24% and is feasible to develop because it is higher than the current loan interest rate in the Bank.

Payback period is the period required to return all costs incurred in investing in a project. From the analysis of available data, it can be observed that the payback period is 3.5, which means that the payback period for investment costs is 3.5 months. Thus, the nutmeg agro-industry business is considered feasible because the payback period for investment costs is shorter than required.

3.5. Factors Influencing the Income Level of Banda's Nutmeg Agroindustry

3.5.1 F Test (Simultaneous)

The F test is used to test all independent variables such as the price of raw materials, the price of supporting materials, production, labor wages and working capital. All of these variables simultaneously affect the income level variable with the calculated F value = 3.156 > F table of 2.62.

3.5.2 Coefficient of Determination (R2)

The percentage of influence of all independent variables on the dependent variable is indicated by the magnitude of the coefficient of determination (Adjusted R2). The coefficient of determination (Adjusted R2) shows the magnitude of the influence of the independent variable on the dependent variable which is expressed in percent (%) [9].

The results of the analysis show that the level of income is 55.7% influenced by variables of raw material prices, prices for supporting materials, production, labors' wages and business capital by and the remaining 44.3% is influenced by other factors outside the model.

3.5.3 T test (partial)

The factors that influence the income level of the nutmeg agro-industry producers are the price of raw materials (X1), the price of supporting materials (X2), production (X3), labor wages (X4), and business capital (X5).

The classical assumption test analysis of the regression model determines whether there are deviations from the model in the income level data of Banda nutmeg agro-industry producers. The results of multiple regression analysis can be seen in Table 7.

Tabel 7. Results of Multiple Regression Analysis on the Income Level of Banda's Nutmeg Agroindustry

	Unstandardized Coefficients		Standardized Coefficients			Collinearity Statistics	
Model	B	Std. Error	Beta	t	Sig	Tolerance	VIF
(Constant)	18.594,16	1.456E6		.0130	0.990		
X1	-784,22	19813.994	-0.010	-2.150	0.045	.468	2.138
X2	-21.277,95	83128.791	-0.061	-2.256	0.024	.499	2.004
X3	34.570,85	79315.395	0.083	2.436	0.016	.793	1.261
X4	-5.995,08	19909.844	-0.064	-2.301	0.019	.637	1.570
X5	1.268,0	0.406	0.549	3.123	0.005	.930	1.075

Table 7 shows several things, such as:

- The regression coefficient of the raw material price variable is 784.22, which means that if the price of raw materials increases, the income will decrease by IDR 784.22. The t-test result of -2,150 > 2.06 indicates that the price of raw materials has a significant and negative influence on income. Fluctuation in the price of raw materials influences the producer's income because the higher the price of raw materials, the lower the producer's income, and vice versa.
- The regression coefficient of the Supporting material price variable is 21,277.95, which means that if the price of Supporting materials increases, the income will decrease by IDR. 21,277.95. The t-test result of -2.256 > 2.06 indicates that the price of Supporting materials have a significant and negative influence on income. Fluctuation in the price of raw materials influences the producer's income because the higher the price of supporting materials, the lower the producer's income, and vice versa
- The regression coefficient for the production variable is 34,570.85, which means that if the number of production increases, the income will increase by IDR 34,570.85. The t-test result of 2.436 > 2.06 indicates that production has a significant and positive influence on income. The number of production influences the producer's income because the more the number of production, the higher the producer's income, and vice versa
- The regression coefficient of the labors' wage variable in the multiple regression equation is 5,995.08, which means that if the labor wage increases, the income will decrease by IDR 5,995.08.

The t-test result of 2.301 > 2.06 indicates that labors' wages have a significant and negative influence on income. Labors' wages influence the producer's income because the higher the labors' wages, the lower the producer's income and vice versa

- The coefficient of the business capital variable in the multiple regression equation is 1.268. which means that if the business capital increases, the income will increase by IDR. 1,268. The t-test result of 3.123 > 2.06 indicates that business capital has a significant and positive influence on income. The higher the business capital, the greater production capacity which leads to bigger sale and higher income, and vise versa.

4. Conclusion

Based on the results of the feasibility analysis of the business, the agro-industry is considered feasible. Nutmeg agro-industry business can be said to be profitable because the payback period of investment costs is shorter than required. The coefficient of determination is 55.7, which means that the income level is influenced by 55.7% by the variables of raw material prices, supporting material prices, production, labors' wages and business capital and the remaining 44.3% is influenced by other factors outside model.

References

[1] Soekartawi 2003 Agribisnis. Teori dan Aplikasi. PT. Raja Grafindo Persada: Jakarta.
[2] Buckle K A, Edwards R A, Flet G H dan Wooton M 1987 Ilmu Pangan. Penerjamah H. Purnomo dan Adiono. Universitas Indonesia Press, Jakarta
[3] Sunanto H 1992 Budidaya Pala Komoditas Ekspor. Penerbit Kanisius: Yokyakarta
[4] Rukmana R 2004 Usahatani Pala. Penerbit Aneka Ilmu: Semarang.
[5] Adrianto T 2007 Menyuling Minyak Atsiri. PT. Citra Aji Parama: Yoyakarta.
[6] Purseglove J W, Brown E G, Green S L, and Robbins S R J 1995 Spices. Longmans *New York. p.* 175–228.
[7] Lawalata M, Thenu S F W, Tamaela M 2017 Kajian Pengembangan Potensi Perkebunan Pala Banda Di Kecamatan Banda Neira Kabupaten Maluku Tengah *Jurnal Agrilan* **5** (2) 132-150.
[8] Timisela N R, Turukay M, Parera W B, Lawalata M 2012 Efisiensi relatif agroindustri pala banda dengan pendekatan Data Envelopment Analysis (DEA) *Jurnal SEPA* **9**(1) 25 – 33.
[9] Gujarati, Damodar N 1995 Basic Econometrics.,Mc Graw Hill Inc *New York*, USA

The effect of organic fertilizers on growth several varieties of soybeans

M Rizwan[1], M Dalimunthe[1], I A Pasaribu [1], H Satriawan[2]

[1] Department of Agrotechnology, Faculty of Agriculture, Universitas Islam Sumatera Utara, Gedung Johor, Medan, Indonesia.
[2] Department Agrotechnology, Faculty of Agriculture, Almuslim University, Bireuen, Aceh, Indonesia 24211

* E-mail: muhammadrizwanagr@gmail.com; satriawan_80@yahoo.co.id

Abstract. This study aims to analyse the effect of three types of organic fertilizers on the vegetative growth several varieties of soybeans. This study used a factorial randomized block design with two treatment factors, namely organic fertilizers and varieties. The first factor is: organic fertilizer from 3 levels, namely: K_1 (1.5 kg of compost), K_2 (1.5 kg of goat manure), and K_3 (1.5 kg of chicken manure). The second factor is: soybean plant varieties consist of 4 levels, namely: V_1 (Dena 1), V_2 (Devon 1), V_3 (Anjasmoro) and V_4 (Derap 1). The parameters observed were plant height, number of branches, number of leaves, amount of chlorophyll, leaf area, and flowering age. The results of our research is organic fertilizers have a significant effect and are able to increase the growth of soybean plants except for leaf area and flowering age. Types of varieties had a significant effect and were able to increase plant height, number of branches, leaf area, and flowering age, but had no significant effect on the parameters of the number of leaves and the amount of chlorophyll. Moreover, there isn't interaction effect between the two treatments for all parameters.

1. Introduction

Soybean is the third most important food commodity in Indonesia after rice and maize, where processed soy products such as tofu, tempeh and soy sauce have become the main source of protein for Indonesian people. This is evidenced by the increase in national soybean consumption published by the Ministry of Agriculture in 2016 which states that national soybean consumption continues to increase every year, in 2015 the total national soybean consumption was 1,563,827.04 tons and increased in 2016 to be as big as 2,486,775.94 tons, from these data, soybean consumption in 2015–2016 has increased by 59%. The increase in consumption is not accompanied by an increase in total soybean production in Indonesia, in 2015 there was a significant deficit between total consumption and total soybean production in Indonesia, there was a deficit of 600,644.04 tons, of the total production in 2015 was 963,183.00 tons, and to meet the national demand for soybeans, in this case the government is still importing soybeans [1].

Based on these facts, efforts aimed at increasing the productivity of soybean plants need to be done with improving the physical properties of the soil. This is because soybean plants are generally planted in paddy fields at the end of the rainy season, so that water is one of the obstacles in achieving the best growth and maximum yield. It is hoped that organic fertilizers from chicken coops and compost will be able to function in improving the physical properties of the soil, especially soil structure through improving the physical properties of the soil. Improvement of soil physical properties can stimulate soil granulation, spur development of better plant roots so that plants can search for water and nutrients more optimally. Thus, the application of inorganic fertilizers in high amounts can be controlled.

Manure is solid and liquid manure from livestock mixed with food scraps and cage mats which can function as a stabilizer for soil aggregates [2; 3]. However, the size of the effect of manure applied to the improvement of soil physical properties will greatly depend on the level of maturity and the dose of manure applied [4]. In soil conditions with low levels of organic matter availability, the application of large amounts of manure is needed. However, if the level of availability of soil organic matter is high, the application of manure is not required. The ideal average content of organic matter is around 2.5% - 5% [5]. The use of manure can improve soil fertility and increase the efficiency of using inorganic fertilizers, thereby accelerating plant growth. The content of N, P, K in manure is not too high [6, 7, 8] but can improve soil permeability, porosity, soil structure, water holding power and soil cation content. This study aims to analyse the response of several soybean varieties to different organic fertilizers.

2. Methods

2.1. Place and Time of Research

The research was conducted in the experimental field at the Faculty of Agriculture, Islamic University of North Sumatra, Medan Johor Sub-district, Medan City. The research field are located at altitude ± 25 meters above sea level, with flat topography. This research was started from March to June 2020.

2.2. Tools and Materials

The tools were used in this study are: a set of SPAD tools, hand sprayer, hoe, tripe machete, analytical scales, cutter blades, buckets, treatment boards, research boards, standard stakes, calculators, documentation tools , writing instruments and others that support the implementation of the research. Whereas, the materials of this research were: Varieties of soybean Dena 1, Devon 1, Anjasmoro, Derap 1, compost, chicken manure, goat manure, and other materials that support the research

2.3. Research Method

This study was used a factorial randomized block design (RBD) with two factors. The firts factor is organic fertilizer consisted of 3 levels, and the second is soybean plant variety type factor consisted of 4 levels. The first factor consists of three levels, namely: K1 = compost fertilizer (8 kg / plot); K2 = goat manure (8 kg / plot); K3 = Fertilizer for the chicken house (8 kg / plot). The second factor consist of four levels, namely: V1 = Dena 1 variety; V2 = Devon variety 1; V3 = Anjasmoro variety; V4 = Derap variety 1.

The plant growth parameters observed were: 1) Plant height (cm). Plant height is measured from the base of the stem at the soil surface to the point of growth. Plant measurements were carried out when the plants were 2 WAP. To prevent measurement errors, each sample plant is marked with a stake. Observation of plant height was carried out at intervals of 1 week until the plants entered the generative growth stage. 2) Number of branches (branches). Observation of the number of branches was carried out when the plants were 2 WAP. With 1 week interval of calculation. The number of branches is done by counting the number of existing branches. Observation of the number of branches is stopped when the plant enters the generative growth phase (flower appears). 3) Number of leaves (strands). Observation of the number of leaves was carried out when the plants were 2 WAP at 1 week intervals. The calculation of the number of leaves is done by counting the number of perfectly open leaves. 4) Amount of Chlorophyll (grain / cm). Observation of the amount of chlorophyll was carried out using a Chlorophyll meter SPAD (Soil Plant Analysis Development) tool. Measuring the amount of leaf chlorophyll, place the leaf sample in the chlorophyll meter head slot then press it down, when the head is closed above the leaf, the meter will sound and the measurement result will appear on the screen. Calculation of the amount of leaf chlorophyll is done once. Calculation of the amount of leaf chlorophyll was only carried out on sample plants. 5) Leaf Area Index (cm^2). Leaf area calculations are carried out at the end of vegetative growth, marked by the appearance of soybean plant flowers. Leaf area calculation is done once. The calculation is only carried out on the sample plants. Leaf area measurement uses a leaf area meter. 6) Flowering Age (days). Observation of flowering age is done by counting the days starting from when the soybean plant is planted until the plants produce flowers.

3. Results and discussion

3.1. Plant Height

The results of analysis of variance showed that the test of several types of organic fertilizers had no significant effect on plant height at 2 and 3 WAP but had a significant effect at 4 WAP. The effect of several types of soybean plant varieties did not significantly affect plant height at 2 and 3 WAP, but had a significant effect at 4 WAP. The interaction between organic fertilizer and variety tests had no significant effect on plant height in each planting (Table 1).

Table 1. Plant height at 4 weeks after planting

Treatment	Organic fertilizers			Average
	K_1 (Compost)	K_2 (Goat Manure)	K_3 (Chicken Manure)	
Varietes				
V_1 (Denal)	33,67	32,83	35,67	34,06 a
V_2 (Devon 1)	31,83	31,83	32,42	32,03 b
V_3 (Anjasmoro)	32,83	36,17	35,67	34,89 a
V_4 (Derap 1)	29,00	33,25	31,75	31,33 b
Average	31,83 b	33,52 a	33,88 a	

Note: Figures followed by different letters in the same treatment group are significantly different at the 5% level based on the DMRT test.

Several types of organic fertilizers had a significant effect on the height of soybean plants with the highest plants obtained in the K3 treatment (chicken manure) resulting in the highest plants, namely 33.88 cm, which was not significantly different from the K2 treatment (goat manure), which was 33.52 cm. and the two treatments were significantly different from the K1 treatment compost), namely 31.83 cm.

The results showed that among the organic fertilizers used, it was seen that chicken manure showed the highest average plant, although not significantly different from goat manure. Manure or organic fertilizer contains different nutrients because each animal has its own characteristics which are determined by the source of organic matter and the type of food and age of the animal. Such as the nutrients found in chicken manure, the nutrients are 3.21% nitrogen, 3.21% phosphor, 1.57% potassium, 1.57% Ca, 1.44% Mg, 250 ppm Mn and 315 ppm Mn[9] (Dwipa, 2020). Nutrients in goat manure N 2.10%, P2O5 0.66%, K2O 1.97%, Ca 1.64%, Mg 0.60%, Mn 233 ppm and Zn 90.8 ppm [10].

Anjasmoro variety produces the highest plants than other varieties. The difference in response shown in the height of the soybean plant was due to different varieties, presumably due to differences in genetic traits of the four varieties tested. This difference in genetic traits causes differences in the response of the four varieties to various environmental conditions, so that the growth activities shown are different. This is in accordance with the statement of [11] that the difference in growth power between varieties is determined by genetic factors.

3.2. Number of Branches

Analysis of variance showed that the organic fertilizer test had no significant effect on the number of branches at 2 WAPs but had a significant effect on 3 and 4 WAPs. The treatment of several varieties had no significant effect on the number of branches at the age of 2 and 3 WAP but had a significant effect on the age of 4 WAP. The average difference test for the application of several types of organic fertilizers and soybean plant varieties on the number of branches at the age of 4 WAP can be seen in table 2.

Table 2. Number of branches soybean at 4 weeks after planting

Treatment	Organic fertilizers			Average
	K_1 (Compost)	K_2 (Goat Manure)	K_3 (Chicken Manure)	
Varieties				
V_1 (Dena1)	19,00	17,75	18,33	18,36 a
V_2 (Devon 1)	14,42	15,25	18,17	15,94 b
V_3 (Anjasmoro)	13,42	15,83	17,25	15,50 b
V_4 (Derap 1)	15,58	15,33	19,33	16,75 ab
Average	15,60 b	16,04 b	18,27 a	

Note: Figures followed by different letters in the same treatment group are significantly different at the 5% level based on the DMRT test.

Table 2 shows that the application of organic fertilizers has a significant effect on the number of branches. The highest number of branches in the K3 treatment (chicken manure), namely 18.27 branches, was significantly different from the K2 treatment (goat manure), namely 16.04 branches and K1 treatment (compost), namely 15.60 branches.

The treatment of chicken manure has the highest number of branches. This condition is suspected that the N content in chicken manure is higher than other organic fertilizers (goat manure and compost). The effect of increasing the number of branches is related to the supply of nutrients to the plant and fixing nitrogen in the air through root nodules. The availability of nitrogen contained in chicken manure affects the number of branches of soybean plants. Nitrogen plays an active role during vegetative growth of plants. The results of this study are in accordance with the statement from [12] that applying fertilizers with higher nitrogen content can accelerate the growth and development of plant organs so that they experience faster increase in the number of leaves, plant height and branch growth.

Table 2 also shows that several types of soybean plant varieties have a significant effect on the number of branches. The highest number of branches was obtained in treatment V1 (Dena 1), namely 18.36 branches, which was not significantly different from treatment V4 (beat 1), namely 16.75 branches but significantly different from treatment V2 (Devon 1), namely 15.94 branches and treatment V3 (Anjasmoro) which is 15.50 branches. Treatment V4 was not significantly different from treatment V2 and V3. These results indicate that the dena-1 variety is more responsive to organic fertilizers.

3.3. Number of leaves

Analysis of variance showed that organic fertilizers had no significant effect on the number of leaves at 2 WAP but had a significant effect at 3 and 4 WAP. The treatment of several varieties had no significant effect on the number of leaves at each observation (Table 3).

Table 3. Number of leaves soybean at 4 weeks after planting

Treatment	Organic fertilizer			Average
	K_1 (Compost)	K_2 (Goat Manure)	K_3 (Chicken Manure)	
Varieties				
V_1 (Dena1)	40,92	47,42	41,00	43,11
V_2 (Devon 1)	33,33	39,08	42,58	38,33
V_3 (Anjasmoro)	32,58	41,58	41,50	38,56
V_4 (Derap 1)	36,08	43,42	43,92	41,14
Average	35,73 b	42,88 a	42,25 a	

Note: Figures followed by different letters in the same treatment group are significantly different at the 5% level based on the DMRT test.

The provision of several types of organic fertilizers had a significant effect on the number of leaves of soybean plants with the highest number of leaves in goat manure treatment (K2), which was 42.88, which was not significantly different from chicken manure (K3), namely 42.25 strands and secondly The treatment was significantly different from compost (K1), namely 35.73 strands.

Macro nutrient elements in organic fertilizers such as nitrogen (N) have an important role in leaf growth, high levels of nitrogen nutrients will produce more and larger leaves. Leaves are an important part of plants, because of the chlorophyll content, the photosynthesis process can take place in the leaves. This concurs with [13] which states that the nutrient nitrogen is very useful for plants for growth and development.

Table 3 also explains that the type of variety has no significant effect on the number of leaves of soybean plants. Treatment V1 (Dena 1) produced the highest number of leaves, namely 43.11, followed by treatment V4 (Derap 1), namely 41.14 strands, then treatment V3 (Anjasmoro), namely 38.56 leaves and treatment V2 (Devon 1), namely 38.33 strands. The four treatments were not significantly different from one another.

3.4. Total Chlorophyll

Analysis of variance showed that the organic fertilizer test had a significant effect on the amount of chlorophyll, while the use of several varieties had no significant effect on the amount of chlorophyll. The average difference test for the application of several types of organic fertilizers and soybean plant varieties on the amount of chlorophyll were shown in table 4.

Table 4. Total chlorohyll several varieties of soybean

Treatment	Organic Fertilizer			Average
	K_1 (Compost)	K_2 (Goat Manure)	K_3 (Chicken Manure)	
Varieties				
V_1 (Dena1)	39,62	42,20	43,79	41,87
V_2 (Devon 1)	41,64	40,51	41,14	41,10
V_3 (Anjasmoro)	38,95	41,64	43,05	41,22
V_4 (Derap 1)	40,84	42,29	43,08	42,07
Average	40,26 b	41,66 a	42,76 a	

Note: Figures followed by different letters in the same treatment group are significantly different at the 5% level based on the DMRT test.

The type of organic fertilizer had a significant effect on the amount of chlorophyll, the highest amount was in the K3 treatment (chicken manure), that is 42.76 grains / cm, not significantly different from the K2 treatment (goat manure) that is 41.66 grains / cm was significantly different from the K1 treatment (compost) which was 40.26 grains / cm3.

The dominant nitrogen element found in manure functions in increasing vegetative growth of plants, especially to spur leaf growth, with more leaf chlorophyll, the higher the photosynthate produced, so that the higher the photosynthate is translocated, the photosynthate is used for plant growth and development including new leaf formation. According to [14] nitrogen can accelerate growth and provide greater yields encouraging the growth of vegetation such as leaves, stems, roots which have an important role in plants. In this case, treatment using chicken manure produced the highest amount of chlorophyll. According to [15, 16] states that providing organic material, especially chicken manure, can increase plant growth and chlorophyll in leaves. Nitrogen is closely related to chlorophyll synthesis and protein and enzyme synthesis [17].

3.5. Leaf Area

Analysis of variance showed that the organic fertilizer test had no significant effect on the leaf area of soybean plants, while the use of several varieties had a significant effect on the leaf area of soybean plants (table 5).

Table 5. Leaf area of several varieties of soybean

Treatment	Organic Fertilizer			Average
	K_1 (Compost)	K_2 (Goat Manure)	K_3 (Chicken Manure)	
Varieties				
V_1 (Denal)	38,77	39,23	39,60	39,20 d
V_2 (Devon 1)	43,00	43,45	43,36	43,27 a
V_3 (Anjasmoro)	41,79	42,58	41,47	41,95 b
V_4 (Derap 1)	32,27	31,99	32,45	32,24 c
Average	38,96	39,32	39,22	

Note: Figures followed by different letters in the same treatment group are significantly different at the 5% level based on the DMRT test.

From Table 5, it can be seen that the type of organic fertilizer has no significant effect on the leaf area index with the widest leaves being in the K2 treatment (goat manure), which is 39.32 cm², which is not significantly different from the K3 treatment (chicken manure), namely 39.22 cm² and K1 (compost) which is 38.96 cm².

In addition, it was known that the V2 treatment (Devon 1) produced the widest leaves, namely 43.27 cm², which was significantly different from other treatments. The narrowest leaf was in treatment V4 (Derap 1), which was 32.24 cm², which was significantly different from treatment V1 (Dena 1) which was 39.20 cm² and treatment V3 (Anjasmoro) which was 41.95 cm².

Each variety shows a significant difference in the mean value between the four varieties. The devon 1 variety has the widest leaves compared to the other three varieties. The difference in leaf area is the result of variations in leaf formation. In addition, genetic factors also affect the leaf area of the varieties planted [18, 19].

3.6. Flowering Age

The analysis of variance showed that the organic fertilizer test had no significant effect on the flowering age of soybean plants while the use of several varieties had a significant effect on the flowering age of soybean plants. The average difference test for the application of several types of organic fertilizers and soybean plant varieties on flowering can be seen in table 6.

Table 6. The flowering age of several varieties of soybean

Treatment	Organic fertilizer			Average
	K_1 (Compost)	K_2 (Goat Manure)	K_3 (Chicken Manure)	
Varieties				
V_1 (Denal)	34,33	34,75	35,08	34,72 a
V_2 (Devon 1)	30,33	30,17	30,00	30,17 b
V_3 (Anjasmoro)	33,75	34,17	34,67	34,19 a
V_4 (Derap 1)	30,33	30,08	30,17	30,19 b
Average	32,19	32,29	32,48	

Note: Figures followed by different letters in the same treatment group are significantly different at the 5% level based on the DMRT test.

From Table 6, it can be seen that several types of organic fertilizers have no significant effect on flowering age with the highest flowering age in the K3 treatment (chicken manure), which is 32.48 days, which is not significantly different from the K2 treatment (goat manure), which is 32.29 days and K1 (compost) which is 32.19 days.

Several types of varieties significantly affected the flowering age of soybean plants with the longest flowering age in treatment V1 (Dena 1), namely 34.72 days, which was not significantly different from treatment V3 (Anjasmoro), namely 34.19 days, both treatments were significantly different from treatment V4 (Derap 1) is 30.19 days and treatment V2 (Devon 1) is 30.17 days. Treatment V4 was not significantly different from treatment V2.

Derap 1 was the variety that had the shortest flowering age followed by Devon 1 variety. While Dena 1 and Anjasmoro varieties had the longest flowering age. [20] state that plants will show a ripe harvest if the total energy adopted has reached a certain level (growing degree day) and a certain level of difference is different for each plant, generally due to genetic factors. Harvest age is influenced by the speed of flowering plants. This is in line with the research of [21], that the length of the reproductive phase and the seed maturation phase for each variety is generally the same. The long life span of rice plants is caused by the long vegetative phase.

4. Conclusions

1. The type of organic fertilizer has a significant effect on the growth of soybean plants, namely plant height, number of branches, number of leaves, and amount of chlorophyll, but does not significantly affect leaf area and flowering age.
2. Varieties have a significant effect on the growth of soybean plants, namely plant height, number of branches, leaf area, and flowering age, but do not significantly affect the number of leaves and the amount of chlorophyll.
3. There is no interaction effect between the treatment of organic fertilizer types and varieties of all observed parameters.

References

[1] Pusat Data dan Sistem Informasi Pertanian 2018 Outlook komoditas pertanian tanaman pangan: Kedelai. Pusat Data dan Sistem Informasi Pertanian Sekretariat Jenderal Kementerian Pertanian.

[2] Whalen J K and Chang C 2002 Macroaggregate Characteristics in Cultivated Soils after 25 Annual Manure Applications. *Soil Sci. Soc. Am. J.* **66** 1637-1647 https://doi.org/10.2136/sssaj2002.1637

[3] Simonetti G, Francioso O, Nardi S, Berti A, Brugnoli E and Francesco Morari E L 2012 Characterization of humic carbon in soil aggregates in a long-term experiment with manure and mineral fertilization. *Soil Science Society of America Journal* **76** 880-890 https://doi.org/10.2136/sssaj2011.0243

[4] Diacono M and Montemurro F 2011 *Long-Term Effects of Organic Amendments on Soil Fertility*. In: Lichtfouse E, Hamelin M, Navarrete M, Debaeke P. (eds) Sustainable Agriculture **2**. Springer, Dordrecht. https://doi.org/10.1007/978-94-007-0394-0_34

[5] Sutanto, 2002. *Dasar Dasar Fisiologi Tumbuhan*. Rajawali Pers. Jakarta.

[6] Martinez E, Domingo F, Rosello A, Serra J, Boixadera J and Lloveras J 2016 The effects of dairy cattle manure and mineral N fertilizer on irrigated maize and soil N and organic C. *Eur. J. Agron*. (2016) http://dx.doi.org/10.1016/j.eja.2016.10.002

[7] Celik I, Gunal H, Budak M and Akpinar C 2010 Effects of long-term organic and mineral fertilizers on bulk density and penetration resistance in semi-arid Mediterranean soil conditions. *Geoderma* **160** (2010) 236–243 doi:10.1016/j.geoderma.2010.09.028

[8] Haynes R and Naidu R 1998 Influence of lime, fertilizer and manure applications on soil organic matter content and soil physical conditions: a review. *Nutrient Cycling in Agroecosystems* **51** 123–137 https://doi.org/10.1023/A:1009738307837

[9] Dwipa I, Warnita and Savitri Y 2020 Effect of mulches types use and chicken manure doses to growth and yield of shallot. *Asian Plant Research Journal* 4 31-38.

[10] Andayani and Sarido L 2013 Uji empat jenis pupuk kandang terhadap pertumbuhan dan hasil tanaman cabai keriting (*Capsicum annum* L.). *Jurnal AGRIFOR* **12** 22-28.

[11] Rizwan M, Rauf A, Rahmawaty and Akub E N 2020 Response of physiology, growth and yield of soybean varieties on fertilization treatment in agroforestry systems. *RJOAS* **6** 145-154. DOI 10.18551/rjoas.2020-06.17

[12] Nasaruddin dan Rosmawati 2010 Pengaruh pupuk organik cair hasil fermentasi daun gamal, batang pisang dan sabut kelapa terhadap pertumbuhan bibit kakao. *Jurnal Agrisistem* **7** 29 – 37.

[13] Dewanto F G, Londok J J M R dan Tuturoong R A V 2013 Pengaruh pemupukan anorganik dan organik terhadap produksi tanaman jagung sebagai sumber pakan. *J. Zootek* **32** 1-8.

[14] Karyati, T. 2004. Pengaruh penggunaan mulsa dan pemupukan urea terhadap pertumbuhan dan produksi cabai merah (*Capsicum annuum* L.). *Jurnal Penelitian Bidang Ilmu Pertanian* **2** 13-16.

[15] Amujoyegbe B J, Opabode J T and Olayinka A 2007 Effect of organic and inorganic fertilizer on yield and chlorophyll content of maize (*Zea mays* L.) and sorghum *Sorghum bicolour* (L.) Moench). *African Journal of Biotechnology* **6** 1869-1873.

[16] Ghosh P K, Ajay, Bandyopadhyay K K, Manna M C, Mandal K G, Misra A K and Hati K M 2004 Comparative effectiveness of cattle manure, poultry manure, phosphocompost and fertilizer-NPK on three cropping systems in vertisols of semi-arid tropics. II. Dry matter yield, nodulation, chlorophyll content and enzyme activity. *Bioresource Technology* **95** 85–93.

[17] Suharja 2009 Biomassa, kandungan klorofil dan nitrogen daun dua varietas cabai (*Capsicum annum*.L) pada berbagai perlakuan pemupukan. *Jurnal Bioteknologi* **6** 11-20.

[18] Muhuria, Tyas KN, Khumaida N, Trikoesoemaningtyas, Sopandie D 2006 Adaptasi tanaman kedelai terhadap intensitas cahaya rendah: Karakter daun untuk efisiensi penangkapan cahaya. *Bul. Agron.* **34** 133-140.

[19] Wirnas D, Widodo I, Sobir, Trikoesoemaningtyas, Sopandie D 2006 Pemilihan karakter agronomi untuk menyusun indeks seleksi pada 11 populasi kedelai generasi F6. *Jurnal Agronomi Indonesia* **34** 19-24.

[20] Masdar M, Karim B, Rusman N, Hakim dan Helmi 2006 Tingkat hasil dan komponen hasil sistem intensifikasi padi (SRI) tanpa pupuk organik di daerah curah hujan tinggi. *Jurnal Ilmu-Ilmu Pertanian Indonesia* **2** 126-131.
[21] Faozi, Khavid dan Rudianto B 2010 Tanggap tanaman padi sawah dari berbagai umur bibit terhadap pemupukan nitrogen. *Jurnal Agronomika* **1** 32-42.

Response of growth and yield of two soybean varieties to seed pre-treatment applications

N Herawati[1,*], A R Aisah[1], I Mardian[1], B N Hidayah[1] and B T R Erawati[1]

[1] Indonesian Agency for Agricultural Research And Development (IAARD) – Institute for Assessment of Agricultural Technology, West Nusa Tenggara Province, Indonesia.

* E-mail: nani.subhan@yahoo.com

Abstract. Plant growth is influenced by seed quality. Seeds need to be treated to prevent pest and disease disorders or to increase seed germination. Soybean planting was carried out in this study by the treatment of varieties and seed pre-treatment application before planting. The study aimed to measure the growth and yield of soybeans by varieties and seed treatment before planting, carried out in the Village of Nggembe, District of Bolo, Bima Regency. The study used factorial randomized block design with two factors. The first factor was soybean varieties consisting of two levels, namely Devon and Dena varieties, and the second factor was the application of pre-planting seed treatments consisting of four levels namely Cruiser, Marshall, Agrisoy, and without seed treatment, each treatment was repeated three times. Observations were made on the performance of agronomic crops and soybean yields. The results showed that the treatment of soybean varieties and the application of seed treatment had a significant influence on plant height and soybean productivity, and both treatments had interactions on parameters of plant height, number of filled pods, weight of 100 seeds, and soybean productivity with the best results successively produced by a combination of Devon-Control, Devon-Marshall, Devon-Agrisoy, and Dena-Cruiser.

1. Introduction

Soya is one of national strategic commodity, consistently enhanced in order to increase national production. This effort is feasible through the use of best varieties and qualified seeds. [1] stated that soybean seeds can have a rapid decrease in viability and vigour. This can lead to low seed germination rates, thereby reducing the number of plant populations. Based on [2], the decline on the soybean population number was not only caused by failed seeds but also by seed flies.

Efforts to get uniform plant growth and to prevent pests attack at the early stage of growth, including seeds treatment or seeds coating. Seeds coating is an application of chemical compound as insecticides, fungicides, growth regulators or nutrients in seeds. This application is expected to increase seeds vigour when planned on field and way of early protection to prevent pests and diseases of plant. The Impact can be traced by plant growing and result achievement [3,4]. Other than chemical compound, biological agents can also be used as substance for seeds treatment to control of seed-borne insects and pathogens[5].

Research related to seeds coating for soybean has been carried out by [6] using biological agents of *Rhizobium* sp and *Azobakter* sp., and insecticides as thiamethoxam. Insecticides with active ingredients thiamethoxam can be applied and absorbed easily, soluble, and can protect seeds equally [3]. The existence of new superior varieties of soybean and seeds coating method thus it is necessary research about the application of seeds treatment to the new superior varieties. This research aimed to measure growth and yield of soybean by treating the soybean varieties and seeds treatment before planting.

2. Methods

This research carried out in drought I (DS I) started in March to May 2019 in Village of Nggembe, District of Bolo, Regency of Bima. This research used factorial randomized block design, consists of two factors: first, seeds treatment before planting and second, varieties of soybean, each treatment repeated for three times. Factor of seeds treatment consist of four levels which are the intake of Cruiser (A), Marshal (B), Agrisoy (C) and without seed treatment as control (D),while seeds treatment have two levels namely variety of Devon I and Dena I. These two varieties obtained from Balai Penelitian Aneka Kacang dan Umbi/Balitkabi, Malang.

This research consists of planting preparation, planting and maintenance, pest and disease control, and harvest. Planting preparation carried out without tillage. The rice straws from previous crop was left on the ground, and then flattened. Before planting, seeds were given a seed coating treatment based on recommended dosage in the package. Planting were carried out by means of hole with a depth of 2-3 cm, each planting hole was filled with 2 seeds, and 15x40 cm spacing. NPK 16:16:16 was applied with dosage of 100 kg/ha and 1 litre/ha of liquid fertilizer with 20 % of active nitrobenzene. NPK (16:16:16) applied by spreading it before spreading of straw mulch, while liquid fertilizers was applied at the age of 20, 30, 45 and 60 days after planting (dap). Watering was carried out during flowering (30 dap), pods formation (45 dap), and pod filling (55 dap). Weeding was conducted 2 times, namely at 21 and 35 dap, while pest control carried out chemically. Harvesting was conducted when 75 % (at a minimum) of leaves of the plants begin to turned yellow, dry and fall out and pods were brown. Harvesting was conducted by cutting stem bases. Sample was taken at 2x5 m sampling plot.

Plants growth observation variables were carried out during vegetative and generative phases included plants height, number of branches, number of leaves, number of nodes, and number of plant per clump. Observations of yield components consisted of number of filled pods, weight of 100 seeds and productivity. Data were taken from 10 sample plants in sampling plot for each treatment replication. The data obtained were then analyzed for variance (Anova) by using SAS 9.0. The treatment that had a significant effect, then followed by Duncan Multiple Ranges Test (DMRT) at α=5%.

3. Results and discussion

3.1. Agronomic performances

3.1.1. Plant height. Based on analysis of variance, soybean varieties and seed treatment gave a significant effect on plants height. Based on single factor of variety, soybean of Devon produced the highest plant height (55.6 cm) and was significantly different from Dena (50.4 cm) (Table 1). Meanwhile for seed treatment, Cruiser treatment (A) resulted in a higher plant height (56.6 cm) and was significantly different than others. There was a significant interaction between soybean varieties and seed treatments in Dena variety using Cruiser (58.4 cm). Different response showed by variety of Devon when seed treatments did not significantly affect plants height in both control and other seed treatments. This result indicated that soybean of Devon had a better adaptation than Dena because eventhough no seeds treatment, plants height is almost the same.

Table 1. Plants height of two soybean varieties (cm) using seeds treatment in the Village of Nggembe, District of Bolo, Regency of Bima in 2019.

Treatment	Cruiser (A)	Marshal (B)	Agrisoy (C)	Control (D)	Average
Devon	54.7 ab	55.5 ab	54.4 ab	57.9 a	55.6 a
Dena	58.4 a	51.1 bc	51.1 bc	45.2 d	50.4 b
Average	56.6 a	53.3 b	50.6 b	51.6 b	

Information: same letters on collums and rows show that there is no real different result based on DMRT α=0.05

Variety of Devon had a better response in parameter of plants height. However when compared to varieties description made by [7], the value obtained was seems low. Application of insecticides resulted in a variety of plant height growth responses. Cruiser which produced the best plant height response contains the active ingredient thiamethoxam. Meanwhile, Marshal contained of active ingredient carbosulfan, and Agrisoy is biological agent consists of *Brady rhizobium*. Thiamethoxam and carbosulfan had a function to protect plants from pest attack. According [8] thiamethoxam was used to control various important pests such as aphid, flies, tryps, leaf borer and species of Lepidoptera.

Other than their function as insecticide, Cruiser also had growth regulator to stimulate plants growth. Therefore, Cruiser treatment resulted the highest growth. Based on [4] showed that Cruiser treatment to Paddy Ciherang variety produced the highest plant and significantly different from control one, especially at 31dap. Meanwhile, Agrisoy treatment which was a Rhizobium biological agent, in this case produced the lowest result even thought it was not different than Marshal treatment and control. [9] stated that commercial Rhizobium inoculation in soybean plant in the field resulted in low nodulation, while the local Rhizobium could increase nitrogen fixation and soybean yields due to its ability to adapt to local conditions.

3.1.2. Number of nodes. The variety of soybean and seed treatment had no significant effect on the number of nodes and there was no interaction between soybean varieties and seed treatment. It showed that both varieties have a good response even though no seed treatment given. [10] stated that genotype of soybean which had a large number of fertile nodes gave higher seed yields and vice versa.

Table 2. Number of nodes on two soybean varieties using the seed treatments in Village of Nggembe, District of Bolo, Regency of Bima, 2019.

Treatment	Cruiser (A)	Marshal (B)	Agrisoy (C)	Control (D)	Average
Devon	11.1	12.5	11.6	12.7	12.0
Dena	12.3	12.1	11.8	11.8	12.0
Average	11.7	12.3	11.7	12.3	

3.1.3. Number of branch. Number of branch per plant was not influenced by soybean varieties or seed treatment. Dena produced the most number of branches (3.5) in the varieties treatment, meanwhile for seed treatment, the highest value obtained by control (3.5) (table 3). This showed that both soybean varieties did not need any application of seed treatment to produce a large number of branches.

Table 3. Number of branches per plan on two varieties of soybean using seed treatment in Village of Nggembe, District of Bolo, Regency of Bima, 2019.

Treatment	Cruiser (A)	Marshal (B)	Agrisoy(C)	Control (D)	Average
Devon	2.4	3.2	2.3	3.2	2.9
Dena	3.6	3.3	3.6	3.4	3.5
Average	3.0	3.3	3.0	3.5	

Different result made by [11] showed that soybean varieties had a significant effect on the number of soybean branches, and the number of branches had a positive correlation with seed weight per plant, plant height, number of pods, number of bunches, and number of seeds.

3.2. Soybean yield component

3.2.1. Number of pods per plant. Soybean varieties and seed treatment had no significant effect on the parameter of the number of filled pods per plant. Varieties treatment produced the highest number of filled pods in variety of Dena (42.9), while the treatment of seeds was produced by the Marshal treatment (45.4). There was interaction between soybean varieties and seed treatment. Untreated soybean plants produced a number of filled pods that din not differ from those of the seed treated plants. Cruiser treatment on the Devon variety actually reduced the number of filled pods per plants. The highest value od filled pods per plant was produced by combination of Devon variety with Marshal treatment (51.5), and the lowest value resulted by the combination of Devon variety with Cruiser treatment (24.4).

Table 4. Number of pods per plant on two varieties of soybean using seed treatment in Village of Nggembe, District of Bolo, Bima Regency 2019.

Treatment	Cruiser (A)	Marshal (B)	Agrisoy (C)	Control (D)	Average
Devon	24.4 c	51.5 a	34.0 bc	44.1 ab	38.5
Dena	49.7 a	39.3 ab	42.8 ab	39.6 ab	42.9
Average	37.1	45.4	38.4	41.4	

Information: same letters given in the columns and rows shows no real different based on DMRT α= 0.05

The result obtained were in line with [12] on the Rhizobium application, which showed that Rhizobium treatment together with manure and PK chemical fertilizer into soybean plants on ultisol soil resulted in the number of filled pods per plant that was not different from control. Meanwhile in rice, [4] showed that insecticide treatment on the seeds of several rice varieties had a significant effect on the number of grains per panicle and number of filled grains per panicle.

3.2.2. Weight of 100 seeds. Soybean varieties and seeds treatment had no significant effect on the weight of 100 seeds. Based on soybean varieties, the weight of 100 seeds of Devon and Dena was not much different (15.6 and 15.7). Meanwhile, based on seeds treatment, Marshal produced the highest value (16.2 gram) (Table 5). There was interaction between soybean varieties and seeds treatment with different value found in the combination of Devon-Marshal (16.9) with Devon-Cruiser (13.8) and Dena-Control (14.4). Application of the Cruiser to the seeds of Devon variety led to a reduction in the size of soybean seed. On the other hand, the application of Cruiser to Dena variety gave a better response when compared to Devon variety.

Table 5. Weight of 100 seeds for on two soybean varieties using seed treatment in the Village of Nggembe, District of Bolo, Regency of Bima, 2019.

Treatment	Cruiser (A)	Marshal (B)	Agrisoy(C)	Control (D)	Average
Devon	13.8 c	16.9a	15.3 abc	16.4 a	15.6
Dena	16.9 a	15.4 abc	15.9 ab	14.4 bc	15.7
Average	15.3	16.2	15.6	15.4	

Information: same letters given in the columns and rows shows no real different based on DMRT α= 0.05

The weight of 100 soybean seeds determined by genetic factor because there were correlations between ability of plants to adapt to the environment in which they were grown[13]. Each soybean variety had a different seed size because it was influenced by genetic factors related to the assimilate translocation process during seed filling [14].

3.2.3. Plants population. Plants population was only affected by soybean varieties treatment, meanwhile seeds treatment had no significant effect. Based on a single variety factor, Devon produced the largest population (108.8) when ompared to Dena variety (84.1) (Table 6). Whereas, the largest plant population based on seeds treatment was produced by the control (106.8). Soybean varieties and seeds treatments had interactions on the total population of soybean plants. The best results were produced by Devon-Control combination (128) but not significantly different from the Devon-Agrisoy combination (112.3).

Table 6. Plant population on two soybean varieties using seeds treatment in Village of Nggembe,District of Bolo, Regency of Bima in 2019.

Treatment	Cruiser (A)	Marshal (B)	Agrisoy (C)	Control (D)	Average
Devon	95.0 bc	100.0 bc	112.3 a	128.0 a	108.8
Dena	91.7 bc	90.0 bcd	69.0 d	85.7 cd	84.1
Average	93.3	95.0	90.7	106.81	

Information: same letters given in the columns and rows shows no real different based on DMRT $\alpha = 0.05$

Plant population can affect the level of soybean production. [13] showed that planting with grain planting tools reduced the population by 27-33.3 %, while the planting method reduced the population by 6.9%.

3.2.4. Productivity. Soybean varieties and seeds treatments had a significant effect on soybean productivity. Based on a single variety factor, Devon had better productivity (2.08 t/ha) than Dena (1.34 t/ha) (Table 7). Meanwhile, the single factor of seed treatment showed that soybean without seeds treatment gave the highest yield (1.90 t/ha) and was different from Agrisoy treatment (1.45 ton/ha) which produced the lowest yield potential. In addition, the factors of varieties and seed treatment had interaction. The best yield was obtained by combination of Devon variety without seeds treatment (2.67 t/ha) and it was different from other combinations.

Table 7. Productivity (t/ha) of two varieties of soybean using seeds treatment in the Village of Nggembe, District of Bolo, Regency of Bima 2019.

Treatment	Cruiser (A)	Marshal (B)	Agrisoy (C)	Control (D)	Average
Devon	1.67 bc	2.06 b	1.97 bc	2.67 a	2.08 a
Dena	1.70 bc	1.53 cd	0.93 e	1.13 de	1.34 b
Average	1.70 ab	1.79 ab	1.45 b	1.90 a	

Information: same letters given in the columns and rows shows no real different based on DMRT $\alpha = 0.05$

The effectiveness of seeds treatment using biological agents depended on the type of inoculant, dosage, technic of intake and adhesive material used [3]. [6] showed that coating soybean seeds with *Rhizobium* sp. and *Azotobacter* sp. did not increase yield components and seeds yield. The differences in yield between varieties was assumed due to differences in yield. Based on description of [7], Devon 1 and Dena 1 varieties had a potential yield of 3.09 t/ha and 2.9 t/ha respectively.

Based on results above, it could be seen that seeds treatment with insecticide or even biological agent was not effective in increasing soybean productivity. Based on [3], thiamethoxam had a positive effect in protecting soybeans from attack by seed flies but it was not consistent with soybean yields. In general, seeds treatment did not affect soybean growth and yield, except for Cruiser that affected plants height.

4. Conclusion

Soybean varieties treatment had a significant effect on the plants height and productivity with the best responses produced by variety of Devon 1. Seeds treatment had a significant effect on the plant height and productivity. The highest plant height value was produced by Cruisers treatment, while the highest productivity resulted by soybean without seeds treatment.

References

[1] Giamerti Y, Yursak Z, Purwantoro 2016 Teknologi invigorasi mendukung ketersediaan benih kedelai bermutu, dalam *Pros. Sem. Nas. Hasil Penelitian Tanaman Aneka Kacang dan Umbi Tahun 2015*, 19 Mei 2015 Malang, 230-236. Bogor: Pusat Penelitian dan Pengembangan Tanaman Pangan, Badan Penelitian dan Pengembangan Pertanian.

[2] Sumadi, Suryatmana P, Sobardini D 2016 Respon benih kedelai terdeteriorasi terhadap aplikasi pelapisan benih, dalam *Pros. Sem. Nas. Hasil Penelitian Tanaman Aneka Kacang dan Umbi Tahun 2015*, 19 Mei 2015 Malang, 204-211. Bogor: Pusat Penelitian dan Pengembangan Tanaman Pangan, Badan Penelitian dan Pengembangan Pertanian.

[3] Sumadi 2015 Prospek pelapisan benih dalam meningkatkan produktivitas kedelai, Dalam *Pros. Sem. Nas. Hasil Penelitian Tanaman Aneka Kacang dan Umbi Tahun 2014*, 5 Juni 2014 Malang, 343-350. Bogor: Pusat Penelitian dan Pengembangan Tanaman Pangan, Badan Penelitian dan Pengembangan Pertanian.

[4] Mahmud Y, Nurlenawati N, Sugiarto 2011 Pengaruh macam perlakuan benih terhadap pertumbuhan dan hasil beberapa varietas unggulan baru tanaman padi (*Oryza sativa* L.) di lahan sawah irigasi Kecamatan Tempuran Kabupaten Karawang *Solusi* **9** (17): 53-63.

[5] Supriadi 2018 Inovasi perlakuan benih dan implementasinya untuk memproduksi benih bermutu tanaman rempah dan obat *Jurnal Litbang Pertanian* 71-80.

[6] Sumadi, Suryatmana P, Sobardini D 2016 Pengaruh aplikasi pelapisan benih terhadap viabilitas benih terdeteriorasi serta pertumbuhan tanaman kedelai *Jurnal Kultivasi* **15**(2): 107-13.

[7] Balai Penelitian Tanaman Aneka Kacang dan Umbi 2016 *Deskripsi Varietas Unggul Aneka Kacang Dan Umbi*. 8thed. Malang: Balai Penelitian Tanaman Aneka Kacang dan Umbi.

[8] Maienfisch P 2006 Synthesis and properties of thiamethoxam and related compounds *Z. Naturforsch* **61b**: 353-59.

[9] Yuan K, Reckling M, Ramirez M D A, Djedidi S, Fukuhara I, Ohyama T, Yokoyama T, Kimura S D B, Halwani M, Egamberdieva D, Ohtsu N O 2020 Characterizationof rhizobia for the improvement of soybean cultivation at cold condition in Central Europe *Microbe Environment* **35**(1): 1-13.

[10] Hakim L 2012 Komponen hasil dan karakter morfologi penentu hasil kedelai *Penelitian Pertanian Tanaman Pangan*, **31**(3): 173-79.

[11] Dwiputra A H, Indradewa D, Susial E T 2015 Hubungan komponen hasil dan hasil tigas belas kultivar kedelai (*Glycine max* (L.) Merr.) *Vegetalika* **4**(3): 14-28.

[12] Lestari A D L, Harsono A 2017 Pengaruh pembenahan tanah dan inokulan rhizobium terhadap hasil kedelai pada tanah ultisol B*uletin Palawija* **15**(1): 8-14.

[13] Somantri R U, Syahri, Thamrin T 2019 Potensi hasil beberapa varietas unggul baru (VUB) kedelai di lahan kering Sumatera Selatan, dalam *Pros. Sem. Nas. Lahan Suboptimal 2018*, 18-19 Oktober 2018 Palembang, 473-481. Palembang: Unsri Press.

[14] Pandiangan D N, Rasyad A 2017 Komponen hasil dan mutu biji beberapa varietas tanaman kedelai (*Glycine max* (L.) Merril) yang ditanam pada empat waktu aplikasi pupuk nitrogen *JOM Faperta* **4**(2): 1 – 14.

Drought stress and K effects on contents of soil water and organic matter, CEC, exchangeable-K, yield and water productivity of sweet corn on Inceptisols

N Nurlaeny[1,*], D Herdiyantoro[1], R M Putra[2], N Ratuliami[1], W N Pratiwi[1] and F S Nurfadilah[1]

[1] Department of Soil Science and Land Resources, Agriculture Faculty, Universitas Padjadjaran, Jl. Raya Bandung - Sumedang Km. 21 Jatinangor 45363 phone: (+62) 227796316; Fax: (+62) 22 7796316 West Java – Indonesia
[2] Department of Agronomy, Agriculture Faculty, Universitas Padjadjaran, Jl. Raya Bandung - Sumedang Km. 21 Jatinangor 45363 phone: (+62) 227796316; Fax: (+62) 22 7796316 West Java – Indonesia

* E-mail: nenny@unpad.ac.id

Abstract. The research was conducted to study the effects of drought stress and K fertilizer on plant growth, contents of soil water (SWC) and organic matter (SOM), cation exchange capacity (CEC), K-exchange, yield and water productivity of sweet corn (*Zea mays* saccharata Sturt.) grown in Inceptisols. The research was conducted on a plastic house in a controlled culture experiment field, Faculty of Agriculture, Universitas Padjadjaran using a factorial randomized block design (RBD). The first factor was the four percentages of drought stress (0%, 20%, 40% and 60%). The second factor was three levels of KCl dose (1, 1.25 and 1.50 times the recommended dose of 50 kg ha^{-1} KCl). The treatment combinations were repeated three times. The interaction between 20-40% drought stress and 1.50 times dose of KCl had a significant effect on stem diameter and independently this KCl dose affected the leaf numbers, but without drought stress. The treatment without drought stress resulted in highest plant height and leaf area index, while drought stress of 40-60% gave higher SOM and CEC, even in 60% drought stress showed a higher exchangeable K. The application of 1.25 times dose of KCl caused higher K-uptake, while the treatment of 40% drought stress gave higher water productivity.

1. Introduction

In Indonesia sweet corn (*Zea mays* saccharata Sturt L.) has developed as a type of young harvested plant that is widely cultivated in the tropical low land. However, the productivity of sweet corn in Indonesia is still relatively lower compared to other tropical countries like Brazil [1] due to the cultivation system which is constrained by environmental factors. Uneven rainfall combined with increasing global temperatures has caused a lack of water availability in dryland farming to be a limiting factor for growth. As a consequent, drastically reduced crop production is occurred by low soil productivity. Highlighting the role of water and fertilizer management for sustainable agriculture therefore must be of particular concern.

Among all the environmental factors, drought is the most important environmental factor which affected growth and development of plants [2]. The availability of plant nutrients which are closely related to soil water content indicates that water is a very influential factor [3]. Therefore, in terrestrial ecosystems, poor plant growth during the severe dry season still can be nourished when soil moisture increases due to water content fluctuations in the soil. Thus, water is a key driver of several ecosystems functions, including biomass and crop yields.

Several studies have shown that the detrimental effects of drought can be mediated by applying a slightly excessive dose of nutrients to mediate this adverse effect, thereby improve drought resistance of plants [4]. As one of the essential macronutrients, Potassium (K) has the potential to develop plant resistance to drought. When K supply is limited, plant growth will be disrupted, especially when there is high light intensity accompanied by the appearance of leaf chlorosis and necrosis. In conditions of minimal soil moisture, K is assumed to play an important role in the plant metabolism process, the regulation of guard cells, flow of water, nutrients, and carbohydrates within the plants, where the opening or closing of stomata depends on the availability of mesophyll water [5]. Therefore, this study aimed to observe the effects of drought stress combined with the application of various doses of KCl on plant growth components, soil physicochemical properties, yield and water productivity of sweet corn (*Zea mays* saccharata Sturt), as well as to know the relationship between net use of consumptive water in increasing the productivity of sweet corn under various drought stress conditions.

2. Methods

2.1. Experiment site

The research was done under a plastic house at the controlled culture experimental field, Faculty of Agriculture Universitas Padjadjaran used factorial Randomized Block Design (RBD). The first factor was the percentage of drought stress which comprised four levels (0%, 20%, 40% and 60% of drought stress), which were equivalent to 100% of soil water availability (SWA) and considered as a control, followed by 80%, 60% and 40%. The second factor was the KCl dose comprised three levels (1, 1.25 and 1.50 times the recommended dose of 50 kg ha^{-1} KCl). The treatment combinations were repeated three times.

The soil was derived from the area around the experimental field. Each polybag containing 37 kg of soil was planted with one seed of sweet corn variety Bonanza F-1 and prepared for the observation of growth parameters in the vegetative and generative stages. The basal rate of compost at 10 t ha^{-1} and recommended doses of N, P fertilizers (300 kg ha^{-1} Urea and 150 kg ha^{-1} SP-36, respectively) were mixed with the soil before planting.

To support the plant growth, the difference between field capacity and the wilting point was considered as the amount of soil water availability (SWA). The Gravimetric method was used before daily irrigation was done by giving an amount of water into each polybag. The amount of water supplied is measured using a cylinder and beaker glass. The formula for water supply was as follow:

$$Irrigation\ water\ (L) = \frac{Wfc\text{-}W/\rho w}{1\text{-}1F} \tag{1}$$

where Wfc = soil weight in each polybag (kg) at field capacity; W = weight of soil before watering (kg); ρw = density of water used (1.0 kg L^{-1}); 1F = the fraction of leaching (lF = 0.20) [6].

To measure the index of leaf area the method used was as follows [7]:

$$A = KLB \tag{2}$$

where A = area of leaf (cm^2), L = length of leaf (cm), B = breath at mid-point, K = factor of reduction for the plant under observation. Along the plant growth stages, the maximum evapotranspiration (ETm) was calculated by the modified equations of [8] and [9]:

$$ET_m = K_c F_c \tag{3}$$

where ET_m = maximum evapotranspiration of the plant (mm), K_c = coefficient of consumptive plant use, Fc = factor of consumptive use which recapitulated from the Hygrothermograph tool and meteorological data. Along the plant growth stages, the net consumptive water use was recorded and related to the mean number of days after planting. The suitability of plant growth phases was calculated by multiplying the period of net consumptive use by the number of days along with the plant growth. To record crop production, water productivity (WP) is calculated from the net amount of water consumed by sweet corn plants per gram of yield in each polybag. The WP was calculated from the total water applied to each treatment (L polybag^{-1}) and the weight of total yield (g plant^{-1}) as shown by [11]:

$$WP\ (L\ g^{-1}\ polybag^{-1}) = \frac{Total\ water\ applied\ (L^{-1}\ polybag^{-1})}{Yield\ (g\ plant^{-1})} \quad (4)$$

2.2. Laboratory analysis

Plant and soil sampling was carried out at the end of the vegetative stages (56 DAS). Soil physicochemical properties were analysed in the laboratory after the soil sample was air-dried and sieved with a size of 2 mm. Different methods adopted for soil properties determinations are listed in Table 1. To calculate the percentage of SOM from organic C used the conversion factor of 1.72 [10].

2.3. Statistical analysis

Data analysis was performed using statistical software packages SPSS v 27.0 and Microsoft Office Excel 2019. Fisher's test at 5% probability [12] was used to determine the mean value with the least significant difference. To determine the relationship between total water supply and crop water production used correlation analysis and significance was tested at 5% probability.

3. Results

3.1. Soil physicochemical characteristics

Initial analysis of soil used in this study was classified as Inceptisols which characterized by the high percentage of clay content (67 %), acidic soil pH (5.40), low contents of total N, available P and exchangeable K^+ (0.20%, 5.61 mg kg^{-1}, 0.22 cmol kg^{-1} respectively) (Table 1). Average daily temperature and relative humidity in the plastic house during the experiment ranged between 24.4 -36.9 °C and 43.0 – 75.5%, respectively.

Table 1. Some characteristics of the soil used in the experiment

Parameter and Analysis methods	Values	*Criteria[13]
pH H_2O (Potentiometric)	5.40	acidic
Total N Kjeldahl [14]	0.20%	low
Organic-C [15]	2.73%	medium
C/N	16	high
Available P [16]	5.61 mg kg^{-1}	low
Cation Exchange Capacity: [17]	17.37 cmol kg^{-1}	medium
K^+	0.22 cmol kg^{-1}	low
Na^+	0.05 cmol kg^{-1}	very low
Ca^{2+}	7.66 cmol kg^{-1}	medium
Mg^{2+}	3.35 cmol kg^{-1}	high
Base saturation	34.14%	medium
Al^{3+}	0.13 cmol kg^{-1}	very low
H^+	0.07 cmol kg^{-1}	low
Texture [18]		
Sand (%)	10	
Silt (%)	23	
Clay (%)	67	clay

IOP Conf. Series: Earth and Environmental Science **883** (2021) 012053 doi:10.1088/1755-1315/883/1/012053

Interaction between 20-40% drought stress and the application of 1.50 times the recommended dose of KCl at 28 DAP significantly resulted in the highest stem diameter of sweet corn compared to other treatments (Table 2). However, in conditions without drought stress (100% SWA) the application of 1 time recommended dose of KCl produced the highest number of sweet corn leaves.

Table 2. Interaction between drought stress and KCl combination on stem diameter and leaves number of sweet corn at vegetative stage (28 DAP*)

Treatments	Stem diameter (cm)			Leaves number		
	Dose of KCl			Dose of KCl		
	k_1	k_2	k_3	k_1	k_2	k_3
w_1: 0% drought stress (100% SWA)	1.47 a A	1.04 a A	1.44 b A	7.33 b A	6.83 a A	6.67 a A
w_2: 20% drought stress (80% SWA)	1.07 a AB	1.01 a A	1.61 b B	6.33 a A	6.50 a A	6.70 a A
w_3: 40% drought stress (60% SWA)	1.18 a A	1.44 c AB	1.49 b B	6.00 a A	6.31 a A	6.64 a A
w_4: 60% drought stress (40% SWA)	0.99 a A	1.24 b A	1.02 a A	6.00 a A	6.33 a A	6.61 a A

Based on Duncan's test at α .05, the numbers followed by the same letter are not significantly different. Horizontally (line) = capital letters; vertically (column) = lowercase *DAP: days after planting; SWA: soil water availability; k_1: 1time the recommended dose of KCl; k_2: 1.25 times the recommended dose of KCl; k_3: 1.50 times the recommended dose of KCl.

The treatment without drought stress (100% SWA) significantly gave higher plant height and leaf area index of sweet corn along the vegetative stages (42-56 DAP) (Table 3). With an increase in the percentage of drought stress (20-60%), it was shown that there was a tendency to decline in the plant height of sweet corn. Meanwhile, the application of various doses of KCl did not affect plant height and leaf area index during the vegetative stage.

Table 3. Effects of drought stress and KCl combination on plant height, and leaf area index of sweet corn along the vegetative stages (42-56 DAP*)

Treatments	Plant height (cm) 42 DAP	Plant height (cm) 56 DAP	Leaf area Index (LAI)
w_1: 0% drought stress (100% SWA)	135.36 c	167.60 c	3.27 c
w_2: 20% drought stress (80% SWA)	112.81 b	129.65 b	2.69 b
w_3: 40% drought stress (60% SWA)	105.60 b	117.54 b	2.49 a
w_4: 60% drought stress (40% SWA)	85.07 a	91.12 a	2.23 a
k_1 : 1 time the recommended dose of KCl	104.83 a	120.77 a	2.25 a
k_2 : 1.25 time the recommended dose of KCl	105.44 a	126.30 a	2.22 a
k_3 : 1.50 time the recommended dose of	118.81 a	132.36 a	2.33 a

KCl				

Based on Duncan's test at α .05, the numbers followed by the same letter are not significantly different.
*DAP: days after planting; SWA: soil water availability

Neither the drought stress nor the KCl treatments had a significant effect on the soil water content (SWC) at 56 DAP, however, significantly higher SOM and CEC levels were found in the 40-60% drought stress treatment (Table 4). The high clay content of this soil type (67%, Table 1) together with higher SOM and CEC values was likely to support the higher concentration of exchangeable K, especially in drought stress of 60% (40% SAW). A higher concentration of exchangeable K also resulted due to the application of the 1 time recommended dose of KCl, whereas increasing the recommended dose of KCl (1.25-1.50 times) did not have any significant effect on SOM and CEC values.

Table 4. Effects of drought stress and KCl combination on soil water content, soil organic matter, cationexchange capacity and exchangeable K at 56 DAP*

Treatments	SWC (%)	SOM (%)	CEC (cmol kg^{-1})	Exchangeable-K (cmol kg^{-1})
w_1: 0% drought stress (100% SWA)	16.93 a	4.84 a	17.88 a	0.36 a
w_2: 20% drought stress (80% SWA)	16.96 a	5.14 ab	18.43 ab	0.52 b
w_3: 40% drought stress (60% SWA)	16.20 a	5.29 b	18.83 b	0.54 b
w_4: 60% drought stress (40% SWA)	15.68 a	5.30 b	18.95 b	0.65 c
k_1: 1 time the recommended dose of KCl	16.59 a	5.02 a	18.50 a	0.59 b
k_2: 1.25 time the recommended dose of KCl	16.40 a	5.05 a	18.17 a	0.50 a
k_3: 1.5 time the recommended dose of KCl	16.33 a	5.06 a	18.90 a	0.49 a

Based on Duncan's test at α .05, the numbers followed by the same letter are not significantly different.
*DAP: days after planting; SWA: soil water availability; SWC: soil water content; SOM: soil organic matter; CEC: cation exchange capacity

Independently, the statistical analysis shows that without drought stress (100% SWA) with 84.87 L of net consumptive water use caused significantly higher K-uptake, cobs diameter and total yield weight of sweet corn. However, the higher water productivity (0.47 L^{-1} g^{-1} $plant^{-1}$) was caused by the treatment of 40% drought stress. Application of various recommended dose of KCl did not have any significant effect on yield component and water productivity, except the application of 1.25 times the recommended dose of KCl significantly caused higher K-uptake sweet corn (Table 5). The relationship between total net consumptive water use and crop water productivity of sweet corn is shown in Figure 1.

IOP Conf. Series: Earth and Environmental Science **883** (2021) 012053 doi:10.1088/1755-1315/883/1/012053

Table 5. Effects of drought stress and KCl combinationon K-uptake, yield component and water productivity of sweet corn at 78 DAP*

Treatments	Net consumptive water use (L plant^{-1})	K-uptake (%)	Yield components: Diameter of corn cobs (cm plant^{-1})	Yield components: Total yield weight of sweet corn (g plant^{-1})	Water productivity (L g^{-1} plant^{-1})
w_1: 0% drought stress (100% SWA)	84.87	1.35 b	4.25 c	319.30 c	0.27 a
w_2: 20% drought stress (80% SWA)	61.62	1.27 ab	4.00 b	211.89 b	0.29 a
w_3: 40% drought stress (60% SWA)	47.19	1.20 a	3.61 a	100.22 a	0.47 b
w_4: 60% drought stress (40% SWA)	36.74	1.14 a	2.97 a	83.78 a	0.44 ab
k_1: 1 time the recommended dose of KCl	54.99	1.14 a	3.55 a	186.25 a	0.30 a
k_2: 1.25 times the recommended dose of KCl	52.39	1.36 b	3.27 a	181.33 a	0.29 a
k_3: 1.50 times the recommended dose of KCl	53.48	1.20 a	3.54 a	169.58 a	0.32 a

Based on Duncan's test at α .05, the numbers followed by the same letter are not significantly different.
*DAP: days after planting; SWA: soil water availability

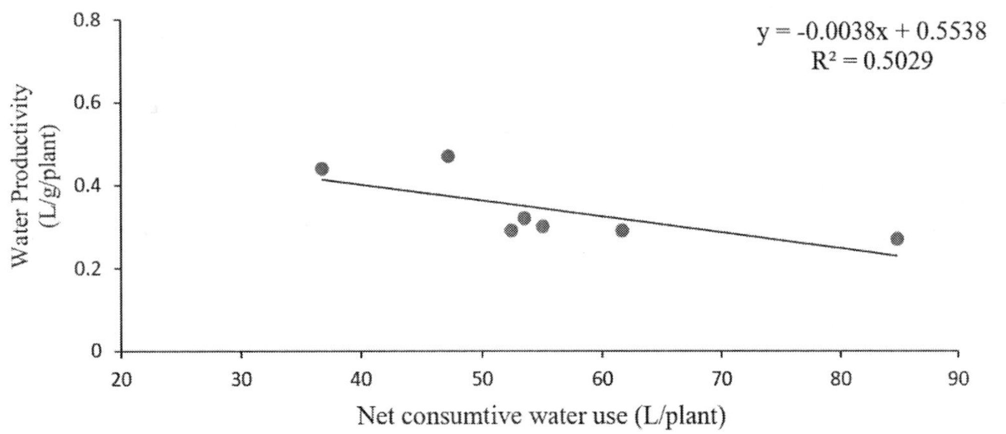

Figure 1. Relationship between total net consumptive water use and water productivity of sweet corn

4. Discussion

Environmental conditions in the plastic house and soil physicochemical properties (Table 1) used in this study supported the growth of sweet corn plants. The average daily temperature (24.4-36.9 °C), relative humidity (43.0–75.5%) and light intensities directly above the tallest canopy (11-12 μmol m^2 s) were found to be adequate for vigorous plant growth [19].

Interaction between drought stress (20-40%) and K fertilizer (1.50 times the recommended dose) at 28 DAP (Table 2) indicated that higher supply of K at soil water availability of 60-80% could support the development of stem diameter due to K function in strengthening the plant cell walls, although the early stage of sweet corn seedling growth was very sensitive to drought. According to [20] the status of water in plants greatly affected stem diameter and if the water supply by the roots was lower than the transpiration from the leaves, the stem diameter can be reduced due to water pressure. This indicates that in supporting maximum plant growth, optimal soil water availability at the vegetative stages must be fulfilled.

The results of this study also revealed that the values of plant height and leaf area index of sweet corn in the vegetative phase were higher (Table 3) due to the treatment without drought stress (100% SWA). In accordance with plant development, water requirements along the growing period of sweet corn were very different, although at the vegetative stage the plant was relatively tolerant of water stress, but very sensitive in the generative phase [21]. It was known that water is used by plants for cell division and enlargement, which was manifested in an increase in plant height and leaf propagation [22]. [23] and [24] stated that decreasing plant canopy during drought caused a decrease in plant water loss, while a decrease in plant height, diameter and weight of sweet corn cobs as a result of stomata closure and high stomatal resistance. Therefore, the barriers to stem elongation, increase in stem diameter and the formation of new leaves (Table 2) and leaf area under drought stress treatment (Table 3) was mainly due to a decrease in plant water content as an inhibitor of plant growth.

According to [25] the number and average leaf area of plants were reduced due to drought stress, which reduced the total leaf area of a plant because enlargement of the leaf size depends on leaf turgidity, temperature and assimilate supply for growth. This study supposed that the lack of water supply at the vegetative growth stage made the leaf area index of sweet corn smaller as mentioned by [26] that when the water content of the plant was limited, less carbon was utilized during the growing period of the plant. The application of various KCl doses did not affect plant height and leaf area index at all presumably because K^+ which was absorbed from the soil will be distributed evenly to all plant organs of sweet corn. Furthermore, [27] found that the distributed K^+ was then converted through metabolic processes in plants to become seeds and husks of sweet corn.

Higher SOM content and CEC levels in the drought stress treatment of 40-60% (Table 4) indicated that the high content of clay in the soil (Table 1) played an important role in maintaining the soil water capacity. [28] mention that fine-textured soil will have a larger amount of surface area and will hold more water as well as cations due to the negative charge on the clay surface. According to [29] soil organic matter played an important role to improve CEC of soils. Lower SOM content in the treatment without drought stress has been recognized as an ineffective condition of improving soil physicochemical conditions, as well as water-holding capacity. It means that the contents of clay and organic matter in this soil studied can be a good indication to assess the soil concentration of exchangeable K in terms of support for plant growth. Meanwhile, the application of 1 time recommended dose of KCl resulted in a higher exchangeable K. It was known that K which came from fertilizer would move to saturate the adsorption complex so that equilibrium with K in the soil solution was achieved [30]. The major contribution of K^+ on CEC might be the cause of a directly proportional relationship between soil organic matter and extractable- K.

The treatment without drought stress (100% SWA) with the amount of 84.87 L of consumptive water resulted in higher K-uptake, cob diameter, and total yield weight of sweet corn significantly (Table 5). Although plant height decreased due to 20-60% drought stress (Table 3), this study found that the cob diameter and yield weight of sweet corn only decreased under severe drought stress (40-60%) compared to without drought stress (100% SWA, Table 5). Surprisingly, the higher water productivity (0.47 L^{-1} g^{-1} $plant^{-1}$) was found due to the treatment of 40% drought stress. This shows that the optimal level of soil water availability at the vegetative stage played an important role in supporting maximum growth continously. On the other hand, the application of various recommended doses of KCl did not significantly affect the yield component, but the uptake of K wa significantly affected due to the application of 1.25 times of KCl (Table 5). According to [31] the availability of water in the soil greatly determined the absorption and transportation of K^+ from the soil to the plant and was involved in various processes of cell regulation, stomatal movement and water transportation [32].

5. Conclusion

The results of this study concluded that plant water productivity (WP) and drought tolerance could be improved through genetic engineering and cultivation techniques (for example, balanced fertilization) so that water use could be utilized effectively and efficiently. There is a relationship between moderate levels of productivity and high yields in the effectiveness of water use.

6. References

[1] Teixeira F F, Pacs M C D, Gama E E G, Filho I A P. Miranda R A, Guimaraes P E O, Parentoni S N, Cotta L V Meirelles W F, Pacheco C S P, Guimaraes L J M, da Silva A R and Machado J R A 2014 BRS Vivi: single-cross super sweet corn hybrid *Crop breeding and Applied Biotechnology* **14** 124-127.

[2] Aslam M, Zamir I, Afzal I, Yaseen M, Mubeen M and Shoaib A 2013 Drought stress, its effect on maize production and development of drought tolerance through potassium application *Cercetari Agronomice În Moldov* **46** 99–114

[3] Clark J S, Campbell J H, Grizzle H, Acosta M V and Zak J C 2009 Soil microbial community response to drought and precipitation variability in the Chihuahuan Desert *Microb. ecol.* **57** 248-260

[4] Raza M A S, Farrukh S M, Mustafa S G, Jamil M and Haider K I 2013 Potassium applied under drought improves physiological and nutrient uptake performances of wheat *(Triticum Aestivun* L.) *J. Soil Sci. Plant Nutr.* **13** 1-8

[5] Anosheh H P, Armin S M, Pakniyat H and Pessarakli M 2016 Stomatal responses to drought stress *Water Stress and Crop Plants: A Sustainable Approach* Volume 1 First Edition. John Wiley & Sons, Ltd.

[6] Bennet D R, Ted E, Harms T E and Entz T 2014 Net irrigation water requirements for major irrigated crops with variation in evaporative demand and precipitation in southern Alberta *Canadian Water Resources Journal* **39** 63-72

[7] Shih S F and Gascho G J 1990 Relationship among stalk length, leaf area and dry biomass of sugarcane *Agron. J.* **72** 309-313

[8] Blaney H F and Criddle W D 1962 Determining consumptive use and irrigation water requirements USDA Technical Bulletin 1275 Beltsville, MD: US Department of Agriculture

[9] Allen R G, Pereira L S, Raes D and Smith M 1998 Crop evapotranspiration: Guidelines for computing crop water requirements. *Irrigation and Drainage Paper 56* Rome, Italy: Food and Agricultural Organization of the United Nations

[10] Ahn C and Jones S 2013 Assessing Organic Matter and Organic Carbon Contents in Soils of Created Mitigation Wetlands in Virginia *Environmental Engineering Research* **18**(3) 151-156

[11] Molden D, Oweis T, Steduto P, Bindraban P, Hanjra M A and Kijne J 2010 Improving agricultural water productivity: Between optimism and caution. *Agric. Water Manag.* **97** 528–35

[12] Gomez K A and Gomez A A 1984 Statistical procedures for agricultural research (2nd ed.) John Wiley and Sons New York 704p ISBN: 978-0-471-87092-0

[13] Indonesian Soil Research Institute (ISRI) 2015 http://en.litbang.pertanian.go.id/unker/one/662/

[14] Bremner J M and Mulvaney C S 1982 Nitrogen total. In: *Methods of soil analysis. Agron. No. 9. Part 2: Chemical and microbiological properties. 2nd ed.* (A. L. Page. ed.) *Am. Soc. Agron*. Madison WI USA 595 – 624

[15] Walkley A J and Black I A 1934 Estimation of soil organic carbon by the chromic acid titration method. *Soil Sci.* **37**: 29-38
[16] Bray R H and Kurtz L T 1945 Determination of total, organic, and available forms of phosphorus in soils *Soil Science* **59** 39-45.
[17] Jackson M L 1973 *Soil chemical analysis* Prentice Hall of India Pvt. Ltd. New Delhi
[18] Berreta A, Silbermann A V, PaladinoL and Lamohte A G 2014 Soil texture analyses using a hydrometer: modification of the Bouyoucos method *Ciencia e investigación agraria* **41**(2)263-271
[19] Eddy R and Hahn D T 2010 Optimizing Greenhouse Corn Productivity: What is the best lighting and plant density https://docs.lib.purdue.edu/cgi/viewcontent.cgi?article=1005&context=pmcg
[20] Ohashi Y, Nakayama N, Saneoka H and Fujita K 2006 Effects of drought stress on photosynthetic gas exchange, chlorophyll fluorescence and stem diameter of soybean plants. *Biol. Plant* **50** (1)138–141
[21] Fischer K S and Palmer F E 1984 Tropical Maize. In *The Physiology of Tropical Field Crops*; Goldsworthy P R, Fischer N M Eds. Wiley: New York NY USA 213–248
[22] Xue R, Shen Y and Marschner P 2017 Soil water content during and after plant growth influence nutrient availability and microbial biomass. *J. Soil Sci. Plant Nutr.* **17**(3) 203-213
[23] Nemeskéri E, Molnár K and Dobos A C S 2017 Csemegekukorica (*Zea mays* L. convar. saccharata) sztóma működése, és hatása a növekedésre és terméskomponensekre eltérő vízellátás alatt [Stomatal behaviour and its influence on the growing and yield components of sweet corn (*Zea mays* L. convar. saccharata]. *Növénytermelés* **66** 75–95
[24] Burton A J, Pregitzer K S, Zogg G P and Zak D R 1998 Drought reduces root respiration in sugar maple forests. *J. App. Ecol* **8**:771–778
[25] Anjum S A, Xie Y X, Wang L C, Salem M F, Man C and Lei W 2011 Morphological Physiological and Biochemical Responses of Plants to Drought Stress *African J. of Agric. Res.* **6**(9) 2026 – 2032
[26] Nilsen E T and Orcutt D M 1996 The physiology of plants under stress: Abiotic Factors U S John Wiley and Sons Inc. 279-357
[27] Khan A A, Hussain A, Ganai M A, Sofi N R and Hussain S T 2018 Yield, nutrient uptake and quality of sweet corn as influenced by transplanting dates and nitrogen levels. *Journal of Pharmacognosy and Phytochemistry* **7**:3567-3571
[28] de Jonge L W, Moldrup P Jacobsen O H and Rolston D E 1996 Relations Between Specific Surface Area and Soil Physical and Chemical Properties Soil Science **161** (1) 9-21
[29] Patel P L, Patel N P, Patel P H and Gharekhan A 2014 Correlation Study of Soil Parameters of Kutch district Agriculture Land. *International Journal of Scientific and Research Publications* **4**(5):1-5.
[30] Singh R P and Mishra S K 2012 Available Macronutrients in the soils of Chiragaon block of district Vanarasi in relations to soil characteristics. *Indian J. Sci. Res.* **3**(1)97-100
[31] Oliveira E M M, Ruiz H A, Alvarez V H, Ferreira P A, Costa F O and Almeida I C C 2010 Nutrient supply by mass flow and diffusion to maize plants in response to soil aggregate size and water potential. *Rev. Brasil. Ciên. Solo* **34**:317–327
[32] Raza M AS, Saleem M F, Shah G M, Khan I H and Raza A 2014. Exogenous application of glycinebetaine and potassium for improving water relations and grain yield of wheat under drought *J. Soil. Sci. Plant Nutr* **14** 348–364

Effect of urea in steamed sago waste on rumen fermentation parameters in vitro tested

N C Tiven[1,*] and T M Simanjorang[2]

[1] Department of Animal Husbandry, Agriculture Faculty, University of Pattimura Jln. Ir. M. Putuhena Kampus Unpatti-Poka, Ambon - Indonesia.
[2] Department of Agricultural Socio-Economics, Agriculture Faculty, University of Pattimura Jln. Ir. M. Putuhena Kampus Unpatti-Poka, Ambon - Indonesia.

* Email: nafly_tiven@yahoo.co.id

Abstract. This study aims to determine urea's effect in steamed sago waste on rumen fermentation parameters in vitro testing. Sago waste was dried for two days, discarded fibre sticks, steamed for 30 minutes, cooled and dried. Weighed 250 g of steamed sago waste, added urea with levels of 0%, 2%, 4% and 6%. Weighed 0.5 g of samples per treatments, inserted in fermentor tubes, added 10 ml buffer and 10 ml of rumen fluid (1:1). Fermentor tubes are inserted in waterbath with a temperature of 39°C, flowed with CO_2 gas and covered with a valved rubber cover. For NH_3 and VFA testing, incubation was carried out for 4 hours, while for DMD and OMD testing, incubation was carried out for 48 hours. The data obtained were analyzed using a completely random design with four urea level treatments (0%, 2%, 4%, 6%), with five replications. The results showed that increase of urea level up to 6%, increasing ($P<0.01$) NH_3. The increase of urea level 2%, increasing ($P<0.01$) VFA, DMD and OMD. The increase of urea levels 4% and 6%, not significant effect on DMD and OMD, while at urea level 6%, decreasing ($P<0.01$) VFA. It can be concluded that urea was added in steamed sago waste, have an optimal effect on the rumen fermentation parameters at level 2–4%.
Keywords: Urea, Steamed sago waste, Fermentation parameters.

1. Introduction

The feed is an important factor in the maintenance and productivity of ruminants [1], so that feed must be available in quantity, quality and continuity. One way to solve this problem is by utilizing industrial waste and processing agricultural products.

One of the byproducts of agricultural product processing that is quite available in Maluku is sago waste, which are the residual products of sago flour processing. In the processing of sago flour, sago waste is usually produced six times that of sago flour, so that sago waste can be obtained as much as 2.7 million tons/year [2]. The limiting factor for the use of sago waste in livestock rations is the high content of crude fibre (especially lignin) which reduces digestibility, due to the presence of lignocellulose bonds that inhibit the penetration of fibre digesting enzymes from rumen microbes. Sago waste contains residues of lignin and cellulose, respectively 21% and 20% [3], containing NDF (Neutral Detergent Fiber) and ADF (Acid Detergent Fiber), respectively 57.27% and 49.54% [4]. In addition, the crude protein content in sago waste is also very low. Crude protein in Tuni sago was 0.92%, while in Ihur sago was 1.01% [5].

To improve the digestibility of the sago waste, try to do the steaming with the aim of loosening the lignocellulose bonds so as to facilitate the penetration of the fibre digestive enzymes from rumen microbial, so that digestibility can be increased. To increase the protein content of sago waste, try to do the adding urea because urea contains 46.7% Nitrogen [6], which is equivalent to 291.88% crude protein. Urea is a source of non-protein nitrogen (NPN) which is most often used as a substitute for true protein feed because it can reduce the cost of animal feed [7].

Urease enzymes will convert urea in the rumen from rumen microbes to CO_2 and NH_3, which are very important for the forming microbial protein if a carbon framework is available from carbohydrate sources. Excessive addition of urea in the ration can cause poisoning in the absence of a source of carbohydrates as the carbon framework. Steaming sago waste is expected to increase the release of the carbon framework, so that urea can be given in large quantities without causing poisoning because the production of NH_3 will be in line with the availability of the carbon skeleton so that protein synthesis and rumen microbial growth are run well so that ration digestibility and protein formation microbes can be increased.

This study aims to determine the effect of urea level, which added in steamed sago waste on rumen fermentation parameters in vitro tested. Through the results of this study, will be obtained a good balance between urea and steamed sago waste so that it becomes the basis for making concentrate feed based on steamed sago waste, which was applied in vivo.

2. Methods

Sago waste was obtained from a processing place for sago flour, which was dried in the sun for two days, then removed the fibre sticks. The sago waste is steamed for 30 minutes, then cooled and dried in the sun. Weighed as much as 250 g of sago waste then added urea according to the treatment level, namely 0%, 2%, 4% and 6%. The rumen fluid was taken from 2 bulls slaughtered at the Slaughterhouse, which was filtered through the cloth into a hot water flask, then added solid CO_2 so that the atmosphere becomes anaerobic.

2.1. Incubation 4 hours

Weighed as much as 0.5 g of the sample, put it in the fermenter tube, then added 10 mL of rumen fluid and 10 mL of Mc buffer. Dougall solution with a pH of 6.9 (1:1). The fermenter tube was supplied with CO_2 gas and covered with a valved rubber cover, and then it was put into a waterbath at 39 °C and incubated for 4 hours. The fermentation process was stopped by adding 0.2 mL of saturated $HgCl_2$ to kill microbes. This fermentative product was centrifuged at 6,000 rpm for 30 minutes to separate the precipitate and the supernatant. The Supernatant was taken to be tested for NH_3 concentration and total VFA. The NH_3 concentration was tested using the micro diffuse Conway technique, while the total VFA was tested by the steam distillation technique [8]. The concentration of NH_3 can be calculated by the formula:

$$\text{mM N-NH}_3 : \frac{\text{vol. titrant (mL)} \times \text{N H}_2\text{SO}_4 \times 1000}{\text{vol. supernatant (mL)}}$$

The concentration of total VFA can be calculated by the formula:

$$\text{mM T-VFA} : \frac{[\text{vol. titer blanco (mL)} - \text{vol. titer sample (mL)}] \times \text{N HCl} \times 1000}{\text{vol. supernatant (mL)}}$$

2.2. Incubation 48 hours

Weighed as much as 0.5 g of the sample, put it in the fermenter tube, then added 8 mL of rumen fluid and 12 mL off buffer Mc. Dougall solution with pH of 6.9 (1:1.5). The fermenter tube was supplied with CO_2 gas and covered with a valved rubber cover, and then it was put into a waterbath at 39°C and incubated for 24 hours. The fermentation process was stopped by adding 0.2 mL of saturated $HgCl_2$ to kill microbes. This fermentative product was centrifuged at 6.000 rpm for 30 minutes to separate the precipitate and the supernatant. The precipitate was taken and incubated with 20 mL of pepsin 0.2 percent solution for 24 hours aerobically. The fermentation product was filtered with Whatman filter paper number 41 with the aid of a vacuum pump and washed with 25 mL hot water. The filter and filter paper was put into a porcelain dish to be analyzed for dry matter digestibility (DMD) and organic matter digestibility (OMD) [8].

For the DMD test, the residue in a porcelain cup was evaporated in an electric oven at 105°C for 24 hours until its weight was constant. The percentage of DMD can be calculated using the formula:

$$\% \text{ DMD: } \frac{\text{DM sample - (DM residue - DM Blanco)}}{\text{DM sample}} \times 100$$

For the (OMD) test, the residue in the porcelain cup was ashed in an electric furnace at 650°C for 8 hours. The percentage of OMD can be calculated using the formula:

$$\% \text{ OMD: } \frac{\text{OM sample - (OM residue - OM Blanco)}}{\text{OM sample}} \times 100$$

2.3. Statistic Analysis

The data obtained were analyzed variance with a completely randomized design, 4 treatments, namely 0% (steamed sago waste + 0% urea); 2% (steamed sago waste + 2% urea); 4% (steamed sago waste + 4% urea); and 6% (steamed sago waste + 6% urea), each with 5 replications. Significant treatment was tested further by Duncan's New Multiple Range Test [9].

3. Results and discussion

3.1. Effect of urea levels in feed with steamed sago waste on NH_3 concentration

The effect of urea level in feed with steamed sago waste on the NH_3 concentration can be seen in Figure 1.

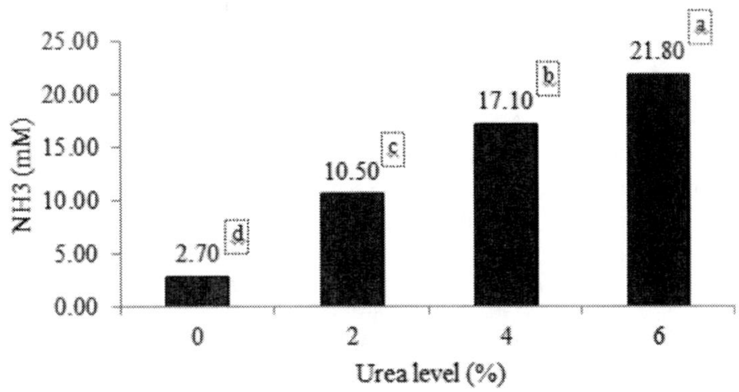

a,b,c,d: significant (P<0,01).

Figure 1. The average of NH_3 concentration (mM) due to the use of urea with different levels in feed with steamed sago waste, in vitro tested.

The results of statistical analysis showed that the addition of urea with different levels in steamed sago waste feed had a significant effect ($P<0.01$) on the NH_3 concentration. These results indicate that changes in the concentration of NH_3 follow the urea level added in steamed sago waste, meaning that the higher the added of urea in the steamed sago waste will increase the NH_3 concentration. This is caused by urea is very easily broken down in the rumen by urease enzyme produced by rumen microbes which convert the available urea to NH_3 continuously in the rumen. According to Kardaya *et al.* [10], urea in the rumen was rapidly hydrolyzed into NH_3 by rumen microbes, then most of it was absorbed rapidly in the bloodstream system, which can have a negative impact on livestock, including decreased consumption and production performance, and death due to poisoning of urea. This shows that the change of N sources (including urea) to NH_3 in the rumen takes place continuously, even though NH_3 has been sufficient or excessive so that the NH_3 concentration continues to increase.

IOP Conf. Series: Earth and Environmental Science **883** (2021) 012054 doi:10.1088/1755-1315/883/1/012054

Excess of NH_3 will be absorbed in the rumen, then converted into urea by the liver, which is finally filtered out by the kidneys and excreted in the urine, so urea breaking down rapidly [11].

3.2. Effect of urea levels in feed with steamed sago waste on T-VFA concentration
The effect of urea level in feed with steamed sago waste on the T-VFA concentration can be seen in Figure 2.

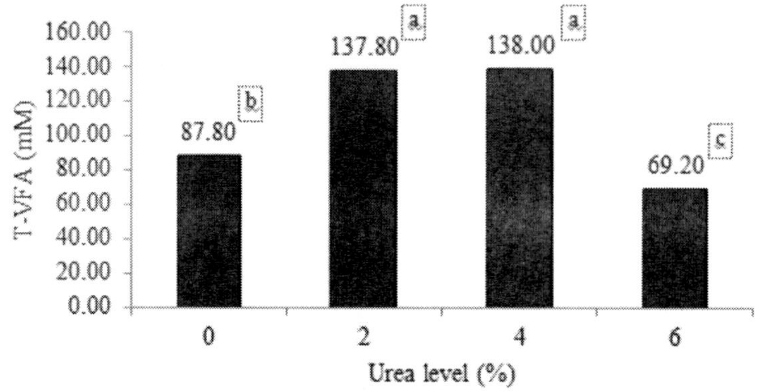

a,b,c: significant (P<0,01).

Figure 2. The average of T-VFA concentration (mM) due to the use of urea with different levels in feed with steamed sago waste, in vitro tested.

The results of statistical analysis showed that the addition of urea with different levels in steamed sago waste feed had a significant effect ($P<0.01$) on the T-VFA concentration. When compared with the un-addition of urea (0%), the addition of urea 2–4% in the steamed sago waste, increased ($P<0.01$) the T-VFA concentration, but the addition of urea by 6% decreased ($P<0.01$) of T-VFA concentration. The increase in urea level up to 6% may negatively affect rumen microbes, which reduces rumen microbes ability to degrade carbohydrates in steamed sago waste feed, thereby reducing T-VFA concentration. According to Sharma *et al* [12], the addition of urea in feed, if not done properly, will cause poisoning and interfere with the rumen's fermentation process. In addition, the decrease in T-VFA concentration may also be caused by the VFA was used for microbial protein synthesis by utilizing the availability of N sources from urea. According to Tiven *et al* [13], the concentration of VFA tends to decrease, it may be used as a source of C framework for microbial protein synthesis when N sources are available, so that microbial protein increases, but the T-VFA concentration decreases. According to McDonald *et al* [14], the optimum VFA content in the rumen is between 10 to 70 mM. It can be said that the decrease in T-VFA is still within the normal range.

3.3. Effect of urea levels in feed with steamed sago waste on dry matter digestibility (DMD) and organic matter digestibility (OMD)
The effect of urea level in feed with steamed sago waste on the DMD and OMD, can be seen in Figure 3 and 4.

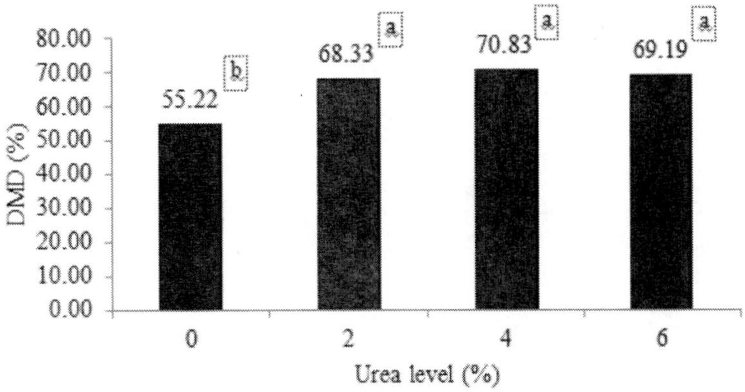

a,b: significant (P<0,01).

Figure 3. The average of DMD (%) due to the use of urea with different levels in feed with steamed sago waste, in vitro tested.

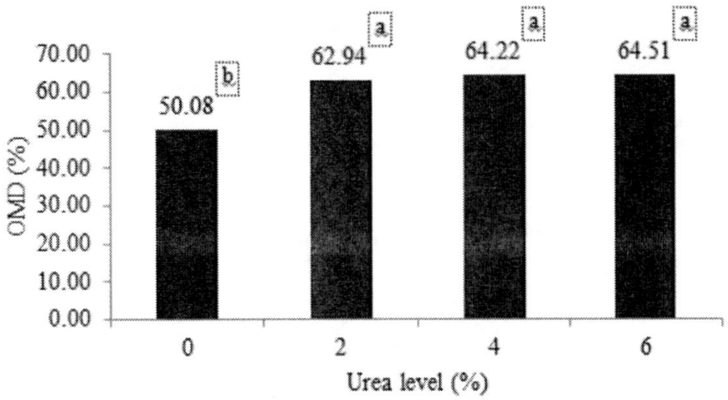

a,b: significant (P<0,01).

Figure 4. The average of OMD (%) due to the use of urea with different levels in feed with steamed sago waste, in vitro tested.

The results of statistical analysis showed that the addition of urea with different levels in steamed sago waste feed had a significant effect ($P<0.01$) on the DMD and OMD. The addition of urea to a level of 2% in the steamed sago waste feed, increased ($P<0.01$) the percentage of DMD and OMD. The addition of urea at the 4% and 6% levels had no significant effect on the percentage of DMD and OMD, even at the 6% level, it tended to reduce DMD. This is probably caused by at the urea level of 6%, the concentration of NH_3 produced is very high so that it has a negative effect on DMD and OMD. The high NH_3 concentration is not in line with the low T-VFA concentration so that the released NH_3 cannot be synthesized by rumen microbes into microbial protein, so the percentage of DMD and OMD tend to decrease. This means that the percentage of DMD and OMD was influenced by rumen microbial activity due to the availability of energy, through the high concentration of T-VFA, but with an increase in urea at the level of 6%, decreases T-VFA (Figure 2), which shows a decrease in microbial activity, so that DMD and OMD also decreased. This shows that the high percentage of DMD and OMD was caused by rumen microbial activity due to the availability of energy.

4. Conclusion

It can be concluded that urea was added in steamed sago waste, have an optimal effect on the rumen fermentation parameters at level 2–4%.

References

[1] Sari A, Liman and Muhtarudin 2016 Supporting of agricultural by product as ruminant feed in District Pringsewu Regency *JIPT* **4** 100-107

[2] Rianza R, Rusmana D and Tanwiriah W 2019 The use of fermented sago pulp as feed for the starter phase super native chicken *Jurnal Ilmu Ternak* **19** 36-44

[3] Haedar, Suardi A, Sapri H and Kasran M 2018 Pemberdayaan masyarakat melalui program pembelajaran pembuatan pakan dari limbah ampas sagu di Desa Buntu Terpedo *Jurnal Dedikasi Masyarakat* **1** 90-97.

[4] Sangadji I, Patty Ch W and Salamena J F 2019 Crude fiber content of sago by product fermented with white root fungi (*Pleurotus ostreatus*) and urea *Agrinimal* **7** 20-25.

[5] Uhi H T 2007 Improvement of nutritive value of sago waste by biofermentation *Jurnal Ilmu Ternak* **7** 26-31

[6] Kurzer F and Sanderson P M 1956 Urea in the history of organic chemistry: Isolation from natural sources *J. Chem. Educ* **33** 452-459

[7] Gonçalves A P, do Nascimento C F M, Ferreira F A, da Costa G R, de Queiroz Monella M., Marino C T, Demarchi J J A A and Rodrigues P H M 2015 Slow-release urea in supplement fed to beef steers *Braz. Arch. Biol. Technol* **58** 22-30

[8] Tilley J M A and Terry R A 1963 A two-stage technique for the *in vitro* digestion of forage crops *J. Brit. Grassland Soc* **18** 104–111

[9] Gupta V K, Parsad R, Bhar L M and Mandal B N 2016 *Statistical Analysis of Agricultural Experiments. Part-I: Single Factor Experiments* (New Delhi: ICAR-IASRILA).

[10] Kardaya D, Wiryawan K G, Parakkasi A and Winugroho H M 2009 Karakteristik urea lepas-lamban pada berbagai kadar molasess dalam ransum berbasis jerami padi secara in vitro *JITV* **14** 177-191

[11] Pratiwi O E, Erwanto and Harris I 2014 The effects of concentrate and amoniated zeolite supplementation on body weight gain of Boerawa Goats *JIPT* **2** 5-11

[12] Sharma S K, Joshi M, Kumar K and Parmjeet 2017 Acute urea poi-soning in buffaloes: case study Research & reviews *J. Vet. Sci* **3** 1-5.

[13] Tiven N C, Yusiati L M, Rusman and Santoso U 2012 Effect of crude palm oil protection on fermentation parameter and rumen microbial activity of male local lamb *Anim. Prod.* **14** 141-146

[14] McDonald P, Edwards R A, Greenhalgh J F D, Morgan C A, Sinclair L A and Wilkinson R G 2010 *Animal Nutrition.* 7th Ed (London: Pearson)

Consumption pattern and hope food pattern in Salahutu District, Maluku Tengah Regency

N R Timisela[1], W Girsang[1,*] and L Tupamahu[2]

[1] Departement Social Economic of Agriculture, Agriculture Faculty, Pattimura University, Jl. Ir. M. Putuhena, Kampus Poka, Ambon 97233 Maluku, Indonesia
[2] Agribusiness Management Post Graduate Program, Kompleks Kampus PGSD, Pattimura University, Jl. dr. Tamaela, Ambon 97113

* E-mail: girsangwardis@yahoo.com

Abstract: This study aims to analyze food consumption patterns, the hope food patterns, and the factors that influence household food consumption patterns. The research was located in Suli and Liang Villages, Central Maluku Regency. The location was determined purposively based on the 2017 Central Maluku Regency Food Security and Vulnerability Atlas (FSVA) results included in Priority 1 and 2 as Food Vulnerable Areas. The research sample was determined by simple random sampling using the Slovin formula with a total sample size of 87 households, 44 (Suli Village) and 43 (Liang Village). The research data were analyzed qualitatively and quantitatively. The results showed that the pattern of household food consumption in the two villages was lower than the ideal weight of food consumption, namely 870 g/capita/day, respectively 783.83 (Liang Village) and 709.81 (Suli Village). The score of the hope food pattern in Liang Village is 75, and Suli Village is 76.9. Analysis of the factors that influence household food consumption patterns in the two villages with a coefficient of determination of 51.56 (Liang Village) and 59.19 (Suli Village). The results of the t-test show that the factors that influence the pattern of household food consumption in Liang Village include income (X_1), the number of family members (X_2), and the education of housewife (X_4), while the factors that influence food consumption patterns Households in Suli Village include income (X_1), age of the housewife (X_3) and education of housewife (X_4).

1. Introduction

Food is everything that comes from biological sources and water, both processed and unprocessed, intended as food and beverage for human consumption, including food additives, food raw materials, and other materials used in preparing, processing food and beverages (Law Number 18 of 2012). Food consumption patterns are related to the type and amount of food consumed by individuals with specific objectives at a particular time. As one indicator of food security, food consumption is strongly influenced by food availability. Law of the Republic of Indonesia Number 18 of 2012 concerning food defines food security as: "The condition of the fulfilment of food for the state to individuals, which is reflected in the availability of sufficient food, both quantity, and quality, safe, diverse, nutritious, equitable and affordable. And not against the community's religion, belief, and culture to live a healthy, active, and productive life sustainably. Food security includes macro aspects, namely the availability of sufficient food, and the micro aspect, namely the fulfilment of food needs for individuals to lead a healthy and active life.

Food's importance in realizing national food security, the government's role, is very much needed to develop excellent and quality food types. The diversity and balance of food consumption at the family level will determine food consumption quality at the regional level, both district/city, province to the national level [1]. Consumption of quality (balanced) food impacts human resources quality and in a region can be determined through the regional expected food pattern score. The Hope Food Pattern Score (HFP) is an indicator of food diversity in food consumption, including value achievement, dietary patterns, and balanced nutrition. HFP score as an indicator in measuring Food Security performance, as stated in the 2015-2019 RPJMN. The achievement of HFP scores in the 2015-2019 RPJMN has been regulated in Government Regulation No. 17 of 2015 concerning Food Security and Nutrition. National and Regional Governments are obliged to realize the diversification of food consumption to meet the community's nutritional needs.

The Maluku region's geographical conditions, most of the oceans, and archipelago areas have various potential sources of local food that are quite diverse, such as sago, corn, cassava, sweet potato. One of the Maluku government's concrete steps towards the potential of local food resources is the Governor's Instruction Number 01 of 2010 concerning Food Diversification in Maluku Province as a form of support for the Indonesian Presidential Regulation Number 22 of 2009 concerning the Policy for Accelerating the Diversity of Local Resource-Based Food Consumption. This instructs all agencies about implementing a diversification program for food consumption from upstream to downstream, increasing foodstuffs' availability to household consumption. The people's food consumption pattern has become increasingly diverse, with the PPH score getting the biggest. However, to get to the hope food patterns, rice consumption must be reduced; on the other hand, for tubers, animal food, and vegetables + fruit, it still needs to be significantly increased [2]. Local food networks as innovations that initially function and develop in local niches within a given food regime. As niche-innovations in local food networks induce socio-ecological changes on the local level, they can foster the dominant food regime [3]. That suggests [4], in their research on local government policies in developing food security, explained that local knowledge has a vital role to play in promoting the effectiveness of food policy.

Based on the Maluku Province Food Security and Vulnerability Map (FSVA) 2017 as one of the instruments used to explain the definition of the concept of food security based on three dimensions of food security (availability, access, and benefits of food) in all conditions, not only in food insecurity situations. It shows that one of the areas in Central Maluku Regency, namely Salahutu District, is included in Priority 4 in a composite manner. This means it is a Food Resistant area, but if viewed per indicator, out of the six (6) villages in the region Salahutu District, Suli Village, and Liang Village are still included in Priority 1 and 2 meaning they are considered vulnerable areas. Based on the Maluku Province Food Security and Vulnerability Map (FSVA) 2017, this study aims to determine the food consumption pattern and HFP score in Salahutu District.

2. Methods

The research was conducted in March-May 2019, located in Suli Village and Liang Village, Salahutu District, Central Maluku Regency. Purposive sampling based on Food Security preparation in Suli Village and Liang Village is included in Priority 1 and 2 as vulnerable food areas. The research sample consisted of 87 households consisting of 43 households in Liang Village and 44 households in Liang Village, randomly selected (simple random sampling). The data collected were analyzed quantitatively and qualitatively.

According to [5], the criteria for household food consumption patterns can be categorized into five categories, namely frequent (if > 1 × a day, 1 × a day, 4-6 × per week), enough (if three times per week), quite often (if < 3x per week, 1-2x per week), rarely (if <once per week), and never. Food consumption can be calculated using the formula:

$$G_{ij} = \frac{BP_j}{100} \times \frac{B_{ddj}}{100} \times KG_{ij} \qquad (1)$$

Information: G_{ij} is nutrient I consumed from food or food j, BP_j is the weight of the food or food j consumed in grams, B_{ddj} is the edible portion j that can be eaten in percent, and KG_{ij} is the content of certain nutrients (i) of food or food j consumed following its unit.

The calculation of the expected food pattern is done by calculating the HFP score. The HFP score is calculated using the formula:

$$Score\ HFP = \%RDA \times weight \qquad (2)$$

% RDA = energy (kcal) food group i: total energy of all food groups × 100 Energy (kcal) = weight of food consumption i: 100 × energy content of food group i × 100/100 (edible portion or BDD)
If the HFP score = 100, then household food consumption is following the expected food pattern (ideal consumption)
If the HFP score is < 100, then household food consumption does not follow the expected food pattern (ideal consumption).

Analysis of the factors that influence food consumption patterns using multiple linear regression analysis.

$$Y = \alpha + \beta_1 X_1 + \beta_2 X_2 + \beta_3 X_3 + \beta_4 X_4 + \beta_5 X_5 + e \qquad (3)$$

3. Results and discussion

3.1. Respondent characteristics

Age affects a person's physical endurance to do work and make decisions. At a productive period, a person can work optimally to meet the needs for quality and nutritious food [6]. The level of education affects food consumption patterns, especially the selection of food types that are good for consumption. Consumption of excellent and quality food can support both physical and non-physical activities. The high level of education is related to the economy, educational facilities and infrastructure in the region, and respondents' motivation and desire to pursue higher levels of education [7]. According to [8], the level of formal education is positively related to nutritional knowledge. Household food security and good nutritional status indicate high healthy understanding and family members' education [9]. A right level of education affects a housewife's knowledge of food and proper nutrition for family consumption. Her ability to manage food becomes more diverse, nutritious, balanced and safe. According to [10], the increasing level of public education affects changing their food consumption perspective, especially nutrition and safety.

The number of family members is related to food fulfilment, namely the amount of food consumed. The majority of Liang Village and Suli Village respondents are in the small family category with 1-4 family members. States that the number of family members determines the level of food security [11]. Household incomes, occupational and educational status of household heads were significant determinants of food security [12]. An increase in the number of household members without being matched by an increase in resources and sources of livelihood to support the economy will cause the allocation of food fulfilment per household member to be low [13].

Table 1. Characteristics of respondents.

Characteristics of Respondents	Category	Village			
		Liang		Suli	
		Amount (Person)	Percentage (%)	Amount (Person)	Percentage (%)
Age Head of RT	Productive (15-64 Years)	34	79.07	40	90.91
	Less Productive (> 64 Years)	9	20.93	4	9.09
	Total	43	100	44	100
Education of the Head of RT	Primary school	19	44.19	15	34.09
	Junior High	10	23.26	8	18.18
	High school	13	30.23	21	47.73
	University	1	2.33	0	0.00
	Total	43	100	44	100
housewife education	Not completed in primary school	1	2.32		
	Primary school	15	34.9	15	34.09
	Junior High	10	23.25	13	18.18
	High school	15	34.9	15	47.73
	University	2	4.65		
	Total	43	100	43	100
Number of Family Members	Small Family (1-4 people)	24	55.81	31	70.45
	Medium Family (5-7 people)	19	44.19	13	29.55
	Total	43	100	44	100

Source: Primary Data 2019 (Processed)

3.2. Household income level

Sources of income for the community come from the agricultural sector, the non-agricultural sector, and a combination of the two sectors. The amount of household income depends on the type of work and the number of family members working. Household income will determine the number of household expenses. The results showed that most of the respondents' income in both locations came from the agricultural sector. Consisting of the food crop sub-sector (cassava commodity), the fruit horticultural sub-sector (rambutan and durian commodities), the plantation sub-sector (cloves and nutmeg commodities), livestock sub-sector (goat husbandry commodity) and capture fisheries sub-sector (fish commodity).

The largest percentage of Liang Village people's income is obtained from the capture fisheries sub-sector, namely 45.19 percent because most of the population work as fishermen. The percentage of income in the plantation sub-sector was 19.71 percent. The percentage is relatively smaller because, during the big wave season, Liang Village people do not go to sea, so they choose to plant gardens. The smallest percentage of income in Liang Village was in the food crop sub-sector by 0.31 percent. Local food that is cultivated in an area of less than 0.25 ha. The harvest is for consumption, and a small portion is sold to buy other food needs.

In contrast to Suli Village, the largest percentage of income is the horticultural sub-sector, such as fruits (rambutan and durian) by 50.57 percent. During the harvest season, rambutan and durian production is very high, so the value of the community's income is relatively large. The percentage of income in the livestock sub-sector was 31.90 percent for pigs and chickens. There is no food crop sub-sector in Suli Village because it does not cultivate food plants for commercialization. Food plants are only grown to fulfil household consumption. Engel's law states that households with low wages or incomes will spend most of their income buying top necessities. On the other hand, high-income households will pay only a small portion of their total income on principal needs.

Table 2. Level of household income.

Source of Income	Village			
	Liang		Suli	
	Average Per Year (IDR)	Percentage (%)	Average Per Year (IDR)	Percentage (%)
Agriculture				
Food Crops:				
Total Food Crops	51,255.81	0.31		
Total Horticultural Plants	862,674.42	5.27	9,757,546.27	50.57
Total Plantation	3,223,859.45	19.71		
Total Livestock	2,079,069.77	12.71	6,155,681.82	31.90
Total Fisheries	7,391,093.02	45.19	2,302,818.18	11.93
Total Agriculture	13,607,952.47	83.21	18,216,046.27	94.40
NonAgricultural				
Total NonAgriculture	2,746,511.63	16.79	1,080,681.82	5.60
Agriculture + NonAgriculture	16,354,464.09	100.00	19,296,728,08	100.00

Source: Primary Data 2019 (Processed)

Table 3. Levels of household expenditure.

Household Expenditure Structure	Village			
	Liang		Liang	
	Average Per Year (IDR)	Average Per Year (IDR)	Average Per Year (IDR)	Average Per Year (IDR)
FOOD				
Total Cereals	4,971,325.58	30.40	4,776,680.71	25.83
Total Tubers	182,790.70	1.12	183,227.27	0.99
Total Animal Food	2,399,627.91	14.68	3,356,500.00	18.22
Total Oil and Fat	572,790.70	3.50	689,072.73	3.74
Total Nuts	867,000.00	5.30	1,038,454.55	5.64
Total Vegetables	273,976.74	1.68	510,136.36	2.77
Total Fruits	141,125.58	0.86	154,100.00	0.84
Total Snacks	1,064,558.14	6.51	1,249,545.45	6.78
Total Others	849,774.19	5.20	1,088,772.73	5.91
TOTAL FOOD	**11,322,939.53**	**69.25**	**13,027,036.36**	**70.72**
NONFOOD				
Total Health	1,060,186.05	6.48	1,106,681.82	6.01
Total Education	1.180.465.12	7.22	1,129,409.09	6.13
Total Kitchen	675,813.95	4.13	833,636.36	4.53
Total Housing	801,000.00	4.90	649,295.45	3.52
Total Outfits	304,883.72	1.86	364,545.45	1.98
Total Others	1,005,325.58	6.15	1,309,227.27	7.11
NONFOOD TOTAL	**5,027,674.42**	**30.75**	**5,392,795.45**	**29.28**
FOOD + NONFOOD	**16,350,613.95**	**100.00**	**18,419,831.82**	**100.00**

Source: Primary Data 2019 (Processed)

3.3. Household expenditure rate

Food expenditure and non-food expenditure depend on the amount of income. Not all households have the same spending, both food, and non-food spending. The results showed that most household expenditure was dominated by food expenditure. The percentage of food expenditure in Liang Village was 69.25 percent, while in Suli Village, it was 70.72 percent. Table 3 shows the level of household food expenditure. Expenditure on cereals is dominated by rice because it is the leading food [14]. The high rice consumption is due to easier access to rice food (available at any time) and relatively easy processing [14]. It is different from local food, especially tubers, whose production is still seasonal, rarely consumed. Local sago food production is still low, so it does not guarantee its availability in the long term. The people's appetite for consumption tends to consume rice as the main carbohydrate food. The convenience to consume local food is highly recommended, meaning that they should go back to local food and increase the consumption of local food for the family. It was necessary to introduce local food products to young generations so that local food will not be neglected [15].

3.4. Household food consumption pattern based on weight of food consumed

Based on the quantity aspect, food consumption is measured by the Nutritional Adequacy Rate (RAN) approach, which includes the Energy Adequacy Rate (RAE) and the Protein Adequacy Rate (RAP). The level of food consumption based on WNPG (2004) is the ideal standard for RAN of at least 2000 kcal/cap/day and RAE of at least 52 grams/capita/day. Simultaneously, the quality aspect or quality of food consumption is seen from the value/score with the HFP approach. For Maluku Province, according to the results of the analysis of food consumption patterns in 2018, it shows that the PPH score is 84.6, with the composition of the grains group being the choice of carbohydrate food sources for consumption, then the animal food group, nuts, fruit, and vegetables.

The results showed that the two villages' household food consumption had not yet reached the ideal weight of food consumption, namely 870 grams/capita/day. The total weight of food consumption in Liang Village is 783.83 grams/capita/day, while in Suli Village is 709.81 grams/capita/day. The two villages' ideal food consumption weight has not reached the standard because there is no balance in consuming the required food types. The type of food consumed goes the perfect weight of food in Liang Village, namely tubers, oils and fats, nuts, and other food types, while Suli Village is animal food, nuts, and other foods. Table 4 shows the pattern of public food consumption based on the standard of consumption (weight) of ideal food.

Table 4. Respondents' food consumption pattern based on ideal food weight.

Food Group	Standard Ideal Food Consumption Weight (gram / capita / day)	Average Weight of Food Consumption (gram / cap / day)	
		Liang Village	Suli Village
Grains	275	239.89	240.46
Tubers	100	164.31	74.56
Animal food	150	111.79	159.93
Oil and fat	20	28.43	37.87
Oily fruit / seeds	10		
Nuts	35	58.84	81.33
Sugar	30	16.75	26.85
Vegetable and fruit	250	101.13	72.62
etc		16.70	16.19
Total	870	737.83	709.81

Source: Primary Data 2019 (Processed)

Table 4 shows that rice as the main carbohydrate food is routinely consumed but has not yet reached the ideal food weight. This is because rice consumption depends on the respondent's income. If income decreases or increases, it will affect the importance of the food consumed. Table 5 shows the percentage of food security levels based on ideal food consumption standards. The results showed that most households in the two villages were classified as food insecure based on the ideal consumption weight standard, namely 90.70 percent (Liang Village) and 86.36 percent (Suli Village). Food consumption per capita per day is still below 850 g or less than 2000 kcal. It is necessary to balance tubers' consumption, animal foods, nuts and vegetables, and fruits, to achieve better food quality. This condition shows that food consumption has not met the recommended balanced nutrition standards: Diverse, Nutritious, Balanced, and Safe Food (DNBSF).

Table 5. Percentage of respondents food security level based on ideal food consumption standarts.

Standard Bert for Food Consumption (gram / capita / day)	Village			
	Liang		Liang	
	Amount (Person)	Amount (Person)	Amount (Person)	Amount (Person)
<850 g or <2000 kcal (Vulnerable to Food)	39.00	90.70	38.00	86.36
> 850 g or> 2000 kcal (Food Resistant)	4.00	9.30	6.00	13.64
Total	43.00	100.00	44.00	100.00

Source: Primary Data 2019 (Processed)

3.5. Household food consumption pattern based on PPH score

The hope food pattern reflects the recommended food consumption pattern for a healthy, active, and productive life. Based on the food score of nine foodstuffs. The availability of food at all times in sufficient quantities, and affordable prices greatly determines food consumption level at the household level [16]. As a guideline for measuring food consumption diversification, including staple food, PPH is an arrangement of various foods based on the main food group's energy contribution from a pattern of availability or consumption pattern. The food consumption pattern of the people is not yet diverse because it is still dominated by grains (56.3%), especially rice (86.3%) [17].

The results showed that the percentage of Nutritional Adequacy Rate (% RDA) was 93 percent (Liang Village) and 92 percent (Suli Village). As measured by the PPH score, the ideal food consumption shows that the PPH score is 75 (Liang Village) and 79.6 (Suli Village). This means that Liang and Suli Villages have not achieved quality food consumption or ideal food consumption because the PPH scores are lower than the PPH standard of 100. The results of research by Hamid et al., 2013 state that the proportion of the grains, animal food, and oil and fat groups has contributed to the actual PPH score of more than 50%, but none of these food groups have achieved the perfect PPH score. There is a need for improvement in quantity as well as the quality of household food consumption by paying attention to the maximum limit so that, in the end, it can help increase the actual PPH score both in villages and in cities. The PPH score is not only influenced by income factors but also by differences in the area of residence. The PPH score of each village is based on the food group (9 food groups); it can be seen that the food groups that achieved the PPH score or achieved ideal consumption for both villages were tubers, nuts, oil, and fat. The results of calculating the PPH score are shown in Table 6.

3.6. Factors affecting household food consumption patterns in Liang Village and Suli Village

The results of the analysis of factors affecting household food consumption patterns in Liang and Suli villages include income (X_1), number of family members (X_2), age of housewives (X_3), education of housewives (X_4), and type of work (X_5). The F test results are 5.67 (Liang Village) and 5.34 (Suli Village), which means that the five factors simultaneously influence household food consumption patterns. The coefficient of determination or R^2 Adjusted was 51.56 percent (Liang Village) and 59.19

percent (Suli Village). This means that 51.56 and 59.19 percent of household food consumption patterns in Liang and Suli villages can be explained by income (X_1), the number of family members (X_2), age of housewives (X_3), education of housewives, stairs (X_4) and type of work (X_5). The remaining 43.49 percent and 40.81 percent are explained by other variables not included in the model. The results of the regression analysis of factors affecting the pattern of food consumption in Liang Village and Suli Village are shown in Table 7.

Table 6. The pattern of Hope Food (PPH) in Suli and Liang Villages, Central Maluku Regency, 2019.

Food Group	Liang Village							Suli Village						
	Consumption (gr / cap / day)	Energy / 100 gr	Energy [kcal]	% RDA	Weight	PPH Score	Standard	Consumption (gr / cap / day)	Energy / 100 gr	Energy [kcal]	% RDA	Weight	PPH Score	Standard
Grains	239.89	361	866	46	0.5	23	25	241	361	868	43	0.5	21.6	25
Tubers	164.31	154	253	14	0.5	7	2.5	75	154	115	6	0.5	2.9	2.5
Animal food	111.79	90	101	5	2	11	24	160	90	144	7	2	14.3	24
Oil and fat	28.43	870	247	13	0.5	7	5	38	870	330	16	0.5	8.2	5
Oily fruit / seeds	0	359	0	0	0.5	0	1	0	359	0.0	0.0	0.5	0.0	1
Nuts	58.84	337	198	11	2	21	10	81	337	274	14	2	27.2	10
Sugar	16.75	364	61	3	0.5	2	2.5	27	364	98	4.9	0.5	2.4	2.5
Vegetable and fruit	101.13	16.8	17	1	5	5	30	73	16.8	12	0.6	5	3.0	30
etc	16.7	6.87	1	0	0	0	0	16	6.87	1.1	0.1	0	0.0	0
Total			1744	93		75	100			1841	92		79.6	100

Source: Primary Data, 2019 (Processed)

Table 7. Results of the regression analysis of factors affecting the pattern of food consumption in Liang Village and Suli Village.

Term	Liang Village					Suli Village				
	Coef	SE Coef	T-Value	P-Value	VIF	Coef	SE Coef	T-Value	P-Value	VIF
Constant										
Income	3.453	0.283	12.20	0.000		4.250	0.303	14.00	0.000	
Number of family	0.1114	0.0360	3.09	0.004	1.18	0.1154	0.0472	2.45	0.019	1.15
members	0.0447	0.0290	2.54	0.031	1.32	0.0506	0.0320	1.58	0.122	1.22
The age of the housewife	0.0221	0.0376	0.59	0.559	1.30	0.1557	0.0503	3.10	0.004	1.77
Housewife education	0.0195	0.0313	2.62	0.036	1.42	0.1245	0.0442	2.82	0.008	1.49
Type of work	0.0230	0.0788	0.29	0.772	1.10	0.0972	0.0692	1.41	0.168	1.16

Source: Primary Data 2019 (Processed)

Factors that influence food consumption patterns in Liang Village include income, number of family members, education of homemakers, and types of work. Simultaneously, the factors that influence food consumption pattern in Suli Village include income, the housewife's age, and the housewife's education. Income is an essential factor in determining household food consumption patterns. If income increases, household food consumption patterns will increase. In the two research locations, it can be seen that the income factor has a positive effect on household food consumption patterns, meaning that if the income increases by one percent, the consumption pattern will also increase by the coefficient value. The results of research by [18][6] stated that the level of income is high, so it is easier for households to access food of higher quality and quantity. High purchasing power also encourages households to consume more diverse, nutritious, and balanced nutrition. The higher the income, the purchasing power of a person also increases. The ability to choose and buy various foods is more elevated, indicating an increasing food consumption pattern [19].

The number of family members affects the household food consumption pattern in Liang Village. If the number of family members increases, the consumption pattern will increase. Many families mean diverse food consumption patterns because family members will choose several food types, especially for purchased food. Hamid et al., 2013 show that the more household members, the more food consumption quantity increases and varies. The burden that is borne by the household is getting bigger, so that the household decides to choose specific types of food that are cheaper and readily available in larger quantities to fulfil the element of satiety, not to meet the nutritional needs of the household. States that the greater the number of household members, the greater the household's burden. There are two ways that households can overcome this, namely increasing household income or reducing food expenditure by choosing cheaper and less diverse food types [20].

Housewife education affects household food consumption patterns in both villages. The housewife regulates the quality of food consumed and the frequency of meals. If the housewife's education level is getting better, knowing the importance of a healthy, safe, and suitable food consumption pattern will increase. Housewives' education is related to the level of knowledge and attitudes they have in meeting their food and nutritional needs. Hopefully, housewives with high education, the higher their knowledge of nutrition, and their awareness will increase in fulfilling nutritious food. This means that the higher the education, the higher the expertise and insight of housewives about nutrition. In processing or cooking daily food it is not only based on habits and the concept of being full. Still, it can choose quality food ingredients by paying attention to the nutritional elements in food to increase the PPH score [18]. The age of a housewife affects household food consumption patterns in Suli Village. If the age increases, the attention to regulating the household's food consumption pattern will get better. This is important because a mother will be more mature and concerned to provide proper and quality food for consumption by all family members.

4. Conclusion

Household food consumption in the two villages has not yet reached the ideal weight of food consumption, namely 870 grams/capita/day (2000 kcal), or has reached PPH standards. The total weight of food consumption in Liang Village is 783.83 gram/capita/day, and Suli Village is 709.81 gram/capita/day. Household food consumption, measured based on the ideal food consumption standard, shows that the two villages have not reached the ideal food consumption. The PPH scores of the two villages are 75 (Liang Village) and 79.6 (Suli Village), respectively, or lower than the ideal food consumption of 100. Factors that affect household food consumption patterns in Liang Village are income (X_1), the number of members family (X_2), and housewife education (X_4). In contrast, the factors that influence household food consumption patterns in Suli Village are income (X_1), age of housewife (X_3) and housewife education (X_4).

References

[1] Badan Ketahanan Pangan 2016 *Laporan Tahunan Badan Ketahanan Pangan Tahun 2015* (Jakarta: Badan Ketahanan Pangan)
[2] Ariani M 2014 Analysis of food consumption at the community level supports the achievement of food diversification *Gizi Indones.* **33** 20–8
[3] Lutz J and Schachinger J 2013 Do local food networks foster socio-ecological transitions towards food sovereignty? Learning from real place experiences *Sustain.* **5** 4778–96
[4] Bastian A and Coveney J 2012 Local evidenced-based policy options to improve food security in South Australia: The use of local knowledge in policy development *Public Health Nutr.* **15** 1497–502
[5] Suhardjo, Hardinsyah and Riyadi H 1988 *Survey Konsumsi Pangan [Food Consumption Survey]* (Bogor: Bogor Agricultural University)
[6] Ningsih M, Suandi and Damayanti Y 2012 The factors that influence the pattern of food consumption and household nutrition of fishermen in Tungkal Ilir District, Tanjung Jabung Barat Regency *Sosio Ekon. Bisnis* **15** 48–56
[7] Imelda 2018 Characteristics and patterns of household food consumption in Pontianak City *Ethos*

J. Penelit. dan Pengabdi. Masy. **6** 250–9
[8] Talahatu A H 2006 *Study of Body Mass Index (BMI) and Weight Gain of Pregnant Women and Their Relationship with the Growth and Development of Babies Born in Ambon City* (Bogor Agricultural University)
[9] Hardinsyah 2007 Review on determinant factors of dietary diversity *J. Gizi dan Pangan* **2** 55–74
[10] Akbar R, Kusrini N and Yurisinthae E 2014 Analysis of food consumption in Pontianak City *J. Soc. Econ. Agric.* **3** 24–36
[11] Rose D 1999 Economic determinants and dietary consequences of food insecurity in the United States *J. Nutr.* **129** 517S-520S
[12] Birhane T, Shiferaw S, Hagos S and Mohindra K S 2014 Urban food insecurity in the context of high food prices: a community based cross sectional study in Addis Ababa, Ethiopia *BMC Public Health* **14** 1–8
[13] Manesa J, Baliwati Y F and Tanziha I 2008 Household food security in village producing resin in Lampung Barat District *J. Gizi dan Pangan* **3** 172–9
[14] Moniharapon G J, Turukay M and Luhukay J M 2013 Changes of household food consumption pattern: From local food to rice. A case in the sub-district of South Leitimur, Ambon City *Agrilan J. Agribisnis Kepul.* **1** 83–93
[15] Timisela N R, Leatemia E D, Luhukay J M, Polnaya F J and Breemer R 2020 An analysis of factors influencing consumers' perception towards the product attributes of sago local food agro-industry *Int. J. Innov. Creat. Chang.* **12** 500–18
[16] Argandi S, Trimo L and Noor T I 2018 Factors that influence the Expected Food Pattern (PPH) in Bandung Regency *J. Agribisnis Terpadu* **11** 126–40
[17] Khomsan A, Baliwati Y F and Dwiriani C M 2004 *Pengantar Pangan dan Gizi [Introduction to Food and Nutrition]* (Depok: Penebar Swadaya)
[18] Hamid Y, Setiawan B and Suhartini 2013 Analysis of household food consumption (Case study in Tarakan Barat Sub District Tarakan City East Borneo Province *Agrise* **13** 175–90
[19] Rachman H P S and Ariani M 2008 Diversifying food consumption in Indonesia: Issues and implications for policies and programs *Anal. Kebijak. Pertan.* **6** 140–54
[20] Suyastiri N M Y P 2008 Diversification of staple food consumption based on local potential in realizing food security for rural households in Semin District, Gunung Kidul Regency *J. Ekon. Pembang.* **13** 51–60

IOP Conf. Series: Earth and Environmental Science **883** (2021) 012056 doi:10.1088/1755-1315/883/1/012056

Test for the antibacterial inhibition of kaffir lime leaf (*Citrus hysteric* D.C) extract against pathogen bacteria in improving food safety

S J Nendissa[1],*, D M Nendissa[2]
[1] Faculty of Agriculture, Pattimura University-Ambon. Indonesia
[2] Faculty of Fisheries and Marine Science, Pattimura University Ambon. Indonesia

* E-mail: sandriananendissa@gmail.com, nendissadm@yahoo.com

Abstract. Kafir lime leaf (*Citrus hystrix*) is a plant from the citrus tribe that has long been known by community as flavor ingredient. To support its use and increase its application in supporting food safety, a test the inhibition of on kaffir lime leaf extract against pathogenic bacteria, namely Gram Negative Bacteria (*Escherichia coli, Salmonella typhimurium*) and Gram Positive bacteria (*Staphylococcus aereus, P. aeroginosa*). Making kaffir lime leaf extract (*Citrus hystrix*) was done by weighing 150g of lime leaf powder, then immersing in 96% ethanol solution and leaving for $\pm$ 3 days. Kaffir lime leaf extract was dissolved with sterile distilled aquades to obtain a concentration of 5%, 10% and 15%. The antibacterial activity of kaffir lime leaf extract was tested by diffusion method using disc paper to determine of the bacterial growth inhibition area. The results showed that kaffir lime extract had antibacterial activity inhibition of 12,78 mm of *S. aereus*, 9 mm of *E.coli*, 7,12 mm of *S. typhimurium* and 9,3 mm of *P. aeroginosa*. Kaffir lime leaf extract has inhibition effectiveness for gram positive bacteria *Staphylococcus aereus* and gram negative bacteria *E. coli, Salmonella typhimurium, P. aeroginosa*. Thus, kaffir lime leaf extract can be used as a decontaminant agait theses 4 type of bacteria, especially *Staphylococcus aereus* which has a strong inhibitory power, so it can maintain quality and increase the safety of mead based foods

1. Introduction

Food safety is a public need in maintaining the quality / preventing food / processed products from being contaminated or contaminated by pathogenic bacteria and not causing or causing foodborne illness and infection. In an effort to prevent food from being contaminated by pathogenic bacteria, spices form plants are used as natural preservatives, namely kaffir lime leaves (*Citrus hystrix DC*).

Kaffir lime leaves (Citrus hystrix DC) is a plant from the citrus tribe that has long been recognized by the public as a natural flavor ingredient in food products (peanut brittle, potato chili sauce, fish meat sauce, etc.) and beverages. The leaves of Citrus stomach (Citrus hystrix DC) contain 1.8% tannin compounds; steroids, triterpenoids, 1-1.5% essential oils, alkaloids, polyphenols, flavonoids so that they can be used as antibacterial agents [1] In addition, kaffir lime leaves (*Citrus hystrix DC*) also have several chemical components, namely 81,49%, citronellal; citronellol 4.22%; linalol 3.69%; geraniol 0.31%; other components 6.29% [2].

Kaffir lime leaf extract with a concentration of 1% and 2% has antibacterial activity against Escherichia coli with MIC and MBC values <0.0625 and has compounds of the terpenes group. According to [3], that the compounds of the terpenes group from kaffir lime leaf extract have antibacterial activity by damaging cell membranes and inhibiting growth and killing pathogenic bacteria by disrupting cell walls [3].

Pathogenic bacteria are usually found in basic ingredients for processed food products, processing equipment and processed products, namely spore-forming bacteria, including *Staphylococcus aureus*, and *Bacillus cereus* (Gram-positive bacteria) and *Salmonella typhimurium*, *Escherichia coli* (Gram-negative bacteria). *Staphylococcus aureus* also plays a role in food safety because it can cause mastitis in dairy cows and has the potential to contaminate cow's milk products [4] as well as flora bacteria found in food products, including poultry, fish, bread, cakes and salads; *Bacillus cereus* is found in meat, fish, milk, vegetables; *Salmonella typhimurium* is found in poultry, raw eggs, vegetables, fruits; *Escherichia coli* is a foodborne bacteria that can cause various lower digestive tract disorders in warm-blooded animals [5], also found in spinach, cucumber, cheese, beef, fish, milk and beverages but in some strains such as O157: H7 can cause food poisoning.

These bacteria often make contamination of food and beverages, causing disease, infection and food poisoning and have an impact on product quality degradation. For this reason, in inhibiting the growth of pathogenic bacteria and increasing food safety, kaffir lime leaf extract is needed as an antibacterial. The research objective was to test the antibacterial inhibition of kaffir lime leaves extract against pathogenic bacteria.

2. Methods

2.1. Material

The materials used are kaffir lime leaves and their extracts. *Staphylococcus aureus* bacteria (FNCC 0047), *Bacillus cereus* (IFO 13690), *Salmonella typhimurium* (IFO 12529), *Escherichia coli* (IFO 3301) obtained by the Food and Nutritional Culture Collection (FNCC) Center for Inter-Study UGM. The medium used was Muller-Hinton Agar (MHA).

2.2. Method

The extract of kaffir lime leaves (Citrus hystrix) was carried out by weighing 150 grams of kaffir lime leaf powder. This powder is then immersed in a 96% ethanol solution and left to stand for ± 3 days. The results obtained are filtered and evaporated for three days. This process aims to evaporate the ethanol to obtain a thick extract from the kaffir lime leaves. Kaffir lime leaves extract was weighed as much as 0.5 g; 1.0 g; 1.5 g was then dissolved with 10 mL of distilled water to obtain a concentration of 5%, 10%, 15%, of the kaffir lime leaf extract.

2.3. Media making

Making Muller-Hinton media according to what is stated on the label, namely by dissolving 38g of agar in 1 liter of distilled water, and then heating it to dissolve completely. The agar solution was sterilized in autoclave at 1210C for 15 minutes. At 450C temperature, then poured into a 20 ml petri dish and allowed to solidify.

2.4. Research methods

In this study, the antibacterial activity of kaffir lime leaf extract was tested against the pathogenic bacteria Staphylococcus aureus, Bacillus cereus, Salmonella typhimurium, Escherichia coli in vitro using the paper disc diffusion method. The parameter measured is the diameter of the inhibition area (DDH) in millimeters (mm).

2.5. *Antibacterial test*

Determination of antibacterial activity was carried out by the paper disc diffusion method. The batteries used were obtained from FNCC PAU UGM-Jogyakarta and had been rejuvenated in Muller-Hinton media and incubated at 37°C for 24 hours. Bacteria that grow from the culture are taken as much as one ose and dissolved in physiological 0.9% NaCl as much as 4.5 ml, so that they have a turbidity according to / close to the turbidity of the main solution of bactreia (4 X 10^9) is diluted three times until a bacterial colony suspension is obtained (4 X 10^6). Bacterial suspension is used as inoculum.

2.6. *Inhibition power diameter*

A total of 0.5 ml of bacterial suspension was poured into a petri dish filled with solid Muller-Hinton agar, the bacterial suspension was leveled and allowed to stand for a while, the excess liquid was piped, and then the media was allowed to dry. On the media was placed disc paper that had been saturated with the test solution for kaffir lime leaf extract at a concentration of 5%, 10%, 15%, respectively. The spelling of each antibacterial activity test was carried out aseptically in a laminar air flow cabinet with three repetitions, then incubated at 37oC for 24 hours. A reading of the total inhibitory power diameter (DDH) when the zone of inhibition formed around the disc paper shows a clear zone. Partial inhibition zone, that is, if the growth of several colonies in the inhibition zone is still visible. Zero zone of inhibition if there is no zone of inhibition around the disc paper. The diameter of the drag zone can be calculated using the caliper in mm.

3. Research design

The research design used a factorial design. The first factor was the concentration of kaffir lime leaf extract (5%, 10%, 15%) while the second factor was the isolates of *Staphylococcus aureus*, and *Bacillus cereus* (Gram Positive bacteria) and *Salmonella typhimurium*, *Escherichia coli* (Gram Negative bacteria). Each treatment has three replications. If there is a significant difference in each treatment, Duncan's multiple distance test is performed [6].

4. Results and discussion

In this study, the antibacterial activity test in kaffir lime leaf extract was carried out in vitro by using the diffusion method to determine the antibacterial activity contained therein. The results of measurements carried out on the diameter of the inhibition area (DDH), using white lime leaf extract against the bacteria Staphylococcus aureus, Bacillus cereus, Salmonella typhimurium, Escherichia coli are presented in Table 1.

Table 1. Effect of the concentration of kaffir lime leaves (Citrus hystrix DC) extract on the Inhibitory Power Diameter (DDH) of the growth of *Staphylococcus aureus, Bacillus cereus, Salmonella typhimurium, Escherichia coli bacteria.*

Kaffir Lime Leaves Extract (*Citrus hystrix DC*) Concentration	DDH
15%	$13{,}75^a$
10%	12.41^b
5%	9.73^c

Note: *Different superscripts in the direction of the column show significant differences (P <0.05)*

The results in Table 1 show that kaffir lime leaves extract is very effective in inhibiting the growth of pathogenic bacteria, namely *Staphylococcus aureus, Bacillus cereus, Salmonella typhimurium, Escherichia coli* at a concentration of 5%, 10% and 15%. The higher the administration of kaffir lime leaves extract, the greater the diameter of the bacterial inhibition (DDH), this means that the antibacterial activity of kaffir lime leaves extract is greater. According to [7], the higher the concentration, the higher the active substance content so that the antibacterial activity will be greater and vice versa, the lower the concentration, the less the active substance content so that the antibacterial activity will decrease.

Table 2 shows the difference in the diameter of the inhibitory power (DDH) which is influenced by several things including the size of the inoculum, the thickness of the agar plate and the diffusion power of the test solution and the sensitivity of bacteria to the lime leaf extract solution. The media used in this study is the Mueller Hinton Agar (MHA) media which is a good medium for the growth of pathogenic bacteria in the antibacterial inhibitory activity test using the disc diffusion method and is a standardization of the Clinical and Laboratory Standards Institute (CLSI) in testing antibacterial activity using the diffusion method. disc. The content of kaffir lime leaves extract as a test solution is known to have antibacterial activity, namely tannin and flavonoid compounds whose working principle is to damage cell walls in order to inhibit bacterial growth. The alcohol groups in flavonoid compounds damage bacterial cells by taking advantage of differences in the polarity of the lipids that make up cells [8].

Table 2. Effect of treatment combinations between concentrations of kaffir lime leaves extract (*Citrus hystrix DC*) and pathogenic bacterial isolates on the diameter of inhibition power (DDH) formed.

Kaffir Lime Leaves Extract (*Citrus hystrix DC*) Concentration (%)	Types of Bacteria	DDH (mm)
15%	*S.aereus*	16.47a
10%	*S.aereus*	14.82^{c}
5%	*S.aereus*	12.78^{d}
15%	*B. cereus*	15.81^{b}
10%	*B. cereus*	13.49^{c}
5%	*B. cereus*	9^{j}
15%	*S. typhimurium*	12.58^{e}
10%	*S. typhimurium*	9.35^{h}
5%	*S. typhimurium*	7.12^{k}
15%	*E. Coli*	11.38^{f}
10%	*E. Coli*	10.63^{g}
5%	*E. Coli*	9.11^{i}

Note: *Different superscripts in the direction of the column show significant differences (P <0.05)*

The antibacterial activity of kaffir lime leaves extract against four pathogenic bacterial isolates consists of two Gram-negative bacteria, namely *Staphylococcus aureus, Bacillus cereus* and two Gram-positive bacteria, namely *Salmonella typhimurium, Escherichia coli* which are presented in Table 3. The antibacterial activity of Gram-positive bacteria Staphylococcus aureus, Bacillus cereus showed the results of measuring the inhibitory power diameter (DDH) were greater than Gram negative bacteria, namely *Salmonella typhimurium, Escherichia coli.* Gram-negative bacteria *Salmonella typhimurium, Escherichia coli* have a high cell wall and lipid content of address [9] 11-22% and a multilayer cell wall structure consisting of lipoproteins,outer membrane phospholipids, and lipopolysaccharides, making it difficult to penetrate the cell walls of Gram negative bacteria. by antibacterial substances compared to Gram-positive bacteria [10];

[11]; [12]. Gram-positive bacteria have a peptidokligan layer on the outside and less act as an effective permeability defense [13].

Table 3. Effect of kaffir lime leaves extract on inhibition diameter of four bacterial isolates of *Staphylococcus aureus, Bacillus cereus, Salmonella typhimurium, Escherichia coli*

Types of Bacteria	DDH
Staphylococcus aureus	16.27[a]
Bacillus cereus	14.09[b]
Salmonella typhimurium	11.54[c]
Escherichia coli	9.51[d]

Note: *Different superscripts in the direction of the column show significant differences (P <0.05)*

5. Conclusion

Extra kaffir lime leaves (Citrus hystrix DC) is effective against pathogenic bacteria and can be used as a decontamination for Gram-positive bacteria, namely *Staphylococcus aureus* which has the largest diameter of inhibition to maintain quality and improve food safety for meat-based food ingredients.

References

[1] Miftahendrawati 2014 Efek Antibakteri Ekstrak Daun Jeruk Purut (*Citrus Hystrix*) Terhadap Bakteri Streptococcus Mutans (In Vitro). *Skripsi. Fakultas Kedokteran Gigi.* Universitas Hasanudin. Makasar.

[2] Warsito, Noorhamdani, Sukardi, Suratmo dan Susanti R 2017 Mikroenkapsulasi Minyak Jeruk Purut *(Citrus Hystrix)* Dan Uji Aktivitasnya Sebagai Antibakteri *Journal of Environmental Engineering & Sustainable Technology* **4**(1) 19-25

[3] Yuliati R, Peni I dan Septi S.R 2011 Aktivitas Antibakteri Minyak Atsiri Daun Jeruk Purut (*Citrus hystrix*) Terhadap *Staphylococcus aureus* dan *Escherichia coli. Jurnal Pharmacom* **12**(2) 50-4

[4] Apriani N J 2015 Uji Aktivitas Antibakteri Ekstrak Daun Jeruk – Pepaya (*Citrus medica* L. Var. *Proper*) Terhadap *Escherichia coli* dan *Staphylococcus aureus* Skripsi Fakultas Farmasi, Universitas Hasanuddin, Makasar

[5] Safithri M, M. Bintang and M. Poeloengan. 2011 Antibacterial Activity of Garlic Extract Against some Pathogenic Animal Bacteria *Media Peternakan* 155-158

[6] Gunawan G S, Setiabudy dan Nafrialdi 2007 Farmakologi dan Terapi, Percetakan Gaya Baru, Jakarta.

[7] Steel R G D and J H Torrie 1980 Principle and Procedures of Statistic.A Biometrical Approach. Edisi Kedua International Student Edition. Mc. Graw Hill International Book Company

[8[Newell D G, Koopmans M, Verhoef L, Duizer E, Aidara-Kane A, Sprong H, Opsteeghb M, Langelaar M, Threfall J, Scheutz, F, Giessen J V D and Kruse H 2010 Foodborne Diseases – The Challenges of 20 Years Ago Still Persist while New Ones Continue to Emerge *J Food Microbiology* **139** S3–S15

[9] Diana M A Dan Tessa S 2014 Efek Ekstrak Jeruk Nipis (*Citrus Aurantifolia,* Swingle) Terhadap Pertumbuhan Bakteri *Salmonella Typhii Jurnal Medika Malahayati* **1**(2) 43 – 46

[10] Silhavy T J, D Kahne and S Walker 2010 The Bacterial Cell Envelope *Cold Spring Harbour Perspectives in Biology* **2**(5) a000414

[11] Pelczar M J 2008 Dasar-Dasar Mikrobiologi Edisi I Ratna Siti Hadioetoemo, penerjemah. Jakarta *Universitas Indonesia Press* 99-119.

[12] Nikaido H and M Vaara 1985 Molecular Basis of Bacterial Outer Membrane Permeability *Journal Microbial Rev* **49** 1-32

[13] Scherrer Re and Philipp G 1971 Molecular sieving by the *Bcaillus megatarium* cell wall protoplast *Jounal of Bacteriological* American Society for Microbiology

Centella asiatica tendril growth of Samosir – Indonesia accession

N S Vinolina[1,]* and R Sigalingging[2]

[1] Department of Agrotechnology, Faculty of Agriculture, Universitas Sumatera Utara, Indonesia
[2] Department of Agricultural Engineering, Faculty of Agriculture, Universitas Sumatera Utara, Indonesia

* E-mail: noverita@usu.ac.id

Abstract. Medicinal plants often used is pegagan (*Centella asiatica*). Due to the benefits of pegagan, it is important to know the growth of this plant in detail, in this case, the growth of tendrils, which is, of course, related to biomass production. Plant materials are needed that are guaranteed production levels and quality by applying recommended cultivation. This study was proposed to understand the influence of phosphorus on the growth of pegagan tendrils. This research was conducted using a single factor with six levels of P2O5 fertilizer dosage with a non-factorial randomized block design repeated 3 times. The 6 levels of fertilizer dosage are 0, 20, 40, 60, 80, 100 kg P2O5/ ha.The treatment of phosphorus (F) at 3-12 WAP did not reveal a prominent influence on the number of the primary tendrils; however, phosphorus's application affects the growth component of the primary tendril length in acid soil conditions. Pegagan can still grow well even though the land conditions are very acidic.

1. Introduction

Centella asiatica or Pegagan is a medicinal plant that has been utilized for generations as medicine of inflammation, rheumatism, asthma, hemorrhoids, tuberculosis, leprosy, dysentery, fever; and as a wound healer and an appetite enhancer. In Japan, medicinal and aromatic plants are highly utilized. It imports the plants from India and China, which China is the largest exporter of the plants [1].

Pegagan contains several saponin compounds, which include asiaticosides [2, 3]. The bioactive compounds of asiaticosides help accelerate the wound healing process and are advantageous to treat leprosy and tuberculosis [4, 5]. The development of pegagan on a large scale needs to be supported by cultivation efforts. To produce quality pegagan products, it requires plant materials with guaranteed production levels and qualities [6]. Plant material is an important thing to pay attention to get simplicia with high centelloside content.

The research objective was to obtain the potential plant material with a high centelloside content to be used as seeds to obtain a biomass production with high centelloside content from North Sumatra. Nowadays, what is meant by organic compounds from natural materials is limited to compounds known as secondary metabolites. The utilization of plants as medicines is related to these plants' chemical content, especially bioactive substances. Without a bioactive compound in plants in general, the plant cannot be used as a medicine. Bioactive compounds found in plants are usually secondary metabolic compounds such as alkaloids, flavonoids, steroids, terpenoids, saponins, etc.

Centella asiatica or Pegagan in the international market has been employed as a medicinal ingredient for various diseases treatment [5, 8, 9]. Its advantages have caused it to be commercial cultivated in several countries, especially in China and India [10, 11]. The increasing demand of Pegagan in national and international is mainly supported by wild populations, which results in the overexploitation of the plant as is the case around the world12.

The studies of the bioactive compounds in pegagan such asasiaticoside, madecasosside, and asiatic acid have focused on the complexes and benefits of biochemical processes [13], their concentration in various parts of the plant [14, 15, 16], and reviews of their growth under controlled conditions [4]. Limited literature or studies, especially in the Indonesian region, deliver information of the harvest time influences on the concentration of important bioactive compounds from pegagan cultivated in fields. The observation has made for the Pegagan field's production vary in terms of quality and quantity throughout the year [2, 3, 6, 17, 18]. Field location, climatic conditions, and type of accession contributes to this inconsistency and deserves further study [13, 17].

The formation of secondary metabolites is influenced by many factors, including temperature, pH, water activity, and light intensity. The thermal (non-photochemical) reaction rate is sensitive to temperature and some reaction rates will rise with an increase in a temperature of 100oC. A relatively dry land, pH, and soil moisture are relevant parameters for forming secondary metabolism.

In this research, planting is carried out with agronomic action by giving phosphorus to see the consistency of the plant's response in biomass production and phytochemicals compounds. In previous studies, it was found that giving phosphorus without methyl jasmonate would result in higher biomass production, and the asiaticoside content in leaves and roots would be affected so that the production of asiaticoside per g of dry matter would be higher [18]. The urgency of this research on pegagan is due to the many benefits of pegagan in various aspects, such as effective for inhibiting or preventing age-related retinal deterioration/ degeneration, useful for developing drugs or functional foods, protecting neuron cells, and useful pegagan herbs [19].

2. Methods

Pegagan Samosir accession was given phosphorus increased from the previously analyzed. The growth and photochemical content of pegagan was analyzed. The material in this research was pegagan accession upland Samosir. The tools used were digital scales, digital cameras, soil processing equipment, meters, and others.

This research was conducted using a single factor with 6 levels of P2O5 fertilizer dosage with a randomized block design repeated 3 times. The 6 levels of fertilizer doses, namely 0, 20, 40, 60, 80, 100 kg P_2O_5/ ha at the Samosir.

2.1. Cultivation Field and Research Preparation

This study used Pegagan seed from the accession of Samosir, North Sumatra, Indonesia. The collected stolons were directly planted after being separated from the mother plant. Each plot contained four seeds planted at a 40 cm distance from each other. The soil fertilization using Urea and KCl was done three times during the cultivation period at 0, 20, and 40 days after planting (DAP). In every fertilization period, Urea's application dose was 30 g/plot, and KCl was 22 g/plot distributed evenly around all of the planting holes.

2.2. Seed Planting and Maintenance

Cultivation maintenance consisted of regular watering, weeding, and replanting crops. Watering was consistently done in the afternoon by considering the weather conditions in the field. Stitching was done two weeks after planting to replace dead plants. Weeding was done manually every day by removing weeds from the soil by hand. Control of pests and diseases was conducted each week to avoid or prevent the distribution of pests and diseases. The plants were harvested at the same time at 12 weeks after planting (WAP) by collecting all parts of the plant.

2.3. Growth characteristics' measurements

Primary tendrils (tendrils that come out from the main plant) and secondary tendrils (tendrils that come out from the primary tendrils) were measured for their number and length every week at 3-12 WAP.

The length of primary tendril data was at 3-12 WAP observations. The stolons were counted every week at 12 WAP from the formed stolons that came out from the tendrils.

2.4. Wet and Dry Weight Measurement
The harvested Pegagan plants were separated into two sample categories: the leave samples containing the leaves and petioles; and the root samples consisting of the roots and tendrils. They were weighed for dry weight.

3. Results and discussion
3.1. Number of primary tendrils (WAP)
Figure 1 shows the number of primary tendrils (tendrils) at 3-12 WAP observation. It reveals that the treatment factor for the dose of phosphorus (F) at 3-12 WAP had no prominent impact on the number of primary tendrils. The highest average number of the primary tendrils was obtained at 8-12 WAP observations in the phosphorus fertilizer treatment at 40 P_2O_5 kg/ha (F2) aged 6-12 WAP and 100 P_2O_5 kg/ha (F5) aged 12 WAP.

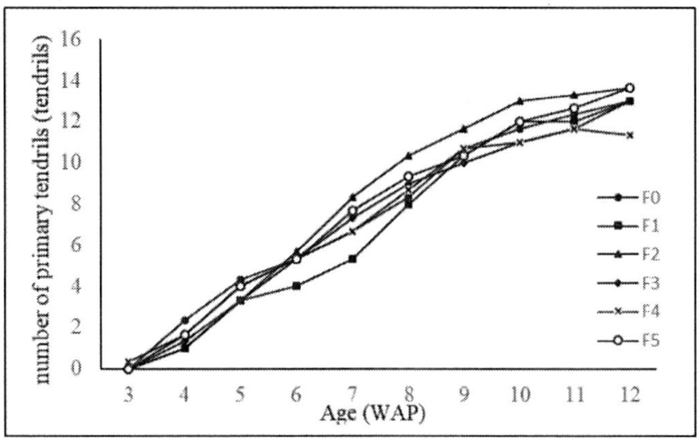

Figure 1. The relationship between the lengths of Primary Tendrils age (WAP) of *Centella asiatica*

3.2. Length of Primary Tendril (cm)
Table 2 lists the length of the primary tendrils at 3-12 WAP. Similarly, the variance analysis demonstrates that the treatment factor for the dose of phosphorus (F) application at the ages of 3-7, 9, and 12 MST also has an insignificant effect on the primary tendrils' length. At the same time, it shows a prominent impact at the ages of 8, 10, and 11 WAP on the length.

The phosphorus treatment at 8 WAP had a notable influence on the length of the primary tendrils. The influence is linear as shown in Figure 2.

The phosphorus fertilization at 10 WAP has significantly affected the length of the primary tendrils. The phosphorus effect on the length of the primary tendrils at 10 WAP is quadratic, as shown in Figure 3.

Figure 4 presents the adding of the phosphorus fertilization at 11 WAP, which is also quadratic in which the higher phosphorus fertilization increases the length of the primary tendril to a certain extent, then a decrease occurs.

Table 1. The average length of Primary Tendrils (cm) in Phosphorus Treatment at 3-12 WAP

Treatment	Primary Tendril Length (cm) (WAP)									
	3	4	5	6	7	8	9	10	11	12
F_0 = 0 kg P_2O_5/ha	0.00	12.67	24.33	34.00	43.00	55.67 b	69.67	80.00 bc	83.33 bc	89.00
F_1 = 20 kg P_2O_5/ha	0.00	12.50	22.83	37.50	47.67	58.67 b	70.67	79.33 bc	82.67 bc	97.33
F_2 = 40 kg P_2O_5/ha	0.00	11.17	34.00	47.17	56.67	69.33 a	82.00	90.00 a	96.00 a	99.00
F_3 = 60 kg P_2O_5/ha	0.00	7.17	20.00	31.00	43.00	56.67 b	73.00	87.33 ab	89.67 abc	91.67
F_4 = 80 kg P_2O_5/ha	4.00	12.67	25.33	39.00	53.67	61.00 ab	80.00	86.00 abc	93.33 ab	104.33
F_5 = 100 kg P_2O_5/ha	0.00	7.00	23.83	33.83	42.00	63.33 ab	73.00	76.67c	82.00 c	100.33

Note: Numbers followed by the same lowercase letter in the same column, show insignificant differences at the 5% level based on Duncan's Multiple Range Test.

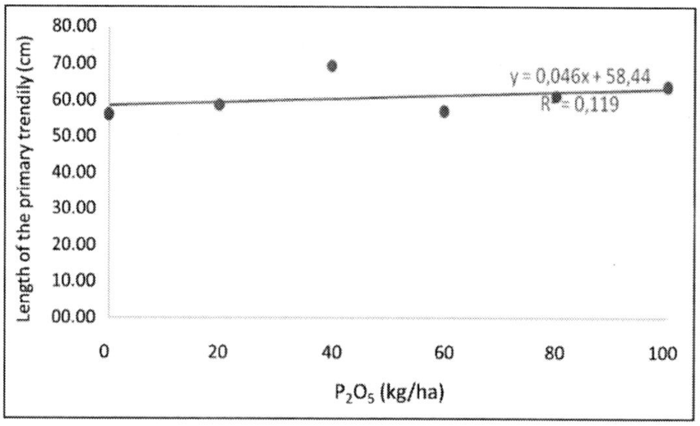

Figure 2. The phosphorus application and the the length of the primary pendrils at 8 WAP

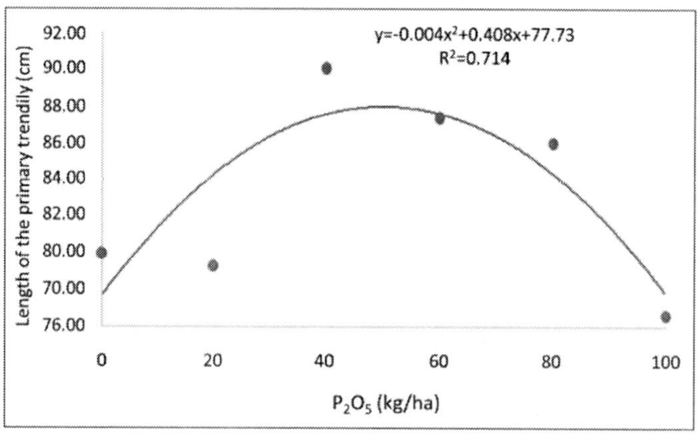

Figure 3. The effect of the phosphorus addition on the length of the primary tendrils at 10 WAP

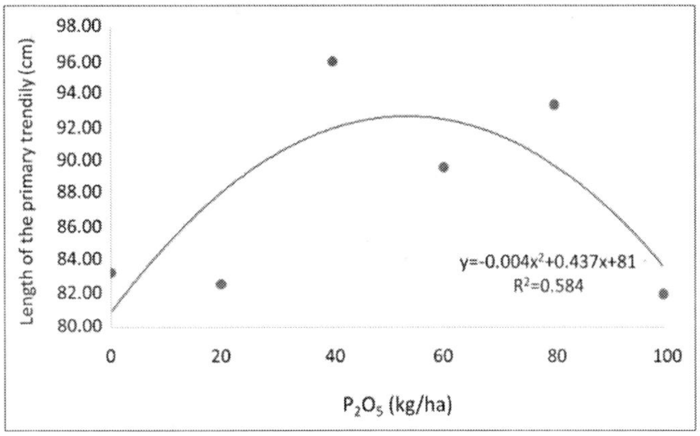

Figure 4. The impact of the phosphorus fertilization on the length of the primary tendrils at 11 WAP

3.3. *Number of Secondary Tendrils (tendrils)*

Table 2. The average number of secondary tendrils at 6-12 WAP in the phosphorus fertilizer treatment

Treatment	Number of secondary tendrils (WAP)						
	6	7	8	9	10	11	12
F_0 = 0 kg P_2O_5/ha	1.33	2.67	7.00	12.33	17.33	25.67	42.33
F_1 = 20 kg P_2O_5/ha	1.67	3.33	6.33	9.67	14.00	23.00	32.33
F_2 = 40 kg P_2O_5/ha	4.00	5.67	10.00	14.67	18.33	25.00	41.00
F_3 = 60 kg P_2O_5/ha	2.33	3.00	9.33	14.00	18.33	20.33	37.33
F_4 = 80 kg P_2O_5/ha	4.33	6.67	11.33	14.33	18.67	24.67	40.67
F_5 = 100 kg P_2O_5/ha	2.67	5.33	10.33	15.00	18.67	26.00	44.33

Table 3 displays the average number of the secondary tendrils at 6-12 WAP observations. It can be seen from the observation of the secondary tendrils aged 6-12 WAP, the number of secondary tendrils is not notably different due to the phosphorus fertilization. The highest number of the secondary tendrils and an increasing rate occur from 8 to 12 WAP observations. The phosphorus fertilization treatment at 5-12 WAP observation on the highest number of secondary tendrils is obtained at 100 P2O5 kg/ha (F5) phosphorus application.

3.4. *Data of Production*

Figure 5 exhibits the average data of the pegagan plant production. The variance analysis reveals that the phosphorus treatment did not prominently impact the dry weight of leaves and petiole, dry weight of roots and tendrils. However, the F2 (40 kg P_2O_5/ ha) treatment was the highest compared to other treatments.

The statistical analysis exhibits that various phosphorus treatments slightly influenced the length of the primary tendrils of 8, 10, and 11 WAP. Simultaneously, the other parameters of the growth and production did not show a significant impact. The phosphorus application escalates the length of the primary tendrils at the age of 10 and 11 WAP in a quadratic response. The length of the primary tendrils rises to a certain extent, and then it declines thereafter. The application of 93 kg P2O5/ha phosphorus obtains the maximum tendril length. Although the phosphorus application, in general, does not reveal a prominent influence, there is an increasing yield trend as a result of the adding of the phosphorus doses given to pegagan plants up to 40 kg P2O5 /ha (F2). Overall, the fertilized plants tend to have better results than those of the unfertilized plants. The growth parameters such as number of leaves, length of petiole, total leaf area of parent plant, number of primary tendrils, number of stolons, wet weight per plot, highest leaf dry weight, and root dry weight, are found in the phosphorus treatment at the level of 40 kg P2O5 /ha. However, the highest parameter of the primary tendrils' length is achieved by adding the phosphorus at the level of 80 kg P2O5 /ha, and the best parameter of the number of secondary tendrils was obtained at the application of phosphorus at the level of 100 kg P2O5 /ha. The adequate nutrients in plants are useful to help the plants grow. Mengel and Kirkby (1982) stated that phosphorus is one of the necessary plant macronutrients which roles in several plants' life processes, such as photosynthesis, carbohydrate metabolism, and energy flow processes.

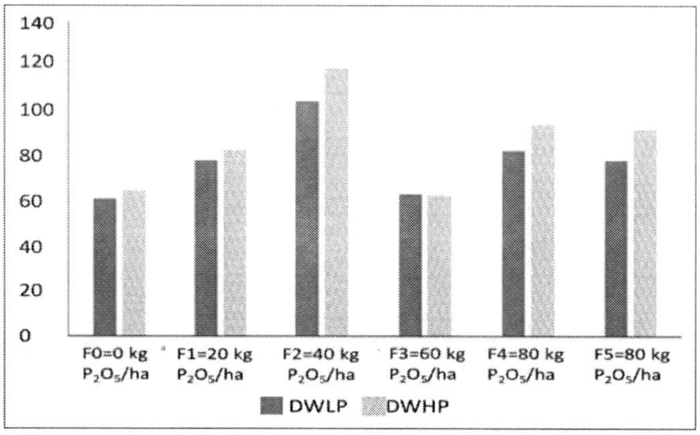

Figure 5. The average of thedry weights of leaves and petiol; androots and tendrils as the phosphorus

This is supported by Vinolina et al (2012) [17] in the earlier study that soil P content had effects on the centelloside content of pegagan. The phosphorus benefit stimulates better root formation to advance the nutrients and water absorption and the amount of leaf chlorophyll (the amount of leaf chlorophyll increases with increasing phosphorus application). The other benefits are that phosphorus can lead to a good photosynthesis process to gain photosynthate. Those benefits are considered to magnify the amount of asiaticoside compounds in plants. Phosphorus cannot be diminished in plants. It remains as phosphates in its free form, but it is esters when bound to organic compounds. Phosphate esters are formed with sugars, alcohols, acids, or other phosphates (polyphosphates). This compound is considered an intermediate for the pentose phosphate pathway of primary metabolites and is derived from precursors to secondary metabolites.Vinolina et al [2, 3, 18] mentioned that the elements of soil phosphorus influence the levels of asiaticoside, madecassoside, and asiatic acid in several accessions observed.

Phosphorus is also a crucial component of various phosphate sugars, which is helpful during reactions in the dark phases of photosynthesis, respiration, and various other metabolic processes. It has a function in energy storage and transfer. Besides, it can elevate the plants' production or dry matter and enhance the yield's quality [20].

4. Conclusion

The addition of phosphorus influences the growth component of the length of the primary tendril in acid soil conditions.

5. Acknowledgments

The authors gratefully acknowledge the research funding support from the Ministry of Research, Technology and Higher Education, Republic of Indonesia with grant No. 11/AMD/E1/KP.PTNBH/2020.

References

[1] Asian Scientist. 2012. Japanese-India Exchange to Promote Sustainable Trade in Medicinal Plants. By admin - Monday, February 13, 2012
[2] Vinolina N S, Nainggolan M and Siregar R 2018 *AGRIVITA J. Agric. Sci.* **40** 304- 312
[3] Vinolina N S 2018*J. Phys. Conf. Ser.* **1116** 052072

[4] Mangas S, Moyano E, Osuna L, Cusido R M, Bonfill M and Palazo J 2008 *Biotechnol. Lett.* **30** 1853-1859
[5] Mangas S, Moyano E, Hernandez V L and Bonfill M 2009*Centella asiatica (L) Urban: An Updated Approach Terpenoids* (Singapore:Research Singpost)
[6] Ghulamahdi M, Sandra A A, Nurliani B dan Octivia T 2010. *Studi Penyiapan Standar Operasional Prosedur Budidaya untuk Produksi Bioaktif Mendukung Standarisasi Mutu Pegagan*. Repository IPB.http://repository.ipb.ac.id/handle/123456789/41772
[7] Aziz Z A, DaveyM R, PowerJ B, Anthony P, SmithR M and LoweK C 2007 *Biol. Plant.* **51** 34-42
[8] Jacinda J and Ian D 2011 *J. Planar Chromat.* **24**(1) 82–87
[9] Adetunji A T, Lewu F B, Mulidzi R and Ncube B 2017 *J. Soil Sci. Plant Nutr.* **17**(3) 794-807
[10] Ariffin F, Heong Chew S.H, Bhupinder K, Karim A A and Huda N 2011 *J Sci Food Agric* **91** 2731–2739
[11] Gallego A, Karla R E, Heriberto R V L, Diego H, Liliana L, Waqas K K, Rosa M C and Javier P 2014*Eng. Life Sci.* **14** 1-10
[12] SinghS, GautamA, SharmaA and Bartra A 2010 *Int. J. Pharm. Sci. Rev. Res.* **4** 9-17
[13] Alqahtani A, Wannit T, Kong M L, Valentina R N, Kelvin C, and George Q L 2015 *Phytochem. Anal.* **26** 436–443
[14] Kim O T, Kim M and Hong M H 2004*Plant Cell Rep* **23**339-344
[15] Lambert E, Ahmad F and Danny G 2011 *Appl. Biochem. Biotechnology* **164** 220-237
[16] Bonfill M, Susana M, Elisabeth M, Rosa M C, Javier P 2011 *Plant Cell Tiss. Organ Cult.* **104** 61–67
[17] Vinolina N S, Luthfi A M, Siregar and Napitupulu J A 2012. Morphology of leaves and content of secondary metabolites asiaticoside in some accession of Pegagan (Centella asiatica l. Urban) in North Sumatera. *Proc. of The 2nd Annual Int. Conf. Syiah Kuala University 2012 & The 8th IMT-GT Uninet Biosciences Conf.* Banda Aceh, 22-24 November 2012, ISSN: 2089 – 208X (200-205)
[18] Vinolina N S 2016 *GJRA Global Journal for Research Analysis* **5**294-297
[19] Wu Z, Weibo L, Jing Z, Xin L, Wang L, Bin C, Mingkui W, Lilian J, Weicheng H and Fu L 2020 *J. Agric. Food Chem.* **68** 6977–6986
[20] Havlin J L, BeatonJ D, TisdaleS L and NelsonW L2005. *Soil Fertility and Fertilizers: An Introduction to Nutrient Management*(New Jersey: Pearson Prentice Hall)

IOP Conf. Series: Earth and Environmental Science **883** (2021) 012058 doi:10.1088/1755-1315/883/1/012058

Evaluation of effectiveness various seed processing combination technique to produce True Shallot Seed (TSS) with good quality

N Waluyo[1,*], A Rahayu[1], R Rosliani[1], T Wikan[2], R Gaswanto[1]

[1] Indonesian Vegetable Research Institute, Jln. Tangkuban Parahu No. 517, Lembang, Bandung Barat, Indonesia 40391
[2] Indonesian Center for Agricultural Engineering Research and Development, Jl. Sinarmas Boulevard, Situ Gadung, Kec. Pagedangan, Tangerang, Banten 15338

* E-mail: nurmalitawaluyo@gmail.com

Abstract. The seed processing technique is essential to maintain quality and suppress seed deterioration rate as long as processing time. This research aims to evaluate various seed processing combination techniques to produce TSS with good quality. The study was conducted at Indonesian Vegetables Research Institute, Lembang (1,250 m sal) from March until December 2018. The research used a Randomized Complete Block Design (RCBD) with three replications. The treatment consisted of 12 combinations of seed processing, including the technique of drying, capsule breaking, and sorting. The research result showed that the best treatment was a combination technique with umbels drying in the room at RH 50 % and 30-35°C for 72 hours, breaking capsule by hand manually, and seed sorting by winnower followed by hand manually. The produced seed quality in this treatment showed the seed germination was 75%, the moisture content was 7.5%, and the physical purity was 99.9%. This research implies that the availability of TSS processing technology can be carried out by massal, but still can produce good seed quality.

1. Introduction

Shallot (*Allium cepa* L. var. ascalonicum) is an essential vegetable with high economic value [1]. Along with the growing population number, shallots domestic consumption demand also increases. Nowadays, shallots are included as the main crop because they have an economic effect on triggering national micro inflation [2].

Indonesia's horticultural statistics informed that the harvested area, production, and shallots' productivity increased in 2015-2019. In 2019 the harvested area was 159,195 ha, or an increase of 1.54% compared to 2018. Shallot production was around 1,580,247 t with a productivity of 9.59 t ha^{-1}. The shallot production and productivity also increased by about 5.11% and 3.55%, respectively. Three provinces in Indonesia that are the main production centers of shallots are Central Java, East Java, and West Nusa Tenggara [3].

All this time, shallot vegetative propagation uses bulbs, but they have some disadvantages. Bulbs can't be kept in the storage for an extended period, and thus after the offseason, the seed availability is usually limited. Sometimes the bulbs also can be a medium for transmitting pathogens [4]. In addition, it is not easy to manage storage and distribution. In terms of genetics, the used same variety continuously of bulbs will decrease diversity, thereby reducing trait improvement through the breeding program [5].

Nowadays be introduced right shallot seed (TSS) as an alternative seed. Some advantages of TSS are easily able to be kept in storage for a long time, free from seed-borne pathogens, more efficient in using the necessity of seed and fertilizers, easier and cheaper for seed distribution, and higher production due to propagation ratio, in this case, 1: 200 compared 1: 10 if using bulbs. However, it takes a long time to maintain and harvest in the field [6] [7].

Agroclimate in Indonesia, as a tropical country, has high temperatures and humidity. The condition is not ideal for producing TSS. It is known that TSS is a conventional seed type with hygroscopic; hence it is straightforward to deteriorate [8]. Many aspects will determine TSS quality, such as plant growth condition in the field, the technique of flowering and pollination, umbel condition at harvesting time, until seed processing technique [9]. Therefore TSS production should be carried out in the highlands (> 900 m asl) during the dry season to produce many umbels and attacks of pollinator insects [10]. Another advantage of plants will be avoided by primary pathogens such as *Colletotrichum* spp, *Alternaria porri,* and *Stemphyllium vesicarium* [11]. The number of umbels can be stimulated and increased by bulbs vernalization technique and application of growth regulators such as benzylaminopurine (BAP) 37.5 ppm [9][6].

Seed quality includes indicators of genetic, physiological, physical purity [12] and its health. A critical part of producing TSS with good quality is seed processing. The processing properly is intended to maintain seed quality at an obtained physiological mature stage and suppress deterioration rate, TSS processing covers drying harvested umbel, breaking capsules, and sorting seeds. The drying process should be done as soon as possible after the umbels are harvested because it is easy to get infected by fungal. The drying process by sunlight is usually for 7-14 days. The capsule drying process correctly will make it easy to separate seeds from shells. The cling shell on seeds can stimulate pathogens to infect it until less the quality.

In Indonesia, TSS processing for only a small scale capacity can be done manually. Seed processing by hand manually usually involves many laborers, high costs, and takes a long time. Indonesia's government had decided on a self-sufficiency policy on shallots; hence it should be supported by sufficient qualified seeds. Therefore TSS processing for large capacity by machine is essential to support the expected production target. The TSS certification minimum standard had already been determined by agriculture ministry No. 131/2015, i.e., physical purity 99.9%, moisture content 8%, and seed germination 70% [13].

Indonesian Center for Agricultural Engineering Research and Development had already assembled some TSS processing machines such as dryer' hallway type', pulper, and winnower machine. It was needed to study the operational effectiveness of these machines and to evaluate the produced seed quality. Finally, the present study's objective was to preliminary research and assessed the efficacy of various seed processing combination techniques to make TSS of good quality.

2. Methods

The research was conducted at the Vegetable Crops Research Institute-Lembang (1,250 m asl) from March to December 2018. The used variety was Trisula. The research activity was carried out in three parts.

2.1. TSS capsule production of Trisula variety

The used bulbs were two months of age after harvesting, then vernalized at a temperature of 6–10°C for four weeks. The growth regulator benzylaminopurine (BAP) was applied with a recommendation dose, then dissolving in 5 ml of 1 M KOH/NaOH. After that, it was diluted in water to obtain a concentration of 37.5 ppm. The vernalized bulbs were soaked in BAP solution for one hour, then drained on a bamboo tray.

The research field was plowed with a depth of 30 cm and cleared of weeds, then loosened. Furthermore, the beds were made with a width of 1 cm, a length of 10 m, or adapted to field conditions. Dolomite lime 1.5 t ha^{-1} was applied on the bed with horse manure at a dose of 20 t ha^{-1}, SP-36 (75-90 kg P_2O_5 ha^{-1}) as basal fertilizer the insecticide for soil insect, then mix it evenly with soil using a hoe. After that, the beds were covered with silver plastic mulch.

After soaking in BAP solution, the drained bulbs have applied a fungicide with active content mancozeb at a dose of 20 g for 1 kg bulbs by spreading evenly. Bulbs were planted in beds with a spacing of 20 x 20 cm. The bamboo shading with transparent U.V. plastic was made after umbels appear (± 3 WAP). NPK 16:16:16 was applied once a week for ten times since 10 DAP at a dose of 60 kg ha-1 per each application. The used boron dose was 3 kg ha^{-1} applied three times since 3, 5, and 7 WAP with 1/3 quantity per each application by dissolving in the water then sprayed to the plants especially head of umbel part.

Pollination can be carried out by some techniques such as planting marry gold plants surrounding of research field for attracting pollinator insects, placing swarms of honey bees in several points of the research field, or swab umbels by hand manually. Watering was done every day if be needed. The dew at the leaves tips was removed by spraying water in the morning. Weeding was done intensively. Selective chemical pesticides control pests and diseases. Harvesting was carried out if the capsule was already cracked and half of the shell was yellow with black seed color. Harvesting TSS was done by cutting the umbel with or without the stalk [7].

Tabel 1. Treatment of TSS processing combination techniques

Treatment	Seed processing technique		
	Drying	Capsule breaking /Pulper	Sorting
1	Drying process by sunlight at the screen house (7-14 days)	Hand manually	Hand manually
2	Drying process by sunlight at the screen house (7-14 days)	Hand manually	Winnower
3	Drying process by sunlight at the screen house (7-14 days)	Pulper	Hand manually
4	Drying process by sunlight at the screen house (7-14 days)	Pulper	Winnower
5	Drying at dehumidifier (R.H. 50%) & heater room (30-35^{0}C for 72 hours)	Hand manually	Hand manually
6	Drying at dehumidifier (R.H. 50%) & heater room (30-35^{0}C for 72 hours)	Hand manually	Winnower
7	Drying at dehumidifier (R.H. 50%) & heater room (30-35^{0}C for 72 hours)	Pulper	Hand manually
8	Drying at dehumidifier (R.H. 50%) & heater room (30-35^{0}C for 72 hours)	Pulper	Winnower
9	Drying by dryer' hallway type' (30-35°C for 5-6 hours)	Hand manually	Hand manually
10	Drying by dryer' hallway type' (30-35°C for 5-6 hours)	Hand manually	Winnower
11	Drying by dryer' hallway type' (30-35°C for 5-6 hours)	Pulper	Hand manually
12	Drying by dryer' hallway type' (30-35°C for 5-6 hours)	Pulper	Winnower

2.2. *TSS capsules processing*

TSS capsule processing used Randomized Complete Block Design (RCBD) with three times replications. There was 12 TSS processing treatment as a technique combination of drying umbels, breaking capsule, and sorting clean seed of TSS. (Table 1). After sorted, all seeds were dried again by dehumidifier room temperature of 30-35°C for two days or by the dryer machine. Then the produced seed was observed on seed weight per treatment.

2.3. Testing of the produced TSS
The produced TSS was conducted quality testing consisted of moisture content, physical purity, viability and vigor, electric conductivity. Also, be observed the time duration of seed processing per treatment. Data were analyzed using the F test and DMRT test at a 5% level

3. Results and discussion
3.1. The produced seed and rendement percentage
Plant conditions in the vegetative and early generative phases were right. Still, in the seeding phase, the stalks began to be infected by the pathogen *Stemphyllium vesicarium*, while the insect Spodoptera exigua infected florets. The situation was due to the research conducted in the rainy season with high humidity, so pest and disease infection was also relatively high. Suppose stalks are infected by *Stemphyllium vesicarium,* so photosynthate and water distribution to umbels were inhibited [11]. In the end, umbels became died. Plant protection was applied by pesticides two times a week.

According to observation could be known that umbels emerging percentage of Trisula variety are around 70%. Naturally, the umbels percentage in the highland (>900 m sal) is high because of low temperature (<18°C). In addition, bulbs vernalization treatment was applied, BAP growth regulators' application stimulated umbels number more than 100%. Umbels emerging began at the age of 2-3 WAP. If without vernalization and BAP application umbels appearing more than 30 DAP with small size and a few numbers [9].

Tabel 2. The produced seed and rendement percentage of harvested umbel per processing treatment

Treatmen t	Weight of harvested umbel (kg)	Capsules containing seed per harvested umbels (%)	Produced clean seed (g)	Seed rendement percentage (%)
1	3.11	91.05	309.27	11.22
2	3.11	92.53	326.28	12.07
3	3.11	92.38	345.02	**12.40**
4	3.67	89.21	441.24	11.61
5	3.13	89.88	368.03	11.86
6	3.47	81.31	426.78	11.82
7	3.96	89.79	**456.10**	11.34
8	3.96	89.39	422.68	10.37
9	3.57	79.75	348.20	9.60
10	3.57	83.28	364.66	10.08
11	3.49	84.17	366.88	10.27
12	3.49	85.59	367.87	10.27
F_{value}	0.92^{ns}	2.04^{ns}	1.47^{ns}	0.26^{ns}
CV (%)	16.08	6.06	17.76	28.31

At the age of 110, DAP was the first time umbel's Trisula variety was harvested. This information is appropriate with Rosliani *et al.* (2017) [7] that mentioned the harvesting time usually at the age of 105-110 DAP identified by the capsules already cracked, and half of the shell was yellow with black seed color. Harvesting was done three times by only selecting the good umbels. The total yield of harvested umbels was around 124.92 kg. Furthermore, these materials were used to research the study, with an average of each treatment per replication range from 3.11-3.96 kg (Table 2).

Prior to processing, the observation was conducted on the percentage of capsules containing seed per harvested umbels distributed to each treatment per replication. The result showed a high percentage in the range 79.75 – 92.53% (Table 2). It indicated that the pollination technique by insect pollinator and swabbed by hand manually were effective enough [10]. These pollination techniques could be applied to produce TSS in the field.

Table 2 could be known that the produced seeds number of each treatment showed no difference significantly. This caused by the processed umbel's weight capacity per each treatment was the same in range 3-4 kg. Treatment No 7 was the best performance in producing TSS (456.10 g). It was possible because the harvested umbel number capacity was also the most a lot (3.96 kg).

Each treatment had the same umbels number capacity. This condition caused the rendement percentage among them also no different significantly. In addition, if the used umbels number that processed only small scale capacity, the possibility of no difference significantly between processed by machine or by hand manually. Treatment No 3 (12.40%) was the highest percentage of seed rendement performance. However, it was suggested that the minimum processed umbel capacity of 10 kg per treatment for the next study could prove a difference significantly between each treatment.

3.2. Preliminary study working effectiveness of TSS processing machine

The results showed that umbels drying by dryer' hallway type' could be used for large capacity per one processing. Umbels as much as 3-4 kg capacity could be dried only within 5-6 hours with a temperature of 30-35^{0}C. Drying by a dehumidifier and heater room needed temperature 30-33°C for three days until the umbels can be processed. Umbel withering treatment should be done at the screen house, either processing by a dryer machine or a dehumidifier & heater room for 1-3 days depending on the harvested umbel humidity to avoid heat shock. At this stage, the stalks and florets must be cut to make easier capsule processing and reduce waste [7]. As for drying processing by sunlight at the screen house is very dependent on time and weather conditions. The processing can't be done in the early morning because the umbel's condition is still chewy due to high humidity in the screen house.

Figure 1. Three kinds of TSS processing machine: (a) Umbels drying 'hallway type'; (b) Pulper; (c) Winnower sorting machine

Umbels drying by dryer machine took the shortest time compared to other treatments because the wither capsules could be processed directly for 4-6 hours depending on the dried capsule capacity per one drying period. But this machine's working was considered inefficient because the processed wither capsule's capacity was only 4-6 kg and needed much more labor than other treatments. In this case, the being dried capsules had to be turned over every 2 hours. Besides that, it took time enough to prepare the machine. Drying by dehumidifier & heater room was more efficient because it could dry umbels in large capacity and not depend on weather conditions. The capsules could be dried for 3.33-4.00 days with controlled temperature and humidity. Hence possibility the maximum viability and vigor achieved at physiological mature could be kept well.

After umbels drying, the capsules were processed by a pulper or by hand manually. The pulper's capsules were considered more effective and efficient than hand manually based on the produced seed amount and speed of time. The processed capsules of 2-3 kg by hand manually needed time for 60 minutes, while the pulper was only for 20 minutes. However, seed processing by the pulper had a risk of damaging seeds. In this case, be indicated by the electrical conductivity value [14].

The winnower is used to sort good seed and dirty waste such as remain capsule, shell, umbels stalks. Based on the produced good seed, the winnower should be evaluated because the physical seed purity was still low. This result was slightly different as reported by [15] mentioned that sorting by winnower more efficient than hand manually with the capacity rate was 11.14 kg/hours and faster than hand manually (only 0.2 kg/hours). The expected seed product causes differences. In this research, the produced seed had to fulfil the minimum standard requirement as a breeder seed class; thus, the physical purity percentage must high.

It is recommended that a winnower still needs to be assisted by hand manually if the expected seeds had to fulfil minimum standard requirements [13]. Unfortunately, sorted by hand manually takes a lot of time, which ± 180 g of seeds took around 9 hours or equivalent with 20 g/hour. Nevertheless, the winnower's seed quality was excellent because it produced a lot of the weighted seed until it predicted that it would affect the germination percentage. Seed sorting by the winnower would be more effective if TSS umbels from good plant conditions would produce a lot of the weighted seed. As high as seed viability at the mature physiological stage so that the winnower working will be useful.

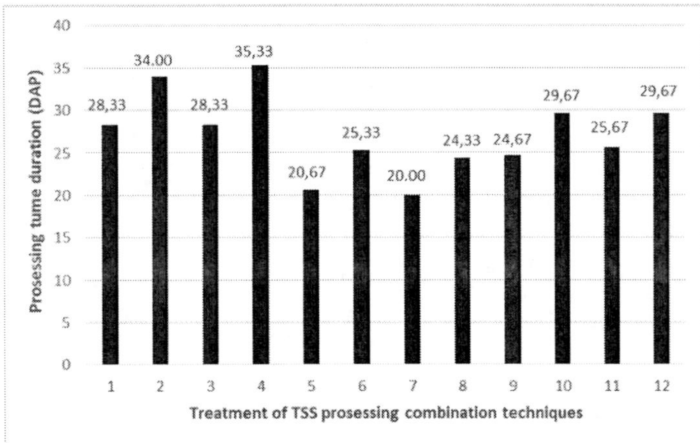

Figure 2. The time duration of TSS processing combination treatment

3.3. Viability, vigor, moisture content, and physical purity percentage of produced TSS

Although the produced TSS is abundant, the processing technique is useless if the quality is not good. Therefore, each treatment's produced TSS must be tested quality as a reference is the minimum standard quality of seed certification [13]. The quality testing results showed that seed germination parameters, seed germination, and electrical conductivity were differences significantly between treatments (Table 3). Still, there was no difference in the parameters of seed moisture content and physical purity (Figure 3). The results showed that the seed processing treatment influenced to quality of the produced seeds, especially viability and vigor. Information on early viability and vigor value will determine the storage ability [12].

The seed germination test represents viability. Germination is one parameter quality with a required minimum percentage of 70% [13]. The result showed that the seed germination percentage of treatment. No. 6 (75.00%) was the highest and differed significantly from other treatments (Table 3). The produced TSS of treatment No 6 was the only one that could fulfill the required minimum standard of seed certification with an average of germination, moisture content, and physical purity were 75.00%, 7.53%, and 99.91%, respectively.

Seed vigor was tested by seed germination (%/etmal) and electrical conductivity (μs/cm/g). It could be known that results in both parameters showed that there were differences significantly among treatments. Treatment No. 6 (12.38%/etmal) was the best performance for speed germination and different from other treatments, whereas treatment No 1 (control) was the worst (6.76%/etmal). This result proved that using the dryer machine positively affected viability and vigor than processed by hand manually. In a recent study, the speed of germination correlated with seed germination directly. As high as germination percentage, so speed germination became higher also. The drying processing technique was predicted to influence seed germination, seed germination, and moisture content. In contrast, the capsule breaking technique affects electrical conductivity as for seed sorting technique influences physical purity.

Tabel 3. Seed quality of the produced TSS

Treatment	Seed germination (%)	Speed of germination (%/etmal)	Electrical conductivity (μs/cm/g)
1	41.58 c	6.76 c	486.99 ab
2	47.92 bc	7.79 bc	411.55 b
3	50.92 bc	8.39 bc	422.70 b
4	50.67 bc	8.35 bc	496.24 ab
5	51.17 bc	8.40 bc	502.81 ab
6	75.00 a	12.38 a	259.88 c
7	57.33 b	9.48 b	561.15 ab
8	55.17 bc	9.11 bc	481.35 b
9	46.33 bc	7.57 bc	528.00 ab
10	56.67 bc	9.31 b	461.52 b
11	44.67 bc	7.42 bc	640.27 a
12	55.33 bc	9.15 bc	401.22 b
CV (%)	14.83	14.92	17.35

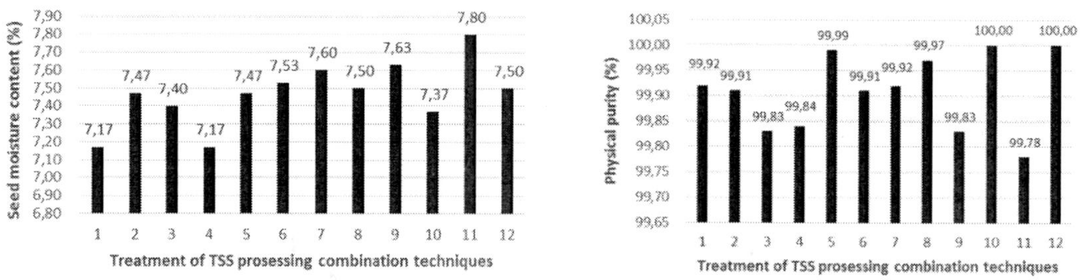

Figure 3. Seed moisture content and physical purity of TSS processing combination techniques

Electrical conductivity is one of the vigor parameters. This parameter is an indicator of seed leakage [14]. It means high electrical conductivity value indicates damage inside a seed and can affect low viability and vigor performance. According to statistical analysis, there was a different significance in electrical conductivity among treatments in the range of 259.88-640.27 μs/cm/g. Treatment No. 6 showed the lowest electrical conductivity performance (259.88 μs/cm/g) and differed significantly from other treatments. It was proved treatment No 6 had the best viability and vigor performance.

Figure 3 shows that seed moisture content value in the range 7.17% (treatment No. 1 and No. 4) until 7.80% (treatment No. 11). The test result showed that the seed moisture content parameter was not significantly different from each treatment because the seed dried after processed manually until minimum reaches moisture content 8.0% as a requirement for all seed classes. Figure 3 could also be known that each treatment's physical purity percentage was not significantly different, in this case in the range of 99.78% (treatment No. 11) until 100.00% (treatment No. 10 and No. 12). This result was caused by a double sortation technique, which the first sorting by winnower, then followed by hand manually;

thus the physical purity level became better. All treatments' physical purity level had fulfilled a minimum requirement as a stock seed (S.S.) class within a limit of 99.5% [13].

4. Conclusion and Suggestion

1. The best treatment was a combination technique with seed capsule drying in the dehumidifier room at 30-35°C for 72 hour - breaking capsule by hand manually – seed sorting by winnower followed by hand manually, then redried in the dehumidifier & heater room at 30-35°C until moisture content was appropriate for breeder seed standard requirement.
2. The produced seed quality in this treatment showed the germination was 75%, the moisture content was 7.5%, and the physical purity was 99.9%.
3. This research implies that the availability of TSS processing technology can be carried out by massal but still can produce good seed quality.

5. Acknowledgments

The authors are thankful to the Indonesia Agency for Agricultural Research and Development (IAARD) for funding research through the IVEGRI DIPA 2018 Project No. 1804.208.053.B.1. Besides that Authors are thankful to the Director of IVEGRI and Dr. Redy Gaswanto, SP., MP, for guidance, motivation encouraging and supporting all this time technically

References

[1] Fanny T, Eliyani and Kurniadinata O F 2020 Can We grow shallot (*Allium ascalonicum* L.) root in hydroponic system with simple growing media?. *J Trop Hortic.* **3** (2) 54-59. DOI: http://dx.doi.org/10.33089/jthort.v3i2.50

[2] Direktorat Jenderal Hortikultura 2017 Petunjuk teknis kegiatan pengembangan sayuran dan tanaman obat tahun 2016. Kementerian Pertanian.

[3] Direktorat Jendral Hortikultura 2020 Statistik hortikultura tahun 2019 (angka tetap). Direktorat Jenderal Hortikultura Kementerian Pertanian.

[4] Sumarni N, Sopha G A and Gaswanto R 2012 Respons tanaman bawang merah asal biji true shallot seeds terhadap kerapatan tanaman pada musim hujan. *J. Hort.* **22** (1) 23–28.

[5] Pangestuti R and Sulistyaningsih E 2011 Potensi Penggunaan true seed shallot (TSS) sebagai sumber benih bawang merah di Indonesia. Prosiding Semiloka Nasional "Dukungan Agro-Inovasi untuk Pemberdayaan Petani, Kerjasama UNDIP, BPTP Jateng, dan Pemprov Jateng, Semarang.

[6] Nurjanani and Djufry F 2018 Test potential for some variety to produce true shallot seed in highland South Sulawesi. *J. Hort.* **28** (2) 201-208.

[7] Rosliani R.,. Hilman Y, Waluyo N and Yufdy M P 2017 Petunjuk teknis teknologi produksi biji botani bawang merah/TSS (*True Seed of Shallot*). Balai Penelitian Tanaman Sayuran. Pusat Penelitian dan Pengembangan Pertanian. Badan penelitian dan Pengembangan Pertanian.

[8] Selvi D T and Saraswathy S 2017 Seed viability, seed deterioration and seed quality improvements in stored onion seeds: a review. *J. of Hort. Scien.and Biotechnology* **93** 1-7.

[9] Prahardini P E R and Sudaryono T 2018 The true seed of shallot (TSS) technology production on Trisula variety in East Java. J Pembang dan Alam Lestari. **9** (1) 27–32.

[10] Sumarni N, Suwandi, Gunaeni, N and Putrasamedja S 2013 Pengaruh varietas dan cara aplikasi GA_3 terhadap pembungaan dan hasil biji bawang merah di dataran tinggi Sulawesi Selatan *J. Hort.* **23** (2): 153–163.

[11] Suheri H and Price T V 2001 Infection of onion leaves by *Alternaria porii* and *Stemphylium vesicarium* and disease development in controlled disease. *Plant Pathol.* **49** (3) 375–82. DOI:10.1046/j.1365-3059.2000.00458x.

[12] Hampton J 2002. What is seed quality?. Seed science and technology **30** (1) 1-10. https://www.researchgate.net/publication/289831470_What_is_seed_quality.

[13] Direktorat Perbenihan Hortikultura 2015 Pedoman Teknis Sertifikasi Benih Bawang Merah Keputusan Menteri Pertanian Nomor : 131/Kpts/Sr.130/D/11/2015. Direktorat Jenderal Hortikultura Kementerian Pertanian.

[14] Lazar S L, Mira S, Pamfil D and Martînez-Laborde J B 2014 Germination and electrical conductivity tests on artificially aged seed lots of 2 Wall-Rocket species. *Turkish J Agric For.* **38** (6) 857–64. DOI: 10.3906/tar-1402-76. https://www.researchgate. net/publication /270684400_

[15] Lestari F and Kusumasari A C 2019 Penerapan teknologi mekanisasi pada proses penanganan pascapanen biji bawang merah. Prosiding Seminar Nasional Kesiapan Sumber Daya Pertanian dan Inovasi Spesifik Lokasi Memasuki Era Industri 4.0. http://repository.pertanian.go.id/handle /123456789/9148

The effect of replacing maize with re-binding banana hump flour on the meat quality, footpad color, and feather or hybrid duck

O Sjofjan[1,*], D N Adli[1], R H Syahputra[1]

[1] Faculty of animal science, University of Brawijaya, 65145, East Java, Indonesia

* E-mail: osofjan@yahoo.com

Abstract. The purpose of this study was to determine the deposition percentage of breast meat, leg colour, and feather weight of hybrid ducks. The research material was 100 hybrid ducks regardless of sex (age 21 days) Peking x Khaki Campbell. The method used was a field experiment using a completely randomized design (CRD) consisting of 5 treatments and four replications. Each replication of 5 hybrid ducks. The treatments used were T0: 20% corn, T1: 15% corn + 5% Banana Hump Flour re-binding, T2: 10% corn + 10% Banana Hump Flour re-binding, T3: 5% corn + 15% Banana Hump Flour re-binding, T4: 20% Banana Hump Flour re-binding. Data were analyzed statistically using Covarian analysis (ANCOVA). If the data results differ significantly between treatments, Duncan's Multiple Range Test (DMRT) will proceed. The results showed that banana hump flour had no significant effect ($P > 0.05$) on the percentage of breast meat deposition and feather weight, but had a very significant effect ($P < 0.01$) on the color of the legs. It can be concluded that the use of re-binding banana hump flour as a substitute for corn up to 20% in feed gives positive results on the percentage of meat and feathers of hybrid.

1. Introduction

The increasing number of population in Indonesia increases the consumption of animal protein that comes from poultry[10]. Ducks are poultry that can be used for meat and eggs [11]. [1] states that duck meat production had increased by 6.21% from 36.4 tonnes in 2017 to 38.7 tonnes in 2018. This increase in duck meat production shows that duck farming is one of the sectors that increase demand and development in society.

Feed-in duck farming contributes 60–70% of the total maintenance costs. One of the energy sources that is widely used in animal husbandry is corn, and this is because corn has ME 3370 Kcal/kg [2]. Currently, to meet the demand for corn, it still depends on imports because the availability of maize in Indonesia is limited, which causes the price of corn to fluctuate very much. Therefore, it is necessary to seek alternative feed ingredients that are good, easy to obtain, relatively inexpensive, and do not compete with humans without neglecting the nutritional value as a substitute for corn on duck farms.

One of the efforts that can be made to replace corn as an energy source is to use local feed ingredients, one of which is banana hump (*Musa paradiciasa L.*) [12]. Banana hump is the underside of a banana stem that is tuber-shaped and is located in the ground. Banana hump is a waste from banana plantations that are rarely used, but in general, plantation waste has a low nutrient content, so it needs processing first to be used as feed material. The advantage of using banana humps is that they have like 76% starch, 20% water, and 4% other ingredients [8]. Banana hump flour is made from fresh banana humps that go through chopping, drying and grinding into flour. Banana hump flour has a crude fibre content of 26.72% [8]. [4] states that banana hump flour has a GE content of 2244.20 Kcal/kg. The presence of high crude fibre content requires technological engineering and nutritional reformulation so that the banana hump can be used as an alternative feed material for energy sources to replace corn.

Re-binding technology is an engineering technology that aims to enrich nutritional content by adding cellulase enzymes to reduce crude fibre and adding amino acids to increase protein levels. Banana hump flour Re-binding can be used as an alternative to ducks as an alternative energy source of feed for corn, which is thought to increase the deposition percentage of breast meat, leg colour, and feather weight of hybrid ducks. Based on the description above, it is necessary to research the effect of using banana hump flour re-binding as a substitute for corn in the feed on the deposition percentage of breast meat, leg colour, and weight of hybrid ducks after being given banana hump flour treatment re-binding.

2. Methods

The materials used in the study were 100non-sexing hybrid ducks with cross-hybrid strains of Peking ducks (male) and Khaki Campbell ducks (females) 21 days old with an average body weight of 421.31±183.90 g with a diversity coefficient of about 44%. Hybrid ducks are obtained from smallholder farms in Blitar Regency for at of IDR. 10,000 per head. The treatment feed was in the form of banana flour from the type of Banana Kepok (*Musa paradiciasa* L.) obtained from Trenggalek, East Java. The number of cages used in this study was 20 flocks with five ducks in each flock. The duck cage is equipped with a place for feeding & drinking, hygrometer and other cage equipment. Feed and drinking water were given ad libitum.

The research method used in this study was a field experiment using a completely randomized design (CRD) consisting of 5 treatments and four replications. Each replication consisted of 5 hybrid broiler ducks. The treatments used were as follows:

T0: Basal feed without banana hump flour re-binding substitution
T1: Feed using substitution of corn with banana hump flour re-binding 5%
T2: Feed using substitute corn with banana hump flour re-binding 10%
T3: Feed using substitution of corn with banana hump flour with re-binding 15%
T4: Feed using substitution of corn with banana hump flour with re-binding 20%

Deposition breast meat is part of breast meat that is taken without the bones then weighed [9]. The percentage of breast meat deposition was obtained by weighing the breast meat of hybrid ducks after slaughtering at the age of 56 days compared to their live weight (%). The nutritional content, especially protein, will affect the deposition of breast meat. The average percentage deposition of hybrid duck breast meat fed with banana hump treatment is re-binding presented in Table 3. The results showed that the percentage of breast meat from the highest sequentially during the study was P4 (8.57 ± 1.66%), P0 (8.44 ± 1.16), P3 (8.07 ± 1.04), P2 (8.06 ± 1.08), P1 (7.89 ± 0.94) use banana hump flour re-binding in the feed had no significant effect (P> 0.05) on the percentage of deposition of hybrid duck breasts. The results of this study were still lower than the results of [2]. The highest analysis results for the deposition of mojosari duck breast meat were in the P3 treatment (30.37+2.8%). In addition from [6] stated the local duck had more breast meat at range 26-29%. This is influenced by the nutritional content of the feed, especially the protein content. The high breast weight allows the ducks to produce a high percentage of the carcass.

3. Results and discussion

Table 1. Nutrient Analysis

Feed Ingredients	CP (%)*	CF (%)**	Fat (%)*	GE (Kcal/g)**
Rice Bran	12.85	1.83	9.66	4104
Corn	9.01	1.73	3.87	3561
Concentrate	38.39	3.91	2.32	3319,52
Banana hump flour	9.22	9.27	1.43	3695, 38
MBM	55.11	12.11	4.97	3901,64

Source: *Analysis Result from Feed Nutrition Laboratory, Faculty of Animal Science, Brawijaya University
**Analysis Result from Centre for Food and Nutrition Studies Laboratory, Gajah Mada University

Table 2. Composition and Nutrient Analysis

Feed Ingredients	Composition (%)				
	T0	T1	T2	T3	T4
Rice Bran	60	60	60	60	60
Concentrate	20	20	20	20	20
Corn	20	15	10	5	0
Banana hump flour	0	4,5	9	13,5	18
MBM	0	0,5	1	1,5	2
Total	100	100	100	100	100
	Nutrition Content				
Energy (Kcal/kg)*	3838,5	3846,3	3854,01	3861,8	4047,5
Crude Protein (%)*	17.19	17.43	17.67	17.91	18.60
Crude Fiber (%)*	2.23	2.62	3.01	3.4	3.88
Crude Fat (%)*	7.03	6.93	6.83	6.72	6.81

Table 3. The effect of replacement to breast meat, leg colour, and feather weight.

Variables	Treatments				
	T0	T1	T2	T3	T4
Deposition of Breast Meat (%)	8.44±1.16	7.89±0.94	8.06±1.08	8.07±1,04	8.57±1.66
Leg Color	8.5 ± 0.58^{cd}	7.0 ± 0.82^{c}	7.5 ± 1.73^{cd}	5 ± 0.82^{b}	2.25 ± 0.96^{a}
Feather Weight (%)	8.32±0.97	9.72±1.04	9.38±0.57	9.26±1.02	8.54±1.15

The average value of the use of banana hump flour re-binding as a substitute for corn in feed to leg colour can be seen in Table 3. The average result of the treatment is P0 (8.5 ± 0.58).), P1 (7.0 ± 0.82), P2 (7.5 ± 1.73), P3 (5 ± 0.82), P4 (2.25 ± 0.96). The variation analysis results showed that the use of banana hump flour re-binding in the feed as a substitute for corn had a significant effect (P <0.01) on the hybrid duck feet colour. This indicates that there is a significant difference in the colour of the legs, this is thought to be due to the carotene content in different feeds. [7] stated that the characteristics of the beak colour and leg colour were found to be colour variations, especially the yellow or orange colour of the duck's beak. The color pigment is carotene. The highest average yield was found in treatment P0 (8.5 ± 0.58) where there was no replacement of corn with a banana hump in the feed. Treatment P4 showed the lowest result, namely (2.25 ± 0.96). This is due to replacing corn as a source of carotene and xanthophyll with banana re-binding humps. [3] explains that the choice of corn is because corn is a source of pigments xanthophyll which causes yellow on the feet and skin.

Feather weight on banana hump flour treatment re-binding as a substitute for corn in feed can be seen in the Table. 3. The average treatment results were P1 (9.72 ± 1.04), P2 (9.38 ± 0.57), P3 (9.26 ± 1.02), P4 (8.54), ± 1.15), P0 (8.32 ± 0.97). Statistical analysis results showed that the use of banana hump flour re-binding as a substitute for corn on feather weight had no significant effect (P> 0.05). The highest average hair weight was in treatment P1 (9.72 ± 1.04). The percentage of feather weight was obtained by weighing the ducks' feathers after plucking in a dry state divided by the live weight (%). According to research by [5] stated that the weight of duck feathers in the treatment showed the results of P1 (7.77%), P2 (6.67%), P3 (13.33%) and P4 (12.23%). This is influenced by the nutritional content, especially protein in the feed. Protein has an important function for poultry, one of which is for feather growth.

4. Conclusion

It can be concluded that the use of re-binding banana hump flour as a substitute for corn up to 20% in feed gives positive results on the percentage of breast meat deposition, as well as the weight of hybrid duck feathers.

References

[1] Agriculture Ministry of Indonesia 2017 *Livestock Animal Health Statistics* (Jakarta-Indonesia)
[2] Putri M, Wiidodo E, and Sjofjan O 2015 *JIIP*. **25** L 2
[3] Royani M 2016 *JIP* **1** L 1
[4] Sjofjan O, Adli DN, and FA Mukflikhien 2020 *JNTTIP* **2** L2 7
[5] Suartiningsih NPM, Dewi GAMK, Utami IAP, and Sutama INS 2016 *JPT* **4** L1
[6] Sjofjan O, Adli D N, Natsir M H, Nuningtyas Y F, Bastomi I, Amalia FR 2021 *J. Indonesian Trop. Anim. Agric.* **46** L 1
[7] Sjofjan O and Adli D N 2021 *LRRD* **33** L 17
[8] Sjofjan O, Adli D N, Sholikin M M, Jayanegara A and Irawan A 2021 *J. of Anim. Feed Sci.* **30**
[9] Widyawati I, Sjofjan O, Adli DN 2020 *JNT* **3** L 1
[10] Sjofjan O, and Adli D N 2020 *IOP Conf. Series: Earth and Env. Sci.* **478**
[11] Sjofjan O, Natsir M H, Adli D N, Adelina D D, and Triana L M. IOP Conf. Series: Earth and Env. Sci. **465**
[12] Adli D N, Sjofjan O, Natsir M H, Nuningtyas Y F, Sholikah N, and Marbun A C 2020 *LRRD* **32** L 120

IOP Conf. Series: Earth and Environmental Science **883** (2021) 012060 doi:10.1088/1755-1315/883/1/012060

Characterization of pectin from tongka langit banana peels with various extraction temperature

P Picauly[1,*] and G Tetelepta[1]

[1] Department of Agricultural Technology, Faculty of Agriculture, Pattimura University, Jl. Ir. M. Putuhena, Ambon, Moluccas, Indonesia.

* E-mail: silyapicauly9806@gmail.com

Abstract. Banana peel waste can be used in the food industry as a source of pectin. Pectin can be obtained from the extraction process tongka langit banana peel. This research's objective was to characterize pectin's properties extracted from banana tongka langit peel with different extraction temperatures. A completely randomized experimental design with 3 level treatment extraction temperature of banana tongka langit peel, i.e., 80°, 85°, and 90°C, with three replicates was applied in this research. The properties evaluated, including equivalent weight, methoxyl content, galacturonate acid, esterification degree, yield, moisture content, and ash content. Result showed that equivalent weight (3104,3 mg, 1084,0 mg, and 699,8 mg), methoxyl content (3,25%, 3,87%, and 5,44%), galacturonate acid (24,56%, 38,06%, and 56,08%), esterification degree (76,02%, 57,45%, and 55,05%), moisture content (11,39%, 10,79%, and 9,85%), ash (4,42%, 5,11%, and 6,47%)), and yield (8,37%, 8,80%, and 9,20%).

1. Introduction

Banana tongka langit (*Musa troglodytarum* L.) is one of the local foods in the Maluku region containing high carbohydrate, fiber, vitamins and caratenoid compounds to be beneficial for health. With this potential, efforts were made to develop tongka langit bananas into several processed products including instant porridge [1], crackers[2], brownies [3], starch [4], and several other products.

Processing the tongka langit banana to produce a product is in the stage of peeling the fruit peel to be taken from the fruit flesh so that the fruit skin will only be disposed of as organic waste or used as animal feed. Waste can cause environmental damage problems, waste elimination, and banana peels can be used as a product with a sale value, one of which is pectin. Banana peels could be used as a food source with high value, such as pectin [5].

Pectin is a polysaccharide compound with a high molecular weight with a functional food value due to its suitable emulsion and stability to be used as a gelling agent and stabilizer, for example, in jelly manufacture jam, marmalade, edible film, and edible coating.

Several research results use banana peels to be used as pectin, including pectin from cavendish banana peels [6], saba banana peels [7], and raw banana peels [8]. Based on the study results, the pectin content in the banana peel tongka langit is 10.41% [9], so that the tongka langit banana is made to produce pectin compounds. Pectin can be obtained from the source through the extraction process. Pectin extraction can be relied on by several factors, including pH, type of solvent, extraction, and extraction temperature. So these factors need to be considered to get the best quality pectin.

This research carried out pectin extraction from tongka langit banana peels with various extraction temperatures based on the description above.

2. Methods

This research's raw material was the tongka langit banana obtained from Soya State, Ambon City, Maluku. The chemicals used include hydrochloric acid (Merck, Germany) and 96% ethanol (Merck, Germany). The equipment consists of a cabinet dryer, analytical scales (Ohaus, USA), and hot plates.

2.1. Pectin extraction

Pectin extraction refers to the method of Ramdja[10] with modifications. 20 g of cleaned tongka langit banana peel is dried using a cabinet dryer. After drying, the banana peel is crushed with a crusher. The powder of the tongka langit banana peel was added with 500 mL of HCl (0.1 N) and extracted based on temperature treatment (80 °C, 85 °C, and 90 °C) for 2 hours. Furthermore, filtering is carried out to separate the precipitate from the filtrate. Pectin phytate was precipitated with 96% ethanol (1: 1). The precipitated pectin is separated from the filtrate using filter paper. The residue was washed with 96% ethanol. The results of the filtration were dried using an oven at 40 ^{oC} for 24 hours. Then pectin analysis was carried out for the equivalent weight[11], methoxyl content [11], galacturonic acid [11], degree of esterification [11], moisture content[12], ash content [12], and yield.

2.2. Equivalent weight

Equivalent weight was defined using weighing 0.5 g of pectin in 100 mL conical flask and moistened with 5 mL of ethanol. 1 g of sodium chloride would be added for the more incisive endpoint. The mixture was then stirred until it reached homogeneity. Titration was then slowly performed (to avoid the possibility of de-esterification) using standard 0.1 N of NaOH until the red indicator turned pink (pH 7.5) and lasted 30 seconds at most. Neutralizing solvent was then used to determine methoxyl content. The equation below was used to calculate the equivalent weight:

$$\text{(mg)Equivalent weight} = \frac{\text{weight of sample x 1000}}{\text{mL titrated NaOH x N NaOH}}$$

2.3. Methoxyl content

In determining the methoxyl content, 25 mL of 0.25 N NaOH was added to titration solvent, shaken, and left for 30 minutes at room temperature in a stoppered flask. And 25 mL of 0.25 N HCl was then added and titrated with 0.1 N NaOH solution until the same previous endpoint.

$$\text{(\%) Methoxyl content} = \frac{\text{mL titrated NaOH x N NaOH x 31}}{\text{pectin weight}}$$

2.4. Galacturonic acid

Galacturonic acid would then be determined using the following:

$$\text{(\%) Galacturonic acid} = \frac{\mu\text{eq (equivalent content+methoxyl content)x 176 x 100}}{\text{pectin weight}}$$

2.5. Degree of esterification

Degree of esterification in pectin was difined as follows:

$$\text{(\%)Degree of esterification} = \frac{\text{Methoxyl content x 176 x 100}}{\text{galacturonic acid x 31}}$$

2.6. Moisture content

Weight about 5 g of sample to the dish. Place the dish with the sample in the oven. Dry for 3 hours at

105 °C. After drying, transfer the dish with a partially covered lid to the desiccator to cool. Reweight the dish and its dried sample.

$$\% \text{ Moisture content} = \frac{\text{weight of sample before drying} - \text{weight of sample after drying}}{\text{weight of sample before drying}} \times 100$$

2.7. Ash content
Weight about 1.5 g sample into the crucible. Heat over low Bunsen flame with the lid half covered. When fumes are no longer produced, place the crucible and lid in the furnace. Heat at 550 °C in the electric furnace until ashes are complete. Transfer the dish with a partially covered lid to the desiccator to cool and then weighed.

$$\% \text{ Ash content} = \frac{\text{weight of ash}}{\text{weight of sample}} \times 100$$

2.8. Pectin yield
Pectin yield was calculated as follows:

$$\% \text{ Yield} = \frac{\text{weight of dried pectin}}{\text{weight dried banana peel taken for extraction}} \times 100\%$$

2.9. Statistical analysis
The research design was carried out using a completely randomized design with three treatment levels of pectin extraction temperature. Each treatment level was repeated three times. Analysis of variance was used to analyze the data obtained and the Tukey test for the avarage treatment difference test using Minitab 17 software.

3. Results and discussion

3.1. Weight equivalent
The average value of the equivalent pectin weight of tongka langit banana peel with temperature treatment can be seen in Table 1. The results of the analysis of variance show that the extraction temperature has a significant effect on the weight equivalent of the pectin tongka langit banana peel.

The pectin extraction of tongka langit banana peels at 80 °C produces an equivalent weight greater than 85 °C and 90 °C. The results showed that the higher the extraction temperature, the smaller the equivalent weight value. This is due to the de-esterification process, namely the change in pectin to pectic acid, which can increase the number of free acid groups so that the amount of pectin equivalent weight will decrease. By the research of Aziz [13], the higher the extraction temperature, the lower the equal weight; this is due to many esterified compounds.

The tongka langit banana peel's pectin equivalent weight at 80 °C is 3104.3 mg, 85 °C is 1084.0 mg, and 90 °C is 699.8 mg. For pectin extracted with a temperature of 90 °C, which meets the quality standards of IPPA (International Pectin Producers Association) [14] because it is in the range between 600-800 mg.

Table 1. Characteristics of banana peel pectin based on extraction temperature.

Pectin characteristics	Treatment extraction temperature (°C)			Quality standard IPPA (*International Pectin Producers Association*)
	80	85	90	
Equivalen Weight (mg)	3104.3 a	1084.0b	699.8c	600-800 mg
Methoxyl content (%)	3.25 a	3.87b	5.44c	< 7% (low methoxyl)
Galacturonic Acid (%)	24.56 a	38.06b	56.08c	Min 35%

Esterifikasi Degree (%)	76.02 a	57.45b	55.07c	>50% (high ester pectin)
Moisture content (%)	11.39 a	10.79b	9.85c	Max 12%
Ash content (%)	4.42 a	5.11b	6.47c	Max 10%
Yield (%)	8.37 a	8.80b	9.20c	

Numbers followed by different letters at the same column are significantly different at 5% of Tukey.

3.2. Methoxyl content

The average methoxyl pectin content of tongka langit banana peels with temperature treatment can be seen in Table 1. The results of the analysis of variance showed that the extraction temperature had a significant effect on the methoxyl pectin content of the tongka langit banana peel.

The pectin extraction of Tongka Langit banana peels at 90 °C resulted in a methoxyl content greater than 80 °C and 85 °C. The results showed that the higher the extraction temperature, the greater the methoxyl content. The increase in methoxyl levels is due to higher temperatures, which will cause a faster protopectin hydrolysis reaction and result in breaking the bonds of the pectin galacturonic component with other compounds, so that there will be an increase in carboxyl groups which will react with alcohol and will produce esters. According to the research results of Triandini[15], the increasing extraction temperature will cause higher methoxyl pectin levels due to the rising esterified free carboxyl groups.

The methoxyl pectin content of tongka langit banana peels at 80 °C was 3.25%, 85 °C was 3.87%, and 90 °C was 5.44%. Based on the quality standards of IPPA [14]. The banana peel pectin of tongka langit is less than 7%, classified as low methoxyl pectin. According to Goycoolea and Adriana[16], pectin is high methoxyl if the methoxyl content is equal to or more than 7%, and low methoxyl if the methoxyl content is less than 7%.

3.3. Galacturonic acid

The mean of galacturonic acid pectin levels of tongka langit banana peels with temperature treatment can be seen in Table 1. The results of the analysis of variance showed that the extraction temperature had a significant effect on the galacturonic acid levels of the pectin peels of Tongka Langit banana peels.

The pectin extraction of tongka langit banana peels at 90 °C resulted in a higher galacturonic acid level than 80 °C and 85 °C. The results showed that the higher the extraction temperature, the higher the galacturonic acid levels. High temperature will cause protopectin's hydrolysis reaction to pectin, which is the basic component of D-galacturonic acid. Galacturonic levels can affect the structure and texture of the gel that is formed. According to Desmawarni & Hamzah [17], the higher the temperature causes the hydrogen ions produced to substitute calcium and magnesium from the more protopectin, the hydrolyzed protopectin produces more pectin.

Galacturonic acid levels of pectin of tongka langit banana peel at 80 °C were 24.56%, 85 °C temperatures were 38.06%, and 90 °C was 56.08%. For the extracted pectin with temperatures of 85 °C and 90°C which meet the quality standards of IPPA [14] because it has a galacturonic acid content of more than 35%.

3.4. Degree of esterification

The average degree of esterification of the tongka langit banana peel pectin with temperature treatment can be seen in Table 1. The results of the analysis of variance showed that the extraction temperature had a significant effect on the degree of esterification of the tongka langit banana peel pectin.

The pectin extraction of Tongka Langit banana peels at a temperature of 80 °C resulted in a greater degree of esterification than 85 °C and 90 °C. The higher the extraction temperature, the higher the degree of esterification. This is due to the presence of a glycosidic bond of the methyl ester group of pectin, which tends to be hydrolyzed to produce galacturonic acid. According to Budiyanto & Yulianingsih[18], the degradation of the methyl ester group in pectin becomes carboxyl acid in the presence of acid. The glycosidic bonds of the methyl ester group of pectin tend to be hydrolyzed to produce galacturonic acid.

The degree of esterification of tongka langit banana peel pectin at 80 °C was 76.02%, 85 °C was 57.45%, and 90 °C was 55.07%. Based on the IPPA [14] quality standard, the banana peel pectin produced in this study belongs to the high ester pectin class because it has an esterification degree of more than 50%.

3.5. Moisture content

The average moisture content of the tongka langit banana peel pectin with temperature treatment can be seen in Table 1. The results of the analysis of variance showed that the extraction temperature had a significant effect on the moisture content of the Tongka Langit banana peel pectin.

The pectin extraction of tongka langit banana peels at a temperature of 80 °C resulted in a greater moisture content than temperatures of 85 °C and 90 °C. The results showed that the higher the temperature, the lower the moisture content. This decrease in moisture content is due to the high temperature, facilitating water evaporation to speed up drying. The results of Roikah[19] showed that the water content of the extracted pectin at a temperature of 60 °C was higher than the extraction temperature of 100 °C. High extraction temperatures can hydrolyze the pectin polymer so that the shorter molecular chains will dry out more quickly.

The tongka langit banana peel pectin's water content at an extraction temperature of 80 °C was 11.39%, a temperature of 85 °C was 10.79%, and a temperature of 90 °C was 9.85%. Based on the IPPA [14] quality standard, pectin's maximum moisture content is 10%, so the moisture content of the banana peel pectin from this study meets the quality standard.

3.6. Ash content

The mean ash content of the tongka langit banana peel pectin with temperature treatment can be seen in Table 1. The results of the analysis of variance showed that the extraction temperature had a significant effect on the ash content of the tongka langit banana peel pectin.

The pectin extraction of tongka langit banana peels at 90 °C resulted in a more significant ash content than 80 °C and 95 °C. The higher the extraction temperature, the higher the ash content. The increase in ash content is due to a heating process, which can result in a protopectin hydrolysis reaction, which will cause an increase in Mg and Ca components in the extract solution so that more magnesium and calcium minerals will increase the ash content [20].

The ash content of the tongka langit banana peel pectin at an extraction temperature of 80 °C was 4.42%, a temperature of 85 °C was 5.11%, and a temperature of 90 °C was 6.47%. Based on the IPPA [14] quality standard, the maximum pectin ash content is 12%, so that the ash content of the banana peel pectin from this study meets the quality standard.

3.7. Yield

The average yield of tongka langit banana peel pectin with temperature treatment can be seen in Table 1. The results of the analysis of variance showed that the extraction temperature had a significant effect on the yield of tongka langit banana peel pectin.

The pectin extraction of tongka langit banana peels at 90 °C resulted in a higher yield than 80 °C and 85 °C. The higher the extraction temperature, the yield will increase. This is because the higher the extraction temperature, the hydrolysis of protopectin to pectin will increase. Protopectin that still binds to cellulose is released and converted into water-soluble pectin by hydrolysis. According to Aziz [13], the higher the temperature caused, the higher the yield because the hydrogen ions produced will substitute more calcium and magnesium from protopectin. The hydrolyzed protopectin produces more pectin.

4. Conclusion

Based on the results of the study, it was concluded that the pectin characteristics of the tongka langit banana peel based on extraction temperatures of 80 °C, 85 °C, and 90 °C resulted in equivalent weight

(3104.3, 1084.0, and 699.8 mg), methoxyl content (3.25, 3.87, and 5.44%), galacturonic acid (24.56, 38.06 and 56.08%), degree of esterification (76.02, 57.45 and 55.05%), moisture content (11.39, 10.79 and 9.85%), ash content (4.42, 5.11 and 6.47%) and yield (8.37, 8.80 and 9.20%).

References

[1] Picauly P and Tetelepta G 2015 Karakteristik kimia bubur instan tersubtitusi tepung pisang tongka langit *Agroforestri* **9** 122-6

[2] Picauly P and Tetelepta G 2016 Uji organoleptik crackers pisang tongka langit *Agritekno* **5** 53-57 DOI: 10.30598/jagritekno.2016.5.2.53

[3] Moniharapon E, Lelmalaya L and Picauly P 2019 Kajian sifat kimia dan organoleptik brownies pisang tongka langit *Agritekno* **8** 60-63 DOI: 10.30598/jagritekno.2018.7.2.60

[4] Palijama S, Singkery M, Breemer R and Polnaya F J 2020 Isolation and characteristics of *Musa troglodytarum* L. starch at different maturity stages *Journal of Physics: Conference Series* 1463: 012015 DOI: 10.1088/1742-6596/1463/1/012015

[5] Happi E T, Ronkart S N, Robert C, Wathelet B and Paquot M 2008 Characterization of pectins extracted from banana peels (Musa AAA) under different conditions using an experimental design *Food Chemistry* **108** 463-471 DOI: 10.1016/j.foodchem.2007.10.078

[6] Oliveira T I S, Rosa M F, Cavalcante F L, Pereira P H F, Moates G K, Wellner N, Mazzetto S E, Waldron K W and Azeredo H M C 2016 Optimization of pectin extraction from banana peels with citric acid by using response surface methodology *Food Chemistry* **198** 113-8 DOI: 10.1016/j.foodchem.2015.08.080

[7] Kamble P B, Gawande S and Patil T S 2017 Extraction of pectin from unripe banana peel *International Research Journal of Engineering and Technology* **4** 2259-2264

[8] Tanaid, R A B and Lauzon R D 2018 Extraction optimization of pectin from unripe banana (*Musa acuminta* balbisina var. cardaba) peel *International Journal of Engineering* **4** 308-315

[9] Sutapa, I W, Siahay V P D and Tanasale M F J D P 2014 Adsorpsi Cu^{2+} pada pektin dari kulit pisang Tongka Langit (Musa speices Van Balbisiana) *Indonesian Journal of Chemical Research* **1** 72-77

[10] Ramdja, A F, Adhitya D P, and Rusman R 2011 Ekstraksi pektin dari kulit pisang kepok dengan pelarut asam klorida dan asam asetat *Jurnal Teknik Kimia* **17** 28-37

[11] Ranggana, S 1997 *Handbook of Analysis and Quality Control for Fruit and Vegetable Products Second Editions Tata McGraw-Hill* Publishing Company Limited: New Delhi

[12] AOAC 2012 *Association of Official Analytical Chemistry*, Washington D.C. Official Method of Analysis

[13] Aziz, T, Johan E G M and Sri D 2018 Pengaruh jenis pelarut, temperatur, waktu terhadap karakterisasi pektin hasil ekstraksi dari kulit buah naga (*Hylocereus polyrhizus*) *Jurnal Teknik Kimia* **24** 17-27

[14] IPPA 2002 *What is Pectin?* https://ippa.info/labelling-safety-regulation-of-pectin/ International Pectin Producers Association Accessed 12th January 2020

[15] Triandini, M M, Aslamiah and Wicakso D R 2014 Pengambilan pektin dari albedo semangka dengan proses ekstraksi asam *Konversi* **3** 1-9

[16] Goycoolea, F M and Cardenas A 2003 Pectins from *Opuntia* spp.: A short review. *Journal of the Professional Association for Cactus Development* **5** 17-29

[17] Desmawarni D and Hamzah F H 2017 Variasi suhu dan waktu ekstraksi terhadap kualitas pektin dari kulit pisang tanduk *JOM Faperta UR* **4** 1-15

[18] Budiyanto, A and Yulianingsih 2008 Pengaruh suhu dan waktu ekstraksi terhadap karakter pektin dari ampas jeruk siam (*Citrus nobilis*) *Jurnal Pascapanen* **5** 37-44

[19] Roikah, S, Rengga W D P, Latifah and Kusumastuti E 2016 Ektraksi dan karakterisasi pektin dari belimbing wuluh (*Averhoa bilimbi* L) *Jurnal Bahan Alam Terbarukan* **5** 29-36 DOI: 10.15294/jbat.v5i1.5432

[20] Hanum, F, Kaban I M D and Tarigan M A 2012 Ekstraksi Pektin dari Kulit Buah Pisang Raja *Jurnal Teknik Kimia USU* **1** 21-26

The impact of el niño southern oscillation and covid-19 on the rice price dynamics in Indonesia: the vector error correction model approach

P Hasudungan[1,*], I Irham[1,2], A W Utami[1]

[1] Department of Agricultural SocioEconomics, Universitas Gadjah Mada, Jl. Flora, 55281, Indonesia
[2] Center of Asia-Pacific Study (PSAP), Universitas Gadjah Mada, Jl. Teknika Utara, 55281, Indonesia

* E-mail: peterhasudungan@mail.ugm.ac.id

Abstract. Half of the world's population makes rice a crop of focus because it has many opportunities in the future. However, there are disruptors for rice development in 2020, which are climate variability and COVID-19 Pandemics. El Nino-Southern Oscillation (ENSO) is a climate variability that can threaten the rice price dynamics. Besides that, COVID-19 also has implications for rice price dynamics. The objectives of this paper are: (1) to empirically study the impact of the El Nino-Southern Oscillation on rice price dynamics and (2) to analyze the effect of COVID-19 pandemics on rice price in Indonesia. The study utilized the vector error correction model (VECM) estimation method and the variables used were sea surface temperature (SST), COVID-19 daily cases, rice price, rice production, and rice price regulation. The result is that El Nino has a positive effect on rice prices, which can increase the price level. The opposite is true also for a La Nina shock. COVID-19 also has a positive impact on the daily cost of rice. The results of this study ought to be of interest to rice producers in Indonesia, as well as processors and intermediaries in the rice markets.

1. Introduction

Rice is a staple food in Asia and nearly half of the world's population makes it a crop of focus because of a lot of opportunities in the future [1]. Rice production is predicted to increase towards 2020 with a growth of 0.76 percent per year [2]. Unfortunately, rice will suffer the most due climate variablitity and COVID-19 because they disrupt high water and labor requirements [3], [4]. Furthermore, the impacts will be discussed in this study.

The impact of climatic factors on production has become a concern of some Indonesian researchers, such as Dhamira & Irham and Mulyasari *et al.* [5], [6]. First, we talk about climate variability. El Niño-Southern Oscillation (ENSO) is a climate variability that fluctuates between a warm phase (El Niño) and a cold phase (La Niña) [7] and it has terrible implications for hydrology and productivity of crops [8]. The worst thing about ENSO is that it has a severe and negative impact on developing countries, especially Indonesia, which has a vast agricultural sector [9].

Rice requires a more significant amount of water than other cereal crops, but it will be optimal if the water availability is stable, especially rainfall [10]. Unfortunately, the ENSO phenomenon makes rainfall fluctuate due to the warm phase (decreasing rainfall) [11] and the cold phase (increasing rainfall) [12]. Furthermore, these fluctuations will reduce rice productivity [3], [13], [14].

The decline in rice productivity can fluctuate rice price dynamics because it cannot meet the increasing demand for rice [15]. Indonesian rice demand is relatively high with an average growth in rice consumption of 0.9% per year [16]. Based on that facts, it can be concluded that rice prices are vulnerable to El Nino and La Nina through high increasing demand.

Several studies have revealed the relationship between ENSO and rice price dynamics. Falcon *et al.* stated that rice price dynamics are closely related to El Niño and La Niña events [17]. Brunner found that rice is a commodity that is more price affected by El Niño and La Niña events than other items [18]. Deng *et al.* indicated that El Nino had a significant effect on international rice prices [14]. In his research, the El Nino event contributed significantly to the increase in domestic rice prices in Indonesia and the Philippines. These studies underline the impact of ENSO on prices between the warm phase (El Nino) and the cold phase (La Nina). As a comparison with other commodities, several studies have found that prices increase during El Niño and decrease during La Niña [19]–[21].

Besides, rice has the most considerable price volatility than other cereal crops [22]. Rice prices also quickly increase from time to time due to the rice market's uncertainty [23]. With these characteristics, coupled with El Nino and La Nina's occurrence, rice prices' volatility will worsen and become the most significant contributor to inflation in a country. Therefore, the El Nino phenomenon is quite dangerous for a country's economy even though 20% of the movement of commodity price inflation is influenced by El Nino [18].

Second, we talk about COVID-19 impact on rice prices. As the number of positive cases continues to rise, agricultural especially rice markets, face disruptions because of labor shortages created by restrictions on movements of people and shifts and income losses [24]. Research has been conducted to study rice prices in China and resulted that no significant impact on rice price [25]. However, it may be different from Indonesia cases until November 2020 because Indonesian COVID-19 cases have increased more than 444,000 cases, while China's growth has decreased sharply and significantly [26].

Therefore, this study was conducted to clarify further the impact of El Nino-Southern Oscillation and COVID-19 on rice prices so that price volatility can be reduced. Because if food prices are stable, poor farmers and consumers will not fall into the poverty trap [27]. This study's objectives are: (1) to empirically study the impact of the El Nino-Southern Oscillation on rice price dynamics and (2) to analyze the effect of COVID-19 pandemics on rice price in Indonesia.

2. Review of Literature

2.1. The Impact of El Nino-Southern Oscillation on Rice Price Dynamics

El Niño has the meaning of' The Little Boy 'or' Christ child 'in Spanish. It was recognized as the appearance of warm water in the Pacific Ocean [28]. Hot ocean temperatures in the Equatorial Pacific are evidence of El Nino's existence. Meanwhile, the opposite of El Niño is called La Niña, which is characterized by freezing ocean temperatures in the Equatorial Pacific. El Niño and La Niña are the opposite phases of the El Niño-Southern Oscillation (ENSO) cycle, a scientific term that describes the fluctuations in temperature between the ocean and atmosphere in the east-central equatorial Pacific [7].

There is a consistent wind called trade winds, blowing from east to west under the normal condition in Tropical Pacific. These winds push warm water near the surface and slowly accumulated on the western side of the ocean (near Asia and Australasia). The warmer water near South and Central America gets pushed away from the coast and replaced by cold water from the deep sea. This process is called upwelling. The upwelling process creates a temperature difference across the tropical Pacific, with warmer water accumulates in the west (Asia and Australasia) and cooler water in the east (South and Central America). Warmer water adds extra heat to the air which causes the air to rise rapidly and this rising air creates a scenario of more unsettled weather with more clouds and rainfall. The rising air in the west sets up an atmospheric circulation across this part of the world with warm moisture rising on the western Pacific Ocean and cooler dry air descending on the other side. This circulation reinforces the easterly winds so that this part of the world sits in a self-perpetuating state until the El Niño event begins [29].

There has been no research that directly addresses the impact of El Niño and La Niña on rice prices in Indonesia. Previous research only focuses on its effect on rice production [3]. However, if El Niño is associated with drought, Jati has proven that the phenomena cause increase rice prices [30].

However, if we look at the broader context than rice, El Niño and La Niña also affect the price of other commodities. For example, Credit Suisse's recent report highlights the spikes in palm oil prices, mostly in the El Niño period (Figure 1). Additionally, several studies have found that El Niño causes an increase in prices, while La Niña causes a decrease in prices [19], [20], [31], [32].

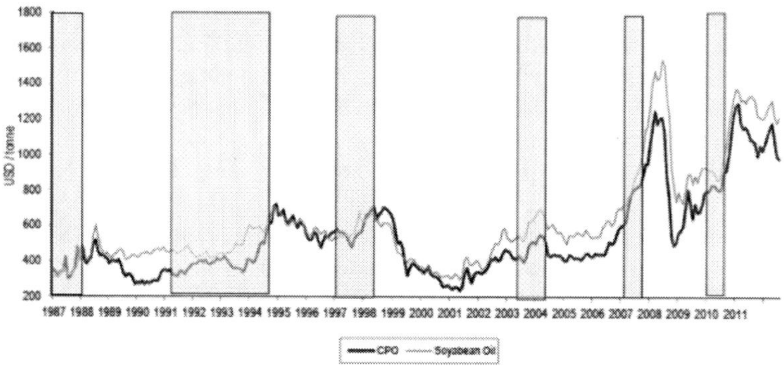

Figure 1. CPO Prices (Line) Soaring During El Niño (Marked in Blue)

Prices also react differently between El Nino and La Nina. Ubilava and Holt stated that prices are more responsive to El Niño shocks than La Nina shocks [19]. This statement means that the response of rice prices is more volatile to the El Nino shock than the La Nina shock. Meanwhile, prices are more sensitive to La Niña shocks than to El Nino shocks. Sensitive means that the La Nina shock variance contribute more to the price of rice than El Nino.

In Indonesia, the price of rice is regulated by state policies, namely *Harga Pembelian Pemerintah* (HPP) and HET (*Harga Eceran Tertinggi*). The HPP policy protects the price of grain or rice at the farm level and the HET policy protects the purchasing power of consumers for rice products as a staple food [33], [34]. Rachman *et al.* stated that the implementation of HET increases the selling price at the farmer level and increases rice farming profitability [35]. HPP implementation's impact is that the purchase price of rice increases at the farm level so that the rice farmer surplus increases [36].

2.2. The Impact of COVID-19 Pandemics on Rice Price

The coronavirus disease 2019 (COVID-19) is an infectious disease caused by the novel coronavirus. Before it has been a global pandemic, the first case was first founded in Wuhan city of China. Almost all the countries in the world have been suffering due to its transmission [37]. Figure 2 shows that how are the changes in COVID-19 cases in Indonesian which shows positive trends.

In the introduction, the study conducted by Yu *et* al. of the COVID-19's impact on rice prices in China was mentioned [25]. There are no enough empirical studies to prove the existence of COVID-19 impact on rice price dynamics. On the contrary, several studies have supported how the COVID-19 pandemic impacted the food agro system to illustrate how the effects on the agricultural sector [24], [38]. In addition, a phenomenon of food crisis could be mirrored by increasing food prices, which are caused by the working limit of low labor [39], [40].

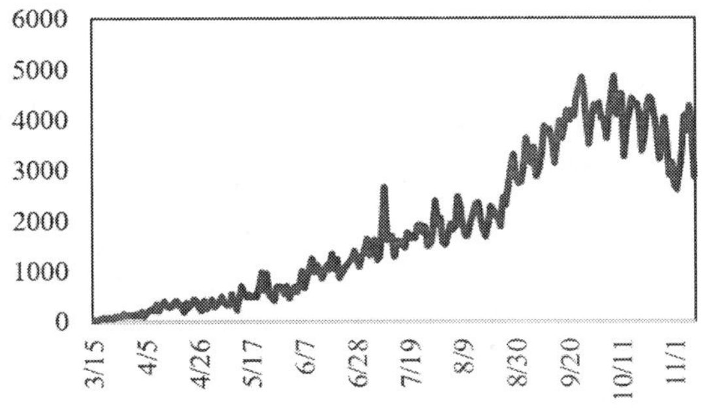

Figure 2. COVID-19 Daily Positive Cases in Indonesia (2020)

3. Data and Methods

3.1. Data

We estimate a Vector Error Correction Model (VECM) model with macroeconomic variables, namely $y_t \equiv [sst_t; prod_t; rpr_t, reg_t]$ and $y_t \equiv [rprd_t, covd_t]$ where sst_t denotes Sea Surface Temperature (SST) anomalies in the Nino 3.4 region", $prod_t$ is the rice production, rpr_t is the rice price in Indonesia, and reg_t is the representative of government regulation, which it dried harvested rice price.

Variables are sampled monthly and cover January 2000 until December 2018, for a total of 228 observations. For the COVID-19 model, $rprd_t$ is the daily rice price in Indonesia and $covd_t$ is the new positive cases reported daily. The variables are sampled daily and cover the period March 15, 2020 (WFH regulation legalized) until November 9, 2020, for a total of 240 observations.

Figure 3 shows that the real price of rice is highly volatile. This practical aspect is coherent with a low price elasticity of supply and low price and income elasticities of demand, whose interaction tends to magnify the price impact of actual and expected supply shortages. Figure 2, the number of positive cases COVID-19, shows higher price volatility than rice even though it shows an increasing trend.

3.2. Identification of the Vector Error Correction Model.

Our VECM model can be written as:

$$RPR_t = \sum_{j=1}^{n} RPR_{t-j} + \sum_{j=1}^{n} SST_{t-j} + \sum_{j=1}^{n} PROD_{t-j} + \sum_{j=1}^{n} REG_{t-j}$$

$$RPRd_t = \sum_{j=1}^{n} RPRd_{t-j} + \sum_{j=1}^{n} COVd_{t-j}$$

The first model is an ENSO model. The following model is a COVID-19 model. Also, an "ENSO shock" is defined as an unpredictable change of the SST index. Positive ENSO shocks identify unexpected El Nino events, while the SST index's unpredictable negative changes represent La Nina episodes. To correctly specify the VECM model, this research followed the standard procedure of time series analyses by following these procedures: (1) Pairwise Granger Causality Test, (2) Impulse Response Function, and (3) Forecast Error Variance Decomposition.

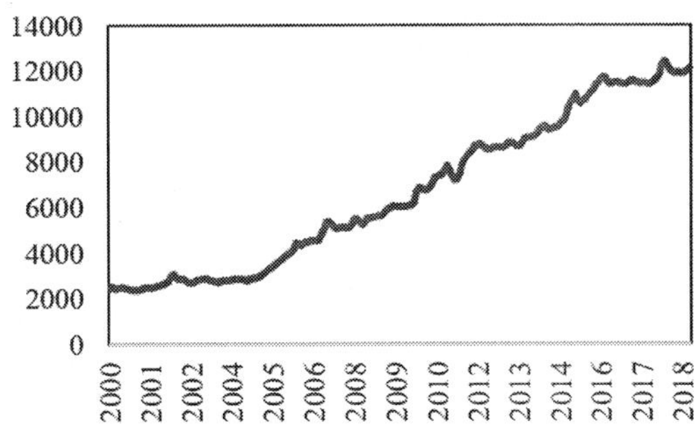

Figure 3. The Real Price of Rice in Indonesia for 2000-2018

4. Results and Discussion

The present empirical analysis starts with checking the time series variables' stationarity as that is the prime requirement for the cointegration and causality test. We found that all the variables do not appear to be stationary in variable levels, which means the variables are non-stationary in their level data and suggest that stationarity be checked at a higher order of differencing (first difference). Following that, the three series are integrated of order one, the cointegration relationship between them is established using Johansen's maximum likelihood (ML) test. The results indicate that the time series variables (rice price, SST, production, GKP, COVID-19 cases) are cointegrated. It gave us Pairwise Granger Causality to testing specific hypothesized causality.

4.1. The Relationship ENSO, COVID-19, and Rice Price Dynamics with Pairwise Granger Causality

Granger causality test is employed to observe whether a causality relationship between each of the variables. In other words, every variable has causality to another variable and can become the exogenous variable and endogenous variable. This research uses $\alpha = 5$ percent as the indicator.

4.1.1. The causality relationship on ENSO Model and COVID-19 Model

The result of the causality test for the ENSO model is shown in Table 1 below. This result also finds between rice price and production, even between rice production and regulation. It shows that climate does not have bidirectional causality, it can be concluded only climatic factor that can affect economic factors, not vice versa [18].

Table 1. Result of Causality Test on El Nino Southern-Oscillation Model

Dependent	RPR	SST	REG	PROD	Inferences
RPR	-	-	0.0023*	0.0136*	RPR ⇄ REG, RPR ⇄ PROD
SST	-	-	-	-	RPR, REG, PROD can not affect SST (climate)
REG	0.0106*	-	-	0.0121*	REG ⇄ RPR, REG ⇄ PROD
PROD	0.0332*	-	0.0004*	-	PROD ⇄ RPR, PROD ⇄ REG

Source: Data Processed

4.1.2. The causality relationship between COVID-19 and Rice Price Dynamics

The result of the causality test is shown in Table 2 below. The result found the existence of bidirectional causality between daily rice prices and COVID cases. The costs can be affected by the COVID pandemic because of its disruption on high labor requirements due to the working restrictions [24]. COVID pandemic can also be affected by rice prices due to demand and consumption during a pandemic [41].

Table 2. Result Pairwise Granger Causality Test on COVID-19 Model

Dependent	RPRd	COVd	Probability
RPRd	-	0.0316**	RPRd ⇄ COVd
COVd	0.0000***	-	COVd ⇄ RPRd, has bidirectional causality

Source: Data Processed

4.2. Impulse Response Function of ENSO model and COVID-19 Model

Impulse Response's function is to track the response of variable overtime after a shock to the VAR system. Kilian & Lütkepohl explains the role of the IRF expectation k-period ahead and the error prediction variable caused by innovations of other variables [42]. Thus, the shock effect duration of one variable to another variable up to the point when the effect of shock disappears or returns to equilibrium can be seen and known.

4.2.1. Impulse Response Function of El Nino-Southern Oscillation Model

The result showed that the positive shocks of ENSO would increase and stabilize in positive areas of rice price dynamics. For La Nina's shock, rice prices tend to response negatively. It supports the statement of Bastianin *et al.* and Ubilava & Holt, which state that El Nino increases prices (positive response) and La Nina decreases costs (negative response) due to the fluctuating demand and supply of rice during El Nino and La Nina [19], [20]. The graph below shows the result of the analysis of impulse response, as follows in Figure 4.

4.2.2. Impulse Response Function of COVID-19 Model

The rice price response shows a high increase from the fifth period until the sixth period. However, the rice price response from the shock of COVID-19 in the last period is highest than in other periods. It can be implied that the more COVID-19 transmitted to other people, the more the price will be increased. Working restrictions caused it for rice farmers and rice supply decreased. The scarcity of rice will increase the price (Figure 5).

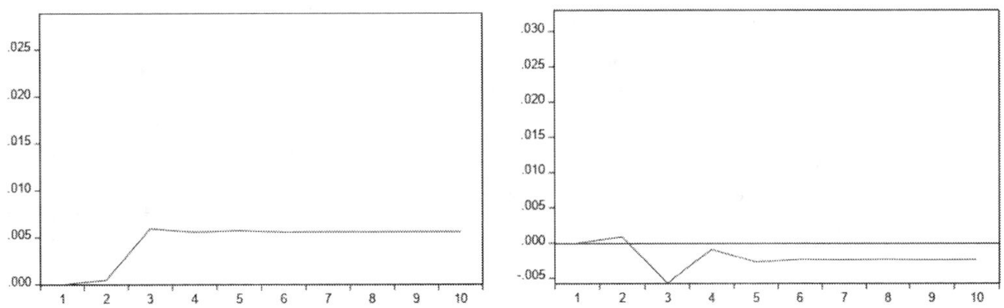

Figure 4. The response of Rice Price to *Cholesky One S.D. (d.f. adjusted) Innovations* of El Nino (left) and La Nina (right).

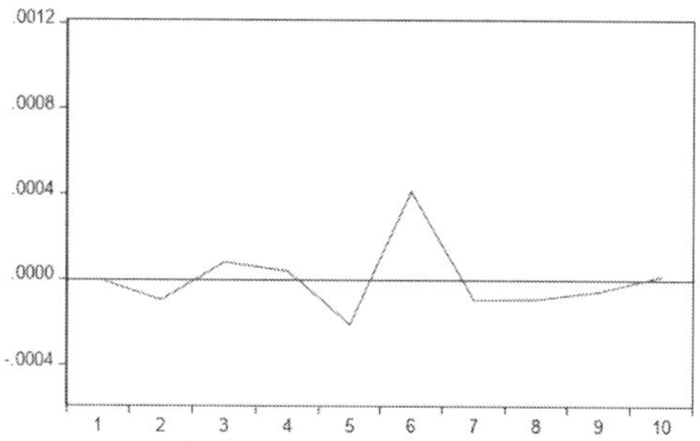

Figure 5. The response of Rice Price to *Cholesky One S.D. (d.f. adjusted) Innovations* of COVID-19

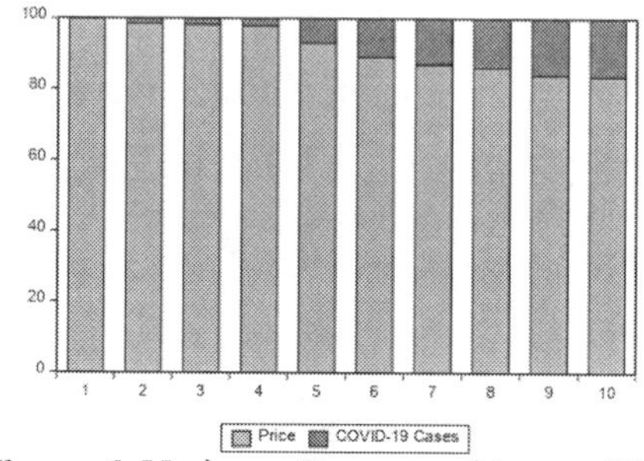

Figure 6. Variance Decomposition on Rice Price Variance (COVID-19 Model)

4.2.3. Variance Decomposition Test of El Nino-Southern Oscillation Model and COVID-19 Model

For the ENSO model, the results of the FEVD indicate that the rice price variance is explained more by itself shock than the ENSO, production, and regulator price shocks (Figure 7). The enormous contribution of rice prices to itself can mean that there is speculative behavior from rice market players considering that several studies have found that market speculators are vulnerable to rice price volatility [22], [23].

Unfortunately, the La Niña (SST negative) shock indicator explains less of the variance of prices. Only the El Nino (SST positive) indicator can explain price variances well. It indicates that prices are more sensitive to El Nino than La Nina, at the same time contradicting what was found by Ubilava & Holt, who found that sensitivity tends to favor La Nina [19]. In Indonesia, drought has a more significant impact on rice prices rather than floods [17].

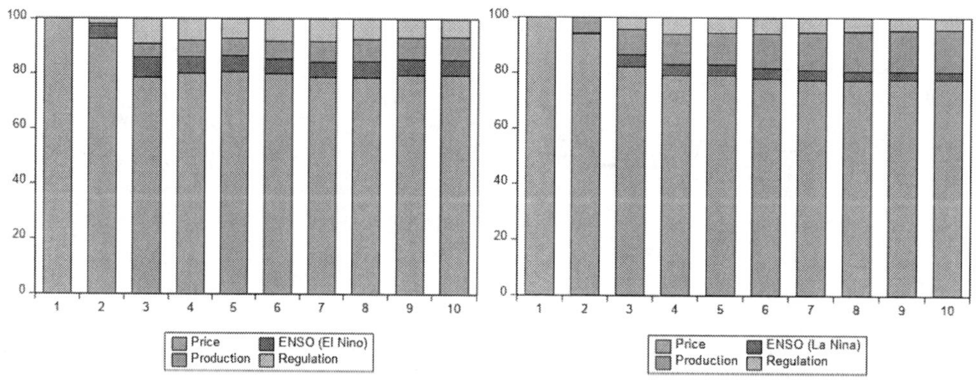

Figure 7. Variance Decomposition on Rice Price Variance (ENSO Model). The left side is an analysis with the El Nino indicator and the right side is La Nina

This figure has the same result as the COVID-19 model (Figure 6). The results of the FEVD indicate that the rice price variance is explained more by itself shock than the COVID-19 model. COVID-19 has started to contribute from the second period until the last period. It can be concluded that COVID-19 has a contribution to rice price fluctuations of 1-16%.

5. Conclusion

This study has examined and modeled a system of rice (price, production, and price-regulation), COVID-19, and the ENSO equation. This study aims to clarify the impact of the El Nino-Southern

Oscillation on rice price dynamics. Consistent with the previous literature [18], [43], evidence supports the hypothesis in the ENSO equation and the system of rice price equations. This study revealed a few features of interest, which can be summarized as follows: a positive ENSO shock—El Nino—has a positive effect on rice price, increasing the cost. The La Nina shock is actual. The second aim of this study is to clarify the impact of COVID-19 on rice prices. Using Granger causality, IRF, and VECD test, this study has concluded the following evidence that COVID-19 has a positive effect on rice prices. This study is the opposite of Yu *et al.* [25] in China that resulted that no significant impact on rice prices. Because China continues to report a lower number of new COVID-19 cases, it differs from Indonesia, which exposes a higher number every day. We hope this analysis is deemed vital for rice producers in Indonesia and processors and intermediaries in the rice markets.

References

[1] Prasad R, Shivay Y.S., and Kumar D 2017 Current status, challenges, and opportunities in rice production, in *Rice production worldwide*, Springer, pp. 1–32.

[2] Gurning I P, Yuprin A D, and Taufik E N 2019 Trend dan Estimasi Produksi Padi dan Konsumsi Beras di Provinsi Kalimantan Tengah *J. Tek. Pertamb.* **14(**1) 48–61.

[3] Naylor R. L, Falcon W P, Rochberg D, and Wada N 2001 Using El Niño/Southern Oscillation climate data to predict rice production in Indonesia *Clim. Change* **50(**3) 255–265, 2001.

[4] Shirsath P B. *et al.* 2020 Agricultural labor, COVID-19, and potential implications for food security and air quality in the breadbasket of India *Agric. Syst.* **185** 102954,

[5] Dhamira A and Irham I 2020 The Impact of Climatic Factors Towars Rice Production in Indonesia *Agro Ekon.* **31(**1).

[6] Mulyasari G, Irham, Waluyati L R, Suryantini A and Prawito P 2019 Climate Change Represented by Seasonal Rainfall, Temperature, and Wind Speed Trends in North Coastal of Java, Indonesia: Implications on Marine Capture Fisheries *Int. J. Adv. Sci. Eng. Inf. Technol.* **9(**6) 2067–2073.

[7] Utami A W, Jamhari J, and Hardyastuti S 2011 El Nino. La Nina, dan Penawaran Pangan di Jawa, Indonesia *J. Ekon. Pembang. Kaji. Masal. Ekon. dan Pembang* **12(**2) 257–271.

[8] Bhuvaneswari K, Geethalakshmi V, Lakshmanan A, Srinivasan R, and Sekhar N U 2013 The impact of El Niño/Southern Oscillation on hydrology and rice productivity in the Cauvery Basin, India: Application of the soil and water assessment tool *Weather Clim. Extrem.* **2** 39–47, 2013.

[9] Thurlow J, Zhu T, and Diao X 2012 Current climate variability and future climate change: estimated growth and poverty impacts for Zambia *Rev. Dev. Econ.* **16(**3) 394–411.

[10] Hafeez M M, Bouman B A M, Van de Giesen N, and Vlek P 2007 Scale effects on water use and water productivity in a rice-based irrigation system (UPRIIS) in the Philippines *Agric. Water Manag.***92(**1–2) 81–89.

[11] D'Arrigo R and Wilson R 2008 El Nino and Indian Ocean influences on Indonesian drought: implications for forecasting rainfall and crop productivity *Int. J. Climatol. A J. R. Meteorol. Soc.* **28(**5) 611–616.

[12] Kripalani R H and Kulkarni A 1997 Rainfall variability over South-east Asia—connections with Indian monsoon and ENSO extremes: new perspectives *Int. J. Climatol. A J. R. Meteorol. Soc.* **17(**11) 1155–1168.

[13] Roberts M G, Dawe D, Falcon W P, and Naylor R L 2009 El Niño-southern oscillation impacts on rice production in Luzon, the Philippines *J. Appl. Meteorol. Climatol.* **48(**8) 1718–1724.

[14] Deng X *et al.* 2010 Impacts of El Nino-Southern Oscillation events on China's rice production *J. Geogr. Sci.* **20(**1) 3–16.

[15] Chen C, McCarl B, and Chang C 2008 Strong El Niño–Southern Oscillation events and the economics of the international rice market *Clim. Res.* **36(**2) 113–122.

[16] Wailes E J and Chavez E C 2012 ASEAN and Global Rice Situation and Outlook *ADB Sustain. Dev. Work. Pap. Ser.* **22** 1–42.

[17] Falcon W P, Naylor R L, Smith W L, Burke M B, and McCullough E B 2004 Using climate models to improve Indonesian food security *Bull. Indones. Econ. Stud.* **40**(3) 355–377.

[18] Brunner A D 2002 El Niño and world primary commodity prices: Warm water or hot air?," *Rev. Econ. Stat.* **84**(1) 176–183.

[19] Ubilava D and Holt M 2013 El Niño southern oscillation and its effects on world vegetable oil prices: Assessing asymmetries using smooth transition models *Aust. J. Agric. Resour. Econ.* **57**(2) 273–297.

[20] Bastianin A, Lanza A, and Manera M 2018 Economic impacts of El Niño southern oscillation: evidence from the Colombian coffee market *Agric. Econ. (United Kingdom)* **49**(5) 623–633.

[21] Lefkovitz Y 2013 Climate and Commodity Prices : An Analysis of the Role of ENSO Forecasts in Agricultural Commodity Markets by Yosef Lefkovitz *Thesis*, May, 2013.

[22] Clarete R L, Adriano L, and Esteban A 2013 Rice trade and price volatility: Implications on ASEAN and global food security *ADB Econ. Work. Pap. Ser.* **368** 1–35.

[23] Wright B 2008 Speculators, storage, and the price of rice *ARE Updat.* **12**(2) 7–10.

[24] Laborde D, Martin W, Swinnen J, and Vos R 2020 COVID-19 risks to global food security *Science (80-.)* **369**(6503) 500–502, doi: 10.1126/science.abc4765.

[25] Yu X, Liu C, Wang, and Feil J H The impact of COVID-19 on food prices in China: evidence of four major food products from Beijing, Shandong and Hubei Provinces *China Agric. Econ. Rev.* **12**(3) 445–458, doi: 10.1108/CAER-04-2020-0054.

[26] Gugus Tugas Percepatan Penanganan COVID-19, "Data," 2020. https://covid19.go.id/.

[27] Dawe D and Timmer C P 2012 Why stable food prices are a good thing: Lessons from stabilizing rice prices in Asia *Glob. Food Sec.* **1**(2) 127–133.

[28] NOAA 2020 Niño 3.4 (anomaly; 1981-2010 mean removed)," 2020. Accessed: Aug. 18, 2020. [Online]. Available: https://psl.noaa.gov/gcos_wgsp/Timeseries/Data/nino34.long.anom.data.

[29] Kamil N N and Omar S F 2016 Climate variability and its impact on the palm oil industry," *Oil Palm Ind. Econ. J.* **16**(1) 18–30.

[30] Jati K 2018 Analisis efek musim hujan dan kemarau terhadap harga beras *J. Manaj. Ind. dan Logistik* **2**(1) 40–51.

[31] Ubilava D 2012 El Niño, La Niña, and world coffee price dynamics *Agric. Econ.* **43**(1) 17–26.

[32] Gutierrez L 2017 Impacts of El Niño-Southern Oscillation on the wheat market: A global dynamic analysis *PLoS One* **12**(6) p. e0179086.

[33] Hermanto S 2017 Kebijakan harga beras ditinjau dari dimensi penentu harga, in *Forum Penelit Agro Ekon.* **35**(1) 31–43.

[34] Suryana A 2015 Dinamika kebijakan harga gabah dan beras dalam mendukung ketahanan pangan nasional *Pengemb. Inov. Pertan.* **7**(4) 155–168, 2015.

[35] Rachman B, Agustian A, Syaifudin A, and K. P. Badan Ketahanan Pangan 2019 Implikasi Kebijakan Harga Eceran Tertinggi Beras terhadap Profitabilitas Usaha Tani Padi dan Harga, Kualitas, serta Serapan Beras *Repos. Publ. Kementeri. Pertan.*

[36] Saputra A, Arifin B, and Kasymir E 2014 Analisis kausalitas harga beras, harga pembelian pemerintah (HPP) dan inflasi serta efektivitas kebijakan HPP di Indonesia *J. Ilmu-Ilmu Agribisnis* **2**(1) 24–31.

[37] Verity R *et al.* 2020 Estimates of the severity of coronavirus disease 2019: a model-based analysis, *Lancet Infect.* **20**(6) 669–677, 2020, doi: 10.1016/S1473-3099(20)30243-7.

[38] Rizou M, Galanakis I M, Aldawoud T M S, and Galanakis C M 2020 Safety of foods, food supply chain and environment within the COVID-19 pandemic *Trends food Sci. Technol* **102** 293–299.

[39] Yu X and Shimokawa S 2016 Nutritional impacts of rising food prices in African countries: a review," *Food Secur.* **8**(5) 985–997.

[40] Yu X 2018 Engel curve, farmer welfare and food consumption in 40 years of rural China," *China Agric. Econ. Rev.* **10**(1) 65–77, 2018, doi: 10.1108/CAER-10-2017-0184.

[41] Baker S R, Farrokhnia R A, Meyer S, Pagel M, and Yannelis C 2020 How Does Household

Spending Respond to an Epidemic? Consumption During the 2020 COVID-19 Pandemic," National Bureau of Economic Research, doi: 10.2139/ssrn.3565521.

[42] Kilian and Lütkepohl 2017 *Structural Vector Autoregressive Analysis*. Cambridge University Press.

[43] Ubilava D and Holt M T 2009 Nonlinearities in the World Vegetable Oil Price System: El Nino Effects *Annu. Meet.* 319-2016–9744, 2009, doi: 10.22004/AG.ECON.49360.

Carrying capacity and environmental capacity of water resources analysis in Maluku Province of Indonesia

P J Kunu[1,*], S Talakua[1], Y Pesulima[2], L Uyara[3], S Laimeheriwa[4], R M Osok[1]

[1] Departemen of Soil Science, Agriculture Faculty of Pattimura University, Ambon, Indonesia
[2] Faculty of Mathematics and Natural Sciences Postgraduate Students, PGRI Indraprastas University, Jakarta, Indonesia
[3] Land Service Officer of East Seram Regency
[4] Departemen of Agrotechnology, Agriculture Faculty of Pattimura University, Ambon, Indonesia

* E-mail: pieterkunu@gmail.com

Abstract. Carrying capacity and environmental capacity of water resources is the ability of the water to support human life, other living creatures, and the balance between the two as well as the ability of the environment to absorb substances, energy and other components that enter or are incorporated into it. The purpose of this research is to analyze the level of carrying capacity and environmental capacity of water resources in several regencies in Maluku Province, Indonesia. The results showed that some water resources were already in a low environmenttal capacity due to limited availability, especially deep groundwater in Wamar Island, Aru Archipelago Regency and the environmental capacity of water resources had exceeded the quality standards set in Indonesia, especially fresh water in Southeast Maluku Regency and Ambon City. Meanwhile, the environmental capacity of fresh water that has exceeded the quality standard is due to the presence of e-colly bacteria. This condition shows the importance of water conservation through regulating human behavior that is not environmentally friendly, building rainwater control and runoff control infrastructure as well as law enforcement efforts related to watershed maintenance and the behavior of communities around water resources.

1. Introduction

The meaning of Carrying Capacity and Environmental Capacity and as stated in Law Number 32 of 2009 concerning Environmental Protection and Management is as follows. The carrying capacity of the environment is the ability of the environment to support the life of humans and other creatures and the balance between the two. Meanwhile, the environmental capacity is the ability of the environment to absorb energy substances and/or other components that are entered or incorporated into it.

In the implementation of national and regional economic development, it will always have an impact, especially the degradation of the environment and natural resources, including forests, land, water and air as a result of an imbalance in the ecosystem. This overall impact will lead to the welfare of human life which is also disturbed and even threatened by the survival of humans and other creatures.

Since the late 1990s, more thorough research has been conducted on the carrying capacity of water resources (WRCC). However, due to the regionality and complexity of influencing factors, the concept of WRCC at home and abroad has not been expressed in a unified way [1]. The water resource carrying capacity (WRCC) can be defined as the maximum capacity of regional water resources to support social and economic development based on a given living standard and production technology without degrading the water environment [2]. The WRCC is an important concept in judging the relationship between regional human activities and water resources [3]. Therefore, studying the water resources carrying capacity (WRCC) is necessary to obtain scientific guidance on water resources allocation andalleviate the conflict between the demand of the society–economy system and water resources supply. WRCC is generally quantified as the ratio between the water demand of social development and water supply under specific conditions [4].

In connection with the existence of environmental problems that have become a global issue, a movement to save the environment was born which resulted in various international policies including SDG's or Sustainable Development Goal's and has been adopted by our country in the form of the National Development Goal's (NDG's) with a set of different indicators starting from the National to Regency / City levels which are implemented in various regulations concerning the need for sustainable development which is of course environmentally sound. This concept in its implementation rests on the limits of the carrying capacity and the environmental capacity of a region's environment. For this purpose, development violations must be preceded by conducting a strategic environmental study with one of the mandatory contents, namely the carrying capacity and environmental capacity (CCEC). If CCEC has been exceeded, the Development Activities, Plans and Programs must be repaired, and all efforts that have exceeded the carrying capacity and environmental capacity are no longer allowed. Referring to these facts, analysis of the carrying capacity and environmental capacity is very important and very much needed in development and must be compiled at every level of government.

The most important components of the environment for life are water and food. The focus in this study is the carrying capacity and environmental capacity of water resources which are common as sources of raw water in the region. [5] state that water is the force powering life, so it is an element crucial to human survival. Nevertheless, the increased number of industries in the area has brought about changes in the land, which, in turn, has made it increasingly more difficult to find clean water in the city of Ambon and the others city and regency in Maluku.

The objective of this research is to analyze the level of environmental capacity and carrying capacity of water resources in Maluku, Indonesia

2. Methods

2.1. Description of Study Site

The Maluku Islands consist of more than 1400 islands. The study was conducted in several regency and city of Maluku Province. The main purpose of the establishment of carrying capacity and enivironmental capacity of water resources. Sample surface water is taken from river bodies, lakes and springs water which are used as raw water sources.

2.2. Data Sources

The data used in this paper were collected from many reports such as Environmental Departement of Southest Maluku Regency (2017) on Environmental Status of Southest Maluku Regency, Environmental Departement of Southeast Maluku Regency (2017) on Environmental Status of Southwest Maluku Regency (2018), Environmental Departement of Aru Archipelago Regency (2018) on and Research in Ambon City (2019) and then compared with the water quality standards according to the ministerial regulation of health and the environment and forestry include Government Regulation No. 82 of 2001 concerning the Management of Water Quality and Water Pollution Control.

2.3. Data Analysis
Water quality data in water resources is obtained from monitoring regularly every year of Environment Agency of Southeast Maluku Regency, Environment Agency of Southwest Maluku Regency and Ambon City. Water quality data is used to determine water quality status, and the capacity of water pollution load or environmental capacity of water resources. Overall water quality is determined by comparing data with criteria from Government Regulation No. 82 of 2001 concerning the Management of Water Quality and Water Pollution Control and Minister of Health Regulation number 492/Menkes/Per/IV/2010.

This regulation is commonly used to determine the water class in nationwide that consists of four water quality criteria based on class, which each of class have different criteria that fit to its purpose. In this case, the designation to be used is the classification of class I-II water quality. First class, water whose designation can be used for raw water drinking, and for other designations that require the quality of water the same as these activities; Second class, the water it is intended for can be used for water recreation infrastructure/facilities, freshwater fish farming, livestock, water to irrigate agriculture, and other designations that require quality the same water for these uses.

2.4. Calculation of Water Demands
Water demand is the total of domestic and non-domestic water demands, and evapotranspiration. Domestic water demands (*Qdomestic*) [m3/month] can be calculated by means of equation 1. The non-domestic water demands (*Qnondomestic*) [m3/month] considered for this calculation only come from the industrial sector as the main non-domestic water user in the region. Non-domestic water demand for the industry is calculated according to equation 2.

$$Q_{domestic} = Water_{consumption} \text{ x population x day} \quad (1)$$

$$Q_{nondomestic} = Water_{employee} \text{ x population x day} \quad (2)$$

The amount of water consumption (*waterconsumption*) for domestic and non-domestic use is 0.12 and 0.5 [m3/person/day], respectively [6]. Population is the total population in area [person]. Employee is the total number of employees in the area [person]. Day is the number of days in a month [day].

2.5. Testing the Capacity of Water Resources
To measure the capacity of water resources is carried out qualitatively by analyzing the data of available (laboratory analysis results) and compared with the quality standard of fresh water resources in accordance with applicable regulations [7].

3. Results and discussion
The result of this research showed that several parameters of water quality have exceeded the standard of water quality according to the minister of health and environment and forestry regulation too.

The results of the analysis of several water resources in Southeast Maluku Regency, Southwest Maluku Regency and Ambon City found that some surface water quality parameters have exceeded the water quality standard.

3.1. Carrying Capacity of Water Resources
3.1.1. Southeast Maluku Regency Water Resources
The following table presents the results of the analysis of the carrying capacity of water resources in Southeast Maluku Regency.

Table 1. The results of the analysis of the carrying capacity of water resources in Southeast Maluku Regency

No.	Water Resources	Annual volume of water (m3)
1	Springs water	179.881.344,00
2	Lakes	2.862.450,00
3	Rivers	1.460.203.524,00
4	Rainfall-runoff	963.593.383,57
	Total	2.606.540.701,57

Source: Environmental Departement of Southeast Maluku Regency (2017) [8] and Kunu (2019) [9]

Remarks : Prediction of Carrying Capasity of Water Resources until 2023 (Under Condition of Forest Area >40%)

Based on the data in the Table 1, it can be seen that total annual available of water in Southeast Maluku Regency is 2,6 billion m3. The results of the analysis of the need for fresh water in the Aru Archipelago Regency, it was found that the amount was around 428 million m3. Compared to the large water availability, the fresh water in Southeast Maluku Regency is in a surplus condition. The following table presents data on water availability and fresh water needs of the people until 2023 in Southeast Maluku Regency. The biggest obstacle in the utilization of these abundant water sources is only the water quality, which is indicated to have been contaminated with e-colly bacteria.

3.1.2. Aru Archipelago Regency of Water Resources

Another research has been carried out in Wamar Island, Aru Islands Regency by Osok, et al (2017). The results showed that on Wamar Island there is a domestic water need of 1,446,145 m3 / year, with Non-domestic water needs are 178,101,520 m3/year, so the total water demand in this area is 1,624,246.52 m3/year. The following table presents the results of the analysis of water resources on Wamar Island under three different conditions. Under normal conditions and conditions of the impact of climate change that cause extreme rainfall. The extreme rainfall that occurs during the rainy season compared to the dry season is very different. Extreme rainfall in the rainy season is 278% greater than the dry season.

To maintain the availability of clean water throughout the year, water conservation efforts need to be put into practice. To increase groundwater recharge naturally, green open space especially in recharge areas must be maintained, meanwhile it is also necessary to harvest rainwater and surface runoff through the construction of water infrastructure such as infiltration wells and reservoirs.

Table 2. Prediction of Total Water Requirements and Avalability until 2023 in Aru Archipelago Regency

Year	Total Water Requirements and Avalability	Total
2018	Total annual water requirements	428.540.111,40 m3
2023		486.391.130,67 m3
2018	Total annual water availability	2.606.540.701,57 m3
2023	The source from the river only	1.460.203.524,00 m3
Water Resources Condition until 2023		SURPLUS
Parameters of Environmental Capacity of Water Resources are Exceeded the Water Quality Standard (Ambon city, Southeast Maluku Regency)		*TDS, E-colybacteria*

Source: Environmental Departement of Aru Archipelago Regency (2017) [10] , *Southeast Maluku Regency (2017)*

Remarks : Prediction of Carrying Capasity of Water Resources until 2023 (Under Condition: Forest Area >40%)

Table 3. The results of the analysis of the carrying capacity of water resources in Aru Archipelago Regency

Rainfall condition	Annual Rainfall (m3)	Lost Total (m3)	Remaining Groundwater (m3)	Domestic and Non Domestic Needs (m3)	Annual Savings (m3)
Normal	118.818.000	116.149.296	2.668.704	1.624.246,52	1,044,457.48
Extreme wet	196.554.000	136.608.720	59.945.280	1.624.246,52	58,321,033.48
Extreme dry	70.454.400	65.273.640	5.180.760	1.624.246,52	3,556,513.48
Total needs				1.624.246,52	

Source : Osok, Kunu and Laimeheriwa (2017) [11]

3.2. Environmental Capacity of Water Resources

3.2.1. Water of Lakes

The results of the analysis of lake water samples in Southeast Maluku Regency showed that all water quality parameters of Abel Lake had not exceeded the fresh water quality standard according to Permenkes 492 / Menks / Per / IV / 2010 except for TDS and fluoride levels.

Table 4. Abel Lake Water Quality as an Indicator of Lake Environmenyal Capacity of Water Resource in Southeast Maluku Regency

Parameters	TDS	pH	DO	NO3-N	NH3-N	Mn	Fe	Zn	Flouride	Nitrit = N
	-----mg/L-----									
SWQ*	500	6,5-8,5	-	50	1,3	0,4	0,3	3	1,5	3
Names of Lake										
Abel	850	7,14	1,14	0,2	0	0	0	0,1	100	0
Witir	-	-	-	-	-	-	-	-	-	-
Nuhuta	-	-	-	-	-	-	-	-	-	-
El	-	-	-	-	-	-	-	-	-	
Ohoidertwaun	-	-	-	-	-	-	-	-	-	-
Nuhu Jeujanan	-	-	-	-	-	-	-	-	-	-

Source: Environmental Departement of Malra Regency (2017)[12]
*Remarks:) - Data not available;) * Standard of Water Quality according to Permenkes 492 / Menkes / Per / IV / 2010 on Drinking Water Quality Requirements*

Fluoride plays a role in the formation of bones and teeth, and helps prevent tooth decay. We know that adjusting the fluoride in our water systems to ideal levels reduces cavities, strengthens teeth, and is an essential part of a healthy life. Fluoride is found naturally in soil, water and food. Minerals such as fluorapatite and fluorite also contain fluoride, such as iron and calcium, fluoride dissolves into the ground water we use for our drinking water. When there is not enough fluoride in drinking water, these elements can be added to optimal levels to protect teeth. Fluoride comes from natural calcium deposits in phosphate rock [13]

Fluorapatite, is a phosphate mineral with the formula $Ca_5 (PO_4)_3F$ (calcium fluorophosphate), has a green, brown, blue, yellow, purple, or colorless color, a pure colorless mineral. Together with hydroxylapatite can be a component of tooth enamel. Fluorapatite crystallizes in the hexagonal crystalline system and combined as a solid solution with hydroxylapatite (Ca5 (PO4) 3OH or Ca10 (PO4) 6 (OH) 2) in a biological matrix. Fluorapatite as a mineral is the most common phosphate mineral, and occurs widely as an accessory mineral in igneous and metamorphic rocks rich in calcium. Based on these facts, the condition of fluoride-rich water exceeded the quality standard in Southeast Maluku Regency, naturally occurring, both from rocks that are rich in calcium as well as from laterite soils which were formed due to lake inundation.

This is according to local geological conditions. According to the Geological Map of Indonesia (1965) Southeast Maluku islands / islands are formed/composed of several types of rock including Alluvium Undak, Coral Reef, Seklis Habluk, Paleogene and Paleozoic Ulagan. Coral reefs or coral reefs are actually organic sedimentary rocks that occur because the deposition process gets help from organisms, namely the remains of houses or carcasses of marine animals buried on the seabed such as shells, coral reefs, bones, Guano bird droppings that mount in Peru and the topsoil in the forest. Examples of organic sedimentary rocks include coral limestone (coral reefs) and shell limestone.

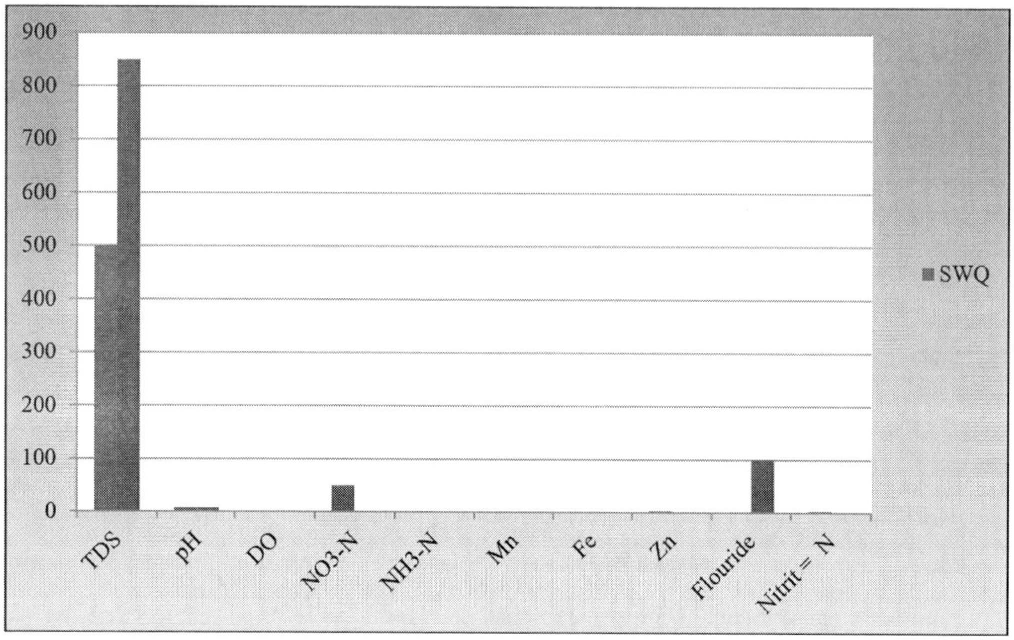

WHO in 1984 and 1993 established a standard for fluoride content in bottled water not to exceed 1.5 mg/liter, because if it exceeds this standard limit can cause the risk of dental fluorosis and in greater levels can cause bone fluorosis. In Indonesia Regulation of the Minister of Health of the Republic of Indonesia No. 492 / Menkes / Per / IV / 2010 regarding requirements for drinking water quality, the fluoride content in drinking water should not exceed 1.5 mg/L and a tighter limit is stipulated by SNI 01-3553-2006 concerning bottled drinking water, it is stated that the content of fluoride in mineral water should not exceed 1 mg/L.

The fluoride level found in lake water bodies in Southeast Maluku Regency is 100 mg/L, so the water is not suitable for use as a source of drinking water for the community unless the fluoride level can be reduced to 1.5 mg/L. Consumption of water with excessive fluoride levels is very dangerous, and can cause problems with teeth, bones and other organs. In addition, the use of drinking water with fluoride levels above 1.5 mg/L has an impact on: (1) The nervous system (decreased IQ) due to drinking water with a fluoride content of 2.5-4 mg/L; (2) The hormone system, which is due to excessive consumption of fluoride, can also cause a decrease in thyroid hormone, an increase in parathyroid hormone and calcitonin, will interfere with glucose tolerance and (3) the reproductive system.

3.2.2. Water Spring

In Southeast Maluku Regency, in addition to the 6 lakes described above, there are also 16 springs which are potential and reliable sources of fresh water if their quality status meets the quality standards of fresh water or drinking water. In Southeast Maluku, there are three parameters in several springs that have exceeded the quality standard of fresh water or are too low from the quality standard value according to Permenkes 492 / Menks / Per / IV / 2010 concerning Requirements for Drinking Water Quality. The results of the analysis found that all springs that have TDS levels have exceeded the quality standard of fresh water, while 8 springs namely Wear Er, Wear But, Wear Matan Manut, Rahareng Atas, Faan Hutan Kota Wells, Watngon 1, Watngon 2, Sumur Tua Warwut , and Faan has too low DO levels, 4 springs that have high fluoride levels are Wear Saai, Wear Er, Wear But and Watngon 1 springs.While the other 2 springs have high Nitrite (N) levels, namely Wear springs. Un and the Warwut Old Well.

In general, the higher the DO value in water, the better quality the water is; on the other hand, if the DO value is low, then it is assumed that the water has been polluted. DO levels indicate the extent

IOP Conf. Series: Earth and Environmental Science **883** (2021) 012062 doi:10.1088/1755-1315/883/1/012062

to which water bodies are able to accommodate aquatic biota such as fish and microorganisms. In addition, the amount of oxygen in water shows the ability of water to fresh water pollution. Nitrate is a form of nitrogen which acts as the main nutrient for plant and algae growth. Nitrate nitrogen is very soluble in water and has relatively stable properties. This compound is produced from a complete oxidation process in the waters. Basically, nitrate is the main source of nitrogen in water, however, plants prefer ammonium to be used in the growth process. The nitrate content in uncontaminated water is usually higher than the ammonium level. Nitrate levels greater than 5 mg / L. describes the state of the waters that have been polluted due to human activities and animal feces. Nitrogen levels greater than 0.2 mg / ltr indicate water eutrophication [14]

Table 5. Quality of Spring Water Resources in Southeast Maluku Regency

Parameters / Names of Springs Water	TDS	pH	DO	NO_3-N	NH_3-N	Zn	Flouri de	Nitri as N
SQW* Class 1-2	1000	6,0-9,0	6,0-4,0	10	0,5-(-)	0,05	0,5-1,5	0,06
SQW** Drinking Water	500	6,5-8,5	-	50	1,3	3	1,5	3
Wear Saai	670	7,67	4,65	0	0,2	0,1	100	0
Sumur Matwair	-	-	-	-	-	-	-	-
Wear Er	880	7,63	2,15	0,2	0	0,2	100	0
Wear But	910	7,71	3,53	0	0	0,3	100	0
Wear Matan manut	1.220	7,6	2,39	0	0	0,2	0	0
Rahareng atas	950	7,71	2,71	0	0	0	0	0
Sumur Faan Hutan Kota	1.130	7,8	2,8	0	0	0	0	0
Watngon 1	860	7,79	2,7	0	0	0	100	0
Watngon 2	990	7,6	2,15	0	0	0	0	
Wear Un	710	8,04	9,48	0	0	0	0,1	100
Wear Matan Rian	860	8,03	6,01	0	0	0	0,1	0
Wear Er	710	7,95	5,13	0	0	0	0,1	0
Sumur Tua Warwut	880	7,75	1,24	0	0	0	0,1	100
Goa hawang	880	8,0	6,68	0	0	0	0,4	0
Faan	1.130	7,8	2,8	0	0	0	0	0

Source: Environmental Departement of Southeast Maluku Regency (2017)
** Standard of Water Quality according to Government Regulation No. 82 of 2001*
*** Standard of Water Quality according to Permenkes 492 / Menkes / Per / IV / 2010*
- Data not available

Nitrite is an intermediate form between ammonia and nitrate (nitrification) and between nitrate and nitrogen gas (denitrification) which is formed under anaerobic conditions. Nitrite sources can be industrial waste and domestic waste. Nitrite levels in water are relatively stable because they are immediately oxidized to nitrate. Natural waters. contains nitrite about 0.001 mg / L. Meanwhile, the allowable nitrite levels do not exceed 0.06-3 mg / L. Nitrate and nitrite levels in water that exceed the quality standard threshold according to Permenkes 492 / Menks / Per / IV / 2010 concerning drinking water quality requirements will have a negative effect on aquatic biota. Pollution is also evident from the results of the table analysis which shows that high TDS levels exceed the quality standard in all rivers in Southeast Maluku Regency. TDS consists of inorganic salts, especially calcium, magnesium, potassium, sodium, bicarbonate, chloride and sulfate and a small amount of water-soluble organic matter. To make it suitable as a drinking water source, treatment must be given in the form of using water purification filter techniques with reverse osmosis technology (RO).

The following table presents the analysis results of water samples from several springs in the city of Ambon.

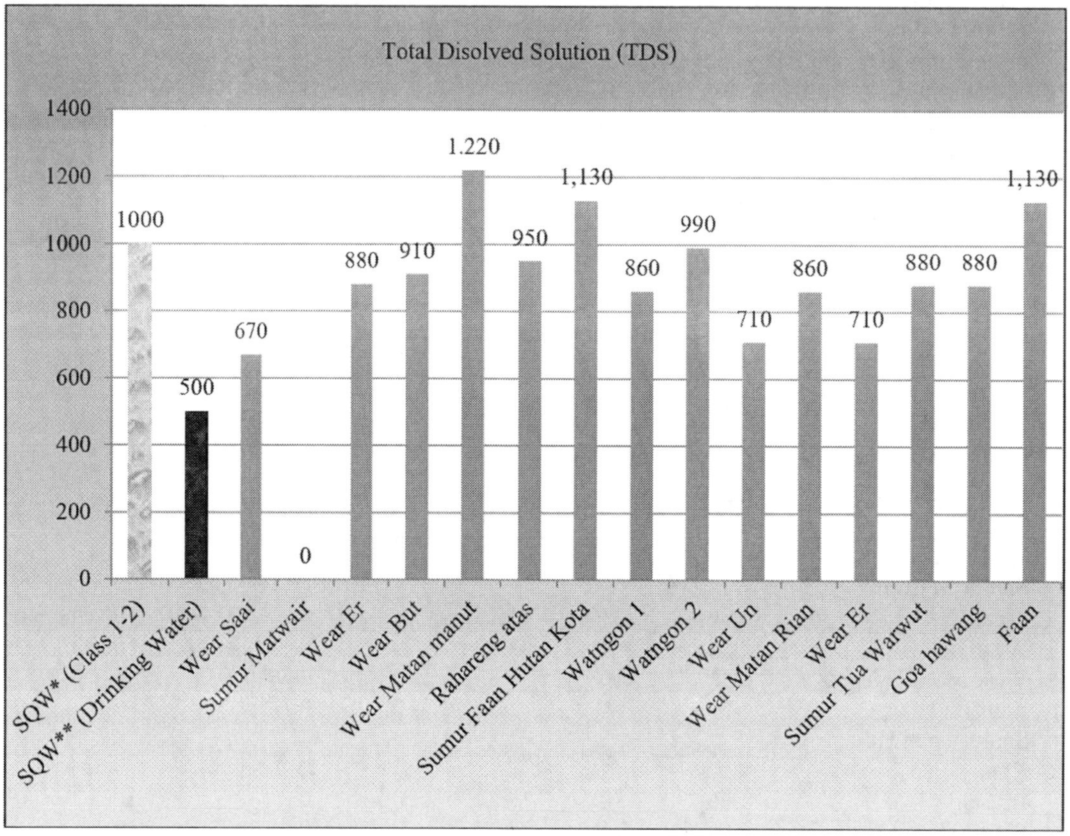

Figure 1. The TDS levels in all water sources exceed the allowable TDS levels

Table 6. Quality of Spring Water Resources in Ambon City

Parameters	Water Quality Standar	Results of Water Quality Analysis			
		Bt Merah	Halong	Wainitu	Amahusu
TDS	500	64	220	191	174
Fluoride	1,5	0,0	0,0	0,0	0,0
Cadmium	0,003	-	0,0	-	-
NO2	3	<0,01	<0,01	<0,01	<0,01
NO3	50	0,51	0,42	0,94	0,09
pH	6,5-8,5	5,64	5,16	5,38	5,52
Amonia	1,5	0,0	0,0	0,0	0,0
E. Colly	0,0	98	0	0	0,0
Total Coliform	0,0	438	0	0	0,0

*The results of the analysis showed that the water was contaminated with bacteria. This result of analysis show that Bt Merah Water is not suitable to use.

Source : Uyara, L., Kunu, P., Talakua, S., 2018 [15]
Remarks : ** Standard of Water Quality according to Government Regulation No. 82 of 2001*
*** Standard of Water Quality according toPermenkes 492/Menkes/Per /IV/2010*
- Data not available

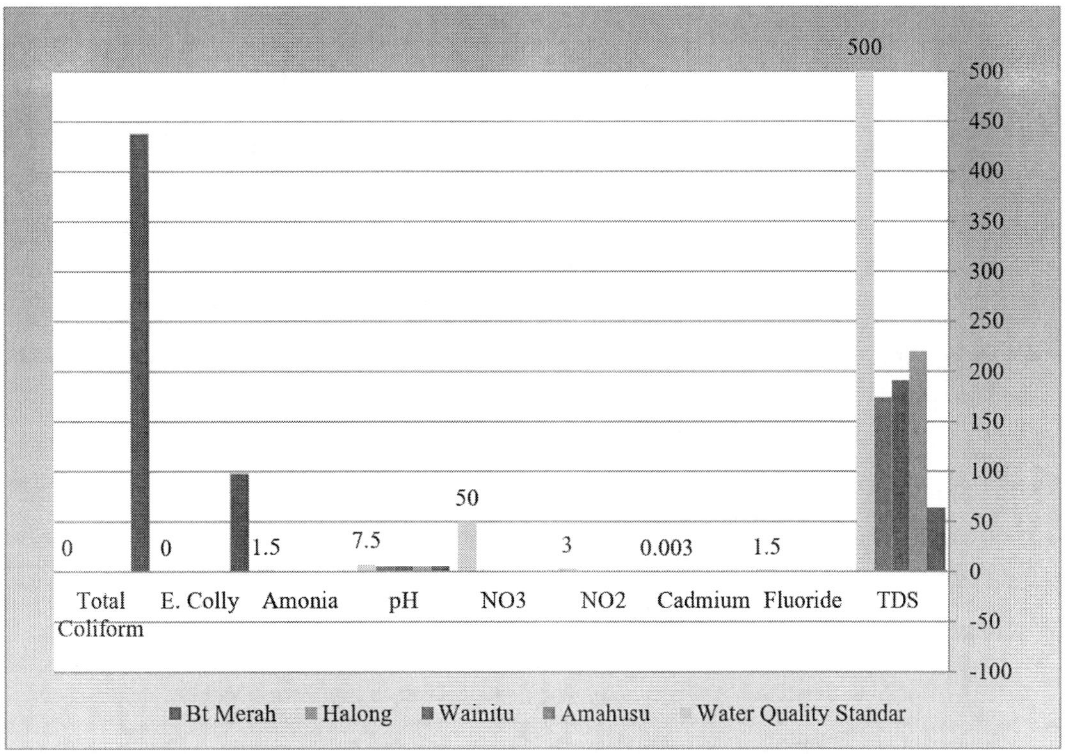

Figure 2. Quality of Spring Water Resources in Ambon City

According to the data in Table 3 above, it shows that in general the water quality in several springs in Ambon City has not exceeded the fresh water quality standard according to applicable regulations in Indonesia. These four springs are the source of fresh water for the people of Ambon

City especially in areas that are difficult to get fresh water or water from daily fresh water supply companies. The only source of fresh water that is no longer suitable for use is Air Besar located in the Batu Merah Village as a result of being contaminated by E-colly bacteria. A similar condition has been found in one of the community's clean water sources in the capital city of Southeast Maluku Regency.

3.2.3. Water of Rivers

The results of the analysis of river water samples in Southeast Maluku Regency and Southwest Maluku Regency are presented in the following tables.

Based on the data in the table below, it shows that the quality of river water in Southeast Maluku Regency, on several parameters does not meet the quality standards of fresh water. TDS levels have exceeded drinking water quality standards, with very low levels of Nitrate and DO and very high levels of fluoride in several rivers. The results of the complete analysis found that some water sources were contaminated with high levels of E-colly bacteria, especially the Evu river, which is the source of raw water for PDAM Southeast Maluku Regency.

Table 7. Analysis of River Water Environmental Capacity in Southeast Maluku Regency

SWQ \ Parameter	TDS	pH	DO	NO3-N	NH3-N	Zn	Flouride	Nitrit as N
Class 2)*	1000	9	4	10	(-)	0.05	1.5	0.06
Dirinking water)**	500	6.5-8.5	--	50	1.5	3	1.5	3
Wear Nen Mas II	750	7.8	1.76	0	0.2	0.1	0	0
Manyeuw Berkat	750	7.8	1.71	0	0.1	0.1	0	0
Wear Imit	720	7.7	1.66	0	0	0.1	0	0
Hoar ten	930	7.67	1.04	0	0	0.2	100	0
Erlarang	950	7.75	3.5	0	0	0.2	0	0
Udar	810	8.56	3.11	0	0	0.1	0	0
Ad wear aur	550	8.92	6.16	0	0	0.1	100	0
Nahak Ohoi	760	7.79	5.58	0	0	0.1	0	0
Wear Un Uwat	710	8.04	9.48	0	0	0.1	0	0
Matan Rian	860	8.03	0	0	0	0.1	100	0
Waer Aur - Ad	550	8.92	6.16	0	0	0.1	100	0
Hoat Sorbay	13.900	7.75	1.24	0	0	0.1	100	0
Wearbaluruk	530	8.16	4.12	0	0	0.1	0	0
Sungai Ur	650	8.8	7.75	0	0	0.2	0	0

Source: Environmental Departement of Southeast Maluku Regency (2017)
** Standard of Water Quality according (SWQ) to Government Regulation No. 82 of 2001*
*** Standard of Water Quality according to Permenkes 492 / Menkes / Per / IV / 2010 on Drinking Water Quality Requirements*

The following table presents the results of the analysis of water resources in Southwest Maluku Regency.

Table 8. Results of water quality analysis of River in Southwest Maluku Regency

Parameters	Water Quality Standard*	Upstream	Downstream
Temperature	Deviasi 3	25,7	27,1
TDS	1000	514	202
TSS	50	5	2

pH	6	7,39	7,19
COD	50	13,21	9,28
BOD	12	4,21	4,94
DO	6	6,35	6,32
NO3 as N	10	0,53	0,17
NH3-N	0,5	0,24	0,12
Cuprum	0,02	0,012	0,008
Nitrit as N	0,06	0,007	0,01

Source: Environmental Departement of Southwest Maluku Regency (2017)
** Standard of Water Quality according (SWQ) to Government Regulation No. 82/2001*

4. Conclusion

Based on the results of this study, it can be concluded as follows:

1. The availability of water from various water resources is generally still sufficient to meet the needs of the population
2. The environmental capacity of water resources in Southeast Maluku Regency, Southwest Maluku Regency and Ambon City generally has exceeded the water quality standard.
3. However, it should be noted that the condition of green open space must be maintained and efforts to conserve water resources so that water availability is guaranteed throughout the year.

References

[1] Ruining J and Xiaohui J, Xingxing S and Chen Wei 2018 Study on the Water Resource Carrying Capacity in the Middle Reaches of the Heihe River Based on Water Resource Allocation *Water* ***10****, 1203; doi:10.3390/w10091203*

[2]. Chi M B, Zhang D S, Fan G W, Zhang W, Liu H L 2019 Prediction of water resource carrying capacity by the analytic hierarchy process-fuzzy discrimination method in a mining area. Ecol. Indic *Ecological Indicators* **96** 1647-655

[3]. Meng C, Wang X L, Li Y 2018 An optimization model for water management based on water resources and environmental carrying capacities: A case study of the Yinma River Basin, Northeast China *Water* **10** 565

[4] Lili W, Zhonggen W and Xiaocong L 2018 Water Resources Carrying Capacity Analysis of YarLung Tsangpo River Basin (I) *Water* **10** 1131; doi:10.3390/w10091131

[5]. Mohammad M H, Ali A, Seyed M R and Seyed M A H 2013 Correlation Study and Regression Analysis of Drinking Water Quality in Kashan City, Iran *Walailak J.Sci. & Tech.* **10** 315-324.

[6] SNI 19-6728.1-2002 Preparation of a Resource Balance - Part 1: Spatial Water Resources (Compilation of Resource Balance - Part 1: Spatial Water Resources, in Indonesian). Institution of National Standardization, Jakarta, 2002.

[7] Deputy for Environmental Governance of the Ministry of Environment 2014 Guidelines For Determining The Carrying Capacity And Capacity Of The Environment.

[8] Environmental Departement of Southeast Maluku Regency 2017 Environmental Status Report.

[9] Kunu P J 2019 Carrying Capacity of Water Resources in Southeast Maluku Regency in Southeast Maluku Regency RPJMD strategic environmental assessment report 2018-2023

[10] Environmental Departement of Aru Archipelago Regency (2017). Environmental Status Report.

[11] Osok R, M., Kunu P J, dan Laimeheriwa S 2017 Climate Impact Assessment on Water Availability in Wamar Island. USAID-Unpati Ambon

[12] Environmental Departement of Malra Regency (2017). Environmental Status Report.

[13] America Academy of Pediatrics, Where Does Fluoride Come From? November 08, 2013 & filed under Factsabout Fluoride. Available online: https://ilikemyteeth.org/fluoride-from/ (Acsessed on November 14, 2020)

[14] Fajar 2010 Kandungan Nitrat/ NO3 dan Nitrit/NO2 Pada Perairan Tawar. Available online : http://illonkjie.blogspot.com/2010/04/kandungan-nitrat-no3-dan-nitritno2-pada.html (Acsessed on April 12 2018)

[15] Uyara L, Kunu P J, and Talakua S 2018 Comparative study of clean water quality in relation to community activities around water sources in Wainitu, Batumerah, AMahusus and Halong villages *Jurnal Budidaya Pertanian* **13** 111-119. http://ojs3.unpatti.ac.id//index.php/bdp

Characteristics of chemical properties of fruit flour of mangrove (*Bruguiera gymnorrhiza* Lamk.) with lower cyanide content

F J Polnaya[1,*], V J Thenu[1], S Palijama[1] and R Breemer[1]

[1] Department of Agricultural Technology, Faculty of Agriculture, Pattimura University, Jl. Ir. M. Putuhena, Ambon, 97233 Maluku, Indonesia.

* E-mail: febby.polnaya@faperta.unpatti.ac.id

Abstract. This study aimed to determine the best method for reducing HCN levels of mangrove (*Bruguiera gymnorrhiza* Lamk.) flour and analyze mangrove flour's chemical properties. This study used a completely randomized design with a treatment level, namely boiling with water for 30 minutes at 100°C, immersion in distilled water, immersion in 0.8 M NaOH solution, immersion in KOH solution 0.8 M, immersion in 0.8 M Na_2CO_3 solution, and immersion in 0.8 M Na_2HPO_4 solution. All immersion treatments were carried out at room temperature for 24 hours. The experimental parameters included HCN, ash, moisture, lipid, carbohydrate and protein contents. The research data were analyzed using analysis of variance and continued with the Tukey test. The results showed that the treatments that were tried to reduce HCN levels could also be used for mangrove fruit flour. The boiling treatment that is usually carried out by the community in Tual, Southeast Maluku is more effective when compared to other treatments in an effort to reduce HCN levels. There were also differences in the chemical characteristics of the mangrove flour with the various techniques that were tried. Boiling treatment for 30 minutes showed the highest reduction (84.14%) with HCN levels of 1.95 mg/kg. The characteristics of mangroves with the lowest levels of HCN were: 14.17% ash content, 12.08% moisture content, 0.51% lipid content, 67.76% carbohydrate content, and 3.64% protein content.

1. Introduction

Mangrove (*Bruguiera* sp.), also known as "*Lindur*" or "*Tancang*" in Central Java [1], thrives in river areas and river estuaries along the muddy coast with low salinity. Mangrove ecosystems provide protection for the beach, stabilize the soil and serve as a habitat for various fauna [2] and has economic benefits [1]. There are several mangrove species, but the one commonly used as a local food source is *B. gymnorrhiza* Lamk. Almost all parts of mangroves can be used, such as "*Pepagan*" (stem bark), leaves [2] and fruit [1,3]. That mangroves are suitable for processing fruit into flour because of their high carbohydrate content which can also be processed into other food products [3].

Mangrove fruit has not been fully utilized because some people still do not know better processing techniques. Mangrove trees that are two years old are already productive [3]. The people in the Tual area, Southeast Maluku Regency, have processed mangroves and produced flour, which is then processed into traditional food ingredients. Mangroves are used as a food source during famine [2].

The research conducted by IPB in collaboration with the East Nusa Tenggara Food Security Agency showed that mangroves' energy content was 371 calories/100 g, higher than rice (360 calories/100 g) and corn (307 calories/100 g). Mangroves' chemical composition includes a water content of 73.76%, a lipid content of 1.25%, a protein of 1.13%, and an ash content of 0.34%. The carbohydrate content of mangrove fruit was 85.1 g/100 g, higher than rice (78.9 g/100 g) and corn (63.6 g/100 g) [4]. Mangroves' nutritional content is complete enough to be processed into cakes and cakes mixed with coconut spices [5].

The limiting factor for mangrove fruit consumption is anti-nutritional content, namely tannins and hydrogen cyanide (HCN). Some other plants that contain HCN include cassava [6–8], sorghum [9], *koro benguk* [10], *koro* nuts [11] and flax [12]. Those containing tannins include betel nut [13]. HCN is a dangerous compound if ingested in doses of 0.5-3.5 mg/kg body weight because it can kill humans [14].

HCN formation activities can be eliminated or reduced through cooking, boiling, or soaking. The soaking method is generally used to remove or reduce the antinutrient content. The immersion medium can be water, salt or alkaline solutions [10]. The treatments that have been carried out to reduce HCN levels are boiling and soaking cassava [7,8], soaking *koro benguk* in Na_2HCO_3 solution [10], soaking *koro* nuts [11], as well as for blanching and soaking in calcium solution [15]. As for mangroves, they are immersed in a solution of husk ash. Soaking mangrove fruit in 30% husk ash solution for 36 hours succeeded in reducing the HCN level to 3.36 mg/kg [16].

Decreasing HCN levels in plant tissues such as cassava can be done by boiling and soaking, and immersion in sodium bicarbonate ($NaHCO_3$). Boiling and soaking can cause cyanide is easily dissolved in water, and its boiling point is 29°C [8]. Immersion in sodium bicarbonate, alkaline, makes it easier to break down cyanide [7,10]. Besides, it can also reduce the hydrolysis of cyanogenic glucoside compounds, namely linamarin. Hydrolysis of linamarine with linamarase produces acetone cyanohydrin and glucose. The cyanohydrin acetone, in turn, has HCN and acetone [17].

Thus, alkaline treatment can reduce the HCN content in a food ingredient. However, so far, the use of alkaline solutions such as NaOH, KOH, Na_2CO_3 has not been carried out to reduce mangroves' HCN levels. When compared to the percentage decrease between the use of the alkaline solution and husk ash, it appears that the alkaline solution gives a higher percentage of HCN reduction (range 24-70%) [7,10] than rub or husk ash (15%) [18]. Thus, the alkaline solution is maximal in reducing the HCN levels of mangroves.

The processing of mangrove fruit into flour with low HCN content is a semi-finished product that can extend its shelf life and be processed further. The processing semi-finished products are one way of preserving crops, especially for commodities with high water content, such as various tubers and fruit [19]. Flour has a low water content with a composition of specific chemical properties. It is more flexible in being applied to various types of food preparations so that it is hoped that it will be easier to introduce to the public. The procedure for making flour is very diverse, differentiated by the properties and chemical components of foodstuffs. However, in broad terms, they can be grouped into two, namely: 1) foodstuffs that do not turn brown quickly when peeled (cereal group); and 2) foodstuffs that promptly turn brown (a group of various tubers and fruits that are rich in carbohydrates) [19]. From this group, mangroves are included in the foodstuff group that turns brown quickly.

In addition to the low HCN content of mangrove flour, it is also necessary to know the chemical properties of flour, such as moisture, ash, protein, lipid, and carbohydrate content. The mangrove fruit flour with low HCN met the requirements as a food ingredient [16]. Thus, research is needed to obtain an appropriate soaking method with an alkaline solution and has the best effectiveness in reducing HCN levels of mangroves to a safe limit for consumption and characterizing the chemical properties of the mangrove flour it produces. This study's objectives were to determine the best method to reduce mangrove fruit flour's cyanide content and describe its chemical properties.

2. Methods

Mangrove fruit used in this study originated from the Tual, Southeast Maluku district, Indonesia. Fresh mangroves fruit are stored properly and then prepared in the laboratory.

2.1. *Preparation of fruit flour of mangrove*

All steps were carried out at room temperature, which ranged from 28 to 33°C. The freshly harvested mangrove fruits were peeled, washed with water and then was cut into pieces with a size of 1 cm. Mangrove pellets were boiled with water for 30 minutes, soaked in distilled water for 24 hours; soaked in 0.8 M NaOH (Merck, Germany) solution for 24 hours; soaked in 0.8 M KOH (Merck, Germany) solution for 24 hours; soaked in 0.8 M Na_2CO_3 (Merck, Germany) for 24 hours; and soaked in 0.8 M Na_2HPO_4 (Merck, Germany) for 24 hours. Furthermore, mangrove pellets were dried using a cabinet dryer at 40°C for 48 hours.

2.2. Total cyanide content
The total cyanide content of mangrove fruit flour was determined by the method of Askhurahman [6]. Five g of mangrove flour is put into a 250 mL measuring flask, then 250 mL of distilled water is added, then shaken, and after that, the distilled water is added again until it is precisely 250 mL, then shaken again. Next, take 1 mL and mix it with 1 mL of 0.1 N NaOH and 5 mL of 0.5% picric acid at pH 11 in a test tube. Then boil for 30 minutes and vortex. The results are then read with a spectrophotometer UV-Vis (Shimadzu, Japan) at a wavelength of 480 nm.

2.3. Ash and moisture contents
Standard AOAC methods [20] were used to measure moisture and ash. Moisture content (%, w/w) was assayed by weight loss on drying at 105°C for five h in a hot air oven (Memmert, Germany).

The ash (%, w/w) was determined by incinerating known sample weights in a muffle furnace (Vulcan A-550 Ney, USA). Starch was burned in a furnace overnight (12 h) at 550°C and cooled down in a desiccator.

2.4. Lipid content
Approximately 5 g of homogenate was weighted and dried in oven 105°C for 2 hours. Removed from the oven, and samples were cooled to room temperature. Cotton has covered the sample and put it into extraction tubes. Dried and weighted extraction balloons were attached to the extraction device. Petroleum ether (Merck, Germany) used as a solvent added into tubes. Extraction completed approximately in 4-5 hours. Balloons were dried in an oven for about 45 min. After cooling to room temperature in a desiccator, balloons were weighted.

2.5. Carbohydrate content
Carbohydrate content was determined by the method of Anthrone. Carbohydrates are dehydrated by conc. H_2SO_4 (Merck, Germany) to form furfural. The reagent's active form is anthranol, the enol tautomer of Anthrone, which reacts by condensing with the carbohydrate furfural derivative to give a green colour in dilute and blue colour in concentrated solutions, which is determined calorimetrically. The blue-green solution shows an absorption maximum at 620 nm, using spectrophotometer UV-Vis 1800 (Shimadzu, Japan).

2.6. Protein content
The sample was weighed as much as 0.5 g, added with 2 g of selenium (Merck, Germany) and 25 mL of concentrated H_2SO_4 (Merck, Germany), then put in a 100 mL Kjeldahl flask. The destruction process is carried out until the solution becomes clear greenish. After cooling, then placed in a 100 ml volumetric flask and diluted with distilled water to the boundary mark. As much as 5 mL of 30% NaOH was put into a distillation flask, then distilled for 5 minutes. The distillate was accommodated as much as 10 mL with 2% boric acid (Merck, Germany) and BCG-MR (Merck, Germany) indicator. The amount of reacted boric acid with ammonia could be identified by 0.01 HCl (Merck, Germany) titration. A change solution color marked the end of the titration from blue to pink. Protein content was calculated by the following formula as mentioned in the previous method [21].

2.7. Statistical analysis
Data were statistically analyzed by analysis of variance test procedure, and significant differences were identified by Tukey's test ($p < 0.05$) using Minitab 18 software (Minitab, Inc.).

3. Results and discussion

3.1. HCN levels for mangrove flour

The results showed that the HCN content of the mangrove flour was 12.3 mg/kg. The levels of HCN in these mangroves are still very high so that if consumed directly without any processing, there is concern that it can cause poisoning. This study showed that HCN levels' value was lower than that of the previous study [16] 31.68 mg/kg for direct powdering, but thorough soaking using a bleach solution yields 12.96 mg/kg.

After boiling or soaking treatment, the HCN levels of mangrove flour decreased to 1.95-3.22 mg/kg or reduced by 73.82–84.14 mg/kg (Table 1). The results of the analysis of variance showed that the immersion treatment had a significant effect on HCN levels ($p < 0.05$). The highest average HCN level of mangrove fruit flour was 3.22 mg/kg. It was indicated by the immersion treatment with KOH 0.8 M. Based on the Tukey test (α = 0,05), it was significant differences from the immersion treatment with distilled water for 24 hours (2.18 mg/kg), immersion treatment with 0.8 M NaOH (3.07 mg/kg) and immersion treatment with 0.8 M Na_2CO_3 (2.82 mg/kg).

The results showed that the boiling treatment for 30 minutes resulted in decreased HCN levels reaching 84.14%, higher than other treatments. A relatively similar effect was also demonstrated for yam tuber [15]. The blanching process of unpeeled yam tuber for 30 minutes combined with immersion in clean water reduced HCN levels to 4.12 mg/kg. This may be due to the higher softening of mangrove fruit tissue at high temperatures. The softer the fruit tissue, the easier the process of removing linamarin from the pulp. The immersion resulted in softening of the tissues and pores so that the transfer of material that could pass through the permeable membrane occurred [15].

The HCN compounds are volatile in the boiling, steaming, and other cooking processes [11]. Compared with immersion in alkaline conditions (pH 12), it can also cause softening of mangrove fruit tissue. Softening the tissue in foodstuffs such as fruit and tubers using an alkaline solution is known as lye peeling; the softening is lower. Also, HCN is water-soluble, so HCN that has been dissolved in water or alkaline solution will be wasted after the immersion is complete. Similar results are also shown by the reseacher [7] on cassava's soaking treatment using an alkaline solution of $NaHCO_3$. The immersion resulted in a 74% reduction in HCN levels. The softer the fruit tissue, the easier the process of removing linamarin and lotaustralin from the fruit.

The decrease in HCN levels is caused by boiling or soaking the linamarin compound will be hydrolyzed (reacting with water) and form cyanide acid soluble in water. The hydrolysis of linamarin will form HCN, which is water-soluble and volatile [7]. This condition causes linamarin levels to be lowered through the immersion process. During the hydrolysis process carried out by β-glucosidase on cyanogenic glucosides, some of the sugar and hydroxyinitrile will be partially separated or enzymatically into cyanide and a mixture of carbonyl (ketose and aldose) [22].

High boiling temperatures also spurred the decrease in HCN levels. In the boiling process, more HCN is dissolved than in the immersion process because it is easy for HCN to evaporate when the temperature rises (HCN boiling point 26.5°C). Also, at high temperatures, it can cause the inactivation of the linamerase enzyme in mangroves to reduce the hydrolysis of linamarin or stop the formation of HCN. Linamarin's hydrolysis with linamarase produces acetone cyanohydrin and glucose. Cyanohydrin acetone spontaneously at pH above 5 has HCN and acetone [17].

Immersion in 0.8 M Na_2HPO_4 for 24 hours showed a lower HCN levels reduction than immersion in Na_2CO_3, NaOH and KOH. This is because the pH value shown in the Na_2HPO_4 solution is around 6.6, which is the optimum pH condition for β-glucosidase enzyme activity. The more effective softening and the inactivation of linamarase can hydrolyze linamarin in alkaline conditions. The optimum activity of β-glucosidase and linamarase enzymes) is pH 5-6 [23]. The linamarase enzyme's optimum pH isolated from cassava tubers is pH 6. Conditions above six can cause enzyme inactivation [6]. The protein denaturation can be caused by enzyme inactivation [23].

Table 1. Cyanide, moisture, ash, protein, lipid and carbohydrate contents of fruit flour of mangrove after treatments.

Treatments	Cyanide content (mg/kg)	Moisture content (%)	Ash content (%)	Lipid Content (%)	Carbohydrate content %)	Protein content (%)
Boiling with water (30 min)	1.95 [c]	12.08 [a]	14.17 [a]	0.51 [bc]	67.76 [c]	3.64 [bc]
Soaked in disttiled water (24 hr)	2.19 [bc]	6.62 [e]	12.95 [b]	0.54 [b]	74.31 [ab]	4.08 [ab]
Soaked in 0.8 M NaOH (24 hr)	3.07 [a]	12.00 [a]	11.72 [c]	0.41 [c]	72.41 [b]	2.78 [d]
Soaked in 0.8 M KOH (24 hr)	3.22 [a]	8.14 [c]	8.84 [e]	0.69 [a]	76.75 [a]	3.83 [abc]
Soaked in 0.8 M Na_2CO_3 (24 hr)	2.83 [ab]	10.09 [b]	11.68 [c]	0.45 [bc]	73.09 [b]	3.59 [c]
Soaked in 0.8 M Na_2HPO_4 (24 hr)	2.01 [bc]	7.68 [d]	10.09 [d]	0.75 [a]	75.98 [a]	4.24 [a]

Note: Means in the same column without the same letter differ ($p < 0.05$) according to the Tukey test ($\alpha = 0.05$)

3.2. Ash content of mangrove flour
The results showed that the mangrove flour's ash content ranged from 14.17–10.09% (Table 1). The results of the analysis of variance showed that the treatment had a very significant effect on the ash content of mangrove fruit flour ($p < 0.01$). The highest average ash content was shown by mangroves boiled for 30 minutes (14.17%), while the lowest ash content was offered by mangroves immersed in 0.8 M KOH for 24 hours (8.84%). The Tukey test results ($\alpha = 0.05$) showed that the boiling treatment for 30 minutes was significantly different from the immersion treatment in distilled water for 24 hours, immersion treatment with 0.8 M NaOH and immersion treatment KOH solution 0.8 M for 24 hours.

Immersion treatment in several alkaline solutions reduced the ash content of mangrove fruit flour by a range of 8.84–11.72%. The dissolving of mineral molecules can cause a decrease in ash content during immersion into the immersion medium. The effect of pH treatment on ash content illustrates that a high pH (KOH 12.9) solution can reduce the ash content significantly compared to other treatments. This is because KOH can break the bonds of complex compounds, lignin, cellulose and pectin and compounds forming cell walls as part of the complex compounds that make up food minerals. The decomposition of these compounds becomes simpler, resulting in a decrease in ash content [24]. Compared with the ash content of some yam flour 3,78%, the ash content of the mangrove flour shows a relatively high value. The lower the ash content of the flour, the better it will be applied in the food industry.

3.3. Moisture content of mangrove flour
The results showed that the mangrove flour's moisture content ranged from 6.62–12.08% (Table 1). The results of the analysis of variance showed that the treatment had a very significant effect on the moisture content of mangrove fruit flour ($p < 0.01$). The results showed that the highest water content was obtained in the boiling treatment for 30 minutes (12.08%), while the lowest water content was obtained in the immersion treatment with distilled water for 24 hours (6.62%). The Tukey test results ($\alpha = 0.05$) showed that the boiling treatment for 30 minutes was not significantly different from the immersion treatment in 0.8 M NaOH for 24 hours but was significantly different from other treatments.

Boiling mangrove fruit for 30 minutes and immersing in 0.8 M NaOH for 24 hours causes the particles to become more porous when compared to other treatments, thereby increasing the ability to absorb water. After drying, the water is trapped in the material. Immersion with a strong alkaline solution and its salts can cause changes in the starch granule structure, making it more porous so that it is easier to absorb water [14]. Meanwhile, immersion in distilled water without heating will not change the starch granule structure so that it provides the lowest water content compared to other treatments. The water content obtained in this study was still included in the water content of commercial flour (11-12%) [25].

3.4. Lipid content of mangrove flour

The results showed that the mangrove fruit flour's lipid content ranged from 0.41–0.75% (Table 1). The analysis results showed that the treatment significantly affected mangroves' lipid content ($p < 0.01$). The highest average lipid content was found in the immersion treatment in 0.8 M Na_2HPO_4 solution for 24 hours (0.75%). The Tukey test results ($\alpha = 0.05$) showed that the treatment was not significantly different from the immersion treatment in the 0.8 M KOH solution for 24 hours but was significantly different from other treatments.

During the immersion process, the mangroves will be hydrolyzed due to contact with the solution, causing the lipid content to decrease due to oxidation. Other researcher showed that soaking and boiling treatments could significantly reduce lipid levels [24]. Treatment of 0.8 M NaOH and 0.8 M Na_2CO_3 solution for 24 hours reduced the lipid content lower than the other treatments. This is because the NaOH solution can break down lipid into water-soluble glycerol and fatty acids to not accumulate in the fruit. Compared with the immersion treatment using KOH with Na_2HPO_4, the value of lipid content is higher, which is because the KOH and Na_2HPO_4 solutions only damage the lipid globules and are not water-soluble but settle in the fruit. Hence, the results of the lipid analysis are high [14].

3.5. Carbohydrate content in mangrove flour

The results showed that the mangrove flour's carbohydrate content ranged from 67.76 to 76.75% (Table 1). The analysis results showed that the treatment significantly affected carbohydrate content ($p < 0.01$). The soaking treatment led to the highest average carbohydrate content with 0.8 M KOH 24 hours (76.75%). The Tukey test result ($\alpha = 0.05$) showed that the treatment was not significantly different from the immersion treatment using 0.8 M Na_2HPO_4 (75.98%) and distilled water for 24 hours (74.31%). Still, it was significantly different compared to other treatments.

The results showed that the boiling treatment showed the lowest value of carbohydrate content (67.76%). This may be due to damage to the starch granules during boiling. The starch in the fruit undergoes gelatinization at boiling temperature (± 100°C) can weaken the hydrogen bonds or breaks the glucoside bonds to cause the dispersion of glucose into the solution. Granule damage caused by immersion in distilled water and alkaline solutions is lower than boiling, indicating higher carbohydrate content. Previous research [26] also suggested that the alkaline solution can break the protein matrix's disulfide bonds that enclose the starch granules to free the starch granules or increase the separation between starch and the protein that wraps the starch granules. This resulted in higher levels of carbohydrates detected in the mangrove flour samples.

3.6. Protein content in mangrove flour

The results showed that the mangrove flour's carbohydrate content ranged from 2.78–4.24% (Table 1). The results of the analysis of diversity showed that the treatment had a very significant effect on protein content ($p < 0.01$). The immersion treatment with Na_2HPO_4 indicated the highest average protein content (4.24%). The Tukey test ($\alpha = 0.05$) results showed that the treatment was not significantly different from the immersion treatment with distilled water and 0.8 M KOH for 24 hours, but it was different from other treatments.

The protein content of mangrove flour in the boiling treatment was 3.64% using an alkaline solution, and the protein content decreased. This is because protein is more difficult to dissolve in acids than in alkaline conditions. The higher the concentration of immersion food, the more acidic the soaking solution is so that the protein does not dissolve easily and not much of it comes out, leaving starch granules. On the other hand, in alkaline Na_2HPO_4 immersion, the protein will form a cross-link with amylose to quickly absorb water. The protein molecules diffuse, leaving the granules dissolving in a soaking solution that will be wasted during the mangroves' washing process. The higher the concentration or strength of the alkaline solution will affect the protein content. The protein content of the ingredients soaked with Na_2HPO_4 is higher than the other solutions.

pH gives varying results on decreasing protein levels where NaOH solution's pH solution with a pH of 12.6 can reduce the largest protein levels. It can be left in mangrove fruit flour is very low. This is by the decrease in HCN levels in mangroves soaked with NaOH compared to KOH solution, both of which have high pH (KOH 12,9). Meanwhile, the Na_2CO_3 solution (pH 8.7) reduced mangrove flour's protein more than the immersion treatment with Na_2HPO_4 solution (pH 6.6). This is because, at high pH, there is damage to protein.

4. Conclusion

All treatments tested were able to reduce levels of HCN mangrove fruit flour. However, the fruit of mangrove treated with boiling water for 30 min showed the highest reduction in the cyanide content, which was 84.14% and were having the final content of 1.95 mg/kg cyanide content. Mangrove fruit flour characteristics with the lowest cyanide content had 4.17% ash, 12.08% moisture, 0.51% lipid, 67.76% carbohydrate, and 3.64% protein contents.

References

[1] Subandriyo and Setianingsih N I 2016 Extraction Process for Reducing Tannin of Mangrove Fruit [*Bruguiera gumnorrhiza* (Lamarck, 1798)] as a Raw Material for Food Flour *Aquat. Procedia* **7** 231–5

[2] Allen J A and Duke N C 2006 *Bruguiera gymnorrhiza* (large-leafed mangrove), ver 2.1 *Species Profiles for Pacific Island Agroforestry* ed C R Elevitch (Hōlualoa, Hawai'i: Permanent Agriculture Resources) pp 1–15

[3] Sadana D 2007 Buah Aibon di Biak Timur Mengandung Karbohidrat Tinggi [Aibon Fruit in East Biak Consist of High Carbohydrat] *Off. Website Local Gov. Biak Numfor*

[4] Fortuna J 2005 Ditemukan Buah Bakau sebagai Makanan Pokok [Mangrove Fruit as Staple Food] *http://www.tempointeraktif.com*

[5] Bayu A 2009 Hutan mangrove sebagai salah satu sumber produk alam laut [Mangrove forests are a source of natural marine products] *J. Oseana* **34** 15–23

[6] Askurrahman 2010 Isolasi dan karakterisasi linamarase hasil isolasi dari umbi singkong (*Manihot esculenta* Crantz) [Isolation and characterization of isolated linamarase from cassava tubers (*Manihot esculenta* Crantz)] *Argointek* **4** 138–45

[7] Hutami F D and Harijono 2014 Effect of solution replacement and $NHCO_3$ concentration on cyanide content reduction in cassava flour processing *J. Pangan dan Agroindustri* **2** 220–30

[8] Suryani C L and Wesniati N 2000 Studi Pembuatan Tepung Kara Benguk [Study of Kara Benguk Flour Making] *Location-Specific Agricultural Technology Seminar in Efforts to Improve Farmer Welfare and Environmental Conservation*

[9] Chrisye 2012 *Sorghum Sebagai Bahan Pangan [Sorghum As a Food Material]* (IPB University)

[10] Sudiyono 2010 Usage of Na_2HCO_3 to lessen hydrogen cyanide content (HCN) at making of fries koro benguk *AGRIKA J. Ilmu-Ilmu Pertan.* **4** 48–53

[11] Suciati A 2012 *Pengaruh Lama Perendaman dan Fermentasi Terhadap Kandungan HCN Pada Tempe Kacang Koro* (Hasanuddin University)

[12] Fieldes M A and Gerhardt K E 2001 Developmental and genetic regulation of ß-glucosidase (linamarase) activity in flax seedlings *J. Plant Physiol.* **158** 977–89

[13] Sulastri T 2009 Analysis of concentration of tannins from ethanol and water extract at the pinang sirih seed (*Areca catechu* L) *J. Chem.* **10** 59–63

[14] Winarno 2004 *Food Chemistry and Nutrition* (Jakarta: PT Gramedia Pustaka Utama)

[15] Djaafar T F, Rahayu S and Gardjito M 2009 The effect of blanching and soaking in calcium solution of yam tuber and chips *Penelit. Pertan. Tanam. Pangan* **28** 192–8

[16] Sulistyawati, Wigyanto and Kumalaningsih S 2012 Low tannins and HCN of lindur fruit flour products as an alternative food *J. Teknol. Pertan.* **13** 187–98

[17] Djazuli M and Bradbury J H 1999 Cyanogen content of cassava roots and flour in Indonesia *Food Chem.* **65** 523–5

[18] Permadi Y B, Sedjati S and Supriyantini E 2012 Effect of ash powder concentration and immersion time on the nutritional content of mangrove fruit flour Avicenia marina *J. Mar. Res.* **1** 39–47
[19] Widowati S, Sukarno L, Suarni and Komalasari O 2003 Labu Kuning: Kegunaan dan Proses Pembuatan Tepung [Yellow Pumpkin: Uses and Process of Making Flour] *Seminar Nasional dan Pertemuan Tahunan Perhimpunan Ahli Teknologi Pangan Indonesia (PATPI)* pp 30–8
[20] AOAC 2012 *Official Methods of Analysis of The Association of Official Analytical Chemists* (Virginia: AOAC Inc. Arlington)
[21] Elfita L 2014 Analysis of profile of protein and amino acids of swallow's nest (*Collocalia fuchiphaga*) from Painan *J. Pharm. Clin. Sci.* **1** 27–37
[22] Frehner M 1995 *The linamarin β-glukosidase in Costa Rican wild lima beans (Phaseoelous lunatus L.) is appoplastics* (University of California)
[23] Ginting E and Widodo Y 2013 Cyanide reduction in cassava root products through processing and selection of cultivars in relation to food safety *Bul. Palawija* **25** 26–36
[24] Yuanita L 2006 The effect of pectic substances, hemicellulose, lignin and cellulose content to the percentage of bound iron by dietary fiber macromolecules: Acidity and length boiling time variation *Indones. J. Chem.* **6** 332–7
[25] Purnobasuki H 2004 Prospect of mangrove as herbal medicine *Biota* **9** 125–6
[26] Chandra A, Inggrid H M and Verawati 2013 *Effect of pH and type of solvent on acquisition and characterization of starch from avocado seeds* (Universitas Katolik Parahyangan)

IOP Conf. Series: Earth and Environmental Science **883** (2021) 012064 doi:10.1088/1755-1315/883/1/012064

Effect of dietary palm kernel meal on laying hens

R D D Prasetya[1], M Rahmadani[2], Nahrowi[3] and A Jayanegara[3,*]

[1] Graduate Study Program of Nutrition and Feed Science, Graduate School of IPB University, Bogor 16680, Indonesia.
[2] Bachelor Degree of Animal Science Faculty, IPB University Bogor 16680, Indonesia.
[3] Department of Nutrition and Feed Technology, Faculty of Animal Science, IPB University, Bogor 16680, Indonesia.

* E-mail: anuraga.jayanegara@gmail.com

Abstract. Indonesia has the potential for palm kernel meal, which has the largest production in the world. Palm kernel meal has been commonly used as an ingredient in ruminant and poultry feeds. However, the various studies conducted have produced very varied results. This article aimed to analyze various research data on the use of palm kernel meal in the chicken feed. The data used were 38 data points from 10 journal articles. The use of palm kernel meal ranged from 0% to 70% in rations. The statistical model used is linear regression. The database was analyzed using SPSS software version 25. Results showed that the use of palm kernel meal in laying hens reduced ($P<0.01$) egg production. Palm kernel meal also increased FCR ($P<0.05$) and tended to decrease Haugh Unit ($P<0.1$). Other parameters such as consumption, egg weight, mortality, egg shell thickness, and yolk color were not influenced by dietary inclusion of palm kernel meal. In conclusion, palm kernel meal cannot be given to poultry directly, it is necessary to process the ingredient to enhance its nutritive value.

1. Introduction

In the new normal era, the food supply chain has to continue to move and ensure the availability of nutritious food as a weapon to maintain immunity to prevent the transmission of COVID-19. Food commodities that have the opportunity to survive in this era are the livestock sector, especially poultry. In from 2019 to 2020, several small-scale layer chicken farmers suffered losses due to falling prices. Egg products are a source of protein and minerals needed in a nutritionally balanced diet and at an affordable price for the wider community. Scientifically, eggs have been shown to contain all the proteins, lipids, vitamins, minerals, and growth factors needed to maintain immunity during the Covid-19 pandemic [1]. [2] also support that supplementation of vitamins A, D, E, and the mineral selenium can be obtained from the consumption of eggs.

Egg production during a pandemic is demanded to be as efficient as possible, one effort is to maximize the use of feed. the feed is the biggest cost in the livestock business. 60%-70% of production costs are determined from the feed. corn and soybean meal are the main ingredients commonly used in poultry rations [3]. The number of maize which depends on the season and the distribution of the location of its wide production makes the price of maize very expensive at certain times. While soybean meal is obtained by importing from other countries, so the price is expensive.

Indonesia is a country with the highest palm oil production in the world. Production of palm oil and palm kernel in 2018 reached 36 594 813 and 7 318 963 tonnes, respectively [4]. According to [5] the proportion of PKM (Palm Kernel Meal) ranges from 45-46% of the palm kernel or 2.0% -2.5% of the weight of the palm kernel meal bunches, so the potential production is around 3,293,533 tonnes of PKM. Therefore palm kernel meal has the potential to be used as a feed ingredient which does not depend on the season also the price is cheap. Optimizing the use of palm kernel meal is expected to push the importation of feed ingredients during the pandemic. The nutritional content contained in the palm kernel meal is 92.1% dry matter, 15.9% crude protein, 9.4% crude fat, and 3.9% ash. The main mineral content in palm kernel meal, namely calcium, phosphorus, and potassium, respectively 3.9, 5.5, 6.4 g / kg [6]. Palm kernel meal has been widely used in ruminant and poultry rations. However, the various studies carried out produced many differences. Therefore, this article aims to analyze various research data regarding the use of palm kernel meal in laying hens.

2. Methods

The database was created from a collection of data from various journals regarding the use of palm kernel meal in poultry. The journals used are sourced from Google Scholar, Research Gate, and Science Direct. 60 journals were obtained and then selected by title. Until finally there were 10 journals with 38 data points. Data inputted into MS. Excel 2016, namely egg production (%), feed intake (grams), Feed Conversion Ratio, egg weight (grams), weight gain (grams), mortality (%), egg shell thickness (mm), egg yolk color, and haugh units. Then it is selected again based on the contents, namely the data selected is the use of PKM in laying hens. PKM usage ranges from 0% to 70%. Regression analysis was performed using SPSS version 25.

Table 1. The study used in assessing the effectiveness of using palm kernel meal on laying hens

No	Reference	PKM level (%)
1	Perez *et al.* [7]	0-50
2	Aderolu *et al.* [8]	20
3	Lee *et al.* [9]	5
4.	Ryu *et al.* [10]	0-8.7
5.	Onwudike *et al.* [11]	0-37.5
6.	Ridla *et al.* [12]	12-28
7.	Wihandoyo [13]	0-36
8.	Zanu *et al.* [14]	0-15
9.	Anya 2014 [15]	15-20
10.	Diarra 2018 [16]	0-20

3. Results and Discussion

Descriptive analysis of the parameters of egg production, feed intake, feed conversion ratio, egg weight, weight gain, mortality, eggshell thickness, yolk color, and haugh unit can be seen in Table 2.

Table 2. Descriptive statistics on the effects of using palm kernel meal on the ration of laying hens.

Parameter	n	Mean	SD	Min	Max
Egg Production (%)	38	68.39	8.85	41.9	81
Feed Intake (gr)	37	116.38	12.02	95.41	140.17
FCR	32	2.98	0.70	2	4.44
Egg Weight (gr)	37	57.76	3.94	50	65.8
Weight gain (gr)	5	522.8	44.09	481	594
Mortality (%)	24	3.5	2.95	0	8.3
Egg Shell Thickness (mm)	16	0.36	0.95	0.18	0.47
Yolk Color	8	5.34	2.68	1.92	8.78
HU	24	84.35	7.45	75.7	97.96

Note: n: the amount of data, SD: standard deviation, Min: minimum, Max: maximum, FCR: Feed Conversion Ratio, HU: Haugh Unit.

The use of PKM had dramatically effect (P <0.01) in reducing egg production. Hen-day egg production is the number of eggs produced from total laying hens and multiplied by 100. The decrease in production was caused by an increase in crude fibre content in the ration. The high fibre content will make it difficult for monogastric animals to digest feed. Poultry as monogastric livestock has a minimal ability to hydrolyze or degrade fibre into simpler compounds; this is because the activity of cellulolytic enzymes in the digestive process is very low [5]. The palm kernel meal contains NSP of 610g / kg [6]. Mannan is the main component of NSP in palm kernel meal (Sathitkowitchai et al., 2018) [17]. The NSP content in PKM is known to bind water in the lumen, so that increasing the viscosity of the digestive tract and reducing its flow rate [18]. Increasing the feed flow rate will decrease feed digestibility. In addition to reducing egg production, the consumption of high-fibre feed makes fecal wet. The variety of shell contamination in PKM also results in various quality. The crude fibre content in palm kernel meal will also be higher following the amount of shell contamination. According to [19] the shell contamination of PKM ranges from 10% to 20% depending on the process of separating the shell from the kernel before extracting palm oil from the palm kernel.

Table 3. Analyzes the regression results of the use of palm kernel meal in laying hens

Parameter	Intercept	Slope	P value	R^2
Egg Production (%)	74.578	-0.304	**	0.364
Feed Intake (gr)	117.17	-0.038	NS	0.003
FCR	2.65	0.15	*	0.152
Egg Weight (gr)	58.99	-0.059	NS	0.071
Weight gain (gr)	553.40	-1.63	NS	0.301
Mortality (%)	2.79	0.023	NS	0.022
Egg Shell Thickness(mm)	0.376	-0.001	NS	0.007
Yolk Color	4.356	0.083	NS	0.078
HU	87.95	-1.66	<0.10	0.173

Note: FCR: Feed Conversion Ratio, HU: Haugh Unit, This model is very significant (**) at P <0.01, (*) significant at P <0.05, tends to be significant at P <0.10, not significant (NS) at P> 0.10.

The use of palm kernel meal has not affected by the feed intake of layer hens (P> 0.05). The amount of feed consumption is the same but the results showed lower egg production in layer hens that are fed with palm kernel meal. This shows that the inefficient use of nutrients to produce eggs. Therefore, given feed containing palm kernel meal could increase (P <0.05) FCR of layer hens. FCR is the amount of feed needed to produce per egg weight in one day [15]. The haugh unit has a downward tendency (P <0.1). Haugh unit is one way to determine the quality of an egg. HU was obtained by measuring albumin height and egg weight. The higher of the HU's value, than the albumen vicosity getting high. Ovomucin albumin which plays a role in binding water to form a gel on albumin which causes albumin to become thick. The more and stronger the ovomusin mesh will increase the albumin viscosity. The higher the HU value shows the better the egg quality [16]. Although the use of palm kernel meal tends to decrease the HU value, this result is still in the normal range above 70. The protein content in feed can affect egg quality. The tendency of PKM to reduce the value of HU was caused by the high NSP (Non-Starch Polysaccharide) in PKM. A diet with high NSP will reduce protein digestibility and increase endogenous amino acid secretion [6].

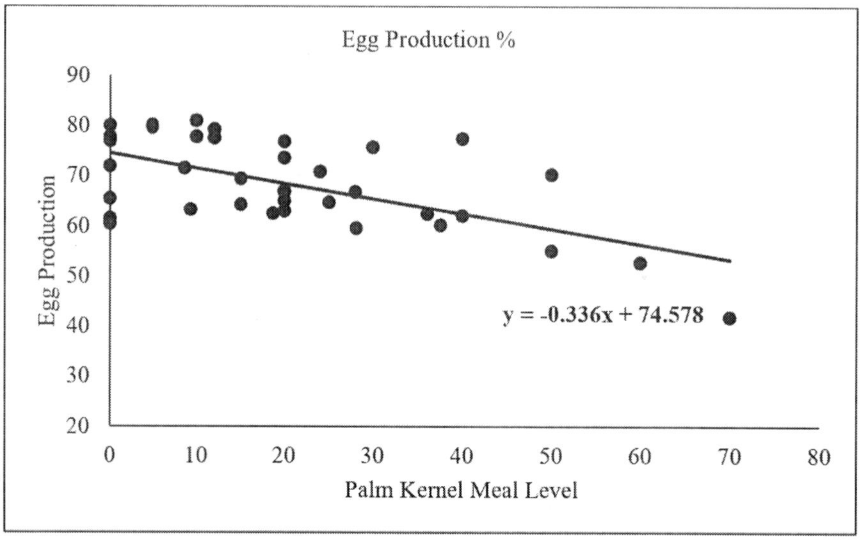

Figure 1. Linear regression graph of the effect of palm kernel meal on egg production.

PKM has low amino acid digestibility. The ileal digestibility coefficient on PKM was 0.42-0.46. Therefore, if the ration formulation based on crude protein will be inaccurate. Formulation based on amino acids will increase its accuracy [6]. Things that can be done to improve the quality of palm kernel meal using physical, chemical, or biological processing methods. Physical processing that can be done is by filtering the palm kernel meal so that contaminated shells can be removed. Based on research [19] filtering with a filter size of 2 mm can reduce shell contamination from 22.8% to 9.92%. Reducing the levels of shells in PKM can increase the proportion of the nutritional content in it. This is because the shells on PKM do not contain nutritional value. Although it can increase the nutrient content in PKM, the filtering process cannot increase metabolic energy, dry matter, and protein digestibility. Palm kernel meal contains 30% -35% β-mannan which is the main component of NSP from palm kernel meal. β-mannan has been shown to have strong anti-nutritional effects in monogastric livestock. Therefore, an enzyme is needed to hydrolyze, namely the enzyme mannanase. Hydrolysis of β-mannan using the enzyme mannase can release sugar, manno-oligosaccharides, and protein [17].

4. Consclusion

The use of palm kernel meal as a feed ingredient without undergoing processing it can reduce egg production tends to reduce the haugh unit value, and increase FCR. However, it does not affect the parameters of weight gain, feed intake, egg weight, mortality, eggshell thickness, and yolk color.

References

[1] Ismoyowati 2020 *Proceedings of the VII Animal Husbandry Technology and Agribusiness Seminar - Webinar: Livestock Prospects in the New Normal Era after the Covid-19 Pandemic (Purwokerto)* vol 7 (Central Java: Indonesia/ Animal Science Faculty of Jenderal Soedirman University p 24-34.

[2] Jayawardena R, Sooriyaarachchi P, Chourdakis M, Jeewandara C, Ranasinghe P 2020 Enhancing immunity in viral infections, with special emphasis on COVID-19: A review, *Diabetes & Metabolic Syndrome: Clinical Research & Reviews*. **14**, 367-382

[3] Oladokun AA, Rahman WA, Suparjo NM 2016 Prospect of maximising palm kernel cake utilization for livestock and poultry in malaysia: a review. *Journal of Biology, Agriculture and Healthcare*. **6**, 107-113.

[4] Statistics Indonesia 2018 *Indonesian Palm Oil Statistics 2018* (Jakarta: Statistics Indonesia).

[5] Pasaribu T 2018. Efforts to Improve the Quality of Palm Kernel Cake through Fermentation Technology and Enzyme Addition for Poultry. *WARTAZOA*. **28**, 119.

[6] Abdollahi MR, Hosking BJ, Ning D, Ravindran V 2016 Influence of Palm Kernel Meal Inclusion and Exogenous Enzyme Supplementation on Growth Performance, Energy Utilization, and Nutrient Digestibility in Young Broilers. Asian Australas. *J. Anim. Sci* **29**, 539–548.

[7] Perez JF, Gernat AG, Murillo JG 2000 Research notes: The effect of different levels of palm kernel meal in layer diets. *Poultry Science*. **79**, 77–79.

[8] Aderolu AZ, Iyayi EA, Onilude AA 2007 Nutrient digestibility and egg production of laying hens fed graded levels of biodegraded palm kernel meal. *Ghana Journal of Agricultural Science*. **39**, 139–146.

[9] Lee JY, Kim SY, Hwan LJ, Heon LJ, Ohh SJ 2013. Effect of Dietary β-Mannanase Supplementation and Palm Kernel Meal Inclusion on Laying Performance and Egg Quality in 73 Weeks Old Hens. *Journal of Animal Science and Technology*. **55**, 115–122.

[10] Ryu MH, Hosseindoust A, Kim JS, Choi YH, Lee SH, Kim MJ, Lee JH, Chae BJ 2017 *β*-Mannanase Derived from *Bacillus Subtilis WL-7* Improves the Performance of Commercial Laying Hens Fed Low or High Mannan-Based Diets. *Jpn. Poult. Sci*. **54**, 212–217.

[11] Onwudike OC 1988. Palm kernel meal as a feed for poultry. 4. Use of palm kernel meal by laying birds. *Animal Feed Science and Technology*. **20**, 279–286.

[12] Ridla M, Imran A, Jayanegara A, Hermana W, Tarman A 2019 Influence of Hemicell® Addition on Diets Containing Different Levels of Crude Fiber on Performance of Laying Hens *Buletin of Animal Science*. **43**, 179-183.

[13] Wihandoyo, Alimon AR, Ismail I, Kassim H 2002 Pemanfaatan bungkil kelapa sawit untuk campuran pakan ayan petelur terhadap kinerja produksi dan karakteristik tinja *Media Peternakan*. **25**, 71-74.

[14] Zanu HK, Abangiba J, Arthur BW, Akparibo AD, Sam R 2012 Laying chickens response to various levels of palm kernel cake in diets. *Int. J. Livest. Prod*. **3**, 12–16.

[15] Anyaegbu BC, Esonu BO, Uchegbu MC and Udedibie ABI 2014 Use of fermented cassava, palm kernel cake and dried brewers' grains to produce maizefree low-cost diets for laying hens. *J. Agric. Sci. Env*. **14**: 42-49.

[16] Purwati D, Djaelani MA, Yuniwarti EYW 2015 Indeks Kuning Telur (IKT), Haugh Unit (HU) dan Bobot Telur pada Berbagai Itik Lokal di Jawa Tengah. *Jurnal Biologi* **4**, 1-9.

[17] Sathitkowitchai, W., Nitisinprasert, S., Keawsompong, S., 2018. Improving palm kernel cake nutrition using enzymatic hydrolysis optimized by Taguchi method. 3 *Biotech*. **8**, 407.

[18] Yaophakdee N, Ruangpanit Y, Attamangkune S 2018 Effects of palm kernel meal level on live performance and gut morphology of broilers. *Agriculture and Natural Resources*. **52**, 75–78.

Vegetation composition in gold post-mining land of PT. Sultra Utama Nickel, Bombana District

F D Tuheteru[1,*], Husna[1], Basrudin[1], A Arif[1], Albasri[1], I Mustafa[1]

[1] Department of Forestry, Faculty of Forestry and Environmental Sciences, Halu Oleo University, Jalan Mayjen S. Parman, Kampus Lama, Kemaraya, Kendari, Southeast Sulawesi Province, Indonesia

* E-mail: fdtuheteru1978@gmail.com

Abstract. This study aimed to determine vegetation composition in the gold post-mining land of PT Sultra Utama Nickel in Bombana District. The method used in this research was the plotted path method by placing transects in accordance with purposive sampling technique. Transects were arranged systematically with a between-transects distance of 50 m The number of transects arranged were 18 transects with 3 plots each with size of each plot was 2 x 2 m, resulted in a total plot of 54 plots. Plant identification results showed that plant species found in the study area were 19 families, 30 genera, 34 species and 4 habitat. Results of vegetation analysis showed that of the 34 plant species found in the research location, plant species with the highest IVI value of 34.54% was *Imperata cylindrica*. The lowest IVI of 0.54% were found for *Lygodium* sp., *Glochidion* sp., *Seleria levis* Retz, *Scoparia dulcis* L., *Sida rhombifolia* L and *Solanum torvum*. The diversity index was 2.59, indicating moderate abundance category.

1. Introduction

Southeast Sulawesi Province is one of provinces having gold mining. One of gold mining activities in Southeast Sulawesi Province is in Bombana Regency. Gold mining activity can severely damage the ecosystem. Post-mining land is marginal land with physical characteristics of compacted soil which can damage the water system and soil aeration, leading to productivity decrease and slow plant growth; hence the need for land revegetation [2].

Revegetation is the activity of replanting post-mining land with plants having the capabilities to grow in a damaged environment. The aim of revegetation is sustainable replanting of native plants to improve biodiversity and restore post-mining land. The restoration directly benefits the environment through improving wildlife habitat, biodiversity, soil productivity and water quality. Revegetation can occur naturally (natural succession), which is indicated by the emergence of new adaptive plant species in the post-mining land [17].

Natural succession in post-mining land has been carried out using several methods. Several studies including [14] reported that there were 23 plant species growing in gold post-mining land in Mandailing Natal. [16] reported that 13 species were found around the tailings of PT Antam, in Pongkor, Bogor Regency. [8] reported that there were 17 plant species with phytoremediation potential in the gold post-mining area of PT Newmont Minahasa Raya. [12] also reported that there were 10 plant species in the Timika gold mine tailings area.

IOP Conf. Series: Earth and Environmental Science **883** (2021) 012065 doi:10.1088/1755-1315/883/1/012065

Research on vegetation composition in gold post-mining land of PT Sultra Utama Nickel (SUN) in Bombana Regency has never been reported. Therefore, this research needs to be done to determine vegetation composition in the gold post-mining land of PT Sultra Utama Nickel in Bombana Regency.

2. Methods

This study was conducted in the gold post-mining area of PT SUN in Bombana Regency. This research was conducted using the plotted path method with transect placement carried out deliberately (purposive sampling). Distance between transects was 50 m. Total number of transects arranged were 18 transects. Each transect had 3 plots with plot size of 2 x 2 m, with an interval of 2 m between plots. The total number of plots arranged were 54 plots. Each plant in the plot was recorded for species name and the number of individuals.
Data collected in the field were analyzed using the following indices:

Important Value Index (IVI) [7]

Density (K) (Indv./ha); $$K = \frac{\text{Number of Individuals}}{\text{The area of the entire plot}}$$

Relative Density (%); $$RD = \frac{\text{Species density}}{\text{The density of all species}}$$

Frequency; $$F = \frac{\text{Number of species found}}{\text{total number of plots}}$$

Relative Frequency (%); $$RF = \frac{\text{Frequency of a type}}{\text{Frequency across species}} \times 100\%$$

Important Value Index: $RD + RF$

Species Diversity Index (H '). Species diversity is determined using the Shannon Index of General Diversity formula (Shannon-Wiener , 1963 in [11]:
$H' = -\sum \{n.i / N\} \ \mathrm{Ln} \{n.i / N\}$

Note:
H '= Shannon-Wiener diversity index
n.i = Number of individuals of type I
N = Total number of individuals
Ln = Natural logarithm
n.i / N = Proportion of samples in species

The criteria for assessing species diversity.
H '> 3: The diversity of a species is high
H '1 ≤ H' ≤ 3: Medium species diversity
H '<1: The diversity of a species is low

3. Results and discussion

3.1.Species Composition Based on Family

Results of plant species identification at the study location showed that there were 34 species from 19 families. Family having the highest number of species was the Poaceae family with a percentage of 17.65%.

Table 1. Vegetation types in PT Sultra Utama Nickel (SUN) based on plant family

No.	Family	Number of types	Percentage (%)
1	Poaceae	6	17.65
2	Fabaceae	4	11.76
3	Cyperaceae	3	8.82
4	Compositae/Asteraseae	2	5.88
5	Malvaceae	2	5.88
6	Phyllanthacae	2	5.88
7	Pteridaceae	2	5.88
8	Schizaeaceae	2	5.88
9	Ericaceae	1	2.94
10	Muntingiaceae	1	2.94
11	Passifloraceae	1	2.94
12	Droseraceae	1	2.94
13	Plantaginaceae	1	2.94
14	Asteraceae	1	2.94
15	Lamiaceae	1	2.94
16	Rubiaceae	1	2.94
17	Linderniaceae	1	2.94
18	Solanaceae	1	2.94
19	Verbenaceae	1	2.94
	Total	34	100

3.2. Species composition based on genera

Plant species identification based on genera observed 34 plant species from 30 genera. The highest number of species belongs to *Imperata, Lygodium, Sida* and *Ceratopteris* genera. Each genus was consisted of approximately 2 species or 5.88%.

Table 2. Vegetation types in PT Sultra Utama Nickel (SUN) based on genera

No.	Genera	Number of types	Percentage (%)
1	Imperata	2	5.88
2	Lygodium	2	5.88
3	Sida	2	5.88
4	Ceratopteris	2	5.88
5	Cheilanthus	1	2.94
6	Chromolaena	1	2.94
7	Cymbopogon	1	2.94
8	Cynodon	1	2.94
9	Golocehidion	1	2.94
10	Grosera	1	2.94
11	Hyptis	1	2.94
12	Acasia	1	2.94
13	Lespedeza	1	2.94
14	Lindernia	1	2.94
15	Actinoscirpus	1	2.94
16	Megathyrsus	1	2.94
17	Oldenlandia	1	2.94
18	Paspalum	1	2.94
19	Passiflora	1	2.94
20	Phyllanthus	1	2.94
21	Polytrias	1	2.94

No.	Genera	Number of types	Percentage (%)
22	Porophyllum	1	2.94
23	Scoparia	1	2.94
24	Selerias	1	2.94
25	Calopogonium	1	2.94
26	Solanum	1	2.94
27	Stachytarpheta	1	2.94
28	Uraira	1	2.94
29	Vaccinium	1	2.94
30	Vernonia	1	2.94
	Total	34	100

3.3. Species composition based on habitat

Plant species identification based on habitat in post-mining land were grouped into 4 habitat, namely shrubs, grasses, shrubs and trees. The highest number of habitat found was grass habitat with 16 species or 47.06%. The lowest habitat was tree habitat, i.e., 2.94%.

Table 3. Vegetation types in PT Sultra Utama Nickel (SUN) based on habitat

No.	Habitus	Number of types	Percentage (%)
1	Grass	16	47.06
2	Shrubs	13	38.24
3	Bush	4	11.76
4	Tree	1	2.94
	Total	34	100

3.4. Vegetation analysis

Vegetation analysis in gold post-mining land of PT Sultra Utama Nickel (SUN) showed that *Imperata cylindrica* had the highest relative density and relative frequency. The lowest IVI were found in *Lygodium* sp., *Glochidion* sp., *Seleria levis* Retz, *Scoparia dulcis* L, *Sida rhombifolia* L, and *Solanum torvum*.

Table 4. Values of Density (D), Relative Density (RD), Frequency (F), Relative Frequency (RF), Important Value Index (IVI), and Plant Diversity Index (H') in Gold Post-Mining of PT Sultra Utama Nickel (SUN)

No	Species	D	RD	F	RF	IVI	H'
1	*Imperata cylindrical*	24722.22	24.20	0.39	10.34	34.54	0.34
2	*Imperata* sp.	10046.30	9.83	0.44	11.82	21.66	0.23
3	*Chromolaena odorata*	5740.74	5.62	0.43	11.33	16.95	0.16
4	*Megathyrsus maximus*	10509.26	10.29	0.24	6.40	16.69	0.23
5	*Phyllanthus* sp.	6250.00	6.12	0.35	9.36	15.48	0.17
6	*Vernonia* sp.	9398.15	9.20	0.19	4.93	14.12	0.22
7	*Polytrias* sp.	8796.30	8.61	0.15	3.94	12.55	0.21
8	*Cynodon dactylon* (L). Pers	6157.41	6.03	0.13	3.45	9.47	0.17
9	*Colopogonium*	3611.11	3.53	0.22	5.91	9.45	0.12
10	*Sida* sp.	1712.96	1.68	0.17	4.43	6.11	0.07
11	*Uraira lagopodoides* (L.)	1805.56	1.77	0.15	3.94	5.71	0.07
12	*Lespedeza* sp.	1944.44	1.90	0.11	2.96	4.86	0.08
13	*Andropogon* sp	1203.70	1.18	0.11	2.96	4.13	0.05
14	*Hyptis brevipes* poit	2129.63	2.08	0.07	1.97	4.05	0.08
15	*Drosera burmanni* Vahl	1990.74	1.95	0.04	0.99	2.93	0.08

No	Species	D	RD	F	RF	IVI	H'
16	*Acasia mangium*	324.07	0.32	0.09	2.46	2.78	0.02
17	*Paspalum* sp.	1481.48	1.45	0.04	0.99	2.44	0.06
18	*Actinas cirpus* sp.	1203.70	1.18	0.04	0.99	2.16	0.05
19	*Vaccinium* sp.	277.78	0.27	0.06	1.48	1.75	0.02
20	*Stachytarpheta cayennensis*	138.89	0.14	0.06	1.48	1.61	0.01
21	*Cyperus rotundus*	416.67	0.41	0.04	0.99	1.39	0.02
22	*Lindernia* sp.	787.04	0.77	0.02	0.49	1.26	0.04
23	*Passiflora foetida* L.	138.89	0.14	0.04	0.99	1.12	0.01
24	*Poropyllum ruderale (jacq) sass*	416.67	0.41	0.02	0.49	0.90	0.02
25	*lygodium flexuosum*	277.78	0.27	0.02	0.49	0.76	0.02
26	*Cymbopogon* sp.	185.19	0.18	0.02	0.49	0.67	0.01
27	*Ceratopteris thalictroides* (L).Brongn	138.89	0.14	0.02	0.49	0.63	0.01
28	*Oldenlandia* sp.	92.59	0.09	0.02	0.49	0.58	0.01
29	*Lygodium* sp.	46.30	0.05	0.02	0.49	0.54	0.00
30	*Glochidion* sp.	46.30	0.05	0.02	0.49	0.54	0.00
31	*Seleria levis* Retz	46.30	0.05	0.02	0.49	0.54	0.00
32	*Scoparia dulcis* L.	46.30	0.05	0.02	0.49	0.54	0.00
33	*Sida rhombifolia* L	46.30	0.05	0.02	0.49	0.54	0.00
34	*Solanum torvum*	46.30	0.05	0.02	0.49	0.54	0.00
	Total	102175.93	100	3.76	100	200	2.59

Results of plants exploratory analysis in the post-mining land of PT Sultra Utama Nickel (SUN) in Bombana Regency showed that there were 34 plant species from 19 families (Table 3). There were six species of the Poaceae family found in post-mining land (17.65%). Results obtained showed that species from the Poaceae family were found in almost all observation plots. The Poaceae family presumably has high adaptation and quick reproduction capabilities. [13] stated that the Poaceae family can grow and adapt well to its environment. This is also in line with research conducted, [5] which showed that species of the Poaceae family can grow in areas having tropical to cold climates with sufficient rainfall; and these plants are also able to live without shade, and can grow at an altitude of 1-2,700 m asl [15]. One species of the Poaceae family found in the study area was *Imperata cylindrica*.

The results showed that grass habitat dominated the post-mining land with 16 species (47.06%), while tree habitat is the least found, having only 1 species of *Acacia mangium*. This situation is presumably due to the open and no shade condition of the post-mining land, leading to the growth difficulties of tree habitat. In post-mining land, such as gold mine, the first vegetation to grow is cover crops [9]. It is assumed that the vegetation community is hyper-tolerant and hyper-accumulator, which can tolerate high metal concentration in the plant tissue [18]. Lower plants can also tolerate various types of environment, including dry, barren and nutrient-poor environment [13]. Therefore, cover crop is widely used as a pioneer crop to rehabilitate marginalized and disturbed land, such as gold post-mining land.

Cover crop found in post-mining land of PT Sultra Utama Nickel consisted of several genera with a variety of species. Several genera, such as *Imperata, Lygodium, Sida*, and *Ceratopteris* were the dominant genera represented by 2 species (5.88) for the respective genera, of the total 30 genera found. *Imperata* is the one genus tolerant to open areas, as well as invasive, and is represented by *Imperata cylindrica* in the study location. This is in line with the study conducted by [1] which showed that *Imperata* is one of the genera that can grow in open areas. The high spread of *Imperata cylindrica* is supported by its propagation method through seeds and rhizobium [4]. *Imperata* seeds are very light and can spread to other places by means of wind, water, animals and human activities [10].

In our study, *Imperata* is found in every plot because of its ability to grow and adapt well to any environment, even adverse environment. This results agrees with the study conducted by [1] which indicated that *Imperata cylindrica* of the *Imperata* genus can wildly grow and are widely distributed in forests, rice fields, gardens or yards, and other open environment types. *Imperata cylindrica* can grow in an environment with direct sun, in wet/moist soil and in dry soil. This ability is supported by the generative (seeds) and vegetative (rhizomes) means of reproduction [4]. The flowering process often occurs during dry season due to stress as a result of burning activity, deforestation, or drought [10].

The number of genera and species found in the study location also had different values of density, frequency and index. From our study, the highest density, frequency, diversity, index of importance and the highest diversity was found in *Imperata cylindrica* of the Poaceae family, with a value of 34.54% (Table 6). The lowest value was found in *Lygodium* sp., *Glochidion* sp., *Seleria levis* Retz, *Scoparia dulcis*, *Sida rhombifolia*, and *Solanum torvum* with respective values of 0.53% (Table 6). The results of vegetation analysis showed that plant diversity found in the post-mining area of PT Sultra Utama Nickel was 2.59 (Table 6), indicating that the species diversity index (H ') of the understory plant in the studied post-mining land was classified as moderate. The index value is influenced by environmental conditions that are not homogeneous, because in the post-mining land there are many mining pits and irregular mounds of soil, as well as high heavy metals content, leading to plant growth difficulties and abnormal plant growth [9].

Plant species with a high frequency have a wide distribution, while plant species with a low frequency have limited distribution [6]. [3] stated that frequency is a parameter that indicates vegetation distribution in an ecosystem.

4. Conclusion

There are 34 species of understory vegetation from 19 families, 30 genera and 4 habitat. The most common species found in the post-mining land was Imperata cylindrica from the genus Imperata. The most common family found in the study location was the Poaceae family. The Important Value Index of Imperata cylindrica was 34.54%, with Diversity Value Index of 2.59 or moderate.

5. Acknowledgment

This research was funded through Fundamental Research National from the Ministry of Research and Technology/BRIN. The authors also thank the Chairman of PT.Sultra Utama Nikel for providing research site.

References

[1]. Balkema A A 1997 Tailings and mine waste Rotterdam. Netherlands. Noordhoff-Groningen

[2]. Sukmana S dan Abujamin S 1986 Pengaruh Pengolahan dalam pada Eutropepts Bekas Sawah Terhadap Sifak Fisik Tanah dan Hasil Tanaman Semusim *Pemberitaan Penelitian Tanah dan Pupuk* **5** 42-48.

[3]. Sipayung J, Delvian, dan Hartini K S 2014 Analisis Vegetasi Tumbuhan Bawah Pada Areal Lahan Bekas Tambang Emas Rakyat (Vegetation Analysis of Ground Cover on Field Area of ex Civillian Gold Mining in KecamatanNaga Juang Kabupaten Mandailing Natal). Program Studi Kehutanan, Fakultas Pertanian, Universitas Sumatera Utara.

[4]. Suharno, Sancayaningsih R P, Soetarto E S, Kasiamdari R S 2014 The presence of arbuscular mycorrhizal fungsi in the tailing of minig gold timika as an attempt of environmentally friendly *Jurnal manusia dan lingkungan* **21** (3) 295-303.

[5]. Juhaeti T F, Syarif dan Hidayati N 2005 Inventarisasi Tumbuhan Potensial untuk Fitoremediasi Lahan dan Air Terdegradasi Penambangan Emas *Jurnal Biodiversitas* **6** (1) 31 -33.

[6]. Purnomo D H, Magandhi M, Helmanto H dan Witono J R 2015 Jenis-Jenis Tumbuhan Reklamasi Potensial Untuk Fitoremediasi Di Kawasan Bekas Tambang Emas.Lembaga Ilmu Pengetahuan Indonesia. Bogor **1**(3) 496-500.

[7]. Indriyanto 2006 *Ekologi hutan* Jakarta: penerbit PT Bumi Aksara.

[8]. Odum E 1993 *Dasar-Dasar Ekologi.*Terjemahan oleh Tjahjono Samingan dari Buku *Fundamentals of Ecology*. UGM. Yogyakarta.

[9]. Setiawan K A, Sutedjo, Matius P 2017 Composition of understorey species in post-coal mine revegetation **1**(2) 182–195.
[10]. Hilu K W 2006 Skewed distribution of species number in grass genera: is it a taxonomic artefact. *In* : Hodikinson, T. R and Parnell, J. A. N. (eds). Reconstructing the Tree of Life ; Taxonomy and Systematics Of Species Rich Taxa. The Systematics Association. New York. p. 165-176.
[11]. Steenis C G G J van. 2013 *Flora Untuk Sekolah di Indonesia*. Jakarta: Terjemahan Moeso Surjowinoto, Soenarto, dan Soerjo. PT Pradnya Paramita.
[12]. Mansur I 2013 *Teknik Silvikultur Untuk Reklamasi Lahan Bekas Tambang*. Southeast Asian Regional Centre for Tropical Biology (SEAMEO BIOTROP). Bogor. Indonesia.
[13]. Widyati E 2011 Potency of Understory as a Heavy Metal Accumulator in Supporting of Ex-mining Site Rehabilitation *Journal Mitra Hutan Tanaman* **6** (2) 37-46
[14]. Atien S 2008 Life Pharmacy for Spices and Wild Plants. Bandung. Yrama Widya.
[15]. Fujiyanto Z, Prihastanti E, Haryanti S 2015 Characteristics of Environmental Conditions, Number of Stomata, Morphometry, Imperata that Grows in the Open Field Areas in Blora and Ungaran Districts *Bulletin of Anatomy and Physiology* **23** (2) 48-53.
[16]. Murniati 2002 From Imperata cylindrical Grasslands To Productive Agroforesty.Ph.D. thesis.Wageningen: Wageningen University.
[17]. Ismaini L, Lailati M, Rustandi dan Sunandar D 2015 Analisis komposisi dan keanekaragaman tumbuhan di Gunung Dempo, Sumatera Selatan *Pros Sem Nas Masy Biodiv Indon*. **1**(6) 1397-1402.
[18]. Fachrul M 2007 *Metode Sampling Bioekologi* Bumi Aksara. Indonesia.

Adaptation test of five introduced cherry tomato lines in Lembang highland

Redy Gaswanto[1,*]

[1] Indonesian Vegetable Research Institute, Jln. Tangkuban Parahu No. 517, Lembang, Bandung Barat, Indonesia 40391

* E-mail: redwanto_1@yahoo.co.id

Abstract. Tomato has some types according to fruit size and shape, one of them is cherry tomato. The purpose of this research was to select and evaluate five introduced cherry tomato lines in Lembang highland that have good potential to be developed as a new variety refer to result of the adaptation and consumer preferences test. The research material consisted of introduced lines IJR 151, IJR 154, IJR 1150 , IJR 1158 , IJR 2170 and one local variety as a check genotype. The adaptation and preference consumer test conducted at Indonesian Vegetables Research Institute, Lembang (1,250 m sal) from March until August 2017. The research used a Ramdomized Complete Block Design (RCBD) with three replications. Research result showed that all cherry tomato lines could adapt well in Lembang. IJR 151 and IJR 154 lines could be developed as a new variety refer to consumer's preference test on fruit shape and taste parameter, eventhought IJR 1158 and IJR 2170 lines had higher fruit weight. Furhermore IJR 154, IJR 1150, IJR 1158 lines could be reccomended to provide and overcome sufficiency vitamine A and C due to their high beta carotene content (> 100 microgram mg^{-1}).

1. Introduction.

Tomato (*Lycopersicon esculentum* Mill.) derived from South America [1] and as one of popular vegetable crop because it has high nutritional content [2]. In Indonesia tomato is widely cultivated in both highland and lowland [3]. Grubben (1990) [4] stated that on Java island mostly tomato cultivation was in highland (60%), whereas in lowland area only 40%. According to Indonesia horticultural statistic data showed that tomato production in 2019 was 1.020.333 ton (18.63 t ha^{-1}) or increase 1.15% with province of West Java, North Sumatra, West Sumatera,East Java, and South Sulawesi as centre of production [5]

There are some tomato types according to fruit size and shape, one of them is cherry tomato [6]. Actually cherry tomato had long been known in Indonesia, but the cultivation still not popular yet among farmers. One of reluctance factor to cultivate it is frequent harvesting and price fluctuation thus considered less profitable economically [7]

Kumari *et al.*(2017) and Diaz (2012) [8] [9] stated that cherry tomato as a main source of vitamin C and B complex in form of beta carotene, thiamin, niacin, riboflavin, and various important minerals for human nutrition i.e. calcium (Ca), phospor (P), iron (Fe), and potassium (K). Unfortunately the added value of cherry tomato has not been utilized yet by Indonesian's consumers. They only consume cherry tomato occacionally as a fresh table fruit. Eventhough Slimestad and Verheul (2005) [10] informed that inside of each 100 g cherry tomato could produce 19 cal.

The opportunity for cherry tomato promotion can be increased in line with arise public awareness on importance nutritional content for health [11], so it is expected the commodity will be economically profitable. Hanson (2003) [12] stated that cherry tomato with high beta carotene content could be developed to overcome vitamin A and C deficiencies that were common in African and Asian countries. Some research institutions such as the Asian Vegetables Research Development Center (AVRDC) have released several cherry tomato lines with beta carotene content of 3-12 times compared to common. Likewise, some introduced cherry tomato lines were reported also to have high nutritional content, but should be verified first by adaptation and preference consumer test.

The purpose of this research was to select and evaluate five introduced cherry tomato lines that have good potential to be developed as a new variety refer to result of the adaptation and consumer preferences test.

2. Methods

The adaptation and preference consumer test conducted at Indonesian Vegetables Research Institute, Lembang (1,250 m sal) from March until August 2017. The research used a Ramdomized Complete Block Design (RCBD) with three replications. The research material consisted of introduced lines IJR 151, IJR 154, IJR 1150 , IJR 1158 , IJR 2170 and one local variety as a check genotype.

The seeds were soaked warm water at 30-40°C for 15-20 minutes before sown. Then the seeds were threated fungicide Benlate with recommendation dosage. The seeds were planted in the plastic pots with mixed media composition soil with chicken manure (1 : 1) and kept at the nursery place. The seedling could be transplanted in the field at 3 weeks age after sown.

Land preparation included of basic fertilizer application with dosage 30 t manure, 35 kg urea, 75 kg Za, 125 kg SP 36 and 100 kg KCl per ha^{-1} [13]. The subsequent fertilizer was applied at age 14 and 30 days after planting (DAP) with dosage 300 kg ha^{-1} NPK (15-15-15) per application. The land was covered by silver plactic mulch with spacing 50 x 70 cm. The population was 20 plants / line / replication until total was 450 plants. Replanting was carried out 7 DAP to change the damaged or died seedling.The plant maintaining included of weeding, watering, subsequent fertilizing, pest and disease protecting, and harvesting.

The observation parameter was carried out to percentage of plant survives in the field (%), plant height (cm), flowering time (DAP), fruit number per clusters (unit), fruit number per plant, fruit weight per plant (g), fruit diameter (cm), and beta carotene content of each line (µg/mg). The parameter data was analized by F test and DMRT test at level 5%.

Preference consumer test was carried out to 50 respondens. The obtained scoring refered to preferred percentage by responden with criteria of shape and taste. The lowest scoring indicated a high preference level.

3. Results and discussion

3.1. Percentage of survive plant in the field

The observation result in the field showed that all lines had survive plant percentage was more than 90.00% (Table 1). This indicated that all the introduced lines could adapt well in Lembang highland. Possibility they all had a genetic background as heat tolerant genotypes. It was not surprisingly because some research institutes such as AVRDC had developed many kind of vegetables including cherry tomato for tropical contries [14]. Recent vegetable cultivation in Indonesia had been focussed to lowland or medium due to environment's damage in higland caused by agriculture activity or residency, eventhough these area actually as a conservation region for water supply. Therefore if availability heat tolerant lines, they could be developed for lowland or medium area which both of them have higher temperature [15]. Thus environmental sustainability in highland area can be maintained.

Table 1. Adaptation test results of five introduced cherry tomato lines in Lembang highland

Lines	Survive plants (%)	Flowering time (DAP)	Plant height (cm)	Flwr number /inflorescence (flowers)	Fruit number/pl (fruits	Fruit weight/pl (g)	Fruit diameter (cm)	Beta carotene content (µg/mg)
IJR 151	100.00	58.67 abc	37.24 a	13.13 ab	770.35 b	4410.65 bc	1.75 d	85.190
IJR 154	93.33	57.60 bc	35.67 a	15.53 a	1010.00 a	4890.65 bc	1.70 d	101.770
IJR1150	96.67	58.20 abc	31.55 ab	14.60 ab	721.25 b	5751.00 b	2.25 c	102.287
IJR 1158	90.00	57.33 c	37.45 a	14.00 ab	602.50 b	8080.65 a	3.00 b	102.166
IJR 2170	100.00	58.67 abc	28.67 bc	11.40 b	355.30 c	7820.00 a	3.80 a	73.648
Local	100.00	60.50 a	23.45 c	8.10 c	197.50 c	1720.33 d	1.65 d	95.959
CV (%)		2.52	9.65	12.75	17.45	15.50	7.50	

Note: Mean followed by the same letters on the same columns are not significantly according to Duncan's multiply range test at 0.05 level

3.2. Flowering time
In the Table 1 could be known that the flowering time of all cherry tomato lines was around 57- 61 days after planting (DAP). IJR 1158 (57.33 DAP) was the earliest eventhought not different significantly with four other lines, i.e. IJR 154 (57,60 DAP), IJR 1150 (58,20 DAP), IJR 2170 (58,67 DAP), IJR 151 (58,67 DAP). These data correlated also with their harvesting time. Generally if one genotype has earlier flowering time, so it will be earlier harvested also. This condition had some advantages i.e. low budget for cost production input and profitable economically. Plant genotypes with early maturity trait also can be implemented to solve pest and disease infection problem because they can stop life cycle of pest-diseae naturally, beside that they can avoid of drought condition with earlier complete biologically life cycle before dry season coming [16].

3.3. Plant height at flowering time
The introduced cherry tomato lines had average of plant height were higher than local line. This caused all introduced lines were indeterminate type. Plant height of IJR 1158 line (37.45 cm) was the highest eventhought not different significantly with three other lines i.e. IJR 151 (37,24 cm), IJR 154 (35,67 cm), IJR 1150 SA (31,55 cm). Commonly indeterminate type has stem is slimmer, but higher because on the main stem will appear and develop upper side branches [17], hence be needed many sticks for their cultivation [18]. The fruit quality of cherry tomato is still good enough as long as maintained well in the cultivation. Thus the prunning technique can be implemented also in this case [19] [20].

3.4, Flower number per inflorescence and fruit number per plant
IJR 154 was the most of flower number per inflorescence (15.53 flowers) eventhought not different significantly with three other lines i.e. IJR 1150 (14,60 flowers), IJR 1158 (14,00 flowers), IJR 151 (13,13 flowers). The consequence of this data was IJR 154 (1010 fruits/plant) also as the most of fruit number per bed compared with other lines. Usually there was a correlation between fruit number parameter with fruit production. As much more as fruit number, it will increase the production also [21].

Generally cherry tomato has more than 10 flowers or fruits per inflorescence. This character can be used also to improve a plant genotype in the breeding program. Using cherry tomato as a male parent crossed with common tomato are expected to be order to inherit trait of flower number per inflorescence or fuit sets [22]. Murti *et al.* (2004) [23] informed that flower number per inflorescence, fruit set, and fruit locule number are qualitatype character that controlled by monogenic, therefore it can be easily inheritted to their offspring.

3.5. Fruit weight per plant and fruit diameter size

In the Table 1 could be seen that IJR 151 was the most of fruit number per plant performance, but in the fact the heaviest of fruit weight was obtained by IJR 1158 (8080.65 g/plant), eventhought not different significantly with IJR 2170 (7820.00 g/plant). This result proved that fruit diameter size of IJR 154 was smaller performance than IJR 1158 and IJR 2170 lines. Thus eventhought both of the lines were fewer of fruit number parameter than IJR 145, but their fruit diameter size were bigger, therefore automatically they were heavier even if compared with other lines. Result of correlation analyze showed that play fruit diameter size (R^2 = 0.450) role was much more than fruit number (R^2 = 0.238) in determining fruit weight performance. The result also showed there was a possitive correlation between fruit weight parameter as a dependent variable with fruit number and diameter size as independend variable (Fig. 1 and 2).

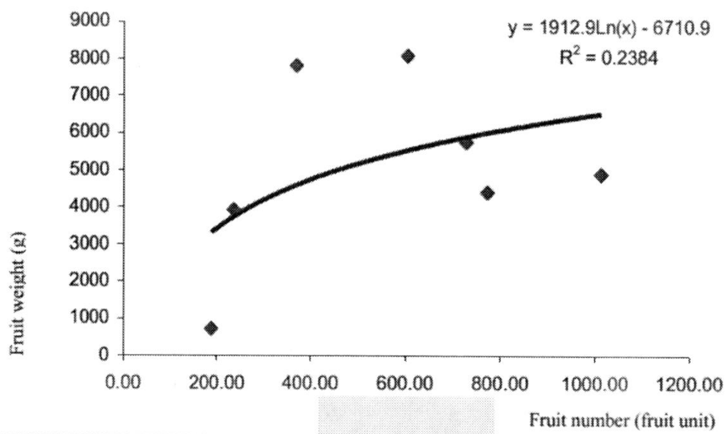

Figure 1. Correlation between fruit number parameter (X) with fruit weight parameter (Y) on cherry tomato lines

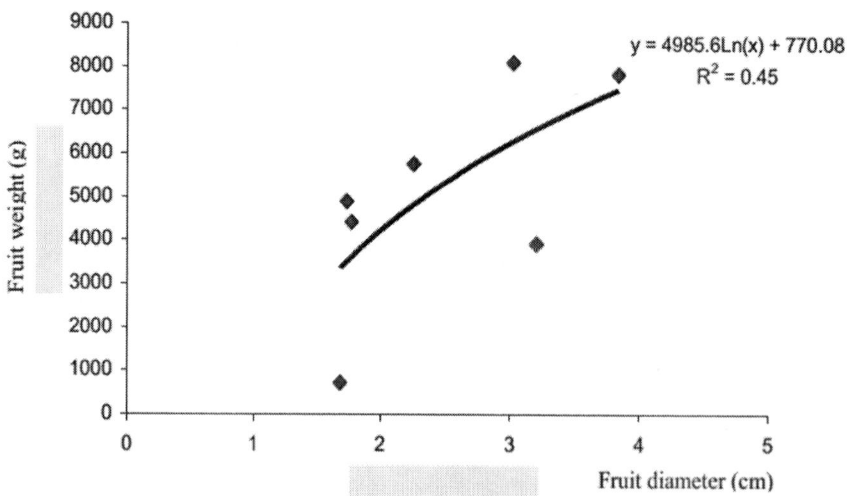

Figure 2. Correlation between fruit diameter size (X) with fruit weight (Y) on cherry tomato lines

3.6. Consumer's preference test

IJR 2170 and IJR 1158 lines were cherry tomato type with bigger diameter size. Nevertheless in the fact fruit diameter size and fruit weight of cherry tomato are not entirely as the prefered characters by consumer. There are some other criteria to be selected by respondents, in this case are fruit shape, fruit taste, fruit hardness, fruit color, etc [24].

Vitamin dan mineral of cherry tomato content also become one of criteria selected by consumers [24]. In the Table 1 could be seen that IJR 154, IJR 1150, IJR 1158 lines had higher beta carotene content (> 100 µg mg^{-1}) than local genotype (46.166 µg mg^{-1}). This data information could recommend the cherry tomato lines which could be consumed to fulfill and overcome sufficiency vitamine A and C. Necessity of vitamine C for daily is around 60 mg [30], but there is a variation of each individual. Some cases of phisical stress such as burns, infection, heavy metal poisoning, smoking, constantly drugs, etc can increase the body's necessity for vitamine C. Furthermore the smookers need vitamin C around 100 mg per daily [25].

Table 2. Preference test according to fruit shape and taste parameter

Parameter	IJR 151	IJR 154	IJR 1150	IJR 1158	IJR 2170	Local	Total
Shape (%)	16 (32,00)	10 (20,00)	7 (14,00)	8 (16,00)	3 (6.00)	6 (12,00)	50
Scoring	1	2	4	3	6	5	
Taste (%)	12 (24,00	14 (28,00)	7 (14,00)	10 (20,00)	2 (4,00)	5 (25.00)	50
Scoring	2	1	4	3	6	5	
Total	3	3	8	6	12	10	

In the Table 2 could be informed that the preferred cherry tomato lines by 50 respondents were IJR 154, IJR 151, IJR 1158, IJR 1150, local genotype, IJR 2170 respectivelly. Certainly this information was very important to be order to select cherry tomato lines which could be developed in the future. At the end the cherry tomato cultivation will be profitable economically.

Table 3. Fruit characterization of six cherry tomato lines

Characters	Cherry tomato lines					
	IJR 151	IJR 154	IJR 1150	IJR 1158	IJR 2170	Local
Immaturity fruit color	Green	Light green	Light green	Green	Dark Green	Light green
Mature fruit color	Red	Orange	Orange	Orange	Red	Orange
Shape of fruit tip	Flat	Intended	Pointed	Flat	Pointed	Flat
Pedicel area	Slightly depressed	Flat	Slightly depressed	Slightly depressed	Slightly depressed	Flat
Fruit shape	Pear	Lengthened cylindrical	Pear	High round	High round	High roind

4. Conclusion and Suggestion

Research result showed that all cherry tomato lines could adapt well in Lembang. IJR 151 and IJR 154 lines could be developed as a new variety refer to consumer's preference test on fruit shape and taste parameter, eventhought IJR 1158 and IJR 2170 lines had higher fruit weight. Furhermore IJR 154, IJR 1150, IJR 1158 lines could be reccomended to provide and overcome sufficiency vitamine A and C due to their high beta carotene content (> 100 µg mg^{-1})

5. Acknowledgments

Authors are thankful to Director of Indonesian Vegetables Research Institute (IVEGRI) for giving motivation and support all this time.

References

[1] Blanca J, Canizares J, Cordero J L, Pascual M J, Diez F and Nuez 2012 Variation revealed by SNP genotyping and morphology provde insight into the origin of the tomato. *Plos One Journals* **7**: 10. https://doi.org/10.1371/journal.pone0048198.

[2] Muhammad A, Hayati R and Aisyah Y 2019 Evaluation of nutrition contents of tomatoes (*Lycopersicum esculentum* Mill.) due to the treatment of plant growth regulators and fertilizers. *Int. J. Biosci.* **14**(2) 134-145. DOI: http://dx.doi.org/10.12692/ijb/14.2.

[3] Pesireron M, Rein E S and Sirappa M P 2014 Adaptation of tomato (*Lycopersicum esculentum* Mill) at dry lowland area in Moluccas, Indonesia. *J. Biol. Chem. Research.* **31** (1): 575-582. https://www.researchgate.net/profile/Marthen_Sirappa /publication /270584177_

[4] Grubben G J H 1990 Timing of vegetable production in Indonesia. *Acta Hortic*: 261-270. DOI: 10.17660/ActaHortic.1990.267.33 https://doi.org/10.17660/ActaHortic.1990.267.33

[5] Direktorat Jenderal Hortikultura 2020 Statistik Hortikultura: produksi, luas panen dan produktivitas buah, sayuran, tanaman hias dan biofarmaka tahun 2019 (Angka tetap). Kementrian Pertanian Indonesia .

[6] Bhattarai K, Sadikshya S and Dilip R P 2018 Diversity among modern tomato genotypes at different levels in fresh-market breeding. *International Journal of Agronomy.* **2018.** https://doi.org/10.1155/2018/4170432

[7] Testa R, Anna M T, Filippo S and Salvatore T 2014 Economic sustainability of Italian greenhouse cherry tomato. *Sustainability 2014.* **6** 7967-7981; DOI: 10.3390/su6117967.

[8] Kumari S, Archana A and Rahul K 2017 Nutritional quality and health benefits of solanaceus vegetables. *Progressive Research – An International Journal.* **12** (Special-II): 1942-1945. https://www.researchgate.net/profile/Archana_Anokhe/publication/320621402_

[9] Dias J S 2012 Nutritional quality and health benefits of vegetables: A Review. *Food and Nutrition Sciences.* **3** 1354-1374 DOI:10.4236/fns.2012.310179 Published Online October 2012. http://www.SciRP.org/journal/fns.

[10] Slimestad R and Verheul M J 2005 Content of chalconaringenin and chlorogenic acid in cherry tomatoes is strongly reduced during postharvest ripening. *J. Agric. Food. Chem.* **53**(18): 7251-7256. https://doi.org/10.1021/jf050737d

[11] Bhowmik D, Kumar K P, Shravan P S and Shweta S 2012 Tomato-a natural medicine and its health benefits. *Journal of Pharmacognosy and Phytochemistry.* **1** (1) https://www. researchgate.net/profil/Shravan_Paswan2/publication/285176270_

[12] Hanson P 2003 AVRDC progress report 2003. *Tomato Unit.* Asian Vegetable Reseach Development Center. Shanhua, Taiwan.

[13] Nurtika N and Hidayat A 1992 Pengaruh waktu pemberian pupuk N dan K terhadap pertumbuhan dan hasil tomat kultivar TW di musim hujan. *Jurnal Hortikultura* **23**(3): 94 – 100.

[14] Hanson P and Chen J T 2005. Introduction to promising AVRDC tomato lines in training workshop on conducting trials of promising vegetable varieties at Shanhua. Taiwan

[15] Islam M T 2011 Effect of temperature on photosynthesis, yield atrributes and yield of tomato genotypes. *Int. J. Expt. Agric.* **2**(1):8-11. https://www.researchgate.net/profileMd_Islam927/ publication/267705727_

[16] Sujinah and Jamil A 2016 Mechanism response of rice under drought stress and tolerant varieties. *Iptek Tanaman Pangan* **11** (1) .http://pangan.litbang. pertanian.go. id/files/ 01-iptek 11012016 Sujinah.pdf

[17] Daryanto A M, Istiqlal R A, Kalsum U and Kurniasih R 2020 Performance of horticultural characteristics of several hybrid tomato varieties in the lowland greenhouse. *J. Agron. Indonesia.* **48**(2):157-164. DOI:https://dx.doi.org/10.24831 /jai.v48i2.30502

[18] Thompson H C and Kelly W C 1979 Vegetable Crop. Mc Graw. Hill Book Co, Inc. New York

[19] Yunita R A and Isnaeni S 2020 Study of growth and production of tomato cultivars in response to fruit thinning at Tamansari, Tasikmalaya, West Java, Indonesia. *Journal of Tropical Crop Science* **7**(2).

[20] Maxa J F J, Schmidt L, Mutwiwa U N and Kahlen K 2016 Effects of shoot pruning and inflorescence thinning on plant growth,yield and fruit quality of greenhouse tomatoes in a tropical climate. *Journal of Agriculture and Rural Development in the Tropics and Subtropics*.**117** (1): 45–56. DOI: 10.15488/589

[21] Souza L M, Melo P C T, Luders R R and Melo A M T 2012 Correlations between yield and fruit quality characteristics of fresh market tomatoes. *Hortic. Bras*. **30** (4). https:// doi.org/10.1590 /S0102-05362012000400011

[22] Raana R 2019 Genetic analysis of fruit quality in tomato. *PhD thesis*, Wageningen University, Wageningen, The Netherlands. DOI https://doi.org/10.18174/503083

[23] Murti R H, Kurniawati T and Nasrullah 2004 Pola Pewarisan Karakter Buah Tomat. *Jurnal Zuriat*. **15** (2). Fakultas Pertanian UNPAD. Jatinangor.

[24] Massaglia S D, Borra, Peano C and Sottile V M.M F 2019 Consumer preference heterogeneity evaluation in fruit and vegetable purchasing decisions using the best–worst approach. *Foods*. **8** (266). DOI:10.3390/foods8070266.

[25] Levine M, Rumsey S C, Daruwata R, Park J B and Wang Y 1999 Criteria and recommendations for vitamin C intake. *JAMA*. **281**(15): 1415-1423 https://www.researchgate.net/profile /Jae_ Park10/publication/13081451_

Organic matter and mixed biofertilizer for plant growth and yield of shallot grown in fertile soil

R Hindersah[1,*], A M Kalay[2], I A Ngabalin[3], M Jamlean and A Talahaturuson[2]

[1] Faculty of Agriculture, Universitas Padjadjaran, Jl. Raya Bandung-Sumedang Km. 21, Jatinanagor Campus, Sumedang 45363, Indonesia
[2] Faculty of Agriculture, University of Pattimura, Jl. Ir. M. Putuhena, Poka Campus, Ambon 97233, Indonesia
[3] Tual Agricultural Office, City of Tual 97611, Indonesia

* E-mail: reginawanti@unpad.ac.id

Abstract. Nowadays, shallot is important horticultural crops in Maluku include in Tual City, Maluku Province. For ensuring the macronutrients level in soil and yield during shallot cultivation, fertilization composed of chemical, organic and bio fertilizer is needed. The objective of field experiment was to evaluate the effect of mixed biofertilizer integrated with organic matter on total nitrogen and available phosphorous in soil as well as shallot yield grown in Vertisols of Tual. The experiment was setup in randomized block design with four treatments of compost and chicken manure amendment with and without biofertilizer containing N-fixing bacteria and P-solubilizing. At the vegetative stage, consortia biofertilizer combined with either compost or chicken manure increased plant height and soil reaction but decreased the available phosphate (P) and didn't change nitrogen (N) total in soil. Chicken manure amendment combined with biofertilizer inoculation increased dry weight of shoots and bulbs even though didn't significantly differ with other treatments. Nonetheles, that treatment resulted in high yield of shallot bulbs up to 14 t/ha.

1. Introduction

Maluku, an archipelago province, has a significant step to self-sufficiency of shallot (*Allium cepa* var. aggregatum) in order to decrease shallot trade from another province and increase the farmer's welfare. Traditionally, shallots have been cultivated in Buru, Tanimbar Island (formerly Maluku Tenggara Barat) and Maluku Barat Daya Regency. Nowadays, shallot is produced in another area include Tual City located in Maluku Tenggara Regency. The contraint of crops production in the island might be low availability of agricultural inputs due to limited transportation when the weather become worst.

In order to reduce chemical fertilizer utilization, another nutrient source such as organic and biofertilizer is considered to fulfill the nutrient sufficiently. In Tual city, some farmers have already produce, commercialized and utilized agricultureal-waste compost but they rarely applied chicken manure in shallot cultivation. Organic fertilizer is considering to contribute in nutrient availability and increase the nutrient efficiency utilization of chemical fertilizer in shallots cultivation [1] and hence decrease the dose of chemical fertilizer up to 50% [2,3]. The effectivity of the organic matter combined with reduced dose of NPK fertilizer to increase soil beneficial microbes and yield of shallot was evidence [4].

Another nutrient source to support shallot production is biofertilizer; the newest eco-friendly technology for supporting the sustainable agriculture. The biofertilizer usually contain the plant growth promoting microbes of nitrogen-fixing bacteria and phosphate-solubilizing mirobes as well. Both microbes enable to enhance the availability of nitrogen (N) and phosphate (P) enzymatically [5,6]. Moreover, they produce significant amount of phytohormones to function on plant growth coordination [7,8,9]. Although the biofertilizer is purchased, actually farmers can proliferate the microbes in biofertilizer by using simple liquid organic matter that locally available such as liquid waste of sago starch production [10].

Intensive agriculture in tropics always face the rapid decrease of available macronutrient in soil after being taken up by the crops. Organic matter amendment is able to support biofertilizer microbes during plant cultivation as well as after the harvest since the microbes utilized organic carbon for their heterotrophic metabolisms [11]. The microbial inoculation combined with organic matter application resulted in the presence of macronutrient and soil beneficial microbes for next crops [12,4].

In general, tropical soil contained low macronutrients mainly N and P due to very intense weathering process [13]. In tropics, high temperature might constantly induce the enzymatic degradation of soil organic matter by soil inhabitants although organic carbon degradation was less sensitive to high moisture [14]. Low soil organic carbon (C) in turn depressed the microbes function for providing plant nutrients. In rainy season, the loss of ionic nutrient is intensive due to leaching; in dry season with high temperature the volatilization of N from N fertilizer is evidence [15,16]. In order to prevent the significant loss of nutrients, the application of balance composition of chemical, organic as well biofertilizer is needed. The objective of this field experiment was to evaluate the positive effect of mixed biofertilizer combined with organic matter and integrated with NPK fertilizer on N and P content as well as yield of shallot grown in Entisols of Tual.

2. Material and Methods

The field experiment was carried out in the agricultural area of Ohoitel Village, Dullah Utara District, Tual City of Maluku Privince (Fig 1) on April to June 2018. The soil was Entisols (Alluvial) has been cultivated for vegetable production since decade and contained high P_2O_5 and K_2O (Table 1) that might be due to intensive chemical fertilization. Agricultural-waste compost was produced by local farmer while the manure was provided by local chicken husbandry with the properties depicted in Table 1.

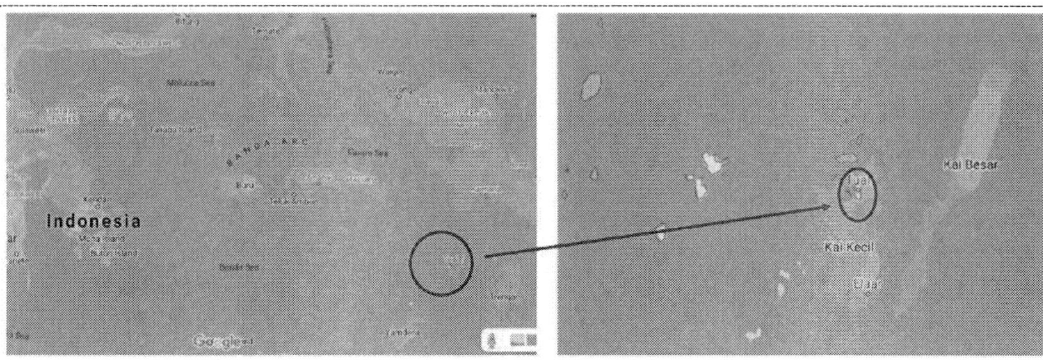

Figure 1. Tual City located in Kai Kecil Island in eastern part of Maluku Province, Indonesia

Mixed biofertilizer was provided by Soil Biology Laboratory, Faculty of Agriculture, Universitas Padjadjaran. Biofertilizer contained N-fixing bacteria *Azotobacter chroococcum*, *A. vinelandii*, *Azospirillum* sp., endophytic *Acinetobacter* sp. and P-solubilizing microbes *Bulkholderia* sp. and *Penicillium* sp. The cell count of bacteria was 10^7 CFU/mL while fungal population was 10^5 CFU/mL.

2.1. *Experimental Establishment*

The experiment was setup in randomized block design to test five treatments each with five replications. The tretments was compost and manure application with and without biofertilizer inoculation. Control treatment was compost application. The rates of organic matter and biofertilizer were 20 t/ha and 3L/ha respectively. The single bulb of shallot cv Bauji were grown in a bed of 100 x 200 cm covered with polyethylene mulch with planting hole of 15 x 20 cm.

Side dress application of biofertilizer was carried out once a week from 10 days to 35 days to maintain their population and activity. To prepare 1% biofertilizer dilution, biofertilizer was mixed with ground water just before inoculation. A total of 20 mL of diluted biofertilizer was applied by soil dressing around the shallot stems. The NPK compound fertilizer at 250 kg/ha was applied at 14 and 30 days after planting (DAP), 0,5 g each in the hole near the stem. During the experiment, purple blotch disease caused by *Alternaria alii* was found at day 27 so that pesticide Dithane M-45 (Mankozep 80 %) was sprayed at the rate of 4 g/L every two weeks. This pesticide enabled repress the Alternaria and also to control fusarium wilt.

Table 1. Propertis of soil, agricultural-waste compost and chicken manure before trial

Nutrients traits	Soil	Agricultural-waste Compost	Chicken Manure
pH	8.2	7.3	6.7
Organic C (%)	1.54	42.1	15.2
N (%)	0.26	2.38	1.1
P2O5 (mg/kg)	30.21	-	-
K2O (mg/kg)	25.32	-	-
P (%)	-	2.1	2.7
K (%)	-	0.12	0.9

2.2. *Experimental parameter and statistical analysis*

Soil and plant samples were collected by composite method from four block consisted of 6 plants. Plant height of sample plants was measured at day 21 and day 49 while the shoot dry weight was only analyzed at day 49. At the harvest time, 56 DAP, the number and fresh weight of bulbs of single plant were measured. All bulbs were store at the room at 24-29°C for 14 days prior to measure their air-dried weight. The average yield in a plot was determined based of fresh and air-dried weight.

Soil samples were taken at harvest time from the samplings blocks. A total of 50 g soil was taken up from each root zone of the six plants. All soil then mixed evenly, put in sealed transparent polyethylene bags and air-dried in the laboratory. The soil reaction, total N and potential P analysis were done by using AOAC proximate analysis methods [17].

All plant and soil data were then subjected to analysis of variance (F test) with $p \leq 0.05$. If the sum square of parameter was significant then Least significant different test ($p \leq 0.05$) was done. Statistical analysis has performed by used of Minitab 18.

3. Results and discussion

The performance of shallot in the experimental plots didn't show either nutrient deficiency nor diseases incidence. The pesticide application was carried out once at 27 DAP when the plants experienced very light attack of purple blotch disease in some plots.

The field experiment results indicated that application mixed non-chemical fertilizer significantly affected plant height but have no effect on shoot dry weight and yield traits as well. Compost with biofertilizer resulted in the highest shoot dry weight over the control and another treatment (Table 2). In general, the shoot dry weight of individual plant was 54 g - 73.33 g. Chicken manure amendment combined with biofertilizer inoculation increased dry weight of shoots and tuber even though didn't significantly differ with other treatments.

Table 2. Effect of organic matter and biofertilizer on Shoot height and dry weight of shallot

Treatments	Shoot height (cm)		Shoot dry weight (g) at day 49
	Day 21	Day 49	
Compost (control)	26.26 ± 0.97 b	35.37 ± 0.59 ab	59.80 ± 7.56 a
Compost and Biofertilizer	28.83 ±1.32 a	38.46 ±1.30 a	61.67 ± 7.31 a
Manure	24.11 ± 0.42 c	32.49 ± 3.71 b	54.00 ± 4.81 a
Manure and Biofertilizer	26.93 ± 0.27 b	36.37 ± 0.42 a	73.22 ± 10.78 a

Number followed by the same letter in a column were not significantly different base on Least significant different at p ≤ 0.05.

Based on analysis of variance, fertilizer treatments affect potential P_2O_5 and soil reaction at harvest time but didn't influence total N in soil (Table 3). At the harvest time organic matter application combined with biofertilizer inoculation surprisingly decreased the available phosphate (P) compared to the control treatment and chicken manure. Slight increase on soil acidity was recorded in soil treated with either compost and chicken manure with biofertilizer.

Table 3. Change in total N, potential P_2O_5 and soil reaction at harvest time after organic matter and biofertilizer application on shallot

Fertilizer Treatments	Total N	Potential P_2O_5 (mg/kg)	Soil reaction
Compost (control)	0.456 ± 0.006 a	39.3 ± 1.3 a	6.26 ± 0.11 c
Compost and Biofertilizer	0.453 ± 0.015 a	26.0 ± 7.3 b	6.53 ± 0.05 b
Manure	0.467 ± 0.040 a	39.2 ±2.6 a	6.46± 0.05 b
Manure and Biofetilizer	0.423 ± 0.015 a	27.8 ± 0.8 b	6.83 ± 0.05 a

Number followed by the same letter in a column were not significantly different base on Least significant difference test at p ≤ 0.05.

Fertilizer treatments have not affected plant yield traits. Regardless of statistical analysis, control treatment resulted in highest number of bulb per plant, but the highest bulbs weight of was showed by shallot treated with compost and manure with biofertilizer (Fig 2). Biofertilizer inoculation followed manure amendment has a potency to enhance air-dried bulbs up to 24.5% compared with the control and compost+biofertilizer (Fig 1).

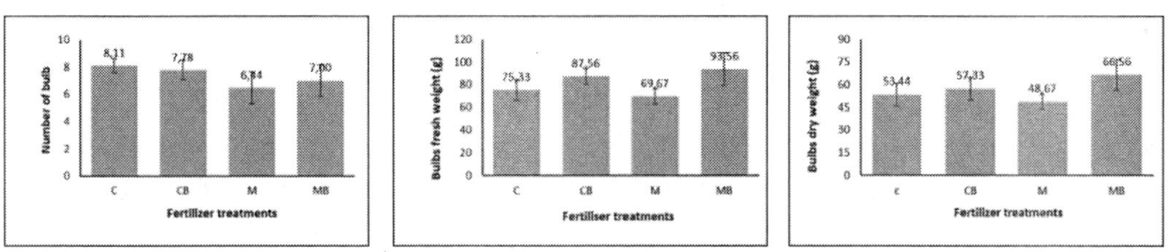

Figure 2. Yield of shallot grown with organic matter amendment with and without biofertilizer. C: compost, CB: compost with biofertilizer; M: manure; MB: manure with biofertilizer

Based on air-dried bulb per plant and the plot dimension, shallot yield in a plot and productivity in a hectare were calculated (Table 4). Manure application with biofertilizer resulted in the highes air-dried shallot weight per plot. The yield per ha was determined for 7,500 m^2 instead of 10,000 since in general some area in the field has been used for access road, space between beds and other utilities. This extrapolated data has not analyzed statistically. The value in Table 4 showed that the plot received chicken manure integrated with biofertilizer has highest yield per ha. Despite of insignificant effect of that treatments on air-dried weight of shallot, the increase of productivity up 24,5% when manure and biofetilizer were used can significantly increase farmer's revenue.

Table 4. Productivity of shallot treated with different organic matter with and without biofertilizer

Fertilizer treatments	Yield of air-dried shallot (kg)	
	per plot (kg)	per ha (ton)
Compost (control)	3.20 b	12.02
Compost and Biofertilizer	3.43 b	12.89
Manure	2.92 c	10.95
Manure and Biofertilizer	3.99 a	14.97

Shoot growth increment following biofertilizer inoculation was caused by the increase of available of N and P in soil. Mixed biofertilizer contained N-fixer and P-solubilizer microbes. Both microbial group also reported elsewhere to produce phytohormone which is important secondary metabolite to regulate plant growth. This results agree with the positive effect of microbial consortia to stimulate root development and shoot growth of maize by 25% compared to the control [18]. However, nutrient content in chicken manure was lower than compost resulted in lower available nutrient for roots uptake and caused delayed growth.

The total N, potential P and pH in soil were measured at the harvest time when most part of available nutrient has been uptake by plant. Biofertilizer treatment caused the decrease in potential P in soil. Phosphate solubilizing bacteria and fungi in biofertilizer provide available P for plant growth and hence bulbs production. Increase in soil microbes due to inoculation might take place but the count of soil microbes at the end of experiment has not been performed. The change of microbial composition might affect the nutrient turnover in which a part of P could be immobilized in the microbial cell.

The treatments only slightly influenced soil reaction. Irrespective of treatments, the soil acidity was neutral which is ideal for plant growth include shallot in Indonesian low land [19]. Neutral soil resulted in high availability of P and optimal proliferation and function of all bacteria in mixed biofertilizer that prefer neutral pH instead of alkaline or acid. The positive effect of biofertilizer was clear even though the soil was rich in N and P before trial. The function of microbes in the mixed biofertilizer didn't only increase the availability of N and P but also produce secondary metabolites phytohormones [20]. It is reported elsewhere that Azotobacter and Pseudomonas synthesize and then excrete the exopolysacharides that is important for soil aggregation and hence nutrient uptake.

At the harvest time, N and P content in soil were average and high respectively since the NPK compound fertilizer was added beside organic matter and biofertilizer. This verified that in fertile soil, decreased doses of NPK is needed. The yield of bulb in this experiment was slightly lower than potential yield of Shallot cv Bauji [21]. However, the yield in plot with chicken manure amendment and biofertilizer inoculation was more than 14 t/ha dried bulbs. Genetically, shallot cv Bauji can produce 13-14 t/ha dried bulbs in the optimal environment. The residue of N and P after the shallots being harvested will be benefit for next plant during entire growth [22]. The nutrient residual effect in our experiment verified that the reduced chemical fertilizer is recommended.

4. Conclusion

The increment of plant height and soil reaction were recorded after consortia biofertilizer inoculation combined with either agricultural-waste compost or chicken manure but both treatments decrease the available phosphate in soil and didn't affect total nitrogen in soil. Chicken manure amendment integrated with biofertilizer inoculation increased dry weight of shoots and yield of shallot bulbs even though didn't significantly differ with the control and other treatments. Application of chicken manure integrated with mixed biofertilizer mostly increase yield of shallot either fresh weight of bulb or air-dried weight. The yield of air-dried bulbs weight following that treatment enhanced up to 24,5 % compared to the control.

5. Acknowledgement

This field experiment was funded by former Maluku Corner of Universitas Padjadjaran and Agricultural Office of Tual City.

References

[1] Lasmini S A, Kusuma Z, Santoso M and Abadi A L 2015 Application of organic and inorganic fertilizer improving the quantity and quality of shallot yield on dry land *Int. J. Sci. Technol. Res.* **4** 43-246.

[2] Trujillo-Tapia M N and Ramírez-Fuentes E 2016 Bio-fertilizer: an alternative to reduce chemical fertilizer in agriculture *J. Global Agric. Ecol.* **4** 99-103

[3] Subedi R, Khanal A, Aryal K, Chhetri L and Kandel B 2019. Response of Azotobacter in cauliflower (*Brassica oleracea* L. var. Botrytis) production at Lamjung, Nepal *Act Sci Malay.* **3** 17–20

[4] Lasmini S A, Nasir B, Hayati N and Edy N 2018 Improvement of soil quality using bokashi composting and NPK fertilizer to increase shallot yield on dry land Aust. *J. Crop Sci.* **12** 1743-1749.

[5] Hoffman B M, Lukoyanov D, Yang Z Y, Dean D R and Seefeldt L C 2014 Mechanism of nitrogen fixation by nitrogenase: the next stage *Chem. Rev.* **114** 4041–4062

[6] Alori E T, Glick B R and Babalola O O. 2017 Microbial phosphorus solubilization and its potential for use in sustainable agriculture *Front. Microbiol.* **8** 971

[7] Patil V 2011 Production of indole acetic acid by *Azotobacter* sp. *Rec. Res. Sci. Technol.* **3** 14-16

[8] Fukami J, Cerezini P and Hungria M 2018 Azospirillum: benefits that go far beyond biological nitrogen fixation. *AMB Expr.* **8** 73

[9] Sharma P P, Verma P P and Kaur M 2017 Phytohormones production and phosphate solubilization capacities of fluorescent *Pseudomonas* sp. isolated from Shimla Dist. of Himachal. *Int. J. Curr. Microbiol. App. Sci.* **6** 2447-2454

[10] Hindersah R, Kalay A M, Jacob A and Talahaturuson A 2014 Sago waste: local waste product for biofertilizer media. *J. Agroekotek.* **6** 12 – 24 Article in Indonesian Abstract in English

[11] Eiler A, Langenheder S, Bertilsson S, Tranvik L 2003 Heterotrophic bacterial growth efficiency and community structure at different natural organic carbon concentrations. *Appl. Environ. Microbiol.* **69** 3701–3709

[12] Berger L R, Stamford N P, Santos C E R S, Freitas A D S, Franco L O and Stamford T C M 2013 Plant and soil characteristics affected by biofertilizers from rocks and organic matter inoculated with diazotrophic bacteria and fungi that produce chitosan *J Soil Sci. Pl. Nutr.* **13** 592-603

[13] Samuel A L and Ebenezer A O 2014 Mineralization rates of soil forms of nitrogen, phosphorus, and potassium as affected by organomineral fertilizer in sandy loam *Adv. Agric.* **2014** 149209

[14] Sierra C A, Trumbore S E, Davidson E A, Vicca S and Janssens I 2015 Sensitivity of decomposition rates of soil organic matter with respect to simultaneous changes in temperature and moisture *J. adv. model. earth sys.* **7** 335-356

[15] Fan X H, Li Y C and Alva A K 2011 Effects of temperature and soil type on ammonia volatilization from slow-release nitrogen fertilizers. *Comm. Soil Sci. Pl. Anal*. **42** 111-1122.

[16] Wang H, Gao J-e, Li X-h, Zhang S-l and Wang H-j 2015 Nitrate accumulation and leaching in surface and ground water based on simulated rainfall experiments *PLoS ONE* **10** e0136274.

[17] AOAC 2012 Official methods of analysis 19th edition *Assoc. official analtical chemist* Washington DC USA.

[18] Bradáčová K, Kandeler E, Berger N, Ludewig U and Neumann G 2020 Microbial consortia inoculants stimulate early growth of maize depending on nitrogen and phosphorus supply *Plant Soil Environ.* **66** 105–112

[19] Syamsiah J, Rahayu R and Binafsih W 2020 Soil properties and shallot yield responses to different salinity levels *Sains Tanah J Soil Sci. Agroclimat.* **17** 30-34

[20] Hindersah R, Setiawati M R, Asmiran P and Fitriatin B N 2020 Formulation of Bacillus and Azotobacter consortia in liquid cultures: preliminary research on microbes-coated urea *Int. J. Agric. Sys.* **8** 1-10.

[21] Balitsa 2018 deskripsi bawang merah varietas Bauji (Description of Shallot var Bauji) *Balai Penelitian Tanaman Sayuran* (*Vegetable Research Institute*). Lembang, Bandung Barat Regency. In Indonesian

[22] Bilkis S, Islam M R, Jahiruddin M, Rahman MM and Hoque T 2019 Residual effects of different manures and fertilizers applied to preceding potato crop on succeeding mung bean (*Vigna radiate* l.) crop in potato-mung bean-rice cropping pattern *SAARC J. Agric*. **16** 67-179

Meat price volatility as implications for food security in Indonesia

R P Destiarni[1,*], A S Jamil[2], and F Septya[3]

[1]Agribusiness Department, Faculty of Agriculture, University of Trunojoyo Madura, Bangkalan
[2]Agricultural Training Centre of Jambi, Muara Jambi, Jambi
[3]Agribusiness Department, Faculty of Agriculture, University of Riau, Pekanbaru

* E-mail: resti.destiarni@trunojoyo.ac.id

Abstract. Indonesian meat consumption has 40 percent deficit which was covered by importing. Meat price in international market tend to fluctuate. The gap between domestic demand and supply meat also the imported price fluctuation causes instability of domestic price. This research is conducted to analyze the volatility of meat price which implicated to food security in Indonesia. ARCH-GARCH model is used to estimate meat price volatility in Indonesia. The Augmented Dickey-Fuller and cointegration test have been used for testing the presence of unit root and cointegration in the series. Langrange multiplier has been utilized to detect the presence of autoregressive conditional effect. Daily meat prices used are national average price which obtained from the Indonesia Ministry of Trade. This study reveals that meat price in Indonesia has high volatility with increasing price over the research period. The empirical model also shows asymetry effect. The results recommend that Indonesia should apply comprehensive managed import such as not only import on fresh meat and ready to cut bovine but also on breeding bovine. By the fulfilling production and stock, meat price can be more stable. By the price stabilization, food security concept will be reached so that every layer society can consume meat.

1. Introduction

Meat is one of the seven main food source commodities that are the focus of the government in realizing food security in Indonesia through self-sufficiency programs. Based on the 2015-2019 Strategic Plan of the Ministry of Agriculture, the seven commodities include rice, corn, soybeans, consumption sugar, beef, fish and salt. The government's focus on the beef commodity is a program to increase beef production to meet household consumption needs [1]. This production increase program is implemented in a self-sufficiency program so that it can meet the needs of consumers from various backgrounds for animal protein. Indonesia's focus on achieving food security is Indonesia's first step towards realizing food sovereignty, which is reflected in the power to overcome problems and fulfill food needs independently [2]. The concept of food security according to Law Number 18 of 2012 is a condition for the fulfillment of food for the state to individuals, which is reflected in the availability of sufficient food, both in quantity and quality, safe, diverse, nutritious, equitable, and affordable and does not conflict with religion, belief, and community culture, to be able to live healthy, active and productive in a sustainable manner. Three important aspects that must be met in order to achieve food security are the availability, access and absorption of food. Beef, which is the product of the livestock sub-sector, is one of the commodities that is the focus for realizing food security.

Changes in the lifestyle of Indonesians who are increasingly aware of the importance of animal protein for body health and hormonal balance have made the demand for beef has an increasing trend every year. The trend of increasing beef consumption in addition to consumers tends to consume beef with consideration of nutrition, social status, culinary, and western cultural influences. Based on data collected by the Center for Agricultural Data and Information Systems, Indonesian beef consumption fluctuates with a tendency to increase annually by 2.89 percent per year with the level of beef consumption in 2017 reaching 2.40 kg / capita / year from various age levels.

2. Methods

Data used in this research are daily observation of meat price. The number of observations was 2737 data observed from September 2010 to January 2017. Data was collected from the Ministry of Trade, Ministry of Agriculture, Kramat Jati Central Market and other sources. This data is used to answer the research objectives, where volatility is analysed using the econometric approach with the ARCH / GARCH model. GARCH ARCH model estimation using Eviews.

ARCH / GARCH model estimation begins with a stationary test. Stationary test was used to prevent spurious regression model. Spurious regression implies that the result of the regression may not be as significant as they seem. In this study, the stationary test was conducted by the augmented Dickey Fuller test. The ADF test is used to detect whether the data being analysed contains unit roots. The data is said to be stationary if it does not contain a unit root. The hypothesis in the ADF test is

H0: data has a unit root
H1: data has no unit root

Testing the hypothesis by comparing the t-statistic and the critical value of MacKinnon. If the t-statistic value in the ADF test is smaller than the critical value MacKinnon means the data is not stationary and it needs to be differentiated or differencing. After the data has been stationary, the tentative ARIMA model estimation is then carried out. Determination of the AR (p) and MA (q) orders is based on correlogram (ACF and PACF patterns), while for the determination of the order d is determined based on the level of data stationarity.

Selection of the best ARIMA model based on the smallest value of Akaike Information Criteria (AIC) and Schwartz Criterion (SC). In addition to the smallest AIC and SC criteria, the selected ARIMA model must also meet the following criteria, namely random forecast residuals, parsimonious models, estimated parameters that are significantly different from zero, invertibility and stationarity conditions must be met as indicated by the number of AR and MA coefficients respectively. each is less than one, the iteration process must be convergence, and the model must have a small MSE.

The selected ARIMA model was tested using the Lagrange Multiplier test to see the presence of the ARCH effect. The ARCH-LM effect test is based on the null hypothesis (Ho), that is, there is no ARCH error. The existence of the ARCH effect is a requirement for estimating the ARCH / GARCH model. After ascertaining that there is an ARCH effect, the process continues by estimating several variance models using the ARIMA model. Several alternative ARCH / GARCH models that have been estimated were then selected based on AIC and SC.

After obtaining the ARCH / GARCH model, the model evaluation is carried out by checking the adequacy of the model. If the model is inadequate, then return to the identification stage to get a better model. Model diagnosis is done by analysing the standardized residuals that include

1. Residual normality
   The residual normality test was carried out to test whether the residuals spread normally. Residual normality was checked by using the Jarque-Bera test. The Jarque-Bera test measures the difference between Skewness (protruding) and kurtosis (tapering) data from a normal distribution, and includes a measure of variability.
2. Residual freedom
   The residual freedom test is carried out to test whether there is autocorrelation in the analyzed data. ARCH-GARCH model shows good performance if it can eliminate the existing autocorrelation in the data, that is, when the residuals are random (white noise). The test was

carried out by examining the residual squared autocorrelation coefficient with the Ljung-Box statistical test.

3. The presence of the ARCH-GARCH effect or the presence of heteroscedasticity.
   A test for the presence of the ARCH effect is carried out to test whether there is still an ARCH effect on the selected ARCH-GARCH model. The presence of the ARCH effect can be detected by the Lagrange Multiplier (ARCH-LM) test.

The ARCH / GARCH model that has been evaluated is then used to estimate the value of beef price volatility. The volatility value of the beef price is described by the standard deviation value which is the square root of the various ARCH / GARCH models estimated.

3. Results and discussion

Figure shows daily beef price trend over the past six years. This figure also shows that the development of meat prices has increased. Meat price growth increased by 66.75% from the beginning of the period to the end of the data period. This significant increase is partly due to the increasing trend of middle-income communities in Indonesia. In addition, according to Nugroho (2016), the increase in population and purchasing power and driven by changes in consumption patterns has an impact on increasing demand for meat.

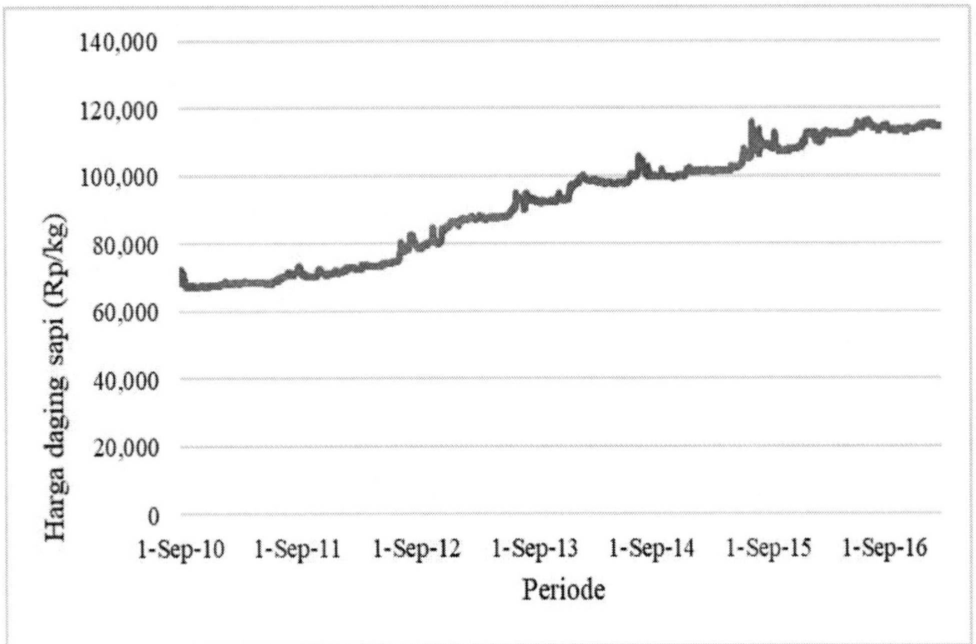

Figure 1. The development of meat price data

An increasing trend in meat prices indicates that there is a change in the average or in other words, the data is not stationary on the average. Based on the Augmented Dicky Fuller test, it shows that at the data level, the p-value is greater than alpha (0.8695> 0.05). This means that the data is not stationary and requires differencing. The 1st differencing shows that the p-value is either tested using intercept or intercept and the trend is stationary.

Tabel 1. The result of Augmented Dicky Fuller test on meat price data

Variables	Intercept		Intercept and Trend	
	t-statistic	Prob	t-statistic	Prob

I (0)	-0.594142	0.8695	-3.539972	0.0354
I (1)	-28.10720	0.0000	-28.10127	0.0000

After the data is stationary, the tentative ARIMA model is identified. The formation of the tentative ARIMA model is carried out by determining the AR order (p) and the MA order (q) [3]. The best ARIMA model selected has met the selection criteria, namely having the smallest Akaike Information Criteria (AIC) and Schwatrz Criterion (SC) values [4], [5]. Each tentative ARIMA model is presented in Table 2. Based on the SIC, SC and R-squared criteria, it shows that ARIMA (3,1,1) is the best model.

Tabel 2. The result of tentative ARIMA models

ARIMA	AIC	SC	R-squared
ARIMA (1,1,0)	15.09488	15.09975	2.5455
ARIMA (2,1,0)	15.10316	15.10803	1.7527
ARIMA (3,1,0)	15.08591	15.09078	3.4663
ARIMA (0,1,1)	15.08683	15.09170	3.2860
ARIMA (0,1,3)	15.08650	15.09137	3.3176
ARIMA (1,1,1)	15.09137	15.08899	3.4566
ARIMA (2,1,1)	15.07739	15.08005	4.3327
ARIMA (3,1,1)	15.06788	15.06323	5.9607
ARIMA (1,1,2)	15.07681	15.07947	4.3708
ARIMA (1,1,3)	15.06673	15.07404	5.3301
ARIMA (2,1,2)	15.10237	15.10503	1.9128
ARIMA (3,1,3)	15.09405	15.08940	3.4673

The ARCH-LM test and the ACF and PACF behaviour patterns were then used to identify the presence of heteroscedasticity elements in the best ARIMA model. Based on the identification of the correlogram pattern, it shows that the autocorrelation value of the squared data analysed is significant in the first 15 lags. This shows the existence of the ARCH effect on the model [3], [6], [7]. In addition, the ARCH-LM test also shows that the p-value is less than alpha, so it can be concluded that there is an ACRH effect on the model.

Tabel 3.The result of heteroscedaticity test

Heteroscedaticity Test	P-value	Conclusion
Correlogram test	Signifikan	Heteroscedasticity
ARCH-LM	0.0000	Heteroscedasticity

The result of tentative ARCH / GARCH model estimation are showed in Table 4. All six models show significant autoregressive and moving average behaviour in the meat price so that the ARIMA model specification is maintained for all models to keep the residuals free from the serial autocorrelation. In addition, the selection of the best model is determined by taking into account the significance of the parameters, the criteria for the smallest AIC and SC [5], [8], [9]. Based on the indicators in Table 4, it can be concluded that the best model is the TARCH model.

Tabel 4. The result of ARCH/GARCH model estimations

Model	Estimation Parameter Significancy		Log Likelihood	AIC	SC
	AR (3)	MA (1)			
ARCH (1,0)	V	V	-17374.81	14.6727	14.68488
ARCH (2,0)	V	V	-17207.88	14.53261	14.54723
GARCH (1,1)	V	V	-16628.47	14.04345	14.05807
EGARCH	V	V	-16571.44	13.99615	14.01321
TARCH	V	V	-16543.69	13.97272	13.98977

ARCH-M	V	V	-17318.84	14.62638	14.64099

The TARCH model is then evaluated for model normality and residual randomness. Testing for normality of errors is done by doing Jarque-Bera. It is concluded that the error is normally distributed if the Jarque-Bera is not statistically significant. Based on the normality test, it shows that the error model is not normally distributed. Although the TARCH model does not spread normally, other ARCH / GARCH models are also not normally distributed. Therefore, the TARCH model remains the best model in this case. In addition, based on the ARCH-LM test, it was found that the model was free from the ARCH effect.

Tabel 5. The result of model evaluation

Model Evaluation	Magnitude	p-value
Jarque-Bera	Signifikan	Heteroscedaticity
ARCH-LM	0.0000	Heteroscedaticity

The optimal result of TARCH estimation is shown in equation below:

$h = 947.7806 + 0.272505\, \varepsilon^2_{t-1} - 0.257075\, d_{t-1} + 0.882690\, \sigma_{t-1}$

$(35.059)^{***}\ (0.008)^{***}\ (0.009)^{***}\ (0.001)^{***}$

Adj.R^2 = 0.051, SSE = 452.363, DW Stat = 2.005, Prob F-Stat = 0.000 (2.0).

This equation shows that all independent variables are significant at the 1% level. This equation also shows that there is an asymmetric effect in the model, which is shown by the significant dt-1 variable. This equation also shows that the volatility of beef prices is influenced by the volatility of the past period and the variance of prices in the previous period. The residual and volatility of the previous period have a positive effect on future volatility.

The ARCH coefficient value in the meat variant model is 0.272505. This value is relatively low, so it indicates that the volatility is low. Meanwhile, the value of the GARCH coefficient is 0.882690. This value is relatively high (close to 1) so that the shock to the price variant occurs for a relatively long time (persistence).

The effect of asymmetry on the model is -0.257075, this indicates that the occurrence of a negative shock has a greater effect than a positive shock. Basically, the dt-1 variable is a dummy variable, where when the dummy coefficient is positive, the effect on volatility is 0.272505. Then, if the dummy coef is negative the effect on volatility is 0.01543 (0.272505 + (- 0.257075)) [10]. In other words, a negative shock event will change the value of future volatility to a greater extent than if a positive shock event occurs. Figure 2 also shows development of the conditional variance over the years. The meat price met the highest conditional variance on 2015 compare to other years.

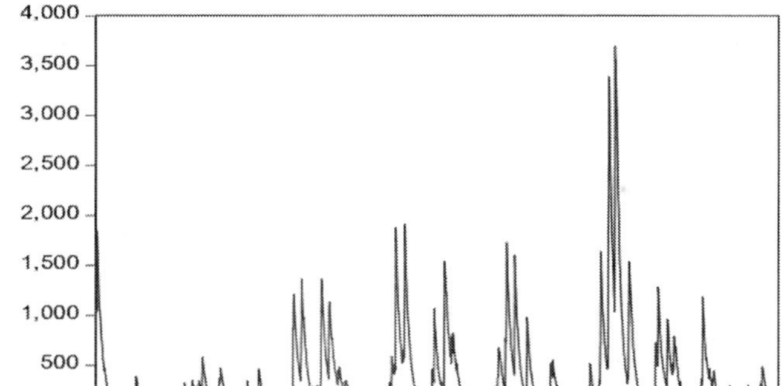

Figure 2. Conditional variance of meat price

4. Conclusion

Meat is one of the main food source commodities which becomes focus on government self-sufficiency program which is written on The Ministry of Agriculture Strategic Planning on 2015 – 2019. Focus on this program is increasing meat production as one of the animal protein sources so that it could fulfil all demand and affordable for all Indonesian. This is related to achievement of food security in Indonesia. Food security concept based on Undang-Undang No. 18 of 2012 have three aspects which should be realized, there are availability, access, and food absorption. Change of lifestyle which is increasingly aware about the importance of animal protein for body healthy and hormonal balance make the meat demand has an increasing trend every year. Based on data from The Ministry of Agriculture, Indonesian meat consumption fluctuates with increasing trend around 2.89 percent per year by the consumption rate of meat on 2017 reaches 2.40 kg/capita/year from various age levels so that meat production still has deficit around 40 percent. That deficit was covered by doing meat importing to keep domestic meat stock. However, meat price in international market tend to fluctuate in increasing trend. The gap availability between demand and supply of domestic meat and the imported price fluctuation causes instability price of domestic meat. That causes price volatility which is a risk to face on meat commodity. Price volatility on meat commodity become fundamental issue especially for Indonesia related to food security. The research about price volatility will increase food security and price problem awareness so that this research is conducted to analyse the volatility of meat price which implicated to food security in Indonesia on meat commodity. The autoregressive conditional heteroscedasticity/generalised autoregressive conditional heteroscedasticity (ARCH-GARCH) regression model is used to estimate meat price volatility in Indonesia. The Augmented Dickey-Fuller (ADF) and cointegration test have been used for testing the presence of unit root and cointegration in the series. Langrange multiplier has been utilized to detect the presence of autoregressive conditional (ARCH) effect. Daily meat prices used are national average price which obtained from the Indonesia Ministry of Trade. This study reveals that meat price in Indonesia has high volatility with increasing price over the research period. The empirical model also shows asymmetry effect. The results recommend that Indonesia should prevent effect of high volatility with applying comprehensive managed import such as not only import on fresh meat and ready to cut bovine but also on breeding bovine which can be bred in Indonesia. Government must be consistent on program implementation to increase domestic production like SIWAB (Sapi Indukan Wajib Bunting), livestock distribution from densely populated to biomass abundant areas but empty cattle, increasing cattle productivity, preventing productive female cattle slaughter, preventing and controlling disease to decrease mortality rate, and also improving genetic quality through selection and crossing. By the fulfilling production and stock, meat price can be more stable. By the price stabilization, food security concept about availability, access, and food absorption will be reached so that every layer society can consume meat.

The results recommend that Indonesia should prevent effect of high volatility with applying comprehensive managed import. Indonesia's Import activity is not only done in one time as the fastest way to stabilize price but also to keep domestic stock. Import is not only on fresh meat and ready to cut bovine but also on breeding bovine which can be bred in Indonesia. Governement must be consistent on program implementation to increase domestic production like SIWAB (Sapi Indukan Wajib Bunting), livestock distribution from densely populated to biomass abundant areas but empty cattle, increasing of cattle productivity, preventing of productive female cattle slaughter, preventing and controlling disease to decrease mortality rate, and also improving of genetic quality through selection and crossing. By the fulfilling production and stock, meat price can be more stable. By the price stabilization, food security concept about availability, access, and food absorption will be reached so that every layer society can consume meat.

References

[1] D. P. dan P. Bappenas, "Penyusunan RPJMN 2015-2019 Bidang Pangan dan Pertanian," 2014, [Online]. Available: ttps://www.bappenas.go.id/files/1914/4299/8417/Penyusunan_RPJMN_2015-2019_Bidang_Pangan_dan_Pertanian.pdf.

[2] Badan Pengkajian dan Pengembangan Kebijakan Perdagangan, "Peluang Ekspor Indonesia di Pasar Negara-Negara Non Tradisional," pp. 2014–2015, 2015, [Online]. Available: http://bppp.kemendag.go.id/leaflet_artikel_perdagangan/view/NDA%3D.
[3] Mittal S, Hariharan V K, and Subash S P 2018 Price volatility trends and price transmission for major staples in India *Agric. Econ. Res. Rev.* **31(**1), p. 65, 2018, doi: 10.5958/0974-0279.2018.00006.x.
[4] Radifan F 2014 Faktor-Faktor Yang Mempengaruhi Ekspor Crude Palm Oil Indonesia Dalam Perdagangan Indonesia *Econ. Dev. Anal. J.* **2(**4) 259–267, 2014.
[5] Saghaian S H 2010 The Impact of the Oil Sector on Commodity Prices: Correlation or Causation? *J. Agric. Appl. Econ.* **42(**3) 477–485, 2010, doi: 10.1017/s1074070800003667.
[6] Apergis N and Rezitis A 2011 Food Price Volatility and Macroeconomic Factors: Evidence from GARCH and GARCH-X Estimates *J. Agric. Appl. Econ.* **43(**1) 95–110, 2011, doi: 10.1017/s1074070800004077.
[7] Nurliza N 2017 The Nature of Food Commodity Prices Volatility in Driving Inflation and Policy *Signifikan J. Ilmu Ekon.* **6(**1) 103–124, 2017, doi: 10.15408/sjie.v6i1.4523.
[8] Ikudayisi A A and Salman K K 2014 Spatial Integration of Maize Market in Nigeria – a Vector Error Correction Model *Int. J. Food Agric. Econ.* **2(**3) 71–80, 2014, [Online]. Available: http://www.foodandagriculturejournal.com/vol2.no3.pp71.pdf.
[9] Rosa F, Vasciaveo M, and Weaver R D 2014 Agricultural and Oil Commodities: Price Transmission and Market Integration between US and Italy *Bio-based Appl. Econ.* **3(**2) 93–117, 2014, doi: 10.13128/BAE-12944.
[10] Ismail A, Ihsan H, Khan S A, and Jabeen M 2017 Price volatility of food and agricultural commodities: A case study of Pakistan *J. Econ. Coop. Dev.* **38(**3) 77–120, 2017.

IOP Conf. Series: Earth and Environmental Science **883** (2021) 012069 doi:10.1088/1755-1315/883/1/012069

Income adaptation of farmers as long covid-19 pandemy on sustainable ub forest management: a case from Indonesia

A Kustanti[1,*]

[1] Socio-Economic Department, Faculty of Agriculture, University of Brawijaya, Malang, Indonesia.

* Email: kustanti@ub.ac.id, asihkustanti@gmail.com
Phone: *+62-83848033998*

Abstract. This study has analysed farmers' income on their adaptation as long Covid-19 pandemic in University of Brawijaya (UB) Forest. UB Forest got a management right of State Forest Management (Ministry of Forestry and Live Environment/MLEF Indonesia). Preliminary, this forest was managed by State Company Forest, namely Perum Perhutani. But, in 2015 it was transferred to the University by MLEF Indonesia for Education and Training Centre. The agricultural plantation has conducted by 824 farmers who were involved in UB Forest management. There are 16 fields of farmers' land use in 514 ha area. After the forest transfer to the UB Forest, there was no clear how farmers right adjusted the vision and mission of UB. This research wants to analyse how the income adaption farmer as long Covid 19, right and obligation as a shape of participation, and UB Forest management with qualitative approach. Selected respondents were in-depth interviewed. The analysis method used Institutional Analysis Design/IAD (Ostrom, 1999) and Bundles of Right (Ostrom, 1990). Farmers adaptation as long Covid-19 on forest management at UB Forest based on bundles of rights as a Claimant. It means that farmers had two right as below: 1)access and withdrawal right (enter the land, cultivate agriculture land, land conservation, harvest the agricultural yield, watch the forest fire along with UB Forest Team, implementation innovation of UB civitas academic, maintain the tree in UB Forest area, including in education and research of UB) and 2)management right on an agricultural plantation under the tree (a decision the species of agricultural plant, manage the plant, and arrange the cycle of the plantation itself). The income of farmers as long as Covid-19 was declined 39 % from the normal condition.

1. Introduction

The University of Brawijaya is a university located in Malang East Java Province. In 2015 this university was given forest by the Ministry of Live Environment and Forestry (MLEF) of Indonesia. The wide was 514 hectares and located on Karangploso District Malang Regency East Java Province. The altimeter was 700-1200 sea above level. This forest was a tropical rain forest that consisted of a production forest and protected forest. Before it transferred to the University of Brawijaya, it was managed by Company State of Forest, namely Perum Perhutani. The process was done in 2015 by Decision Letter of MLEF with number: 676/MenLHK-Setjen/2015 about Establishment Forest Area as Specific Purpose on Protected (42.72 ha) and Production (471.28 ha) Forest in Karangploso District, Malang Regency, East Java Province widely 514 hectares as an Education and Training

Forest [3-4]. Furthermore, this forest has managed as a Tridharma of the university, i.e., as an education, research, and community service development.

The University of Brawijaya managed the forest, namely UB Forest. The vision and mission of UB Forest were different from Perum Perhutani. Perum Perhutani worked the forest as a profit-oriented company while UB Forest was an education and training forest. The Strategies of UB Forest management were: 1) optimising the authority and bargaining position of UB Forest in forest and environmental management, 2) human resources of UB Forest should focus on strengthening national and international cooperation targeting education, research, and community development (i.e. Tridharma of University), besides developing the industry of forest yields and environment, 3) increased support from regional and central government, and from other stakeholders for fundraising and sustainable management, 4) enhancing the technology and information, 5) management planning on forest area by exploring the role of community in forest management, and 6) developing entrepreneurship in forest and environment management by a) minimising the monopoly of forest yields and coffee marketing, b) enhancing the sustainability of forest ecotourism, wood and non wood productions, c) local community recruitment as a forest guider, and d) regularity of community services in forest institution management, and law enforcement to maintain the forest regulation [4].

According to the fifth strategy, UB Forest management included community roles in and around the forest. It was named a farmer. The number of a farmer who worked in UB Forest land was 824 farmers. Their activities as long as working in the forest were different between State Forest Company/SFC (Perum Perhutani) and UB Forest time [6]. In the Perum Perhutani period, farmers who were involved in the forest worked on the forest land. They must maintain the primary plantation, e.g., Forest's standing stock (mahoni and pine trees). They also must give the 3 kilograms of pine latex dairy to the Perum Perhutani, besides space utility of land among the trees. The transfer of ownership between Perum Perhutani and UB Forest implied the difference of vision and mission. But, the farmers' involvement was still going as UB Forest as a new manager. As long as the covid-19 pandemic, it influenced agricultural activities. There is no data about how it affected the income of farmers. This research was essential to analyse how the farmers' income was affected by this condition.

2. Methods

This research was conducted in June-July 2020. The location was in Dusun Sumbersari Tawangargo Village, Dusun Sumberwangi Donowarih Village, and Dusun Buntoro Ngenep Village. The type of research was qualitative. Data used primary and secondary data. The primer data were collected by field observations, in-depth interviews with farmers, and government selected by snowball sampling. Respondent of farmers selected by snowball sampling based on age class standing stock in UB Forest. There were four age classes (3, 5, 6, and 7) of standing inventory. In series of age class classification were: 1)Third (3) was the standing stock with 15 years old; 2)Fifth (5) was 25 years old; 3)Sixth (6) was 30 years old, and 4)Seventh (7) was 35 years old. Then the existing farmers for each class of age were selected by snowball sampling. Analysis data used descriptive analysis of farmers' roles based on their right stratum as property rights bundles (Ostrom, 1999).

3. Results and discussion

Since 2015, UB Forest's management has had bundles of rights as followed: 1)access and withdrawal rights, 2)management rights; 3)exclude non-right people; and 4)without tradable rights. Access and withdrawal rights mean that UB Forest can do all activities into the forest and take advantage of its yields. It was like a non-wood product, limited wood yields suitable with the regulation, services of forest, ecotourism, education and training, research, and community services.

The topography of UB Forest area varied as followed: (a) Slope of 0-8% was (40.97 ha); (b) > slope of 8-15% was (484.89 ha); and (c) slope > 15% was (23.81 ha). The average rainfall per years was 250 mm, and an average temperate was 27°C. The accessibility comes into the forest was become three gates (Table 1). Gate A is located in Zona 1 through wonokoyo area, gate B located in

Zona 2 through sumberwangi, gate C located in Zona 2 turn right before wonokoyo, and gate D in Zona 3 through Bocek [10].

Table 1. Area of each category of slope gradient

Slope (degree)	Area (Ha)	Proportion (%)
0-8%	40,97	7,45
8-15%	484,89	88,21
>15%	23,81	4,33
TOTAL	549,67	100

The vegetation in forest protected area of UB Forest consisted as trees and shrub, as followed: 1)trees group gintungan [(*Bischoffia javanica*), dadap (*Erythirma lithosperma*), anggrung (*Trema orientalis*), ringin (*Ficus benjamina*), kesek (*Muntingia calabura*), gondang (*Ficus variegata*), tepus (*Etlingera solaris*), puspa (*Schima wallicii*), eukaliptus (*Eucalyptus spp*), durian (Durio zibethinus), avocado (Perse Americana), breadfruit (*Artocarpus altilis*); and shrup group 2) bamboo (*Bambosa spp.*), fern (*Cycas* spp.), and kaliandra (*Calliandra callothyrsus*). Hereafter the protected forest, the production forest had a different formation of vegetation. In this area consisted of the industrial woody and agricultural plantation. Management of two kinds of plantation-based on the regulation of MLEF that woody was managed by UB Forest and agricultural plantation was managed by farmers (Table 2).

Table 2. Potential of Woody and Agricultural Plantation in UB Forest

No.	Species	Age Class (AC) of Standing Stock	Agricultural Plants
1.	Pine (*Pinus merkusii*)	AC 2-7	Talas (*Colocasea esculenta*), coffee (*C. robusta* & *Arabica*), carrot (*Daucus carota*), and chilli.
2.	Mahoni (*Swietenia mahagoni*)	1977, 1979, 1982	Talas (*Colocasea esculenta*), curcuma (*Curcuma longa*), cassava (*Manihot esculenta*), coffee (*C. robusta* & *Arabica*), and chilli.
3.	Suren (*Toona surena*)	2007	Talas (*Colocasea esculenta*), cassava (*Manihot esculenta*)

The space among the pines plantation in the field was 3 x 3 meters. At the area, farmers utilised by growing agriculture plants as following: coffee, carrot, ginger, curcuma, chilli, etc. The species of agriculture plant was defined by the farmers. They had an obligation to manage the land and maintain the standing stock of trees. It had been held since the 1980s when State Forest Company/SFC (Perum Perhutani).

Some set of regulations were used as management guide: 1) National Education System Law No. 20 year 2003 (*Staatsblad* year 2003 number 78, *Staatsblad* number 2301); 2) National Development Planning System Law No. 25 year 2004 (*Staatsblad* No. 24 year 2004, *Staatsblad* number 4421); 3) Indonesia Teachers and Lectures Law No. 14 year 2005 (*Staatsblad* number 4586); 4) Indonesia National Long-term Development Plan Law 2005-2025 No. 17 year 2007 (*Staatsblad* number 33 year 2007, *Staatsblad* number 4700); 5) University Education Law No. 12 year 2012 (*Staatsblas* No 5336);

6) Government Regulation of National Education Standards No. 19 year 2005 (*Staatsblad* No. 14 year 2005, *Staatsblad* No. 5339) as amended for the last time by Government Regulation No 32 year 2013 of Amandment Government Regulation No. 9 year 2005; 7) Lectures Government Regulation No. 37 year 2009 (*Staatsblad* No. 76 year 2009, *Staatsblad* No 5007); 8) Government regulation of Fund for Education No. 48 year 2008 (*Staatsblad* No. 91 year 2008, *Staatsblad* No. 4864); 9) President Regulation of Indonesian National Qualifications Framework No. 24 year 2012; 10) Ministry of National Education Regulation of Strategic Plan Ministry of National Educational year 2010-2014, as amended for the last time by Ministry of Nastional Education No. 44 year 2010; 11) Ministry of National Education Regulation of Statute University of Brawijaya No. 34 year 2011; 12) Minister of Administrative reform and Bureaucratic Reform of Lecturer Functional Position and Credit Score No. 17 year 2013 as amended for the last time by Minister of Administrative Reform and Bureaucratic Reform No. 46 year 2013 Amandment of Minister of Administrative Reform and Bureaucratic and Reform regulation No. 17 year 2013 of Lecturer Functional Position and Credit Score; 13) Indonesia Forestry law No. 41; and 14). The Special Purpose Forest Area rule is explained on P15/MENLHK/SETJEN/KUM.1/5/2018 the year 2018.

UB Forest management's strategy was suitable with its characteristic, the community around the forest, and finally proper with Tridharma of University, e.g. education, research, and community development [3]. Priorities strategy were: 1) optimisation UB Forest management; 2)Human resources development in Tridharma implementation; 3)National and International support for the forest sustainability, and 4)Increasing Information and Technology.

4. Farmers characteristic

UB Forest was surrounded by three villages, e.g. Tawang Argo, Donowarih, and Ngenep. The primary livelihood was a farmer. There was a farmer group, namely Kelompok Tani Sumber Makmur. This group was formed to organise the member. All member planted coffee species in the forest land. The size of the land cultivated under the trees was varied. There was 0.5 ha until 3 ha. That land scattered on different age class of UB Forest. The existing of land and agricultural each farmer laid in different of age class standing stock.

It can found on the land two kinds of farmer types. The wide land of each farmer was varied. It was from 2000 meter per sea till 10.000 m per sea. One was as a Magersaren. It means that the farmer who lived in the forest area as a worker [8] and the have land agricultural in the UB Forest. They didn't have a house. They depended on State Forest Company. Since 1970 State Forest Company had a program to make their worker's housing, namely *Magersaren base-camp*. The obligation of Magersaren should do anything of forestry work. It was lax tapping, wood harvesting, pruning, skidding, and log transporting to the wood store location.

Magersaren communities are people who do not own their cultivation land so that their lives are highly dependent on the intercropping land granted by Perum Perum Perhutani for reforestation activities. It was as well as relying on Perum Perum Perhutani forestry work, such as sap tapping, logging, thinning, skidding.

The second was a Pesanggem. Pesanggem was a farmer who was not living in the forest, but they have agricultural land in the UB Forest. The number of farmers was more than Magersaren. Sometimes they did the ground by themselves, and sometimes they did not do. They only gave the financial land operation by another farmer. The land of pesanggem was not fixed, often it happened illegal transfer to another person. So the handover of the land, it caused the messed upland cultivation. It was forbidden to an illicit transaction of state forest area. It happened since Perum Perhutani operated to this location. The settlement of Magersaren consisted of huts made of wood branches, roofing with alang-alang grass or dry leaves in the forest area and continue to move around following the work rotation of Perum Perhutani. Magersaren tends to be rated as a poor community. In Pakistan, another occasion based on [1] stated that forest officials provided timber and fuelwood was just 14% and 10%, respectively. The kinds of agricultural plants depended on the family, owners/tenants, and fellow-farmers' opinions. It was likely too with the farmer behavior in UB Forest agricultural plants. They decided the kinds of plants based on economic section, environmental benefit, hereditary planting habits, and market demand.

Since the 1970s, Perum Perum Perhutani (SFC) started to launch a program to build magersaren settlement units in the forest area with the building of semi-permanent houses. It provided for farmers working on intercropping land called magersaren base-camp. This base camp was not an improvement from the long-standing magersaren community but is a new settlement unit for intercropping farmers and Perum Perhutani supervisors in temporary forest areas. It was occupied temporarily during the planting work contract [8]. In the other farmer context, [9] have categorized the assets of community forest farmers in the following terms: (i) human capital; (ii) physical capital; (iii) natural capital; (iv) social capital; and (v) financial capital. Human capital refers to the number of persons in the community categorized as of productive age: all of the magersaren and pesanggem farmers sampled in this study fell within the productive age group. The activities performed by all farmers comprised both agricultural and non-agricultural endeavors. The community-based forest (HKm) farmers had 14 sources of income, while the farmers of the privately-owned forest had 12.

Magersaren and Pesanggem were different farmers. UB Forest continued SFC pattern with the new management. UB Forest tried to in good arrangement between the right of the land and its capability. Firstly, activity was conducted by mapping the land based on the age of a class of trees and the farmers' rights. It founded that many farmers had many lands in the other area. Contrary to this condition, not all the land was right in good cultivation. Some were left unused. UB Forest tried to manage the disorder of this right by identifying, mapping, and communicating with the farmers to get well administration of the land.

UB Forest should make it well-managed the land and the farmers. They were driven by the effectiveness and efficiency of land use. The purpose was the equity and welfare can increase. In the future, It could not find unused land. The farmers should adapt to the policy of UB Forest management. Magersaren and Pesanggem will manage base on their capability and responsible for land management. Land and farmers were taken as a sample based on stratified sampling and administration region (Table 3).

Table 3. Farmers, administration region, and plantation

No	Dusun	Village	Swath	∑ Farmers	Plantation
1.	Sumbersari	Tawangargo	85	32	Pine, Coffee
			90	6	Pine, Coffee
2.	Sumberwangi	Donowarih	91	30	Pine, Coffee
			93	16	Pine, Coffee
			94	14	Pine, Coffee, Vegetable
			97	8	Pine, Coffee
			100	3	Mahogani, Coffee
3.	Buntoro	Ngenep	98	40	Mahogani, Coffee, Talas
			99	44	Mahogani, Coffee, Talas
	Total Farmers			193	

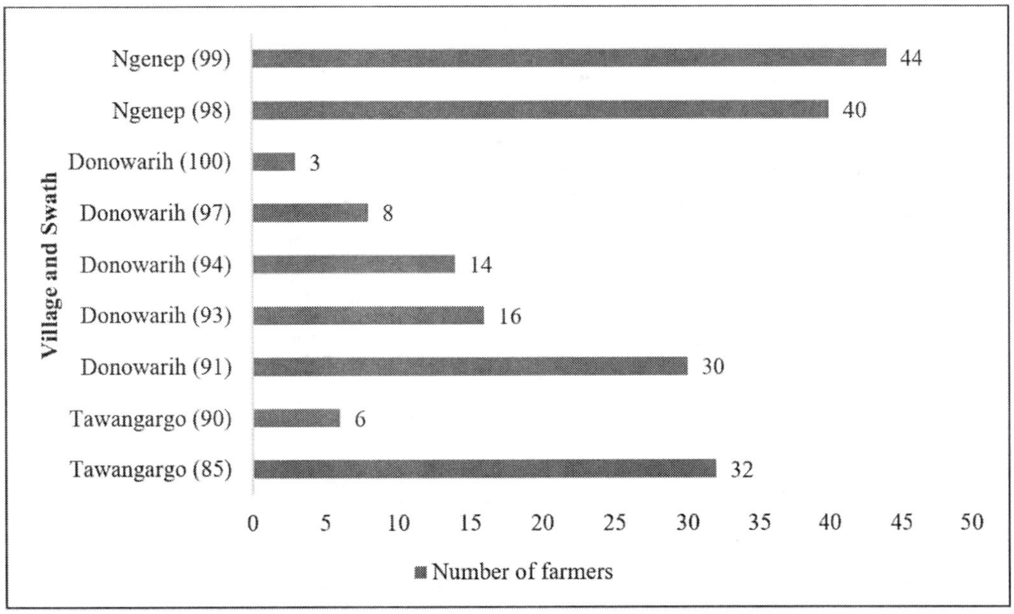

Figure 1. Number of a farmer of each village and swath category

The administration region in UB Forest has founded 3 Dusun (Sumbersari, Sumberwangi, and Buntoro) and 3 Villages (Tawangargo, Donowarih, and Ngenep). There were nine swaths (85, 90, 91, 93, 94, 97, 100, 98, and 99). Farmers in the sample were 193. Meanwhile, the plantation in that area was pine, coffee, and talas. Based on data is obtained from the interviews, the age of respondents ranging from 32 years to 58 years. The level of education of all informants is only up to elementary school. The income of each informant varies from 1,500,000 to 5,000.00. It is because the area of agricultural land owned by informants also varies. But according to the respondent, the income earned every month as a farmer still lacks to meet daily needs. So some respondents also have other jobs as traders, so they get an extra income.

The existence of COVID-19 affects the habits of farmers in UB Forest. Moreover, at the same time, Covid-19 coincided with the primary harvest season and calls for a government lockdown due to the pandemic causing harvest prices to drop dramatically. One of the changing habits of farmers is in labor, which is usually for harvesting; they employ workers from outside the family, but because of the pandemic and the concurrent harvest, they only use workers from within the family to reduce costs incurred are not too high. They can still set aside for their daily lives. Many of the farmers interviewed did not grow vegetables due to pandemics, and they only depend on the coffee plants even though the production of this year's coffee is not as good as in previous years.

The selected respondents begun with recommendations from predetermined key informants. Researchers took information from key informants aimed at validating the data from interviews with respondents. The head of the farmer group was chosen as the key informant because he coordinated directly with UB Forest and the farmers. Director of UB Forest had an authority to regulate the management and decisions regarding the management of UB Forest.

Respondents selected lived in the state forest area. Formerly, the term used by SFC in people who lived in forest areas is Magersaren. The Magersaren community was a farm employee who worked for Perum Perhutani, or an organization is living in a state-owned forest area. The kinds of work were land clearing, planting, maintaining plants, and protecting the forest.

The existence of the Magersaren community has existed since SFC era. Perum Perhutani (SFC) had permitted them to live in a forest area. The objective was to integrate reforestation with activities to increase the income and welfare of forest village communities. In this context, farmer groups are formed around agroforestry on forest land and activities conducted on farmers' private land, animal husbandry, small-scale industry, and public character activities such as building facilities for clean water, bridges, and village roads [11]. They could not make a permanent residence. Besides, the Magersaren community was also required to maintain the staple plants owned by SFC. The Magersaren community has a variety of agricultural activities under the main stand of pine trees. They plant crops that are seasonal - chili, corn, cassava, turmeric, and taro. In comparison, the annual plants cultivated by the Magersaren community were coffee plants. Most of them also worked as farms worker whose owners are capital owners and come from outside the village. Every day they were paid Rp. 30,000 for female workers (approximately 8 hours of work/day) and Rp.50,000 for male workers (around 8 hours of work/day)[5].

Table 4. Farmers Income before and during Covid-19

No	Before Covid-19				As long as Covid-19			
	Plants	Total Cost (Rp)	Revenue (Rp)	Income (Rp)	Plants	Total cost (Rp)	Revenue (Rp)	Income (Rp)
1	Corn	2.455.000	3,450.000	995.000	Chili	2.900.000	4.800.000	1.894.000
2	Chili	12.530.000	24.000.000	11.740.000	Turmeric	770.000	2.000.000	1.230.000
3	Turmeric	600.000	2.200.000	1.600.000	Carrot	2.364.747	5.500.000	3.135.263
4	Cassava	1.160.000	2.800.000	1.640.000				
5	Taro	2.730.000	4.500.000	1.720.000				
6	Average revenue			3.539.000	Average revenue			2.086.421

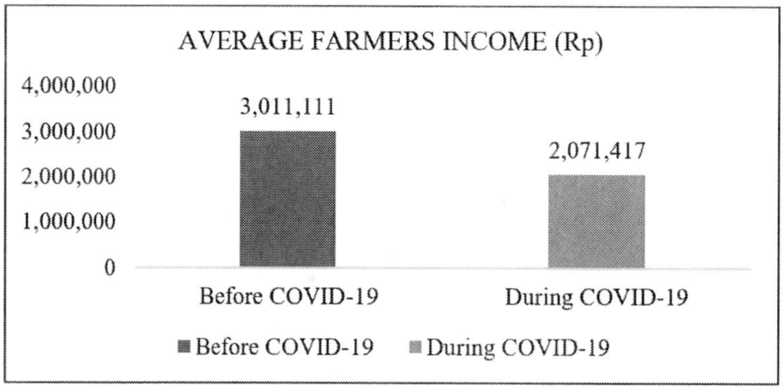

Figure 2. Average farmers income (IDR)--before and during COVID-19

Even though food is always needed by the community, due to PSBB (Large-Scale Social Restriction), which is carried out in several areas, UB Forest magersaren farmers experience problems during a pandemic. For example, farmers cannot freely sell their crops. Suddenly the Coronavirus made people's purchasing power decrease, accompanied by a reduction in commodity prices because supply increased, but purchasing power decreased. Some farmers continue to harvest and sell their crops even though lower prices and not a few farmers leave their crops not harvested until they die. To keep earning income during a pandemic, they rely on coffee plants that they have planted for a long time to be harvested. According to the farmers interviewed, the price of coffee is not too affected by this pandemic. The income of farmers as long as Covid-19 was declined 39 % from the normal condition. Before Covid-19, farmers' income was IDR 3.539.000, then after IDR 2.086.421 (Table 4 and Figure 2).

The farmers' existence in this forest area will be optimized according to their abilities and capacities to assist in the planning and management of UB Forest. The community is expected to help the UB Forest carry out supervision in social, economic, and environmental settings. Besides, the location of Sumbersari Magersaren is usually used by UB Forest to facilitate students to conduct research, practicum, and other activities that support education. The Magersaren community of Sumbersari Magersaren helped to meet the adequacy of facilities provided in the forest area, such as water, tools or working tools, etc.

5. Conclusion

UB Forest management's sustainability depended on knowing the real characteristic itself, how the actors involved, and the regulation implemented. Priorities strategy were: 1) Optimize UB Forest management; 2) Human resources development in Tridharma implementation; 3) National and International support for the forest sustainability, and 4) Increasing Information and Technology. The handover farmers consisted of Magersaren and Pesanggem managed by Tridharma University implementation-education, research, and community service.

Farmers adaption as long Covid-19 on forest management at UB Forest based on bundles of rights as a Claimant. It means that farmers had two right as below: 1)access and withdrawal right (enter the land, cultivate agriculture land, land conservation, harvest the agricultural yield, watch the forest fire along with UB Forest Team, implementation innovation of UB civitas academica, maintain the tree in UB Forest area, including in education and research of UB) and 2)management right on an agricultural plantation under the tree (a decision the species of agricultural plant, manage the plant, and arrange the cycle of the plantation itself). The income of farmers as long as Covid-19 was declined 39 % from the normal condition.

References

[1] Hussain T GS Kahn SA Khan N Masood M Ashfaq and N Sarwer 2010 Farmers' Agroforestry in Pakistan Farmers' Role-Trends and Attitudes Current Research *Journal of Social Sciences* **4**(1) 29-35, 2012 ISSN: 2041-3246.

[2] Jatana G and Jegeye P 2017 Farmer's Participation in Participatory Forest Management and Factors Affecting its Performance (The Case of Sodo Zuriya District Wolaita Zone Ethiopia) *Journal of Economics and Sustainable Development* www.iiste.org ISSN 2222 1700 (Paper) ISSN 2222-2855 (Online) **8**(9),

[3] Kustanti A Eko GS Yayuk Y and Syed AR 2017 *Forest Management Planning in Education and Training Forest at The University of Brawijaya (UB Forest) Malang East Java: A Case From Indonesia* The 14TH HISAS Hokkaido Indonesian Student Association Scientific Meeting Sapporo March 18-19 2017 ISSN-2597-8640

[4] Kustanti A 2017 *The Strategy on Education Forest of UB Forest Management and Mangrove Forest of LMC* Prosiding Seminar Nasional Pembangunan Pertanian II (Arah dan Tantangan Pembangunan Pertqanian dalam Era SDG's) Fakultas Pertanian Universitas Brawijaya Malang 25 November 2017 ISBN: 978-602-60456-5-2 p: 285-289

[5] Meliala D P 2017 *Participation Of Farmer Group in the Management of Forest Area with Special Purpose (KHDTK) UB Forest (In Sumbersari Village Tawangargo Village Karangploso Sub-district Malang Regency)* Thesis Not publication

[6] Nurdiansah G B 2017 *Transfer of Ownership Rights of Forest Management from Perhutani to Brawijaya University in Karangploso Subdistrict Malang Regency* Ostrom E 1999 *Institutional Rational Choice: An Assessment of the IAD Framework* In Paul Sabatier ed Theories of the Policy Process Boulder CO: Westview Press

[7] Peluso N L 2006 *Hutan Kaya, Rakyat Melarat: Penguasaan Sumberdaya dan Perlawanan di Jawa* (N. Fauzi, Penyunt.) Jakarta: Konphalindo atas dukungan Ford Foundation

[8] Saraswati Y and Dharmawan A H 2014 *Resiliensi Nafkah Rumahtangga Petani Hutan Rakyat di Kecamatan Giriwoyo Wonogiri* Sodality: Jurnal Sosiologi Pedesaan **2**(1) 63–75 DOI: http://dx.doi.org/10.22500/sodality.v2i1.9413

[9] Sugiharto B, Asihing K Eko GS Sudarto 2016 *Masterplan Kawasan Hutan dengan Tujuan Khusus (KHDTK) UB Forest Tahun 2016-2030* Universitas Brawijaya Malang

[10] Sunito S and Saharudin 2001 *Local Farmer Organisations In Upland Natural Resource Management In Indonesia* Center of Agrarian Study of the Agricultural University of Bogor Bogor, Indonesia page 7 line 10

The feasibility of environmentally friendly and conventional shallot farming in Bantul Regency

D R Kamardiani[1,*], M Fauzan[1], K S Riansyah[1]

[1] Department of Agribusiness, Faculty of Agriculture, Universitas Muhammadiyah Yogyakarta

* E-mail: kamardiani@umy.ac.id

Abstract.This study analyses the differences in costs, profit, and feasibility of environmentally friendly and conventional shallot farming in Bantul Regency. Data were collected from 128 farmers taken by cluster sampling and analyzed quantitatively using profit analysis and R/C ratio. The results showed that the total production cost of environmentally friendly shallot farming was 12.29% higher than conventional shallot farming, but conventional shallot farming had a higher profit of 41.94%. Both environmentally friendly and conventional farming is feasible, but the R/C ratio of conventional shallot farming is higher by 0.7.

1. Introduction

Shallots (*Allium oscalonicum* L) are strategic commodities that drive the national development of horticultural crops. Shallot production in Indonesia in 2015 was in four production center areas, namely Central Java, East Java, West Java, and West Nusa Tenggara. Shallot production in production centers applies a conventional agricultural system with intensive use of fertilizers and pesticides. Data from the Indonesian Organic Agriculture System in 2017 shows that there is no organic shallot production in production center areas.

The conventional agricultural system is an agricultural system that still uses chemical fertilizers or pesticides, and the organic farming system is a holistic production management system to improve and develop agroecosystem health, including biodiversity, biological cycles, and soil biological activities [1]. An environmentally-friendly agricultural system is a sustainable farming system that aims to increase and maintain high productivity by keeping the supply of nutrients from organic matter, minimizing dependence on inorganic fertilizers, improving soil biota, and controlling plant pests based on ecological conditions and diversifying crops.

Yogyakarta Special Region is not the largest shallot production center in Indonesia, but Bantul Regency, especially Selopamioro Village, Imogiri District, has implemented an environmentally friendly agricultural system. In addition, production growth was higher than the production centres of Central Java, West Java, and East Java, reaching 39.12% in 2016 [2]. Selopamioro village farmers do not use toxic chemical pesticides that are harmful to human health or the environment. In June 2018, shallot farmers from Nawungan Hamlet, Selopamioro Village, had their first harvest of environmentally friendly shallot farming. Farmers in Selopamioro Village, especially Nawungan Hamlet, are different from other farmers in Bantul Regency. They have used chemical pesticides excessively, which causes lindan organochlorine residues to exceed the maximum residue limits.

The term "environmentally friendly system" is deemed more appropriate than "organic system" because farmers have not fully implemented the organic farming system. It is only limited to the use of organic fertilizers and vegetable pesticides, but the land is not yet free from chemical residues from fertilizers and pesticides. This is because farmers still apply conventional farming systems to rice plants' cultivation, which are planted before shallots. The cropping pattern implemented by farmers in Selopamioro Village is rice-shallots-shallots. Shallots are planted in April / May for the first planting season and July / August for the second planting season. An environmentally friendly or conventional agricultural system implemented by farmers in the Bantul Regency affects production facilities, costs, and shallots' production.

From the above explanation, it is important to know the differences in costs, profits, and feasibility of environmentally friendly shallot farming with conventional shallot farming in Bantul Regency. The results of this study will be very useful as a material for consideration in making decisions about the use of shallot farming technology, given that there are not many onion cultivation systems that are organic or environmentally friendly, especially in Indonesia. In addition, there are not many scientific publications related to shallot farming that do not use chemical pesticides.

2. Methods

This research was conducted in Bantul Regency, Yogyakarta Special Region, Indonesia. The location of this research was determined purposively by the research objectives. An environmentally-friendly system is found in Selopamioro Village, Imogiri Districts, a hilly area, while the conventional method is in Parangtritis Village, Kretek District, which is in the lowlands. The two locations are shallot production centers in Bantul Regency. The determination of the sample of farmers was carried out using the cluster sampling technique. Data collection was conducted on 128 respondents using a survey method in August-October 2020 during the harvest season. Data analysis using descriptive analysis.

The differences in costs, revenues, profits, and feasibility of green and conventional shallot farming, using the cost, profit, and feasibility formula (R / C Ratio) as follows:

$$TC = TEC + TIC$$

TC is the total cost, TEC is the total explicit cost, and TIC is the total implicit cost.

$$\pi = TR - \mathrm{TC}$$
$$TR = P.Q$$

Where π is profit, TR is total revenue, TC is the total cost, P is product price, and Q is the amount of production (kg) calculated on a land area of 1,000 m^2.

$$R/C\ Ratio = \frac{TR}{TC}$$

Where TR is total revenue, and TC is the total cost. With the following criteria: if R / C> 1, then shallot farming is feasible to cultivate, if R / C <1, then shallot farming is not feasible to cultivate, and if R / C = 1, then red onion farming is not profitable or not loss [3].

3. Results and discussion

3.1. Characteristics of Shallot Farming in Bantul Regency

Shallot farming in Bantul Regency is generally cultivated in narrow land measuring 1,000-2,000 m^2 by farmers over 50 years of age with more than 20 years of experience running their farms. About 31% of conventional shallot farmers are elementary school graduates for education level, and 46.88% for environmentally friendly shallot farmers. Most of the farmers are landowners, and a small proportion of them are farmers who rent other people's land.

Table 1. Characteristics of shallot farming in Bantul Regency

Items	Environmentally Friendly Shallot Farming	Conventional Shallot Farming
Age (year)	50,91	52,58
Farming experience (year)	20,77	23,42
Land area (m²)	1.453	1.824
Seeds (kg/1.000 m²)	105	99
Manure (kg/1.000 m²)	1.844	-
Chemical fertilizer (kg/1.000m²)	108	245,51
Liquid fertilizer (ml/1.000 m²)	119,9	163,34
Solid fertilizer (kg/1.000 m²)	0,51	1,019
Labor (HKO/1.000 m²)	46,04	31

The average land tenure for environmentally friendly shallot farmers in Bantul Regency is 1,453 m^2, narrower than the land tenure for conventional shallot farmers which is 1,824 m^2 (Table 1). However, the average use of environmentally friendly shallot seeds is higher than traditional because the spacing is tighter, which is about 15 x 15 cm. In comparison, the spacing of conventional shallots is 15 x 20 cm. The main characteristic of environmentally friendly shallot farming is the use of manure. The high use of manure on environmentally friendly shallots is 1,844 kg/1.000 m^2, which is 84.4% more than ecologically friendly shallots in Tegal Regency [4].

Meanwhile, conventional shallot farming does not use manure. The use of chemical fertilizers in environmentally friendly shallot farming is 43.99% lower than traditional shallot farming. Less use of chemical fertilizers because farmers have used manure. However, the use of labor is higher than conventional shallot farming.

3.2. *Shallot farming costs*

Farming costs are all expenses or sacrifices used in farming activities, including labor value in the family and interest on own capital. Two kinds of prices in the production process must be incurred, namely explicit and implicit costs [5]. The explicit cost is the expenditure incurred by farmers to buy production factors, which refers to real spending regarding the purchase or procurement of input needs. Meanwhile, implicit costs are the opportunity costs of using production factors owned by farmers in the production process.

Table 2. The cost of shallot farming in Bantul Regency

Cost Items (Rp/1.000 m²)	Environmentally Friendly Shallot Farming		Conventional Shallot Farming	
	Cost (Rp)	Percentage (%)	Cost (Rp)	Percentage (%)
Seeds	4.798.844,30	40,42	4.009.741,63	37,93
Fertilizer	1.188.645,30	10,01	956.706,49	9,05
Pesticide	48.272,17	0,41	218.065,28	2,06
Irrigation	1.011.431,33	8,52	829.074,87	7,84
Labor	3.246.551,20	27,35	2.797.879,28	26,47
Deprication	316.944,64	2,67	553.523,89	5,24
Land rent	508.398	4,28	694.119	6,57
Other cost	752.052,16	6,34	512.191,26	4,85
Total cost	11.871.139,53	100	10.571.301,84	100

The results showed that the largest proportion of costs in environmentally friendly shallot farming in Bantul Regency was the seed production factor of 40.42%. The same results can also be seen in conventional shallot farming, where the seeds occupy the largest portion of production costs, which is

37.93%. According to many other research results, this finding states that the cost of sources generally occupies the largest amount of the fees of farming shallots [6,7,8]. However, this is different from the Central Java Province Agricultural Technology Research Center's findings, which states that the cost of fertilizer occupies the largest proportion, which about 42.77%, followed by seed costs of 35.25% for environmentally friendly shallots in Tegal Regency [4].

The second-largest cost is the labour cost for both environmentally friendly and conventional, but the proportion is greater for environmentally friendly shallots. This result is different from the labour cost in shallot farming in Batu City occupies the largest position, which is 47.31% [9]. It also happens for environmentally friendly shallot in Tegal Regency, where labour cost is in the third position with 23,68% in number [4] and other shallot research [10]. It shows that shallot farming is quite labour-intensive.

The average cost of fertilizers used in environmentally friendly shallot farming is 0.06% higher than conventional shallot farming. It is quite interesting because environmentally friendly shallot farming only uses organic fertilizers and does not use chemical fertilizers. However, it turns out that the cost value is higher than conventional shallot farming that uses chemical fertilizers. It is the price that farmers with environmentally friendly systems must pay because their agricultural land quality will continue to improve in the long run. Meanwhile, conventional systems that use chemical fertilizers in large doses, in the long run, will reduce the quality of the soil and its environment.

The field findings show that environmentally friendly shallot farming still uses chemical pesticides, although in small amounts. It is seen in the very small proportion of pesticide costs, only 22.14% of the price of conventional shallot pesticides. Overall, the total cost of environmentally friendly shallot farming was 12.29% higher than traditional farming of shallot, and the results of t-test analysis showed a significant difference.

3.3. Profits of shallot farming

Profits are generally an indicator of the success of a farm [11]. Farming profits are closely related to several factors, including technology application, land tenure, input prices, and output prices [12]. If the increase in output prices received by farmers is not proportional to the rise in input prices that must be paid and is accompanied by inefficient farm management, it will result in low farm profits [13].

Table 3. Profits of shallot farming in Bantul Regency

Items	Environmentally Friendly Shallot Farming	Conventional Shallot Farming
Production (kg)	1.116,97	927,02
Price (Rp)	23.335	34.086
Total revenue (Rp/1.000 m^2)	26.627.866	31.516.592
Total implicit cost (Rp/1.000 m^2)	3.018.457	1.646.518
Total explicit cost (Rp/1.000 m^2)	8.852.683	8.924.784
Total profit (Rp/1.000 m^2)	14.756.726	20.945.290

The results showed that the profit of conventional shallot farming in Bantul Regency is 41.94% higher than environmentally friendly shallot farming, as shown in Table 3. The low profitability of environmentally friendly shallot farming is due to the low selling price and the relatively high implicit costs, although environmentally friendly shallots production are higher. The low selling price of ecologically friendly shallot farming happens because the harvest season coincides with the national shallot production center's main harvest season in Brebes Regency in June-August. Meanwhile, the conventional shallot harvest season is after the national shallot production center's main harvest in Brebes Regency in October 2020, so the selling price is higher.

The profit of environmentally friendly shallot in Bantul Regency is lower than the ecologically friendly shallot in Tegal Regency [4] but is higher than the profit of conventional shallot in Batu City [14]. The result shows that environmentally friendly shallot farmers are quite successful in managing their farming and can still be improved. The difference in the profit value of shallot farming in various regions in Indonesia is generally caused by differences in planting seasons, seed varieties, allocation of fertilizers and pesticides, selling prices, and technology application by farmers.

3.4. Feasibility of shallot farming

The results showed that the R/C ratio for environmentally friendly and conventional shallot farming were 2.22 and 2.92, respectively, as shown in Table 4.

Table 4. Feasibility of shallot farming in Bantul Regency

Items	Environmentally Friendly Shallot Farming	Conventional Shallot Farming
Total revenue (Rp/1.000 m^2)	26.627.866,02	31.516.591,71
Total cost (Rp/1.000 m^2)	11.871.139,53	10.571.301,84
R/C ratio	2,22	2,92

Based on the criteria for the R / C ratio value, a value greater than one is obtained so that it can be interpreted that shallot farming is environmentally friendly and conventional is feasible to be cultivated, although conventional farming is 0.7 greater than environmentally friendly shallot farming. This result is different from environmentally friendly shallot farming in Tegal Regency [4], which stated that the R/C ratio of environmentally friendly shallot farming is higher than conventional ones. It can be explained that with the capital of Rp. 11,871,139.53 (environmentally friendly) and Rp. 10,571,301.84 (conventional) can get sales results of 2.22 times and 2.92 times the total capital. The value of this R/C ratio is higher than the shallot production areas in Junrejo District [9], Cirebon Regency [10], and Majalengka Regency [15].

4. Conclusion

The results showed that the total production cost of environmentally friendly shallot farming was 12,29% higher than conventional shallot farming, but conventional shallot farming had a higher profit of 40.48. Both of two farming methods are feasible, but the R/C ratio of conventional shallot farming is higher by 0.7. Environmentally friendly shallot farming can be developed in other regions by observing the planting season, which does not coincide with the planting season at the national shallot production centers for a better selling price so that higher profits can be obtained.

References

[1] Wihardjaka A 2018 Penerapan model pertanian ramah lingkungan sebagai jaminan perbaikan kuantitas dan kualitas hasil tanaman pangan *Jurnal Pangan* **27**(2) pp 155-164

[2] Badan Pusat Statistik 2018 *Produksi Sayur-Sayuran Menurut Jenisnya dan Kecamatan Tahun 2013-2017* Bantul Badan Pusat Statistik Kabupaten Bantul

[3] Suratiyah K 2015 *Ilmu Usahatani Edisi Revisi* Yogyakarta Penebar Swadaya

[4] Mardiyanto T C, Prastuti T R, and Pangestuti R 2017. Analisa kelayakan usaha budidaya bawang merah ramah lingkungan di Kabupaten Tegal. *Agronomika* **12**(01) pp 34-39

[5] Sitorus S F J and Ramli R 2013 Analisis efisiensi faktor produksi padi sawah dalam rangka ketahanan pangan di Desa Tumpatan Kec Beringin Kab Deli Serdang *Jurnal Ekonomi dan Keuangan* **1**(10) pp 35-48

[6] Azmi C, Hidayat I M, and Wiguna G 2016. Pengaruh varietas dan ukuran umbi terhadap produktivitas bawang merah *Jurnal Hortikultura* **21**(3) pp 206-213

[7] Aldila H F, Fariyanti A, and Tinaprilla N 2015 Analisis profitabilitas usahatani bawang merah berdasarkan musim di tiga kabupaten sentra produksi di Indonesia. *SEPA: Jurnal Sosial Ekonomi Pertanian dan Agribisnis* **11**(2) pp 249-260

[8] Lawalata M, Darwanto D H, and Hartono S 2015. Efisiensi relatif usahatani bawang merah di Kabupaten Bantul dengan pendekatan data envelopment analysis. *Ilmu Pertanian*

(Agricultural Science) **18**(1) pp 1-8

[9] Eka H A, Ilmi P I, and Sugeng R 2019 Implementation of farm management and financial feasibility evaluation of shallot (Allium Ascalonicum L.). *Russian Journal of Agricultural and Socio-Economic Sciences*, **92**(8)

[10] Hamdani K K, Nurawan A, Rachman A, and Dianawati M 2020. Kajian usahatani bawang merah di Kecamatan Pabuaran Kabupaten Cirebon *Prosiding Seminar Nasional Teknologi Pertanian Kesiapan Sumber Daya Pertanian dan Inovasi Spesifik Lokasi Memasuki Era Industri 4.0* pp 339–346

[11] Fauzan M 2016 Pendapatan, risiko, dan efisiensi ekonomi usahatani bawang merah di Kabupaten Bantul. *AGRARIS: Journal of Agribusiness and Rural Development Research* **2**(2) pp 107-117

[12] Kay R D, Edwards W M, and Duffy P A 2008 *Farm Management (Sixth Edition)* New York McGraw-Hill International Edition

[13] Fauzan M 2020. Pendapatan rumah tangga petani bawang merah lahan pasir pantai di Kabupaten Bantul. *JAS (Jurnal Agri Sains)* **4**(1) pp 60-66

[14] Maharani N 2019. Pendapatan usahatani bawang merah di Kecamatan Junrejo Kota Batu *Jurnal Ilmiah Hijau Cendekia* **4**(2) pp 70-73.

[15] Rahmadona L, Fariyanti A, and Burhanuddin B 2016. Analisis pendapatan usahatani bawang merah di Kabupaten Majalengka *Agricultural Socio-Economics Journal* **15**(2) p 72.

IOP Conf. Series: Earth and Environmental Science **883** (2021) 012071 doi:10.1088/1755-1315/883/1/012071

Contribution of agrosilvopastoral pattern to sustainable communities economic improvement (study in Namaa Hamlet, Pelauw Village, Haruku Island Sub-District, Central Maluku Regency)

M Sahureka[1,*], M Hadijah[1], H Marasabessy[1]

[1] Forestry Department, Agriculture Faculty, Pattimura University, Ambon, 97232

* E-mail: mersisahu@gmail

Abstract. The agrosilvopastoral management pattern can provide multiple benefits to integrating the agriculture sector with the interrelated forestry and livestock sectors. This research in Namaa sub-village, Pelauw Village.The research aimed to : 1) Knowing the respondent's characteristics in agrosilvopastoral management, and 2) The contribution of the agrosilvopastoral pattern to improve community's economy. This research was conducted by using the purposive sampling method, with a representation of the 20 households of farmers. Primary and secondary data were taken by observation and interviews related to respondents' income and the agrosilvopastoral pattern. The data obtained were analyzed with a quantitative approach and qualitative descriptions. The results showed that the respondents' characteristics were the factors that influenced the management of agrosilvopastoral, namely the level of education, age level, number of family members, land area, and land ownership status, while the contribution of an agrosilvopastoral for respondents was 274,882,000 / year.

1. Introduction

Agrosilvopastoral is a combination of components or agricultural activities with forestry and livestock/animal. This system's purpose is an integrated process between sectors in the farming industry that mutually utilizes the residue of an industry's management process, reused in other sectors to produce other useful benefits. Many things must be considered in the agrosilvopastoral application to support its sustainability, namely market conditions, climate, soil types, and types selection. The selection of varieties is intended for high value, including fast-growing species, deep roots, capable of providing products, and providers of environmental services [1].

The agrosilvopastoral management practice in Maluku is carried out at *dusung* locations, a form of traditional agroforestry in Maluku [2]. The *dusung* management activities usually begin with the preparation stage. This stage consists of determining the location, burning the land, clearing the land, planting, maintaining, and harvesting. It requires a long time, extra effort, patience to wait for the harvest, and harvest management skills. The agrosilvopastoral pattern can be used to continue farmer activity because farmers' activities in *dusung* will be more numerous and varied. Apart from farming activities such as hoeing, planting, cleaning lianas, and harvesting, farmers can also raise livestock by feeding, foraging for livestock, or cleaning pens.

Agrosilvopastoral or hybrid systems, namely agroforestry systems, combine tree crops (forest), agricultural crops, and forage crops and/or livestock [3]. (The application of the agrosilvopastoral management pattern in this *dusung* is very beneficial for improving the community's economy because the results can guarantee short-term, medium-term, and long-term needs. Besides, the system produces various foodstuffs for consumption, namely vegetable food needs and animal food needs to ensure the community's food needs. In general, the dusung farming pattern's contribution to the community has 3 benefits for the community's food security: 1) providing for family food, 2) providing cash, and 3) increasing family income. The *community widely practices the dusung farming pattern* because of several main reasons a) the *dusung* has various types of plants with different harvest seasons, b) the *dusung* provides cash to its owner like a bank, c) the location of the *dusung* is usually not far from the residential area so that it is easier to maintain [4]. It means that the agrosilvopastoral pattern as a component of traditional agroforestry provides a positive contribution from an economic perspective to improve farmers' welfare [5].

Namaa Hamlet is a part of Pelauw village. It is located in the Central Maluku Regency, where most of the people are farmers. Farmers in the Namaa hamlet still apply a forest and land management pattern based on the agrosilvopastoral design. This management pattern is a legacy and tradition passed down from one generation to the next. [4] [6] [7]. The *dusung* owned by farmers is an inheritance and is located not too far from the settlement [4] to allow farmers to have plenty of time to manage the *dusung* [6]. If This management pattern is made seriously and on a large scale, it will significantly impact the family economy and sustainable family food availability.

Based on the description above, what needs to be done to improve the community's economy is the maximum application of agrosilvopastoral in terms of resources and management to ensure food security, safety, and quality. It is to strengthen food availability and provide an increase in farmers' income and welfare. Therefore, it was necessary to research the title: Contribution of Agrosilvopastoral Pattern to Sustainable Communities Economic Improvement (Study in Nama Hamlet, Pelauw village, Haruku Island Sub-District, Central Maluku Regency). The research objectives were to know: 1) the characteristics of farmers in Agrosilvopastoral management, and 2) the contribution of the agrosilvopastoral pattern to the increasing of the community's economy.

2. Methods

This research was conducted in the Namaa hamlet Pelauw village, which took place from May to June 2018. The sample selection of the research location was carried out using the purposive sampling method, namely the representation of farmer elements, namely as many as 20 families. Also, several community leaders who could provide information on the application of the agrosilvopastoral pattern were taken. Data collected includes primary data obtained from respondents through interviews and observations and secondary data obtained from related parties.

The data obtained were analyzed using a qualitative approach, namely an analytical technique to describe a phenomenon, explain relationships, test hypotheses, and determine the implications of a problem under study. To get the amount of contribution of the agrosilvopastoral pattern to income, the researchers used the following formula [8]:

Formulas used

$$P = \frac{X}{Y}\ 100\%$$

Where :

P = income precentage

X = Income derived from the Agrosilvopastoral pattern

Y = Total of respondents' income

3. Results and discussion

3.1. Characteristics of Respondents in the Management of Agrosilvopastoral

3.1.1. Respondent Education

The social characteristics of the community greatly influence activities in the management of agrosilvopastoral. These characteristics include education level, income, age, number of dependents of family members, and distance between settlement (house) to the location of the *dusung*. The characteristics of the respondents based on their education level are as follows :

Table 1. Characteristics of Respondents Based on Education Level

Level of education	Number of Respondents	Percentage (%)
Elementary School	2	10
Junior High School	4	20
Senior High School	14	70
Bachelor degree	-	-
Total	20	100

Based on the table above, it can be seen that the education level of the respondents is the highest school in general with 14 respondents, and the lowest is an elementary school with one respondent. The high and low level of a person's education significantly affects their agrosilvopastoral management mindset and behavior. The attitude and behavior are about what must be planted and maintained, as well as the application of new technology/innovation in the fields of agriculture, forestry, and animal husbandry.

3.1.2. Respondents' Income

Family income dramatically affects the level of fulfillment of basic family needs. Family income is the total income that the family receives from various sources of income, both agricultural and non-agricultural. The respondents' income based on the following income level:

Table 2. Characteristics of Respondents Based on Level of Income

Income level/month	Number of Respondents	Percentage (%)
≤ 500.000	1	5
500.000- 1.000.000	9	45
1.000.000 – 2.000.000	8	40
2.000.000 – 3.000.000	2	10

The table above shows the highest income of the respondents, namely Rp. 2,000,000 to Rp. 3,000,000 / month for two respondents. Meanwhile, the income range of Rp. 1,000,000 to Rp. 2,000,000 / month is eight respondents, followed by nine respondents with income in the range of Rp. 500,000 to Rp. IDR 1,000,000 / month and only one respondent with the lowest income, namely ≤ 500,000. The respondents' income depends on the type and number of plants, livestock being raised, and side jobs.

3.1.3. Respondents' Age Level

The age level of a person also affects the activities and productivity in managing the *dusung*. Age level also influences the decision/policy-making process, and the absorption of technology/innovation developed. The respondents' age level also affects the division of tasks in agrosilvopastoral management, and their involvement can be measured in each process. Age levels of respondents in agrosilvopastoral management in *dusung* Namaa generally range from 26 years to 70 years. The characteristics of the respondents are as follows:

Table 3. Characteristics of Respondents Based on Age Levels

Age Level (Years)	Number of Respondents	Percentage (%)
19 - 25	-	
26 – 50	16	80
51 – 70	4	20
Total	20	100

Table 3 shows that respondents aged 26-50 years are the most dominant in terms of quantity and strength of work because 26-50 years is a productive age where a person is very busy with routine daily work to meet their daily needs. Generally, a person is married at this age and has the responsibility to support his family in farming and caring for livestock in the *dusung*.

3.1.4. Number of Family Members

Family members are internal workers in the family who can help complete daily tasks. Likewise, the involvement of family members in agrosilvopastoral management activities in the Namaa hamlet, Pelauw village, is very beneficial. In addition to fulfilling the needs of family members, working-age family members can also help work in *dusung*, both men and women. The following is a table regarding the characteristics of the respondents based on the number of family members

Table 4. Characteristics of Respondents Based on Number of Family Members

The number of dependents	Number of Respondents	Percentage (%)
≤ 4 people	2	10
5-6 people	12	60
≥ 7 people	6	30
Total	20	100

Based on Table 4, it can be seen that the respondents with the most number of family members were seven people, *i.e.*, six respondents and the least number were four people with the number of 2 respondents. Most of the respondents, namely 60% or 12 respondents, had 12 family members as dependents. The number of family members can help work in sharing tasks and roles. Usually, women do light work such as pulling weeds, planting, and looking for firewood, while men usually hoe, make fences, harvest, and feed livestock. Thus the number of family members and gender in a family also dramatically affects the capacity and type of work done by a person in managing *dusung* with an agrosilvopastoral pattern.

3.1.5. The distance from the house to the respondent's Dusung

The distance between settlement and the *dusung* significantly affects the activities and types of plants planted in the *dusung*, as shown in Table 9 below:

Table 5. Characteristics of Respondents Based on the Distance from Home to Hamlet

Distance from Home	Number of Respondents	Percentage (%)
≤ 1 km	7	35
1 – 2km	13	65
≥ 2 km	-	-
Total	20	100

Table 5 shows that most of the respondents have a *dusung* that is not far from the settlement. About 35% of the respondents or seven respondents distance from the house to the agrosilvopastoral *dusung* as far as ≤ 1 km. It can be seen that most of the respondents also use the yard to carry out farming and raising livestock. Meanwhile, 65% of respondents, or 13 respondents, had a distance of 1-2 km from the house to the *dusung*. The road to the *dusung* location is still a path with flat conditions so that farmers were comfortable to go to the *dusung* with a travel time of about 30-60 minutes. It can be concluded that the distance and travel time will also affect supervision and activities in agrosilvopastoral management.

3.2. Application of Agrosilvopastoral *Pattern in Namaa Hamlet*

Agrosilvopastoral is a land management system that combines agriculture components with forestry and livestock/animals to solve land availability problems and increase land productivity, especially on marginal lands. The limitations of the agrosilvopastoral pattern in the Namaa hamlet, Pelauw village, are raising livestock without making a patent cage, lack of knowledge, management system, and limited capital. Therefore, the agrosilvopastoral management pattern is still carried out at a small scale level to fulfill daily life necessities and the celebration of major religious holidays so that the results given are not maximal [5]. On the other hand, farmers are not interested in doing business in the forestry sector since they must fulfill administrative problems to sell wood so that farmers are more likely to plant secondary crops.

In the management of agrosilvopastoral, the farmers' activities in *dusung* were varying, among others, as follows:

1. Planting: Before planting in the *dusung*, land preparation is usually carried out, namely cutting down trees and shrubs, cleaning the lianas, and burning them. After that, planting holes and planting with a combination of various types of long-life and short-life plants, such as forestry/timber, agricultural and plantation crops such as secondary crops.
2. Plant nurturing: at this stage, the activities carried out fertilizing, weeding, and clearing the land by removing grass/lianas and shrubs that grow between staple plants to provide space for plants to grow.
3. Harvesting; each plant has a different harvest age. Harvesting is carried out when the plants reach maturity, indicated by changes in the plants' physical properties, namely color and smell. When the harvest season comes, the farmers choose to stay in the *dusung* to collect the *dusung* products.
4. Raising livestock: in addition to carrying out activities related to the agricultural sector, in an agrosilvopastoral land, farmers also carry out livestock raising activities by looking for grass/animal feed, feeding livestock, and cleaning cages.

3.3. Contribution of the Agrisilvopastura Pattern

An area's ecological conditions will affect the composition of plants that make up agroforestry [3]. Land with an agroforestry pattern has a layered canopy structure that functions to reduce rainwater's kinetic energy. Based on observations, it was found that the main components of Agrosilvopastoral were mango (*Mangifera indica*), langsat (*Lansium domesticum*), durian (*Durio zibethinus*), jackfruit (*Artocarpus heterophyllus*) and teak (*Tectona grandis*), ironwood (*Instia bijuga*), lenggua wood. (*Petrocarpus indicus*). Agricultural and plantation crops are dominated by seasonal crops of chili

(*Capsicum frutescens*), eggplant (*Solanum melongena*), long beans (*Vigna unguiculata ssp*), and tomato (*Solanum lycopersicum*). Meanwhile, annual plants or longevity plants are dominated by cloves (*Eugenia aromatic*), nutmeg (*Myristica fragrans*), bananas (*Musa paradisiacal*), and coconut (*Cocos nucifera*). The types of livestock that are kept are cows, chickens, goats, and ducks.

From the results of the analysis, it can be seen that the agrosilvopastoral pattern carried out by farmers in Namaa hamlet provides a significant contribution to the community's economy, which guarantees a long and sustainable period. Furthermore, the contribution of the agrosilvopastoral pattern can be seen in the following table.

Table 6. Contribution of Agrosilvopastoral Patterns from Respondents in Each Sector

No	Agriculture	%	Livestock	%	Forestry	%	Total
1	8,504,000	84.50	1,200,000	11.92	360,000	3.58	10,064,000
2	1,200,000	4.43	25000000	92.37	864,000	3.19	27,064,000
3	4,285,000	39.58	6,300,000	58.20	240,000	2.22	10,825,000
4	4,985,000	31.26	10.360.000	64.97	600,000	3.76	15,945,000
5	4,350,000	43.11	5,500,000	54.51	240,000	2.38	10,090,000
6	1,100,000	8.03	12.300.000	89.78	300,000	2.19	13,700,000
7	1,330,000	18.83	5,300,000	75.05	432,000	6.12	7,062,000
8	5,425,000	24.87	16.100.000	73.81	288,000	1.32	21,813,000
9	1,250,000	9.35	11,400,000	85.27	720,000	5.39	13,370,000
10	920,000	9.89	8,200,000	88.17	180,000	1.94	9,300,000
11	900,000	13.98	5,300,000	82.30	240,000	3.73	6,440,000
12	6,670,000	38.40	10,300,000	59.30	400,000	2.30	17,370,000
13	1,805,000	17.86	8,000,000	79.17	300,000	2.97	10,105,000
14	1,725,000	78.95	340,000	15.56	120,000	5.49	2,185,000
15	1,475,000	15.79	7,000,000	74.95	864,000	9.25	9,339,000
16	950,000	3.01	30,000,000	95.16	576,000	1.83	31,526,000
17	2,510,000	12.89	16,000,000	82.18	960,000	4.93	19,470,000
18	7,755,000	53.05	6,000,000	41.04	864,000	5.91	14,619,000
19	2,515,000	23.43	8,100,000	75.45	120,000	1.12	10,735,000
20	3,000,000	21.77	10,540,000	76.49	240,000	1.74	13,780,000
Total	62,654,000	22.80	203,240,000	73.96	8,908,000	3.24	274,802,000

Based on Table 6 above, it can be seen that the livestock sector provides a very high contribution of IDR 203,240,000 / year or 73.96%. The agricultural and plantation sectors contribute IDR 62,654,000 / year or 22.80%, and the forestry sector gave the lowest contribution of IDR 8,908,000 / year or 3.24%. Therefore, the total contribution of the agrosilvopastoral pattern is IDR 274,802,000 / year.

The results showed that the livestock sector gave the largest contribution because farmers were more interested in raising cows and goats. After all, the price was quite high. For example, cows were sold for around Rp. 5,000,000 - Rp. 10,000,000 / tails adjusted to body size and market prices.

On the other hand, eggs and meat can be taken from chickens and ducks for sale or consumption. Meanwhile, for agriculture and plantations, farmers are more interested in planting short-lived plants to provide quick yields than long-life crops or timber plants. The forestry sector contributes to the sale of firewood for IDR 12,000/ inch. At the same time, commercial wood such as teak (*Tectona grandis*) has not yet entered the cutting period as well as lenggua wood (*Petrocarpus Indicus*) and ironwood (*Instia bijuga*). The farmers only do the cutting according to need/cutting to hardly provide nominal value within a year for community income. One of the efforts made to increase the interest and awareness of farmers in managing agrosilvopastoral. Therefore, the Pelauw village government in 2016 carried out various types of plants in *dusung* by planting teak wood (*Tektona grandis*), procuring agricultural equipment as well as counseling and socialization activities on the management of agrosilvopastoral.

In general, the application of the agrosilvopastoral pattern in the Namaa hamlet has several advantages. From an ecological perspective, land productivity with high yields is due to plant species' diversity (polyculture). It has a sustainable harvest season compared to monoculture crops. With a mixed crop pattern, the productivity of forest land can be increased optimally and sustainably. Besides that, it is also possible that the regeneration of nature and wildlife also supports plants' combination. The natural interactions of agroforestry components are easy to identify. The role of stands as a provider of wild animal feed (for example, fruits for various bird species) and vice versa is a simple example. Also, wildlife's function can be the process of pollination or stand regeneration and a source of animal protein for farmers [10].

Suppose the practice of agrosilvopastoral management is taken seriously. In that case, it will provide several benefits, namely: 1. the needs of the plant and animal food as well as shelter, 2. to ensuring long-term, medium-term and short-term needs, and 3. ensuring the stability of community income. It means that the agrosilvopastoral pattern provides sustainable benefits for the economic improvement and welfare of farmers [5] [6] [11]. Thus, it can be said that agrosilvopastoral is a very complicated land management pattern because it combines the agricultural, forestry, and livestock sectors. For this reason, for the development of agrosilvopastoral practices in the future, it is necessary to have intensive supervision and guidance from stakeholders to farmers. Farmers need to be guided through training to get better and mastering agrosilvopastoral techniques. The farmers also need counseling and advice regarding forest management to produce high-quality timber plants and make a more significant contribution to farmers' income [11].

4. Conclusion

(a). Characteristics of the respondents, namely education level, income level, age level, number of family members, distance from house to *dusung,* also influence community activities in agrosilvopastoral management.

(b). The agrosilvopastoral pattern contribution is Rp. 274,802,000 / year, with details of the livestock sector giving significant influence with Rp. 203,240,000 / year or 73.96%, then the agricultural and plantation sectors contribute Rp 62,654,000 or 22.80% and the forestry sector give the lowest contribution, namely IDR 8,908,000 / year or 3.24%.

References

[1] Clason T 1998 An On Going Study To Understand Tree, Forage, and Livestock Systems Inside Agroforestry, *USDA National Agroforestry Center. Nebraska.*

[2] Ajawailla J W 1996 Potensi Sosial Budaya Agroforestry Dusung [The Socio-Cultural Potential of Dusung Agroforestry.] *Pusat Studi Maluku, Universitas Pattimura Ambon*

[3] Abdul Rauf 2017 System agroforestry,Upaya Pemberdayaan Lahan Secara berkelanjutan [Agroforestry systems, sustainable land empowerment efforts.] *Edisi Revisi, USU Press*

[4] Matinahoru J 2007. Kontribusi Pola Pertanian *Dusung* bagi Ketahanan Pangan Masyarakat Maluku [Contribution of *Dusung* Agricultural Patterns for Food Security of the Maluku People.] *Jurnal Agroforestry* **2**(2).

[5] Sahureka M, Marassabessy M, Hadijah M 2019 Agrosilvopastoral Sebagai Model Pemanfaatan Lahan Bagi Peningkatan Pendapatan Masyarakat di Dusun Namaa Negeri Pelauw Kecamatan Pulau Haruku [Agrosilvopastura As a Model of Land Use for Increasing Community Income In Dusun Namaa Hamlet Pelauw Village Haruku Island Sub Distric.] *Prosiding Seminar Nasional Perhutanan Sosial Dalam Pengelolaan Hutan Berkelanjutan di Wilayah Kepulauan Fakultas Pertanian Universitas Pattimura Ambon*

[6] Sahureka M, Talaohu M 2018 Pengelolaan Agroforstry Tradisional *Dusung* Berbasis kearifan Lokal *Masohi* [Dusung Traditional Agroforestry Management Based on Masohi Local Wisdom.] *Jurnal Hutan Pulau-pulau Kecil* **2**(2).

[7] Silaya Th M 2015 Kearifan Masyarakat Lokal Dalam Pengelolaan Sumber Daya Hutan di Wilayah Pegunungan Masela Seram Utara [Wisdom of Local Communities in Managing Forest Resources in the Masela Seram Mountains Region of Utara*], Jurnal Makila Vol IX Nomor 1 Jurusan Kehutanan univeritas Pattimura*

[8] P Iskandar 2005 Teory Mikro *M itra Wacana Media, Jakarta,*

[9] Prayogo, Hamdani Fauzi , D Naemah 2020 Analisis Sosial Dan Ekonomi Masyarakat Dalam Penerapatan Pola Agroforestry Pada Hutan Kemasyarakatan (Studi Kasus Desa Tebing Siring, Kabupaten Tanah Laut) [Social and Economic Analysis of Communities in the Application of Agroforestry Patterns to Community Forests (Case Study of Tebing Siring Village, Tanah Laut District)] *Jurnal Sylva Scienteae* **03**(4) ISSN 2622-8963.

[10] Henny M, Azhari 2011 Pengembangan Agroforestry untuk Mendukung Ketahanan Pangan dan Pemberdayaan Petani di Sekitar Hutan [Agroforestry Development to Support Food Security and Empowerment of Farmers Around the Fores.t] *Forum Penelitian Agro Ekonomi* **29**(22).

[11] Ma'ruf A 2017 Agrosilvopastoral Sebagai Sistem Pertanian Terencana Menuju Pertanian Berkelanjutan [Agrosilvopastoral as a planned agricultural system towards sustainable agriculture.*] Jurnal Penelitian Pertanian Bernas* **13** (2) ISSN 0216-7689 Fakultas Pertanian Universitas Asahan . tanggal akses 18 November 2020

Correlation of normalized difference water index and baseflow index in small island watershed landscapes

B Latuamury[1,*], M Talaohu[1], F Sahusilawane[1], W N Imlabla[1]

[1] Department of Forestry, Faculty of Agriculture, Pattimura University, Ambon 97123, Indonesia

* E-mail: okky.environmentalscience@gmail.com

Abstract. The utilization of remote sensing data in the field of environmental hydrology is experiencing rapid progress. The Normalized Difference Water Index (NDWI) approach to transforming the water content of various land cover types and its implications for small island watersheds' hydrological characteristics is essential. NDWI is an algorithm used to detect water bodies, with the capacity to absorb visible and infrared wavelengths strongly. This study aims to analyze the correlation between the NDWI water index and the BFI baseflow index in the small island landscape of Ambon City. The Landsat 7 ETM + and Landsat 8 OLI image processing methods use ENVI 5.3 software to transform the NDWI algorithm and the BFI + 3.0 digital recursive filtering (RDF) method for hydrological characterization. The results showed that there was a strong correlation between the NDWI water index and the baseflow index (BFI) for the small island watershed of Ambon city. This result is relevant to the geographic area of Ambon City, which is dominated by the ocean 95% and land area 5%, so the application of the NDWI water index and the hydrological conditions of small island watersheds are significant.

1. Introduction

Ambon City has 2/5 of Ambon Island's total area with a land area of 359.45 km^2 and an ocean area of 17.55 km^2 with a coastline length of 98 Km. Ambon City area's topographical conditions cover hilly land areas to steep slopes with a slope of >20° as much as 73%. The land area tends to be flat or sloping with a slope of <20° by 17%, and about 10% are beaches, coasts, and bays. The unique topographical conditions characterize the specific river characteristics of narrow watersheds (DAS) and the main rivers' relatively short lengths. The longest river is Way Sikula in Laha Village with a length of 15.5 km, while the shortest river is Way Tomu and Way Batu Gajah with a length of 1.5 km, which flows in the center of Ambon City [1].

According to the hydrometeorological disaster risk study, Ambon Island is at a score of 156, which is in the high-risk category. In the last five years, Ambon Island's flood hazard's vast potential has experienced floods and landslides, impacting infrastructure damage and casualties and deaths. The possibility of the population exposed to flooding includes the five watersheds of Ambon City due to the river border protection area's condition, crowded with human settlements, and many vulnerable groups.

The water index using the Normalized Differences Water Index (NDWI) is an index modified from the NDVI [2]–[4]. The water content index shows the water content of land recorded by the image [3]–[7]. NDWI and NDVI values are directly proportional. The denser the vegetation density in an area, the higher the recorded NDWI value. NDWI is an algorithm used for the identification of water bodies. Water bodies strongly absorb visible and infrared wavelengths of light. An NDWI value that is more than zero is assumed to be a water body, and if the NDWI is less than zero, it is considered land [8]–[11]. The water index transformation research by [12]–[16] proves that the water index method has advantages in terms of ease of use. Use and short processing time for coastline data acquisition. Some of the water indices used in this study are NDWI [3], [17], MNDWI [15], [16], [18], and AWEI [11], [15]. The use of the water index algorithm can adjust to specific research objectives. The advantages of the water index algorithm at a particular location can be optimally useful in other sites.

The Normalized Difference Water Index (NDWI), according to [2], [17], [19], [20], is a popular drought index for vegetation humidity. NDWI has been widely used for humidity and dryness identification. According to [9], [21]–[23], NDWI effectively monitors vegetation water content in various studies. [21], [24] conducted a comparison of NDVI and NDWI to monitor vegetation moisture and grassland drought, which resulted in NDWI being more sensitive to vegetation drought. According to [18], [21], [24], ISODATA performs the clustering and initial classes based on computers' statistical computations, then classifies pixels and modifies the criteria for each category. This process is repeated until the spectral interval (range) between classes is maximum.

The Landsat 8 satellite, known as the Landsat Data Continuity Mission (LDCM), was launched by NASA on February 11, 2013, at Vandenberg Air Force Base, California. This project is a continuation of the previous Landsat satellite mission known as the Continuity Mission. Landsat 8 is a refinement of Landsat 7 with improvements related to the spatial, temporal, and spectral resolution. The advantages of Landsat 8 are especially the specifications of the bands and electromagnetic wave spectrum. The Landsat 8 specifications are mainly in bands 1, 3, 10, and 11. Band 1 (ultra blue) has a lower electromagnetic wavelength than blue waves at a wavelength of 0.43 - 0.45 μm. Other researchers also find the superiority of this band in distinguishing aerosol concentrations in the atmosphere and identifying the characteristics of the appearance of seawater at different depths [3], [15], [25]–[27].

Cirrus cloud detection is also optimal in band 9 with an OLI (Operational Land Imager) sensor at a wavelength of 1.36 - 1.38 μm. Landsat 8 with a thermal sensor has two thermal bands: band 10 (wavelength of 10.6 - 11.19 μm) and band 11 (wavelength 11.5 - 12.51 μm). The spatial resolution of the TIRS (Thermal Infrared Sensor) sensor has a spatial resolution of 100 meters. However, it is resampled to 30 meters to facilitate multispectral data analysis (OLI). Besides, sensor calibration also produces noise, called noise-equivalent-change-in-temperature (NEΔT), as the maximum limit of the resulting error [11], [28]–[31].

The water index characteristics of various land cover types can accommodate the water index's detection in different land cover classes. This study analyzes the water index and hydrology characteristics of small island watersheds unique and vulnerable to disasters' hydrometeorological aspects, based on the advantages of satellite imagery and the water index in various environmental studies. Therefore, this study investigates the Water Index Correlation using the Normalized Difference Water Index (NDWI) approach and the baseflow index for five small island watersheds in Ambon City, Maluku Province.

2. Methods

2.1 Research Area

The research was conducted in five small island watersheds in Ambon City, Wae Batu Merah, Wae Ruhu, Wae Batu Gajah, Wae Batu Gantung, and Wae Tomu watersheds. These five watersheds were chosen considering that the River Flow Measurement Station (SPAS) availability from the Ministry of Public Works of Maluku Province and the spatial dynamics of land changes in the riverbank area are rapid and have implications hydrological conditions of the watershed. The morphometric characteristics of the five watersheds are presented in Table 1.

Table 1. Morphometric characteristic of watersheds in the current study

Watershed Name	SPAS Coordinates		Watershed Length (km)	Watershed Width (km)	Circumference (km)
	X	Y			
Wae Batu Merah	409415.73	9592511.44	4,69	2,65	23,12
Wae Ruhu	410922.34	9594695.91	6,46	3,29	47,96
Wae Batu Gajah	408291.73	9591503.334	5,22	2,25	18,49
Wae Batu Gantung	408291.73	9591235.01	4,92	4,24	25,32
Wae Tomu	409111.96	9592214.09	5,10	1,55	12,90

Source: Visual Map of Indonesia, BPDASHL Maluku Province

2.2 Research Procedure

2.2.1 Normalized Difference Water Index

Landsat 7 ETM + and Landsat 8 OLI image processing begins with atmospheric correction using ENVI 5.3 software, especially the FLAASH module [32]–[34]. Water Index, Normalized Difference Water Index (NDWI) is a transformation of the reflectance of the band to extract the water's brightness level, including two algorithms with mathematical equations for NDWI are [2], [19], [35]:

$$NDWI = \frac{NIR - SWIR}{NIR + SWIR} \quad \text{.......................................} (1)$$

Table 2. Multispectral characteristics according to LAPAN

Sensor	Spectral	Wavelength	Spatial Resolution	Plot Width Measure	Temporal Resolution
LISA	B1-Blue	0.41-0.49μm	18 m		21 days
	B2-Green	0.51-0.58 μm	18 m		
	B3-Red	0.63-0.70 μm	18 m		
	B4-NIR	0.77-0.99 μm	18 m		

In this study, the red, green, and blue bands were tested on satellite images with a modified formula into the model equation:

$$NDWI = \frac{Blue - \mathrm{Re}\,d}{Blue + \mathrm{Re}\,d} \quad \text{.......................................} (2)$$

RNir is the NIR band's reflectance, while R SWIR6 reflects the SWIR band 6 in Landsat 8.

The NDWI index is influenced by leaf moisture content, vegetation types, and land cover [2]. The high moisture content of vegetation and a high fraction of vegetation cover will have a high NDWI value. Low NDWI values mean that the vegetation's moisture content is low, and the vegetation cover fraction is also low (Table 2).

Table 3.NDWI class according to Gulacsi and Kovacs, 2015.

NDWI Category	Description
0,6 ≤ NDWI ≤ 0,7	High water content
0,5 ≤ NDWI ≤ 0,6	Medium water content
0,4 ≤ NDWI ≤ 0,5	Low water content

Source: Gulacsi dan Kovacs, 2015

2.2.2 Hydrological characteristics of small island watersheds

The hydrological characteristics of the five watersheds of the small island of Ambon City use the Recursive Digital Filter (RDF) method from the hydro office 12.0 software module BFI + 3.0 program and, more specifically, using filtering algorithm [36]–[38] as presented in Table 3.

Table 4. Recursive digital filter method using the BFI + 3.0 program

Filter Name	Mathematical Equation	Remarks
Algoritma Lynie & Hollick	$q_{f(i)} = \alpha q_{f(i-1)} + (q_{(i)} - q_{(i01)} \frac{1-\alpha}{2}$	$q_{f(i)} \geq 0$, if α 0,925, it is recommended for river flow data filter and implemented in three phases of basic flow, qb = q - qf.

Source: hydrooffice 12.0 module BFI^{+}*3.0 [37]*

2.3 Data Analysis

Correlation analysis between the two parameters of the NDWI index and the BFI base flow index is carried out to determine the direction of the relationship between the two variables, whether positive or negative and to predict the value of the dependent variable if the value of the independent variable has increased or decreased. The data used is usually an interval or ratio scale. The simple linear regression correlation formula is as follows:

Y' = a + bXn .... (3)

3. Results and discussion

3.1 Water content using the NDWI algorithm for the 2015 and 2019 periods

The water content analysis in the Landsat imagery in 2015 and 2019 for the five Ambon City watersheds showed water content variation for the NDWI class. The variation in water content for the period 2015 shows that the five watersheds have a low water content of NDWI -0.949655 to -0.581242 (2636.21 hectares), followed by the moderate water content of NDWI -0.581242 to -0.212828 (1250.13 hectares), and high water content of NDWI - 0.212828 to 0.155585 (458.15 hectares). The highest to lowest NDWI water content in a row is the high NDWI class, followed by medium and lastly low, as shown in Table 4 and Figure 1.

Table 5. NDWI water content level in 2015 (in hectares)

Level of water content	NDWI Class	Area (Hectares)
High water content	-0.212828 to 0.155585	458.15
Medium water content	-0.581242 to -0.212828	1250.13
Low water content	-0.949655 to -0.581242	2636.21

Source: Landsat 8 Oil image analysis in 2015

The spatial pattern of water content for 2015 shows that the water content class is relatively high in coastal areas with sloping topography. The movement pattern of non-vegetation land, such as settlements and built-up areas, is centered in the coastal zone, the center of the economy and government of the capital city of Maluku Province and Ambon City. The development concentration is centered in the coastal area, the estuary for the five research watersheds, as shown in Figure 1.

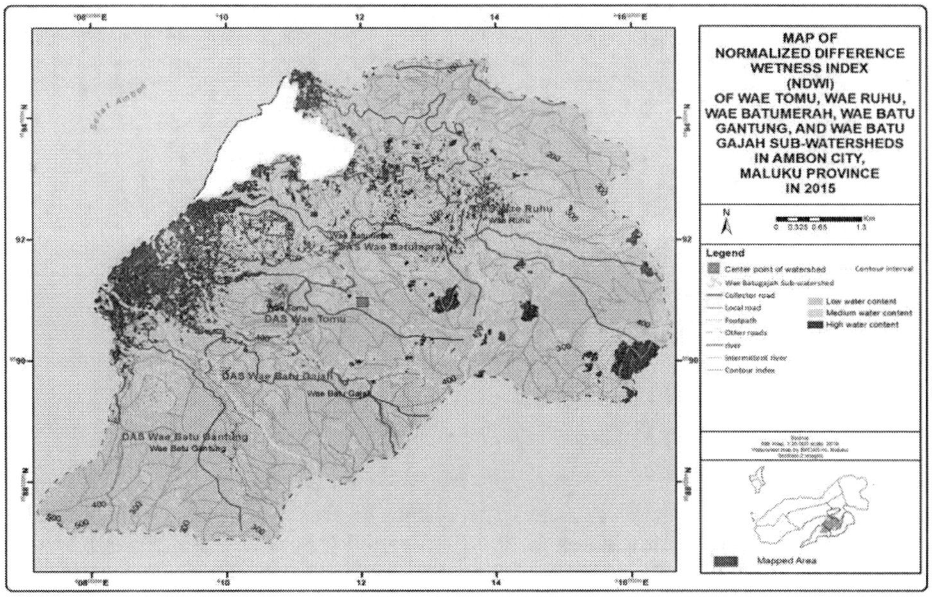

Figure 1. The spatial pattern of water content in the five watersheds of Ambon City in 2015

The trend of water content variation for the 2019 period also shows a relatively uniform pattern for the three NDWI classes. The movement pattern of water content in the five research watersheds shows water content for the Batu Merah, Wae Ruhu, Batu Gajah, Batu Gantung, and Wae Tomu watersheds same relative spatial pattern. The variation of all water content classes for 2019 has a more dominant water content in the moderate to high NDWI class, as presented in Table 5 and Figure 2.

Table 6.NDWI water content level in 2019 (in hectares)

Level of water content	NDWI Class	Area (Hectares)
High water content	-0.212828 to 0.155585	396.06
Medium water content	-0.581242 to -0.212828	892.81
Low water content	-0.949655 to -0.581242	3055.60

Source: Landsat 8 Oil image analysis in 2015

Water content in the 2015 and 2019 period shows that the water content's movement pattern has changed for all study watersheds. The water content experienced a shift in the high water content class, followed by the medium and low water content classes. It can be traced from the direction of changes in the movement of non-vegetation areas, especially the distribution of community settlements, which have begun to shift towards hilly areas. The rate of population growth causes the rate of change in land cover to increase sharply so that the spatial pattern of changes in water content changes significantly.

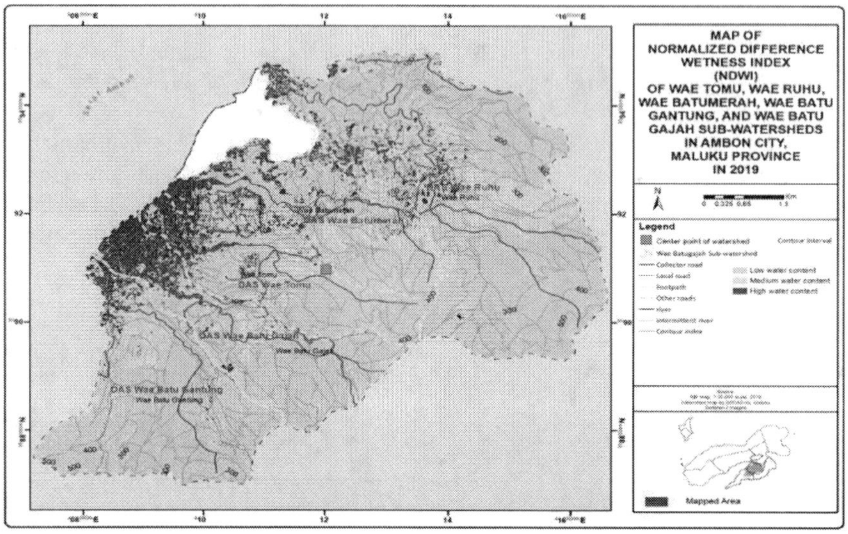

Figure 2. The spatial pattern of the NDWI water content in the five Ambon City watersheds in 2019

3.2 Hydrological characteristics of small island watersheds

The discharge and basic flow calculation results for 2015 and 2019 use the BFLOW hydrooffice12.0 filter module and set the α value of 0.85. The results of the calculation of the average monthly discharge, base flow, and baseflow index for the 2015 period for the Wae Batu Merah watershed are a discharge of 0.39 m^3/sec to 0.60 m^3/sec; base flow 0.32 m^3/sec to 0.43 m^3/sec, and a BFI index of 0.79 to 0.94. The Wae Ruhu watershed's hydrological characteristics have an average monthly discharge of 0.139 m^3/sec to 1.92 m^3/sec; base flow 0.11 m^3/sec to 1.15 m^3/sec, and BFI index 0.76 to 0.93. The monthly average discharge value for the Wae Batu Gajah watershed is 0.54 m^3/sec to 1.02 m^3/sec; base flow 0.45 m^3/sec to 0.89 m^3/sec s, and BFI index 0.79 to 0.92. The hydrological condition of the Wae Batu Gantung watershed has an average monthly discharge of 0.39 m^3/sec to 0.75 m^3/sec; base flow 0.34 m^3/sec to 0.57 m^3/sec, and BFI index 0.83 to 0.90; and finally the Wae Tomu watershed has an average monthly discharge of 0.40 m^3/sec to 0.77 m^3/sec; base flow 0.29 m^3/sec to 0.58 m^3/sec, and BFI index 0.78 to 0.86. The average monthly discharge variation, baseflow, and baseflow index for the five watersheds were relatively uniform in the 2015 period, as presented in Table 6 and Figure 3.

Table 7. Characteristics of average monthly debit, base flow, and BFI index for the period 2015

Month	Wae Batu Merah			Wae Ruhu			Wae Batu Gajah			Wae Batu Gantung			Wae Tomu		
	Discharge	Base flow	BFI	Discharge	Base flow	BFI	Discharge	Base flow	BFI	Discharge	Base flow	BFI	Discharge	Base flow	BFI
January	0.44	0.32	0.79	0.25	0.16	0.82	0.83	0.59	0.79	0.44	0.35	0.83	0.59	0.45	0.80
February	0.41	0.34	0.87	0.21	0.13	0.82	0.80	0.72	0.92	0.41	0.35	0.89	0.48	0.38	0.80
March	0.39	0.35	0.94	0.13	0.11	0.87	0.77	0.70	0.92	0.39	0.34	0.88	0.40	0.29	0.78
April	0.59	0.42	0.88	0.23	0.17	0.83	0.81	0.67	0.88	0.59	0.45	0.87	0.52	0.41	0.84
May	0.44	0.37	0.89	0.92	0.45	0.78	0.91	0.77	0.89	0.53	0.40	0.85	0.44	0.34	0.84
June	0.41	0.35	0.89	1.92	1.15	0.80	0.93	0.79	0.89	0.41	0.35	0.90	0.77	0.58	0.82
July	0.39	0.34	0.90	0.70	0.41	0.80	0.98	0.82	0.89	0.43	0.35	0.89	0.52	0.39	0.79
August	0.60	0.43	0.88	0.27	0.23	0.93	0.99	0.88	0.92	0.65	0.45	0.86	0.59	0.45	0.79
September	0.42	0.36	0.90	0.31	0.16	0.76	1.02	0.84	0.89	0.44	0.38	0.90	0.48	0.34	0.78
October	0.39	0.34	0.91	0.22	0.16	0.81	0.99	0.89	0.92	0.48	0.39	0.88	0.56	0.45	0.86

November	0.46	0.37	0.91	0.22	0.19	0.88	0.82	0.67	0.88	0.52	0.41	0.87	0.51	0.42	0.84
December	0.58	0.42	0.89	0.85	0.48	0.80	0.54	0.45	0.88	0.75	0.57	0.84	0.48	0.37	0.84

Source: Hydrological data analysis, 2015

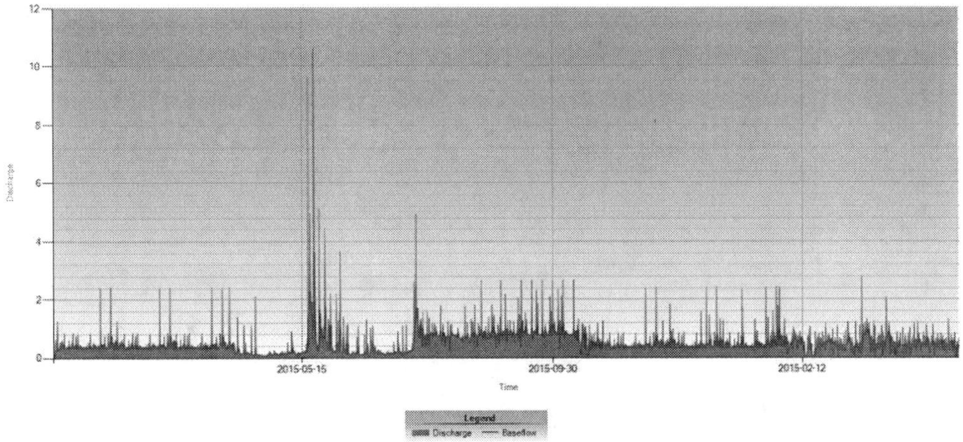

Figure 3. The hydrological characteristics of the discharge and baseflow of the five Ambon City watersheds for the 2015 period

The calculation of hydrological characteristics for 2019 shows changes in the five watersheds' three flow components. The Wae Batu Merah watershed's hydrological characteristics have decreased from the previous period with an average monthly discharge of 0.23 m^3/sec to 0.38 m^3/sec; baseflow 0.21 m^3/sec to 0.29 m^3/sec, and BFI index 0.83 to 0.92. The Wae Ruhu watershed characteristics also decreased the three flow components, successively discharged from 0.23 m^3/sec to 0.41 m^3/sec; baseflow was 0.19 m^3/sec to 0.35 m^3/sec, and BFI index 0.81 to 0.90. The Wae Batu Gajah watershed's hydrological characteristics are the monthly average discharge of 0.53 m^3/sec to 0.74 m^3/sec; base flow 0.44 m^3/sec to 0.58 m^3/sec, and BFI index 0.84 to 0.92. The three flow components for the Wae Batu Gantung Watershed are the monthly average discharge of 0.26 m^3/sec to 0.43 m^3/sec; base flow 0.21 m^3/sec to 0.35 m^3/sec, and BFI index 0.83 to 0.91; and finally the Wae Tomu watershed has an average monthly discharge of 0.05 m^3/sec to 0.86 m^3/sec; base flow 0.03 m^3/sec to 0.53 m^3/sec, and BFI index 0.70 to 0.84. The variation of the three flow components has decreased, as shown in Table 7 and Figure 4.

Table 8. Characteristics of average monthly discharge, base flow, and BFI index for the period 2019

Month	Wae Batu Merah			Wae Ruhu			Wae Batu Gajah			Wae Batu Gantung			Wae Tomu		
	Discharge	Base flow	BFI	Discharge	Base flow	BFI	Discharge	Baseflow	Discharge	Discharge	Base flow	BFI	Discharge	Base flow	BFI
January	0.26	0.23	0.92	0.25	0.2	0.89	0.59	0.51	0.9	0.26	0.22	0.85	0.64	0.49	0.83
February	0.26	0.23	0.9	0.23	0.2	0.89	0.66	0.51	0.85	0.33	0.23	0.83	0.47	0.36	0.81
March	0.23	0.21	0.92	0.23	0.19	0.85	0.61	0.49	0.86	0.26	0.21	0.9	0.05	0.04	0.79
April	0.38	0.27	0.83	0.38	0.28	0.81	0.54	0.46	0.88	0.41	0.32	0.85	0.05	0.03	0.77
May	0.33	0.27	0.91	0.33	0.26	0.88	0.74	0.58	0.84	0.43	0.34	0.9	0.13	0.07	0.7
June	0.32	0.29	0.92	0.33	0.29	0.89	0.6	0.52	0.92	0.38	0.33	0.91	0.14	0.1	0.8
July	0.28	0.24	0.88	0.31	0.25	0.88	0.58	0.51	0.9	0.38	0.28	0.87	0.29	0.21	0.82
August	0.35	0.29	0.87	0.41	0.35	0.89	0.6	0.51	0.89	0.41	0.35	0.88	0.86	0.53	0.81

September	0.29	0.24	0.9	0.36	0.28	0.89	0.73	0.58	0.87	0.29	0.25	0.89	0.8	0.53	0.79
October	0.27	0.24	0.9	0.34	0.27	0.87	0.53	0.44	0.89	0.27	0.23	0.87	0.59	0.37	0.76
November	0.3	0.25	0.88	0.31	0.27	0.9	0.6	0.51	0.89	0.3	0.25	0.86	0.16	0.12	0.84
December	0.29	0.26	0.9	0.33	0.27	0.89	0.66	0.51	0.84	0.29	0.25	0.87	0.08	0.07	0.84

Source: Hydrological data analysis, 2019

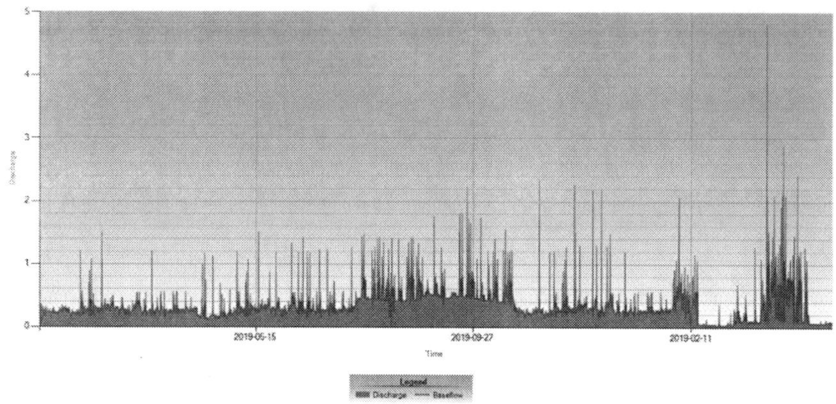

Figure 4. The hydrological characteristics of the discharge and baseflow of the five Ambon City watersheds for the period 2019

The hydrological characteristics (discharge and baseflow) between 2015 and 2019 show a significant change trend. The visualization of hydrological changes is relatively fluctuating from April to June 2015, and there is an increasing flow from July to September 2015. However, from November to December 2015, it is relatively uniform. While the 2019 period's hydrological characteristics show that the visualization of discharge and base flow is relatively consistent in January to June, then in July to September, the flow increases and is relatively uniform in October. However, in November, the flow conditions fluctuate somewhat with decreasing relative flow. Flow conditions in December were relatively fluctuating and increasing. Land cover changes strongly influence the visualization of small island watersheds' hydrological characteristics in the watershed ecosystem. Small island watershed morphometry is a locus that is hydro-morphometrically very vulnerable to changes in the global environment. It is in line with the hydro-morphometric conditions of small island watersheds that affect water systems, especially flood vulnerability and riverbed flow storage and groundwater drainage [39]–[41].

Drainage density is an essential factor related to flood vulnerability and groundwater drainage. Likewise, the river branching ratio affects watershed conditions in high peak flood discharge and short recession times. This condition tends to have a high flood susceptibility characterized by high surface runoff but low permeability and infiltration. A bifurcation ratio factor determines the watershed's shape concerning peak flood discharge, fast-rising times, and fast recession times. This condition creates a high vulnerability to small island watersheds [42].

3.3 Correlation of NDWI water content and BFI base flow index

The NDWI water content's spatial pattern shows that the water content shift in the five watersheds during the study period. Land change factor is one of the critical dynamic factors in influencing small island watersheds' hydrologic characteristics. The NDWI water content's spatial pattern correlates with the watershed's hydrological characteristics with a coefficient of determination R^2-adjusted 0.382, namely, the NDWI water content index and hydrological characteristics correlate (38.8%). The smaller the SEE, the more precise the regression model will be in predicting the dependent variable.

Table 9. Regression models for the relationship between spatial patterns of water content and hydrological characteristics

Model	R	R Square	Adjusted R Square	Std. Error of the Estimate	Durbin-Watson
1	0.645^a	0.416	0.382	0.103	1.271

Remark : predictors: (constant), NDWI algorithm, watershed area (km2), river gradient (m), elevation (m dpal), drainage density (km / km^2); dependent variable: flowrate and base flow (m3 / sec)

The simultaneous correlation value using the ANOVA test obtained an F-count value = 4.234, more significant than the F-table 2.367 at a significance of 0.001 <0.05, which means that there is an influence between the NDWI water content on the hydrological characteristics (discharge and baseflow) significantly for the research watershed.

4. Conclusion

The spatial pattern of water content changes in the study period shows that water content changes tend to experience significant coastal area changes. The spatial pattern of shifting the water index has developed, especially in dense settlements in coastal areas. In 2019, the spatial pattern moved to the mountainous area, the water catchment area in Ambon City, and became a concern to affect the small island watershed's hydrological characteristics. The visualization of hydrological characteristics showing trends that have fluctuated in the period of the study. The correlation analysis results show that variations in the spatial pattern of water content and hydrological characteristics of small island watersheds have a strong negative relationship. These results suggest that if the spatial pattern of water content in areas with massive vegetation cover changes will affect the watershed ecosystem's hydrology. Thus, mitigation efforts and adaptation to changes in small island watershed ecosystems need to be accommodated in planning and managing land and water resources in small island areas.

5. Acknowledgment

The authors would like to thank the head and staff of the River Basin Station at Maluku Province for providing the authors with the opportunity to utilize the data recorded at the stream gauge stations installed at the researched watersheds.

References

[1] Latuamury B, Aponno H S E S, Marasabessy H, Hadijah M H, and Imlabla W N 2020 The spatial dynamics of land cover change along the Wallacea corridor in the key biodiversity area Buano Island Maluku, Indonesia, *J. Degrad. Min. Lands Manag.* 7(4) doi: 10.15243/JDMLM.2020.074.2373.

[2] Gao B C 1996 NDWI - A normalized difference water index for remote sensing of vegetation liquid water from space, *Remote Sens. Environ.*, doi: 10.1016/S0034-4257(96)00067-3.

[3] Ji L, Zhang L, and Wylie B 2009 Analysis of dynamic thresholds for the normalized difference water index, *Photogramm. Eng. Remote Sensing* doi: 10.14358/PERS.75.11.1307.

[4] Zarco P J-Tejada, Rueda C A, and Ustin S L 2003 Water content estimation in vegetation with MODIS reflectance data and model inversion methods *Remote Sens. Environ.* doi: 10.1016/S0034-4257(02)00197-9.

[5] Estallo E L 2012 Effectiveness of normalized difference water index in modelling Aedes

aegypti house index, *Int. J. Remote Sens.* doi: 10.1080/01431161.2011.640962.

[6] Jackson T J 2004 Vegetation water content mapping using Landsat data derived normalized difference water index for corn and soybeans, in *Remote Sensing of Environment* doi: 10.1016/j.rse.2003.10.021.

[7] Wibowo A, Sukojo B M, Harianto T, and Djajadihardja Y S 2015 Ekstraksi Kandungan Air Kanopi Daun Padi dari Data Ground Field Spectrometer dan Airborne-Hyperspectral, pp. 12–25.

[8] Delbart N., Kergoat L, Le Toan T, Lhermitte J, and Picard G 2005 Determination of phenological dates in boreal regions using normalized difference water index, *Remote Sens. Environ.* doi: 10.1016/j.rse.2005.03.011.

[9] Kaplan G and Avdan U 2017 MAPPING and MONITORING WETLANDS USING SENTINEL-2 SATELLITE IMAGERY, in *ISPRS Annals of the Photogrammetry, Remote Sensing and Spatial Information Sciences* doi: 10.5194/isprs-annals-IV-4-W4-271-2017.

[10] Suprayogi A and Sasmito B 2018 Kajian Ekstraksi Unsur Dalam Identifikasi Tutupan Lahan Berbasis Layer Stacking Indeks Citra (Studi Kasus : Kecamatan Wedarijaksa, Kabupaten Pati) *J. Elipsoida* **01**(01) 26–32.

[11] Zhou Y 2017 Open surface water mapping algorithms: A comparison of water-related spectral indices and sensors, *Water (Switzerland)* doi: 10.3390/w9040256.

[12] Mallinis G, Emmanoloudis D, Giannakopoulos V, Maris F., and Koutsias N 2011 Mapping and interpreting historical land cover/land use changes in a Natura 2000 site using earth observational data: The case of Nestos delta, Greece, *Appl. Geogr.* **31**(1) 312–320 doi: 10.1016/j.apgeog.2010.07.002.

[13] Schmitt K, Albers T, Pham T T, and Dinh S C 2013 Site-specific and integrated adaptation to climate change in the coastal mangrove zone of Soc Trang Province, Viet Nam, *J. Coast. Conserv.* doi: 10.1007/s11852-013-0253-4.

[14] Van De Griend A A and Owe M 1993 On the relationship between thermal emissivity and the normalized difference vegetation index for natural surfaces, *Int. J. Remote Sens.* doi: 10.1080/01431169308904400.

[15] Rokni K, Ahmad A, Selamat A, and Hazini S 2014 Water feature extraction and change detection using multitemporal landsat imagery *Remote Sens.* doi: 10.3390/rs6054173.

[16] H. Xu 2006 Modification of normalised difference water index (NDWI) to enhance open water features in remotely sensed imagery, *Int. J. Remote Sens.* doi: 10.1080/01431160600589179.

[17] Acharya T D, Subedi A, and Lee D H 2018 Evaluation of water indices for surface water extraction in a landsat 8 scene of Nepal, *Sensors (Switzerland)* doi: 10.3390/s18082580.

[18] Yang X, Zhao S, Qin X, Zhao N, and Liang L 2017 Mapping of urban surface water bodies from sentinel-2 MSI imagery at 10 m resolution via NDWI-based image sharpening, *Remote Sens.* doi: 10.3390/rs9060596.

[19] Zha Y, Gao J, and Ni S 2003 Use of normalized difference built-up index in automatically mapping urban areas from TM imagery, *Int. J. Remote Sens.* doi: 10.1080/01431160304987.

[20] Yang H, Wang Z, Zhao H, and Guo Y 2011 Water body extraction methods study based on RS and GIS, in *Procedia Environmental Sciences* doi: 10.1016/j.proenv.2011.09.407.

[21] Chen X L, Zhao H M, Li P X, and Yin Z Y 2006 Remote sensing image-based analysis of the relationship between urban heat island and land use/cover changes," *Remote Sens. Environ.* doi: 10.1016/j.rse.2005.11.016.

[22] McFeeters S K 1996 The use of the Normalized Difference Water Index (NDWI) in the delineation of open water features, *Int. J. Remote Sens.* doi: 10.1080/01431169608948714.

[23] Parsa M, Dirgahayu D, and Harini S 2019 Pengembangan Metode Klasifikasi Lahan Sawah Berbasis Indeks Citra Landsat Multiwaktu (Development of Paddy Field Classification Method Based on Multi-Temporal Indeces of Landsat Images), *J. Penginderaan Jauh dan Pengolah. Data Citra Digit.* **1**(1) 35–44.

[24] Huete A, Didan K, Miura T, Rodriguez E P, Gao X, and Ferreira L G 2002 Overview of the radiometric and biophysical performance of the MODIS vegetation indices, *Remote Sens. Environ.* doi: 10.1016/S0034-4257(02)00096-2.

[25] Li W 2013 A comparison of land surface water mapping using the normalized difference water index from TM, ETM+ and ALI, *Remote Sens.* doi: 10.3390/rs5115530.

[26] Maria Octarina T, Dewa Nyoman I, Putra N., and Kadek Ayu Wirdiani N2019 Penginderaan Jauh Pemrosesan Data Satelit Landsat 8 Untuk Deteksi Genangan, *J. Ilm. Merpati (Menara Penelit. Akad. Teknol. Informasi)*, **7**(1) 77 doi: 10.24843/jim.2019.v07.i01.p09.

[27] Roy D P, Boschetti L, and Trigg S N 2006 Remote sensing of fire severity: Assessing the performance of the normalized burn ratio, *IEEE Geosci. Remote Sens. Lett.* doi: 10.1109/LGRS.2005.858485.

[28] Myneni R B 1997 Estimation of global leaf area index and absorbed par using radiative transfer models, *IEEE Trans. Geosci. Remote Sens.* doi: 10.1109/36.649788.

[29] She X, Zhang L, Cen Y, Wu T, Huang C, and Baig M H A 2015 Comparison of the continuity of vegetation indices derived from Landsat 8 OLI and Landsat 7 ETM+ data among different vegetation types, *Remote Sens.* doi: 10.3390/rs71013485.

[30] Villar R G, Pelayo J L, Bantugan J, and Opiso E 2017 Algorithm for modeling agricultural land cover classification and land surface temperature," in *GISTAM 2017 - Proceedings of the 3rd International Conference on Geographical Information Systems Theory, Applications and Management.*

[31] Zhai K, Wu X, Qin Y, and Du P 2015 Comparison of surface water extraction performances of different classic water indices using OLI and TM imageries in different situations, *Geo-Spatial Inf. Sci.* doi: 10.1080/10095020.2015.1017911.

[32] Ahn K H and Merwade V 2016 Role of watershed geomorphic characteristics on flooding in Indiana, United States, *J. Hydrol. Eng.* **21**(2) doi: 10.1061/(ASCE)HE.1943-5584.0001289.

[33] Dutta D, Das S N, Sharma J R, and Radhakrishnan K 2001 Watershed Morphometry and Land Characteristics for Watershed Hydrological Response Analysis: A Remote Sensing and GIS-based Approach, *Ann. Arid Zone*, **40(1**).

[34] Mangan P, Haq M A, and Baral P 2019 Morphometric analysis of watershed using remote sensing and GIS—a case study of Nanganji River Basin in Tamil Nadu, India, *Arab. J. Geosci.* **12**(6) doi: 10.1007/s12517-019-4382-4.

[35] JRC 2011 NDWI (Normalized Difference Water Index), *Product Fact Sheet.*

[36] Eckhardt K 2008 A comparison of baseflow indices, which were calculated with seven different baseflow separation methods," *J. Hydrol.* **352**(1–2) 168–173 doi: 10.1016/j.jhydrol.2008.01.005.

[37] Gregor M 2010 Manual: Bfi+ 3.0." Department of Hydrogeology, Faculty of Natural Science, Comenius University Bratislava, Slovakia, Slovakia.

[38] Wang L and Qu J J 2007 NMDI: A normalized multi-band drought index for monitoring soil and vegetation moisture with satellite remote sensing, *Geophys. Res. Lett.* doi: 10.1029/2007GL031021.

[39] Balzan M V, Potschin-Young M, and Haines-Young R 2018 Island ecosystem services: Insights from a literature review on case-study island ecosystem services and future prospects," *International Journal of Biodiversity Science, Ecosystem Services and Management* **14**(1) doi: 10.1080/21513732.2018.1439103.

[40] Scandurra G, Romano A A, Ronghi M, and Carfora A 2018 On the vulnerability of Small Island Developing States: A dynamic analysis, *Ecol. Indic.* **84** doi: 10.1016/j.ecolind.2017.09.016.

[41] Sith R, Watanabe A, Nakamura T, Yamamoto T, and Nadaoka K 2019 Assessment of water quality and evaluation of best management practices in a small agricultural watershed adjacent to Coral Reef area in Japan, *Agric. Water Manag.* **213** doi: 10.1016/j.agwat.2018.11.014.

[42] Jazwa C S, Duffy C J, Leonard L, and Kennett D J 2016 Hydrological Modeling and Prehistoric Settlement on Santa Rosa Island, California, USA, *Geoarchaeology*, **31**(2), doi: 10.1002/gea.21532.

Physiological characters of four lowland chilli varieties (*Capsicum annum* L.) with root cutting

S Kusumaningrum[1], E Sulistyaningsih[1,*], R Harimurti[1] and K Dewi[2]

[1] Faculty of Agriculture, Universitas Gadjah Mada
[2] Faculty of Biology, Universitas Gadjah Mada
Bulaksumur, Sleman 55281, Yogyakarta, Indonesia

* E-mail: endangsih@yahoo.com

Abstract. The root system plays some vital roles in overall plant development, promoting plant anchorage, absorption of nutrients and water, and hormone production. Cutting is an attempt to root regeneration and enhance the plant's metabolism capability, including canopies and roots for high yielding. Non-hybrid chilli can be used as a model plant because Indonesia's chilli yield has not been maximal yet. This research was aimed to determine the physiological characters of four lowland chilli varieties with root cutting. This study was designed using a randomized, complete block design (RCBD) with two factors: root cutting and varieties. The root cutting factor consisted of 4 levels: root cutting in seeding, root cutting in ridging, root cutting in seedling and ridging, and non-cutting. The varieties factor consisted of 4 levels, namely Lembang, Kencana, Tanjung, dan Ungu. The results showed that root cutting in ridging for Kencana, Lembang, and Ungu increased stomatal conductance, transpiration rate, CO_2 intercellular, chlorophyll accumulation, photosynthesis rate, and improved fruit yield per plant. However, Tanjung did not respond to root cutting treatment because it did not improve plant physiology characters and fruit yield than non-cutting treatment. It is concluded that root cutting in ridging can be improved plant physiology, which contributes to increasing yield on Kencana, Lembang, and Ungu.

1. Introduction

The root is an organ that functions to absorb water and nutrients to support vegetative and reproductive organ growth. A roots system with a wide range can access a large soil volume so it can contribute to the anthesis and fruit formation process [1]. Root could change the rizosfer zone by metabolites on the exudate affected by water, nutrient, and environment. Upper soil generally tends to higher in nutrient and dryer and prone to extreme weather compared to subsoil layers [1]. Root architecture is a spatial root system distribution [2] whose role is to support resource distribution evenly and contribute to plant productivity [3].

Root architecture can be improved by cutting roots. Cutting is a technique to reduce vegetative plant growth classified as practical and economical for stunting but can induce new roots to maintain growth [4]. Root cutting can change the growth balance, assimilation, nutrient distribution, and hormones. Increased hormone production in the meristem can initiate lateral roots [5], and the small diameter roots from where they cut can increase the water and nutrient absorption.

Cell division and cell enlargement during root formation need a massive quantity of energy and carbon. The primary carbon source comes from sucrose, which is translocated to the sink. The translocated sucrose by basipetal is used as energy for cell division and cell enlargement and a carbon source for roots growth [6]. The acceleration of the physiology plant process can help accumulate photosynthetic products used in organ regeneration. Stomata activity, CO_2 content at the leaf, transpiration rate, and photosynthetic rate in the canopy can affect plants to produce triose phosphate, which will be used as a basic component in the complex carbohydrate molecule formation [7].

Chilli plants can act as model plants that can be used to measure success in root-cutting treatment. Chilli is an economic crop with a high demand in Indonesia that can trigger inflation [8]. National chilli productivity has reached optimal productivity that can be produced, yet [9]. This research was aimed to determine the physiological characters of four lowland chilli varieties with root cutting.

2. Methods

2.1. Experimental details

The research was conducted at Greenhouse of Agriculture Faculty, Universitas Gadjah Mada for nursery, Kebun Tridharmaof Agriculture Faculty, Banguntapan, Bantul, Yogyakarta for planting local curly red chillies. The research was conducted from August 2019 – March 2020.

The material used was four non-hybrid chilli varieties, namely Lembang, Kencana, Tanjung, Ungu. Other materials used were media for nursery from soil: fertilizer: sand: cockpit with composition 1:1:1:1, media for transplanting is from the soil and manure composition, polybag, nail polish, acetone, NED, SA. The tools used were 1). Measuring instruments: analytic scale, gauge, ruler, nailed board; 2). Cultivation tools: plastic rope, sprayer, hoe, measuring cup, and a bucket; 3). Plant physiology observation tools: microscope, optical lab, photosynthetic analyzer type Li COR 6400.

This study was designed using a two-factor factorial design (4 x 4) arranged in a randomized completely block design (RCBD) with three blocks replication. The first factor is four non-hybrid chilli varieties, namely Kencana, Lembang, Tanjung, dan Ungu. The second factor is root cutting consisted of 4 levels, namely root cutting in seeding, root cutting in ridging, root cutting in seedling and ridging, and non-cutting. Each block contains 16 treatment combinations. Each block consisted of 10 plants, so in total, there were 480 chilli plants each block. Three plants selected were observed periodically and harvested sample plants; 2 plants were observed destructively. Root cutting treatment was given by lateral root cutting by 50% when seedlings were ready to be transplanted, cutting by 50% of the sides plant when planting, and transplanting them into bigger polybags.

2.2. *Sampling and measurement*

2.2.1. *Environmental Data.* Leaf environmental data include leaf temperature and light intensity observed by a leaf thermometer and lux meter.

2.2.2. *Dry Weight.* The root, stem, and leaf of the plant, before being oven-dried, is first let dry for 6 hours. The oven was carried out at 80 °C for three days until the weight was constant.

2.2.3. *Character physiology.* Measurements on sample plants were determined by one maintenance leaf as a random sample to measure photosynthesis rate, transpiration rate, leaf CO_2, photosyntetic radiation, and H_2O using the photosynthetic analyzer model Licor 6400. The photosynthetic analyzer can be used after the battery, and the chemical tube is filled, and the chamber tube clamped to leaves connected to a photosynthetic analyzer cable connection. The tool is then turned on by pressing the button and waiting for a few minutes until AC=0 (warning up). Then the sample leaves are clamped by chamber d for 5 minutes, and after the measurement number is stabilized, the chamber is removed. The monitors displayed on the monitor are the photosynthetic measurement rate, transpiration rate, leaf CO2, PAR, and dan H_2O.

2.2.4. *Chlorophyll and carotenoid contents.* Leaf chlorophyll content was analyzed by weighing 1 gram of leaves ground with a mortar and then adding 20 ml acetone 80%. The mixture is filtered using a paper filter. Next, I turned on the spectrophotometer and left for 10 minutes. The 80% pure acetone solution was put into cuvet until the limit was a blank. With the button set at a wavelength of 645 nm, the absorbance is set to zero. The pigment solution sample was poured into another cuvet until the limit, and the absorbance number was recorded (A645). The step can be repeated to get an absorbance number

at a wavelength of 663 nm (A663). Chlorophyll a and chlorophyll b on leaves content were calculated using the equation [10] with units of mg g leaf fresh weight $^{-1}$.
Chlorophyll content a = (12,7 x A663 – 2, 69 x A645) x 0,02
Chlorophyll content b = (22,9 x A645 – 4,68 x A663)x 0,02
Carotenoid concentration (Cx+c) was calculated with a spectrophotometer at a wavelength of 470 nm and acetone 95.5% used as a blank.
Carotenoid concentration = A470 + (0,144 A663 – 0,638 A645)
2.2.5. *Length and Width of Stomatal.* Stomata characters were observed in the form of height, and width stomatal carried of by printed stomata using transparent nail polish smeared evenly under the leaves across 4 cm. Dried it for 5 minutes, the stained part fixed with transparent insulation. Then the insulation part is carefully peeled to get the stomata print. The isolation was affixed to the glass side under a microscope binocular Olympus (Germany) with magnification to 40x [11]. After that, take a picture of stomata used using the OptiLab Mikroskop Digital Advanced Type and the software OptilAB Viewer version 2.1. Furthermore, the number of stomata and the length and width of the stomata opening were measured using OptiLab ImageRaster software version 2.1. Measurement of the number of stomata is carried out per unit square millimetre (mm^2), and the width and length of the stomata opening are expressed in units of micrometres (μm).

2.3. Statistical analysis
The observational data obtained were tested using variance (ANOVA) for factorial design arranged in a completely randomized block design. The linear model that follows this design appears in equation (1).

$$Yijk = \mu + \alpha_i + \beta_j + (\alpha\beta)_{ij} + \varepsilon_{ijk} \quad (1)$$

By i = 1, 2, 3, 4 shows the first factor levels; j= 1, 2, 3, 4, shows the second factor levels and k = 1, 2, 3 shows block. Yijk shows the observed data; α shows the first-factor effect; β shows the second-factor effect; αβ shows the interaction between the first and second factor; dan ε shows error (Gomez & Gomez, 1976). If there is an interaction, it is further tested with a simple effect analysis (*simple effect)* by *post hoc honestly significant difference* visualized using a table [12]. Data analyzed using software SAS version x64 3.1.2

3. Results and discussion
Environmental factors were one of the factors that affect plant growth. The temperature and light intensities received by plants can be metabolized inside the plant's body. Temperature did affect the physiological process and the biochemical reaction rate. The optimum temperature did contribute to physiological reaction, so the simple formed sugar becomes the raw ingredients for optimum regeneration cells to happen [13]. The plants' high temperature can inhibit plant growth by suppressing enzyme activities and increasing the transpiration rate.

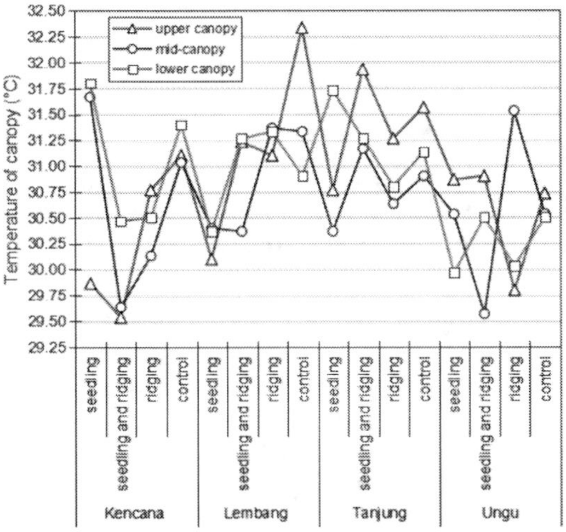

Figure 1. Leaf temperature by the leaves at the upper caonpy, midle canopy, and lower canopy, n=6

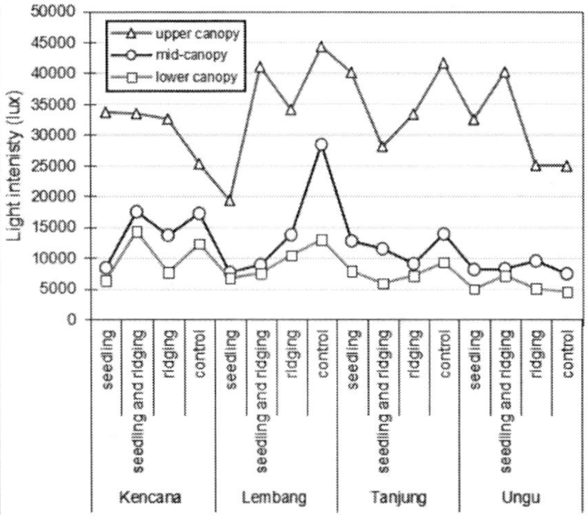

Figure 2. Light intensity by the leaves at the upper caonpy, midle canopy, and lower canopy, n=6

The characters of temperature and light intensity received by plants canopy shown in Figure 1. The plant canopy temperature between the upper and lower one of all varieties and root cutting treatment ranged from 29,5 °C - 32.25 °C. The top, middle, and bottom canopy temperature did not show any significant symptoms in the whole treatment. However, the plant temperature over 35 °C can cause symptoms that aggravate physiological conditions and plant growth [14]. Unlike the temperature case, the light intensity received by upper part plants is higher than the light intensity reached the middle and lower canopy. The middle and lower canopy showed no significant effect on the overall plant treatment. The light intensity that decreased to the canopy located at the bottom shows that the composition of the middle and lower leaves position is tighter than at the top. The light at the bottom is lower than that of the top of the canopy.

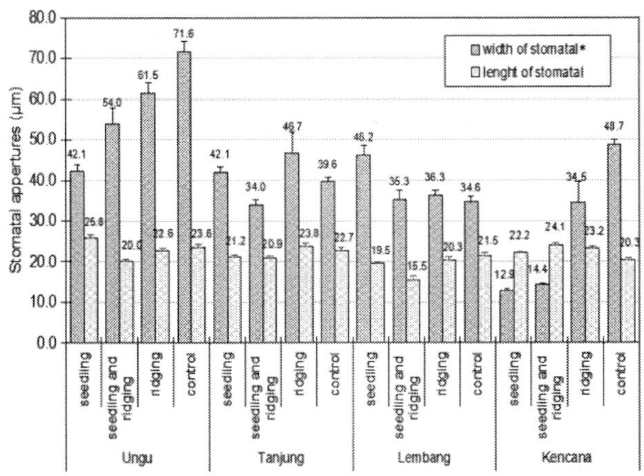

Figure 3. Organel stomata on leaves from root cutting treatment of 4 non-hibrid chilli varieties. * : significant P < 0.05

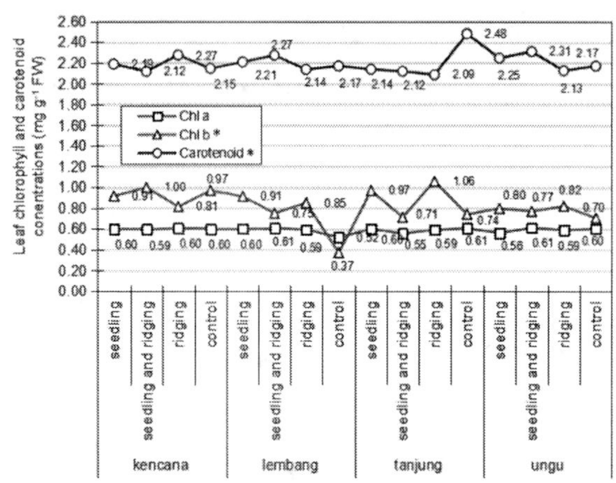

Figure 4. Chlorophyll and caretenoid content on leaves from root cutting treatment of 4 non-hibrid chilli varieties. * : significant P < 0.05

Stomata are organelle where CO_2, H_2O, and O_2 are exchanged, abundant at lower leaves epidermis. Stomata openings can be affected by light, temperature, and water content in leaves [15]. The stomata length of the longest diameter and the width, which is the shortest diameter, can describe the extent of gas exchange in the leaves. Based on observation, Ungu, Tanjung, and Lembang varieties had a wider opening than the length opening. Width opening, which is wider than the length opening, shows that the guard cells have turgor to open wide [16]. Ungu offers a control treatment that gives the most wider width opening. While in Tanjung and Lembang, the widest wide opening from ridging treatment and seedling. In Kencana, seedling and seedling ridging shows a much longer length opening than width opening. This indicates that the guard cells condition is not turgor making the stomata opening not maximal. Ridging and controlling Kencana had a much wider width opening than length opening, showing stomata had urgency. The stomatal width also indicated that water availability in the soil is sufficient so that the transpiration process occurred to maintain optimal body temperature. The stomata activity in chilies belongs to type C3, which is an indicator of optimal metabolism. This thing can spur the physiological processes that occurred.

The involvement of stomata in photosynthesis is supported by the mineral and water content on the leaves; also, the leaf's thickness is related to the chlorophyll content. The chlorophyll content is a measurement for the pigment photosynthesis quantity. It is located on leaves palisade tissue. The analyzed result (Figure 4.) shows that chlorophyll b content is higher than chlorophyll a. Chlorophyll b on Lembang and Tanjung shows a significant difference compared to Kencana and Ungu. In Lembang and Tanjung, root cutting treatment during ridging can increase chlorophyll b content. Chlorophyll B is found in the pigment antenna system that makes it function as a light catcher. And chlorophyll A is on the photosystem I and Photosystem II and the system in the antenna pigment [16]. The pigment-protein light harvester LHC-I from PS1 has a ratio of a/b ~3, while LHC-II from PSII has a rate of a/b 1.1-1.3. The balance a/b LHC-II from PSII, which is variate, shows an adaptation response toward the light. Shade plants have a much higher LHC-II than an exposed plant, resulting in a lower a/b ratio than the exposed plant. Thus, a decreased a/b can indicate there is an enlargement antenna system on PSII. When the a/b rate is low, it also shows a perfect leaves development, classified as shading leaves and C3 plants [18]. The chlorophyll a and b ratio toward total carotenoid also indicates a greenish presence of the plants. The chlorophyll a and b with a lower carotenoid shows aging, stress, and damage to photosynthesis equipment, which means chlorophyll break down is faster than carotenoid. However, it can also happen when the ratio < 1, the plant's professional chromoplast development on ripe fruits, which changes from green to red, so the ratio keeps decreasing [19]. This indicates a chromoplast development on fruit so that the leaves' chlorophyll content is lower than carotenoids. The a/b rate is also lower due to the shading effect on the canopy shown in the light intensity (Figure 2.).

Table 1. Nett activity of photosynthesis (Pn), stomatal conductance (g_s), CO2 intercellular (Ci), Transpiration rate (Tr), H_2O, Photosynthetic active radiation (PAR) on chili canopy

Variety	Cutting root	Pn (µmol CO_2 m^{-2} s^{-1})	g_s (mol H_2O m^{-2} s^{-1})	Ci (µmol CO_2 mol^{-1})	Tr (mmol H_2O m^{-2} s^{-1})	H_2O Daun (mol H_2O mol/air)	PAR ($mol/m^2/s$)
Kencana	seedling	17.34^b	0.0062^b	1129.3^b	1.36^b	2.56^b	3922.5^a
	seedling and ridging	20.90^b	0.0267^{ab}	1990.8^b	5.42^{ab}	6.79^{ab}	3983.3^a
	ridging	35.48^a	0.0315^a	3101.5^a	7.01^a	9.07^a	4274.0^a
	control	22.05^b	0.0285^{ab}	1868.3^b	7.27^a	5.85^{ab}	3506.3^a
Lembang	seedling	15.86^b	0.0140^{ab}	1389.0^b	3.78^a	5.30^a	3141.5^a
	seedling and ridging	20.23^b	0.0094^b	2348.0^b	2.82^b	3.57^a	4625.8^a
	ridging	26.40^a	0.0284^a	4702.7^a	4.50^a	5.82^a	4729.8^a
	control	15.29^b	0.0091^b	2559.8^b	2.03^b	5.51^a	3387.7^a
Tanjung	seedling	23.96^a	0.0194^a	2206.8^a	4.96^a	6.13^b	4799.7^a

	seedling and ridging	6.31 c	0.0287^{a}	585.7^{b}	4.01^{a}	24.01^{a}	3126.5^{a}
	ridging	16.05^{b}	0.0113^{a}	1916.8^{ab}	2.78^{b}	4.24^{b}	3602.5^{a}
	control	11.01 bc	0.0111^{a}	1462.2^{ab}	2.52^{b}	3.84^{b}	3372.7^{a}
Ungu	seedling	13.97^{b}	0.0231^{a}	1086.7^{b}	5.21^{ab}	6.88^{a}	3993.2^{a}
	seedling and ridging	11.57^{b}	0.0084^{b}	1901.2^{ab}	2.94^{b}	4.10^{a}	3747.2^{a}
	ridging	26.42^{a}	0.0275^{a}	2511.4 a	7.11^{a}	7.58^{a}	4054.5^{a}
	control	18.37^{b}	0.0182^{ab}	1131.1^{b}	5.33^{ab}	8.19^{a}	3214.5^{a}

Notes: Means different letters were significantly different at same variety, P < 0.05 (n=9) using the Honestly significant difference

Active photosynthesis, water content on leaves, transpiration, stomata conductance, co2 leaf concentration, and photosynthesis is a leaf physiology activity that can be observed. Photosynthesis active radiation is visible radiation used in the photosynthesis process. Where PAR is half of the total radiation received by the earth's surface. Of the total radiation, it is assumed half of it is absorbed by plant canopy, and the other half is reflected and absorbed by the earth's surface. Based on the PAR experiment received from the treatments, the plants had the same result of high PAR from the leaf that can affect the plant's stomata conductance activity. Based on the results show that the increase in stomata conductance in the root cutter is offset by the availability of H_2O and CO_2 in the leaves. Transpiration is the loss of water from the plant in water vapor through stomata, cuticle, and lenticel to the outside, which is not saturated with water. Transpiration functions for the plant in water absorption, nutrient transportation, and photosynthate distribution. The decrease in the transpiration rate is a physiological mechanism for the plant to maintain its water status to do the physiology mechanism. The existence of CO_2 and water on cell affect the photosynthesis rate. Photosynthesis is an anabolism reaction that produced the final product, glucose. Plant production depends on the size and efficiency of the photosynthesis system. Based on the analysis result, it is shown that Kencana, Lembang, and Purple have the same photosynthesis level as the root cuts at planting time.

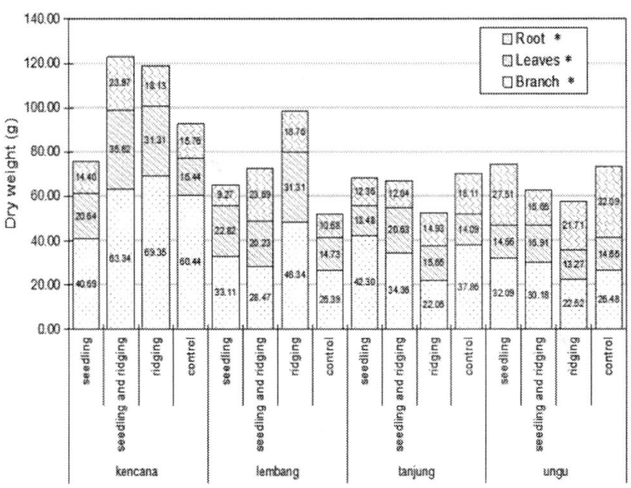

Figure 5. Dry weight of plant organs from root cutting treatment of 4 non-hibrid chilli varieties. * : significant P < 0.05

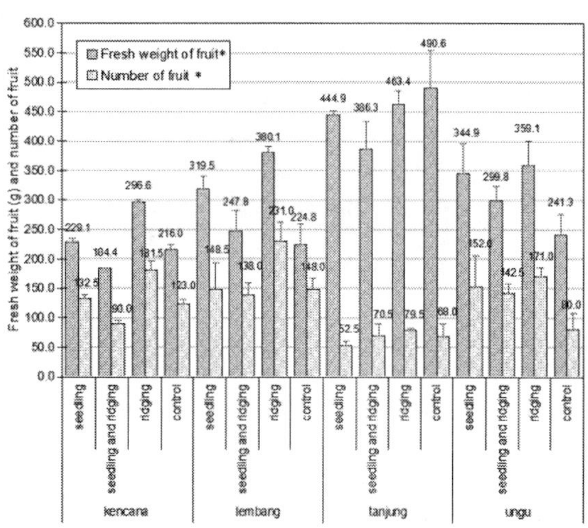

Figure 6. fresh weight and number of fruit per plant from root cutting treatment of 4 non-hibrid chilli varieties. * : significant P < 0.05

Dry weight as a variable can describe the accumulation of assimilating, which translocated during the growth phase referred to as biologic plant weight. In this case, the dry weight plant is divided into root, stem, and leaf. The root dry weight is a parameter that explains the growth and development of the plant root system. [20] explain the growth and development of the plant root system affected by division rate and enlargement cell, in addition to the accumulation rate of dry weight in the root. This assimilated accumulation results from the absorption of solar energy used for the photosynthesis process [20]. Based on the analysis results, Kencana dry weight at the time of ridging and seedling and ridging had the growth of root, stem, and the other stem that dominates the other root cutting treatment. The same thing can be seen on the Lembang variety given root cutting at ridging. Root cutting at ridging had a positive effect on plant dry weight by the improved physiology process. Also, plant dry weight as a variable, fresh fruit weight, and whole fruit in the ridging treatment of Kencana, Lembang, and Ungu varieties showed a positive result. However, this does not happen on the Tanjung type.

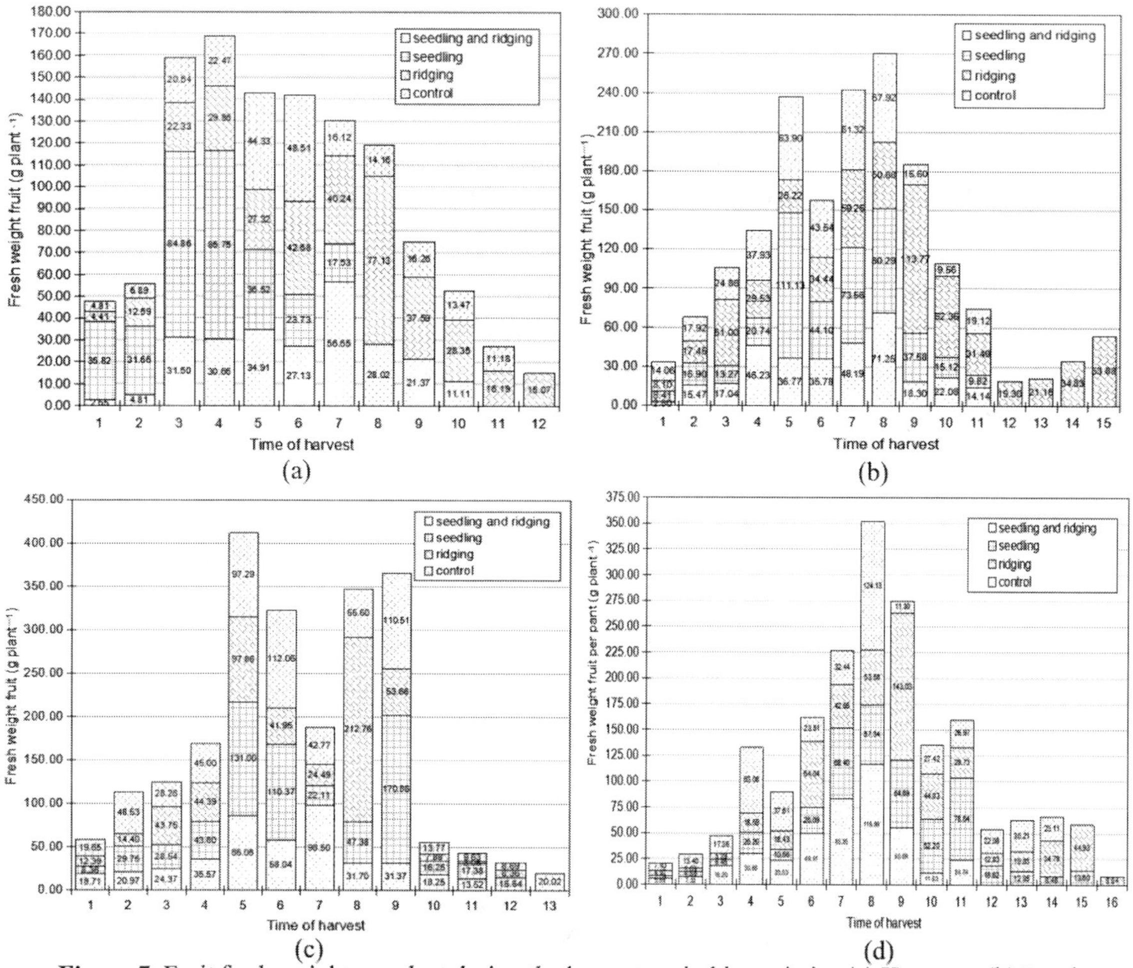

Figure 7. Fruit fresh weight per plant during the harvest period in varieties (a) Kencana, (b) Lembang, (c) Tanjung, and (d) Ungu in root cutting treatment

When observed in the harvesting chili process, the Kencana variety was harvested up to 12 times, while Lembang up to 15 times. The 3rd to 8th harvest period indicates too high. It indicates at the 3rd to 8th harvest period ridging treatment the maximum results. In contrast, the Lembang variety had an optimum harvest period from the 5th, 7th, and 8th. The harvest period shows that root cutting has a maximum harvest time. Tanjung has a harvest period of up to 13th, as an Ungu until 16th. Tanjung, the optimal harvest occurred on 5th to 6th and 8th up to the 9th period. In that period, care provides the most optimal value. While in Ungu, the peak crop harvest happens in the 8th period. In this period, the yield provides the most optimal value in treatments. In the treatment of control, optimal results occur in the

7th-8th period, in the ridging treatment happened at 7th up to 9th period, in seedling happened at 9th period, while in seedling and ridging happened at 9th period.

4. Conclusion

Root cutting treatment during the planting can increase the fruit's fresh weight, and the number of fruit by increasing the plant's physiologic processes make the growth increased, with the increase of dry weight as the variable. In addition, the ridging also accelerates the harvesting time so that the fruit yields are optimal.

5. Acknowledgements

The authors would like to acknowledge the Directorate General of Higher Education's financial support under Universitas Gadjah Mada's management based on contract No. 5712/UN1.DITLIT/DIT-LIT/LT/2018

References

[1] Lynch J 1995 Root architecture and plant productivity *J Plant Physiol* **109** 7
[2] Kochian L V 2016 Root architecture Journal of Integrative J *Plant Biol* **58** 203.
[3] Beidler K V, Taylor B N, Strand A E, Cooper E R, Schonhlz M, and Pritchard S G 2014 Changes in root architecture under elevated concentrations of CO_2 and nitrogen refletc alternate soil exploration strategies *J New Phytol* **205** 1153
[4] Gao K, Zhu T, Wang L and Gao Y 2017 Effects of root pruning radius and time on yield of tuberous roots and resource allocation in a crop of Helianthus tuberosus L *Sci Rep* **9** 7
[5] Xing S J, Ma B Y, Liu F C, Ma H L and Wang Q H 2012 Effects of root pruning on the growth and rhizosphere soil characcteristics of short-rotation closed canopy poplar *J For Syst* **21** 236
[6] Matsuda R, Suzuki K, Nakano, Higashide A and Takaichi M 2011 Responses of leaf photosynthesis and plant growth to altered source-sink balance in a Japanese and Dutch tomato cultivar *J Sci Hortic* **127** 520
[7] Zhu G X, Long S P and Ort D R 2010 Improving photosinthetic efficiency for greater yield *Ann Rev Plant Biol* **61** 235
[8] Anwarudin M J, Sayekti A L, Marendra A K and Hilman Y 2015 Dynamics production and price volatility of chili: anticipation of development strategies and policie *J Agri Inn Dev* **8** 33
[9] Farid M and Subekti N A 2012 Overview of the production, consumption, distribution and price dynamics of chili in Indonesia *J Trade Res Dev Scient* **6** 211
[10] Arnon D I 1949 Copper enzymes in isolated chloroplast polypphenoloxidase in Beta vulgaris *J Plant Physiol* **24** 4
[11] Haryanti S 2010 The number and distribution of stomata on the leaves of several species of dicot and monocot plants *J Anat and physiol* **2** 21
[12] Gomez K A and Gomez A A 1976 *Statistical Procedures for Agricultural Research With Emphasis on Rice* International Rice Research Institute Los Banos (Philippiñes) p 244
[13] Ratna M, Sarker R, Ara R,Hossain M M and Kamruzzaman M M 2018 Performance of chilli (Capsicum annum L.) lines with dfferent plantation time during rainy season *J Agri Environ Sci* **3** 240
[14] Jackson R D, Idso S B, Reginato R J and Pinter P J 1981 Canopy temperature as a crop water stress indicator *J Water Resour Res* **4** 1133
[15] Smirnov O E, Kosyan A M, Kosyk O I and Taran N Y 2014 Buckwheat stomatal traits under aluminium toxicity Modern Phytomorphology **6** 15
[16] Kumekawa Y, Miyata H, Ohga K, Hayakawa H, Yokoyama J, Ito K, Tebayashi S, Arakawa R and Fukuda T 2013 Comparative analysis of stomatal size and density among ecotypes of Aster hipidus (Asteraceae) *Am J Plant Sci* **4** 524
[17] Taiz L and Zeiger E 2010 Plant physiology 5thed Sinauer Accociates Sunderland (USA)

[18] Kume A, Akitsu T and Nasahara K N 2018 Why is choropyll b only used light-harvesting systems? *J Plant Res* **131** 961
[19] Mostofa K M G, Liu C, Pan X, Vione D, Hayakawa K, Yoshioka T and Kornissarov G G 2013 Chorophylls and their degradation in nature *J Environ Sci Eng* **13** 687
[20] Gardner F, Pearce R B and Mitchell R L 1991 *Cultivated Plant Physiology* University of Indonesia Publisher (Jakarta) p 55

Development of home-based ready-to-use supplementary food (RUSF-HB) to overcome nutrition-related problems among children under five during the covid-19 pandemic

S Sastrawan[1,*], M Menap[1], L Sulaiman[2] and H Hendrayani[3]

[1] Faculty of Health Sciences, Universitas Qamarul Huda Badaruddin, Indonesia
[2] Central for Research and Community Engagement, Universitas Qamarul Huda Badaruddin, Indonesia
[3] Nutrition Services Department, Sengkol Public Health Centre, Indonesia

* E-mail: sastrawan@gmail.com

Abstract. The most direct causes of malnutrition are food consumption and infection. Yet the prolonged Covid-19 pandemic has limited low-income families' ability to fulfil the need for nutrition consumption, particularly for children under five. Responding to this situation, we sought to develop home-based ready-to-use supplementary foods (RUSF-HB) from local ingredients that are energy-protein-dense, affordable, simple and easy to produce at home. We created three milk-free formulas (MFFs) and six standard formulas (STFs). Three cheap and abundant local foods: soybeans, cowpeas, and mung beans were combined with rice flour, refined sugar, and coconut oil. A cross-over study design was used to assess food organoleptic, which showed that the products were comparable in several sensory aspects except for the odour and the taste. Soybeans-based formulas contain slightly more energy and protein compared to mung beans or cowpeas-based ones. However, the mung beans-based formulas were more favourable to caregivers and children, particularly their taste and smell. The products contain slightly less energy than the recommended ones but high enough to supply macronutrient for those in need on a regular basis. The caregivers considered the total price for the products was affordable, and the formula was easy to follow.

1. Introduction

The Covid-19 pandemic has been affecting all aspects of life. Those of low socioeconomic classes have been reported to suffer the most from the global economic downturn following the pandemic [1]. With the prolonged pandemic, nutrition problems, particularly among children from such social classes, along with other vulnerable groups, are becoming more imminent [2]. As the risk of intra-hospital infection cannot be omitted [3], many families avoid seeking help from the public health centres, clinics or hospitals, due to the fear of the Covid-19 infection in those places. This implies that their access to health services has been limited, which possibly leave the malnourished children untreated.

Severe acute malnutrition children are treated using ready-to-use Therapeutic Foods (RUTFs) [4]. On the other hand, mild acute malnutrition children usually are given Ready-to-use Supplementary Foods (RUSF) [5]. Both RUSF and RUTF are very similar in their requirements, food characteristics and composition except that RUTF should be fed exclusively. Also, the dose of RUTF should be suited

IOP Conf. Series: Earth and Environmental Science **883** (2021) 012074 doi:10.1088/1755-1315/883/1/012074

to the children's weight while the quantity of RUSF is normally fixed [5,6]. RUSF/RUTF is a mixture of highly nutritious foods intended for emergency and supplementary feeding for either malnourished children or individuals with special nutrition requirements.

Although The effectiveness of RUTF for improving individual nutrition status has been proved [7,8], many people cannot afford industry-grade RUTF. Commercial food supplements are considered costly, especially for those of lower-to-middle class families in developing countries [9] which limit their access to the food. Hence, to improve accessibility, some countries created their version of RUSF/RUTF from local ingredients to reduce the overall price [10]. In many places, RUTF is distributed to those in need free of charge. However, in some other locations, people do not have access to RUTF. With the spread of covid-19, the distribution of free RUSF/RUTF was almost interrupted entirely, leaving the vulnerable group in latent danger of malnutrition.

It is paramount to stay alert of potential malnourished issues among the vulnerable individuals during this economically challenging time [11]. It is necessary to find appropriate solutions to this issue amidst the pandemic and the economic downturn. The development of alternative RUSF/RUTF formulas from local foods potentially provides solutions for both health and economy issues during the pandemic Covid-19. Indeed, several crucial points should be taken into consideration when developing home-based RUSF/RUTF. These considerations include the principles of health, affordability, practicality, simplicity, and locality. The resulting foods should be able to reproduce at household without a sophisticate cooking utensil.

2. Objectives

To develop energy-dense home-based Ready-to-use Supplementary Food (RUSF-HB) from local produces that are easy to make at the household level to fight malnutrition issues among children under five during the Covid-19 Pandemic. Although we focused on developing supplementary foods, the finished products have the potential to use as therapeutic food in an emergency situation.

3. Methods

This research was done in two phases: Development phase and Evaluation phase

3.1. Development phase.

Development phase consisted of two steps: identifying typical local foods and creating formulas and recipes

3.1.1. Identifying typical foods. The first step was finding typical foods of the local community, which was done through observation in two local markets. To ensure the validity of the derived data, we also interviewed four locals.

3.1.2. Creating and analysing formulas. We created several combinations of ingredients. The formulas development was done based on the international practice of RUTF [6] with additional considerations such as affordability, practicality (easiness), simplicity, and locality of the process and the products.

We developed two groups of formulas: the first group was milk-free formulas (MFFs), which was created for mothers who have limited access to milk, either due to economic or geographic factors. For this target group, milk powder was dropped from the formulas. The other group was Standard Formulas (SFs) created by following standard practice of RUSF/RUTF development, which made use milk powder. The detailed combination of the formulas is presented in table 1.

To determine the macronutrient content of each combination, we referred to and consulted the latest Indonesian Food Composition database [12]. The primary method of processing involved steps: roasting, grinding, mixing serving/packaging.

1. Roast rice flour for approximately 30 minutes over low heat
2. Roast the beans over low heat for approximately 30 minutes. Ground the beans.

3. Mix all the ingredients according to the prescribed formula, stir evenly. Any available utensils (modern such as a food processor, or traditional/simple such as a spoon) can be used to mix all ingredients as long as they are evenly mixed.
4. The mothers are encouraged to use a beautiful and eye-catcher cake mould to attract children's attention. However, when a cake mould is unavailable, the mother can use their creation using any available utensils.

3.2 Evaluation Phase

Evaluation phase consisted of two steps: a food test (organoleptic test) and real setting test.

3.2.1. Organoleptic test. The organoleptic/sensory test was performed to ensure the acceptability of the products among the target groups. We invited caregivers (mothers) as the proxy panellists for their children because using children panellists tends to be problematic. We asked the panellists to score each formula according to their best knowledge of their children's preferences. It is understandable that mothers know best about their children's characteristics; hence we assumed mothers' response to the food test represents their children's. Twelve semi-trained mother panellists participated in the test. They were given a brief training before the test commenced. The inclusion criteria for the panellist was that mothers should be a caregiver of an underweight or severe underweight child (a child who has Weight for Age Z Score of less than minus 2 (WAZ <= -2).

The panellists were randomly assigned into two groups: group A, who commenced the first organoleptic test with milk-free formulas, and group B, who started with the standard formulas. We used cross-over design, hence for the second test, the panellists were assigned to the other group (see figure 1).

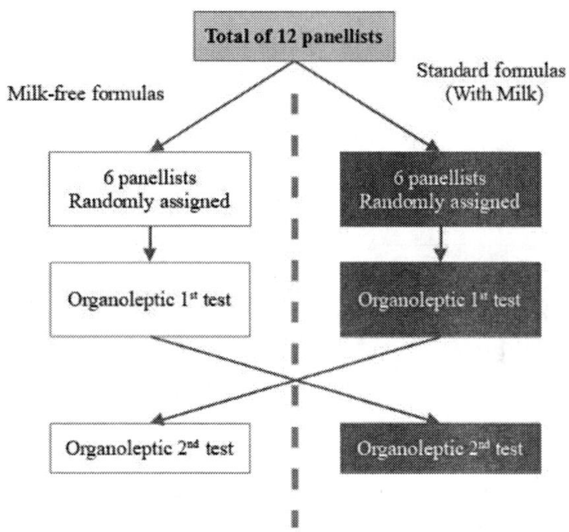

Figure 1. The study design.

3.2.2. Qualitative analysis of food acceptability. We chose the best formula (based on the organoleptic test result) from both groups. We provided raw materials based on the chosen formula and distributed it along with the recipe to 15 families with children under five who has nutrition problem or requiring special nutrition treatment. We asked the mother to reproduce the recipe and give the finished product to their children. After three days, we went on interviewing the caregivers to collect qualitative data to assess the acceptability and practicality of the products.

4. Statistical analysis

With regard to the organoleptic tests, both groups were analysed separately. Friedman's two-way analysis of variance by rank was performed to determine differences among formulas in those two groups. As significant differences were observed, the test was continued using the Wilcoxon signed-rank test to identify differences between formulas. [13]

5. Ethics

The ethical clearance for this research was obtained from the Universitas Qamarul Huda Badaruddin (UNIQHBA) Ethics Committee, approval number: 0025.

6. Results and discussions

We created a total of nine formulas: three milk-free formulas (Group MFFs) and other six standard formulas (group SFs). Total weight for each formula was 100 grams.

6.1. The formula

The composition of each formula can be seen in the following tables. The food codes shown in the tables are used as a reference to the Indonesian Food Composition database [12].

Table 1. Group MFFs: Milk-free formulas.

No	Food (Food Code)	Formula 1	Formula 2	Formula 3
1	Rice flour (AP004)	15 gr	15 gr	15 gr
2	Soybeans (CR017)	40 gr	-	-
3	Mung beans (CR014)	-	40 gr	-
4	Cowpeas (CR033)	-	-	40 gr
5	Refined sugar (MP007)	20 gr	20 gr	20 gr
6	Coconut Oil (KR011)	25 gr	25 gr	25 gr

Table 2. Group SFs: Standard formulas (with milk).

No	Food (Food Code)	For-mula 4	For-mula 5	For-mula 6	For-mula 7	For-mula 8	For-mula 9
1	Rice flour (AP004)	15 gr	-	15 gr	-	15 gr	-
2	Soybeans (CR017)	20 gr	40 gr	-	-	-	-
3	Mung beans (CR014)	-	-	20 gr	40 gr	-	-
4	Cowpeas (CR033)	-	-	-	-	20 gr	40 gr
5	Skim milk (JP009)	20 gr	15 gr	20 gr	15 gr	20 gr	20 gr
6	Refined sugar (MP007)	20 gr	20 gr	20 gr	20 gr	20 gr	20 gr
7	Coconut oil (KR011)	25 gr	25 gr	25 gr	25 gr	25 gr	20 gr

The main ingredients of the product were rice flour, soy flour, mung beans, cowpeas, fine granulated sugar, and coconut oil. An optional ingredient: skim milk, may be added when possible. Each formula contains energy that varies from 457.00 kcal to 502.55 kcal, and protein ranges from 10.46 – 21.75 grams, which is a good amount of macronutrient for those requiring nutrition supplement. Details of the macronutrient content of the formula are illustrated in table 3.

Formula one and formula five scores the best for macronutrient contents as it has the densest energy-protein component in their group. Formula one and five contains 501.65 and 502.22 kcal of energy, 17.46 and 21.75 grams protein respectively which are needed to quickly restore body mass for those in need of it. Formula five also has more fat than other formulas on the list and has the least amount of carbohydrate. Nonetheless, it has slightly lower kcal than international standard RUSF macronutrient content. The recommended energy value for RUSF is between 510 – 560 kcal, protein: 11 – 16 grams, fat 26 – 36 grams [5]. The recommended value for RUSF is almost identical to that of RUTF Energy: 520 – 550 kcal, Protein 10-12 % total energy, lipid 45-60% total energy [4].

Table 3. Macronutrient contents of each formula (100 grams).

	Energi (kcal)	Protein (grams)	Fat (grams)	Carbohydrate (grams)	Total cost/100 gr (US$)
Group 1 (MFFs)					
Formula 1	501.65	17.46	31.26	40.76	0.09
Formula 2	478.45	10.46	25.18	53.52	0.10
Formula 3	481.65	11.06	25.34	53.44	0.11
Group 2 (SFs)					
Formula 4	497.25	16.50	28.12	46.18	0.20
Formula 5	502.55	21.75	31.33	36.56	0.17
Formula 6	485.65	13.00	25.08	52.56	0.21
Formula 7	479.35	14.75	25.25	49.32	0.19
Formula 8	487.25	13.30	25.16	52.52	0.22
Formula 9	457.00	17.08	20.56	51.84	0.23
Maximum	*502.55*	*21.75*	*31.33*	*53.52*	*0.23*
Average	*485.64*	*15.04*	*26.36*	*48.52*	*0.17*
Minimum	*457.00*	*10.46*	*20.56*	*36.56*	*0.09*

With regard to affordability, we were able to minimise the total cost for each formula. The average price for the MFFs group was US$ 0.10/100 gram, halved the SFs group with the average cost of US$ 0.2/100 grams. This difference was affected by the addition of skim milk powder. We opted to use skim milk rather than full cream milk to minimise the product price. The price components in these products were only affected by the local market prices of the ingredients. The HB-RUSF price fluctuation was more predictable than commercial RUTF/RUSF in which price was affected by not only the core prices of the components but also the global competition of RUTF market [14].

Price analysis of the product shows that the cheapest formula was formula one, with only US$ 0.09 per 100 grams final product while the most expensive one was formula nine that cost US$ 0.23. From interviews with mothers who participated in the product testing, it emerged that even for the most expensive product on the list, the price was considered affordable. This signified that our price-wise principles were accomplished. Despite the lowest price, formula one contained total energy of 501.65, which was the most energy-dense formula in MFFs group. This was due to the use of soybean, which was the cheapest beans of the beans selected in this research. The price of soybeans in the local market was the cheapest among other beans. Yet soybeans contains the highest energy (381 kcal/100 gr) and protein (40.4 gr/100 gr) compared to mung beans and cowpeas [12].

Given the high price of milk, we provided the option milk-free formulas for mothers to allow them to choose the best formulas for their financial situation. In term of effectiveness as assessed through amino acid metabolism, however, a study showed that standard RUTF (with milk) and localised ones (milk-free with improved amino acid profile) had a comparable effect on amino acid metabolism to support nutritional recovery among severely malnutrition children under five [15].

6.2. Organoleptic analysis

Organoleptic analysis of group MFFs reveals that the formulas were similar in some sensory aspects including colour (p=0.971), texture (p=0.799), sweetness (p=0.273), saltiness (p=0.590), bitterness (p=0.125), rancid (p=0.572), and burnt (p=0.572). However, the difference was observed in the odour (p=0.000) and the taste (p=0.001). Further analysis showed that all three formulas in this group were all different in the odour. However, formula 2 appeared to be most favoured both in the odour and the taste among this group. The panellists were able to distinguish the odour and the taste of each formula. This was due to the distinct odour of each bean that can be perceived quite objectively by the panellists.

Among group SFs, we found differences in three aspects of sensory, including the odour (p=0.000), the sweetness (p=0.000), and the taste (p=0.000). All other sensory aspects were comparable statistically ($p > 0.05$), meaning that the panellists did not have any particular preferences in those aspects except for the odour, sweetness and taste. The more advanced statistical analysis showed that formula 7 was the most favourable formula both in the odour and the taste among other formulas in group SFs. Yet, formula 7 was not of the highest energy and protein contents. On the other hand, formula 5 that contains more energy was not as much acceptable as formula 2 or 7.

An interesting finding was that the products containing mung beans seemed to be superior with respect to the acceptability, odour, and taste. The panellist showed preferences toward formulas with mung beans both in group MFFs (formula 2) and SFs (formula 7). The taste and the odour of both formulas statistically more acceptable to the panellists. The interviews following the organoleptic test revealed that caregivers were confident that their children would prefer the product with mung bean better than the ones with cowpeas or soybeans. This was because generally their children had been introduced to mung bean flavour as weaning commenced, that was about six months old. Furthermore, children often are given mung bean porridge in *Posyandu* (integrated health post) as part of the *Posyandu* program, which helped to develop their preference and familiarity toward mung beans.

Another issue that worth considering is the colour of the formulas. As described by the caregivers, the colour of the products might be unattractive for children. The colour indeed is one of the crucial factors to consider when preparing food for children as research has found that food colour influences children choice of food [16]. However, as we focused on practicality, simplicity, and affordability, we dropped the option for colour enhancement from our RUSF-HB as we tried to create the most economical composition of ingredients, although we were aware that children generally prefer coloured food [17,18]. Besides, we did not prescribe colour agent in our formula for the fear that the less educated mothers might mistakenly use inappropriate colouring agents such as textile dyes rather than food colouring, which was quite common mistakes in villages. On top of all, the new RUSF/RUTF standards that endorsed by the UNICEF prefer cream to light colour or orange-brown colour [6], which exactly was the colour of this finished products (see figure 2 and 3).

The original texture of the product was generally a relatively smooth sticky flour. However, the final texture that served to children is dependent upon the process of grinding, printing / pressing into a cake mould. This means that the texture of the finished product can be adjusted to mothers' wishes. During the printing, we also found that adding a little fresh water to the mixture helps to improve the final texture. However, adding water to the product that was intended to store for days is not recommended as it increases the chances of bacterial multiplication in the product [5,19]. The texture of this RUSF was a little differ from the common texture of the standard RUSF such as thick paste that easy to squeeze out of a sachet [6]. The prescribed texture of therapeutic food was regarded as less important because the product was not intended to be mass-produced, nor to be stored for a long time.

With regard to odour, fresh odour of the beans used in this formula, particularly mung beans, was not pleasing for the panellists. Yet, we found that the addition of skim milk to the formula helps to neutralise the discernible odour of the beans. This explains why the odour score of the formulas from group SFs scored better than those in group MFFs. The odour of the finished product appears to comply with the UNICEF new standard such as free from foreign odours and flavours [6].

6.3. Food acceptability: qualitative analysis

After the organoleptic test was done, we distributed samples of formula two and formula seven to fifteen families with underweight children in order to access the children acceptability of the formulas in their natural setting. We perform interviews with the families three days after formulas distribution. We found that participating children and family enjoy the products as expressed by the mothers "my kid like it so much that she asked for it multiple time in the first day" and "he likes playing with it, smell and then eating it". Also, some mothers claimed to have tasted the food themselves and found it "familiar in the tongue". All respondents have agreed that the food was easy to make. Yet few mothers do not like the idea of using skim milk as it was thought to be costly. When the option for non-milk was given, however, participants split with more than half preferred to use skim milk while others dropped it for economic reason.

We also asked the mothers to explain their opinion regarding these formulas and how they view the formulas in regard to the family's need for food and nutrition, particularly during the pandemic Covid-19. As they had been informed that the formulas were of high energy food, the majority of the mothers saw these formulas as a doable alternative that can help them to fulfil their children's need for extra nutrition. With 'doable' we mean that the formulas were easy and affordable enough for mothers to make, even during this tough situation of the pandemic. Mothers also viewed these nutritious supplementary foods as simply a 'snack' that can be given to not only underweight children but also to normal children. Many of them served the formulas to normal adults as alternative foods to be consumed in the afternoon tea. Hence, as part of ethics consideration, we gave education for these families that the use of the formulas as a regular snack without proper control possibly create other problems such as the overweight due to the density of energy and protein of the food [20]. More importantly, the formulas were designed as supplementary foods for mildly malnourished children. But in case of emergency, the products can also be used as a treatment for severely malnourished children. However, the exclusive and prolonged usage of the formulas to treat severely malnourished children is not recommended as the products contain slightly less energy than the recommended ones, and the texture is not in the form of a paste which is a requirement for standard therapeutic food.

Figure 2. The RUSF-HB ingredients.

Figure 3. The finished product.

The strength of these formulas lies in the use of local materials that are affordable and can be obtained around the community, which allow home-based care for those with nutrition problems. Home-based care of malnourished children allows saving financial resources [21] from the reduced cost of the overall treatment. Moreover, the selection of local ingredients has an impact on the acceptance of many sensory aspects that effortlessly suit the locals' taste. Yet, the selection of local foods as the main component of RUSF/RUTF should proceed carefully as not all local food be equally accepted as demonstrated in a Cambodian study of Fish-based RUTF [22].

7. Limitations

There was no laboratory test performed to determine nutrient loss during the processing. However, as all ingredients used were of standard and typical food with no sophisticated processing, we were confident to assume that there would be insignificant loss of nutritional content. There was no product storage stability test performed because the initial intention of developing HB-RUSF was to be produced at the household level and be consumed immediately or within a short time. This means that we could not determine how long the product can be safely stored before expired although we were also aware that bean-based RUSF/RUTF commonly regarded safe to store the product for twelve months and some products were proved to be stable over 18 months [23]. For a scale-up development, however, thorough laboratory test, including nutrient content after processing, metal contamination, and food storage stability, would be a necessity.

8. Conclusions

We were able to create simple RUSF-HB formulas using common local foods that needed only simple processing with available cooking utensils. The RUSF-HB is a viable, doable and affordable alternative to the more expensive industrial-grade mass-produced supplementary foods as it uses only locally-produced ingredients. The formulas were easy enough to reproduce, and the caregivers (mothers) were provided with a flexible option: to use or drop milk powder from the formula dependent upon their circumstances. Except for the odour and taste, all developed formulas were comparable in most sensory aspects. Soybeans-based formulas tended to be less expensive and contain more macronutrients but were poorly rated for their organoleptic test. On the other hand, mung-beans formulas scored better in their organoleptic test, although containing slightly less energy and protein compared to soybeans-based formulas. Formula 2 and formula 7, both were mung beans-based formulas, emerged as the most acceptable formulas for children and their caregivers. These formulas can supply extra nutrient to mild acute malnutrition children or used as treatment foods in the event of an emergency for severely malnourished children, particularly during the Covid-19 pandemic.

9. Recommendations

The further clinical test is warranted to access the real contribution of the developed RUSF-HB formulas to the improvement of general nutrition status of malnourished children. In addition, as we notice soybean-based formula have more energy and protein content but were scored down from its odour, further research might need to find ways to eliminate the strong odour of soybeans while improving colour and texture of the final products. This helps to provide a wider range of RUSF-HB for mothers and children to choose from.

References

[1] Bong C-L, Brasher C, Chikumba E, McDougall R, Mellin-Olsen J, Enright A 2020 The COVID-19 Pandemic: Effects on Low-and Middle-Income Countries. *Anesth Analg.*

[2] Kalea A Z, Klimis-Zacas D 2020 Challenges and Insights amidst the Covid-19 pandemic: Nutrition, the immune system and disease risk. *Curr Opin Clin Nutr Metab Care* **23**(4):233–5.

[3] Gan W H, Lim J W, David K O H 2020 Preventing intra-hospital infection and transmission of COVID-19 in healthcare workers. *Saf Health Work.*

[4] WHO 2017 Application for inclusion of ready-to-use therapeutic food (RUTF) on the WHO model list of essential medicines (EML) and model list of essential medicines for children (EMLc) [Internet]. New York, USA: World Health Organisation; Available from: https://www.who.int/selection_medicines/committees/expert/21/applications/s26_therapeutic_food_add.pdf

[5] WFP 2019 Technical specification for ready-to-use supplementary food (RUSF) [Internet]. New York, USA. Available from: https://docs.wfp.org/api/documents/WFP-0000104979/download/

[6] Fleet A 2019 Expert meeting on ready-to-use-therapeutic foods (RUTF) [Internet]. Available from: https://www.unicef.org/supply/media/3316/file/RUTF-technical-expert-meeting-report-

Sept-2019.pdf
[7] Tandon S, Arora B, Prasad M, Narula G, Banavali S, Gadge M, et al. 2018 G360(P) A randomised controlled trial of ready to use therapeutic food (RUTF) for moderate/severe acute malnourished indian children with cancer. *Arch Dis Child* [Internet] **103**(1):A146 LP-A147. Available from: http://adc.bmj.com/content/103/Suppl_1/A146.2.abstract
[8] Hendrixson DT, Godbout C, Los A, Callaghan-Gillespie M, Mui M, Wegner D, et al. 2020 Treatment of severe acute malnutrition with oat or standard ready-to-use therapeutic food: a triple-blind, randomised controlled clinical trial. *Gut* [Internet];gutjnl-2020-320769. Available from: http://gut.bmj.com/content/early/2020/03/16/gutjnl-2020-320769.abstract
[9] Sandeep K, Vikas K, Satish K, Sheenam S, Jaspreet K 2020 Considerations for development of low-cost supplementary foods for lactating women in India – a review [Internet]. Vol. ahead-of-p, Nutrition & Food Science. Available from: https://doi.org/10.1108/NFS-02-2020-0036
[10] Kohlmann K, Callaghan-Gillespie M, Gauglitz JM, Steiner-Asiedu M, Saalia K, Edwards C, et al. 2019 Alternative ready-to-use therapeutic food yields less recovery than the standard for treating acute malnutrition in children from Ghana. *Glob Heal Sci Pract* [Internet]. **7**(2):203 LP – 214. Available from: http://www.ghspjournal.org/content/7/2/203.abstract
[11] Ozili PK, Arun T 2020 Spillover of COVID-19: impact on the global economy. *Available SSRN 3562570.*
[12] Kementrian Kesehatan Republik Indonesia 2017 Data komposisi pangan Indonesia [Internet]. 2017 [cited 2020 Nov 4]. Available from: https://www.panganku.org/id-ID/beranda
[13] R Core Team. R 2020 A language and environment for statistical computing [Internet]. Vienna, Austria; Available from: https://www.r-project.org/
[14] Kakietek J 2018 Determinants of the global prices of ready-to-use therapeutic foods. *Food Nutr Bull.* **39**(3):435–48.
[15] Sato W, Furuta C, Matsunaga K, Bahwere P, Collins S, Sadler K, et al. 2018 Amino-acid-enriched cereals ready-to-use therapeutic foods (RUTF) are as effective as milk-based RUTF in recovering essential amino acid during the treatment of severe acute malnutrition in children: An individually randomised control trial in Malawi. *PLoS One* **3**(8):e0201686.
[16] Mura Paroche M, Caton SJ, Vereijken CMJL, Weenen H, Houston-Price C 2017 How infants and young children learn about food: A systematic review [Internet]. Vol. 8, Frontiers in Psychology. 2017. p. 1046. Available from: https://www.frontiersin.org/article/10.3389/fpsyg.2017.01046
[17] Rioux C, Picard D, Lafraire J 2016 Food rejection and the development of food categorisation in young children. *Cogn Dev.* **40**:163–77.
[18] Carstairs S A, Caton S J, Hetherington M M, Rolls B J, Cecil J E 2020 Colour as a cue to eat: Effects of plate colour on snack intake in pre-school children. *Food Qual Prefer* **83**:103862.
[19] UNICEF. Ready-to-use therapeutic foods scale-Up [Internet]. New York, USA; 2019. Available from:https://www.unicef.org/evaldatabase/files/8._EOI_Case_Study_Ready-to Use_Therapeutic_Foods_Scale-Up.pdf
[20] Garrido-Miguel M, Oliveira A, Cavero-Redondo I, Álvarez-Bueno C, Pozuelo-Carrascosa DP, Soriano-Cano A, et al. 2019 Prevalence of overweight and obesity among European preschool children: A systematic review and meta-regression by food group consumption. *Nutrients* **11**(7):1698.
[21] Manary M, Callaghan-Gillespie M 2018 Ready-to-use therapeutic food (RUTF) and ready-to-use supplementary food (RUSF) *Sight Live* **32**:40.
[22] Sigh S, Roos N, Sok D, Borg B, Chamnan C, Laillou A, et al. 2018 Development and acceptability of locally made fish-based, ready-to-use products for the prevention and treatment of malnutrition in Cambodia. *Food Nutr Bull.* **39**(3):420–34.
[23] Owino VO, Irena AH, Dibari F, Collins S 2014 Development and acceptability of a novel milk-free soybean–maise–sorghum ready-to-use therapeutic food (SMS-RUTF) based on industrial extrusion cooking process. *Matern Child Nutr.* **10**(1):126–34.

In vitro propagation of six selected sugarcane mutant clones through leaf explants

R Purnamaningsih[1,*], D Sukmadjaja[1], S Suhesti[2], and S Rahayu[1]

[1] Balai Besar Penelitian dan Pengembangan Bioteknologi dan Sumberdaya Genetik Pertanian, Jl. Tentara pelajar no.3A Bogor, Indonesia
[2] Pusat Penelitian dan Pengembangan Perkebunan, Jl. Tentara Pelajar No. 1 Bogor, Indonesia

* E-mail: raga_padmi@yahoo.com

Abstract. Six mutant clones of sugarcane with high productivity have been produced through tissue culture techniques combined with mutations using gamma-ray irradiation and Ethyl Methane Sulfonate. The six mutant clones have been tested for stability in the field. They are proven to have high productivity and yields, so that they are very potential to be developed as superior varieties. To support the planting material sufficiency of these clones, an efficient propagation method was needed. Media formulations with different physical properties and composition of growth regulators were tested to obtain high seedling propagation rates. The media formulation for callus induction was Murashige dan Skoog (MS) + 3 mg/l 2,4-D + 3 g/l casein hydrolysate + 3% sucrose and for shoot regeneration was MS + 0,5 mg/l BA + 0,1 mg/l IBA + 100 mg/l PVP and 2% sucrose. Shoot proliferation was carried out on MS liquid (1, ½) + (0.3; 0.5 mg/l) BA + 0.1 mg/l IBA + 1 mg/l Kinetin + (0; 0.5 mg/l) GA3+ sucrose 2%. The results showed that callus induction, callus regeneration, and shoot proliferation of sugarcane mutant clones were influenced by the genotype and medium composition. The fastest callus induction was obtained from the MSP-4 clone (5.82 days), and the longest was MSB-7 (8.82 days). The largest callus diameter was obtained from MSB-6 clone on MS medium containing 1 mg/l BA, 100 mg/l PVP, and 2% sucrose. The highest number of shoots was obtained from the MSB-6 clone, while the least number of shoots conducted from the MSB-8 clone. The MSB-8 clones were more difficult to regenerate compared to the others. The best media formulation for shoot proliferation was ½ MS containing 0.5 mg/l BA, 1 mg/l Kinetin, and 0.1 mg/l IBA, while the best formulation for rooting was ½ MS.

1. Introduction

Sugarcane (*Saccharum officinarum* L.) is a monocotyledonous crop plant that belongs to the family Poaceae. It is the main crop that supplies sugar and ethanol production. The main byproducts of the sugar industry are bagasse and molasses. Molasses is the main raw material for alcohol and alcohol-based industries. Excess bagasse is also used as a raw material in the paper industry. Indonesia's sugar needs exceed domestic production capacity, so part of it must be met from imports. To achieve this self-sufficiency, providing superior seeds is one of the priorities.

Increasing sugarcane production is influenced by the provision of quality superior seeds. The use of improved varieties has a contribution of 30-35% to yield. Releasing new sugarcane varieties that have superior characteristics must be supported by the propagation of seeds in large quantities to anticipate demand for seeds from newly released varieties. The propagation of sugarcane seeds, which will be exploited on a large scale in a fast time, will be difficult to achieve through conventional techniques because it suffered from low propagation rates, expensive labor, time-consuming and potential transmission of pathogens through seed cane from generation to generation [1].

Six mutant clones of sugarcane with high productivity have been produced through tissue culture techniques combined with mutations using physical mutagens (gamma-ray irradiation) and chemical mutagens (Ethyl Methane Sulfonate). The six mutant clones have been tested for stability in the field. They are proven to have high productivity and yields, so that they are very potential to be developed as superior varieties. To support the planting material sufficiency of these clones, an efficient propagation method was needed.

Plant tissue culture offers the best methodology through micropropagation of sugarcane for quality and phytosanitary planting material at a faster rate in a shorter period. Propagation of seeds through tissue culture could be scheduled according to the needs and the desired amount, and the level of uniformity of plant material is quite high. According to [2], sugarcane from tissue culture has a better growth rate, tillers, stem production, and yields compared to conventional plants.

The multiplication and proliferation of shoots are a crucial stage in the propagation of any species for commercial exploitation, and the most rapid rates are required. The most common additives to standard media are cytokines, usually BA and Kinetin alone or with the combination of low amount auxins like NAA, IBA, and IAA [1]. However, seed propagation through tissue culture influenced many factors, including type and combination of plant growth regulators, explants type, culture medium composition, and genotype [3]. Therefore, it is necessary to optimize the method for each genotype to achieved the seed production target.

The last stage in shoot propagation through tissue culture is transferring plantlets from the laboratory to the greenhouse (acclimatization). The factor that determines the success of plantlet acclimatization is the root conditions of the plantlets. Plantlets that have perfect roots will greatly support the success of obtaining quality seeds.

2. Methods

2.1 Induction of indirect adventitious shoots and callus regeneration

Plant material (explants) used in this study were young leaves from 6 mutant clones. The young leaves used were rolled leaves isolated from the shoots of sugarcane. The leaves washed using a detergent solution mixed with 30% NaOCl. Then sugarcane shoots were put into Laminar Air Flow (LAF) for further sterilization. The sugarcane shoots are then dipped in 96% alcohol, burned over a fire, and cut into ± 2 mm in a sterile petri dish. After that, the slices of sugarcane leaves are inserted into the growing medium.

The first stage of this activity is callus induction from young leaves. The media formulations used for callus induction were Murashige and Skoog (MS) basic media + 3 mg / l 2,4-D + 3 g / l Casein Hydrolyzate + 3% sucrose. Each bottle contains 2 explants with 20 repetitions. Callus was induced in dark conditions in the culture room at a temperature of 22 ± 2°C...

The callus obtained was transferred to shoot regeneration media, namely MS with the addition of BA (0.5 and 1 mg/l), thidiazuron (0 and 0.1 mg/l), IBA 0.1 mg / l, PVP 100 mg / l, and 2% sucrose) . The cultures were incubated in the culture room at 22°C. Irradiation is carried out for 16 hours in light conditions and 8 hours in dark conditions with a radiation intensity of 1000 lux. The variables observed were callus initiation time, callus percentage formed, callus diameter, and callus color.

2.2 Shoot proliferation using liquid media

Clones with a good growth response were transferred to the media for induce shoot proliferation using two (2) media formulations, namely: 1) MS (liquid) with the addition of 3 mg/l BA, 1 mg/l kinetin, 0.5 mg/l GA_3, 2% sucrose, and 200 mg l PVP, and 2) ½ MS (liquid) with the addition of 0.5 mg/l BA, 0.1 mg/l IBA, 100 mg/l PVP and 2%.sucrose.

The experimental design used was a completely randomized design with 20 replications. The bottles were placed in a culture rack using a TL lamp with an irradiation intensity of 1000 lux for 16 hours a day at a temperature of 22°C. Culture was incubated statically. Irradiation carried out for 16 hours. Observations were made on the number of shoots produced.

2.3 Rooting induction

Root induction were performed on MSP-2 and MSB-6 clones using MS1/2 + IBA (0 and 0.1 mg / l) media. The experimental design used was a completely randomized design with 20 replications.

The experimental design used was a completely randomized design with 20 replications. The bottles were placed in a culture rack using a TL lamp with an irradiation intensity of 1000 lux for 16 hours a day at a temperature of 22°C.. Observations were made on the number of roots formed.

2.4 Data analysis

Data were analyzed using ANOVA at the 95% confidence level. If there is a significant difference between treatments, then a further test is carried out using the Duncan test at the 95% confidence level.

3. Resulst and discussion

3.1 Induction of indirect adventitious shoots and callus regeneration

The use of MS + 3 mg/l 2,4-D + 3 g/l casein hydrolyzate + 3% sucrose induced callus formation in all clones, with varying responses in all observed variables (Table 1). 2,4-D is a type of auxin with high activity, effective for sugarcane callus induction [4]; [5]. Growth regulators from the auxin group play a role in the callus formation process. At the same time, the selection of types and concentrations is determined by the type of growth and development of the explants, and the genotype used [5], [6]; [7]). The research of [2] showed that 2.4-D with a concentration of 1-6 mg/l was effective in inducing sugarcane callus formation. The addition of casein hydrolyzate to the media containing auxins seems to increase embryogenic callus formation [8]. Casein hydrolyzate is a precursor in the formation of nucleic acids and other cellular processes.

The observations showed that the MSP-4 clones could form callus in the fastest time (5.82 days) than other mutant clones, while the callus from MSB-7 and MSB-8 clones were formed the latest at 8.82 and 8.03 days, respectively. MSP-4 clones were also able to produce the highest callus formation, namely 86.13%, while the callus diameter of the six clones was almost the same (Table 1). The callus color of each clone was different. The callus, which is black in color, indicates that the ability of callus proliferation is low (Figure 1).

Table 1. Callus induction and the percentage callus formation of six sugarcane mutant clones.

Mutant clones	Callus induction (days)	Callus formation (%)	Diameter of callus (cm)	Color
MSP-2	6,49	47,50	1,08	Reddish white
MSP-3	6,18	50,74	1,28	Brown
MSP-4	5,82	86,13	1,73	Reddish
MSB-6	6,69	65,92	1,78	Yellowish white
MSB-7	8,82	36,43	1,62	Black
MSB-8	8,03	37,94	1,64	Black

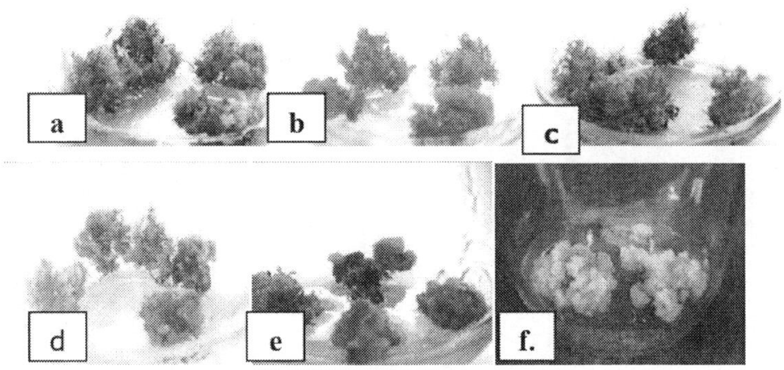

Figure 1. Callus appearance of six clones of mutant sugarcane.
a = MSP-2, b= MSP-3, c=MSP-4,
d= MSB-6, e= MSB-7, f =MSB-8

TP1, TP2, TP3, and TP4 media influenced callus development until callus regeneration formed shoots. The statistical analysis results showed an interaction between the media formulations and the mutant clones on callus diameter (Table 2). The MSB-6 mutant clones produced the largest callus diameter in all media formulations. In contrast, in the MSB-8 clones, the best callus development was obtained using TP2 and TP3 media, indicated by callus diameters of 2.40 cm and 2.0 cm. The smallest callus diameter resulted from MSP-3 clones.

Table 2 . Diameter of callus on media containing cytokinins, 5 weeks after planting

Media / **Mutant clones**	**TP1**	**TP2**	**TP3**	**TP4**
MSP-2	1,92 cdefg	1,92 cdefg	2,05 bcde	1,63 ghi
MSP-3	1,40 ij	1,25 j	1,44 hij	1,42 ij
MSP-4	1,63 ghi	2,02 bcdef	1,83 defg	1,83 defg
MSB-6	2,22 abc	2,40 a	2,19 abc	2,12 abcd
MSB-7	1,69 fghi	1,61 ghi	1,47 hij	1,76 efgh
MSB-8	1,94 cdefg	2,30 ab	2,00 bcdef	1,94 cdefg

Means within a colomn having the same lettter are not statistically significant (p=0,05) according to Duncan's Multiple Range Test.
TP 1 = MS+ BA 0.5 mg/l + PVP 100 + Sukrosa 2%
TP 2 = MS+ BA 1 mg/l + PVP 100 + Sukrosa 2%
TP 3 = MS+ BA 0.5 mg/l+ TDZ 0.1 mg/l + PVP 100 + Sukrosa 2%
TP 4 = MS+ BA 1 mg/l +TDZ 0.1 mg/l+ PVP 100 + Sukrosa 2%

The TP2 media formulation appears to be the best medium for callus development. Increasing BA concentration from 0.5 mg/l (TP1) to 1 mg/l (TP2) could increase callus development. However, BA 1 mg/l combined with thidiazuron 0.1 mg/l (TP4) suppresses callus development. The use of BA as a growth regulator in the in vitro propagation of sugarcane has also been reported by [9].

The ability of shoot regeneration varied depending on the plant genotype. MSB-6 clones produced the highest number of shoots on TP2 media (21.6 shoots), followed by MSP-2 on TP1 media (16.50 shoots), while MSP-3 and MSB-8 clones produced the least number of shoots. In all media formulations (Table 3). The results indicated that each clone had a different response to the same media formulation

[10]. This condition indicates that each mutant clone has changed its genetic trait even though all the mutant clones come from the same parent (variety). Changes in genetic traits are due to the treatment given at the time of culture, which causes changes in the DNA arrangement that makes the gene expression is different. These various responses are in accordance with the statement of [11] that genotype affects the growth and development of a plant species.

Table 3. The number of shoots of six mutant clones in four media formulations, 5 weeks after planting

Mutant clones \ Media	TP1	TP2	TP3	TP4
MSP-2	16,50 ab	11,50 bcde	12,83 bc	6,67 cdef
MSP-3	3,83 ef	2,00 f	3,60 ef	2,20 f
MSP-4	1,00 f	1,30 f	3,80 ef	5,00 def
MSB-6	12,22 bcd	21,60 a	16,60 ab	13,00 bc
MSB-7	5,78 cdef	4,67 def	1,22 f	2,30 f
MSB-8	1,10 f	1,10 f	1,10 f	1,60 f

Means within a colomn having the same lettter are not statistically significant (p=0,05) according to Duncan's Multiple Range Test
TP 1 = MS+ BA 0.5 mg/l + PVP 100 + Sukrosa 2%
TP 2 = MS+ BA 1 mg/l + PVP 100 + Sukrosa 2%
TP 3 = MS+ BA 0.5 mg/l+ TDZ 0.1 mg/l + PVP 100 + Sukrosa 2%
TP 4 = MS+ BA 1 mg/l +TDZ 0.1 mg/l+ PVP 100 + Sukrosa 2%

Sugarcane callus production and regeneration depend on the plant genotype, the source of the explants used, and the media formulation to regenerate it [5]; [9]; [12]; [13]. Plant regeneration through tissue culture is usually specific, which means that media formulations that can be used to regenerate certain plant varieties may not necessarily be used for other varieties [8].

The callus regeneration ability to form shoots was not affected by the callus induction rate of each mutant clone. MSP-4 clones were able to form a callus at the fastest time than MSP-2 and MSB-6 clones (Table 1), but callus was not able to divide and form shoots rapidly (Table 2 and 3). On the other hand, MSP-2 and MSB-6 clones, although initially, the callus formation was rather slow, the callus could develop well, even the fastest shoot formation. The results obtained indicated that the regeneration ability of each clone was different.

Table 4. The number of shoots from two mutant clones in shoot proliferation, 4 Weeks After Planting

Media (liquid)	MSP-2	MSB-6
MS + BA 0.3 + K1 + IBA0.1 +GA_3 0.5+S2% (P1)	9,80 abc	7,00 abc
½ MS + BA 0.5 + K1+ IBA0.1+ S2% (P2)	11,00 ab	14,67 a

Means within a colomn having the same lettter are not statistically significant (p=0,05) according to Duncan's Multiple Range Test

Statistical analysis showed no significant effect of the two media formulations used on the number of shoots produced. The highest number of shoots was produced from MSB-6 clones on P2 media (14.67 shoots), as well as in MSB-2 clones (11 shoots) (Table 4). The use of liquid media does not appear to increase the proliferation of shoots of sugarcane. This is different from the research by[14], which showed that the use of MS media containing BA 0.5 mg / l and IBA 0.1 mg / l could increase the number of shoots produced.

Figure 2. Shotts proliferation on liquid media.

3.2 Root induction

Rooting is one of the stages in plant propagation that determines the transfer of plant tissue culture products (plantlets) to the greenhouse (acclimatization). Plants with many roots with root hairs are needed for plantlet adaption in the greenhouses.

The statistical analysis results showed no interaction between clones and the formulations used on sugarcane rooting. The number of roots produced was not significantly different between R1 and R2 media, namely 13.50 and 16.20, respectively (Table 5). Similar results were obtained from the number of roots in the MSP-2 and MSB-6 clones (Table 6).

The use of auxin growth regulators (NAA and IBA) is effective in accelerating the initiation of the root formation of sugarcane. The use of IBA 1 mg/l was able to induce rooting of sugarcane by 90% [10]. However, in this study, the use of ½ MS medium without the addition of IBA was able to induce rooting in sugarcane clones to reduce production costs. Different results were also shown from the study of [15]., which showed that the use of 2.5 mg/l NAA on MS medium stimulated root formation.

Table 5. Effect of media formulation on the number of roots, 4 weeks after planting.

Media formulation	2 MST	3 MST	4 MST
½ MS + IBA 0.1 (R 1)	4,6 a	9,9 a	13,50 a
½ MS (R 2)	5,0 a	12,6 a	16,20 a

Means within a colomn having the same lettter are not statistically significant (p=0,05) according to Duncan's Multiple Range Test

Table 6. Roots number of two mutant clones, 4 weeks after planting.

Mutant clones	2 MST	3 MST	4 MST
MSP-2	5,25 a	12,4 a	17,45 a
MSB-6	4,60 a	10,63 a	14,58 a

Means within a colomn having the same lettter are not statistically significant (p=0,05) according to Duncan's Multiple Range Test

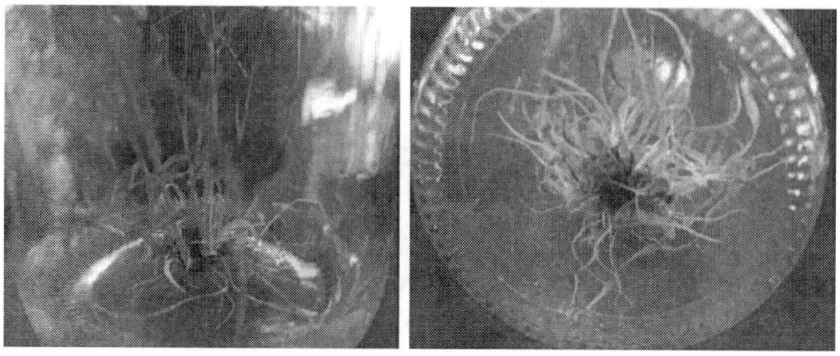

Figure 3. Root performances of mutant clones.

4. Conclusion

Callus induction, callus regeneration, and shoot proliferation of sugarcane mutant clones in vitro were influenced by the media's genotype and composition. The fastest callus induction came from the MSP-4 mutant clone, which was 5.82 days, and the longest was MSB-7, which was 8.82 days. The largest callus diameter was obtained from MSB-6 clones on MS medium containing BA 1 mg / l, PVP 100 mg / l, and sucrose 2%. The highest number of shoots was obtained from MSB-6 clones, while the least number of shoots were conducted from MSB-8 clones. MSB-8 clones tend to have a difficult response to regenerate than other clones. The best media formulation for shoot proliferation was ½ MS containing 0.5 mg / l BA, 1 mg / l Kinetin and 0.1 mg / l IBA, while the formulation to induce rooting was ½ MS.

References

[1] Getnet B 2017 Review On In Vitro Propagation of Sugarcane to Advance The Value of Tissue Culture. *Agric. Res, and Technol.* **54** 99-104

[2] Jalaja N C, Neelamathi D and Sreenivasan T V 2008 *Micropropagation for Quality Seed Production in Sugarcane in Asia and The Pacific* (FAO, APCoAB and APAARI) 46

[3] Hailu M, Chimdessa M and Muthswamy M 2018 In Vitro Propagation of Selected Sugarcane (*Saccharum Officinarum* L.) Varieties (C 86-165 and C 86-12) Through Shoot Apical Meristem. *Int. J Hort & Agric.* www.symbiosisonline.com

[4] Gandonou C, Errabii T, Abrini J, Idaomari M, Chibi F and Senhaji N S 2005 Effect of Genotype on Callus Induction and Plant Regeneration from Leaf Explants of Sugarcane (*Saccharum* sp.). *African J. Biotechnol* **411** 1250-55

[5] Ali A, Naz S, Siddiqui F A and Iqbal J 2008 Rapid Clonal Multiplication of Sugarcane (*Saccharum officinarum*) Through Callogenesis and Organogenesis. *Pak. J. Bot.* **411** 123-138

[6] George F E, Hall M A and Geert-Jan De Klerk 2008 *Plant Propagation by Tissue Culture. The Background.* (Springer Publihser. Dordrecht, Netherlands). 501 p

[7] Jamil S, Shahzad R, Talha G M, Sakhawat G, Rahman S U, Sultana R dan Iqbal M Z 2017 Optimization of Protocols for In Vitro Regeneration of Sugarcane (*Saccharum officinarum*). *Int. J. Agronomy* **1** 1-8

[8] Purnamaningsih R 2006 Induksi Kalus dan Optimasi Regenerasi Empat Varietas Padi Melalui Kultur In Vitro J. AgroBiogen **22** 74-80

[9] Sukmadjaja D, Supriati Y dan Pardal S J 2014 Kultur Apeks untuk Penyediaan Bibit Unggul Tebu Varietas PS864 dan PS881. *J AgroBiogen* **102** 45-52

[10] Azizi AAA, Roostika I dan Efendi D 2017 Multiplikasi Tunas In Vitro Berdasarkan Jenis Eksplan pada Enam Genotipe Tebu (*Saccharum officinarum* L.). *J LITTRI.* **232** 90-7

[11] Ali S, Iqbal J and Khan M S 2010 Genotype Independent In Vitro Regeneration System in Elite

Varieties of Sugarcane *Pakistan J of Bot* **426** 3783-90
[12] Behera K K and Sahoo S 2009 Rapid In Vitro Micropropgation of Sugarcane (*Saccharum officinarum* L. cvNayana) Through Callus Culture *Nat. Sci* **74** 1-10
[13] Sukmadjaja D dan Mulyana A 2011 Regenerasi dan Pertumbuhan Beberapa Varietas Tebu (*Saccarum officinarum* L.) secara *in vitro*. *J Agro Biogen* **72** 106-18
[14] Roostika I, Rahayu S, Sukmadjaja D dan Mariska I 2015 Aplikasi Teknik Kultur Cair Untuk Efisiensi Perbanyakan Tunas Tebu Secara Kultur Jaringan. *Pros., Sem., Nas., Tebu Inovasi Teknologi Budi Daya Tebu Mendukung Swasembada Gula*, Malang 22 April 2014 Balittas. 248 hal
[15] Dinesh P, Thirunavukkarasu P, Saraniya A R and Ramanathan T 2015 In Vitro Studies of Sugarcane Variety Co-91017 Through Micropopagation of Shoot Tip Culture. *Adv Plants Agric Res* **26**:263–68

IOP Conf. Series: Earth and Environmental Science **883** (2021) 012076 doi:10.1088/1755-1315/883/1/012076

Utilization of herbal supplements on feed on growth and survival rate of catfish (*Clarias* sp.)

S N Pratama[1] and R C Mukti[1,*]

[1] Department of Aquaculture, Faculty of Agriculture, Universitas Sriwijaya

* E-mail: retnocahyamukti@unsri.ac.id

Abstract. The feed is one of the most important factors in the growth and survival of fish. One of the efforts made to increase the growth and survival of fish is the provision of herbal supplements in fish feed. The research aim is to provide information about the utilization of herbal supplements on the growth and survival of catfish (*Clarias* sp). This research has been carried out from July to August 2020 at Bandar Agung Lahat Village, South Sumatera. Catfish rearing containers use two waring units placed in concrete ponds with two treatment that P0: control (without herbal supplements) and P1: utilization of herbal supplements and. The data taken includes absolute growth and absolute length, survival, feed efficiency, and water quality. The utilization of herbal supplements has a good effect on fish growth, feed efficiency, and survival rate. Maintenance of P0 (control) resulted in an absolute weight of 9.79 grams, an absolute length of 6.3 cm, the survival of 85.15%, feed efficiency of 107.89% while maintenance of P1 with the utilization of herbal supplements resulted in an absolute weight of 14.17 grams, an absolute length 7.8 cm, survival 98.57%, feed efficiency 161.170%.

1. Introduction

Catfish (*Clarias* sp.) is a type of freshwater fish that has economic value and is very easy to cultivate. Catfish has several advantages, such as being able to be maintained more easily, being able to be kept in a narrow space, and being able to live in an unfavorable environment. Besides, catfish has a taste of meat that is quite tasty and contains high nutrition so that it is very much in demand by the community so that catfish farming continues to increase along with the increasing public demand [1]. The growth and high survival rate of catfish is an indicator of success in cultivation so that catfish production increases. One of the factors supporting this success is to feed.

The feed is one of the most important factors in the growth and survival of fish. The utilization of feed by fish is strongly influenced by the quality of feed both in terms of nutritional content and feed digestibility. Quality feed, apart from acting as the main energy source, is also expected to increase the digestibility of fish so that growth can be optimal [2]. The provision of herbal supplements in feed can increase the body's resistance to disease, save on feed use and increase the appetite for fish and improve the digestive system [3].

Herbal supplements are made by utilizing ingredients of natural origin so that production costs are affordable. Herbs are plants or plants that have high value in medicine. The utilization of herbs is now widely used, because this natural ingredient is more economical, easy to obtain, and has no side effects. Herbal supplements for fish are made from plants such as turmeric, kencur, and ginger which contain the main active substance called curcumin. The curcumin content in turmeric functions as an antioxidant and can increase palatability [4].

Information about the use of the herbal supplements has been widely published in the public, such as the addition of ginger, turmeric, kencur, and ginger extracts to catfish [5]. According to Syawal [6], giving herbal supplements in the form of kencur, ginger, turmeric with a fermentation process in feed can stimulate the appetite of catfish, increase fish immunity to disease and reduce stress levels of fish against environmental changes. Samsundari [7] reports that curcuma in feed can accelerate gastric emptying, causing hunger and stimulating appetite so that it can accelerate the growth of koi fish. Therefore it is necessary to research the utilization of herbal supplements on the growth and survival of catfish (*Clarias* sp.).

2. Experimental

2.1. Place and Time

This research was conducted in Bandar Agung, Lahat District, Lahat Regency, South Sumatra from July to August 2020.

2.2. Tools and Materials

The tools used include a concrete pool, jerry cans, a scale, a ruler, a thermometer, and styrofoam. While the materials used include catfish seeds 5-7 cm in size, 30% commercial protein pellets, turmeric, tamarind, kencur, rice, ginger, molasses, commercial probiotics, and water.

2.3 Method

2.3.1. Preparation of maintenance containers

The maintenance container uses a concrete pool measuring 3.5x2.5x0.8 m^3 which is divided into 2 parts with a waring barrier. Where one part is used as a control fish rearing container and the other part is used as treatment fish. Before using the pool, it is cleaned first by brushing and then rinsing with water, then filled with water as high as 0.5 m and left for 24 hours before use.

2.3.2. Making Herbal Supplements

The herbal ingredients are turmeric, kencur, ginger, tamarind each 100 g which have been cleaned then boiled with 1L of water to boil, then remove and chill. Herbal supplements are made by mixing all the ingredients consisting of boiled turmeric water with tamarind, kencur boiled water with 100 g of refined rice (rice flour), boiled water for ginger, 175 ml of molasses then stirring so that they are mixed. After that, it is fermented by adding 50 ml of commercial probiotics containing bacteria from the Lactobacillus genus and homogenized. The herbal supplement mixture is then stored in a jerry can and tightly closed. The herbal supplements are stored for 10 days, and once a day the lid of the jerry can be opened so that the steam inside the jerry can escape. After 10 days, the herbal supplement is ready for use. Herbal supplements can be stored for up to a month.

2.3.3. Addition of Herbal Supplements to Feed

There are two treatments, P0: control (without the addition of herbal supplements) and P1: treatment with the addition of herbal supplements. The addition of herbal supplements by mixing herbal supplements is done by mixing 200 ml of herbal supplements into a basin containing 1 kg of pellets and stirring until homogeneous. The result of the mixing is then aerated until it dries and then the pellets are stored in a tightly closed container. The addition of herbal supplements to feed is only done at the beginning of maintenance and can be done again if the feed runs out.

2.3.4. Fish Maintenance

The stocking density of fish used is 350 fish/m^3. Before stocking, catfish were acclimatized and then adapted to commercial feed for 10 days. After that, the fish were weighed and measured as many as 30 fish in length. Before being given treated feed, the fish have fast first. The feed that has been added with herbal supplements is given to catfish on the 11th day [8]. The amount of feed given is 5% of the biomass. The time of feeding is 3 times, that is 07.00 am, 12.00 am and 5.00 pm. Fish rearing will be carried out for 30 days and sampling is every week. Fish weights and lengths are weighed at baseline after adaptation, at sampling, and harvest.

2.4. Parameters

2.4.1. Growth

Absolute Length Growth

The absolute length growth during maintenance is calculated using the formula as follows:

$$L = Lt - Lo$$

Information:
L = absolute length growth of fish (cm)
Lt = Length of fish at the end of maintenance (cm)
L0 = Length of fish at the beginning of rearing (cm)

Absolute Weight Growth

The absolute weight growth during maintenance is calculated using the formula, as follows:

$$W = Wt - Wo$$

Information:
W = absolute growth of reared fish (g)
Wt = weight of fish at the end of rearing (g)
W0 = Fish weight at the beginning of rearing (g)

2.4.2. Specific Growth Rate

The specific growth rate during maintenance is calculated using the formula, as follows:

$$SGR = \frac{\text{In Wt} - \text{In W0}}{t} x\ 100\%$$

Information:
Wt = weight of fish at the end of rearing (g)
W0 = Fish weight at the beginning of rearing (g)
t = Time (days)

2.4.3. Feed Efficiency

Feed efficiency is calculated using the following formula:

$$EP = \frac{(Wt - Wd) - Wo}{F} x\ 100\%$$

Information:
EP = Feed efficiency (%)
Wt = final weight of fish biomass (g)
Wo = initial weight of fish biomass (g)
Wd = Weight of dead fish (g)
F = weight of feed given (g)

2.4.4. Survival Rate

The percentage of survival rate can be calculated using the formula, as follows:

$$SR = \frac{Nt}{No} x100\%$$

Information;
SR = Survival Rate (%)
N0 = Number of fish at the beginning of maintenance (tail)
Nt = Number of fish at the end of maintenance (tail)

2.4.5. Water quality

Water quality parameters measured are temperature and pH. Measurement of temperature and pH of maintenance water is carried out every morning.

2.4.6. Data analysis

Growth data, feed efficiency, survival, and water quality obtained will then be processed using Microsoft Excel and analyzed descriptively and supported by literature.

3. Results and discussion

3.1 Results

3.1.1. Absolute length growth

The absolute length growth of the catfish in 28 days culture is shown in figure 1.

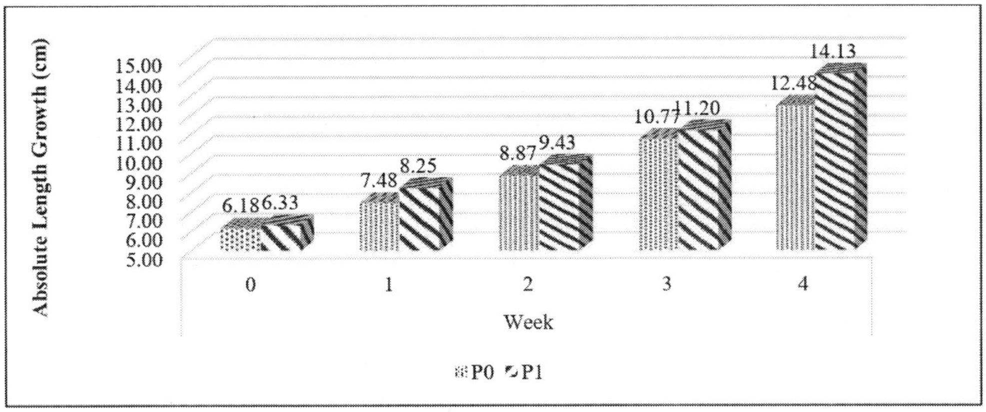

Figure 1. The absolute length growth rate of catfish in 28 days of culture.

The absolute length growth of P0 reaches 6.3 cm while P1 7.8 cm. The results showed that the fish treated with herbal supplements had a high absolute length growth than the fish that were not treated with herbal supplements.

3.1.2. Absolute weight growth

The absolute weight growth of the catfish in 28 days culture is shown in figure 2.

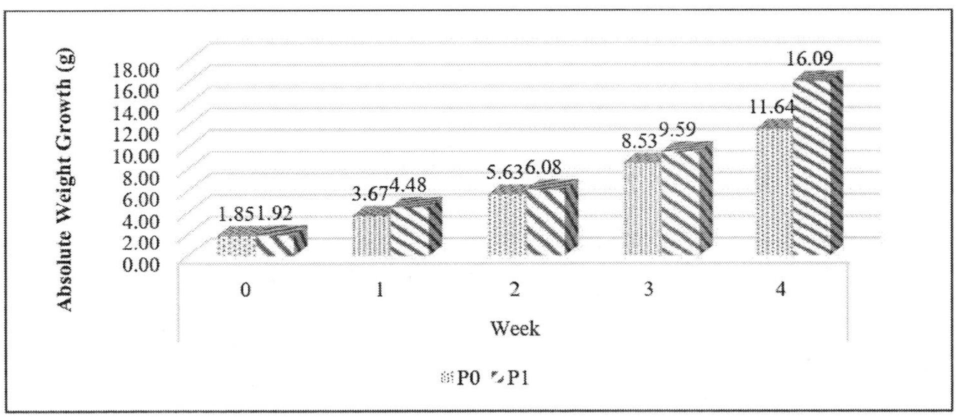

Figure 2. The absolute weight growth rate of catfish in 28 days of culture

The absolute weight growth of P0 reaches 9.79 g while P1 14.17 g. The results showed that the fish treated with herbal supplements had a high absolute weight growth than the fish that were not treated with herbal supplements.

3.1.3. Specific Growth Rate

The specific growth rate of the catfish in 28 days culture is shown in figure 3.

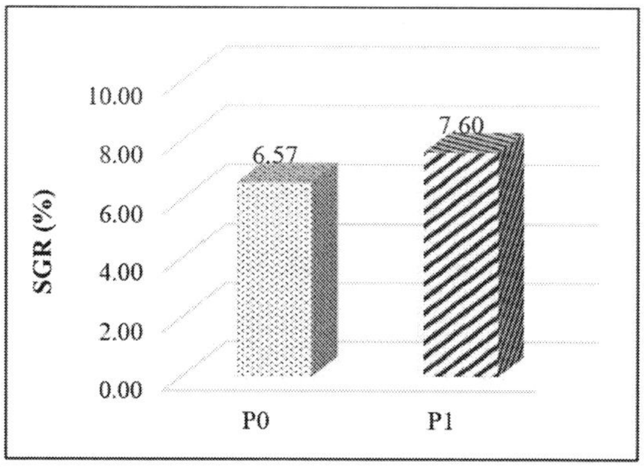

Figure 3. The specific growth rate of catfish in 28 days of culture

The specific growth rate of P0 reaches 6.57% while P1 7.60%. The results showed that the fish treated with herbal supplements had a high specific growth rate than the fish that were not treated with herbal supplements.

3.1.4 Feed Efficiency

Feed Efficiency of the catfish in 28 days culture is shown in figure 4.

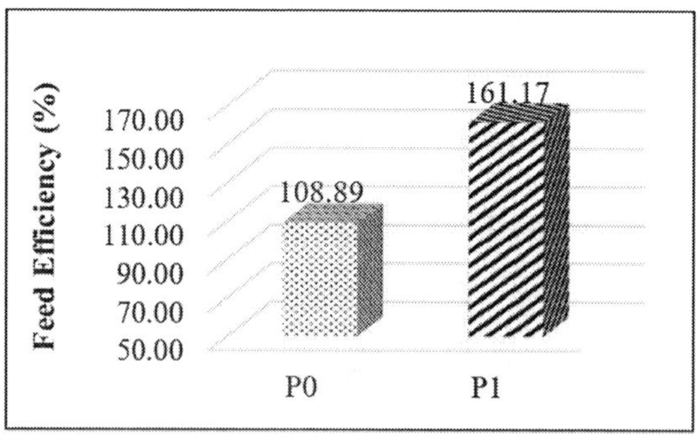

Figure 4. Feed efficiency of catfish in 28 days of culture

The feed efficiency of P0 reaches 108.89% while P1 161.17 %. The results showed that the fish treated with herbal supplements had a high feed efficiency than the fish that were not treated with herbal supplements.

3.1.5. Survival Rate

The survival rate of the catfish in 28 days culture is shown in figure 5.

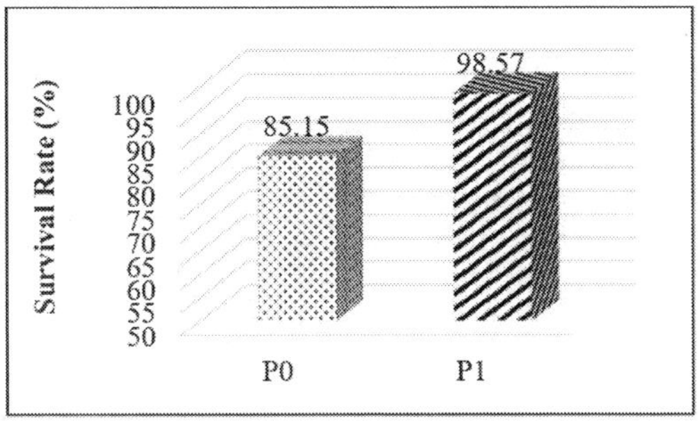

Figure 5. The survival rate of catfish in 28 days of culture

The survival rate of P0 reaches 85.15% while P1 98.57%. The results showed that the fish treated with herbal supplements had a high survival rate than the fish that were not treated with herbal supplements.

3.1.6. Water quality
Water temperature and pH of catfish culture in 28 days are shown in table 1. The water temperature of all treatments is in the normal range.

Table 1. The water temperature and pH of the catfish culture

Water quality	Result
Temp (°C)	26,4 -29,7
pH	6,3-7,2

3.2. Discussion
The results showed that the fish treated with herbal supplements had high growth over the fish that were not treated with herbal supplements. According to Yuwono [9], increased feed consumption provides an adequate supply of nutrients to meet increased metabolic needs during periods of rapid growth. According to Effendi [10], growth is an increase in length and weight over time. The factors that affect growth are heredity, age, parasites, food, and water temperature. The absolute length growth of 7.8 cm and the absolute weight growth of 14.17 g. The resulting growth value is high. Taqwa [11] reported that the growth of catfish fed with local raw materials resulted in an absolute length growth of 3.83 cm and an absolute weight growth of 9.94 g.

The high growth in the use of herbal supplements in feed is due to the feed added with herbal supplements consisting of turmeric, ginger, kencur, commercial probiotics, molasses, rice, tamarind contains many metabolites that act as antioxidants and maintain fish health by increasing immunity and appetite. fish [12]. The herbal ingredients used in this field practice are fermented using probiotics that contain lactobacillus bacteria which can change the compounds in the herbal ingredients, making them easier to absorb. The fermentation process causes the hydrolysis of complex compounds to make them easier to absorb [12]. Not only that, the fermentation process also produces a distinctive aroma that can increase the palatability of fish to feed [13]. The feed efficiency of those given herbal supplements was 161.17%, while the control commercial feed without the addition of herbal supplements was 108.89%. According to Winarti [14], the factors that influence the high value of feed efficiency are thought to increase the digestibility of feed protein. Herbal supplements fermented with probiotics are thought to still contain active bacteria and yeast so that when added to the feed, herbal supplements can remodel complex compounds in the feed into simple compounds. The process of breaking down bacteria and yeast may make the feed which is given additional herbal supplements can be utilized by the fish body efficiently and effectively for growth [15].

The feed efficiency value reaching 100% indicates that the addition of herbal supplements in feed can result in increased feed efficiency than no addition of herbal supplements. According to Purnomo [16], herbal sources added to feed can be converted as a source of energy to produce large amounts of protein biomass and can be utilized by fish with high protein. The survival percentage of catfish with the addition of herbal supplements is higher than control. Fish survival is influenced by various factors including water quality (dissolved oxygen, ammonia, temperature, and pH), age of fish, environment, health condition of fish, and feed [17]. The provision of herbal supplements can increase the immunity of catfish so that the fish become healthier, as shown by the high consumption of feed each day. According to Syawal [6] giving herbal supplements mixed with feed can stimulate fish development and reduce mortality. Death in control was caused by stocking density because it affected energy utilization and was a stress factor due to increased interaction between fish.

Water quality is one of the main environmental factors that can optimize the growth of fish in aquaculture containers. Temperature can affect the life activities of organisms such as fish appetite. If the temperature increases, it will increase the intake of food by the fish and the decrease in temperature causes the digestive and metabolic processes to run slowly and pH values that can interfere with fish life are too low (very acidic) and too high (very alkaline) pH values [18]. Based on the water quality that has been measured during field practice activities, it is known that the temperature obtained during maintenance reaches 29.7°C and the pH reaches 7.2. Based on BSN [19], the optimal temperature for catfish is 25-30°C and according to BSN [20], the optimum pH range for catfish growth ranges from 6-8. So, during temperature maintenance activities and within catfish tolerance, catfish grow well.

4. Conclusion

The utilization of herbal supplements has a good effect on fish growth, feed efficiency, and survival rate. Maintenance of P0 (control) resulted in an absolute weight of 9.79 grams, an absolute length of 6.3 cm, the survival of 85.15%, feed efficiency of 107.89% while maintenance with the addition of herbal supplements resulted in an absolute weight of 14.17 grams, absolute length 7.8 cm, survival 98.57%, feed efficiency 161.170%.

5. References

[1] Hasim dan Tuiyo 2014 *Pendederan ikan lele secara intensif* (Bogor: Balai Riset Perikanan Budidaya Air Tawar)

[2] Ahmadi H, Iskandar and Nia K 2012 Additopn of probiotic in comercial feed to growth of sangkuriang catfish (*Clarias gariepinus*) on nursery II *Jurnal Perikanan Dan Kelautan* **3**(4) 99-107

[3] Suseno dan Haetami 2004 *Pemeliharaan Ikan Lele dengan Menggunakan Probiotik* (Jakarta: Penebar Swadaya)

[4] Ariyanto A N, Iriyanti N and Mufti M 2013 Utilization of turmeric (*Curcuma domestica* Val) and bitter (*Andrographis paniculata* Nees) meal in feed comsumption and weight gain *Jurnal Ilmiah Peternakan* **1**(2): 471-478

[5] Gupta A, Mahajan S and Sharma R 2015 Evaluation of antimicrobial activity of *Curcuma longa* rhizome extract againts *Sthapylococcus aureus*. *Biotechnology Reports* **6**: 51-55

[6] Syawal H M, Riauwaty, Nuraini dan Hasibuan S 2019 Pemanfaatan pakan herbal (jamu) untuk meningkatkan produksi ikan budidaya *Dinamisia-Jurnal Pengabdian Kepada Masyarakat* **3**: 188- 193

[7] Samsundari S 2006 Pengujian ekstrak temulawak dan kunyit terhadap resistensi bakteri *Aeromonas hydrophila*. *Buletin Teknik Pertanian*. *Badang Litbang Pertanian* **7**(1): 33-36

[8] Puspitasari D 2017 Efektivitas suplemen herbal terhadap pertumbuhan dan kululushidupan benih

ikan lele (*Clarias* sp.) *Jurnal Ilman: Jurnal Ilmu Manajemen* **5**(1): 53-59

[9] Yuwono E, Sukardi P dan Sulistyo I 2005 Konsumsi dan efisiensi pakan pada ikan kerapu bebek (*Cromileptes altivelis*) yang dipuasakan secara periodik *Berk. Penel. Hayati* **10**(2) : 129-132

[10] Effendi M I 2002 *Biologi perikanan* (Yogyakarya: Yayasan Pustaka Nustama)

[11] Taqwa F H, Yulisman Y, Mukti R C, Rarassari M A, dan Antika R M 2020 Efektivitas pemanfaatan bahan baku lokal sebagai pakan ikan terhadap peningkatan produktivitas budidaya ikan lele (*Clarias* sp.) di Desa Sakatiga, Kecamatan Indralaya, Kabupaten Ogan Ilir, Sumatera Selatan. *Journal of Aquaculture and Fish Health* **9**(3), 222-231

[12] Meilina 2012 *Mudah dan Praktis Membuat Tahu Tempe. Teknologi Tepat Guna* (Jakarta: Wahyu Media)

[13] Simanjuntak D H, Herpandi and Lestari S D 2016 Chemical characteristics and antioxidant activity of water lettuce (*Pistia straiotes*) leaves kombucha during fermentation *Fishtech* **5**(2): 123-133

[14] Winarti, Subandiyono and Sudaryono A 2017 Utilization of fermentes wheat *Lemna* sp. in artificial feed on the growth of common carp fish (*Cyprinus carpio*) *Jurnal Sains Teknologi Akuakultur* **1**(2), 88-94

[15] Asriyanti I N, Johannes H dan Herawati V E 2018 Pengaruh penggunaan tepung *lemna* sp. terfermentasi pada pakan buatan terhadap tingkat pemanfaatan pakan, pertumbuhan dan kelulushidupan benih ikan lele dumbo (*Clarias gariepinus*) *Jurnal Rekayasa dan Teknologi Budidaya Perairan* **7** (1), 783-798

[16] Purnomo P D 2012 The effect of carbohydrate addition in aquaculture media towards production of intensive tilapia culture *(Oreochromis niloticus). Journal of Aquaculture Management and Technology* **1**(1): 161- 179

[17] Adewolu M A, Adenji C A and Adejobi A B 2008 Feed utilization , growth and survival of *Clarias gariepinus* (Burchell 1882) fingerlings cultured under different photoperiods *Aquaculture* **283**, 64-67

[18] Effendi H 2003 *Telaah Kualitas Air Bagi Pengelolaan Sumber Daya Dan Lingkungan Perairan* (Jakarta: Kanisius)

[19] BSN (Badan Standardisasi Nasional) 2014 6484.4: *Ikan lele dumbo (Clarias sp.) produksi benih.* (Jakarta: Badan Standardisasi Nasional)

[20] BSN (Badan Standardisasi Nasional) 2000 SNI : 01-6484.3.2000 *Produksi benih ikan lele dumbo (Clarias gaperiepinus x C. fuscus) kelas induk pokok (parent stock)* (Jakarta : Badan Standar Nasional)

IOP Conf. Series: Earth and Environmental Science **883** (2021) 012077 doi:10.1088/1755-1315/883/1/012077

Collaborative and sustainable management strategy of the Wallacea Key biodiversity area (KBA) in Kasa Island

D V Pattimahu[1,*], **A Kastanya**[1]

[1]Forestry Department, Pattimura University, Jl. Ir. M. Putuhena, Kampus Poka, Ambon, Maluku, Indonesia

* E-mail: debbypattimahu@yahoo.co.id ; aguskastanya@yahoo.com

Abstract. Kasa island, with an area of 52.80 Ha, was declared a KBA in the Wallacea region with the staut of wildlife sanctuary and Ocean Tourist Reserve because of its particular biodiversity and its service to the environment. This research aimed to determine the social, biophysical potential and the threats to the region, as well as to explore a collaborative management. The methods used included survey, spatial analysis of the vegetation and Focus Discussion Groups (FDG). The results showed a variety index level of 1.37%. The dominant tree species were belo hitam, salamuli daun besar and kayu besi pantai. The belo hitam species had the highest INP. The wildlife sanctuary is the habitat of the *Gosong Maluku* bird (*Eulipoa* and *Megapodis),* an endemic species called Gosong Maluku (*Euliopa wallacei*). Other species include Gosong Kelam (*megapodius freycinet*). The coastline with its white sand and natural ocean is a tourist attraction with development potential. The strategy which needs to be developed is by using strengths to take advantage of opportunities so they can overcome weaknesses and threats. the strategies developed are as follows: Collaborative Management; Strengthening The Capacity of Local Community and Formulation of regional policy regarding Kasa Island management

1. Introduction

Nature Reserve is an area with certain characteristics, both on land and in waters which functions mainly as a preservation area for the reservation of wild plants and animals diversity as well as their ecosystem which also functions as life support system area. In contrast, Wildlife Reserve is a part of nature reserve area which characteristics are in the form of diversity and or uniqueness of animals which for their survival in which their habitat must be maintained (Anonymous, 1990 [1]).

Kasa Island was designated as a conservation area according to Decree of the Minister of Agriculture Number: 653/Kpts/Um/10/1978 dated 25 October 1978 as a Nature Reserve which functions as a Wildlife Reserve in 900 Ha area and a Marine Park in 1,100 Ha area. The designation of Kasa Island forest and the surrounding sea as a conservation area was due to the existence of the island as a habitat for Maluku Gosong bird (Eulipoa wallacei), Gosong Kelam bird (Megapodius freycinet), Soa-soa or Ambon monitor lizard (Hidrosaurus amboinensis). On the other hand, it has a natural beauty with the presence of coral reefs and aquatic animals which need to be maintained for its sustainability for the benefit of science, education, culture, recreation, and tourism. (Anonymous,2016 [2]); (Anonymous, 2016.[3]).

IOP Conf. Series: Earth and Environmental Science **883** (2021) 012077 doi:10.1088/1755-1315/883/1/012077

The management of Kasa Island conservation area face various threats from ecological aspects like the decrease of environmental quality, such as the destruction of coral reef ecosystems due to illegal fishing or overfishing, damage to the habitat of Gosong birds, as well as from local aspects like a lack of elders community involvement in its management. Thus, the management of Kasa island must be done comprehensively and integrated to anticipate and overcome the changes and threats.

Besides designated as Wildlife Reserve and Marine Park, Kasa Island is also one of the hotspots or key biodiversity area of Wallacea region. Considering the important role of the Kasa island area, a scientific study has been carried out to determine the bio-ecological potential of the conservation area so that the direction of its management can be designed. This study aimed to: (1) identify the biophysical potential of Kasa Island conservation area; (2) identify the social potential of Kasa island; and (3) design the direction of Kasa Island management as one of the conservation areas in Maluku.

2. Methods

2.1. Research Location

This study was carried out on the conservation are of Kasa Island, which is located on West Seram regency, Maluku.

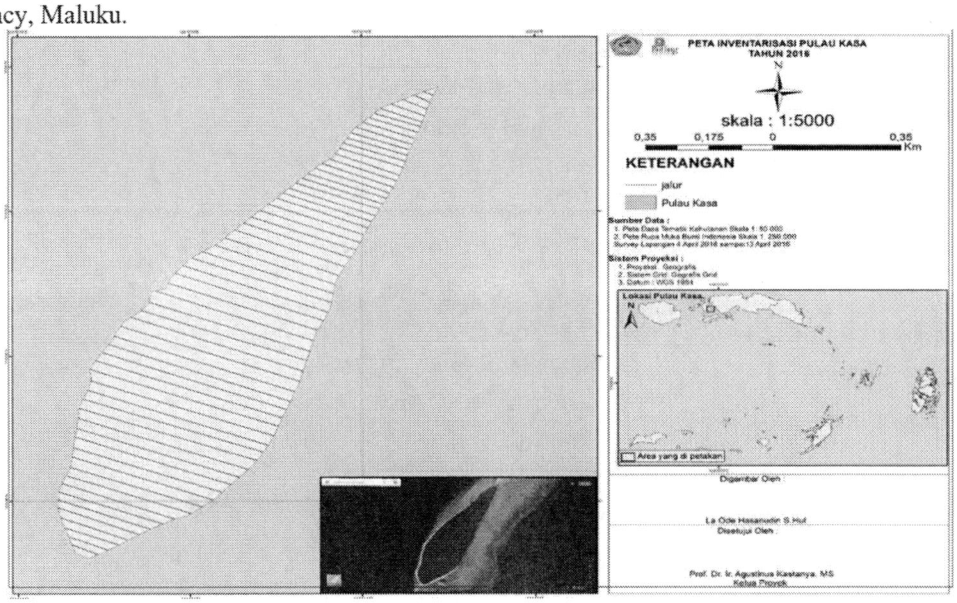

Figure 1. Map of Kasa Island reservation area

2.2. Analysis of the Vegetation

Data collection of the potential of vegetation in Kasa Island conservation area was done by making observation track and measurement plot on the filed. The observation tracks was made with the width of 20 m, and with some criteria of vegetation observations on the field as follows:

a) Three category (with diameter above 20 cm), the size of measurement plot was 20 m x 20 m.
b) Pole category (with diameter 10 cm to 20 cm), the size of measurement plot was 10 m x 10 m.
c) Sapling category (diameter < 10 cm and height > 1,5 cm), the size of measurement plot was 5 m x 5 m.
d) Seedling category (height < 1,5 cm), the size of measurement plot was 2 m x 2 m.

Measurement plots which were made to measure the density, frequency and vegetation dominance were as follows:

Explanation :
T : Trees
P : Poles
Sp : Sapling
Sd : Seedling

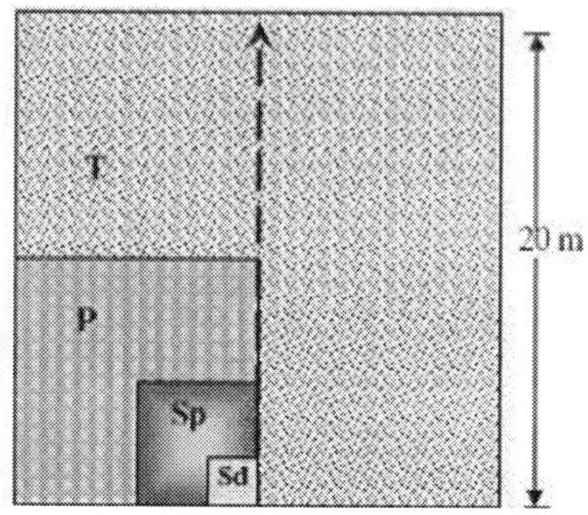

Figure 2. Measurement Plot of Vegetation Sampling

From the data of vegetation, it was used to find INP which was then analyzed descriptively. To find INP, the following formulas were used:

$$\text{Density} = K = \frac{\text{Number of Individual}}{\text{Total Sample}}$$

$$\text{Relative Density} = KR = \frac{\text{Density of A Variety}}{\text{Density of All Varieties}} \times 100$$

$$\text{Frequency} = F = \frac{\text{Total Plot Where A Variety Is Found}}{\text{Total Plot}}$$

$$\text{Relative Frequency} = FR = \frac{\text{Frequency of A Variety}}{\text{Frequency of All Varieties}} x100$$

$$\text{Dominance} = D = \frac{\text{Number of Basic Area}}{\text{Width of the Sample}}$$

$$\text{Relative Dominance} = DR = \frac{\text{Dominance of A Variety}}{\text{Dominance of All Varieties}}$$

$$\text{Index of Significance Value} = INP = KR + FR + DR$$

2.3. *Analysis of Diversity Index*

Diversity index is used to estimate the variety of species in the community analysis, which is (Sugianto, 1994 [9]).

Shannon index of general diversity (H)

$$H = -\sum\left\{\left(\frac{n.i}{N}\right)\log\left(\frac{n.i}{N}\right)\right\}$$

Explanation :
H = Shannon index of general diversity
n.i = Significant value of each species
N = Total of significant value

2.4. Analysis of the Wildlife

The collected data of the wildlife potential covered mammals, aves, reptile, and amphibian. The data collection of wildlife in the inventory and identification of wildlife potential in Kasa Island was carried out by rapid assessment. The observations did not have to be done in a special lane or location. Observations were made directly in the field to the footprints, dirt, traces in plants, feathers or other body parts which left behind, sounds, and nest marks. Other clues were also used such as guidelines for introduction of species, habitat, animals active period and morphological forms.

Additional data of the animal presence were also obtained by finding information from the community who have seen the presence of animals in the area. Information from the community was then verified or reviewed to ensure its validity.

2.5. Descriptive Analysis

Data of environmental service potential were obtained based on the direct observation on the field, which then analyzed by descriptive analysis. Later, the data of social community was also analyzed by using descriptive analysis. The management model of Kasa Island was done based on the result of Focus Group Discussion (FGD) as well as expert judgment regarding their expertise.

2.6. SWOT Analysis

Data analysis in this study was focused on the strategic environment analysis based on the SWOT matrix at the macro level, both on internal and external environments which involved community, Non-Government Organizations (NGO), and related agencies. The SWOT Analysis Approach (Strengths, Weaknesses, Opportunities and Threats) was used to obtain strategic priorities of Kasa Island Area Management.(Rangkuty,2014 [7]).

3. Results and discussion

Biophysical Condition of Kasa Island Conservation Area

3.1. Vegetation Potential Condition

Kasa Island conservation area has high vegetation potential, in which some of them are endemic to this area. Tree-category forest vegetation on Kasa Island is dominated by various types of wood. The overall potential inventory which has been done on Kasa Island showed that the category of trees found on Kasa Island consists of forty two species; forty-five species of pole category; forty species of sapling and seedling category. The tree category is dominated by black belo wood species (black ebony or charcoal wood: *Diospyros celebica*), ironwood (ironwood: *Eusideroxylon zwageri*), large leaf Salamuli wood (Kendal Wood: *Cordia spp*) and beach ironwood (ironwood: *Eusideroxylon zwageri*). The pole category is dominated by black belo wood (black ebony or charcoal wood: *Diospyros celebica*), hanet wood (puspa wood: *Schima wallichii*), sea guava (Malapari pods (*Pongamia pinnata*) and ironwood (ulin wood: *Eusideroxylon zwageri*). The sapling category is dominated by black belo wood (black ebony or charcoal wood: *Diospyros celebica*), key lime wood (Citrus aurantifolia), airal wood and totrain wood. Whereas, the seedling category is dominated by black belo wood, sea guava, hanet wood and airal wood.

The density of tree category was 26.34, the pole category was 130.68, the sapling category was 823.14 and seedling category was 17603.71. Black belo wood (black ebony or charcoal: *Diospyros celebica*) dominated the Kasa Island area with the highest density for all categories of vegetation.

The role of one tree species against another can be seen from the Importance Value Index (INP). If a species shows a high INP, the role of the species is very large towards other tree species in the ecosystem. Based on the analysis result, black belo wood showed a fairly high INP of around 48.96% for the tree category; 122.23% for the pole category; 62.21% for the sapling category and 98.41% for seedling category. Black Belo wood also showed a significant Index Value for all types of categories.

The diversity level of natural resources in Kasa Island was quite diverse based on the results from Shannon diversity index analysis. Based on the results of the analysis, the highest diversity index value was 1.37%. It means that the diversity of vegetation on Kasa Island in the West Seram Regency was moderate. This result was consistent with the classification of Odum diversity index (1971) which stated that the diversity index value between 1-3 has moderate diversity, and the diversity index value of more than 3 is high diversity. Based on the data above, it can be stated that Kasa Islandhas received ecological pressure.

Flora biodiversity in this area is a biotic resource which functions as genetic bank that is very potential to be developed. The high biodiversity needs serious effort to maintain its sustainable management, although this are is vulnerable to a higher extinction.

3.2. Wildlife Potential Condition

Kasa Island Wildlife Reserve is a habitat for Gosong Maluku bird (Eulipoa and Megapodis) known as endemic fauna. Gosong/ maleo bird is a type of bird found in Maluku, known as Gosong Maluku (Eulipoa wallacei). The result showed that there were two types of Gosong/ maleo bird that belongs to the genus of Eulipoa and Megapodius. Another type of Gosong bird found was a Gosong kelam bird (Megapodius freycinet). Resence the bird is the environmental balance in the ecosystem components, (Maya Adelina, Sugeng P. Harianto, dan Nuning Nurcahyani. 2016 [6]). The results showed that both types of Gosong birds were threatened by population decline due to habitat degradation and irresponsible hunt of maleo egg. This degradation occurred due to natural factors and human activities. The natural factor caused high abrasion, while human activity which created new paths by not paying attention to maleo nest would eventually threat maleo habitat. The hunting of maleo eggs and sea turtles by the community also greatly threatened the availability of these biological resources. Besides, that there was also the destruction of coral reefs due to illegal fishing done by the people outside Kasa island, especially from several village nearby such as from Telaga, Ketapang and its surroundings.

Various other types of aves were also found on Kasa Island such as Pombo (*Ducula bicolor*), Sea eagle (*Haliastur indus*), Trinil pantai (*Actitis hipoleucos*), Seagull (*Stercorarius pomarinus*), Dara laut (*Sterna hirundo*), Pekaka (*Halcyon spp*), Walet (*Collocalia infuscata*), and types of sea eagles (*Haliaeetus leucogaster*).

On the other hands, there are some types of reptiles found on Kasa Island, such as snakes, monitor lizards (*Varanus sp*), Soa (*Hydrosaurus sp*), chameleon and gecko. Monitor lizards found were with the average length of 60-100 cm. These lizards are also predators of Gosong/ maleo bird eggs and predators of these birds. Green turtles were also found, in which according to the community, this species has begun to decrease.

3.3. Environmental Services Potential

Kassa Island has various tourism objects which are potential to be developed, especially on the sea both on the surface and under water. Some attractions which are worth developing are in the form of beautiful natural panorama, biodiversity of flora and fauna, animal attractions and distinctive ecosystem characteristics. There are other ecotourism potentials in the form of beaches filled with white sand, coral reefs and various other marine life. By considering Kasa Island based on its function as a nature reserve, the type of tourism that can be done is special interest tourism. However, the development of ecotourism must prioritize the principles of sustainable ecotourism that are ecologically still maintaining the supporting and carrying capacity of the region, and can economically improve the welfare of the community (Scheyvens, 2000 [8]).

3.4. Social Condition of the Village Community

There were 75 people (100%) of Kaibobu villagers who became the respondents of this study. The age distribution were dominantly came from age group of 36-55 years old (45%), 17-35 years old (25%), and the rest fro above 56 years old (30%). Age had an effect to the participation level. Generally, when people reach certain age, there is a possibility of deeper involvement in community activities due to the awareness towards environmental issue.

Their education level was primary school graduates (49%), secondary school graduates (27%), senior high school graduates (23%), and university graduates (1%). Regarding the low education level, the people were given understanding of the importance to maintain the environment especially in Kasa Island.

The profession of the people were mostly farmer (40%), fishermen and farmer (37%), fishermen (9%), private worker and builder (4%) and others such as civil servant and speed boat driver by 1% for each. These types of work were related to the level of education of the respondents that mostly worked as farmers and fishermen.

The people of Kaibobu Village in general were very supportive to the management plan of the Kasa Island area as a Nature Reserve as well as its development as a special interest tourism object. The village community expected that collaborative management of the Kasa island by involving the community, both directly and indirectly, would mean that the community also participated in maintaining the existence of the Kasa Island area so that it would not be degraded. The degradation will possibly threaten Maluku endemic of Gosong bird. It will also protects the area from the outsiders whose activities will possibly damage the area, so that the management of the Kasa Island can be sustained.

3.5. Management Strategy of Kasa Island

In formulating the management strategy of Kasa Island in West Seram regency, SWOT analysis was conducted by selecting the relations or interactions between internal elements (strengths and weaknesses) against external elements (opportunities and threats) based on natural factors and characteristics of the community and other facts at the research setting. The SWOT components of the Kasa Island Management Strategy were as follows:

Strengths	Weaknesses
• High potential of biodiversity in Kasa Island • High diversity of vegetation • The existence of Maleo as endemic bird • Has a potential environmental service	• Institutional management of Kasa Island which was not optimum • Low supervision from relevant agencies • Limited knowledge of the people in the community about the management of Kasa Island • No local government regulation of Kasa Island management • Limited capital of local community to access Kasa Island
Opportunities	**Threats**
• Increase of the economy level of coastal community • Potential of special natural destination • Allocation of government's budget to accelerate the management of Kasa Island • Presence of RPJP in Kasa Island • Increase of fishery products commodity	• Maleo bird extinction • Construction of cottage • Coral reefs damage • Weak law enforcement to illegal fishing

Table 1. Weighing of Strategy Factor

Internal Factor	Weight	Rate	Score
Strengths (S)			
• High potential of biodiversity in Kasa Island • High diversity of vegetation • The existence of Maleo as endemic bird • Has a potential environmental service	0,20 0.10 0.15 0.09	3,50 3.40 3.65 3.50	0,70 0.34 0.54 0.32
Total			1.90
Weaknesses (W)			
• Institutional management of Kasa Island which was not optimum	00,5	2,10	0,10
• Low supervision from relevant agencies	0.10	2.15	0.22
• Limited knowledge of the people in the community about the management of Kasa Island	0.05	2.15	0.11
• No local government regulation of Kasa Island management	0.05	1.15	0.06
• Limited capital of local community to access Kasa Island	0.10	1.18	0.11
Total			0,60

External Factor	Weight	Rate	Score
Opportunities (O)			
• Increase of the economy level of coastal community	0,17	3,30	0,56
• Potential of special natural destination	0.17	3.25	0.55
• Allocation of government's budget to accelerate the management of Kasa Island	0.15	3.30	0.46
• Presence of RPJP in Kasa Island	0.10	3.40	0.34
• Increase of fishery products commodity	0.10	2.50	0.25
Total			1.91
Threat (T)			
• Maleo bird extinction • Construction of cottage • Coral reefs damage • Weak law enforcement to illegal fishing	0,17 0.12 0.09 0.12	2,20 2.10 2.10 1.10	0.38 0.25 0.19 0.13
Total			0,95

From table of SWOT above, it can be illustrated that the management strategy of Kasa Island is located in Quadrant I as follows:

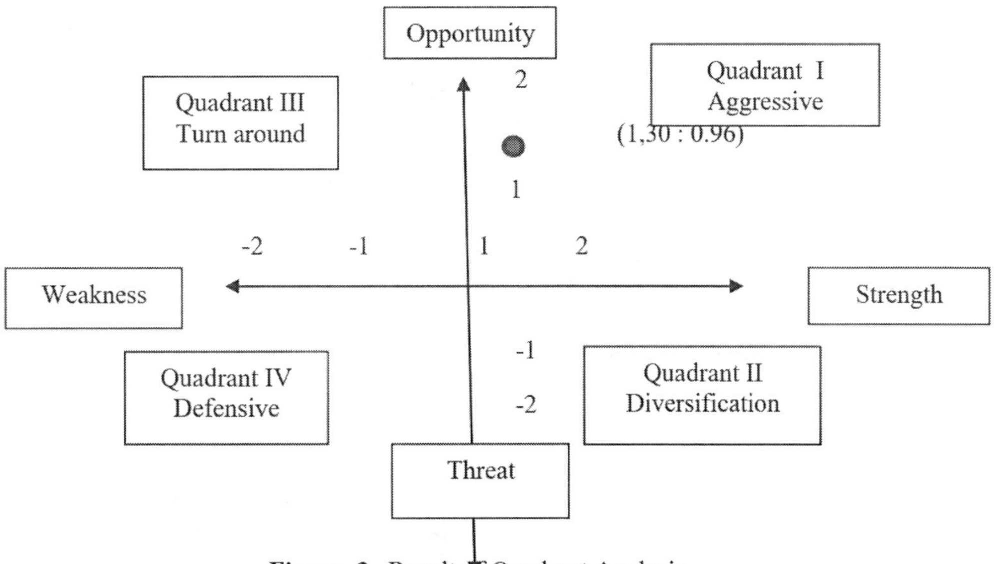

Figure 3. Result of Quadrant Analysis

Based on Figure 3, the management position of Kasa Island was located in Quadrant I. It showed that the management of Kasa Island had strengths that must be optimized by utilizing the opportunities existed to maintain its sustainability. The strategy which needs to be developed is by using strengths to take advantage of opportunities so they can overcome weaknesses and threats.

The management of the Kasa Island in the quadrant above showed a favorable situation. The strategy which must be applied in this condition is to support growth oriented strategy (Rangkuty, 2014 [7]). Based on quadrant I, the strategies developed are as follows:

1. Collaborative Management

The legal basis for collaborative management of Kasa Island and its implications is Law No. 5 of 1990 about Conservation of Biological Resources and Its Ecosystems, as a legal basis which allows Kasa Island to be managed together and accommodate various interests for research, science, education, support cultivation and recreation. Besides that, it is also strengthened by *Permenhut* No. P.19 of 2004 about Management of Nature Reserves and Nature Conservation Areas, which included guidelines for preparation, implementation, monitoring and evaluation in collaborative management of conservation areas, which involves various stakeholders. The implication is that collaborative institution of multi parties needs to be made for the management of Kasa Island with clear guideline from planning to evaluation (Falah F,2013 [5]).

Collaborative management can be done by promoting integrated environmental sustainability. Coordination among stakeholders needs to be improved to solve various fundamental problems in the management of Kasa Island, especially in infrastructure and socio-economic issues. (Carlsson L, Berkes F. 2005 [4]). Besides that, DKP stakeholder authority through the enactment of Law No. 23 of 2013 which was transferred to the province level also greatly influenced the management of Kasa Island, so that collaborative management needs to be implemented in various forms of innovative programs and activities. Several collaborative management can be effective in achieving sustainability of the functions and benefits of conservation areas, including: (1) Holistic principles, which is the management of the area which must consider all ecological, economic and social functions in the ecosystem; (2) Integrative principles, which includes management based on cooperation between all parties.

2. Strengthening The Capacity of Local Community

Strengthening the capacity of local community needs to be increased through training. Types of training that can be provided to local communities to support Kasa Island management starts with the formation of a guided group (increasing the sense of group)based on the social and cultural conditions of the local communities around Kasa Island.

3. Formulation of regional policy regarding Kasa Island management

The existence of regional policy in the form of regional or domestic regulation will be the basis for the legality of Kasa Island management.

4. Conclusion

1. Vegetation potential on Kasa Island is more dominated by black belo wood species, with the highest density level at all vegetation category. The density of the tree category was 26.34, the pole category was 130.68, the sapling category was 823.14 and the seedling category was 17603.71.
2. Kasa island has fairly diverse natural resources , with the highest diversity index value of 1.37%, which includes in the medium category.
3. The wildlife potential is more dominated by two types of Gosong/ maleo birds from the genus of Eulipoa and Megapodius. Another type of Gosong bird found was Gosong kelam (*Megapodius freycinet*).
4. Based on its function, Kasa Island is a nature reserve, so that the type of tourism that can be done is special interest tourism for research and education purposes.
5. Kasa Island management strategies which can be followed up in collaborative and sustainable management are as follows: (1) Collaborative management; (2) Strengthening the capacity of local communities; (3). Formulation of regional policies regarding the management of Kasa Island.

References

[1] Anonymous 1990 Undang-Undang No. 5 tahun 1990.

[2] Anonymous, 2016. Laporan Penataan Kawasan Konservasi Pulau Kasa. Balai Konservasi Sumberdaya Alam Wilayah IX Maluku.

[3] Anonymous 2016 Pengelolaan Sumberdaya Alam Pulau Kasa secara Kolaboratif dan Berkelanjutan. Laporan Penelitian. Lembaga Penelitian Universitas Pattimura Ambon.

[4] Carlsson L, Berkes F 2005 Co-management: concepts and methodological implication *Journal of Environmental Management* **75**(1) 65-76.

[5] Falah F 2013 Kajian Efektivitas Pengelolaan Kolaboratif Taman Nasional Kutai *Jurnal Analisis Kebijakan Kehutanan* **10**(1) 37-57.

[6]. Maya Adelina, Sugeng P. Harianto, dan Nuning Nurcahyani. 2016. Bird Diversity in Community forest Kelungu Village Kota Agung Sub District Tanggamus Distric *Slyva Lestari Journal* **4**(2) 51-60. ISSN 2339-0913 p

[7] Rangkuti F 2014 *Analisis SWOT: teknik membedah kasus bisnis (reorientasi konsep perencanaan strategis untuk menghadapi abad 21)*. Buku. PT.Gramedia Pustaka Utama. Jakarta. 246 p

[8] Scheyvens R 2000 Promoting Women's Empowerment Through Involvement in Ecotourism: Experiences from the Third World. Journal of Sustainable Tourism **8**(3) 232 – 249..

[9] Sugianto 1994 Ekologi Kuantitatif. Penerbit Usaha Nasional. Surabaya

IOP Conf. Series: Earth and Environmental Science **883** (2021) 012078 doi:10.1088/1755-1315/883/1/012078

Study of changes in carbon stock in primary and secondary forest in protection forest of Soya Village, Ambon City

L Siahaya[1,*], L Pelupessy[1], Y D Komul[1], A Sahupala[1], T E Siahaya[1], B B Seipala[1]

[1]Department of Forestry, Faculty of Agriculture, Pattimura University

* E-mail: ledysiahaya@gmail.com

Abstract. This study aimed to determine changes in litter biomass and the effect of changes in litter biomass on carbon stock changes in Permanent Sample Plots of primary and secondary forest in the Protection Forest of Soya Village, Ambon City, for three years of measurement. This study used a sampling method with harvesting/collection and using laboratory analysis, where the working procedure used the SNI: 7724 guidelines. The data taken includes the water content of litter, total wet weight, dry weight of the sample, and wet weight of the sample to calculate the range of biomass or organic matter and carbon content of litter organic matter. The results showed that the average value of water content in both primary and secondary forests decreased. The depletion degree was higher in the first year than the second and third year. The biomass content of litter in primary forests decreased from year to year for three years of measurement. It causes a decrease in the carbon content in the primary forest. In contrast, the biomass content in secondary forests increased from year to year and increased carbon content from year to year for three years of measurement.

1. Introduction

The current climate change is evidenced by the long dry season and relatively short rainy season with high rain intensity. The result is prolonged drought, crop failure, food crisis, lack of clean water, warming sea levels, floods, and landslides [1] and [2].

The cause of this climate change is an increase in the concentration of Greenhouse Gases (GHG), namely Carbon Dioxide (CO2), Methane (CH4), Nitrous Oxide (N2O), Hydro Fluoro Carbon (HFC), Perfluoro Carbon (PFC), and Sulfur Hexafluoride (SF6) in the atmosphere. According to [2], the largest source of emissions in the world comes from the energy sector (24% power generation, 14% industry, 14% transportation, 8% construction, and other energy sources 5%) and from the non-energy sector (land change including forestry 18%, agriculture 14% and waste 3%).

Forest biomass plays a critical role in the biogeochemical cycle, especially for the carbon cycle [3]. Photosynthesis causes the binding of CO2 to biomass, while decomposition and combustion cause the release of CO2 into the atmosphere. The natural forest acts as the highest sink (carbon sink or storage) compared to other land-use systems due to the high diversity of trees, dense undergrowth, and litter covering the soil surface. If the forest's function is changed or the forest density decreases, then the stored C will be reduced, and it could be lost entirely [4].

According to [5], forest plants or trees are considered a carbon storage place or deposition. The amount of carbon and tree biomass content varies based on the part of the plant being measured, growth stage, plant level, and environmental conditions. The carbon content and biomass of understorey are influenced by the species of constituent plants [6]. The litter layer or forest floor is all dead organic material that is above the soil surface. Some of this organic material can still be recognized or is still slightly decomposed [7].

Therefore measuring the amount of carbon in biomass inland can describe the amount of carbon dioxide (CO2) in the atmosphere that plants absorb, and measuring carbon in dead plant parts (necromass) can define the carbon dioxide (CO2) that is not released into the air through combustion.

Mount Sirimau Protection Forest is a forest ecosystem that functions to support life, namely to regulate water management, prevent floods and erosion, maintain global climate stability and maintain soil durability and fertility. Mount Sirimau Protection Forest also has a high diversity of flora and fauna as well as potential tourist attractions. Therefore, research was conducted with the title "Study of Changes in Carbon Stock in Primary and Secondary Forest in Protected Forest of Soya Village, Ambon City." The purpose of this study was to determine the changes in litter biomass and the effect of changes in litter biomass on changes in carbon stocks in primary forest and secondary forest PSP in the Protection Forest of Soya Village, Ambon City, for three years of measurement.

2. Methods

The research was conducted on a Permanent Sample Plot (PSP) built in the Soya Village Protected Forest Management Unit, Sirimau District, Ambon City, and continued in the Silviculture Laboratory Forestry Department, Faculty of Agriculture, Pattimura University. This research was conducted for three years, and data were taken every year.

There were several tools used in this research. GPS to determine the coordinates of sample plots. Digital scale (Ohaus Balance) with 0.5% accuracy. Other tools include meter roll to make a 2 m x 2 m plot, knife and cuttings to cut the litter into small pieces, brown envelope for laying litter, plastic bags for storing wet samples, and oven to dry the sample. The material or object used was litter on PSP floors in Protected Forest of Soya Village, Sirimau District, Ambon City.

2.1. Procedure

Measurement of litter biomass using the steps in [8], where the stages of measuring litter biomass were carried out as follows: a) collect litter in the measurement plot; b) weigh the total litter weight; c) take approximately 300 grams to consider the sample weight; d) oven-dry the litter sample at a temperature range of 70 ° C to 85 ° C until it reaches a constant weight; e) weigh the dry weight of the litter; f. do an organic carbon analysis in the laboratory to see the carbon content.

2.2. Data Analysis

The analysis used in this method was quantitative analysis by comparing the re-measurement data with the data from the previous measurement results. Calculating the water litter content using the formula :

$WC = ((WWs\text{-}DWs)) / WWs \times 100$ .... [9]

Information : WC = Water Content (%), WWs = Sample Wet Weight (gr), DWs = Sample Dry Weight (gr).

The estimate of litter biomass was calculated using the following formula: Calculation of litter organic matter :

$Wo = (DWs \times WWt) / WWs$ ....[8]

Information : Wo = weight of organic matter, expressed in kilograms (kg); DWs = dry weight of the sample, expressed in kilograms (kg); WWt = total wet weight, expressed in kilograms (kg); WWs = sample wet weight, expressed in (kg).

Carbon accounting of dead organic matter (litter), Calculation of carbon from litter organic matter using the following formula :

$Cm = Wo \times \%\ C\ organic$ .... [8]

Information: Cm is the carbon content of dead organic matter, expressed in kilograms (tonnes/acres); Wo is the total biomass/organic matter, represented in kilograms (tonnes/acres); % C organic is the percentage value of carbon content, amounting to 0.47 or using the percent carbon value obtained from measurement results in the laboratory measurement of litter biomass using the steps in [8], where the stages of measuring litter.

3. Results and discussion

3.1. Water Content of Litter

The amount of water content affects the amount of biomass contained in an object. Water content is an important indicator that must be known, where the greater the value of water content, the smaller the biomass value will be. Based on the research results, it was found that the water content of the litter in the Primary Forest ranged from 43.13 to 46.60% in year I; 63.34-65.50% in year II; and 52.86-64.33% in year III. Meanwhile, the range of water content in Secondary Forest is between 33.50-43.37% in the year I, 60.33-66.67% in year II, and 53.33-66.83% in year III.

The average water content in Primary Forest for the first year reached 45.13%; in year II, it reached 64.63%, and in year III reached 60.40%. Meanwhile, the average water content in secondary forest in the first year was 39.87%; in year II amounted to 63.27%, and in year III amounted to 59.89% (see Table 1 and Figure 1).

Table 1. Interval of the water content of primary and secondary forest in protection forest of soya village

Location	Interval Year I (%)	Interval Year II (%)	Interval Year III (%)
Primary Forest	43.13-46.60	63.34-65.50	52.86-64.33
Secondary Forest	33.50- 43.37	60.33-66.67	53.33-66.83

Source: *Research Data for 3 Years*

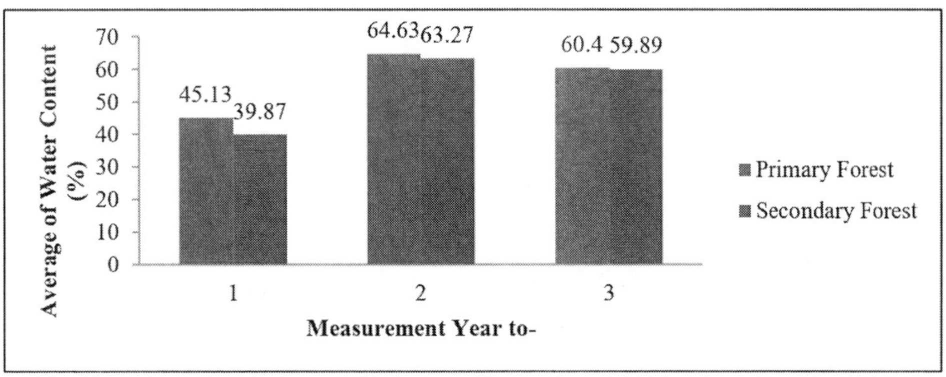

Figure 1. The average litter water content of primary and secondary forest at protection forest of soya village

From the graph above, it can be seen that the average water content in the two forest types shows that in the year I the water content is low, in the second year it has increased (by 19.50% in Primary Forest and 23.40% in Secondary Forest). Then in the 3rd year, it decreases again (by 4.23% in Primary Forest and 3.38% in Secondary Forest). It happens because of changes in rainfall from year to year. The rain in the first year is 181.5 mm / day; in year II amounting to 420.1 mm/day, whereas in year III, it was 296 mm / day.

The graph shows that the average water content in each measurement year in the primary forest is higher than in the secondary forest. Due to the sampling plots' condition in the primary forest, it has a higher stand density than the secondary forest density. In the primary forest, treestands dominate over seedling sand poles, but many have died and fell, forming more gaps. According to its characteristics, the primary forest contains large, long-lived trees, alternating with dead tree trunks that are still upright, stumps, and fallen wood. The collapse of these logs can form gaps or gaps in the stands, allowing rainwater to fall onto the floors o that the forest makes the litter water content higher.

While the sampling plots in the secondary forest were many stands at the level of seedlings to poles found in each measurement plot in the secondary forest. It is consistent with the statement put forward by [10] that secondary forest is the forest that grows and develops naturally after damage/change occurs in the first forest. Secondary forest is a phase of forest growth from a bare footprint, due to natural or anthropogenic, to a climax again.

The physical condition of the litter tends to be different each year. Since the litter's potential water has evaporated due to temperature and sunlight's influence, this condition causes the litter's water content to be less.

3.2. Biomass Content of Litter

The biomass content in primary forest was 3.6713-13.0720 tones/acres with 8.6614 tonnes/acres on average at Year I. The result was 4.0533-7.7475 tonnes/acres with an average of 5.5089 tones/acres in next year. For Year III, the range was between 3.9-5.8742 tones/acres with an average of 4.802 tones/acres at Year III. Meanwhile, secondary forest ranged from 0.7481 to 4.2950 tones/acres with an average of 2.2710 tones/acres for Year I; 2,7133-4,1058 tonnes/acres with an average of 3.2840 tonnes/acres for Year II; and 2.2950-5.3850 tones/acres with an average of 3.808 tones/acres for Year III. To more clearly see the biomass content interval in Table 2 and the average biomass content in Figure 2.

Tabel 2. Interval of biomass content of litter of primary and secondary forest in protectin forest of soya village

Location	Interval Year I (tonnes/acres)	Interval Year II (tonnes/acres)	Interval Year III (tonnes/acres)
Primary Forest	3.671-13.072	4.053-7.748	3.9-5.874
Secondary Forest	0.748-4.295	2.713-4.105	2.295-5.385

Source: *Research data for 3 Years*

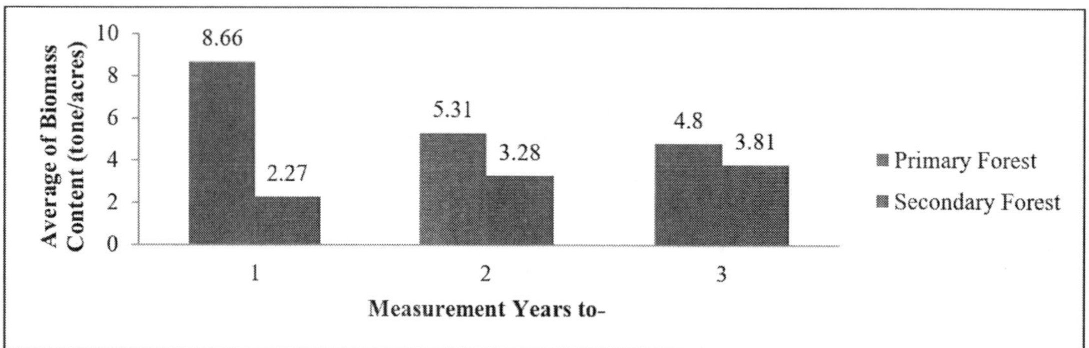

Figure 2. Average of biomass content of litter of primary and secondary forest at protectin forest of soya village

From the results of biomass calculations, it can be seen that the average biomass content in primary forests has decreased from year to year, while in secondary forests, it is increasing. The value of biomass was higher in the primary forest than in the secondary forest. The result has a similar finding with the previous study [11], where biomass content in secondary forest (6.656) was higher than grove (4.624) and shrubs (1.574). It is due to many plants with a diameter of 10 cm and above in the plot.

Although the secondary forest has a higher vegetation density in water content, it is at the seedling to pole level. In contrast, the primary forest has lower viscosity at a seedling to pole level, but more large trees have a large crown density in the primary forest.

Forests with a high canopy density will get less sunlight so that the decomposition process that occurs will be slower than forests with a lower canopy density [12]. The primary forest has a higher canopy density resulting in a slower decomposition process. The slow decomposition process results in a higher biomass content in the litter compared to the secondary forest.

The factors causing the amount of litter vary because it generally occurs due to litter biomass and the increase in tree age and crown density. Canopy or stand density is a factor that affects the fall of forest litter due to competition for sunlight. The tighter a stand or canopy will produce an enormous amount of litter because the trees that grow in a relatively dense forest quickly release branches and leaves starting from below because there is not enough light for them for the photosynthesis process. In addition to the factor of increasing tree age and the density of the litter canopy, it is also influenced by the dropping of litter both in quantity and quality, influenced by environmental factors (climate, altitude, soil fertility), species of plant, and time (season and age of stands).

Forests with a high vegetation density will get less sunlight so that the decomposition process that will occur will be slower than forests with a lower canopy density [12]. Litter productivity is also influenced by vegetation and rainfall. Rainfall affects vegetation's physiology because the higher the rainfall, the lower the leaves' leaves, twigs, flowers, and fruit. When the rainfall is high, the humidity will increase, so the leaves' evaporation will decrease so that the leaves remain fresh and do not fall easily.

3.3. *Carbon Content of Litter*

In the calculations carried out, the carbon in primary forest for the year I ranged from 1.7225-6.1438 tonnes/acres with an average of 4.0708 tonnes/acres; for Year II ranged from 1.9051 to 3.6413 tones/acres, with an average of 2.4952 tones/acres; and for Year III ranged from 1.833-2.761 tonne/acres with an average of 2.2571 tonne/acres. Whereas in the secondary forest, the stored carbon content is in the range of 0.3516-2.01833 tones/acres with an average of 1.0674 tones/acres for Year I; 1.2753-1.9297 tones/acres with an average of 1.5435 tones/acres for Year II; and 1.0787-2.5310 tones/acres with an average of 1.7899 tones/acres for Year III. For more details, see Table 3 and Figure 3.

Table 3. Interval of the carbon content of litter at primary and secondary forest in protection forest of soya village

Location	Interval Year I (tonnes/acres)	Interval Year II (tonnes/acres)	Interval Year III (tonnes/acres)
Primary Forest	1.7225-6.144	1.905-3.641	1.833-2.761
Secondary Forest	0.351-2.019	1.275-1.9297	1.0787-2.531

Source: *Research data for 3 Years*

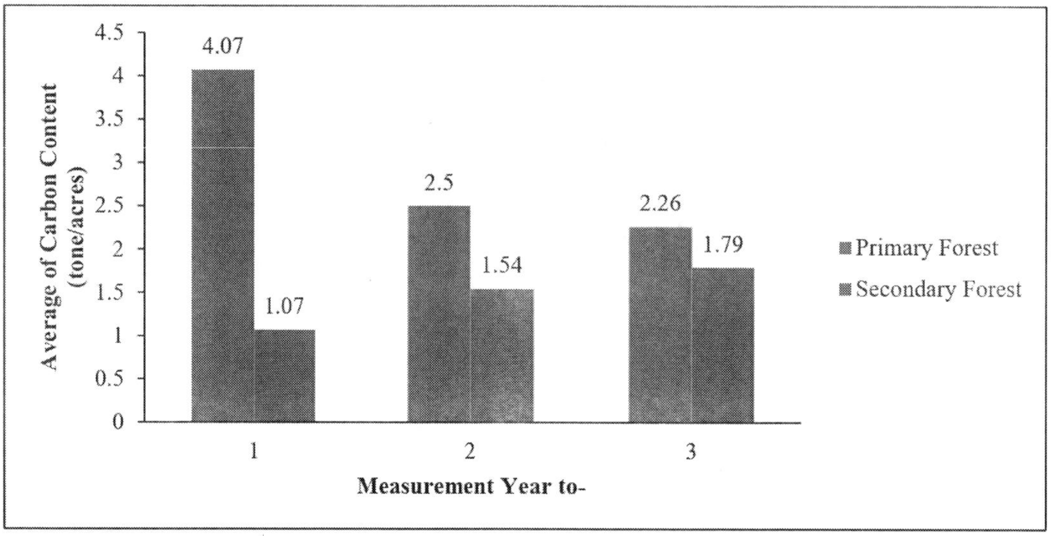

Figure 3. Average of the carbon content of litter of the primary and secondary forest at protection forest of soya village

From the graph above, it can be seen that the average carbon content of litter in primary and secondary forests is different. This difference is influenced by the amount of biomass contained in the litter. Plantations' ability to absorb CO2 from the atmosphere varies according to type, age level, and plant density, so that forests with high plant densities contain more carbon.

Based on these results, it was found that the carbon content stored in litter changes and the biomass in the litter. This change occurs in primary and secondary forests from year to year. Changes in the litter's carbon content are highly dependent on the litter's biomass, so if the biomass content changes, inevitably, the carbon content will also change. At this stage, the primary forest's carbon content is found to be higher than in the secondary forest. It can be compared with the results of research from [11], where carbon in secondary forest (amounted to 3.129) was higher than that in grove (amounting to 2.173) and shrubs (amounting to 0.739). It is due to a large amount of woody vegetation in the plot.

According to [4], the amount of carbon stored between land varies depending on the diversity and density of existing plants, soil types, and management methods. The carbon storage of land becomes greater if soil fertility conditions are perfect. In other words, the amount of carbon stored above the soil (plant biomass) is determined by the amount of carbon stored in the soil (soil organic matter).

When conducting the research, 21 species of trees were found in primary forest and 15 species in secondary forest in the Protection Forest of Soya Village. Vegetation species in the primary forest consist of Bintanggur Hutan (*Callophyllum soulatri*), Siki (*Palaqium javense*), NanTi (*Meterosideros vera*), Redwood (*Eugenia cumini*), Mangosteen (*Garcinia mangostana*), Forest mango (*Mangifera* sp), Loasing, Cempedak Hutan (*Artocarphus* sp), Sageru wood (*Gurania cydrocarpa*), Damar (*Agathis alba*), Iron wood (*Intsia bijuga*), Ketapang Hutan (*Terminalia copelandi*), Taheru (*Eugenia* sp), Pala Hutan (*Myristica* sp), Tawang, Samar (*Homalium foetidum*), Kayu Sisir (*Cudrania* sp), Makila (*Litsea angulata*), Forest Cloves (*Syzgium aromaticum*), Gayam (*Inocarphus fagiferus*), and Bamboo (*Bambusa* sp).

Some species also found in secondary forests, e.g., Bintanggur Hutan (*Callophyllum soulatri*), Siki (*Palaqium javense*), forest jack fruit (*Artocarphus* sp), forest mango (*Mangifera* sp), Nani (*Meterosideros vera*), Ketapang Hutan (*Terminalia.* sp), Pala Hutan (*Myristica* sp), Red Wood (*Eugenia cumini*), Sageru Wood (*Gurania cydrocarpa*), Taheru (*Eugenia* sp), Damar (*Agathis alba*), Guava, Makila (*Litsea angulata*), Nanari (*Cannarium commune*), and Kayu Sisir (*Cudrania* sp).

Several species were only found in plots of primary or secondary forest, namely Mangosteen (*Garcinia mangostana*), Loasing, Ironwood (*Intsia bijuga*), Tawang, Samar (*Homalium foetidum*), forest cloves (*Syzigium aromaticum*), and Gayam (*Inocarphus fagiferus*). In contrast, the Guava species (*Eugenia* sp) and Nanari (*Cannarium commune*) were only found in secondary forest plots.

Litter is organic material in the form of fallen leaves and twigs from woody vegetation. A dense canopy or stand will affect the fall of forest litter due to competition from sunlight. Trees that grow in a slightly dense forest are faster to release branches and leaves from below because there is not enough light for them to photosynthesize [13]. Litter production is a loss of vegetative and reproductive structures caused by aging, stress, or a combination of both [14].

A large amount of litter that affects increasing the amount of biomass and carbon content in primary forests also occurs due to standing age, crown density, and stand density.

4. Conclusion

The water content of litter at the primary and secondary forest in the protected forest of Soya village increased from year I of measurement to year II of measurement but decreased in year III of measurement.

The Biomass content of litter in the primary forest from year I to year III decreased, while the secondary forest's biomass content has increased from year to year. However, the content of litter biomass in the primary forest is higher than in the secondary forest.

The content of litter carbon in the primary forest from year I to year III decreased, while the Carbon content in the secondary forest has increased from year to year. But the Carbon content of litter in the primary forest is higher than in the secondary forest.

References

[1] IPCC 2006 *IPCC Guidelines for National Greenhouse Gas Inventories* IPCC National Greenhouse Gas Inventories Programme IGES Japan

[2] Stern N 2007 *The Stern Review: The Economics of Climate Change* Cambridge University Press Bogor

[3] Sutaryo D 2009 *Penghitungan Biomassa* Wetland International Indonesia Programme Bogor

[4] Hairiah K dan Rahayu S 2007 *Pengukuran Karbon Tersimpan di Berbagai Macam Penggunaan Lahan* Bogor World Agroforestry Centre – ICRAF SEA Regional Office University of Brawijaya, Unibraw Indonesia p 77

[5] Center for International Forestry Research 2003 *Perdagangan Karbon* Warta Kebijakan No.8 Februari 2003.

[6] Yuanita Windusari Nur A P Sari Indra Yustian dan Hilda Zulkifli 2012 Dugaan Cadangan Karbon Biomassa Tumbuhan Bawah dan Serasah di Kawasan Suksesi Alami pada Area Pengendapan Tailing PT Freeport Indonesia *Jurnal Biospecies* **5(**1) 22-28

[7] Pearson T R H Brown S Ravindranat N H 2005 *Integrating Carbon Benefit Estimates into GEF Project* UNDP Global Environment Facility Dikutip dalam Windusari et. al. 2012 Dugaan cadangan karbon biomassa tumbuhan bawah dan serasah di Kawasan Suksesi Alami pada area pengendapan Tailing PT Freeport Indonesia *Jurnal Biospecies* **5** (1) 22- 28. (Akses November 11,2015).

[8] Standar Nasional Indonesia (2011 a) SN1 No:7724 *Pengukuran dan penghitungan karbon tersimpan. Pengukuran lapangan untuk penaksiran karbon tersimpan hutan (ground based forest carbon accounting)* Jakarta.

[9] Haygreen John G dan Jim L Bowyer 1993 *Hasil Hutan dan Ilmu Kayu, Suatu Pengantar* Cetakan Kedua Gadjah Mada University Press Yogjakarta

[10] Lamprecht H 1986 *Waldbau in den tropen* Parey Hamburg

[11] Zikri Azham 2015 Estimasi Cadangan Karbon pada Tutupan Lahan Hutan Sekunder, Semak dan Belukar di Kota Samarinda *Jurnal Agrifor* **14**(2)

[12] Alrasjid H 1998 Pelepasan Unsur Hara C Organik dan Unsur Hara Mineral lainnya selama pelapukan serasah di areal tegakan sisa hutan alam Mangrove, sungai sepada Kalimantan Barat *Buletin penelitian Hutan* **503** 29-44.

[13] Budiman M Gusti H Herlina D 2010 Estimasi Biomassa Serasah dan Tanah pada Basal Area Tegakan Meranti (Shoreamacrophylla) di Kalimantan Bara *Jurnal Hutan Lestari* **3**(1).

[14] Hanafi N R Biroum B 2012 *Pendugaan Cadangan Karbon pada Sistem Penggunaan Lahan di Areal PT. Sikatan Wana Raya* Universitas Palangkaraya.

IOP Conf. Series: Earth and Environmental Science **883** (2021) 012079 doi:10.1088/1755-1315/883/1/012079

The impact of climate change on the resin productivity of dammar tree (*Agathis alba*) in Inamosol subdistrict, West Seram district, Maluku-Indonesia

J M Matinahoru[1,*]

[1] Department of Forestry, Faculty of Agriculture, Pattimura University, Jl. Ir. M. Putuhena, Kampus Poka, Ambon 97233, Maluku Indonesia

*E-mail: johan.matinahoru@faperta.unpatti.ac.id

Abstract. This research was aimed to determine the impact of climate change on the resin productivity of dammar tree. This research will be useful as data and information for farmers and government to maintain the resin of dammar tree to be optimal and sustainable in production. This research was conducted in Inamosol Sub-district, West Seram District, Maluku Indonesia, during September-October 2020. Village and farmer samples were determined by purposive sampling technique. The selected villages were Honitetu, Hukuanakota and Rambatu. Furthermore, from each village, It was ten farmers to select for interviews and filling the questionnaire. The results showed that the average resin production of farmers in 2019 was 904.2 kg/farmer, while in 2020 was 523.7 kg/farmer. This means that it occurred a decline in resin production in 2020 about 42.08 % for each farmer—the leading cause of the decreased production as climate change factors, namely rainfall, temperature and humidity. Based on climate data of West Seram District in 2019 indicated that rainfall has occurred during six months with an average temperature of 27 °C and relative humidity of 82 %. Meanwhile, in 2020 the rainfall occurs for nine months with an average temperature of 26.5 °C, and relative humidity of 85 %.

1. Introduction

Dammar tree (*Agathis alba*) is a type of plant from the *Araucariaceae* family, with a cylindrical trunk, diameter at breast height up to 2 m, a tree height up to 60 m the bark always emits sap if injured. In Indonesia, the dammar tree grows and spreads from Sumatra, Kalimantan, Sulawesi to Maluku. In Maluku, the dammar tree is found growing naturally on the island of Seram, in the mountainous area of Sub-districts of Taniwel, West Seram, and Inamosol at an altitude of 700 m above sea level. Dammar tree is a highland and humid tropical plant, found at an altitude of 500-1,200 meters above sea level. The observation showed that in Inamosol Sub-district, West Seram District, dammar forest grows and develops well on humus cambisol soils in mountain peaks with sufficient sun exposure and temperature condition of 24-28 °C and relative humidity of 76-90 %. The dammar tree that is naturally growing in Maluku is a type *of Agathis alba*, and it can produce sap well at the age of over 25 years [1].

The benefits of resin are industrial raw material, especially as a base material for the paint, turpentine, varnish, ink and cosmetics [2]. In related to this utility, farmers considered the dammar forest to be a bank that can provide cash for farmers at any time to meet their various needs.

In Inamosol Sub-district, West Seram District, there are three marketing system channels for the resin products, namely: 1) Farmers sell to collectors in the village at a selling price of IDR 8,000-

12,000/kg; 2) Farmer's market to collectors in the sub-district city with a selling price of Rp.10,000-13,000/kg; and 3) Farmers sell to collectors in Provincial City at a selling price of IDR 14,000-15,000/kg. In general, farmers prefer to use the first marketing chain because it will require quite high transport costs if they choose another marketing channel.

Because the resin is quite economically valuable, there is often a theft by people who do not have dammar forest. To overcome tree resin's theft, farmers usually take precautions according to the custom, namely the Sasi system. Sasi is a customary prohibition to take other people's belongings without the owner's permission. If someone takes another person's belongings, they will receive punishment or expected sanctions [3]. The resin sap is usually harvested by tapping with a machete. The sap that is tapped or treated with a machete will fall into a container made merely of wood bark or sago midrib.

This study aims to determine the impact of climate change on the productivity of resin. It is hoped that this research results can be useful as data and information for farmers and the government to maintain the productivity of resin.

2. Methods

This research was conducted in resin production villages in Inamosol Sub-district, West Seram District, Maluku Indonesia from September to October 2020. The sample of village and farmer in this study was determined by using a technique of purposive sampling. The selected villages were a village that still relies on resin sap as cash. Meanwhile, the sample of farmers is a farmer who actively manages the resin forest. Therefore, the selected villages were Honitetu, Hukuanakota, and Rambatu. In each village, ten farmers were chosen to conduct an interview and fill out the questionnaire lists.

3. Results and discussion

3.1. Resin production

The resin tree is found to grow naturally in Inamosol Sub-district, especially in Honitetu, Hukuanakota and Rambatu. In these villages, some of the farmers have a dammar forest that can produce sap every year to collect for sale to collectors. In the management of the dammar forest, farmers usually had problems related to maintaining or increasing resin production. Farmers always strive to do this, so their income can increase. The resin production is determined by tree diameter, tree height, bark thickness, and tree crown density [1]. Furthermore, it was concluded that the increasing in diameter, tree height, bark thickness and tree crown would increase resin sap production. On the other hand, the resin production is also determined by the shape and location of the tapping, where the shape of the letter (W) and the directional tapping to sunlight will be better in the production of sap [4] [5]. The trees number of production age, maintenance intensity, habitat conditions and season are strongly determining the resin production of dammar tree [6][7]. When the injury of tree bark is applied properly, the tree age of 25-60 years can produce an average resin of 1-4 kg/tree, the age of the tree 61-120 years can give sap of 4-6 kg/tree, and trees over 120 years old can have resin of 6-8 kg/tree [8].

The research results on the amount of resin that harvested and sold by farmers during 2019 and 2020 were as follows.

Table 1. Resin production in 2019 and 2020

Sample of Farmers	Average of resin production in 2019 and 2020 (kg/year)					
	Honitetu Village		Hukuanakota Village		Rambatu Village	
	2019	2020	2019	2020	2019	2020
1	1.200	720	1.300	754	1.450	841
2	560	364	890	534	1.210	701
3	700	420	900	513	1.010	606
4	450	284	560	308	676	419
5	670	396	1.100	583	490	290
6	1.300	637	701	421	674	425

7	811	446	1.123	641	1.430	858
8	457	274	615	376	1.600	820
9	912	529	847	458	917	550
10	818	491	914	548	840	504
Total	7.878	4561	8.959	5136	10.297	6014
Average	787.8	456.1	895.9	513.6	1.029	601.4

The research results in table 1 show that resin production in 2019 and 2020 was very varied, with the highest average sap production in Rambatu with 1029 kg/farmer in 2019 and 601,4 kg/farmer in 2020. This variation is caused by the number of trees in production age and the maintenance intensity that carried out by the farmers themselves. Interviews with farmers showed that farmers in Rambatu carried out more intensive maintenance activities on their resin trees than farmers in other villages. In Honitetu village, the intensity of maintenance is only twice a year due to the long-distance access to the dammar forest, the lack of workers in the family, and the existence of other business alternatives that can generate income.

In these three research villages, the frequency of applying the injury to the tree bark to encourage sap production and thinning the disturbance trees to create space for sunlight to exposure the tree trunks was the maintenance activity that carried out. It can increase the temperature and reduce humidity, facilitating the discharge of sap from tree trunks that have been injured. The practice of wounding or applying injury to the tree bark is usually carried out 2-3 times/year to each production age dammar tree. First, the bark stem injury is generally done in early summer, namely in October or November, and the second time in January or February. Thus, the resin harvest can only be done after six months from the second application of the bark injury.

However, farmers' main complaint is the problem of seasonality or climatic factors, wherein 2020 the sap production is deficient because the rainy season is longer than usual, which only occurs for three hot months and nine rainy months.

3.2. Climate of 2019 and 2020

In general, the West Seram District's normal climatic conditions are six months of summer and six months of the rainy season. The average resin production per tree ranges from 4-6 kg/tree/year. In this condition but in abnormal season conditions, such as in 2020, resin production only reaches an average production of 1-3 kg/tree/year.

Table 2. Climate conditions in 2019 and 2020

Months	Average of Rainfall (mm)		Average of Temperature (C°)		Average of Relative Humidity (%)		Average of Solar intensity (lux)		Average of Solar Duration (hour)	
	2019	2020	2019	2020	2019	2020	2019	2020	2019	2020
January	75	54.0	28.5	29.0	76	76	16.000	15.000	10	12
February	125	57.0	28.3	28.8	77	79	15.500	14.600	8	9
March	198	220	28.0	27.0	79	80	15.200	14.100	7	10
April	149	98.0	28.4	27.2	77	80	15.000	14.300	7	7
May	304	298	27.0	26.8	81	82	13.600	13.200	6	7
June	504	601	24.5	24.0	90	93	9.000	8.500	4	1
July	481	500	25.0	24.5	88	92	9.500	8.100	5	1
August	406	497	25.5	25.0	88	91	10.000	9.300	5	3
September	310	450	26.5	25.7	85	90	11.000	10.251	6	4
October	188	248	27.2	26.8	83	88	12.500	11.912	7	6
November	156	240	27.8	27.0	82	85	13.000	12.675	8	6
December	120	235	28.3	28.0	80	85	13.500	12.122	8	7
Average	251.3	291.5	27.0	26.5	82	85	12.816	12.005	6.75	6.08

The research results in table 2 show that the average rainfall in 2019 is 251.3 mm with an average temperature 27 °C and relative humidity 82 %. Meanwhile, the year 2020 has an average rain 291.5 mm with an average temperature 26.5 °C and relative humidity 85 %. This shows that the average rainfall intensity in 2020 is higher and longer, namely during the nine months of the rainy season. Also, it can be seen that the average intensity of radiation and the duration of solar radiation in the Inamosol Sub-district in 2020 is lower than in 2019. In general, West Seram District, Maluku Indonesia has two seasons, namely the dry season, namely October-April, and the rainy season from April-October. However, in 2019 there was a slight change, namely five wet months and seven dry months, and in 2020 there were nine wet months and only three dry months. This change has a significant impact on the production of resin as a non-timber forest product.

Plants use various methods to protect themselves from pests and diseases, such as what dammar plant species do by removing sap to cover wounds on the trunk or roots, and branches of trees. Resin sap is produced in the sap tubes found in the parenchyma in the xylem and stem cortex and then channelled to epithelial cells in tree bark. Resin formation occurs due to injury to tree bark due to friction or damage by animals and humans. Resins are plant products insoluble in water, harden when exposed to air, do not play a role in fundamental plant growth processes, and are generally produced by woody plants. The epithelial cells secrete resin into large central resin ducts, where the resin is deposited until an injury occurs. One or more layers of axial parenchyma cells (sheaths) envelop the duct structure of the resin. The resin channel or resin channel extended in the space between tubular cells and surrounded by epithelial cells secrete the resin into the canal. These canals are oriented longitudinally and radially between the fusiform ray. They are usually found in denser woods-growing late in the season such as *Pinus merkusii* [9].

The epidermis of wood contains a layer of cells called keratinocytes. Keratinocytes are the stem cells for epithelial cells. The cells are arranged neatly to form the epidermis layer of the bark or known as the epidermis's main cells. Epithelial tissue is widely distributed throughout the epidermis, forming the covering of all body surfaces, body cavities and orifice organs, and is the gland's main network. They perform various functions that include protection, secretion, absorption, excretion, filtration, diffusion, and sensory reception [10][9].

Cortex resin channels are important resin reservoirs during the early stages of plant development, whereas xylem resin channels are the main source of resin in mature trees. Axial resin channels are present in the xylem and the cortex of roots and stems. Pines produce resin acids, turpentine, essential oils, resene (a non-volatile neutral organic mixture) and water. Common resin compounds are toxic to protect plants from various herbivores, insects and fungi. Resins are plant products that are insoluble in water and harden when exposed to air. Some woody plants generally produce resins although they do not play a role in fundamental plant growth. To increase the purity and density of resin, and the essential oils as a whole, which are then converted into trichomes and turpentine, so plants need nutrients in high concentrations, especially phosphorus (P), potassium (K), and silicic acid (Si) [11].

4. Conclusion

The average resin production of farmers in 2019 and 2020 is 904.2 kg/farmer and 523.7 kg/farmer, respectively. There was a decrease in resin production by about 41.5 % in 2020 if compared to 2019. The leading cause of farmers' declining average resin production is climatic factors, especially the duration of rainfall, temperature, humidity, intensity, and solar radiation duration. Farmers have considered the dammar forest as a bank in providing cash for their daily needs.

References

[1]. Manuputty D.N. 1955. *Agathis* family in Indonesia. Keluarga *Agathis* di Indonesia *Rimba Indonesia* **3** (4) 132-188.
[2] Rahayu N 2009 Utilization of Resin. Pemanfaatan Damar. http: //id. utilization of resin.

[3] Souhaly R 2016 Study of the Implementation of Adat Sasi and Its Implications. Kajian terhadap Pelaksanaan Sasi Adat dan Implikasinya *Jurnal Kajian Teologi* **2**(2).

[4] Lempang M 1997 Test several tapping patterns to predict copal production from Agathis trees (*Agathis hamii*). Uji beberapa pola sadap untuk menduga produksi kopal dari pohon Agathis (*Agathis hamii*) *Forestry Research Bulletin* **2** (1) 15-25.

[5] Santosa G 2006 Development of copal tapping method through the application of incision techniques. Pengembangan metode penyadapan kopal melalui penerapan teknik sayatan [disertasi]. Bogor Agricultural Institute.

[6] Melmambessy Y 2020 Study of resin forest management in West Seram District. Kajian pengelolaan hutan damar di Kabupaten Seram Bagian Barat, Tesis Pascasarjana, Program Studi Managemen Hutan, Universitas Pattimura, Ambon

[7] Setiawan H 1997 Effect of shape, tapping location and black plastic cover on the sap production of *Agathis loranthifolia*. Pengaruh bentuk, letak sadapan dan pemberian tutup plastik hitam terhadap produksi getah pohon *Agathis loranthifolia* [skripsi]. Bogor Agricultural Institute.

[8] Matinahoru J M 2011 Potential of Dammar Tree (*Agathis alba*) in Honitetu Village, West Seram Regency. Potensi Pohon Dammar (*Agathis alba*) di Desa Honitetu, Kabupaten Seram Barat. *Agroforestry Journal* **7**(2).

[9] Kramer P J dan T T Kozlowski 1960 Physiology of Trees. New York: Mc Graw-Hill Book Co. inc.

[10] Franceschi V R, Krokene P, Christiansen E, and Krekling T 2005 Anatomical and chemical defences of conifer bark against bark beetles and other pests *Journal of New Physiologist Foundation*. https://doi.org/10.1111/j.1469-8137.2005.01436.x

[11] Pratiwi A 2008 Effect of potassium fertilizer application on the production of *Agathis spp* sap in the Gunung Walat Education Forest, Sukabumi. Pengaruh pemberian pupuk kalium terhadap produksi getah *Agathis spp* di Hutan Pendidikan Gunung Walat Sukabumi [skripsi]. Sukabumi [thesis]. Bogor Agricultural Institute

Institutional Characteristics of Supporting Potato Farming in South Sulawesi

R Indrasti[1,*], J B M Rawung[2], N R Sudolar[3], S Handoko[1] and Nurjanani[4]

[1] Indonesian Center for Agricultural Technology Assessment and Development, Indonesia.
[2] North Sulawesi Assessment Institute for Agricultural Technology, Indonesia.
[3] Jakarta Assessment Institute for Agricultural Technology, Indonesia.
[4] South Sulawesi Assessment Institute for Agricultural Technology, Indonesia.

E-mail: ritaindrasti@yahoo.com

Abstract. Potatoes are one of the commodities that have been prioritized for development which considered as a source of carbohydrates, highly nutritious and have the potential for food diversification. In this case the institutional activities of potato farmers may contribute to support their agribusiness aspects. The purpose of the study was to analyze the feasibility of potato farming as well as to characterize the farmer institutions. Data collection was carried out through FGD (Focus Group Discussion) with 25 respondents of potato farmer group members and secondary data as supporting data. The survey was conducted in 2017 in South Sulawesi Province. Data were analyzed descriptively and R/C ratio was measured to determine the potato farming feasibility. The results showed that potato farming was feasible economically. The farmers were mostly active in supporting institutional activities, including group meeting activities, paying fees and extension activities.

1. Introduction

The horticulture sub-sector in agricultural development is a strategic position. Horticulture's contribution to agricultural development continues to increase as reflected in several growth indicators, including Gross Domestic Product (GDP), export volume of horticulture, labor absorption, Farmer Exchange Rate, nutrition and aesthetics. According to [1], the strategic role of the horticulture sub-sector can still be improved due to the high diversity of horticultural genetic resources (SDG), large land area, abundant human resources, and a conducive agro-climate.

Apart from being a center for rice production in Eastern Indonesia, South Sulawesi also produces vegetables including tomatoes, potatoes, cabbage and mustard greens. Potatoes are a plant that is cultivated intensively, especially in highland areas, one of which is Gowa Regency. Potato farming in the Gowa area requires relatively high organic matter, which is around 10-30 tonnes/ha. Potato farmers still feel the need for large enough inputs, both organic materials and energy. Thus, farmers only use small amounts of organic fertilizers, even not using fertilizers at all [2]. Potato production centers in South Sulawesi Province are located in 4 districts, including the regencies of Gowa, Bantaeng, Enrekang and Tana Toraja, where Gowa district ranks as the largest potato producer with an area of about 6.9 thousand hectares per year. The largest contribution of potato production to total production is Gowa Regency at 50%, followed by Bantaeng Regency 34%, Enrekang Regency 8% and Tana Toraja Regency and other districts only 4% [3].

Potato (*Solanum tuberosum* L.) is one of the vegetable crops that gets priority in development, because potatoes are considered to have high competitiveness compared to other vegetables. The demand for potatoes in Indonesia is increasing, both as a fresh product and as a processed product. In addition, the position of potatoes as food for the future is increasingly important, because they can be used as vegetables and for diversification of carbohydrate sources that substitute rice and are a reliable source of food security [4]. Potato is well adaptive under varieties of soil, even under saline stress, especially treated with suitable combination of fertilizers [5].

According to [6], the production, acreage and productivity of potatoes in Indonesia from 2014 to 2017 have always fluctuated. In 2014, a land area of 76,291 ha resulted in a production of 1,347,815.1 tonnes with a productivity of 17.66 tonnes/ha. In 2015, the land area of 66,983 ha resulted in a production of 1,219,269.7 tonnes with a productivity of 18.20 tonnes/ha. In 2016, the land area of 66,450 ha resulted in a production of 1,213,038.4 tonnes with a productivity of 18.25 tonnes/ha. In 2017, the land area of 76,611 ha resulted in a production of 1,164,738.1 tonnes with a productivity of 15.40 tonnes/ha. According to General Hortikultura (2018), production, acreage and productivity of potatoes in Indonesia from 2014 - 2017 always fluctuated. In 2014 the land area of 76,291 ha resulted in a production of 1,347,815.1 tonnes with a productivity of 17.66 tonnes/ha. In 2015, the land area of 66,983 ha resulted in a production of 1,219,269.7 tonnes with a productivity of 18.20 tonnes/ha. In 2016 the land area of 66,450 ha resulted in a production of 1,213,038.4 tonnes with a productivity of 18.25 tonnes/ha. Meanwhile, in 2017 the land area of 76,611 ha can produce a production of 1,164,738.1 tonnes with a productivity of 15.40 tonnes/ha.

In agricultural development, the farming community has an important role. Therefore, it is necessary to empower the farming community, so that the farmers are able to solve the problems faced in their farming activities. One of the government's efforts for farming communities is to build farmer groups in rural areas. Farmer groups are farmers formed on the basis of common interests and similar environmental conditions (social, economic, resources), familiarity and harmony led by the chaiman [7]. With the existence of farmers groups, the goal will be easier to achieve than individual farmers, because farmers can exchange experiences, knowledge and farming abilities to increase their productivity. This study aims to determine the feasibility of potato farming as well as to characterize the potato farmer institutions in South Sulawesi.

2. Methods

The research location was determined purposively at Patapang Village, Tinggi Moncong District, South Sulawesi Province, with the consideration that this place is a potato producer. Data collection was conducted through FGD (Focus Group Discussion) with 25 respondents and secondary data using data from BPS South Sulawesi in Figures and other supporting data. Data collection methods, using polymer data obtained from interviews, questionnaires, documentation, and secondary data obtained from library data. The research method used is descriptive analysis. The survey was conducted in 2017. Data were analyzed descriptively using R/C for farming feasibility tests.

3. Results and discussion

3.1. Institutional characteristic

Institutional characteristics of Mitra Bandengngia farmer group and several other farmer groups in Patapang Village, Tinggi Moncong District are farmer groups that are active in activities that support the institution, including actively participating in farmer group meetings, paying contribution for group needs, participating in extension programs and other activities that support the institution. The culture of mutual cooperation is still reflected in the people there.

Table 1. Farmer institutional characteristics

Activity	Active (respondent)	Middle (respondent)	As needed (respondent)
Meeting activity	10	7	8
Farmer group saving	16	5	4
Extension program	11	8	6

3.2. Meeting activities
The meeting activities were carried out with the agreement of each farmer group according to the agreed agenda. The number of group meetings held by the farmer groups in Patapang Village was twice a month. The highest result with 10 respondents shows that respondents attend meetings regularly. Meanwhile, seven (7) respondents stated that sometimes this is because the respondent has other activities. There were eight (8) respondents who attended meetings as needed. The respondent was allegedly passive in conducting meeting activities.

3.3. Farmer group saving
Paying fees for the farmer group saving is an obligation for the farmer. This fee can be used as a loan fund if a member experiences difficulties, but the member is still obliged to repay the loan. 16 respondents have a high awareness of paying fees because they were considered that the fees will be used for their own interests as well. However, five respondents do pay dues sometimes; it is possible that the respondents in this category think this is not too important. Four respondents paid non-routine contributions, only as needed. The respondent may have other activities that involve making contributions as well.

3.4. Extension program
The extension program has several agendas, one of which is learning activities or non-formal education carried out by extension agents to farmers with the aim of providing information so that farmers can be independent in dealing with problems. This is consistent with the statement of [8] which states that there is a change in behavior among farmers who participate in extension activities because there will be more useful information for farmers, new skills, as well as stronger mental attitudes and motivation as expected. Extension activities are carried out twice per month. Eleven respondents routinely participate in extension programs because they consider this program very important and support institutional activities which are expected to be sustainable. Eight (8) respondents stated that they sometimes participate in extension programs, which generally depend on time availability. If the respondent has free time to take part in the extension program, the respondent will attend, and vice versa. Such respondents are not too serious in supporting these activities. There were six (6) respondents who participated in the extension program according to their needs, such respondents were considered respondents who had other activities.

3.5. Potato Farmer Performance
One of the potato production centers is in Gowa Regency, with a total potato production averaging 6.9 thousand tons per year. One of the potato production centers is in Tinggi Moncong District, Patapang Village. In this village there are 37 farmer groups working on horticultural crops. The horticultural crops commonly cultivated by farmers in Patapang Village are potatoes, cabbage, carrots, tomatoes and mustard greens. About 75 percent of farmers in this village cultivate potato crops with an average group area of around 20 hectares with a number of farmer members per group ranging from 20 to 25 people. The average area planted with potatoes per farmer ranges from 0.5 to 1 hectare. Potato plants are cultivated 2 times a year, first planted in March and harvesting in June, the second planting of potatoes is planted in July and harvest time around October. Farmers in this village have been cultivating potato crops for the past 20 years, with an average age of farmers ranged from 30 to 54 years.

Farmers in this village know the potato plant from farmers who come from West Java and live in this village. Until now, the potato seeds planted by farmers come from West Java. Through this intermediary farmer from West Java, they order a number of seeds that are needed; the potato seeds that are planted generally come from West Java Canning. After this sub-district became known as a center for potato production, the local government through the Agriculture Office helped increase community potato production.

3.6. Farming analysis

Prior the technology is adopted by the farmers, the technology have to be feasible either economically, technically and socio-culturally, because the technology is considered to provide more profitable income. Farmers are always faced with costs. [9] stated that there are three (3) ways to maximize farm income, namely: 1) technical efficiency, 2) input efficiency, and 3) production efficiency. Achieving high technical efficiency is very important to increase the competitiveness and profitability of farming. Profitable farming is also supported by environmental factors [10]. However, when the existing problems can be solved and the inputs are available and used properly, then the production and income can be optimized [11].

In principle, production costs are the sum of all costs (fixed costs and variable costs) that have been incurred to obtain production. Fixed costs that are calculated are the rental of land, equipment and other supporting materials. Meanwhile, variable costs include seeds, fertilizers, pesticides and labor.

Table 2. Analysis of potato crop farming using NPK 300 kg/ha + 0.15% HA/ha per planting season

No	Detail	Volume	Price/unit (Rp)	Total (Rp)
I.	**Production Cost**			
1.	Potato seed	1.500 Kg	15.000	22.500.000
2.	Fertilizer			
	• NPK Super	300 Kg + HA 0.15%	2.800	840.000
3.	Pesticide			
	• Fungicide	13 kg	240.000	3.120.000
	• Insecticide	10 l	260.000	2.600.000
	Total Input cost			**29.060.000**
II.	**Labor Cost**			
1.	Soil Processing	50 HOK	30.000	1.500.000
2.	Planting	60 HOK	30.000	1.800.000
3.	Weeding I	50 HOK	30.000	1.500.000
4.	Weeding II	50 HOK	30.000	1.500.000
5.	Plant Pest Control	40 HOK	30.000	1.200.000
6.	Harvesting	60 HOK	30.000	1.800.000
7.	Transportation of Harvest	16.506 Kg	50	810.500
	Total labor cost			**10.125.333**
III.	**Other cost**			
1.	Land Rent Per Season			1.500.000
2.	Hand Sprayer	2 pcs	400.000	800.000
3.	Plastic Drum 200 l	3 pcs	60.000	180.000
4.	**Sack**	200 pcs	2.000	400.000
	Total other fees			**2.880.000**
	Total production cost (I+II+III)			**42.065.333**
IV	**Production/Revenue**	16.470 Kg	7.000	**115.546.672,44**
V	**Income**			**73.481.330,07**
VI	**R/C ratio**			**2,75**

Source: analyzed primer data, 2017

Assumption:

Potato seed tubers prices = IDR 15.000/kg
Consumption tubers prices = IDR 7.000/kg

NPK Super + HA 0,075 = IDR 2.600 /kg
NPK Super + HA 0,15 = IDR 2.800/kg
NPK Super = IDR 2.300/kg

According to the types of costs used, the production facilities and labor were the components of the highest cost, reached IDR 29.06 million and 10.1 million, respectively, and followed by other cost components of IDR 2.88 million. Then the total cost of potato farming is IDR 42.065 million per hectare per planting season. In accordance with the treatment (NPK super 300 Kg / h + 0.15% HA) can produce high production which reached 16,506 kg/ha consisting of 1,351 kg of consumption potato and 15,154 kg of seed potato tubers. The current price of consumption potatoes is IDR 7,000, and the price of seed tubers is IDR 12,000. Thus with this production assumption, an income of IDR 181,848,000 can be obtained. If it is reduced by the total production costs, the net income received by the farmers is IDR 139,782,667. However, if all production is sold as consumption potatoes, the net income received by farmers is IDR 73,481,339.07. The results of potato farming analysis can be seen in Table 1. From the table, it can be seen that the R/C ratio analysis reaches 2.75 if the production is sold in the form of consumption potatoes. This illustrates that the treatment is assumed to be economically viable. Nevertheless, [12] revealed that the potato feasibility depends on differences in yield, N fertilizer source and dose, as well as the selling price. Besides cultivated in the field, potato are feasible to be cultivated aeroponically [13]. However, [14] stated that there are some major constraints on potato agribussiness which should be taken into account, namely price instability, high storages charges, uncertainty of water, inadequate marketing facility and spoilage of potato, even though the problems may varies by location.

4. Conclusion

The institution in Patapang Village is a group that is active in supporting institutional activities, including group meeting activities, paying fees and extension activities. The meeting activities carried out by the farmer groups are considered important to advance the group's institution. Paying dues is very beneficial for the group and for themselves because they can help personally when experiencing difficulties. Extension program Conducted so that farmer groups can learn from extension agents who demonstrate through extension programs, including demonstration plots and technical assistance activities.

Efforts to increase potato productivity should have strong support from the local government, including a commitment to always demonstrate technology for potato farming through a demonstration plot (demonstration plot). Demonstration activities must be carried out more frequently so that farmers are interested so that farmers can imitate it. Increased productivity in line with sustainable institutions, the farmer groups will always improve their standard of living. The increase in income has been seen with a profitable analysis of potato farming.

References

[1] Puslitbanghorti 2010 *Pedoman umum program dukungan pengembangan kawasan agribisnis hortikultura* (PDPKAH). (Indonesia: Puslitbanghort Balitbangtan Deptan)

[2] Arinong A R, Nane M N, and Arsyad M 2013 *Peranan Kelompok Tani Abulosibatang Dalam Pemenuhan Kebutuhan Usahatani Padi Sawah Di Kelurahan Lanna Kecamatan Parangloe Kabupaten Gowa* [Jurnal]

[3] Badan Pusat Statistik Sulawesi Selatan 2012 *Sulawesi Selatan Dalam Angka. Badan Pusat Statistik Sulawesi Selatan. Makasar.*

[4] Hamdani J S 2009 Pengaruh Jenis Mulsa terhadap Pertumbuhan dan Hasil Tiga Kultivar Kentang (Solanum tuberosum L.) yang Ditanam di Dataran Medium *J. Agron. Indonesia* **37 (1)** 14 – 20.

[5] Al-taey D K A Al-Naely I J C and Kshash B H 2019 A study on the effects of water quality, cultivars, organic and chemical fertilizers on potato (Solanum tuberosum L.) growth and yield to calculate the economic feasibility *Bulgarian J. of Agricultural Sci.* **25** (6) 1239-45

[6] Direktorat Jenderal Hortikultura 2018 *Laporan kinerja 2018* Kementerian Pertanian Direktoral Jenderal Hortikultura. p 122

[7] Trimo 2006 *Evaluasi penyuluhan pertanian permasalahan dan upaya pemecahannya di Kecamatan Banyudono Kabupaten Boyolali* Unpublished

[8] Yustina I and Sudrajat A 2003 *Membentuk pola prilaku manusia pembangunan* (Bogor: IPB press) p 292

[9] Kumbhakar S C and Lovell C A K 2000 *Stochastik Frontier Analysis*. (Cambridge: Cambridge University Press) p 333

[10] Rizk A H Mashhour A M A Abd-Elhadyand E S E and El-Ashri K M A 2010 The rote of some humic acid products in reducing of use mineral fertilizer and improving soil properties and nutrient uptake *J. Soil Sci. and Agri. Engineering* **1** (8) 765-74

[11] Sapkota S C Rokaya P R Acharya H and Uprety S 2019 An economic analysis of potato production in Achham District of Nepal *Int J of Hort & Agr* **4** (2) 1-9

[12] Elisabeth D A A Ginting E and Restuono J 2018 The economic feasibility of sweet potatos farming by using selected N fertilization *J. Socioecon. and Development* **1** (2) 100-5

[13] Demo P Chiipanthenga M Mwenye O J Chipungu F Chimwala L Chinoko G Mbewe and W Kumar N 2013 Assessment of technical and financial feasibility of seed potato multiplication using aeroponics and field techniques in Malawi. Transforming potato and sweetpotato value chains for food and nutrition security. 9. *Triennial Congress of the African Potato Association Naivasha* (Kenya) 30 Jun-9 Jul 2013 (Kenya).

[14] Honaryar G 2019 An economic analysis of production marketing and value chain of potato in Bamyan Province *Int J. of Research – Granthaalayah* **7** (10) 1-25.

Organoleptic characteristics of nutmeg tea with variations in the roast time

S Palijama[1,*], L O Kakisina[2], N R Timisela[2], J M Luhukay[2], R Breemer[1], S Laterissa[1], E Kaplale[1] and F J Polnaya[1]

[1] Department of Agricultural Product Technology, Faculty of Agriculture, Pattimura University, Jl. Ir. M. Putuhena, Kampus Poka, Ambon 97233 Indonesia
[2] Department of Agricultural Socio-Economic, Faculty of Agriculture, Pattimura University, Jl. Ir. M. Putuhena, Kampus Poka, Ambon 97233 Indonesia

* E-mail: annie_jeane@gmail.com

Abstract. This study aims to determine the effect of roast time on the organoleptic characteristics of nutmeg tea. The experimental design carried out in this study was a randomized block design (RBD), which is arranged in a factorial consisting of 2 factors with three replications. The factor used was the roast time (S), which consisted of 4 levels, namely S1 (20 minutes), S2 (25 minutes), S3 (30 minutes), S4 (35 minutes), and part of the nutmeg (P) which consisted of 2 levels, namely P1 (fruit skin) and P2 (pulp) to obtain 24 experimental units. The response variables analyzed included organoleptic responses (color, taste, and flavour), both hedonic and hedonic quality. The results showed that the roast time, the nutmeg part, and the interaction between the two affected the color, taste, and flavour of both hedonic and hedonic quality. The nutmeg skin treatment and roasting time were 25 minutes; on average, the panellists responded to 3.96 (very like) color, 3.32 (like), flavour and 2.80 (like) taste hedonic. In hedonic quality, the treatment took 25 minutes, and the skin nutmeg, on average, the panellists responded to 3.35 color (tea-colored), 3.24 flavour (nutmeg flavor) and 2.56 taste (slightly nutmeg taste).

1. Introduction

Tea is one of the favourite drinks and has become a habit of Indonesian society because it has a distinctive taste. Today, the tea consumed has been modified with various ingredients for specific purposes so that the use of herbs and spices has begun to be studied and used. Many spice plants in Maluku have not been fully utilized, one of which is nutmeg.

There are many nutmeg plants (*Myristica fragrans* Houtt) in the Maluku islands where the potential for a nutmeg commodity in Maluku was 31,624.10 ha, which a smallholder plantation with a total production of 5,512.10 tons/year [1]. The nutmeg plant is known as a spice with economic value and is multipurpose. Apart from functioning as an essential oil-producing plant, nutmeg is also used in the food, beverage, medicine, perfume, and cosmetic industries.

The chemical composition of nutmeg was 1.37% flavonoids, 22.14 mg oxalate, 49.32% saponins, 8.42% alkaloids, and 16.00% phytate [2]. The active ingredients in nutmeg are minerals, Vitamin A, Vitamin B, vitamin C, folic acid, riboflavin, niacin, and many flavonoids. These compounds can interfere with the fungal cell membranes' permeability and change the organic components and nutrient transport, ultimately resulting in a toxic effect on fungi [3].

In making tea, the drying process significantly affects the quality of the tea produced, both chemically, physically, microbiologically, and consumer acceptance. Drying is one of the techniques used to extend storage capacity because it can reduce the water content of food, affecting the appearance, texture, and taste of food [4] and can inhibit the growth of unwanted microbes that can cause damage. Foodstuffs during storage [5][6].

The process of making herbal tea is drying with heat energy. Drying is the process of reducing water from a material using heat energy. [7] added that drying is one way of preservation to reduce water content in the material to inhibit unwanted microbes' growth. Based on previous research conducted by [8], the best temperature and drying time for soursop leaf tea processing is at a temperature of 50˚C with a drying time of 150 minutes. [9] added that medicinal plants' best drying temperatures are 50˚C and 60˚C. Various processes of drying food ingredients have been carried out, one of which is roasting.

The traditional roasting process uses a temperature of 100-200 ° C, where heat transfer occurs by conduction. The roasted material is stirred rapidly and continuously so that water vapor is quickly carried out and the heat is uniformly and thoroughly distributed [10] [11]. This roasting process is crucial in tea processing, but currently, there is no data on the right roasting time to produce tea according to consumer desires. Nutmeg so that harvesting is usually wasted so that tea was made using the nutmeg's skin and pulp in this study. This study aimed to determine the effect of the part of the fruit used and the roasting time on the organoleptic characteristics of the nutmeg tea.

2. Methods

2.1. Materials

2.2. Materials preparation

Material preparation is done by sorting the nutmeg after being harvested. Then wash it with running water to remove dirt and dust that stick. Then peel/peel using a slicer and do the pulp so that the skin and pulp used to have the same thickness. It is weighing the skin and pulp of the fruit before roasting.

2.3. Withering

The nutmeg skin and flesh are roasted, and the roasting time is adjusted according to the treatment used. The factor used is the roasting time (S), which consists of 4 levels, namely S1 (20 minutes), S2 (25 minutes), S3 (30 minutes), S4 (35 minutes), and the part of the nutmeg used (P) which consists of 2 levels, namely P1 (fruit skin) and P2 (pulp). This withering process aims to reduce the moisture content of the nutmeg's skin and pulp before use.

2.4. Reduction in size

The next process is reducing the size of the skin and flesh of the nutmeg using a cruiser. The aim is to facilitate the release of the nutmeg tea components during brewing.

2.5. Packaging

The skin and pulp of the fruit that has been mashed are then weighed at 2 g; 2 g of stevia leaf powder are added and packed using the prepared tea bag.

2.6. Research variable

This study's response variable was the organoleptic analysis carried out on color, taste, and flavour using the hedonic and hedonic-quality test (Table 1).

Table 1. Hedonic test and hedonic quality of color, taste and flavour

Scoring	Hedonic Test			Hedonic Quality Test		
	Color	Taste	Flavor	Color	Taste	Flavor
4	Very like	Very like	Very like	Very tea color	Very nutmeg taste	Very nutmeg flavor
3	Like	Like	Like	Tea colored	Nutmeg taste	Nutmeg flavor
2	Slightly like	Slightly like	Slightly like			

1	Not like	Not like	Not like	Slightly tea colored	Slightly nutmeg taste	Slightly nutmeg flavor
				Not tea colored	Not nutmeg taste	Not nutmeg flavour

2.7. Statistical analysis

This research was conducted experimentally using a randomized block design arranged in a factorial consisting of two factors and three replications.

3. Results and discussions

Organoleptic testing, including color, taste, and aroma, showed the panellist's response to the resulting nutmeg tea. Color is also crucial in processed and unprocessed food or drinks because color can also provide clues about its chemical changes. Taste is influenced by several factors, namely chemical compounds, temperature, concentration, and interactions with other taste components. Various chemical compounds cause different tastes due to organic compounds found in food ingredients. The effect of temperature can also result in sensitivity to taste, while interactions with other components can affect the taste value. The assessment of smells is influenced by psychological and physiological factors that give rise to different opinions. The flavor of a food or beverage product has a vital role in the assessment and appearance because if it has a distinctive flavor, the product is said to be good. The organoleptic test was carried out using the hedonic test method and the hedonic quality test to determine the level of preference or acceptance of the panelists for nutmeg tea to see whether the product was liked or not.

The results showed that the nutmeg used and the length of roasting time affected the resulting nutmeg tea. The results of further tests showed that the nutmeg parts' treatment and the interaction had a significant effect on the color, taste, and flavor of nutmeg tea, both hedonic and hedonic quality test. The nutmeg rind has the highest panelist acceptance value for the color, taste, and flavor of nutmeg tea both by hedonic test and hedonic quality test. They were compared with the nutmeg pulp (Table 2).

Table 2. The effect of the part of the fruit used on the color, taste, and flavor of nutmeg tea.

The part of the fruit used	Response variable					
	Hedonic test			Hedonic quality test		
	color	taste	flavor	color	taste	flavor
Nutmeg skin	3.62 a	2.95 a	3.28 a	3.32 a	2.69 a	3.32 a
Nutmeg pulp	2.52 b	2.79 b	2.54 b	2.63 b	2.07 b	2.51 b

Note: Figures followed by different letters show significant differences according to the Tukey test (α = 0.05).

Table 2 shows that the average hedonic assessment of the color, taste, and flavor of high nutmeg rind tea was 3.62 (very like), 2.95 (like), and 3.32 (like) compared to the flesh of the nutmeg 2.52, 2.79, and 2.54. (like). For the hedonic quality test, the average hedonic assessment of the color, taste, and flavor of high nutmeg rind tea was 3.32 (tea color), 2.69 (slightly nutmeg flavor), and 3.32 (nutmeg flavor) compared to 2.63 nutmeg pulp. (slightly tea-colored), 2.07 (slightly nutmeg flavored) and 2.51 (slightly nutmeg flavored). It is suspected that the flavonoid compounds and enzymes found in the nutmeg skin are oxidized by air to produce a specific color, taste, and flavor. Herbal teas rich in phenolic and flavonoid compounds have high antioxidants that can inhibit free radical activity [12].

The results showed that the treatment of roasting time had no significant effect on the color, taste, and flavor of nutmeg tea, both hedonic and hedonic quality (Table 3). Still, the interaction between the treatment of the nutmeg parts and roasting duration had a significant effect.

Table 3. Effect of roasting time on the color, taste, and aroma of nutmeg tea.

Length of roast time (minutes)	Response variable					
	Hedonic test			Hedonic quality test		
	color	taste	flavour	taste	color	flavour
20	3.04 a	2.96 a	2.30 a	2.94 a	2.52 a	2.72 a
25	3.22 a	2.94 a	2.34 a	2.92 a	2.28 a	2.92 a
30	3.08 a	2.88 a	2.40 a	3.08 a	2.42 a	2.94 a
40	2.94 a	2.86 a	2.44 a	2.94 a	2.30 a	2.88 a

Note: Figures followed by different letters show significant differences according to the BNJ test at the 5% level.

Table 3 shows that the interaction of nutmeg skin treatment and roasting time is 25 minutes. The average hedonic assessment of the color, taste, and flavor of high nutmeg tea is 3.96 (really like), 2.80 (like), and 3.32 (like). They compared it to other treatment interactions. For the hedonic quality test, the interaction of nutmeg skin treatment and roasting time was 30 minutes, the average hedonic quality of the assessment of the high nutmeg tea color, taste, and flavor was 3.52 (very tea-colored), 2.72 (slightly nutmeg flavor) and 3.32 (nutmeg flavored) compared to other treatment interactions.

Table 4. Effect of interaction of parts of the nutmeg used and roasting time on the color, taste and aroma of nutmeg tea.

Response variable	Roast Time (minutes)			
	20	25	30	35
	Hedonic test			
color				
- Nutmeg skin	3.36 bc	3.96 a	3.44 ab	3.32 c
- Nutmeg pulp	2.72 d	2.48 d	2.32 d	2.56 d
taste				
- Nutmeg skin	2.80 a	2.80 a	2.96 a	2.74 a
- Nutmeg pulp	1.80 b	1.88 b	1.84 b	1.64
flavor				
- Nutmeg skin	3.32 a	3.32 a	3.40 a	3.08 ab
- Nutmeg pulp	2.40 b	2.56 b	1.84 b	1.64 b
	Hedonic quality test			
color				
- Nutmeg skin	3.24 ab	3.36 a	3.52 a	3.16 ab
- Nutmeg pulp	2.64 c	2.48 c	2.60 c	2.86 bc
taste				
- Nutmeg skin	2.68 ab	2.56 ab	2.72 ab	2.80 a
- Nutmeg pulp	2.36 bc	2.00 cd	2.12 cd	1.80 d
flavor				
- Nutmeg skin	3.24 a	3.24 a	3.32 a	3.08 ab
- Nutmeg pulp	2.20 c	2.60 bc	2.56 c	2.68 bc

Note: Figures followed by different letters show significant differences according to the BNJ test at the 5% level.

4. Conclusion

The results of the study, it can be concluded that the treatment of nutmeg skin and roasting time of 25 minutes produced tea with organoleptic characteristics including hedonic test, namely 3.96 color (very like), 3.32 flavor (like) and 2.80 taste (like) while hedonic quality test is 3.35 color. (tea color), flavor 3.24 (nutmeg flavor) and 2.56 taste (slightly nutmeg flavor)

References

[1] Maluku Central Statistics Agancy 2017 The area and production of nutmeg plantations in Maluku
[2] Okukpe K M, Adeloye M A, Alli O I, Adeyina O A, Annongu A 2012 Investigation of pytohormonal potential of some selected tropical plants *Res j med plants* **6**(6) 425-432.
[3] Jupriadi L 2011. Activity Test of Ethanol Extract of Waru Leaf (*Hibicustilaceus* L.) against Malassezia furfur. Malang: Veterinary Medicine Study Program, Brawijaya University
[4] Ananda A D 2009 Antioxidant Activity and Organoleptic Characteristics of Functional Drinks Green Tea (*Camellia sinensis*) Instant Spice. Essay. Faculty of Agriculture. Bogor Agricultural Institute.
.[5] Angraiyati D and Hamzah F 2017 Dried long time in making herbal tea fragrant pandan leaves (*Pandanus amarilifolius* Roxb) *Journal Online Agriculture Faculty Riau University* **4** (1).
[6] Jamaluddin J, Molenaar R, dan Tooy D 2014 Study of Water Sorption Isotherm and Water Fraction Bonded with Green Bean Pia Cake from Gorontalo City *Journal of Food Science and Technology* **2** (1) 27.
[7] Sribudiani E, Parlindungan and Volliadi 2011 Study of Temperature and Drying Time on Organoleptic Quality of Rosella Herbal Tea (*Hibiscus sabdariffa* Linn.) *Sago* **10** (2) 9-15.
[8] Adri D, and Wikanastri H 2013 Antioxidant Activity and Organoleptic Properties of Soursop Leaf Tea (*Annona muricata* Linn.) Based on Variations of Drying Time *Journal of Food and Nutrition* **04** (07).
[9] Rocha R P, Melo E C and Radunz L L 2011 Influence of Drying Process on the Quality of Medical Plants *Journal of Medical Plants Research* **5** (22) 7076-7084.
[10] Karyadi J N W, Lumbanbatu J, and Sri R 2009 Effect of Temperature and Roasting Time on the Physical-Mechanical of Robusta Coffee Beans. National Seminar and Holds PERTETA Technology. Faculty of Agricultural Technology. Gajah Mada University.
[11] Panjaitan V 2018 The effect of roasting and drying process towards physical characteristics and sensory quality of green tea from leaves of torbangun (*Coleus amboinicus* L.) Agricultural Engineering Study Program, Department of Agricultural Technology, Faculty of Agriculture, Sriwijaya University Thesis
[12] Komes D, Horzik D H, Belscak A, Ganik K K, and Vulic I 2010 Green tea preparation and its influence on the content of bioactive compounds *Food Research International* **43** 167–176.

IOP Conf. Series: Earth and Environmental Science **883** (2021) 012082 doi:10.1088/1755-1315/883/1/012082

Corn farming analysis in Babar Island, District of Babar Islands, Southwest Maluku Regency

S F W Thenu[1,*]

[1]Lecturer at Agribussines Study Program, Faculty of Agriculture, Pattimura University.

* E-mail: stevethenu@gmail.com

Abstract. Babar Island is corn production center in District of Babar Islands. Maize is the dominant food commodities and a major food source for the community. Corn is the dominant crop commodity in Barbar Island and become staple food for the community. The farming are traditional and subsystems to meet farmers household needs and the excess (if any) will be sold but in very limited quantities. However the sales purpose is not commercial but just to meet the households needs that can not be substituted by other goods and only by few farmers if necessary. The farming is carried out on slash and burn dry land by polyculture cropping pattern that is combination of annual and perrenial crops. This study aimed to examine the feasibility and profitability of corn farming in Barbar island, using survey methods with stratified random sampling technique. The results showed that the cost of corn production in Barbar Island, district of Barbar Islands, South West Maluku Regency is consisted of fixed cost and variable cost. The fixed cost components is including : equipment depreciation costs and land rent costs. The variable costs includes : inputs costs (seeds, fertilizers, chemical) and labor costs. Variable costs is the largest cost component for corn farming. Corn farming is profitable and viable, as indicated by the average income Rp.2.647.396 compared to the average cost incurred by farmers Rp.2.520.934 and BC ratio of 1.05. This means that the production costs of Rp.2.520.934 will generate benefit 1.05 times of the cost.

1. Introduction

Babar island is corn production center in District of Babar Islands, Southwest Maluku Regency. The farming is carried out on dry land, traditional and subsystems. The cultivation is using slash and burn technique, located around and outside (forest) residential areas and still practice organic farming. The inputs are local with the use of family labor.

Farmers as much as possible reduce / avoid cash costs, except for equipment and post harvest tools (processing tool). Corn production is used as food stocks and not for commercial use.

Dryland farming systems are simple, use miniimum cash costs, depending on nature, using local varieties without chemicals, no-tillage (TOT) or minimum tillage (OTS) and still hold strong local wisdom. This farming character is appeal to be observed whether the activities is profitable or economically viable if all production factors calculated.

In this study, all costs such as labor, seeds, fertilizer, chemical and marketing are taken into account as cash costs (incurred by farmers), although not exist in real terms. This calculation aims to provide an overview of all economic inputs and outputs that should be taken into account in corn farming analysis such as the cost of production, revenue and income even the feasibility. This will give more relevant result than the real condition that can cause bias in interpreting the results. This study aims to analyse profit and feasibility of corn farming in Babar island.

2. Methods

The study carried out from October to December 2018, in 3 villages namely Village, Tela and Letsiara Village. The number of farmers sample is 120 persons (40 persons each village), using survey methods with multi-stage sampling technique. The villages is selected purposively while respondents selected by Stratified Random Sampling. Costs, revenues and income can be analyzed using the following formula.

Total cost can be calculated by adding fixed costs and variable costs by formula :

$$TC = FC + VC$$

Where :
TC = *total cost*
FC = *fixed cost)*
VC = *variable cost*

Farming revenue can be calculated by multiplication of production (yield) with selling price which can be written by the following formula [1]

$$TR = Q \times P$$

where :
TR : *total revenue*
Q : production *(quantity)*
P : Selling Price *(price)*

$$B = TR - TC$$

where :
B : net income
TR : total revenue
TC : total ciost

B/C ratio (Benefit Cost Ratio) used to determine the economic feasibility of corn farming. B/C ratio is the ratio between total income and total cost of farm production [2]. If the BCR > 1, then the farming is feasible, conversely, if the BCR < 1 then the farming is not feasible.

3. Results and discussion

3.1. Production Costs

Production cost is the summation of fixed cost and variable cost component that used in the production processes [3] Farming activities can not be separated from the use of various costs in the production process. The greater the revenue or the smaller the costs, then the farming will be more profitable or generated greater income [4].

3.1.1. Fixed Cost

Fixed costs are costs that are not influenced by the amount of production [2]. Fixed costs are also defined as costs that are not used up in the production process. These costs are formed due to the use of various agricultural inputs such as farm equipment. Fixed costs on corn farming can be seen at the following table.

Tabel 1. Average cost of equipment depreciation and land rent for corn farming in Babar Island, District of Babar Islands, Southwest Maluku Regency 2018.

No	Fixed Cost Components	Value (Rp)
1.	Chopping knife	42.933
2.	Hoe	24.008
3.	Shovel	25.000
4.	Harrow	16.313
5.	*Pakwel*	24.392
6.	Ax	43.725
7.	Crowbar	16.263
8.	Hand Sprayer	69.075
9.	Mixer	278.958
10.	Land Rent	264.083
	Total	804.750

Source: Primary Data, 2018 (Proceed)

Table 1 shows that the average total fixed cost is Rp. 804.750, consisted of equipment depreciation costs Rp. 540.667 and land rent costs Rp. 264.083. Equipment depreciation value is the highest cost component of fixed costs while the highest share of equipment cost is mixer namely Rp.278.958. The high cost of mixer compared to the other equipment is due to high price per unit namely Rp. 500,000 (manual mixer) and Rp. 2.500.000 (machine mixer) and owned by farmers as a means for corn processing.

3.1.2. Variable Cost

Variable costs are costs that are affected by the amount of production [2]. This cost is also known as consumables cost in a production process. This costs is one of the important cost component in the farming operation, because it deals with the means of production used by farmers. In practice, there are several components of variable costs that commonly used such as: seeds, fertilizers, chemical and labor. The average variable cost of corn farming is as follow:

Table 2. Average variable cost of corn farming in Babar Island, District of Babar Islands, Southwest Maluku Regency 2018.

No	Variable Cost Components	Value (Rp)
1.	Seeds	234.392
2.	Fertilizer	627.938
3.	Chemical	52.479
4.	Labor	801.376
	Total	1.716.184

Source: Primary Data, 2018 (Proceed)

Table 2 shows that, on average, total variable cost is Rp.1.716.184 consisted of inputs costs (Rp.914.809) and labor costs (Rp.801.376). The high of inputs cost is due to the need for fertilizer, Rp.627.938, seeds, Rp.234.392, and chemical Rp.52.479. The seeds produced locally (local variety) while fertilizer usually use organic fertilizer (ash from grass, weeds, corn stubbles and other forage) or manure.

Production costs as the summation of fixed costs and variable costs is shown in Table 3 below.

Table 3. Average Production Cost for Corn Farming in Babar Island, District of Babar Islands, Southwest Maluku Regency 2018.

No	Cost Components	Values (Rp)
1.	Fix Cost	804.750
2.	Variable Cost	1.716.184
	Production Cost	2.520.934

Source: Primary Data, 2018 (Proceed)

Table 3 shows that the average total production cost for corn farming is Rp.2.520.934, consisted variable costs (Rp.1.716.184) and fixed costs (Rp.804.750). Variable costs contributed most for the production process due to the needs of production input used in the production process and labor used.

3.2. Farm Revenue

Revenue is the multiplication between the production and the selling price. Farmer revenue depends on the size of production and selling price of the commodity on the market [5] Average farm revenue in Babar Island shown in Table 4 below.

Table 4. Average production, Selling price and Farm Revenue for Corn Farming in Babar Island, District of Babar Islands, Southwest Maluku Regency 2018.

No	Description	Values (Rp)
1.	Production	1.124
2.	Selling Price	4.600
	Revenue	5.168.330

Source: Primary Data, 2018 (Proceed)

Table 4 shows that the average revenue of corn farming is Rp.5.168.330 with average production (dry shelled) is 1.124 kg and selling price Rp.4.600. Prevailing selling price at farmers level ranged between Rp.4000-5000/kg.

3.3. Farm Income

The success of a farm can be seen from the level of income received by farmers from their farm operation. Or in other words, the profit depends on total revenue and total production costs incurred by the farmers [6]. Table 5 shows the average income of corn farming.

Tabel 5. The average of revenue, Production cost and Income for Corn Farming in Babar Island, District of Babar Islands, Southwest Maluku Regency 2018.

No	Decription	Values (Rp)
1.	Revenue	5.168.30
2.	Production Cost	2.520.934
	Income	2.647.396

Source: Primary Data, 2018 (Proceed)

Table 5 shows that corn farming income is Rp.2.647.396, that obtained from the revenue of Rp.5.168.330 less the production cost of Rp.2.520.934. These results indicate that farmers earn profit from corn farming, because the revenue is greater than the costs incurred during the production process.

3.4. Feasibility Analysis

Benefit Cost Ratio (BCR) is a comparison between the income earned by farmers and costs incurred during the production process, and these values are used to analyze the feasibility of corn farming [7] The average value of the B/C ratio of corn farming shown in table 6 below.

Table 6. The average value of the B/C ratio of corn farming in Babar Island, District of Babar Islands, Southwest Maluku Regency 2018.

No	Description	Value (Rp)
1.	Income	2.647.396
2.	Production Cost	2.520.934
	B/C Ratio	1,05

Source: Primary Data, 2018 (Proceed)

Table 6 shows that the Benefit Cost Ratio (BCR) of corn farming is higher than one (BCR> 1). This indicated that corn farming on Babar island is profitable and feasible. The value of B/C ratio is 1.05, indicates that the incured cost of Rp.2.520.934 will generate profit 1.05 times the cost.

4. Conclusion and Recommendation

4.1. Conclusion

1. Production cost of corn farming in Babar island, District of Babar Islands, Southwest Maluku regency consists of fixed cost and variable costs. Fixed costs include: equipment depreciation expense and land rental costs and variable costs include inputs cost (seeds, fertilizers, pesticides) and labor costs. Variable costs contributed most to the production cost of corn farming.
2. The analysis results of cost, revenue and income showed that corn farming is profitable with income of Rp.2.647.396 as the difference between farm revenue (Rp.5.168.330) with production costs (Rp.2.520.934)
3. Corn farming in Babar island is feasible because the value of BCR (benefit cost ratio) is greater than one (BCR> 1). The value of BC ratio is 1.05 illustrates that corn farming will make a profit of 1.05 times the production cost of Rp.2.520.934.

4.2. Recommendation

Farmers need to make production planning such as: efficient use of inputs, anticipated planting time due to growing season and rainfall as well as maintain environmental balance to make the farming sustainable in Babar island.

Refferences

[1] Prajinanta 1999 Agriobusiness of Hybrid Red pepper. Penebar Swadaya. Jakarta
[2] Suratiyah K 2006 Farm Science. Published by Swadaya, Jakarta
[3] Soeharjo 1975 The Principles of Farming. Department of Socio-economics, Agriculture Faculty, IPB. Bogor
[4] Soekartawi 1995 Farm Analysis. Published by Universitas Indonesia. Jakarta
[5] Suherman O, Burhanuddin, Faesal, M. Dahlan, and F. Kasim 2002 Development of National free pollen and Hybrid corn HYV. Proceeding of Corn and Other Cereal
[6] Sarasutha I G P 2002 Farm Performance and Corn Marketing in the production center. Journal of Agriculture Research and Development **21**(2) 39–47. Research
[7] Nugroho R A 2008 Soybean (*Glycine max* (L) merril) Farming Analysis for *Foundation Seed* (FS) level to produce certified seeds. Minithesis. Major in Agribusiness. UGM. Yogyakarta. Not published.

Moler disease and cultivation practiced by shallot farmers in Brebes Central Java

Supriyadi[1,2,3,*], Supyani[1,2,3], SH. Poromarto[1,2,3], Hadiwiyono[1,2,3]

[1] Agrotechnology Department, Agricultural Faculty of Universitas Sebelas Maret (UNS), Surakarta,
[2] Magister Program of Agronomy, Agricultural Faculty of Universitas Sebelas Maret (UNS), Surakarta,
[3] Doctoral Program of Agricultural Sciences, Agricultural Faculty of Universitas Sebelas Maret (UNS), Surakarta,

* E-mail: supriyadi58@staff.uns.ac.id; hadiwiyono@staff.uns.ac.id

Abstract. Shallot is one of the leading commodities for farmers in Brebes, Central Java. Brebes is the largest supplier of Indonesia's national shallot production. This is what makes Brebes famous as a center for shallot production. However, recently farmers have been faced with the attack of pathogens that cause Moler disease, which recently increased their attacks so that certain conditions cause significant yield losses of up to 40%. The relationship between increasing the intensity of moler and cultivation practised by shallot farmers in Brebes is discussed in this text. The research was conducted through a survey by taking seven districts with the Brebes district's highest production area. Five farmers were selected for each speed. The data was excavated from farmers with a questionnaire about the shallot cultivation practice carried out by farmers. The results showed that the cultivation practices carried out by farmers had the potential to increase the intensity of moler disease. The practice that is meant is without crop rotation or varieties, without organic fertilizers, intensive application of fungicide, and balanced fertilization.

1. Introduction

Shallots are a mainstay commodity for the Brebes district of Central Java. Central Java province contributes around 30% of Indonesia's national production, and Brebes is the largest supplier [1]. Economically, shallot farming is very attractive and promising, with a B / C ratio of 1.41 [2]. The high-yielding varieties of shallots can be harvested two months after planting. Therefore, many Brebes farmers are very enthusiastic about depending on their livelihoods on shallot production. The farmer's hope cannot be avoided from the obstacles they have to face. Recently, some farmers in Brebes have complained about a decline in land productivity and an attack by the pathogen *Fusarium oxysporum* f.sp. *cepae,* which in the initial attack causes leaf twisted symptoms, so farmers call it moler disease. In the last three years, the attack of molecular pathogens has increased significantly, especially in the rainy season, and can cause up to 40% [3]. Some farmers think that this disease is the cause of the decline in land productivity. Many factors can induce the increasing incidence of Moler disease in Brebes. One of the factors that need attention is the practice of shallot cultivation carried out by local farmers. This paper discusses the increased incidence of moler disease in Brebes and its relationship to farmers' cultivation.

2. Methods

The research was carried out in Brebes Regency, Central Java, Indonesia, from August to September 2020. The data was collected using a survey method of shallot farmers and their land. Brebes Regency has 17 sub-districts that have shallot land. Seven of the 17 districts, seven districts were selected as samples. The seven samples are the ones that have the largest area of shallot land. The seven sub-districts are Larangan, Tanjung, Bulakamba, Wanasari, Songgom, Jatibarang, and Brebes. Five farmers were selected in each district as a sample of respondents. Using a questionnaire, the respondent farmers were asked about everything related to shallot cultivation, including varieties planted, crop rotation, variety rotation, organic fertilizers, N, P, K fertilizers, pesticides, and the origin of seeds. The data obtained were presented in tabular form and analyzed descriptively.

3. Result and discussion

3.1. Moler conducive soil

The attack of moler pathogens in Brebes' shallot cultivation has shown an increasing trend in the last three years (Table 1). In general, molecular pathogens' attack varies greatly from the lowest 0.2% to the highest 60%. This shows that not all land shallot lands are conducive to moler pathogens, while some still show moler disease suppression. This suppressive soil area can be used as an approach model in the development of land conservation that has been degraded by disease pathogens [4, 5, 6, 7, 8, 9]. The supresivance of soil to plant disease can be explained in depth [6, 9, 10, 11].

The essential character factors in conducive or suppressive soils are biological factors [8, 9, 10, 11, 12]. Many studies have shown that these physical factors' involvement in suppressive soils is that these soils' suppressive properties can be transferred to conducive soils to convert them to suppressive soils [13, 14]. Suppressive soils have a community structure with higher diversity and population and are complicated than conducive soils [4, 8, 10, 11, 12]. Mazzola M.J [15] explained that rhizosphere bacteria could be manipulated to develop suppressive soils against certain diseases. The results of previous studies on garlic, basal rot suppressive soil indicated that the bacterial community was more diverse than that of conducive soil [16, 17]. In other suppressive soils, fungi' role can be more significant than bacteria, as reported by Xiong et al. [18], which occurred in Fusarium vanilla wilt. The soil's low organic matter can cause the low population of conducive soil microbes in Brebes.

Table 1. Disease intensity of Moler caused by *Fusarium oxysporum* f.sp. *cepae* on shallots in 2017 to 2020 at Brebes, Central Java

Year	Season	Highest	Lowest	Average	Standard Deviation
2017/2018	Dry	2.00	0.20	1.49	0.58
	Rainy	25.00	15.00	20.00	2.43
2018/2019	Dry	4.00	0.50	1.84	0.57
	Rainy	30.00	20.00	26.63	3.18
2019/2010	Dry	4.00	0.20	2.21	0.82
	Rainy	60.00	25.00	32.14	5.76

3.2. Impact of shallot cultivation without organic matter

Soil organic matter is one of the health indicators where healthy soils generally have a high organic matter content [18]. The survey results showed that almost none of the respondent farmers planted garlic by providing organic fertilizers (Table 2). Of the 35 respondents, only one farmer said that he had used organic fertilizers. The soil analysis of the diseased plant samples showed that the soil organic matter content from the shallots' conducive soil was 1.04% (complete data not presented). This soil organic matter content is included in the very low criteria [20].

The low level of organic matter in the conducive soil of shallots in Brebes can be considered one of the factors causing this conducive character. Organic amendments such as organic fertilizers are essential in developing disease-suppressive soils [21, 22]. And soil health [23, 24]. Soil containing high organic matter provides a suitable environment for functional microbes in the soil, including antagonistic and other microbes involved in plant health [9, 10, 11, 14, 22, 23, 25]. Low organic matter content and soil microbial community structure will affect soil nutrients' quality [22, 26].

Table 2. Shallot cultivation practised by farmers and disease intensity of moler at Brebes Central Java

Culture Practices	Number of respondent(%)	Disease Intensity of Moler (%)	Explanation
CV. Bima	35(100)	32.14	No other variety are planted
Own seed	35(100)	32.14	Own farmer cluster is included
Plant rotation	0(0)	-	
No Plant Rotation	35(100)	32.14	
Variety rotation	0(0)	-	
No Variety Rotation	35(100)	32.14	
No Organic fertilizer	35(100)	32.14	
Nitrogen Fertilizer	14(40,00)	31.43	
Phosphate Fertilizer	11(31,42)	29.55	
Potassium Fertilizer	1(2,86)	50.00	
NPK compound fertilizer	0(0)	-	
Synthetic chemical fungicides	35(100)	32.14	3 or 5 or 7 days starting at the age of 20-30 days after the plant until just before harvest
Bio pesticide	0 (0)	-	

3.3. Impact of shallot cultivation without balanced fertilization

Plant nutrients are an essential factor in managing plant diseases [22, 27, 28]. The results of the survey provide information that the Brebes farmers are not wise in managing plant health. Generally, this shows that most farmers (40%) only use nitrogen fertilizer, and only one farmer out of 35 respondents uses potassium fertilizer (Table 2). The results of the analysis showed that in conducive land, the macronutrient content of N, P, K was low, namely 0.25%, (moderate) 13.07 ppm (very low), and 0.33 me% (low) (complete data no presented) [20]. Therefore, the low soil fertility status of the shallot cultivation area in Brebes should be assumed to be a contributing factor to the increased attack of molecular pathogens. In addition to not having farmers use organic fertilizers, unbalanced inorganic fertilization can be a factor in the shallots' low soil fertility in Brebes.

This phenomenon is consistent with the field's facts conducive to basal rot of garlic in Tawangmangu, Central Java, which indicates that the intensity of garlic basal rot tends to be high on land with low soil macronutrient content of Phosphorus and Potassium [29]. *F.o.* f.sp. *cepae* on garlic performs as a weak parasite, whose virulence is increased in plants that grow less fertile or lack nutrients [29, 30]. F.o shows the same. f.sp. *cube* in banana [31].

3.4. Impact of intensive fungicide application

The survey results showed that the application of fungicides by shallot farmers in Brebes was very intensive. The application was made periodically for 3, 5, or 7 days (Table 2). Intensive application of fungicides can negatively impact non-target microorganisms, including functional ones [32]. The most widely used fungicide is propined. Too intensive application of functions can promote pathogen resistance [33, 34]. Therefore, it is essential to conduct an empirical study of the impact of propine on non-target microorganisms and moler pathogens' resistance.

3.5. Acknowledgements

This paper is one of the outputs of research activities supported by PNBP-UNS funding sources from the Mandatory Research Scheme for the 2020 fiscal year.

References

[1] [BPS] 2019. Distribusi perdagangan bawang merah Indonesia tahun 2019. (Jakarta: Badan Pusat Statistik RI).

[2] Agrowindo. 2020. Peluang usaha budidaya bawang merah dan analisa usahanya. Available from:http://www.agrowindo.com/peluang-usaha-budidaya-bawang-merah-dan-analisa-usahanya.htm

[3] Supyani, SH. Poromarto S H, Supriyadi, Hadiwiyono. 2020. Moler disease of shallot in brebes central java in the last three years: the intensity and resulting yields losses is increasing. Presented online in the 2nd Sriwidjaja International Conference of Environmental Issues Conduted on Oct.21st. 2020.

[4] Hadiwiyono. 2013. Tanah Supresif dalam Praktik Pengelolaan Penyakit Tumbuhan. *Sains Tanah-J. Soil Sci. Agroclim.* **7** 31-40

[5] Hadiwiyono. 2008. Tanah supresif: Terminologi, sejarah, karakteristik, dan mekanisme. *J. Perlind. Tan. Indones* **14** 47-54.

[6] Chandrashekara C, BhatRavinder J C, and Chandrashekara Kn K. 2012. Supressive soils in plant disease management. In book: *Eco-friendly Innovative Approaches in Plant Disease Management*. Chapter: 14 (Dehradun: International Book Distributors). pp 242-255

[7] Cook R J. 2014. Plant health management: Pathogen suppressive soils. *Encyclopedia of Agriculture and Food Systems*. (Elsevier) pp 441-455. Avalaible from: https://doi.org/10.1016/B978-0-444-52512-3.00182-0.

[8] Song C, Zhu F, Carrión VJ, Cordovez V. 2020. Beyond plant microbiome composition: Exploiting microbial functions and plant traits via integrated approaches. *Front Bioeng Biotechnol.* **8** 896.

[9] Mazzola M. 2002. Mechanisms of natural soil suppressiveness to soilborne diseases. *Antonie Van Leeuwenhoek.* **81** 557-64.

[10] Weller DM, Raaijmakers JM, Gardener BB, Thomashow LS. 2002. Microbial populations responsible for specific soil suppressiveness to plant pathogens. *Annu Rev Phytopathol.* **40** 309-48.

[11] Schlatter D, Kinkel L, Thomashow L, Weller D, Paulitz T. 2017. New insights from the Soil Microbiome. *Phytopathology*. **107** 1284-1297.

[12] Mazzola M, Freilich S. 2017. Prospects for Biological Soilborne Disease Control: Application of Indigenous Versus Synthetic Microbiomes. *Phytopathology.***107** 256-263.

[13] Raaijmakers J M, and Mazzola M 2016. Soil immune responses. *Science* **352** 1392–1393.

[14] Gómez Expósito R, De Bruijn I, Postma J, and Raaijmakers J M 2017. Current insights into the role of rhizosphere bacteria in disease suppressive soils. *Front. Microbiol.* **8** 2529.

[15] Mazzola M.J 2007. Manipulation of rhizosphere bacterial communities to induce suppressive soils. *Nematol.* **39** 213-220.

[16] Hadiwiyono, R.D. Wuspada, S. Widono, S.H. Poromarto, Z.D. Fatawi. 2009. Kesupresifan tanah terhadap busuk pangkal (*Fusarium oxysporum* f. sp. *cepae*) bawang putih di Tawangmangu, Karanganyar. *Sains Tanah* **6** 1-6.

[17] Zainal D. Fatawi, Hadiwiyono, Salim Widono 2012. Analysis of rhizosphere bacterial community in suppressive and conducive soils to basal rot of garlic based on PCR-RISA. *Biomirror* **3** 9-12.

[18] Xiong W, Li R, Ren Y, Liu C, Zhao Q, Wu H, Jousset A, Shena Q. 2017. Distinct roles for soil fungal and bacterial communities associated with the suppression of vanilla Fusarium wilt disease. *Soil Biol. Biochem.* **107** 198-207.

[19] Fonte S J, Yeboah E, Ofori P, Quansah G W, Vanlauwe B, Six J. 2009. Fertilizer and residue quality effects on organic matter stabilization in soil aggregates. *Soil Sci. Soc. Am. J.* **73** 961–966.

[20] Balai Penelitian Tanah, 2009. *Analisis Kimia Tanah, Tanaman, Air dan Pupuk.* Balai Penelitian Tanah. Balai Penelitian dan Pengembangan Pertanian RI (Bogor: Balai Penelitian Tanah).

[21] Baileya KL and G Lazarovits G. 2003. Suppressing soil-borne diseases with residue management and organic amendments. *Soil Tillage Res.***72** 169-180.

[22] Wingate. E. 2020. Building Disease-Suppressive Soils for Organic Agriculture. Organic Farmer. Available form: http://organicfarmermag.com/2020/11/building-disease-suppressive-soils-for-organic-agriculture/

[23] Titiek Yulianti T. 2010. Bahan Organik: Perannya dalam Pengelolaan Kesehatan Tanah dan Pengendalian Patogen Tular Tanah Menuju Pertanian Tembakau Organik. *Bul. Tan. Tembakakau, Serat, Minyak Industri.* **2** 26-32

[24] Bonilla N, Gutiérrez-Barranquero J A, Antonio de Vicente A. and Cazorla, F M.2012. Enhancing soil quality and plant health through suppressive organic amendments. *Diversity* **4** 475-491.

[25] Bonilla N, Cazorla F M, Martínez-Alonso M, Hermoso J M, González-Fernández J, Gaju N, Landa B B, de Vicente A. 2012. Organic amendments and land management affect bacterial community composition, diversity and biomass in avocado crop soils. *Plant Soil.* **357** 215–226.

[26] Lin W, Lin M, Zhou H, Wu H, Li Z, Lin W. 2019. The effects of chemical and organic fertilizer usage on rhizosphere soil in tea orchards. *PLoS ONE* **14** e0217018.

[27] Dordas C. 2008, Role of nutrients in controlling plant diseases in sustainable agriculture. *Agron. Sustain. Develop.* **28** 33-46.

[28] Gupta N, Debnath S, Sharma S, Sharma P, Purohit J. 2017. Role of Nutrients in Controlling the Plant Diseases in Sustainable Agriculture. ed Meena V, Mishra P, Bisht J, Pattanayak A. *Agriculturally Important Microbes for Sustainable Agriculture*. (Singapore: Springer).

[29] Hadiwiyono, Widono S. 2008. Hubungan faktor lingkungan tanah terhadap intensitas busuk pangkal bawang putih di Tawangmangu. *Agrin* **12** 15-22.

[30] Cahyani VR, Hadiwiyono, Rahayu D P, Siswanto A, Prasetyo D. 2014. Succession of fungal community structure in degradative land caused by basal rot pathogen of garlic. *Sains Tanah.* **11** 47-51.

[31] Widyantoro A, Hadiwiyono, Subagiya. 2019. Biological control of Fusarium wilt on banana plants using biofertilizers. *Biodiverstas*. **21** 2119-2123

[32] Yang C, Hamel C, Vujanovic V, Yantai GanY. 2011. Fungicide: Modes of action and possible impact on non-target microorganisms. *ISRN Ecol.* **2011** 130289

[33] Hobbelen PH, Paveley ND, van den Bosch F. 2014. The emergence of resistance to fungicides. *PLoS One.* **9** e91910.

[34] Lucas J A, Hawkins N J, Fraaije B A. 2015. The evolution of fungicide resistance. *Adv. Appl. Microbiol.* **90** 29-92.

IOP Conf. Series: Earth and Environmental Science **883** (2021) 012084 doi:10.1088/1755-1315/883/1/012084

Performance of native chicken fed on ration containing fermented sago waste

T N Ralahalu[1,*], CH C E Latupeirissa[1], S CH H Hehanussa[1] and S. Fredriksz[1]

[1]Animal Science and Technology Laboratory, Animal Nutrition; Animal Science Departement, Faculty of Agriculture, Pattimura University, Jl. Ir. M. Putuhena, Kampus Poka – Ambon 97233

* Email: ipopralahalu@gmail.com

Abstract. The use of fermented sago waste (9 days incubation time) in the ration of native chicken is studied. 100 chicks of 7 days old native chicken were used. The experiment is being conducted in a completely randomized design in which the animals were divided into four treatments of ration. Each treatment was replicated five times. The experimental rations are: R0 (0% fermented sago waste), R1 (5% fermented sago waste), R2 (10% fermented sago waste) and R3 (15% fermented sago waste). Parameters measured were feed intake, feed conversion, body weight gain and carcass percentage. It appeared from this experiment that feeds intake were similar between birds. In grower birds, the use of fermented sago waste in the rations had no significant effect on feed intake. In both, starter and grower birds body weight gain was higher for birds consuming control ration than for those consuming ration with fermented sago waste. The use of 5 up to 15 percent fermented sago waste in the rations significantly affects feed conversion and carcass percentage in both, starter and grower birds. Keywords: sago waste, fermentation, aspergillus niger, native chicken.

1. Introduction

Native chicken is raised for meat, eggs and display. They are raised in extensive traditional semi-intensive and intensive system. The development of native chicken in Indonesia is important. As one of animal source that contributes largely to providing animal protein to the community.

Compared to other chicken species, native chicken is more resistant to diseases, more adaptable to the environment, meat and eggs tastier [1]. The population of native chicken in Maluku province was about 2.871.000, in 2019 [2], this shows that the chicken has potentially been improved in the province.

Nutrition is one of the important factors that influence native chicken productivity. To achieve optimal production of native chicken, both quantity and quality of nutrition are necessary. On the other hand, feed cost is the highest cost of cost production, about 65–70% in chicken farming or poultry business. Therefore feed cost is a crucial factor determining whether or not a business can be profitable [3].

Commercial poultry/chicken feed factory is not available in Maluku province so, suppliers purchase poultry/chicken feed from outside the province. Generally, native chicken is raised in an extensive system in which farmers feed the chicken by offering rice, corn and chopped coconut. Such feedstuffs are not providing adequate nutrients, both in quality and quantity to the chicken, and this result in low production. In addition, those feedstuffs are also human food, so they become more expensive.

Therefore it is important to substitute or replace those kinds of feedstuffs with unconventional one which is cheap, contains enough nutrients and does not compete with human needs. Sago waste is a solid waste generated from sago processing that can be used in native chicken ration. [4], stated that the ratio between sago mash and the waste is 1:6. The production of sago meal from a mature sago tree is about 220 kg, so that the waste from the tree could reach about 1320 kg. This means that sago waste is available abundantly in the province and could be potentially used as local feedstuffs. The waste still contains high starch. However, its protein content is low and high in fibre, so to be used in the chicken ration, it needs to be treated. Nutrients content of sago waste is crude protein 2.30% [5] and s starch 52.98% [6]..

Microbiology fermentation treatment can increase protein content, digestibility, and form many kinds of amino acids, to produce enzyme and vitamin. [7] reported that the protein content of sago waste increased from 1.2% (dry unfermented sago waste) to 12.31% (steam fermented sago waste). Fermentation mixture of chicken excreta and cassava meal with *Aspergillus niger* resulted in increasing protein digestibility from 56.45% to 65.90% [8]. Giving fermented sago waste 7.5% in swine ration at grower phase result in increase weight gain compare to ration without fermented sago waste [9]. Fermented mixture of sago waste and chicken excreta could be used in broiler ration up to 12% not affecting feed intake, abdominal fat, water and protein content of meat [10]. To improve native chicken production in Maluku province, using local feedstuffs in formulating the ration is meaningful.

The objective of this study was to determine the effect of using different level of fermented sago waste in ration of native chicken at starter and grower period on feed intake, live weight gain, feed conversion and carcass percentage.

2. Methods

2.1. Materials

The experiment was conducted for five months at Agricultural Poultry Research Centre, Passo, Ambon, six weeks for preliminary and fermentation period, eight weeks for a starter period and six weeks for the grower. Sago waste of Ihur sago (*Metroxylon rumpii*) collected from Tawiri and Hutumuri villages Ambon Island, Maluku province. *Aspergillus niger* van Tieghan MC 482 LIPI (average colony was 51.500.000) and minerals (ammonium sulphate, urea, sodium hydrogen phosphate, magnesium sulphate, potassium chloride, fero sulphate and calcium sulphate) were used to make fermented sago waste.

Fermented sago waste made using dry sago waste, hot water with a ratio of 2:1. The minerals and 12 g of inoculum were added, mixed thoroughly and then incubated for 6, 9, 12, 15 days. After the incubation period, the fermented sago waste was dried at 60^{0}C for two days. Proximate analysis was done, and based on the nutrient content of fermented sago waste incubated for nine days was selected to be used in the ration for the native chicken on starter and grower period.

100 native chicken aged one week with average live weight about 14.6–21.6 g were used and fed on homemade pellet. The animals randomly placed in battery cages, 100 × 75 × 60 cm, equipped with through, drinking bottles, lamp, thermometer and weighing scale.

The animals were divided into four treatments according to the level of fermented sago waste used in the tested rations, namely R0 (ration with 0% fermented sago waste), R1 (ration with 5% fermented sago waste), R2 (ration with 10% fermented sago waste) and R3 (ration with 15% fermented sago waste). An adaptation period 0f six days allowed the animal to accustom to the tested feed, followed by seven weeks and five weeks measurements period for starter and grower phase, respectively. During the trials, the animals were fed twice daily at 07.00 a.m and 16.00 p.m, and they had free access to drinking water. Fed offered was recorded daily, feed refusal was weighed every morning before the next feeding. Rations for starter and grower chicken are shown in Table 1 and 2.

Table 1. Composition and nutrient content of starter ration

Feedstuffs (%)	R0	R1	R2	R3
Yellow corn	59.00	58.50	57.50	55.50
Rice bran	18.25	15.75	13.75	10.50
Coconut oil by-product	9.00	8.00	7.00	9.00
Fermented sago waste	0.00	5.00	10.00	15.00
Fish meal	13.00	12.00	11.00	9.50
Coconut oil	0.25	0.25	0.25	0.25
Premix D	0.25	0.25	0.25	0.25
Salt	0.25	0.25	0.25	0.25
Total	100.00	100.00	100.00	100.00
Nutrient content:				
Crude Protein (%)	18.08	18.05	18.03	18.07
ME ration (Kcal/kg)	2903.61	2903.66	2904.82	2908.12
Crude fibre (%)	3.47	3.41	3.40	3.37
Fat (%)	4.07	3.91	3.73	3.83

Table 2. Composition and nutrient content of grower ration

Feedstuffs (%)	R0	R1	R2	R3
Yellow corn	59.00	59.75	59.00	59.00
Rice bran	17.50	16.00	14.00	12.00
Coconut oil by-product	13.00	9.75	8.25	6.25
Fermented sago waste	0.00	5.00	10.00	15.00
Fish meal	9.25	8.75	8.00	7.00
Coconut oil	0.25	0.25	0.25	0.25
Premix D	0.25	0.25	0.25	0.25
Salt	0.25	0.25	0.25	0.25
Total	100.00	100.00	100.00	100.00
Nutrient content:				
Crude Protein (%)	16.07	16.09	16.17	16.02
ME ration (Kcal/kg)	2900.14	2901.39	2900.16	2903.36
Crude fibre (%)	3.56	3.50	3.47	3.42
Fat (%)	4.34	3.98	3.77	3.51

2.2. Methods

Variables measured for starter were voluntary feed intake, live weight gain and feed conversion, besides that carcass percentage was also measured for the grower. A complete randomized design was used, four treatments and five replicates of each treatment with five chicken. Voluntary feed intake was calculated by subtracting feed refusal from feed offered. Live weight gain was determined by weighing the animal at the end of the week, then subtracting final with initial body weight. Feed conversion was calculated by dividing voluntary intake and live weight gain. Slaughtered weight was determined by weighing the live animal after 10 hours fasting, carcass weight was determined by subtracting non-carcass weight from the slaughtered weight.

The effect of treatments on voluntary feed intake, live weight gain, feed conversion and carcass percentage were examined analysing of variance using the General Linear Model. Main effect differences were detected using LSD.

3. Results and discussion

3.1. Starter period

3.1.1. Voluntary feed intake

The result shows that average voluntary intake decrease as the level of fermented sago waste increase in the ration. Statistically, there is a significant effect on voluntary feed intake ($P<0.05$), as shown in Table 3. The energy content of the ration is one of the factors that contribute largely to voluntary feed intake. According to [11], chicken eats to meet their energy requirement, as feed energy is high, feed intake will decrease.

Tabel 3. Average feed intake, live weight gain and feed conversion of native chicken feed on different level of fermented sago waste in the ration

Treatments	Feed intake (gr)	Live weight gain (gr)	Feed conversion
R0	23.38^{a}	6.78^{a}	3.48a
R1	21.90^{ab}	6.47^{ab}	3.40^{a}
R2	20.82^{b}	5.69^{b}	3.67^{a}
R3	21.07^{b}	5.70^{b}	3.74^{a}

^{a,b} within categories of response, means with different superscript are significantly different ($P<0.05$)

It appears from Table 1 that Energy metabolism of R0 was the lowest, although the differences among the treatments are not much. This result in high voluntary feed intake for R0 compares to R2 and R3. The result of the present study is in line with [12], who states that animal will consume a large amount of feed if the energy content of ration is low.

3.1.2. Live Weight gain

Using fermented sago waste in the ration of starter native chicken has a significant effect ($P<0.05$) on live weight gain. Data from Table 3 indicates that as the level of fermented sago waste increase in the ration, live weight gain decrease. BNJ test result shows no significant difference between R0 and R1, but differed significantly with R2 and R3. [13] states that, poultry which starts to grow to require high protein because their body needs to form new cells. The difference in live weight gain between R0 and R2, R3 is in line with voluntary feed intake and protein intake. Protein intake is 4.22 g/h/d, 3.95 g/h/d, 3.75 g/h/d and 3.80 g/h/d for R0, R1, R2 and R3, respectively. In addition, the quality of feedstuffs protein is another factor that could affect the difference in live weight gain. It can be seen from Table 1, increasing the amount of fermented sago waste leads to decreasing in a fish meal as a source of protein. Therefore protein from the fish meal was substituted by fermented sago waste protein. If the protein qualities are the same, then it is expected that live weight gain would be the same. However, using fermented sago waste above 5 % in the ration gives less live weight gain than that of control ration (R0). This indicates that the quality of protein in fermented sago waste is not as good as in fish meal. [14] points out that fish meal contains complete amino acid. Fermented sago waste protein is a single-cell protein of *Aspergillus niger*. It has been known that single-cell protein contains low sulphur amino acid such as [15]. It is suggested that increasing fermented sago waste in the ration may increase methionine deficiency. However, there it needs to be proved by amino acid analysis and digestibility of fermented sago waste.

3.1.3. Feed Conversion

Statistical analysis shows that using fermented sago waste in the ration does not significantly affect feed conversion ($P>0.05$). However, using 5% fermented sago waste decreased feed conversion, numerically. This means that chicken utilises ration containing 5% fermented sago waste and convert it to be meat more efficiently compare to control ration (R0) those of 10 and 15 % fermented sago waste. Low feed convcersion ration shows that add some more feed result in increase live weight gain of broiler with big proportion [16]. This current study shows that voluntary feed intake and live weight gain differed significantly, in which control ration (R0) was consumed more and also results in high live weight gain, compare to ration with 5%, 10 % and 15%, but it gives similar feed conversion. This may be caused by high digestibility of fermented sago waste. [17], reported that the digestibility of cassava meal fermented with *Aspergillus niger* increase, enzyme formed during the fermentation process could digest certain substituents in dry matter of the cassava to be digested. [8] also reported that fermentation of mixed chicken excreta and cassava meal with *Aspergillus niger* increased protein digestibility from 56.45% to 65.90%. Moreover, [18] states that a nutrient balance influences feed conversion in feed and animal ability to digest feed nutrients. feedstuffs.

3.2. Grower period

3.2.1. Voluntary feed intake

Although statistical analysis has shown that using fermented sago waste in the grower ration has no significant (P>0.05) effect, in general, voluntary feed intake tends to decrease as the level of fermented sago waste in the ration increase. This shows that fermented sago waste could be used up to 15% in the grower ration. [19] stated that protein and energy had been related to voluntary feed intake, live weight gain and feed conversion. Protein and energy content of tested ration were relatively similar, therefore no significant effect was found in voluntary feed intake among the treatments. Both energy and protein levels of the diet should be taken into account when formulating diets aimed at achieving optimal feed intake in growing chicken [20]. [11] and [21], pointed out that, poultry eat to fulfil the requirement of energy, high level of feed energy will lead to low voluntary feed intake. A similar response in voluntary feed intake also shows that using fermented sago waste in the ration does not affect the ration's palatability.

Tabel 4. Average feed intake, live weight gain and feed conversion and percentage carcass of native chicken feed on a different level of fermented sago waste in the ration

Treatments	Feed intake (gr)	Live weight gain (gr)	Feed conversion	Carcass percentage (%)
R0	41.55^{a}	8.48^{a}	4.90^{a}	66.37^{a}
R1	38.78^{a}	8.03^{ab}	4.84^{a}	65.62^{a}
R2	38.44^{a}	7.72^{b}	4.99^{a}	63.26^{a}
R3	38.61^{a}	7.65^{b}	5.06^{a}	66,.06^{a}

$^{a, b}$ within categories of response, means with different superscript are significantly different (P<0.05)

3.2.2. Live Weight Gain

Data in Table 4 show that using fermented sago waste in the grower ration has a significant effect (P<0.05) on live weight gain. Giving 5% (R1), 10% (R2) and (15%) fermented sago waste in grower ration gives low live weight gain compared to control ration (R0). It is suggested that the difference in live weight among the treatments is because of amino acid unbalance in the tested rations. [22] state that amino acid balance which adequates to support animal growth is needed. As stated earlier, fish meal contains complete and balanced amino acid compared to those in the fermented sago waste, therefore, increasing the level of fermented sago waste in the ration up to 10 and 15% results in high unbalanced amino acid. This means that the quality of protein ration R0 and R1 is better than R2 and R3. [23] state that intake nutrient has related to chicken performance. Nutrients deficiency could not support the optimum performance of chicken. This is in line with [24] who point out that nutrient has related directly to growth rate and body composition during growth. Nutrients deficiency is also slower the peak of growth and fat deposit.

3.2.3. Feed Conversion

Statistical analysis shows that voluntary feed intake on grower native chicken was similar among the treatments, but numerically R0 tend to be higher, resulting in higher live weight gain significantly. However, feed conversion was high for ration, which contains 5% fermented sago waste, numerically. It is suggested that ration R1 (5% fermented sago waste) is more digestible so that the chicken uses it more efficiently.

3.2.4. Carcass Percentage
Statistical analysis shows that using fermented sago waste up to 15% in grower ration has no significant effect (P>0.05) on carcass percentage. This determines that fermented sago waste can be used up to 15% in the grower ration. Carcass percentage in the present experiment is ranging from 63.26–66.37 %. This value is lower than that reported by [25], who feed the chicken with pollard and duckweed in the ration, which was 66.49–69.33 %. [26] used fermented sago waste in the grower duck ration reported that carcass percentage is about 63.22–68.35%.

4. Conclusion
It can be concluded from the present study that using fermented sago waste in starter ration at 5% has no significant effect on voluntary feed intake, meanwhile using 10% and above decreased voluntary fed intake significantly compare to control ration. Control ration gives high live weight gain, while using fermented sago waste 15% results in lower live weight gain on both period, starter and grower and has no significant effect on feed conversion on starter and grower chicken also carcass percentage.

References
[1] Dudung A M 1990 *Memelihara Ayam Kampung Sistim Battery.* Kanisius, Yogyakarta.
[2] Anonimous 2019. *Provinsi Maluku dalam Angka 2019.* Badan Pusat Statistik (BPS) Provinsi Maluku. 16 Agustus 2019.
[3] Prawirokusumo S 2005 *Masalah dan Prospek Bisnis Peternakan di Indonesia. Makalah Seminar Nasional Perunggasan Indonesia Sekarang dan Masa Depan.* Fakultas Peternakan UGM, Yogyakarta.
[4] Rumalatu F J 1981 *Distribusi dan Potensi Pati Beberapa Sagu (Metroxylon , sp) di Daerah Seram Barat.* Karya Ilmiah. Fakultas Pertanian/Kehutanan yang Berafiliasi dengan Fateta IPB, Bogor.
[5] Nurkurnia E 1989 *Hasil fermentasi rumen kambing kacang betina dengan pemberian beberapa tingkat ampas sagu (Metroxylon sp) dalam ransum* [skripsi]. Bogor: Fakultas Peternakan. Institut Pertanian Bogor.
[6] Ralahalu T N 2014 *Potensi Ampas Sagu sebagai Pakan Babi.* Prosiding Seminar dan Lokakarya Nasional Ternak Babi. Denpasar Bali. 5 Agustus 2014. Hal.143-153.
[7] Ralahalu T N 2002 Kualitas Ampas Sagu yang Difermentasi dengan Aspergillus niger pada taraf Pemberian Urea dan Lama Fermentasi yang *Berbeda Jurnal pertanian kepulauan Fakultas Pertanian Universitas Pattimura, Ambon.*
[8] Djunaedi I, Surisdiarto O, Sjofjan, 1999 Biokonversi Kotoran Ayam dan Onggok oleh Jamur Aspergillus niger dan Manfaatnya pada Ayam Petelur *Jurnal Ilmiah Ilmu-Ilmu Peternakan.*
[9] Ralahalu T N 2005 *Penggunaan ampas sagu yang difermentasi dengan Aspergillus niger dalam ransum ternak babi periode pertumbuhan.* Seminar Nasional Inovasi Teknologi Pertanian Berwawasan Agribisnis Mendukung Pembangunan Pertanian Wilayah Kepulauan Ambon. Tanggal 22-23 November 20005. Hal. 487-490.
[10] Hehanussa S, Supadmo dan Wibowo A 2006 Pengaruh penggunaan Campuran Ampas Sagu dan Ekskreta Ayam Fermentasi dalam Ransum Terhadap Penampilan, Kualitas Karkas dan Daging Ayam Broiler *Jurnal Agrosains* **19**(2).
[11] Anggorodi 1985 *Kemajuan Mutaktir Dalam Ilmu Makanan Ternak Unggas.* Universitas Indonesia. Jakarta.
[12] Wahyu J 2004 *Ilmu Nutrisi Unggas.* Cetakan ke-5. Gadjah Mada University Press. Yogyakarta.
[13] Sterling K G, Costa E F, Henry M H, Petsi G M and Bakalli R I 2002 Respons of Broiler Chickens to Cottonseed-andSoybean Meal-Based Diets Fed Several Protein Levels *Poult.Sci.***81**271-226.
[14] Anggorodi 1995 *Ilmu Makanan Aneka Ternak Unggas.* Gramedia. Jakarta.
[15] Kosaka K 1990 Feed Grain Subtitutes and Nonconventional Feedstuff for Poultry and Livestock in Japan , Food and Fertilizer Technology Centre *Extention Bulletin* **308** 1-13.

[16] Suwarta 2014 *Feed Conversion Ratio (FCR)*. Usaha Ternak Ayam Broiler di Kabupaten Sleman. Publishing Universitas Widygama. Malang.
[17] Aryogi dan Umiyasih U 2001 *Kandungan dan Nilai Kecernaan In Vitro Bahan Kering, Bahan Organik dan Protein Kasar Cassapro dengan Lama Fermentasi yang Berbeda*. Prosiding Seminar Nasional Teknologi Peternakan dan Veteriner. Pusat Penelitian dan Pengembangan Peternakan. Badan Penelitian dan Pengembangan pertanian. Bogor.
[18] Kamal M 1994 *Nutrisi Ternak I*. Fakultas Peternakan Universitas Gadjah Mada, Yogyalarta.
[19] Nesheim M C, Austic R E and Card L E 1979 *Poultry Production 12th Ed*. Lea and Febiger, Philadepia.
[20] Mbajiorgu C A, Ngambi J and Danna N 2011 Voluntary Feed Intake and Nutrient Composition in Chickens *Asian Journal of Animal and Veterinary Advances* **6**(1) 20-28.
[21] Zuprizal 1998 *Nutrisi Unggas lanjut*. Hand Out. Jurusan Nutrisi dan Makanan Ternak. Fakultas Peternakan. Universitas Gadjah Mada, Yogyakarta
[22] Tillman A D, Hartadi H, Reksohadiprodjo S, Prawirokusumo S dan Lebdosoekojo S *1991 Ilmu Makanan Ternak Dasar*. Gadjah Mada University.
[23] Malheiros R D, Moraes M B, Collin A, Janssens P J, Decuypere E and Buyse J 2003 Dietary Macronutrient , Endocrine Functioning and Intermediary Metabolism in Broiler Chickens *Nutr. Res.* **23** 567-578.
[24] Mc. Donald P, Edwards R A and Greenhalgh J F D 1987 *Animal Nutrition 4th Ed*.English Language Book Society. Longman Groups. England.
[25] Arief D 2002 *Evaluasi Ransum yang Menggunakan Terhadap Persentase Berat Karkas, Bulu, Organ Dalam, Lemak Abdominal, Panjang Usus dan Sekum Ayam Kampung*. Skripsi Fakultas Peternakan IPB. Bogor.
[26] Antawidjaja T, Bintang I A K, Supriyati, Sinurat A P dan Kompiang I P 1997 Penggunaan Ampas Kirai (Metroxylon sagu) dan Hasil Fermentasinya sebagai Bahan Pakan Itik yang Sedang Tumbuh *Jurnal Ilmu Ternak dan Veteriner*. Balai Penelitian Ternak. Bogor.

Identification of endophytic and rhizosphere bacteria in rice (*Oryza sativa* L.) in the experimental field at Payakumbuh State Agriculture Polytechnic, West Sumatra, Indonesia

Y Sondang[1,*], R Siregar[1] and K Anty[1]

[1] Department of Food Crop Cultivation, Payakumbuh State Agriculture Polytechnic. Jl. Raya Negara Km 7 Tanjung Pati, Limapuluh Kota Regency, 26271, West Sumatera, Indonesia

* E-mail: silitongayun27@gmail.com

Abstract. The decrease of soil fertility and fewer soil microorganisms will lower crop production, particularly rice, thus threatening the national food security program. This study is (a) to isolate and identify the bacteria in the endophytic and rhizosphere of rice plants (b) to study the bacteria from the endophytic and rhizosphere of rice plants which potentially stimulate plant growth. The experiment was carried out at the Laboratory of Food Crop Cultivation at Payakumbuh State Agriculture Polytechnic, Limapuluh Kota Regency, West Sumatra for four months. The sampling method was carried out by random sampling at rice planting in the Payakumbuh State Agriculture Polytechnic Experimental Field. Endophytic bacteria were taken from the root tissue of rice plants, and rhizosphere bacteria were taken from a layer of soil around rice roots. Isolation of bacteria was carried out by using the pour plate and scratchplate methods. Four bacteria were identified using the 16S rRNA sequencing method. The identification results showed that in the rice root tissue found the bacteria *Chromobacterium rhizoryzae* and *Brevibacillus brevis*. In the rice rhizosphere, *Bacillus pseudomycoides* and *Bacillus thuringiensis* are found. Bacteria are dominated by the Bacillus genera which can stimulate plant growth.

1. Introduction

Lowland rice production in Limapuluh Kota Regency from 2012 to 2016 decrease continuously, from 23,771.28 ha, 21,440.00 ha, 23,392.00 ha, 23,388.00 ha, and 22,585.00 ha. This decrease is caused by the transfer of land functions to non-agricultural functions [1]. Besides the decrease in land area, there was also a drastic decrease in paddy fields' fertility. One cause is the continuous use of inorganic fertilizers with excessive doses, resulting in reduced soil organic matter that makes soil microorganisms reduced or dead. Microorganisms play an important role in controlling soil and plant productivity. The existence of eco-friendly microorganisms, especially bacteria, guides sustainable agricultural management [2]. Bacteria closely related to plants and the soil environment are endophytic bacteria and rhizosphere bacteria around plant roots. Experiments on identifying and characterizing bacteria in plant tissue and rice rhizosphere are needed to determine the types of bacteria beneficial for plants and survive in soil fertility conditions.

The nature of the bacteria influences endophytic bacteria found in plant tissue, the host plant's conditions, and the environment [3, 4]. Pathogenic bacteria are not included in endophytic bacteria, because the group of bacteria found in plant tissue does not cause damage to plants. Endophytic bacteria are plant-beneficial bacteria that grow in plants and increase growth in normal and suboptimal conditions [4]. Endophytic bacterial colonization is more commonly found in plant roots than in plant stems. [5]

found that endophytic actinomycetes are more diverse in colonized roots than stems of rice plants. Endophytic bacteria that live in plant tissue originate from the root rhizosphere, which colonizes the roots and penetrates other hosts' inner tissues [6]. Endophytic bacteria colonize the host plant tissue through gaps or wounds formed when roots emerge in the lateral or root elongation zone and root differentiation and then spread to other plant parts such as stems and leaves.

Bacteria on plant roots have a positive effect through direct and indirect effect mechanisms. The direct mechanism is to contribute nutrients to plants and indirectly increase plant resistance by producing growth-promoting hormones such as auxins and gibberellins [7]. Plant roots are dominated by growth-promoting bacteria, which can support plant health by increasing nutrient availability and absorption [8]. Bacteria have a direct effect on plants through their role as plant growth stimulants, with nitrogen binding mechanisms, P and K dissolution, siderophore production, production of phytohormones in the form of IAA [9].

The following mechanism is that plants through their roots will release exudate in the form of organic acids. Bacteria around the roots will metabolize some organic acids as a source of carbon and nitrogen. Bacteria will produce several organic acids and minerals and be taken back by plants for growth and development [10]. The rice rhizosphere bacteria can produce organic acids such as succinic acid, propionic acid, oxalic acid, malic acid, and mineralize insoluble P to make it available for rice plants [2]. [11] stated that plant growth-promoting rhizobacteria (PGPR) in the soil can support plant growth and development by secreting different metabolites in the rhizosphere and available to plants.

Exploration, isolation, and characterization of bacteria will provide information on the potential of beneficial bacteria from endophytes and the rhizosphere of lowland rice plants as growth and abundance promoters. This study aims to identify, characterize, and calculate the abundance of bacteria from the endophytic and rhizosphere of rice plants in the Payakumbuh State Agriculture Polytechnic Experimental Field.

2. Methods

The research was conducted at Payakumbuh State Agricultural Polytechnic, Limapuluh Kota Regency, West Sumatra for four months. The source of bacteria comes from lowland rice cultivation at the Payakumbuh State Agricultural Polytechnic Experiment Field. Bacteria were isolated from the endophytic and rhizosphere of lowland rice plants. Isolation and identification of bacteria were carried out in the Food Crops laboratory of Payakumbuh State Agricultural Polytechnic and continued with identification using molecular analysis based on 16S rRNA fragments at LIPI Bogor.

Materials needed include Nutrient Agar, Nutrient Broth, and Pikovskaya's media, litmus paper, wrapping, alcohol, distilled water, and chemicals for molecular identification of bacteria. The tools used are beaker glass, Erlenmeyer, petri dish, test tube, one use hose needle, Bunsen lamp, analytical scales, oven, and laminar airflow

2.1 Isolation of Endophytic Bacteria

Endophytic samples were taken from rice plants' roots aged two months with criteria for healthy plants, green color, and normal growth. Root samples were washed with flowing water until clean and dried on filter paper. Sterilize the root surface by immersing in 70% ethanol solution for 2 minutes, 0.52% sodium hypochlorite solution for 5 minutes, rinse three times with sterile water, and dry on sterile filter paper. After dry, weigh 1 g and gently grind the roots in a sterile mortar, then put in the Erlenmeyer and add up to 10 ml of sterile distilled water. Stir by turning the Erlenmeyer. Take 1 ml of extract using a micropipette, put it in a tube, and add 9 ml of sterile distilled water. The same goes for a 10-7 dilution. Then take a 1 ml solution and grow it in a petri dish containing NA media. See the development of bacteria on the third day. The growing bacteria are separated and purified. The morphological characteristics of pure isolates of endophytic bacteria were based on shape, edge, elevation, surface, color, halo zone, and gram stain.

2.2 Isolation of Rhizosphere Bacteria

The soil was taken from a layer of soil around the same roots of rice plants. Dried the soil, take 1 g, then put it into Erlenmeyer and add sterile water to 10 ml. Then dilute it until 10-7. Next, take 1 ml of solution and grown it on a Petri dish containing NA media. Look at the development of bacteria on the third day. Bacteria that grow are separated and purified. Pure isolates of endophytic bacteria are characterized by their morphology-based on the shape, edge, elevate surface, color, halo zone, and gram staining.

2.3 Identification of bacterial isolates

After that, the identification using molecular analysis based on 16S rRNA gene fragments. The extraction of bacterial genomic DNA using the GES method. Amplification of the 16S rDNA fragment was done using GoTaq (Promega) with a pair of general 27F primer (5-AGAGTTTGATCCTGGCTCAG-3 ') and 1492R (5-GGTTACCTTGTTACGACTT-3') [11]. Then, sequencing data is processed with the Bioedit program. Isolates were identified using the EzTaxon server based on the 16S rRNA gene sequence data.

2.4 Calculation of Bacterial Populations

The bacterial population on NB media was calculated using a serial dilution method from dilution 10^{-1}, 10^{-2}, 10^{-3} until a dilution of 10^{-6}. Each dilution was piped 0.1 ml and put into NB media, then incubated for 3 × 24 hours. To calculate the population used formula:

$$Bacterial\ density = Colony\ count\ x \frac{1}{Dilution\ factor}\ x\ (\frac{CFU}{ml})$$

3. Results and Discussion

The results of bacterial isolation from root endophytes and rhizosphere of rice plants in Payakumbuh State Agriculture Polytechnic Experiment Field totalled 16 isolates. Still, four bacterial isolates were selected, which were dominant and potentially promoted growth. The morphological characteristics of endophytic bacteria and rhizosphere of rice plants are shown in Table 1.

Table 1. The morphological characteristics of endophytic bacteria and rhizosphere of rice plants

Source of bacteria	Isolate Code	Colony shape	Colony edge	Elevate	Surface	Color	Halo zone	Gram
Endophyte	AP3A	Irregular	lobate	raised	smooth	violet	3 mm	-
	AP3B	circular	entire	raised	smooth	cream	4 mm	+
Rhizosphere	TP1D	rhizoid	lobate	flat	smooth	white	3 mm	+
	TP3A	circular	wave	raised	smooth	cream	5 mm	+

The four bacteria's morphological characteristics have various shapes, edges, elevations, colors, and halo zones. The colony's part, which is surrounded by clear areas (halo zone), indicates that the bacteria can dissolve P and suspected have properties that stimulate plant growth. These beneficial characteristics will open opportunities for bacteria to work together and can be used as biofertilizers. [12] state that close interactions between bacteria and plants, and between bacterias will form colonization as phytopathogens, so endophytic bacteria have great potential as biocontrol agents and biofertilizers.

Table 2. Sources of bacteria, bacterial species, strains, homology, and TPC of bacteria.

Source of Bacteria	Bacterial Species	Strain	Homology	TPC CFU/ml
Endophyte	*Chromobacterium rhizoryzae*	LAM1188	99%	5,96 × 10^7
	Brevibacillus brevis	NBRC 15304(T)	99,86%	6,78 × 10^8
Rhizosphere	*Bacillus pseudomycoides*	FJAT-40028	99%	4,6 ×10^6
	Bacillus thuringiensis	BT	99%	7,96 × 10^8

Bacteria originating from the endophytic and rhizosphere of rice plants were dominated by the genera Bacillus from the *Bacillus cereus* group with 95% genetic relationship (Figure 1). This group of bacteria can dissolve phosphates called Phosphate Solubilizing Bacteria (PSB). According to [2], PSB strains isolated from endophytes can dissolve P, produce organic acids, enzymes, IAA, siderophore and are competent for antagonistic effects against pathogens. P availability will increase in waterlogged soils such as lowland rice. Likewise, soils with acidity (pH) close to normal will increase the availability of P [13].

The abundance of bacteria in the endophytic and rhizosphere is quite high, and this is due to rice plants that grow in water which can accelerate the spread of bacteria. The factors influencing population density are plant age, type of bacteria, biotic, and non-biotic [14]. The types and abundance of microbes in rice plants' rhizosphere are mostly influenced by root secretions or exudates such as organic acids, amino acids, and sugars [15].

The composition of the nucleotide bases tested by molecular analysis based on the 16S rRNA gene fragment resulted in a similarity of bases (homology) between 99% - 99.86%. Phylogenetic tree based on the 16S rRNA gene for endophytic bacteria and rhizosphere processed using Mega 06 software (Figure 1)

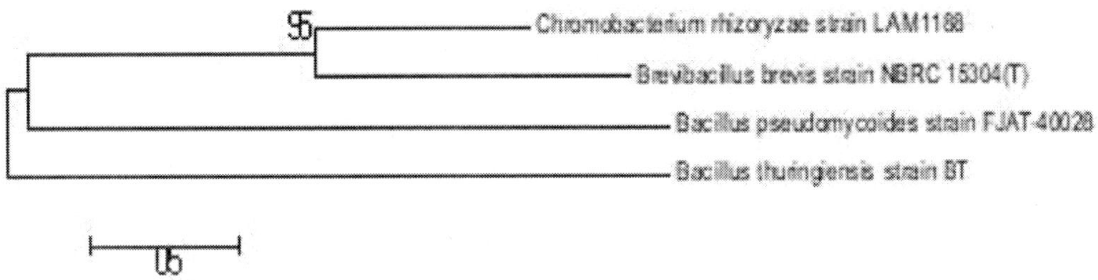

Figure 1. Phylogenetic joining between endophytic bacteria and rice rhizosphere based on the 16S rRNA gene sequence

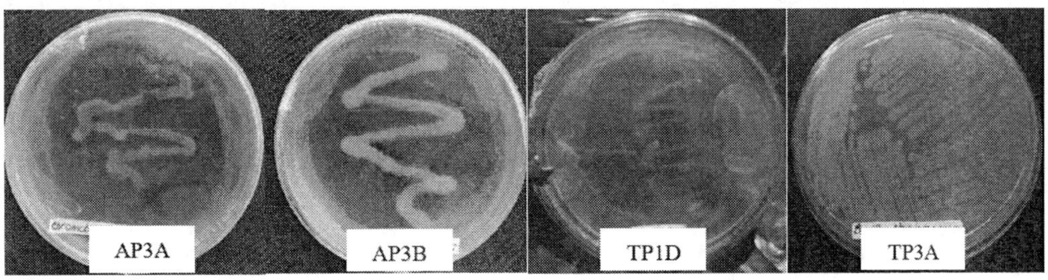

Figure 2. Bacterial isolates with codes AP3A, AP3B, TP1D, and TP3A

In this study, rice plant endophytic bacteria the *Chromobacterium rhizoryzae* strain of LAM 1188T is the dominant bacteria found in rice root endophytes. Seeing the name of the species, it is suspected that bacterium originated from the rhizosphere which succeeded in colonizing plant roots and living in the endosphere of rice plants. [16] Explain bacterial colonization's interaction on the root surface through the phases of bacterial movement, attachment, recognition, and penetration of bacteria in the root epidermis. Characteristics of the genera Chromobacterium is the production of violins and pigments, whose synthesis is regulated by quorum sensing and acts as a biocontrol against certain diseases [17]. Most of the genera Chromobacterium are found to live freely in the environment of water, soil, and the rhizosphere. Some of these bacterial species are found in plant roots and can suppress plant diseases [17]. Previous research by [18] that these bacteria have been found in rice plants' roots at The Experimental Demonstration Base at Huazhong Agricultural University, Wuhan City, China.

Some plant exudates that out through the roots can act as repellents against microorganisms, while others act as attractants to accommodate microbes. This exudate composition depends on the physiological status and species of plants and microorganisms [10]. [19] reported that four isolates from the endophytic origin and rhizosphere of rice plants, including Chromobacterium sp. MWU328 which potentially can produce bioactive compounds. Natural bioactive compounds are found in antimicrobial, anti-fungal, or antibiotic forms [20] stated that fifty percent of all endophytic bacteria isolated were antagonistic to rice pathogens.

Brevibacillus brevis NBRC 15304(T) in the endophyte of rice plants have the highest abundance because the bacteria *B. brevis* is multifunctional. According to [8] *B. brevis* is a PGP bacteria capable of producing growth stimulants on cotton plants. *B. brevis*, which is co-inoculated to the seeds, will reduce the disease caused by Fusarium oxysporum and Fusarium licopersici in tomato plants, increase plant and root height growth [21]. *B. brevis* bacterial isolate is very effective in reducing nickel toxicity, increasing shoots, and plant root biomass on nickel-isolated soil [22]. Plant the IAA phytohormone influences growth and development. *B. brevis* bacteria can produce IAA closely related to improving root growth and nodule production [23]. The antimicrobial activity of ethylparaben from *B. brevis* can inhibit harmful bacteria's growth and fungi [24], so that plant health is maintained and plant growth is getting better. Some of the benefits of endophytic bacteria to their host plants are increasing plant growth-promoting activities, modulation of plant metabolism, and phytohormones signalling [14] ; [25].

Rice plant rhizosphere bacteria the *Bacillus pseudomycoides* are found in the rhizosphere of lowland rice plants in the Payakumbuh State Agricultural Polytechnic Experimental Field. Based on the 16S rRNA gene sequence analysis and its relationship, *B. pseudomycoides* is one of the 12 species of *B. cereus* members [26]. These bacteria are rod-shaped, facultatively anaerobic, gram-positive stain. Several types of *B. pseudomycoides* from plant rhizosphere that have been reported can produce cytotoxic compounds. [27] found that B. pseudomycoides species with Pseudomycoicidin production as a new antibiotic with antimicrobial activity produced heterological in E. coli. These bacteria are heterotrophic bacteria that can inhibit the growth of pathogenic bacteria with the antimicrobials they produce.

Bacillus thuringiensis is well known in the biological insecticide industry (bioinsecticide). The main products of these bacteria are compounds with antimicrobial activity in the form of several bacteriocins. The by-product of B. thuringiensis can synthesize Cry, Vip, and Cyt proteins and active against insects and parasites [28]. The results of previous research [29] found *B. thuringiensis* in the rhizosphere of cotton plants. Bacterial isolates are inoculated into plants and can dissolve phosphate, IAA production, and siderophore so that these bacteria have the potential as biological fertilizers. Bacterial inoculation experiments on several plants show maximum plant growth [29]. *B. thuringiensis* bacteria belong to the *B. cereus* group [26]. *B. thuringiensis* is one of the bacteria used as an insecticide because this bacterial gene can form spores, has toxic parasporal crystal inclusions for insect larvae. As an entomopathogenic bacterium, *B. thuringiensis* also has the potential to produce growth stimulants [8]. Previous researchers reported that some Bacillus species in the rhizosphere could mineralize insoluble P to become available to plants.

The mechanism for bacteria's presence starts from the roots that release metabolic secretions in the form of exudates that become food for microbes around the plant's rhizosphere. It was reported by [30] that root exudates provide an abundant source of energy and nutrients to microbes. Root and microbial exudates are important components of the rhizosphere's ecology and so important in changing the bioavailability of nutrients. Many factors determine bacteria's ability to live in the rhizosphere (Rhizocompetence), including bacterial strains, bacterial population, and soil fertility [31].

The Bacillus genera dominate bacterial species found in the endophytic and rhizosphere of rice in the Payakumbuh State Agricultural Polytechnic experimental garden. Likewise, the endophytic and rhizosphere bacteria of maize plant found by [32] at the same location (Kandang Lamo, Limapuluh Kota Region) were more dominated by the genera Bacillus. Testing bacteria on seeds by [33] that bacteria from the genera Bacillus and Pseudomonas inoculated on corn seeds will affect maize seeds' germination metabolism. The mechanism is bacteria with the help of carbohydrates in the endosperm to multiply, dissolve phosphate in the endosperm and produce IAA as a growth-promoting substance. The availability of phosphate will stimulate embryo growth in forming root and leaf growth of corn.

Previous studies indicated that differences in soil chemical composition could affect the endophytic microbial community of rice plants. The production of siderophore by endophytic bacteria will improve plant growth by binding to available iron [8]. Plant growth promoter bacteria can increase plant growth and protect plants from disease through various mechanisms [9]. Exudate organic acid compounds will regulate the soil microbial community in the root zone [11]. Utilization of endophytic bacteria from plants to obtain phytohormones or growth stimulating substances can be inoculated into agricultural products. [34] stated that endophytes affect plant growth through N fixation, phytohormone production, nutrient acquisition, and tolerance for abiotic and biotic stresses. All bacteria have the potential to dissolve phosphate, produce IAA and siderophores, all of which can stimulate plant growth.

4. Conclusion

The identification results showed that in the rice root tissue found the bacteria *Chromobacterium rhizoryzae* 5.96 × 10^7 CFU/ml and *Brevibacillus brevis* 6.78 × 10^8 CFU/ml. Bacillus pseudomycoides 4.6 × 106 CFU/ml and Bacillus thuringiensis 7.96 × 108 CFU/ml are found in the rice rhizosphere. Bacteria are dominated by the Bacillus genera which can stimulate plant growth.

5. Acknowledgement

The author is grateful to DRPM Kemenristek Dikti for funding this research. The article entitled "Identification of Endophytic and Rhizosphere Bacteria in Rice (*Oryza sativa* L.) in The Experimental Field at Payakumbuh State Agriculture Polytechnic, West Sumatra, Indonesia " is part of the Superior Basic Research for Higher Education. Thanks to Payakumbuh State Agriculture Polytechnic, especially the Food Crop Cultivation Laboratory, to facilitate this research.

References

[1] Badan Pusat Statistika 2018 Statistika data lahan pertanian tahun 2012-2016 Pusat Data dan Sistem Informasi Pertanian 20 Juli 2018.

[2] Panhwar Q A, Othman R, Rahman Z A, Meon S and Ismail M R 2012 Isolation and characterization of phosphate-solubilizing bacteria from aerobic rice. *Afr. J. Biotechnol.* **11** 11 pp 2711-19 DOI: 10.5897/AJB10.2218 ISSN 1684–5315 © 2012 Academic Journals.

[3] Mano H and Morisaki H 2008 Endophytic bacteria in the rice plant. *Microbes environ*, **23** 2 pp 109-17. DOI: 10.1264/jsme2.23.109

[4] Afzal I, Shinwaria Z K, Sikandar S, and Shahzad S 2019. Plant beneficial endophytic bacteria: Mechanisms, diversity, host range, and genetic determinants. *Microbiological Research* **221** pp 36-49

[5] Tian X L, Cao L X, Tan H M, Zeng Q G, Jia Y Y, Han W Q and Zhou S N 2004 Study on the communities of endophytic fungi and endophytic actinomycetes from rice and their antipathogenic activities in vitro. *World Journal of Microbiology & Biotechnology* **20** pp 303–09.

[6] Reinhold-Hurek B and Hurek T 2011 Living inside plants: bacterial endophytes *Current Opinion in Plant Biology* **14** 4 pp 435-43 DOI: 10.1016/j.pbi.2011.04.004

[7] Ek-Ramos M J, Gomes-Flores R, Orozco-Flores A A, Rodriguez-Padilla C, Gonzalez-Ochoa G, & Tamez-Guerra P 2019 Bioactive Products from Plant-Endophytic Gram-Positive Bacteria *Frontiers in Microbiology* https://doi.org/10.3389/fmicb.2019.00463

[8] Nehra V, Saharan B S and Choudhary M 2016 Evaluation of Brevibacillus brevis as a potential plant growth-promoting rhizobacteria for cotton (*Gossypium hirsutum*) crop. *SpringerPlus* **5** p 948. DOI: 10.1186/s40064-016-2584-8.

[9] Souza R, Ambrosini A and Passaqlia L M P 2015 Plant growth-promoting bacteria as inoculants in agricultural soils *J. Genet Mol. Biol.* **38** 4 pp 401–419

[10] Kang B G, Kim W T, Yun H S and Chang S C 2010 Use of plant growth-promoting rhizobacteria to control stress responses of plant roots. *Plant Biotechnol.* Rep. **4** pp 179–183.

[11] Ahemad M and Kibret M 2013 Mechanisms and applications of plant growth-promoting rhizobacteria: a current perspective. *J King Saud Univ Sci* **26** pp 1–20.

[12] Liu H, Carvalhais L C, Crawford M, Singh E, Dennis P G, Pieterse C M J and Schenk P M 2017 Inner plant values: Diversity, colonization, and benefits from endophytic bacteria. *Journal Frontiers in Microbiology*. https://doi.org/10.3389/fmicb.2017.02552

[13] Morales L A, Vázquez E V, Paz-Ferreiro J 2014 Spatial distribution and temporal variability of ammonium-nitrogen, phosphorus, and potassium in a rice field in Corrientes, Argentina *Sci World J* 135906 pp 1–12.

[14] Miliute I, Buzaite O, Baniulis D and Stanys, V 2015 Bacterial endophytes in crops and their role in stress tolerance: a review. *Zemdirbyste Agriculture*, **102** 4 pp 465–478. DOI 10.13080/Z-A.2015.102.060. ISSN 1392-3196/e-ISSN 2335-8947.

[15] Adhya T K, Kumar N, Reddy G, Podile A R, Bee H, Samantaray B 2015 Microbial mobilization of soil phosphorus and sustainable P management in agricultural soils. *Curr Sci* 108 7 pp 1280–87.

[16] Tuson H H and Weibel D B 2013 Bacteria-surface interactions. *Soft Matter RSC Publishing* **9** pp 4368–80. DOI: 10.1039/c3sm27705d.

[17] Batista J H, and da Silva Neto J F 2017 *Chromobacterium violaceum* Pathogenicity: Updates and Insights from Genome Sequencing of Novel *Chromobacterium* Species. *Front. Microbiol.* https://doi.org/10.3389/fmicb.2017.02213.

[18] Zhou S, Guo X, Wang H, Kong D, Wang Y, Zhu J, Dong W, He M, Hu G, Zhao B, Zhao B and Ruan Z 2016 *Chromobacterium rhizoryzae* sp. nov., isolated from rice roots. *Int. J. of Systematic and Evolutionary Microbiology* **66** 3890–96. DOI 10.1099/ijsem.0.001284.

[19] Kurniawati S, Mutaqin K H and Giyanto 2015 Eksplorasi dan uji senyawa bioaktif bakteri agensia hayati untuk pengendalian penyakit kresek pada padi *J. HPT Tropika,* **15** 2 pp 170–179. ISSN 1411-7525

[20] Liarzi O, Bucki P, Braun Miyara S, Ezra D 2016 Bioactive volatiles from an endophytic *daldinia cf. concentric* isolate affect the viability of the plant-parasitic nematode *Meloidogyne javanica. PLoS ONE* **11** 12 e0168437. https://doi.org/10.1371/journal.pone.0168437

[21] Chandel S, Allan E J and Woodward S 2010 Biological control of *Fusarium oxysporum* and *F. sp. licopersici* on tomato by *Brevibacillus brevis*. *J. Phytopathology* **158** pp 7-8.

[22] Vivas B, Biro T, Nemeth, Barea J M and Azcon R 2006. Nickel tolerant *Brevibacillus brevis* and arbuscular mycorrhizal fungus can reduce metal acquisition and nickel toxicity effects in plant growing in nickel supplemented soil. *Soil Biology and Biochemistry*. ISSN 0038-0717 **38** 9 2694-2704.

[23] Vivas A, Barea J M and Azcón R 2005 Interactive effect of *Brevibacillus brevis* and *Glomus mosseae*, both isolated from Cd contaminated soil, on plant growth, physiological mycorrhizal fungal characteristics and soil enzymatic activities in Cd polluted soil. *Environmental Pollution* **134** 2, 257-266

[24] Jianmei C, Bo L, Zheng C, Huai S, Guohong L and Cibin Ge 2015 Identification of ethylparaben as the antimicrobial substance produced by *Brevibacillus brevis* FJAT-0809-GLX. *Microbiological Research* 172 pp 48–56.

[25] Tamosiune I, Baniulis D and Stanys V 2017 Role of endophytic bacteria in stress tolerance of agricultural plants: Diversity of Microorganisms and Molecular Mechanisms. In Book *Probiotics in Agroecosystem*, Chapter 1 pp 1-29. Springer Nature Singapore Pte Ltd. 2017 DOI: 10.1007/978-981-10-4059-7_1. ISBN 978-981-10-4059-7.

[26] Liu Y, Du J, Lai Q, Zeng R, Ye D, Xu J and Shao Z 2017 Proposal of nine novel species of the *Bacillus cereus* group. *International J. of Syst. Evolution Microbiol* **67** 2499–2508. DOI 10.1099/ijsem.0.001821

[27] Basi-Chipalu S, Dischinger J, Josten M, Szekat C, Zweynert A, Hans-Georg S, Bierbaum G 2015 Pseudomycoicidin, a Class II Lantibiotic from *Bacillus pseudomycoides*. *Applied and Environmental Microbiology* **81** 10 pp 3419-29.

[28] Salazar-Marroquın E L, Galan-Wong L J, Moreno-Medina V R, Reyes-Lopez M A and Pereyra-Alferez B 2016 Bacteriocins synthesized by *Bacillus thuringiensis*: generalities and potential applications *Reviews in Medical Microbiology* **27** pp 95–101

[29] Pindi P K, Sultana T, and Vootla P K 2014 Plant growth regulation of Bt-cotton through Bacillus species. *3 Biotech*, **4 pp** 305–15. DOI 10.1007/s13205-013-0154-0

[30] Ma Y, Oliveira R S, Freitas H, and Zhang C. 2016. Biochemical and molecular mechanisms of plant-microbe-metal interactions: Relevance for phytoremediation *Front. Plant Sci.* https://doi.org/10.3389/fpls.2016.00918.

[31] Albareda M S, Dardanelli C, Sousa M, Meg´ıas F, Temprano and Rodr´ıguez-Navarro D N 2006 Factors affecting the attachment of rhizospheric bacteria to bean and soybean roots," *FEMS Microbiology Letters*, **259** 1, pp. 67–73.

[32] Sondang Y, Anty K and Siregar R 2019 Identification of endophytic and rhizosphere bacteria in maize (*Zea mays* L.) in Limapuluh Kota Region, West Sumatra, Indonesia. *IOP Conference Series: Earth and Environmental Science* **347** (2019) 012002. DOI: 10.1088/1755-1315/347/1/012002.

[33] Sondang Y, Anty K, Siregar R, and Hayatunufus 2018 Application of corn endofit bacteria (*Pseudomonas sp* and *Bacillus sp*) to the physiological quality of corn seed. *Pros. Seminar Nasional Peranan Teknologi Pembenihan Berbasis Sumberdaya Lokal dalam Mendukung Ketahanan Pangan di Era Industri 4.0* Politeknik Pertanian Negeri Payakumbuh 26 September 2018, pp 101-108 ISBN: 978-602-51262-2-2

[34] Kandel S L, Joubert P M and Goty S L 2017 Bacterial endophyte colonization and distribution within Plants. *J. Microorganisms*, **5** 4 p 77. doi:10.3390/microorganisms504007

The physicochemical characteristics of gandaria (*Bouea macrophylla*) leather with sugar concentration treatment

V N Lawalata[1,*]

[1] Department of Agricultural Product Technology, Faculty of Agriculture, Pattimura University, Maluku, Indonesia

* E-mail: vitalawalata@yahoo.com

Abstract. Gandaria leather is a kind of fruit leather processed by gandaria fruit in the form of thin sheets with a thickness of 2-3 mm and has a specific consistency and taste according to gandaria fruit. The purpose of this study was to determine the effect of added sugar on the physicochemical characteristics of gandaria leather. This study used a single-factor completely randomized design, with 0%, 5%, 10%, and 15% sugar concentrations. The results showed that the sugar concentration had a very significant effect on the parameters of water content, total sugar, total acid, vitamin C, and water activity (a_w), and had a significant effect on the tensile strength of gandaria leather. The physicochemical characteristics of gandaria leather from various sugar concentration treatments are as follows: water content (14.05% -20.13%), total sugar (80.56% - 84.90%), total acid (3.10 - 12.37 mg/100g), vitamin C (59.44 - 78.48 mg/ 100g), water activity (0.48 - 0.62), and tensile strength (0.20-0.37 N). The resulting gandaria leather qualifies as good fruit leather, and the treatment with a 10% sugar concentration is recommended for gandaria leather which has the potential to be developed and commercialized.

1. Introduction

Gandaria (*Bouea macrophylla*) is one of the typical fruit types in Maluku Province, which is usually consumed directly as table fruit or made for juice. When the fruit is fully ripe, it is indicated by all parts of the fruit turn orange. The nutritional content of fresh gandaria fruit is very good to support the fulfillment of nutrition for the human body. The composition of gandaria fruit per 100 g of the edible portion is 85 g of water, 12 mg of protein, 600 mg of fiber, 230 mg of ash, 6 mg of calcium, 10.8 mg of phosphorus, 0.31 mg of iron, 0.043 mg of carotene, 0.031 mg of thiamine, 0.025 mg riboflavin, 0.286 mg niacin, and 75 mg vitamins [1].

Diversification of gandaria fruit is needed to increase the economic value of the fruit, given its very high nutritional content. One of the diversified products from the gandaria fruit is gandaria leather. Gandaria leather is a type of fruit leather and a snack product made from dried fruit puree to form a thin sheet that can be rolled.

Gandaria leather is similar to nori which is a traditional Japanese food made from *Porphyra* sea algae (Bangiales, Rhodophyta). Nori is in the form of thin sheets (0.2 mm in size, composed of 10-20 layers), finely cut to uniform sizes - kizaminori or aonori), dried or seasoned or toasted (nori - ajitsuke nori or okazunori [2] . Research on gandaria fruit leather reported by [3] is only limited to its organoleptic properties, therefor further testing of the nutritional composition of the gandaria leather is needed so since the product has the potential to be developed. Based on this gap, this study aims to determine the effect of adding sugar on the physicochemical characteristics of gandaria leather.

2. Methods

2.1. Materials

The raw material used in this study was the optimal ripe gandaria fruit marked with all parts of the fruit orange. Additional ingredients used are refined sugar, kappa carrageenan, and water.

2.2. Methods

The study was designed using a single-factor completely randomized design with sugar concentration as a treatment, namely control (0%), 5%, 10%, and 15% of sugar. The research stages of making gandaria leather are as follows: gandaria fruit is sorted, washed and dried, then peeled and taken from the flesh. The pulp is then blended while adding 1% carrageenan and sugar according to the treatment (0%, 5%, 10%, and 15%) so that it becomes fruit pulp.

The pulp is then poured on a 24 x 24 cm tray which has been smeared with margarine with a thickness of ± 4 mm, after which it is dried in an oven using a kerosene stove at 60°C for 1 hour. The next drying phase uses sunlight (effective at 12.00 noon) for 3 hours. The dried fruit leather is then cut and packaged and ready for analysis.

Analysis of physicochemical characteristics includes water content with a thermogravimetric method [4], total acid and vitamin C by titration method [4], total sugar by phenol method [5], water activity (a_w) [4, tensile strength with a universal testing machine.

The research data will be statistically tested according to the design used, if the data is significant and highly significant then it can be proceeded with the Tukey test at confidence level 95%.

3. Results and discussions

Physicochemical characteristics reflect the nutritional value of a food product. Data on the nutritional value of a food product is very important to determine whether the food product conforms to standardized quality requirements or not before the food product is commercialized. The average test results for the physicochemical characteristics of gandaria leather from various sugar concentration treatments are presented in Table 1.

Tabel 1. Average test results for the physicochemical characteristics of Gandaria leather

Physicochemical characteristics	Sugar concentration treatment (%)			
	0 %	5%	10%	15%
Water content (%)	20.13_a	19.35_b	16.54_c	14.05_d
Total acid (mg/100g)	12.37_a	7.27_b	3.60_c	3.10_d
Vitamin C (mg/100g)	78.48_a	67.73_b	62.38_c	59.44_c
Total sugar (%)	80.56_c	81.42_c	82.79_b	84.90_a
a_w	0.62_a	0.61_b	0.53_c	0.48_d
Tensile strength (N)	0.37_a	0.22_b	0.20_b	0.20_b

Note: The numbers followed by different letters on the same line are different on the Tukey test with 95% confidence level.

3.1. Water content

The water content value of gandaria leather from various sugar concentration treatments ranged from 14.05 - 20.13% which is shown in table 1. The sugar concentration treatment had a very significant effect on the water content of gandaria leather produced. The water content of gandaria leather decreased with increasing sugar concentration treatment. The decrease in water content of the gandaria leather product is due to the nature of sugar as a hydrating agent (water-binding) so that sugar can easily absorb and hold water so that sugar can reduce the water content in the added food [6], [7] .

The water content of the Gandaria leather produced has met the requirements for good fruit leather because according to [8] good fruit leather has a water content of 10 - 20%. The effect of water content in food is very important in the formation of the durability and shelf life of these food products.

3.2. *Total acid*

The total amount of acid in the gandaria leather produced ranged from 3.10 - 12.37 mg/100g which can be seen in table 1. The sugar concentration treatment had a very significant effect on the total acid produced. The total acid of gandaria leather decreased along with the increase in sugar concentration treatment. The decrease in total acid is due to the addition of sugar so that it will reduce the taste or the total acid value of the product. This is due to the nature of sugar which gives a sweet taste so that with increasing sugar concentration treatment, the total acid value will also decrease.

Most of the gandaria fruits when ripe still have a relatively high sour taste due to the high content of organic acids, such as citric acid which is the main organic acid in these acids. The addition of sugar to the gandaria fruit pulp can reduce the taste and acid content, so that the produced gandaria leather has a low total acid content because it is covered by the addition of sugar.

3.3. *Vitamin C*

Vitamin C or ascorbic acid has a molecular formula of $C_6H_8O_6$ with a molecular weight of 178, in its pure form, it is a colorless and odorless crystal [9] The content of vitamin C gandaria leather produced ranged from 59.44 - 78.48 mg/100g which can be seen in table 1. The sugar concentration treatment had a very significant effect on vitamin C in gandaria leather. The value of vitamin C in gandaria leather decreases with increasing sugar concentration treatment. The decrease of vitamin C in gandaria leather is because of the addition of sugar. This is due to the nature of sugar which gives a sweet taste so that with increasing sugar concentration treatment, the vitamin C content will also decrease. Vitamin C is one of the elements forming organic acids in gandaria fruit, so it is equal to total acid, with the addition of sugar, there is a decrease in vitamin C in gandaria leather products.

The content of vitamin C in food products is very important for humans, because it can maintain immunity and as an antioxidant that can inhibit the oxidation process. According to [9] , the amount of vitamin C intake needed by adults is 10 mg/day, so that gandaria leather in this study can be an alternative healthy snack that can meet the needs of vitamin C.

3.4. *Total sugar*

The total sugar in the gandaria leather produced ranges from 80.56 - 84.90% which is shown in table 1. The treatment of sugar concentration has a very significant effect on the total sugar in gandaria leather produced. The total sugar in gandaria leather increases with increasing sugar concentration treatment. The increase in total sugar was due to the addition of sugar concentration treatment.

The increased total sugar in Gandaria leather is due to the addition of sugar which causes glucose and fructose resulting from the inversion of sucrose to also increase [10]. This is because sugar only consists of sucrose, while the reducing sugar found in the pulp contains simple sugars, namely glucose and fructose.

3.5. *Water activity (a_w)*

Water activity (a_w) in gandaria leather from various treatments, the concentration of sugar is presented in Table 1 ranged from 0.48 - 0.62 mg/100g. Water activity is free water that can be used by microbes for their growth. A_w results in gandaria leather obtained in this study are quite good because the value is less than 0.7. As [8] suggests that a_w for good fruit leather is below 0.7.

The results showed that the gandaria leather produced had met the safe limit from microbial attack. According to [9], various microorganisms have a minimum a_w to grow, such as bacteria: 0.9, yeast: 0.8 - 0.9, and mold: 0.6 - 0.7.

3.6. Tensile strength

The tensile strength value is the maximum tensile strength that can be achieved before the product breaks or tears. The tensile strength value shows the force required to achieve maximum tensile on each unit area of the product [11] . The tensile strength in this study is related to the plastic texture of the fruit leather so that it can be rolled and does not break easily. The results showed that the tensile strength value of the gandaria leather with various sugar concentrations ranged from 0.20 - 0.37 mg/100g (Table 1). These results illustrate that gandaria leather has good tensile strength because it can be rolled and does not break easily.

4. Conclusion

The physicochemical characteristics of gandaria leather from various sugar concentration treatments are as follows: water content (14.05% - 20.13%), total sugar (80.56% - 84.90%), total acid (3.10 - 12.37 mg/100g), vitamin C (59.44 - 78.48 mg/100g), water activity (0.48 - 0.62), and tensile strength (0.20 - 0.37 N). The resulting gandaria leather qualifies as a good fruit leather, and the treatment with a sugar concentration of 10% is recommended as Gandaria leather which has the potential to be developed and commercialized.

References

[1] Tanasale V L 2011 Kajian Agronomi Dan Pemanfaatan Buah Gandaria (*Bouea Macrophylla.Griff*). *Jurnal Ilmiah agribisnis dan Perikanan (agrikan UMMU-Ternate)* **4** (2) 69-74.
[2] Levine I A and Sahoo D 2010 Porphyra: harvesting gold from the Sea. [India] Banglore: IK International Pvt Ltd.
[3] Lawalata V N, Ega L, Sipahelut S G, Tetelepta G 2015 Mutu Organoleptik *Fruit Leather* Gandaria (*Bouea Macrophylla*) *Jurnal Agroforestri* **10** (2) 127-130.
[4] AOAC 1999 Officials Methods of Analysis Associaton Official Agricultural Chemistry, Washington D.C.
[5] Sudarmadji S, Bambang H, and Suhardi 1997 *Prosedur Analisa untuk Bahan Makanan dan Pertanian.* Liberty, Yogyakarta.
[6] Sulandari S, Zein M S A, Paryanti S, Sartike T, Astuti M, Widjastuti T, Sujana E, Darana S, Setiawan I, and Garnida D 2007 *Keanekaragaman Sumber Daya Hayati Ayam Lokal Indonesia: Sumbergenetik ayam lokal Indonesia.* Penerbit Swadaya, Jakarta.
[7] Engka 2016 Pengaruh Konsentrasi Sukrosa Dan Sirup Glukosa terhadap Sifat Kimia Dan Sensoris Permen Kerasbelimbing Wuluh (*Averrhoa Bilimbi* L.) *Jurnal Teknologi Pertanian*, Fakultas Pertanian Universitas Sam Ratulangi **7**(3) 49-58.
[8] Nurlaely E 2002 Pemanfaatan Buah Jambu Mete untuk Pembuatan Leather Kajian dari Proporsi Buah Pencampur. *Skripsi*. Teknologi Hasil Pertanian. Universitas Brawijaya, Malang.
[9] Winarno F G 2008 Kimia Pangan dan Gizi. PT Gramedia Pustaka Utama, Jakarta
[10] Buckle KA, Edwars R A,Flett G H, Wootton M 1989 Ilmu Pangan. Penerbit UI Press, Jakarta.
[11] Krochta J M and Mudler-Johnson C D 2002 Edible and Biodegradabel Polymer Films : Challenges and Opportunities *Food and Technology* **51** (2) 62-74.

IOP Conf. Series: Earth and Environmental Science **883** (2021) 012087 doi:10.1088/1755-1315/883/1/012087

The utilization of agro-industrial waste for soil amendment and liquid biofertilizer mixed bacterial antagonist in eggplant production

S Boonraeng[1],* and N Punyoyai[1]

[1] Department of Agro-industry, Faculty of Agricultural Technology
Chiangmai Rajabhat University, Thailand

* E-mail: boonraengs@gmail.com

Abstract. This research aimed to improve soil fertility with soil amendment, prepared from agro-industrial waste, and application of antagonistic bacteria for controlling bacterial wilt disease. The experiments were tested on sandy-to-sandy loam soil with low organic matter and nutrients for planting eggplants. Besides the soil improvement with spent mushroom waste and mango peel mixed kernel, the results showed that soil pH, soil organic matter, nitrogen, phosphorus, and potassium increased as available levels. Also, two species of bacterial antagonist named *Bacillus subtilis* and *B. amyloliquifaciens* were cultured and mixed in liquid biofertilizer. It was spiked for soil preparation before planting and during crop growth every week, which was affected by eggplant survival from bacterial wilt by 100% and 97% in plot and field-tested, respectively. The results from testing on sandy loam with the use of liquid biofertilizer three times/week showed that *Ralstonia solanacearum* in soil suppressed to the reduction of 1,000-10,000 times. The yield increased by 25.9%. The optimum harvesting time was 12-15 days of fruit growth with the antioxidant activity. Finally, this study has excellent potential to be extended for farmers who organically grown.

1. Introduction

Eggplant is one of the most important vegetables for a food processing factory, and it is usually grown in the northern part of Thailand. Although the demand has increased. But the yield has not been attained due to the eggplant had infected by bacterial wilt disease. It had caused by *Ralstonia solanacearum* that was low yield and productivity. Besides of *R. solanacearum* is a soil-borne disease and survive from day to year. That is very hard to manage by the use of chemicals on the soil cannot control. Thus, the use of bacterial antagonists is being emerged method of interest. Due to disease suppression, the antagonists have some advantages like not harmful to humans, animals, and the environment. Among the useful bacterial strains in *Bacillus subtilis* and *B. amyloliquefaciens* had been reported as effective species for the biological control eggplant pathogen.

B. amyloliquefaciens showed significantly control bacterial wilt of eggplant in glasshouse and field studies by the reduced rate of 92.1% and 73.7%, respectively [1]. However, biological control methods have been not too easy for farmers to prepared and cultured. And the production of eggplant was relevant to many factors to manage with building soil fertility, irrigation, and agricultural practices. For example, applying agricultural wastes for soil amendment directly impacts plant health and crop productivity. The degradation of organic matter in the soil can directly affect pathogens' vitality and survival by restricting available nutrients and releasing natural chemical substances with varying inhibitory properties. Also, enhanced soil microbial communities are indirectly involved in disease suppression.

Pre-treatment for growing eggplant is the best prerequisite for effective wilt prevention by the antagonistic bacteria. That was the importance of plant colonization ability of the antagonist bacteria for bacterial wilt control [2]. Therefore, this study had selected rhizosphere bacteria that have inhibition properties for *R. solanacearum*. And the antagonist bacteria were cultured with mango residues and others as biofertilizer. It had applied for eggplant production under organically grown by using spent mushroom waste as a soil amendment in sandy-to-sandy loam soil.

2. Methods

2.1. Isolation of Ralstonia solanacearum

Eggplants had collected from infectious bacterial wilt. For a quick field diagnostic, identification of *R. solanacearum* can be used by the section method. The stems were cut from a wilting plant with a blade, and the samples were placed against the inside wall of a water-filled clear conical flask. The end of the section had touched the water surface. Milky white strands containing bacteria and extracellular polysaccharide will stream from the cut ends of the xylem. Then, the prospective samples had kept for further testing as follows. Stem segments of length 10 cm from wilted plants were rinsed with sterilized distilled water containing 1% Clorox and filled in a tube. The tubes were stirred on Vortex mixture for bacterial exudation were obtained. Then a loop full of turbid suspension was streaked on 2,3,5-triphenyltetrazolium chloride media (TTC) and incubated at 35-37 °C for 2 days, reddish fluidal colonies were again streaked on TTC plates and the process was repeated till purified bacterial cultures were obtained with the homogeneity of colony morphology with pinkish center. The colony was picked up and maintained with 20% glycerol at -40 °C in the freezer.

2.2. Isolation of rhizosphere bacteria

The rhizosphere soil samples were collected from 5 sources; bamboo, forest tree, lettuce, pepper and kale. Bacillus species were isolated by using trypticase soy agar (TSA) medium. Soil samples (10 g) were suspended in 90 mL of 0.85% normal saline (pH 7.0) and shaken well at 150 rpm for one h. An appropriate dilution of this suspension was spread on TSA plate. The cultures were incubated at 37 °C for 48 h. Every single colony was assayed for morphological and Gramm's staining, endospore.

2.3. Invitro screening for antagonistic activity

Bacillus isolates were tested for ability to inhibit *R. solanacearum* by using the paper disc method, and each bacterial were cultured on Tryptic Soy Broth (TSB) at 35 °C for 24 h. for the cell concentration 10^{12} CFU/mL *R. solanacearum* was cultured with tetrazolium chloride broth (TTCB) at 35 °C for 24 h. and smeared on nutrient agar (NA). The paper with 5 mm. of diameter and sterilized. Bacillus's suspensions for 20 μL with 2×10^9 CFU were dropped on paper discs and placed on NA plates. On the other control plates consisted *R. solanacearum* only. The culture plates were incubated at 35 °C for 48 h. The zone of inhibition around paper discs was observed and recorded.

2.4. Identification of Bacillus

Two bacillus isolates showing antagonistic activity to *R. solanacearum* were identified by 16S rRNA gene sequence analysis at the National Center for Genetic Engineering and Biotechnology (BIOTEC).

2.5. PCR amplification of 16S rDNA

DNA templates for PCR amplification were prepared by using "Genomic DNA mini kit (Blood/culture cell)" (Geneaid Biotech Ltd., Taiwan). DNA coding for 16S rRNA regions was amplified by means of PCR with Taq polymerase [3–5]. A PCR product for sequencing 16S rDNA regions was prepared by using the following two primers, 20F (5'-GAG TTT GAT CCT GGC TCA G-3', positions 9-27 on 16S rDNA by the *E. coli* numbering system [6]) and 1500R (5'-GTT ACC TTG TTA CGA CTT-3', position 1509-1492 on 16S rDNA by the *E. coli* numbering system [6]). The PCR amplification was carried out with DNA Engine Dyad® Thermal Cycler (Bio-Rad Laboratories). One hundred μL of a reaction mixture contained 15-20 ng of template DNA, 2.0 μmoles each of the two primers, 2.5 units of Taq polymerase, 2.0 mM $MgCl_2$, 0.2 mM dNTP and 10 μL of 10xTaq buffer, pH 8.8, containing $(NH_4)_2SO_4$, which was comprised of 750 mM Tris-HCl, 200 mM $(NH_4)_2SO_4$ and 0.1% Tween 20. The PCR amplification was programmed to carry out an initial denaturation step at 94 °C for 3 min, 25 cycles of denaturation at 94 °C for 1 min, annealing at 50 °C for 1 min and elongation at 72 °C for 2 min, followed by a final amplification step at 72°C for 3 min. The PCR product was analyzed by 0.8% (w/v) agarose gel electrophoresis and purified with a QIAquick® PCR purification kit (QIAGEN GmbH, Hilden, Germany). The purified PCR product was stored at -20 °C for further step.

2.6. Direct sequencing of 16S rDNA

Direct sequencing of the single-banded and purified PCR products (ca. 1500 bases, on 16S rDNA by the *E. coli* numbering system; [6]) was carried out. Sequencing of the purified PCR products was carried out with an ABI PRISM® BigDye™ Terminator Ready Reaction Cycle Sequencing Kit (version 3.1, Applied Biosystems, Foster City, California, USA). The primers 27F (5'-AGA GTT TGA TCM TGG CTC AG-3') and 518F (5'-CCA GCA GCC GCG GTA ATA CG-3') for partial sequencing, and additional 1492R (5'-TAC GGY TAC CTT GTT ACG ACT T-3') and 800R (5'-TAC CAG GGT ATC TAA TCC-3') for full-length sequencing were used for sequencing of 16S rDNA. Ten μL of a sequencing reaction mixture contained 5–20 ng of template DNAs, 2.0 μL of BigDye™ terminator ready reaction mixture, 5–20 ng of DNA template, 1.6 pmole of the sequencing primer, 1.5 μl of 5xBigDyeTM sequencing buffer and deionized water. The PCR reactions were carried out as follows: an initial denaturation step at 96°C for 30 sec, 25 cycles of denaturation at 96 °C for 10 sec, annealing at 50 °C for 5 sec and elongation at 60 °C for 4 min. Eighty μL of freshly prepared ethanol/acetate solution was added to the sequencing reaction mixture in a 1.5 mL microcentrifuge tube and mixed well with a brief vortex. The mixture was left to stand at room temperature for 15 min and centrifuged at the maximum speed of 14,500 rpm for 20 min at room temperature. The ethanol solution was immediately removed carefully from the tube with an aspirator equipped with a fine tip. The resulting DNA pellets were washed by adding 250 μL of 70% ethanol to the tube, and vortexed briefly. The precipitated DNA was collected by centrifugation for 5 min at the maximum speed. The remaining ethanol was carefully removed from the tube with an aspirator equipped with a fine tip. The DNA obtained was dried in a heat box at 90 °C for 1 min, and the dried DNA was stored at either 4 °C or -20 °C. The DNA pellets were suspended in 20 μL of a terminator sequencing reagent, mixed on a vortex and spun down. The double-stranded DNA was completely separated by heating at 95 °C for 2 min and immediately placed on ice until ready to load on the instrument. The DNA sequencing was performed on an ABI Prism® 3730XL DNA Sequence (Applied Biosystems, Foster City, California, USA).

2.7. Sequence analyses

The nucleotide sequences obtained from all primers were assembled using the Cap contig assembly program, an accessory application in BioEdit (Biological sequence alignment editor) Program. The identification of phylogenetic neighbors was initially carried out by the BLAST [7] and mega BLAST [7] programs against the database of type strains with validly published prokaryotic names [8]. The 50 sequences with the highest scores were then selected to calculate pairwise sequence similarity using a global alignment algorithm, which was implemented at the EzTaxon server (http://www.eztaxon.org/; [8].

2.8. Biocontrol efficacy in pot trials

This experiment was used completely randomized design (CRD) with 5 treatments and 5 replications. The 5 treatments were used of: *R.* solanacearum, *R. solanacearum* + *B. amyloliquefaciens*, *R. solanacearum* + *B. subtilis*, *R. solanacearum* + *B. amyloliquefaciens* + *B. subtilis*, *B. amyloliquefaciens* + *B. subtilis* and a control in which there was no application. Thirty days old of 45 plants were performed for each treatment and transferred to pots with 2 kg. sterilized mixture (sandy-loam soil:compost:spent mushroom media, 3:1:1), left for 7 days. The biocontrol bacterial suspension of 50 mL (10^{10} CFU/mL) was poured into the soil at each pot's base. After three days of inoculation, a

50 mL of *R. solanacearum* suspension (10^{10} CFU/mL) grown in TTCB medium was poured into the soil. The eggplants were irrigated two times/day and left for wilt development for up to four weeks. Wilt exhibit was recorded at weekly intervals, and the wilting percentage was also calculated, the height was measured.

2.9. Field evaluation of eggplant for wilt disease incidence
Field treatment was carried out for six months from February to July 2019. The plot size was 54 × 20 m on sandy-loam soil.

2.10. Liquid biofertilizer production
The liquid biofertilizer was produced from fruit-based waste material obtained by Princess Food Co., Ltd. and using a commercial method practised by [9]. Three independent batches of liquid biofertilizer were produced for 200 L and mixed by two bacterial antagonists. Liquid biofertilizer was packed in 10 L plastic (PP) bottles. The samples were then delivered for dilution 100 times with water and poured into the eggplant every week.

2.11. Soil preparation
The spent mushroom media treated soil amendments, mango peel mixed kernel and broiler litters for 2 kg/m^2. The microbial activator PD-1 obtained by the Land Development Department of Thailand was dissolved and sprayed on soil. Then, the soil was plowed and left for 30 days. The soil was turned over for 25 plots, each measuring 1.2 × 20 m and were laid out, with 1.0 m alley between 2 plots. Soil samples were collected for physicochemical property analysis by comparing before and after preparation.

2.12. Planting of eggplant
The experiment was conducted in a CRD, with five treatments and five replications. The treatments were the same pot experiments. Transplanting of eggplant seedlings was done four weeks after seed germination. On each plot there were two columns with 20 plants in each column. Within each column, the planting interval was 80 cm. The interval between 2 columns was 1 m. Two weeks after transplanting, application of compost was done for 100 g/plant around the stem. Sampling and data collection were carried out at weekly intervals for wilt infection, and heights were measured.

2.13. Scaling up of eggplant production
Eggplant production at farmers' scale was conducted in three fields and each using area of 0.25 hectares. The efficacy of liquid biofertilizer on eggplant yields, quality and harvesting index, the experiments were prepared soil as the same methods at previous studies. The treatments were compared the frequencies between 1-3 times/week of application for liquid biofertilizer.

3. Results and discussion
3.1. Screening of Bacillus antagonistic potential
The wilt disease was infected on eggplant, and it could exhibit symptom in the host for 5-10 days. It was isolated and purified for antagonistic testing of *Bacillus*. A total of 34 Bacillus isolates were exhibited antagonistic activity for seven isolates. All of them were two isolates that high activity, as shown in Table 1. The in-vitro test showed that the lettuce soil's isolate exhibited the highest clear zone and included the isolate from kale soil. Two strains were identified through the 16S rRNA gene sequencing followed by ABI Prism® 3730XL DNA sequence. They were *Bacillus amyloliquefaciens* and *B. subtilis*. Both strains showed 99-100% similarity with the matching isolates at the BLAST and mega BLAST database. The *B. amyloliquefaciens* was showed the highest inhibitory zone for 1.8 cm. *Bacillus* is the most potential microorganism that used for biocontrol of plant. It was secreted various peptides, which have shown distinct capacities to inhibit plant pathogen [1]. However, *B. amyloliquefaciens* DSBA-11 was the best among other species of *Bacillus* and indicated maximum inhibited the growth of *R. solanacearum* 4.91 cm [9].

Table 1. Screening of *Bacillus* antagonistic potential for wilt disease in eggplant

Sources of Bacillus	Inhibition test	Clear zones (cm)	PCR identification
Bamboo soil	+++	$1.2^{b} \pm 0.1$	-
Forest soil	++	$0.9^{bc} \pm 0.2$	-
Lettuce soil 1	+++++	$1.8^{a} \pm 0.1$	*Bacillus amyloliquefaciens*
Lettuce soil 2	++	$0.8^{c} \pm 0.1$	-
Lettuce soil 3	++	$0.8^{c} \pm 0.2$	-
Pepper soil	++	$0.9^{bc} \pm 0.1$	-
Kale soil	+++++	$1.6^{a} \pm 0.2$	*Bacillus subtilis*

3.2. Biocontrol efficacy in pot trials

The comparative bio-efficacy to control of wilt as well as growth promoting for eggplant in pot trials which was designed 5 treatments. The results (Table 2) showed that on the 1st week after inoculation of the pathogen, eggplants were infected by *R. solanocearum* and disease incidence of 17.78%.

Table 2. Biocontrol efficacy in pot trials of eggplant

Treatments	Height (cm.) (1 day)	Height (cm.) (30 days)	Disease incidence (%) of eggplant		*R. solanacearum* (CFU/g of soil)
			1 week	4 weeks	
Control	16.83 ± 0.35^{ns}	41.03 ± 0.55^{c}	4.45 ± 3.8^{a}	4.45 ± 3.8^{a}	$7.0 \times 10^{1} \pm 1.41^{c}$
R. solanacearum	16.90 ± 0.82^{ns}	21.03 ± 0.23^{d}	17.78 ± 3.8^{b}	80.0 ± 1.7^{c}	$1.91 \times 10^{7} \pm 2.12^{a}$
R. solanacearum + BL	17.13 ± 0.25^{ns}	52.03 ± 0.68^{b}	0.0 ± 0.0^{a}	0.0 ± 0.0^{a}	$6.7 \times 10^{4} \pm 4.95^{b}$
R. solanacearum + BS	16.87 ± 0.31^{ns}	56.27 ± 0.27^{a}	0.0 ± 0.0^{a}	0.0 ± 0.0^{a}	$9.6 \times 10^{4} \pm 3.54^{b}$
R. solanacearum + BL+BS	16.77 ± 0.51^{ns}	55.83 ± 0.40^{a}	0.0 ± 0.0^{a}	0.0 ± 0.0^{a}	$4.4 \times 10^{4} \pm 2.83^{b}$

Note: BL = *Bacillus amyloliquefaciens*, BS = *Bacillus subtilis*
Means (a-d) with a column followed by different letter are significantly different at 95% confidence

Also, the *R. solanocearum* treatment has contained the soil's disease at the highest concentration in 1.9×10^{7} CFU/g. The treatment with treated by two strains of Bacillus antagonists and wilt pathogen remained survival for 100% of four weeks. The wilt disease in soil was decreased by *B. amyloliquefaciens* and *B. subtilis* nearly 1,000 times. After 30 days of disease inoculation, the maximum height of eggplants for three treatments was not significant; *R. solanacearum* + BS and *R. solanacearum* + BL+BS. *Bacillus* spp. showed plant growth promotion expression attributes such as siderophores and IAA 10]. That was indicated the *Bacillus* spp. which rhizobacterial properties of their antagonistic activity and plant growth promotion.

3.3. *Field evaluation of eggplant for wilt disease incidence*

According to eggplant inoculated with *R. solanacearum* showed wilt symptoms after seven days of inoculation. And the results from pot trials that indicated the two strains of *Bacillus* were suppressed pathogen and no wilt incidence for 4 weeks. Field trials were conducted the same treatments, but soil was improved fertility due to sandy – loam soil and low organic matter. The use of spending mushroom media, mango peel mixed kernel, broiler litters for 2 kg/m^2 and included the microbial activator PD-1. The soil properties were better than before treating (Table 3), and especially organic matter, micronutrients were also at optimum levels.

Table 3. Soil properties for field evaluation of eggplant for wilt disease incidence

Soil Parameters	Before	After	Optimum*
pH	5.3	6.9	6.0-7.0
Bulk density (g/cm^3)	1.72	1.45	-
Organic matter (%)	0.12	2.9	2.5-3.0
Total Nitrogen (mg/kg)	Trace	74	50-80
Phosphorus (mg/kg)	8	409	26-42

Potassium (mg/kg)	36	790	130
Calcium (mg/kg)	Trace	1008	1,040
Magnesium (mg/kg)	Trace	183	135

* Reference from the manual of eggplant production, Princess Food Co., Ltd., Chiang Mai, Thailand

From Table 4, the field evaluation of wilt incidence showed that at after 4 weeks of disease inoculation was 67.0%, whereas the treatments of *Bacillus* antagonists were exhibited below 5%. The results of *R. solanacearum* in rhizosphere soil and soil were indicated that suppression for 10,000-1,000 times, respectively. The factor was associated with wilt controlling that was soil preparation by using organic materials and microorganism. The saprophytic microorganism that was obtained from spent mushroom media. This method was the same Bokashi technique, were introduced as the best performer in the suppression of *R. solanacearum* [11]. The application of *Bacillus amyloliquefaciens* strain S20 alone could control eggplant wilt with an efficacy of 25.3% during a 40 days experiment. If strain S20 was used with organic fertilizer, the control efficacy against eggplant wilt reached as high as 70.7% [12].

The production results at farmers' scale showed that the soil for planting contained *R. solanacearum* 2.5×10^3 CFU/g. So, the method for soil preparation was the same method for field evaluation. Then the use of two strains of *Bacillus* antagonists mixed with liquid biofertilizer for the ratio of 1:100. The experiments were conducted with four treatments and five replications in 0.25 hectares. They were a controlled treatment (no use of pathogen and bacterial antagonists), liquid bio-fertilizer and antagonist bacteria for 1, 2 and 3 times/week. The results shown in Table 5 indicated that disease incidence of control treatment for 13.84%, whereas other treatments bacterial wilts were reduced with the survival rate to 97%. The height of eggplants with the Bacillus sp. was better for high yield by an average of 0.74 kg/plant/week that yield was increased about 25%. However, the frequencies of liquid biofertilizer mixing *Bacillus* antagonists were not significant for the height and growth rate of fruit.

Table 4. Field evaluation of eggplant for wilt disease incidence

Treatments	Height (cm.)		Disease incidence (%) at 4 weeks	*R. solanacearum* (CFU/g)	
	(1 day)	60 days		Rhizosphere soil	Soil
Control	15.73 ± 1.54	80.25 ± 1.84^b	15.0 ± 2.8^b	4.0×10^3±4.84^b	1.4×10^5±5.05^b
R. solanacearum	16.12 ± 1.37	58.54 ± 1.39^b	67.0 ± 4.2^c	3.9×10^7±6.38^a	2.4×10^7±12.52^a
R. solanacearum+BL	15.89 ± 1.61	125.62 ± 0.61^a	3.3 ± 1.1^a	2.9×10^3±3.00^b	8.4×10^4±6.28^c
R. solanacearum+BS	15.65 ± 1.72	130.35 ± 0.45^a	3.2 ± 1.2^a	6.6×10^3±0.66^b	7.5×10^4±9.08^c
R.solanacearum+BL+BS	16.05 ± 1.58	128.80 ± 2.70^a	3.0 ± 1.3^a	3.4×10^3±2.40^b	4.7×10^2±5.57^e

Means (a-c) with a column followed by different letter are significantly different at 95% confidence

Table 5. Scaling production up of farmers' field

Treatments	Height (cm)		Disease incidence (%) at 4 weeks	Growth rate of fruit (cm/day)^{ns}	Yields Kg/plant /week)
	1day ^{ns}	60 days ^{ns}			
Control	28.00 ± 7.54	121.17 ± 14.57	13.84 ± 2.33^b	0.58 ± 0.18	2.84 ± 0.23 ^{b}
LF 1 time/week	32.67 ± 4.93	132.50 ± 17.18	2.11 ± 0.64^a	0.69 ± 0.15	3.57 ± 0.12^a
LF 2 times/week	33.17 ± 6.91	135.00 ± 13.16	2.42 ± 1.02^a	0.73 ± 0.16	3.56 ± 0.15 ^{a}
LF 3 times/week	29.33 ± 5.43	133.84 ± 19.02	2.32 ± 0.95^a	0.71 ± 0.15	3.60 ± 0.19^a

ns = not significant at 95% confidence
Means (a-b) with a column followed by different letter are significantly different at 95% confidence

So, farmers could use the antagonists at a minimum 1 time/week due to colonization, formation of biofilm by Bacillus strains on root surface may provide an additional advantage for the prevention of wilt pathogen [2]. The eggplant production under field condition for harvesting time can be investigated of anthocyanin, antioxidant capacity and total phenolic acid content. The optimum was 12-15 days of fruit growth for highly bioactive ingredients.

4. Conclusion

Eggplant production was succeeded by the use of agro-industrial wastes for soil amendments and soil fertility improvement. And the application *Bacillus* antagonists that indicated wilt disease suppression by the reduced rate 1,000-10,000 time and wilt incidence was exhibited 3%. The most effective biocontrol technique was inoculated *Bacillus* antagonists before planting for seven days, and it was used as liquid biofertilizer for eggplant every week. So, this method was suited for organically grown eggplant.

References

[1] Sakthivel K, Manigundan K, Gautam R K, Singh P K, Nakkeeran S and Sharma S K 2019 Bacillus spp. for suppression of eggplant bacterial wilt pathogen in Andaman Islands: Isolation and characterization *Indian J. Exp. Biol.* **57** 131–7

[2] Achari G A and Ramesh R 2019 Colonization of eggplant by endophytic bacteria antagonistic to Ralstonia solanacearum, the bacterial wilt pathogen *Proceedings of the National Academy of Sciences, India Section B: Biological Sciences* vol 89 pp 585–93

[3] Kawasaki H, Hoshino Y, Hirata A and Yamasato K 1993 Is intracytoplasmic membrane structure a generic criterion? It does not coincide with phylogenetic interrelationships among phototrophic purple nonsulfur bacteria *Arch. Microbiol.* **160** 358–62

[4] Yamada Y, Katsura K, Kawasaki H, Widyastuti Y, Saono S, Seki T, Uchimura T and Komagata K 2000 Asaia bogorensis gen. nov., sp. nov., an unusual acetic acid bacterium in the α-Proteobacteria *Int. J. Syst. Evol. Microbiol.* **50** 823–9

[5] Katsura K, Kawasaki H, Potacharoen W, Saono S, Seki T, Yamada Y, Uchimura T and Komagata K 2001 Asaia siamensis sp. nov., an acetic acid bacterium in the α-Proteobacteria *Int. J. Syst. Evol. Microbiol.* **51** 559–63

[6] Brosius J, Dull T J, Sleeter D D and Noller H F 1981 Gene organization and primary structure of a ribosomal RNA operon from Escherichia coli *J. Mol. Biol.* **148** 107–27

[7] Altschul S F, Madden T L, Schaffer A A, Zhang J, Zhang Z, Miller W and Lipman D J 1997 Gapped BLAST and PSI-BLAST: a new generation of protein database search programs *Nucleic Acids Res.* **25** 3389–402

[8] Chun J, Lee J H, Jung Y, Kim M, Kim S, Kim B K and Lim Y W 2007 EzTaxon: A web-based tool for the identification of prokaryotes based on 16S ribosomal RNA gene sequences *Int. J. Syst. Evol. Microbiol.* **57** 2259–61

[9] Boonraeng S, 2016 Quality and Food Safety of Minimally Processed Organic Vegetables. Faculty of Agricultural Technology Chiang Mai Rajabhat University 325 pp.

[10] Singh D, Yadav K D, Chaudhary G, Rana V S and Sharma R K 2016 Potential of Bacillus amyloliquefaciens for biocontrol of bacterial wilt of tomato incited by Ralstonia solanacearum *J. Plant Pathol. Microbiol.* **7** 1–6

[11] Lwin M and Ranamukhaarachchi S L 2006 Development of biological control of Ralstonia solanacearum through antagonistic microbial populations *Annu. Rev. Earth Planet. Sci.* **8** 657–60

[12] Chen D, Liu X, Li C, Tian W, Shen Q and Shen B 2014 Isolation of Bacillus amyloliquefaciens S20 and its application in control of eggplant bacterial wilt *J. Environ. Manage.* **137** 120–7

IOP Conf. Series: Earth and Environmental Science **883** (2021) 012088 doi:10.1088/1755-1315/883/1/012088

In-situ kuskus conservation strategy in Wakasihu Village, West Leihitu District, Central Maluku District

Y Th Latupapua[1,*], F F Tetelay[1], F Soselisa[1], C K Pattinasarany[1], L Latupapua[1], A Sanduan[1]

[1] Forestry Study Program, Faculty of Agriculture, Pattimura University, Ambon, Indonesia

* E-mail: vthaforester@gmail.com

Abstract. Kuskus, a marsupial species belonging to the Phalangeridae family that is vulnerable in the IUCN Redlist, is one of the endemic animals whose survival is currently very concerning due to habitat destruction and poaching, so conservation activities need to be carried out immediately. This study aims to identify the characteristics of the habitat, type of feed, and in situ conservation strategies in the Wakasiu forest, Leihitu Barat sub-district, Central Maluku district. This research was conducted in October 2020. Habitat characteristics were identified using vegetation analysis, forage types were identified using direct and indirect observations, and in-situ conservation strategies using SWOT analysis. The highest index of importance at the tree level in the line of observation is the type of gayam (Inocarpus fagifer) (20.9%), at the level of weaning is the langsat species (70.32%) the level of the pole type of brown (69.27%). The diversity level of plant species is high and the level of evenness of plant species is even. There are 24 types of plants recorded as potential for special feed. The in-situ conservation strategy that is carried out is planting the type of feed preferred by the specialists, so as to provide a stable feed availability during the successive fruiting period. Apart from that, the restriction of hunting for specialties is through written regulations and the imposition of sanctions.

1. Introduction

Wakasihu Village, which is located in the coastal area of Ambon city, West Leihitu sub-district, Central Maluku district, is a representative of the tropical rainforest ecosystem and is the habitat for one of the endemic animals, namely Kuskus (Phalanger sp). The Kuskus, which is a member of the Phalangeridae family, is a type of marsupial mammal which distribution in the Maluku region. Kuskus is one of the endemic species of Maluku which is protected based on Kuskus in Indonesia which has been protected since 1990 through the Wild Animal Hunting Regulation (PPBL) No. 226/1931, Law no. 5/1990 concerning the conservation of living natural resources and their ecosystems, and Law no. 7/1999 on the preservation of plant and animal species [1-3].

Kuskus is in the vulnerable status on the IUCN Red List because currently, its population has decreased due to deforestation and illegal hunting. Kuskus is a type of mammal that is nocturnal (active at night), and has a very slow movement so that it can be a threat to the existence of Kuskus because it is easily caught by humans. One of the important factors that must be considered in ensuring the survival and management of specialities is the availability of information about the types

of feed consumed by these animals. The function of food is as a constituent of the body, an energy source, and a regulator of metabolic processes [4,5].
Information related to Kuskus feed carried out in previous studies showed that around 31 plant species were identified as sources of Kuskus 'food in the Tangkoko Nature Reserve, North Sulawesi [6], 10 species of Kuskus' food in the Tangkoko Batu Nature Reserve. Angus, North Sulawesi [7], 4 types of food sources for bear Kuskus in the Educational Forest of Hasanuddin University, South Sulawesi [8], and 44 types of food sources in Manusela National Park [9].

The loss of Kuskus habitat due to changes in land use into small-scale community agricultural areas, cutting down of feed sources for housing wood and household furniture, and rampant illegal hunting are the main threats to the survival of cuscu[5]. For this reason, it is necessary to make in-situ conservation efforts in supporting the preservation of the specialities in their habitat. Supporting the conservation of Kuskus in situ, one of the important factors that must be considered in ensuring the survival and management of wild animals is the availability of information about the types of feed consumed by these animals. For this reason, the purpose of this study is to determine the condition of the Kuskus' feed source in Wakasihu Village, and the form of in-situ conservation strategies to support the preservation of this type of marsupial.

2. Methods

The research was conducted in Wakasihu forest, West Leihitu District, Central Maluku Regency. The research was conducted in October - November 2020. The tools used were binoculars, measuring tape, tape measure, raffia rope, stakes, compasses, cameras, GPS, area maps, plant species identification manuals, tally sheets, and stationery. The materials used were questionnaires and vegetation in the research location.

Data collection is generally carried out using direct observation methods. The data collected were the types of feed consumed in utilizing feed, as well as the potential types of feed (density, dominance and frequency) in the study area. There are three ways of collecting data in observing the types and potential of Kuskus feed. 1) Collecting data on the types of food eaten, is carried out by direct observation methods concerning the species of feed eaten by Kuskus. 2) Data collection of preferred feed frequency is carried out by observing the types and parts of plants eaten during observation of feeding activities. 3) Collecting data on the ecological status of the type of feed in the movement area. Observation of the type of Kuskus' feed, was carried out by recording the type of plant selected to be eaten and recording the part of the plant consumed (leaves, fruit, flowers and branch bark) using the scan sample method. The data collection methods are presented in Table 1.

Table 1. Data collection methods

No	Type of Data	Data Collected	Data Collection Methods
1.	Habitat Characteristic	▪ Altitude ▪ Temperate and humidity ▪ Source of Water ▪ Composition and structure of Vegetation	• Measurenment by GPS • Measurenment by termometer dry-wet • Measurtenment by GPS • Vegetation Analysis(Line plot method)
2.	Food Potential	▪ Type of food ▪ Parts eaten	• Direct and indirect observation (discovery of feed traces), interviews
3.	Conservation Strategy	▪ EFAS ▪ IFAS	Indepth interview

Vegetation analysis was carried out to determine the composition and structure of the habitat of bear Kuskus. The method used is the checkered line method [10]. Plots are created based on. the location of the discovery of Kuskus and the location of the discovery of traces of eating bear Kuskus. The stages of the vegetation analysis are carried out by:

IOP Conf. Series: Earth and Environmental Science **883** (2021) 012088 doi:10.1088/1755-1315/883/1/012088

- The encounter with Kuskus and the discovery of traces of eating bear Kuskus.
- Making a checkered line, with a total of 3 plots at each location where bear Kuskus encounters and the location where bear Kuskus feeds are found. The distance between the plots is 10 meters.
- Collecting data in plots in the form of plant levels in it, namely seedlings, saplings, poles and trees. The line plot method design is shown in Figure 1.

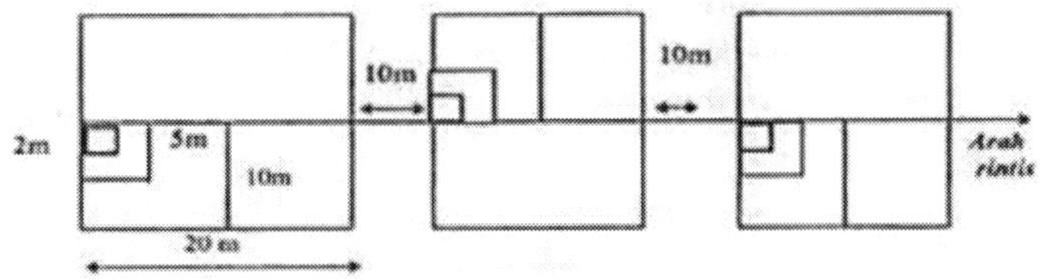

Figure1. Line plot method

The vegetation plots made were 10 plots, with 3 plots each at 3 points where the Kuskus was encountered and 5 points where the Kuskus' food was found. Data were collected in plots measuring (20 m × 20 m) for the tree level, with each sub-plot (10 m × 10 m) for the pile level, (5 m × 5 m) for the sapling level, and (2 m × 2 m) for the tree level. seedling level. Identification of plant species was carried out using the Tree Types book, compiled by Regional Name and Botanical Name Book 10.

Identification of the types of Kuskus 'feed was carried out by direct and indirect observation (finding food traces), as well as conducting inirviews with redundant to determine the types of plants that could potentially serve as Kuskus' feed. The data collected included plant species and plant parts eaten by Kuskus. The line of observation was determined based on the results of a field survey, namely the location which was thought to be the location of the meeting of the Kuskus and was thought to represent the availability of the habitat function as a place to eat, shelter, and rest the Kuskus. The sample plots were made in size (20 m × 20 m) at the location where Kuskus was found. The data from the analysis of vegetation in the field have calculated the values of density, relative density, frequency, relative frequency, dominance, relative dominance, and index of the importance of a plant species. In addition, the Shannon-Wiener species diversity index and the evenness index of a plant were calculated. The important value index according to has the following equation [11]:

Density = number of individuals of each species/area
Relative Density = (Density of Species) / (Density of All Species) × 100%
Dominance = the total area of the base area of a type / area
Relative dominance = dominance of a species / dominance of all species x 100%
Frequency = (Number of plots found in species) / (total number of plots)
Relative frequency = (Frequency of a species) / (frequency of all species) × 100%
Importance Value Index = KR + FR + DR (Trees and poles)
Importance Value Index = KR + FR (Seedlings and saplings)
Shannon Wiener species diversity index with the formula :

$$H' = -\sum pi \ln pi \; ni = 1 \; \text{pi} = \text{ni} / \text{N}$$

Information:

H '= Shannon-Wiener diversity index
ni = number of individuals or the importance of type i
N = total of individuals or the importance of all species
Evenness index: E ' = H' / ln (S)

Information:

E ' = evenness index
H '= Species diversity index
S = Number of species

The observation data of bear Kuskus' feed in the field were made a table in the form of plant species eaten by bear Kuskus, and a graph was made of the part of the plant eaten by Kuskus. The results were analyzed by descriptively qualitatively. Data from observations of daily activity of bear Kuskus in the field are in the form of daily activity duration of bear Kuskus that appears during observation and each activity recorded is calculated as a value.

3. Results and discussion

3.1. Habitat characteristics kuskus

3.1.1. Physical components

Research conditions when collecting data in October 2020 in the rainy season conditions. The average temperature during the study ranged from 25°C-28.5°C an air humidity of 86%-92%. During data collection, rain rarely occurs, but when it rains the rainfall is very heavy but with a short duration. The research was conducted at an altitude of 8-10 m above sea level in lowland rain forests. According to [12], Kuskus can live up to an altitude of 600 m above sea level. Based on the results of the observations, there was no direct use of water by Kuskus at the observation location. This is because Kuskus is a nocturnal and arboreal animal that rarely descends to the surface of the soil, besides that Kuskus gets a source of water from the contents of the shoots and young leaves on the tree which is used as a tree for feed, nest, and play.

3.1.2. Biotic components

The vegetation composition and structure of the Kuskus' habitat were obtained from the results of vegetation analysis while in the field. There are 5 measuring squares on the left lane and 5 measuring squares on the right lane. Plant species that dominate other species are seen from the results of the Importance Value Index (IVI), this index serves to describe the ecological position of a species in a community [10]. The highest vegetation importance value index (IVI) at the tree, weaning, sapling and seedling levels in the left and right lanes is presented in Table 2, 3, and 4.

Table 2. Tree-level importance value index (IVI) on the path of observation

No	Species (local name)	INP	Scientific Name	Family	English name
1	Durian	96.621	*Durio zibethinus*	Bombacaceae	Durian
2	Langsa	21.284	*Lansium parasiticum*	Meliaceae	Duku var. aqua
3	Pala	19.414	*Myristica fragans*	Myristicaceae	Nutmeg

Table 3. Pole-level importance value index (IVI) on the path of observation

No	Species (local Name)	INP	Scientific Name	Family	English name
1	Cokelat	69.278	*Theobroma cacao*	Sterculiaceae	Cocoa
2	Duku	58.687	*Lansium domesticum*	Meliaceae	Duku
3	Langsa	58.354	*Lansium parasiticum*	Meliaceae	Duku var. aqua
4	Pala	16.101	*Myristica fragans*	Myristicaceae	Nutmeg

Table 4. Seedling-Level Importance Value Index (IVI) on the Path of Observation

No	Jenis (local name)	INP	Scientific Name	Famili	English Name
1	Gayam	20.1872	*Inocarpus fagifer*	Fabaceae	Gayam
2	Kopi	17.2460	*Coffea spp*	Rubiaceae	Coffe plant
3	Langsa	17.2460	*Lansium parasiticum*	Meliaceae	Duku var.aqua

Based on the results in Table 2, the highest vegetation significance index (IVI) in the observation path at the tree level is durian (*Durio zibethinus*) (96.62%), duku var. aqua (*Lansium parasiticum*) (21.28%), and nutmeg (*Myristica fragans*) (19.41%). This shows that these species dominate and are important species and are used by Kuskus as a source of food, shelter and rest areas.

3.1.3. Components of Kuskus' feed in the Wakasihu Village area

The results of observations and interviews with communities around the Wakasiu forest area show that there are 24 (twenty four) types of vegetation used as food for Kuskus which includes forest vegetation types and agricultural and plantation plants. The parts of plants that are preferred and function as Kuskus' feed are young plants. This result is also supported in the research results of [1], [9], which state that in general, consumers prefer the type of feed that is still young, has a fresh aroma and has a sour taste. The selection of this type of feed must of course be adapted to the condition of the Kuskus' organs which are classified as small mammals. This is supported by the results of the study by [13], that the choice of feed type in marsupials is related to the morphological and histological structures of the gastrointestinal tract.

Of the 24 types of feed found in the study location, 16 types of young leaves were eaten (67%), 20 types of fruit (83%), 6 types of flowers (25%). The choice of feed type in Kuskus is related to the morphological and histological structure of the gastrointestinal tract. This is the reason for preferring young plant organs as a source of food.

Table 5. Type of Kuskus' Feed

No	Type (local name)	Scientific name	Family	English Name
1	Cengkeh	*Syzygium aromaticum*	Myrtaceae	Clove
2	Coklat	*Theobroma cacao*	Sterculiaceae	Cocoa
3	Duku	*Lansium domesticum*	Meliaceae	Duku
4	Gandaria	*Bouea macrophylla*	Anacardiaceae	Gandaria
5	Ganemu	*Gnetum gnemon*	Gnetaceae	Gnetum
6	Gayam	*Inocarpus fagifer*	Fabaceae	Gayam
7	Gondal	*Ficus septica*	Moraceae	
8	Gujawas	*Psidium guajava*	Myrtaceae	Guava
9	Jambu Hutan	*Eugenia spp*	Myrtaceae	
10	Ketapang Hutan	*Terminalia copelandii*	Combretaceae	Ketapang
11	Kayu Burung	*Eleocarpus ganitrus*	Eleocarpaceae	Rudraksa/Ceylon Olive
12	Kayu Samar			
13	Kemiri	*Aleurites moluccana*	Euphorbiaceae	Candlenut
14	Kenari	*Canarium amboinensis*	Burseraceae	Canary/Walnut
15	Kopi	*Coffea spp*	Rubiaceae	Coffe plant
16	Langsa	*Lansium parasiticum*	Meliaceae	Duku var.aqua
17	Lenggua	*Pterocarpus indicus*	Fabaceae	Red Sandalwood Amboyna
18	Linat			
19	Mangga	*Mangifera indica*	Anacardiaceae	Manggo
20	Matoa	*Pometia pinnata*	Sapindaceae	Fijian Longan
21	Pala	*Myristica fragans*	Myristicaceae	Nutmeg
22	Rambutan	*Nephelium lappaceum*	Sapindaceae	Rambutan
23	Siripopar	*Ficus spp*	Moraceae	
24	Tomi-Tomi	Flacourtia inermis	Salicaceae	batoko plum

Source : Primary data, 2020

[14] states that in the Table Mountain Tourism Park there are 26 types of plants used as food for Kuskus. The high potential of vegetation types found in this area provides opportunities for the production of Kuskus' feed ingredients. The types of vegetation consumed by Kuskus in the forest area of Wakasiu Village are presented in Table 6. The comparison of the types of feed consumed by the two types of Kuskus shows that they both consume the same plant species but based on their home

range, P. orientalis and S.maculatus are often found in addition to forests are also in plantation or agricultural areas, because of the solitary nature of large populations [15,16]. The types of feed consumed include forest vegetation, agricultural or plantation crops. The parts consumed by Kuskus are fruit (ripe), young leaves (shoots /shoots) and flowers. The same opinion was expressed in the research results of [8,9], which state that the vegetation parts of the preferred feed are fruit, shoot, and flowers which have a sour aroma and taste. The same opinion was expressed in [17] which stated that apart from fruit, young leaf shoots were also consumed. Perhaps this is due to the fact that when the plant has not produced fruit, the young shoots are consumed. This phenomenon is natural, because [18] explains that when the preferred food is not available, wild animals usually shift their consumption to alternative feed or "emergency foods". Based on the results in Table 2, the highest vegetation importance index in the line of observation at the tree level is durian (*Durio zibethinus*) (96.62%), langsat (*Lansium parasiticum*) (21.28%), and nutmeg (*Myristica fragans*) (19.41%). This shows that these species dominate and are important species and are used by Kuskus as a source of food, shelter and rest areas. However, the density value shows the opposite result.

3.1.4. Alternative Kuskus conservation strategy

In managing the habitat of Kuskus, it is necessary to formulate a strategy based on the carrying capacity of the habitat. SWOT analysis (Strength, Weakness, Opportunity, Threat) is one of the analysis methods used to get the best strategy by measuring the strengths, weaknesses in the existing sector and simultaneously measuring the opportunities and challenges /threats that will be faced later after determining the strengths, weaknesses, opportunities, threats, and challenges in every sector. The method of determining the internal and external strategic factors is as follows: 1. In the Strength (S) box, identify some of the strengths that exist in the Wakasiu Village forest area in achieving Kuskus preservation; 2. In the Weakness box (W), identify several existing weaknesses, namely weaknesses in achieving success in Kuskus conservation 3. In the Opportunity (O) box, identify some external opportunities that will be obtained in efforts to preserve Kuskus 4. In the Threat (T) box, identify some of the challenges that will be faced in efforts to manage Kuskus, inbox 5. Identify strategic possibilities of conservation and management of Kuskus based on the consideration of a combination of four strategic factors, namely SO, ST, WO, and WT strategies.

The identification of each factor was obtained from the results of direct observation in the field, the results of interviews with key actors of Raja Wakasiu, Saniri Negeri, and the community around the area, as well as a literature study on activity reports at the Maluku BKSDA. Based on the carrying capacity of the habitat in the form of altitude-related to the behaviour and movement of Kuskus which is relatively easy at medium altitude, location slope, water availability, area boundaries in the form of natural boundaries in the form of rivers, the condition of the community around the area, availability of food can be used as an internal factor divided into strength factors and weakness factors.

Weaknesses can be seen from the current condition that has not been developed and is a factor causing damage to the habitat of Kuskus, so it does not support in-situ conservation efforts in habitat locations. This can be identified from the physical condition of the habitat that is not supportive, the condition of human resources and management documents. External factors in the form of threats and opportunities, threats can be seen from the disturbance of the area in the form of hunting and logging, the ease of access to the area which is supported by roads so that community activities around the area are high, including hunting. The opportunity factor is an external condition outside the existing or non-existing areas so that it can be developed as a supporting factor for Kuskus conservation. The opportunities that already exist are laws and regulations that state Kuskus as protected animals, cooperation with stakeholders who focus on the preservation of Kuskus. Meanwhile, the opportunity that does not exist at this time is the development of ecotourism in the form of observing the behaviour of Kuskuses as unique and protected marsupials.

a. Identification of IFAS (Internal Strategic Factors Analysis Summary) and EFAS (External Strategic Factors Analysis Summary)

After identifying each of the internal and external factors above, to formulate a strategy for Kuskus conservation, an analysis of each of these factors was carried out through scanning between IFAS and

IOP Conf. Series: Earth and Environmental Science **883** (2021) 012088 doi:10.1088/1755-1315/883/1/012088

EFAS. The stages of the analysis using SWOT, giving weight to each element of the internal and external factors, are the key to the success of strategy making (key success factor). The weighting is given based on the relative advantage of a factor. The weight value in this study is the result of observation and in-depth interviews with resource persons. Give a weighted value for each factor ranging from 1.0 (very important) to 0.0 (not important). The key success factors are then given a rating indicating the value of support for each factor in achieving the goal. The granting of a rating starts with number 4 (very influential) to number 1 (very weak).

Table 6. Identification of internal factors (strengths and threats)

No.	*Strength*	*Weakness*
1	The amount of food availability is large and varied	Season affects the availability of feed Kuskus
2	Flat slopes so that the proper use factor is quite high	Forage, playing and sleeping trees become trees that are often cut down to meet community household needs
3	Temperature and humidity suitable for Kuskus' habitat	Low understanding of the community about the role and function of Kuskus for the environment
4	Clear area boundaries in the form of a river	Kuskus meat is often a need for animal protein for the community so it is often hunted
5	Shade conditions of vegetation that meet the life requirements for Kuskus as cover, play and sleep.	The low level of community knowledge in supporting reforestation in Wakasiu Village in supporting the availability of food sources for Kuskus' trees

b. Identification of external factors (opportunities and threats)

Table 7. Identification of external factors (opportunities and threats)

No.	*Opportunity*	*Threats*
1	Laws and regulations that define Kuskuses as protected animals	High hunting of Kuskus
2	The high level of community participation in supporting the conservation of Kuskus' habitat in the forest	Easy access to Kuskus habitat in Wakasiu village
3	There are regulations in the village regarding animals that are allowed to be eaten and which cannot be eaten	Having a high economic value as animal protein
4	Has attractiveness as an object of ecotourism (unique and rare animals)	There is no written prohibition on hunting Kuskus
5	The population can still support captive activities	Low monitoring of the Kuskus population

Sources : Primary data, 2020

c. Identification of IFAS (Internal Strategic Factors Analysis Summary) and EFAS (External Strategic Factors Analysis Summary).

After identifying each of the internal and external factors above, to formulate a conservation strategy for Kuskus in Wakasiu village, an analysis of each of these factors was carried out through scanning between IFAS and EFAS. The stages of the analysis using SWOT, giving weight to each element of the internal and external factors, are the key to the success of strategy making (key success factor). Weights are given between 0-1, where the number 0 indicates unimportance and the number 1 indicates the most important. The key success factors are then given a rating indicating the value of support for each factor in achieving the goal. The value is between 1-5, where, 1 = very little, 2 = little effect, 3 = moderate effect, 4 = influential, and 5 = very influential.

d. Weighting for IFAS and EFAS

Table 8. Weighting for strength

No. Strength	Strategies Factor	Quality				
		1	2	3	4	5
1	The amount of food availability is large and varied					√
2	Flat slopes so that the proper use factor is quite high				√	
3	Temperature and humidity suitable for Kuskus' habitat				√	
4	Clear area boundaries in the form of a river				√	
5	Shade conditions of vegetation that meet the life requirements for Kuskus as cover, play and sleep.					√

Sources : Primary data, 2020

Table 9 . Weighting for *Weakness*

No. Weakness	Strategies Factor	Quality				
		1	2	3	4	5
1	Season affects the availability of feed Kuskus					√
2	Forage, playing and sleeping trees become trees that are often cut down to meet community household needs					√
3	Low understanding of the community about the role and function of Kuskus for the environment					√
4	Kuskus meat is often a need for animal protein for the community so it is often hunted				√	
5	The low level of community knowledge in supporting reforestation in Wakasiu Village in supporting the availability of food sources for Kuskus' trees					√

Sources : Primary data, 2020

Table 10. Weighting for *opportunity*

No. Opportunity	Strategies Factor	Quality				
		1	2	3	4	5
1	Laws and regulations that define Kuskuses as protected animals					√
2	The high level of community participation in supporting the conservation of Kuskus' habitat in the forest					√
3	There are regulations in the village regarding animals that are allowed to be eaten and which cannot be eaten					√
4	Has attractiveness as an object of ecotourism				√	

	(unique and rare animals)					
5	The population can still support captive activities					√

Source : Primary data, 2020

Table 11 . Weighting for *threats*

No. Threats	Strategies Factor	Quality				
		1	2	3	4	5
1	High hunting of Kuskus					√
2	Easy access to Kuskus habitat in Wakasiu village					√
3	Having a high economic value as animal protein					√
4	There is no written prohibition on hunting Kuskus				√	
5	Low monitoring of the Kuskus population					√

Source : Primary data, 2020

Table 12. Rangking for *strength*

No. Strenght	Strategies Factor	Rating				
		1	2	3	4	5
1	The amount of food availability is large and varied					√
2	Flat slopes so that the proper use factor is quite high					√
3	Temperature and humidity suitable for Kuskus' habitat				√	
4	Clear area boundaries in the form of a river				√	
5	Shade conditions of vegetation that meet the life requirements for Kuskus as cover, play and sleep.					√

Source : Primary data, 2020

Table 13. Rangking for *weakness*

No. Weakness	Strategies Factor	Rating				
		1	2	3	4	5
1	Season affects the availability of feed Kuskus					√
2	Forage, playing and sleeping trees become trees that are often cut down to meet community household needs					√
3	Low understanding of the community about the role and function of Kuskus for the environment				√	
4	Kuskus meat is often a need for animal protein for the community so it is often hunted				√	
5	The low level of community knowledge in supporting reforestation in Wakasiu Village in supporting the availability of food sources for Kuskus' trees				√	

Source : Primary data, 2020

IOP Conf. Series: Earth and Environmental Science **883** (2021) 012088 doi:10.1088/1755-1315/883/1/012088

Table 14. Rangking for *opportunity*

No. Opportunity	Strategies Factor	Rating 1	2	3	4	5
1	Laws and regulations that define Kuskuses as protected animals					√
2	The high level of community participation in supporting the conservation of Kuskus' habitat in the forest					√
3	There are regulations in the village regarding animals that are allowed to be eaten and which cannot be eaten					√
4	Has attractiveness as an object of ecotourism (unique and rare animals)				√	
5	The population can still support captive activities					√

Source : Primary data, 2020

Table 15. Rangking for *threats*

No. Threats	Strategies Factor	Rating 1	2	3	4	5
1	High hunting of Kuskus					√
2	Easy access to Kuskus habitat in Wakasiu village					√
3	Having a high economic value as animal protein				√	
4	There is no written prohibition on hunting Kuskus				√	
5	Low monitoring of the Kuskus population				√	

Source : Primary data, 2020

After determining the weight and rating of internal and external factors, the next stage of analysis in identifying the in-situ conservation strategy is to determine the value of importance, which previously determined the total score of each internal and external factor in Tables 17 and 18.

Table 16. Internal strategies

No	Strategies Factor	Quality	Rating	Influence Value
Strength				
1	The amount of food availability is large and varied	0,3	5	1,5
2	Flat slopes so that the proper use factor is quite high	0,2	4	0,8
3	Temperature and humidity suitable for Kuskus' habitat	0,2	4	0,8
4	Clear area boundaries in the form of a river	0,1	3	0,3
5	Shade conditions of vegetation that meet the life requirements for Kuskus as cover, play and sleep.	0,2	4	0,8
		1,0		4,2
Weakness				

1	Season affects the availability of feed Kuskus	0,3	4	1,2
2	Forage, playing and sleeping trees become trees that are often cut down to meet community household needs	0,2	3	0,6
3	Low understanding of the community about the role and function of Kuskus for the environment	0,2	3	0,6
4	Kuskus meat is often a need for animal protein for the community so it is often hunted	0,1	3	0,3
5	The low level of community knowledge in supporting reforestation in Wakasiu Village in supporting the availability of food sources for Kuskus' trees	0,2	3	0,6
		1,0		3,3

Table 17. External strategies

No	Factor Strategies	Quality	Rating	Influence Value
Opportunity				
1	Laws and regulations that define Kuskuses as protected animals	0,2	4	0,8
2	The high level of community participation in supporting the conservation of Kuskus' habitat in the forest	0,3	3	0,9
3	There are regulations in the village regarding animals that are allowed to be eaten and which cannot be eaten	0,2	3	0,6
4	Has attractiveness as an object of ecotourism (unique and rare animals)	0,2	3	0,6
5	The population can still support captive activities	0,1	3	0,3
		1.0		3,2
Threats				
1	High hunting of Kuskus	0,3	4	1,2
2	Easy access to Kuskus habitat in Wakasiu village	0,1	3	0,3
3	Having a high economic value as animal protein	0,1	2	0,2
4	There is no written prohibition on hunting Kuskus	0,3	3	0,9
5	Low monitoring of the Kuskus population	0,2	4	0,8
		1,0		3,4

Based on the analysis of the influence value of the Tables and Tables, then the IFAS value is calculated which is the total difference in the value of the internal influence of strengths and weaknesses, namely 4.20-3.30 = 0.9, while the EFAS value which is the difference in the total value of the influence of external factors, opportunities and threats, which is equal to 3, 20-3.40 = - 0.2. The results obtained indicate that the IFAS value (+) means that the cumulative strength factor is greater when compared to the weakness factor. While the EFAS value (-) means the cumulative opportunity factor is smaller than the threat factor.

4. Conclusion

1. Potential for 24 species of Kuskus' feed-in Wakasiu village, with the highest IVI values for the tree level, are durian (*Durio zibethinus*), duku var. aqua (*Lansium parasiticum*), and nutmeg (*Myristica fragrans*)
2. Insitu Conservation Strategies that need to be carried out in support of Kuskus conservation are:
   a. Developing Kuskus' tourist attraction objects,
   b. Optimizing the role of the surrounding community in planning and developing objects of ecotourism that are focused on developing endemic and community-based flora and fauna, Improve the protection of the preservation of the area through law enforcement.

5. Acknowledgment

In the end, we, as a team, would like to thank the head of the Wakasiu village and the Saniri Negeri staff, who were involved in providing input regarding data needs during the research.

References

[1] Saragih E W, Sadsoetoeboen M J, and Pattiselanno F 2010 The Diet Spotted Kuskus (*Spilokuskus maculates*) in natural and captivity Habitat J. *Bioscience* **2** 78- 83.
[2] Pattiselano F 2007 Hunting of Kuskus (Phalangeridae) by the Community Napan di Pulau Ratewi, Nabire, Papua *J. Biodiversitas* **8** 274-278.
[3] Usmany M, Tuaputty H, Kakisina P 2015 Kajian Fenotip Kuskus (*Famili Phalangeridae*) di Penangkaran Desa Lumoli, Kecamatan Piru, Maluku *Jurnal Sain Veteriner* **33** ISSN : 0126 – 0421.
[4] Soemarwoto O 1991 Environmental Ecology and Development. Jakarta (ID): Publisher Djambatan.
[5] Nugraha R dan Mustari A H 2017 Habitat Characteristics and Feed Types of Bear Kuskus (*Ailurops ursinus*) in Tanjung Peropa Wildlife Reserve, Southeast Sulawesi *Journal Wasian* **4** 55-68.
[6] Dwiyahreni A A, Kinnaird M F, O'Brien T G, Supriatna J, and Andayani N 1999 Diet and activity of the bear Kuskus (*Ailurops ursinus*) in North Sulawesi, Indonesia *Journal of Mammalogy* **80** 905 - 913.
[7] Wowor F M, Kiroh H J, Rawung V, and Wunguw R H 2016 Density estimation of bear Kuskus (*Ailurops ursinus*) in Nature Reserve Tangkoko Batu Angus Kota Bitung *Jurnal Zootek* **36** 395 – 404.
[8] Achmad A N, Maulani N P, Asrianny R I 2016 Potential of Feed and Preference of Bear Kuskus (*Ailurops ursinus*) in the UNHAS Educational Forest. Proceedings of the International Seminar on Biologists 2016. ISBN 978-602-72198-3-0.
[9] Handayani S dan Kunda R 2019 Identification of Plant Types as Feed for Kuskus (Phalangeridae) from Maluku in the North Manusela National Park, Central Maluku Regency *Journal of Mathematics, Science and Technology* **20** 9-19.
[10] Indriyanto 2006 Forest Ecology. Jakarta: PT Aksara Bumi.
[11] Soerianegara I and Indrawan A 2005 Indonesian Forest Ecology. Bogor: Faculty of Forestry, Bogor Agricultural University.
[12] Flannery T 1990 Mammals of New Guinea. Australia: Robert Brown and Associates.
[13] Stannard H A, McAllan, B M, & Old, J M 2014 Dietary composition and nutritional outcomes in two marsupials, *Sminthopsis macroura* and *S. crassicaudata Journal. of mammal* **99** 1512 5141.
[14] Linthin N 2000 Identification of Vegetation Types as Feed for Kuskus on Moor Island, Napan Weinami District, Nabire Regency. Unpublished thesis, Department of Forestry, Faculty of Agriculture, University of Cenderawasih, Manokwari.

[15] Fatem S, and Sawen D 2007 Types of Kuskus on the North Coast of Manokwari, Papua. Journal Biodiversitas ISSN: 1412-033X. **8**, No.2 Juli 2007 pp.: 233-237.

[16] Pratiwi A A. Talumepa R S H, Wungow Z, Poli S C, Rimbing 2016 Daily Behavior of Bear Kuskus (*Ailurops ursinus*) in Nature Reserve Tangkoko Batu Angus *Zootek Journal* **36** 174 - 183 (Januari 2016).

[17] Duwiri F, Nunaki J H, Pattiselanno F 2010 Identification of Natural Cushes (Phalangeridae) Types of Feed in Hariti Island, Weinami District, Nabire Regency *Natural Journal* **9** 24-28.

[18] Bailey J A 1984 Principles of wildlife management. John Wiley & Sons, Inc. USA. Collins, L.R.1973. Monotremes and Marsupials: A Reference for Zoologi Institution. Washington: Smithsonian Institution.

The influence of agricultural diversification on community income in East Taniwel District

La Ega[1], Muspida[2], E Kembauw[3,*], S J Ferdinandus[4], L J Sinay[5], M Sangadji[6], M Faisal[7], H S E S Aponno[8], M Wenno[9], E Regar[10], R Apituley[11]

[1]Agricultural Product Technology Study Program, Faculty of Agriculture, Pattimura University
[2,6]Development Economics Department, Faculty of Economics and Business, Pattimura University
[3]Social Economics Department, Agribusiness Study Program, Faculty of Agriculture, Pattimura University
[4,7,9,10]Department of Management, Faculty of Economics and Business, Pattimura University
[5]Mathematics Departments, Faculty of Mathematics, Pattimura University
[8]Department of Forestry, Faculty of Agriculture, Pattimura University
[11]Patimura University Planning Department

* E-mail: ekembauw@yahoo.co.id

Abstract. This study entitled "The Effect of Agricultural Diversification on Community Income in the East Taniwel District". The research objective was to determine the relationship and influence of agricultural diversification on people's income in the East Taniwel District.
The data used in this study are primary data from eight villages in East Taniwel District, West Seram District, Maluku Province. This data is the result of a baseline survey at the household level, where the sample is determined using the Slovin formula. The data analysis technique used is the application of the backward method in multiple linear regression modeling based on OLS. The results showed that there was a relationship and influence between agricultural diversification and the income of rural communities in East Taniwel District.

1. Introduction

Indonesia is an agricultural country, because most of the population works as farmers. Agricultural activities require land as agricultural land. Soil in Indonesia is known to be very fertile, so it is suitable for use as agricultural land. Humans in maintaining life always need clothing, food and housing, to fulfill their living needs they do a job (livelihood), because by working they will get results to make ends meet.

The Indonesian population is one of the basic assets for development which is quite influential in increasing national development, especially in the agricultural sector. Development in the agricultural sector is directed at increasing the income and level of welfare of the Indonesian people. Achieving agricultural objectives requires efforts to increase agricultural production. Increasing agricultural production can be done by intensifying and extending agriculture. Agricultural intensification is an effort to increase agricultural production through post-farming programs, consisting of land processing, use of superior seeds, irrigation, fertilization and pest control. Agricultural extensification is an effort to increase agricultural production by expanding agricultural land [1].

Land area as one of the factors of production which is the factory of agricultural products which has a large enough contribution to farming, the size of the production from farming is influenced by, among other things, the narrow area of land used [2].

[3] The economic development of the agricultural sector has the following roles:

1. Agriculture is the backbone of the economic development process and functions as an equalization effort from all its aspects in accordance with historical factors and opportunities for development.
2. Agricultural development supports people's businesses in the field of cultivation and processing technology as well as services in the processing and marketing of their products.
3. Agricultural development is a support capable of accommodating the entrepreneurial development of farmers in a rational direction

The importance of agricultural sector development, especially farmers. The agricultural sector as a macro livelihood sector, namely farmers must be able to create prosperity, economic growth and new job opportunities. This success mainly depends on the opportunity for farmers to implement new methods and sell their produce at a fair price.

Most of the income they earn comes from cultivated agricultural products, but sometimes this income is not sufficient for their living needs, so it needs intensive handling of agricultural management so that the income they get can reach maximum results. The income of farmers is very dependent on the success of the plants they cultivate, so this research is entitled "The Effect of Agricultural Diversification on Community Income in East Taniwel District".

The area of East Taniwel District is in Seram Island Region, where in the north it is bordered by the Seram Sea, in the south by Inamosol and Elpaputih Districts, in the east by Central Maluku Regency, and in the west by Taniwel District [4]. In the District of East Taniwel, there are villages that are in lagging condition (80%) and around 20% are developing. There are some leading sectors that became one of the factors supporting the increase of economic activities in the province of Maluku. The rate of economic growth is an indicator of the success of the development of an area that can be seen in the GDP and per capita income, [5].

2. Literature review

2.1. Definition of agriculture

Agriculture is the activity of utilizing biological resources carried out by humans to produce food, industrial raw materials or energy sources, as well as to manage the environment, [6]. Agriculture is divided into two, namely smallholder agriculture and agricultural enterprises.

a. Smallholder farming is a family farming business where the main foodstuffs are produced, such as rice, secondary crops (corn, beans and tubers) and horticultural gardens, namely vegetables and fruits. Farming for the people is cultivated in rice fields, fields and yards.
b. Agricultural companies are agricultural enterprises that produce certain products with a uniform agricultural system under a centralized management system using various scientific methods and efficient processing techniques to obtain the maximum profit.

Farming is an organization of nature (land), labor and capital aimed at production in the agricultural field. The organization stands alone and is deliberately managed by a person or group of people as its manager. [7] Mention the types of agriculture, namely:

a. Traditional Agriculture (Subsystem)

Traditional agricultural production, if agricultural production and consumption are the same and only one or two kinds of plants (rice or maize) are the main source of foodstuffs. Low production and productivity due to the use of very simple equipment. The use of capital is small, while land and human labor are the dominant factors of production.

b. Traditional to Modern Agriculture

Agricultural diversification is a major logical step in the transition from traditional agriculture to modern agriculture. Staple crops no longer dominate agricultural production, because new cash crops such as; fruits, coffee, tea and others have started to be run together with simple livestock businesses. The use of simple tools such as small tractors and plow-pulling animals can be used to increase agricultural productivity. The use of superior seeds, fertilizers and good irrigation also increases agricultural production, so that farmers can obtain a better production surplus so that they can be sold to the market.

Agricultural diversification will minimize the impact of the failure to harvest staple crops and guarantee income security that never existed before.

c. Modern Agriculture
Modern agriculture or known as specialized agriculture describes the most advanced level of agriculture. This situation can be seen in more developed countries. Modern farms can vary in size and function. Nearly all types of wheat and maize farming in North America use highly labor-efficient mechanical equipment, from the largest types of tractors and modern types of harvesters to air-spraying techniques that allow a family to cultivate and plant thousands of hectare's farmlands.

2.2. Agricultural diversification

Agricultural diversification is an effort to diversify the types of business or agricultural crops to avoid dependence on one agricultural product. Agricultural diversification can be done in two ways, namely:

1. Increase the types of agricultural activities, for example, besides farming, farmers also raise chickens and raise fish.
2. Increasing the types of crops in a certain area, for example, in addition to planting corn, fields are also planted.

[8] Diversified agriculture is also known as mixed farming. Food availability is one of the subsystems in food security besides distribution and consumption. Therefore, it must be anticipated that availability is never less than demand.

2.3. Income

Income is the result of work or effort received by someone in the form of money or goods. Income is the result in the form of money or other material, which is achieved from the use of wealth or human services, [9].

[10] Income becomes two definitions, namely:

1. Income in the form of money, namely: Salary or wages obtained from main work, part time work, overtime work and occasional work. Own business which includes net proceeds, commissions on sales and home crafts and from investors' proceeds, namely income obtained from the rights of the land owner.
2. Income in the form of goods, namely in the form of: Payment of wages and salaries in the form of rice, medical treatment, transportation and housing, goods produced and commissions at home.

[10] Income, namely: "Revenue is an increase in assets resulting in an increase in owner's equity, but not because of the increase in new capital from the owner and not an increase in assets due to increased liabilities" explains that an increase in assets can be called revenue if the increase in assets is derived from contra performance received by the company for services provided to other parties. The increase or increase in assets will result in an increase in the owner's equity.

3. Methodology

3.1 Location and data

The data used in this study are primary data from eight villages in East Taniwel District, West Seram District, Maluku Province. This data is the result of a baseline survey at the household level, where the sample is determined using the Slovin formula

$$n = \frac{N}{1 + Ne^2}$$

Where n is the number of sample villages, N is the number of villages, and e is the tolerance limit (5%). The number of respondents sampled were 116 household heads spread across 8 villages, namely Makububui Village, Solea Village, Hatunuru Village, Maloang Village, Sohuwe Village, Uwen Pante Village, Tounusa Village, and Walakone Village.

3.2 Models and data analysis

The method used in this research is multiple linear regression modeling method. In general, multiple linear regression models are defined as follows:

$$Y_i = \beta_0 + \beta_1 X_{i1} + \beta_2 X_{i2} + \cdots + \beta_k X_{ik} + \varepsilon_i$$

with:

Y_i is the value of the dependent variable, the *to-i* observation;
β_0 and β_1 ... βk are the parameters of the regression coefficient;
X_{ik} is the value of the k-th independent variable in the *to-i* observation;
ε_i is the random error value assuming that

$$\varepsilon_i \sim N(0, \sigma^2)$$

which means that the mean E (εi) = 0 and the variance $Var(\varepsilon_i) = \sigma^2$; *εi and εj* are not correlated so that the covariance value $Cov(\varepsilon_i, \varepsilon_j) = 0$ for all values of i and j, i ≠ j, i, j = 1,2, ..., n.

3.3. Procedure

1. Determine the dependent variable and the independent variable in the model.
2. Preprocessing data, the aim is to determine the characteristics of the data (variables) used in the study.
3. Build a linear regression model by estimating parameters using the Ordinary Least Square (OLS) method.
4. Selecting significant independent variables in the model (backward elimination)
   This process is usually carried out using the t test, F test
5. Examination of the goodness of the model uses the coefficient of determination
6. Perform model diagnostic checks.
   This stage is often called the classic assumption test (from the ordinary least square / OLS estimation method or least squares method), which is checking for the absence of multi collinearity between independent variables, then checking assumptions on model residuals such as normally distributed residuals, no serial correlation to the residuals, and residual variance that is constant (homoscedasticity).
7. Model interpretation.

4. Results and discussion

4.1. Data characteristic

The sample in this study were household heads in 8 villages in East Taniwel District, Maluku Tengah Regency, Maluku Province. To facilitate the analysis, the eight villages are structured as follows: Village 1 is Makububui Village; Village 2 is Solea Village; Village 3 is Hatunuru Village; Village 4 is Maloang Village; Village 5 is Sohuwe Village; Village 6 is the village of Uwen Pante; Village 7 is Tounusa Village; and Village 8 is Walakone Village. Visually, the conditions of income and agricultural diversification in the eight villages are shown in Figures 1 and 2, where Figures 1 - 8 on the (horizontal) axis shown in both figures are a symbol of Village 1 - Village 8.

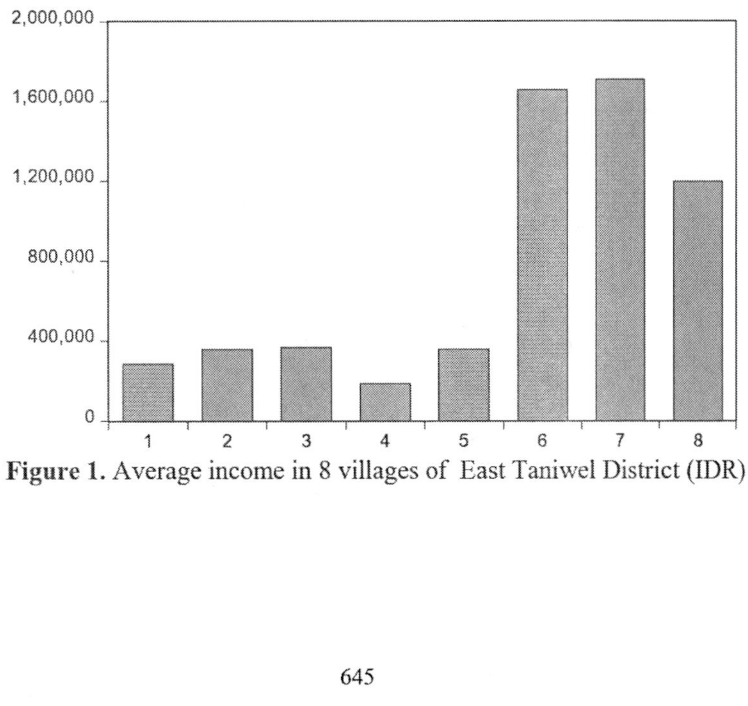

Figure 1. Average income in 8 villages of East Taniwel District (IDR)

IOP Conf. Series: Earth and Environmental Science **883** (2021) 012089 doi:10.1088/1755-1315/883/1/012089

Figure 1 shows that the income condition in Village 1 - Village 5 in 1 sales period is very small, namely less than (IDR) 400,000, while the average income in Village 6 - Village 8 is more than (IDR) 1,000,000. The lowest average income was found in Village 4 (Maloang village), while the highest was in Village 7 (Tounussa village). Statistical figures on the average income in the eight villages are shown in Table 1.

Agricultural diversification was carried out by the heads of households in the eight villages. This business includes the use of available agricultural land. One of the efforts made by the head of the household (community) in the eight villages is the diversification of food plants, namely increasing the types of plants on the available land. Types of cultivated crops are tubers, coconut, oil palm, nutmeg, cloves, cocoa, and bananas. Visually, the agricultural diversification undertaken by the eight villages is shown in Figure 2.

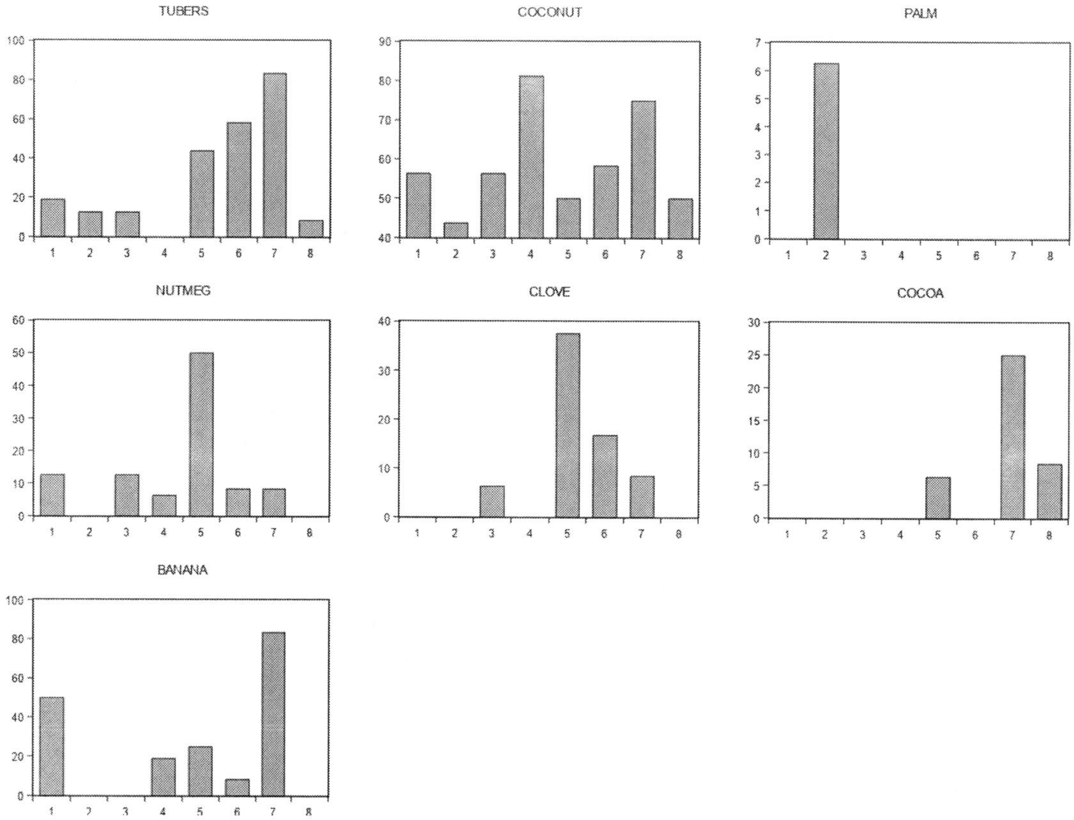

Figure 2. Agricultural diversification in East Taniwel District (percent)

Figure 2 shows that the plant found in all villages is coconut, while the oil palm plant is only found in 1 village, Solea Village. Villages that have a diversity of plants are Tounussa Village and Sohuwe Village. The statistical description of the seven variables is shown in Table 1. Based on the results of graphic analysis, Tounussa Village is the village that has the most diversity of plants and is also the village with the highest average income in East Taniwel District. On the other hand, Maloang Village is the village that has the lowest average income and has relatively little plant diversity. Based on this analysis, there is a relationship between agricultural diversification and income in the East Taniwel District.

Table 1. Descriptive statistics for each variable

	Variable							
	Dependent	Independent						
	INCOME	TUBERS	COCONUT	PALM	NUTMEG	CLOVE	COCOA	BANANA
Mean	764,240.8	29.69	58.85	0.78	12.24	8.59	4.95	23.18
Maximum	1,710,000.0	83.33	81.25	6.25	50.00	37.50	25.00	83.33
Minimum	184,078.1	0.00	43.75	0.00	0.00	0.00	0.00	0.00
Std. Dev.	647,049.1	29.14	12.88	2.21	16.00	13.10	8.76	29.74
Obs.	8	8	8	8	8	8	8	8

4.2. Analysis of the effect of agricultural diversification on income

Analysis of the effect of agricultural diversification on income uses a statistical approach, namely applying multiple linear regression methods to model data. The dependent variable used is average income (INCOME), while the independent variable is the percentage of household heads who plant tubers, coconut, oil palm, nutmeg, cloves, cocoa and banana (the notation is adjusted to the type of plant). The estimation results of the linear regression model are shown in Table 2.

Table 2. Estimated model parameters and VIF value

Variable	Coefficient	Std. Error	t-Statistic	Prob.	VIF
C	729613.9	169533	4.303669	0.0126	
NUTMEG	-82290.9	17879	-4.602658	0.01	5.456159
CLOVE	93000	21644.25	4.297414	0.0127	5.362388
BANANA	10459.99	4212.143	2.483294	0.068	1.046502

Table 3. Simultaneous test and model determination coefficient

	F-statistic	Prob (F-statistic)	R-squared	Adjusted R-squared
Model	7.971382	0.03661	0.856703	0.749231

Table 2 is the multiple linear regression model obtained based on the backward elimination results. These results indicate that there are only 3 independent variables that significantly affect the average income (INCOME), namely NUTMEG, CLOVE, and BANANA. Thus, other crops did not significantly affect INCOME. Based on Table 2, NUTMEG and CLOVE significantly affect INCOME at the significance level $\alpha = 5\%$, while BANANA significantly affects INCOME at the significance level $\alpha = 10\%$. This is supported by the simultaneous test results using the F test in Table 3 which shows that there is at least 1 independent variable in the model that affects the INCOME variable at the significance level $\alpha = 5\%$. The R-squared and adjusted R-squared values obtained are more than 50% and close to 100%. This shows that the model obtained is good. Based on the adjusted R-squared value, the INCOME variable is influenced by the NUTMEG, CLOVE, and BANANA variables by 75%, while 25% is influenced by other factors outside the model.

Table 4. Residual analysis

	Normality (Jarque-Bera Test)		Serial Correlation (Breusch-Godfrey LM Test)		Heteroscedasticity (Breusch-Pagan-Godfrey Test)	
	JB Stats.	Prob.	BGLM Stats.	Prob.	BPG Stats.	Prob.
Model	0.233	0.89	1.076	0.58	0.307	0.31

The results of the multi collinearity assumption test on the model show that the VIF value of each coefficient is less than 10 (Table 2). This shows that there is no correlation between the coefficients in the model (multi collinearity does not occur). The results of the assumption test on the residual model are shown in Table 4. The table shows that the residuals are normally distributed, because the probability value of JB stats is more than α = 5%. In addition, there is no serial correlation between residuals (no auto relation occurs), because the BGLM Stats probability value is more than α = 5%. Table 4 also shows that the residual model is free from heteroscedasticity, because the probability value of BPG stats is more than α = 5%. Based on the results of the model assumption test, the model obtained is an appropriate and good model in describing the characteristics of the data.

4.3. Model interpretation

The results discussed in this section are based on the model coefficients obtained in Table 2. Based on Table 2, the NUTMEG coefficient is negative. This means that, if the number of household heads who plant nutmeg has increased, it can reduce the average income. This shows that there is a cost burden experienced by the head of the household who grows the nutmeg plant in the East Taniwel District, so that it can cause losses. These costs can be in the form of operational costs for planting, harvesting, production, and even sales. On the other hand, the CLOVE and BANANA coefficients are positive. This shows that planting cloves and bananas can increase the average income for household heads in East Taniwel District. Based on Table 2, planting clove plants provides greater benefits than banana plants.

5. Conclusion

Agricultural diversification has been carried out by the community in East Taniwel District with the aim of increasing household income. The process of agricultural diversification carried out by the village community in East Taniwel District is an effort to plant several types of plants which are the main commodity. Based on the results of data analysis, there is a relationship between agricultural diversification and income. This is supported by the model obtained based on the backward elimination method in multiple linear regression modeling based on OLS. The model is an appropriate model and is good at describing the characteristics of the data. In the regression model, there are 3 types of plants planted by the community in East Taniwel District which significantly affect income, namely nutmeg, cloves, and bananas. Based on the coefficient of determination, income can be influenced by the cultivation of nutmeg, cloves and bananas by 75%, while 25% by other factors outside the model. These results indicate that the recommended plants for planting in the East Taniwel District are cloves and bananas, because they can increase their income compared to nutmeg.

6. Acknowledgment

Thanks to the International Fund for Agricultural Development (IFAD) Regional Director for Asia-Pacific: Ivan Cossio Cortez and the International Fund for Agricultural Development (IFAD) Indonesia Program Manager: Anissa and also thanks to the Ministry of Villages, PDT, Transmigration especially the Director of Development Leading Products for Villages, Disadvantaged Areas, and Transmigration Ir. Leroy Samy Uguy, MA, Ph.D. along with the team that has entrusted us to carry out a survey of Integrated Village Transformation in West Seram Regency and Central Maluku Regency in Maluku Province. We also thank the chancellor of the Pattimura university who have entrusted us to conduct studies and input for future development.

References

[1] Rukmana R 1997 Kacang Hijau, Budidaya dan Paska Panen. Penerbit Kanisius, Cet.I. Yogyakarta.
[2] Suhartini T 2012 Hubungan Faktor Sosial Ekonomi dengan Produksi Padi di Kecamatan Tabunganen Kabupaten Barito Kuala. Tugas Akhir tiak diterbitkan, Banjarmasin: Pendidikan Geografi FKIP Unlam Banjarmasin.
[3] Soeratno 1986 Ekonomi Pertanian. Karunika Jakarta Universitas Terbuka. Jakarta
[4] Badan Pusat Statistik (BPS) 2018 Seram Bagian Barat Dalam Angka, 2018.
[5] Kembauw E, Sahusilawane A M, Sinay L J, and Batlolona J R 2018 The Role of the Economic Sector Growth in the Province Maluku *Journal of Advanced Research in Law and Economics*, IX **6**:36 2035-2041

IOP Conf. Series: Earth and Environmental Science **883** (2021) 012089 doi:10.1088/1755-1315/883/1/012089

[6] Firdaus, Rachmat H 2009 Manajemen Perkreditan Bank Umum-Teori, Masalah, Kebijakan, dan Aplikasinya-Lengkap dengan Analisis Kredit. Alfabeta, Bandung.
[7] Soekartawi 2010 Agribisnis Teori dan Aplikasinya. Jakarta: PT. Rajawali Persada.
[8] Handayani E A S 2005 Usaha Diversifikasi Dalam MeningkatkanVolume Penjualan Produk Olahan Apel. Skripsi. Malang: Fakultas PertanianUniversitas Brawijaya.
[9] Winardi 1986 Evaluasi Kerja Sumber Daya Manusia. Jakarta: Salemba Empat.
[10] Rifani A 2013 Pengaruh Good Corporate Governance terhadap Hubungan Manajemen Laba dan Kualitas Laba. Skripsi Sarjana Program Studi Akuntansi pada Fakultas Ekonomi Universitas Negeri Padang, Padang.

Moringa oleifera milk powder as a supplementary food for malnutrition children (SUSUKE)

S W Djoko[1], E Kembauw[2], I B D Kapelle[3]

[1] Faculty of Medical, Pattimura University, Ambon, 97232
[2] Faculty of Agriculture, Pattimura University, Ambon, 97232
[3] Faculty of Mathematics and Natural Sciences, Pattimura University, Ambon, 97232

E-mail: wahyunisd2017@gmail.com; ekembauw@yahoo.co.id;
berly_molluca@yahoo.com

Abstract. Moringa oleifera are known throughout the world as a nutrition plants, WHO has introduced moringa as an alternative food to overcome malnutrition. In Africa and Asia Moringa leaves are recommended as a nutrient-rich supplements for nursing mothers and children during infancy phase. Moluccas is one of the top 5 regions with quite high rates of malnutrition, while rich in natural products, the lack of knowledge and poor processing method results in nutrients unable to be properly absorbed. Moringa milk-powder can be the answers to malnutrition problems. Methods: This Study used a Nested case-control design , subject were all children aged 1 month-12 years who were diagnosed with malnutrition using WHO and CDC chart during reseach. Result: Sample size was 60 children's 30 subjects are using malnutrition treatment and moringa milk powder and 30 subjects are using only malnutrition treatment. Conclusion: Moringa milk powder have a benefit for malnutrition children.

1. Introduction

Moringa oleifera L. is a tropical plant that is easy to grow in Indonesia. Moringa plants are shrubs with an altitude of 7-11 m and thrive from the lowlands to an altitude of 700 m above sea level (asl). Moringa plants can grow on all types of soil and are resistant to dry seasons with drought tolerance of up to 6 months (Mendieta et al., 2013) [1]. WHO has introduced Moringa as an alternative food to overcome malnutrition (Broin, 2010) [2]. In Africa and Asia Moringa leaves are recommended as a nutrient-rich supplement for nursing mothers and children during their infancy. One part that can be used from the moringa plant is the leaves. Moringa leaves are very rich in nutrients, including calcium, potassium, iron, vitamin A, vitamin C, and vitamin B (Misra and Misra, 2014 [3]; Oluduro, 2012 [4]; Ramachandran et al., 1980 [5]; Krisnadi, 2015 [6]). Moringa leaves contain higher iron than other vegetables, namely 17.2 mg / 100 g (Yameogo et al., 2011) [7]. The potassium content in Moringa leaves is 259 mg / 100 g (Bey, 2010) [8]. People rarely use plants directly for treatment, so people do not recognize plants that are beneficial for health. Therefore, plants with medicinal properties that are around the community need to be explored and developed. One of them is the Moringa oleifera Lam. Plant, which is rich in benefits. Malnutrition remains a difficult problem to face in Indonesia. WHO data states that Indonesia ranks in the top ten children with the most malnutrition, so the issue of malnutrition is still a priority for handling in Indonesia. The result of the joint agreement between Indonesia and 188 other UN countries in the Millennium Development Goals (MDGs) declared the Millennium Declaration which is committed to eradicating poverty and prospering the people. One of the 8 points is to reduce child mortality, and reduce poverty and hunger. Until now, many Indonesian people still live below the poverty line and still lack nutrition. Maluku is one of the top 5 regions with quite high rates of malnutrition. While Maluku is rich in natural products, the lack of knowledge and poor processing methods results in nutrients unable to be properly absorbed. Moringa leaves, which thrive on the mainland of Maluku, can be one of the answers to problems that often occur. The purpose of this study was to determine the effectiveness of Moringa leaves as a powdered milk product which is used as a supplementary food for malnourished children.

2. Methods

The research design used a nested case-control design. Malnourished children aged 1 month to 12 years as evidenced by anthropometric examinations in the form of stunting, stunted, and wasted. Anthropometric status data were taken first, then entered in the sample, in the data, and made the research subject. The children were then divided into two groups, namely the control group using WHO malnutrition management, and the same case group using the WHO malnutrition management group and added by giving Moringa milk powder 4 times a day alternating with other malnutrition therapies. For three months followed and continued anthropometric measurements every month. At the end of the third month, see whether there is a difference between these cases and controls. The minimum sample in each group is 30 people. So that the minimum total sample required is 60 people.

3. Results and discussion

Sample size was 60 children's 30 subjects are using malnutrition treatment and moringa milk powder and 30 subjects are using only malnutrition treatment. In this study, using concoctions that took data only three months, so this study did not evaluate the data when the patient experienced outcome. In this study, did not see primary disease as one of the factors causing malnutrition.

4. Conclusion

One of the benefits of Moringa milk is that it helps increase weight in children with poor nutrition. These benefits can be obtained by consuming Moringa powder for two months regularly.

Acknowledgements

We also thank the chancellor of the university who have entrusted us to conduct studies and input for future development.

References

[1] Mendieta A B, Spörndly E R, Sánchez N, Salmerón M F, Halling M 2013 Biomass production and chemical composition of Moringa oleifera under different planting densities and levels of nitrogen fertilization *Agroforest. Syst.* **87** 81-92
[2] Broin 2010 *Growing and processing moringa leaves* France: Imprimerie Horizon
[3] Misra S dan Misra M K 2014 Nutritional Evaluation of Some Leafy Vegetable Used by the Tribal and Rural People of South Odisha, India *Journal of Natural Product and Plant Resources* **4** 23-28
[4] Oluduro A A 2012 Evaluation of Antimicrobial Properties and Nutritional Potentials of Moringa oleifera Lam. Leaf in South-Western Nigeria *Malaysian J. Microbiol* **8** 59-67
[5] Ramachandran C, Peter K V, Gopalakrishnan P K 1980 Drumstick (*Moringa oleifera*): a Multipurpose Indian Vegetable *J. Econ. Bot* **34** 276-283
[6] Krisnadi A D 2015 Kelor Super Nutrisi. Blora : Pusat Informasi dan Pengembangan Tanaman Kelor Indonesia Lembaga Swadaya MasyarakatMedia Peduli Lingkungan (LSM-MEPELING) [Tersedia di: http://kelorina.com/ebook.pdf] [Diakses pada tanggal 28 Agustus 2018].
[7] Yameogo, W. C., Bengaly, D. M., Savadogo, A., Nikiema, P. A., Traore, S. A. 2011. Determination of Chemical Composition and Nutritional Values of Moringa oleifere Leaves. Pakistan Journal of Nutrition, 10 (3), 264-268.

[8] Bey, H. 2010. All Things Moringa. [Online]. www.allthingsmoringa.com. [Diakses pada tanggal 11 September 2018].

IOP Conf. Series: Earth and Environmental Science **883** (2021) 012091 doi:10.1088/1755-1315/883/1/012091

Clove processing as a source of increasing business income in Ambon City

E Kembauw[1]*, A S Mahulette[2], A P Kakisina[3], M T F Tuhumury[1], M C B Umanailo[4], M G I Kembauw[3]

[1]Department Social Economic, Faculty of Agriculture, Pattimura University, Ambon Indonesia
[2]Agrotechnology Study Program, Faculty of Agriculture, Pattimura University, Ambon Indonesia
[3]Center for Training and Empowerment of Village Communities, Disadvantaged Regions and Transmigration Ambon, Ministry of Villages, Development of Disadvantaged Regions and Transmigration Indonesia
[4]Department of Agricultural and Forestry, University of Iqra Buru Indonesia

E-mail: ekembauw@yahoo.co.id

Abstract. Clove is one of the important plantation commodities in Maluku, especially in Ambon City. Clove processing diversification will provide opportunities to increase business income. This study aims to determine the factors that support the diversification of clove processing on the income of entrepreneurs in the city of Ambon. The method used was purposive sampling. The results showed that the diversification of clove processing gave a large enough contribution to income, but the percentage contribution of processing diversification was not visible because the income outside of processing was relatively large. The results showed that the greater the number of species produced, the greater the capital. Based on the analysis, it is known that the level of raw materials available to entrepreneurs to support diversification activities.

1. Introduction

Clove is a native Indonesian plant originating from the Maluku Islands, Indonesia [1,2] and is one of the most important plantation crop commodities because 99% is cultivated by farmers. According to data [3], Maluku is the largest clove-producing province in Indonesia from 2015 with an average contribution of 15.37%. According to BPS Maluku data [4], the area of clove plantations in Maluku in 2016 was recorded at 43 620.3 ha then increased to 43 780.1 in 2017 with total production increasing from 20 805.6 tons in 2016 to 21 159.6 in 2017. not as wide as other provinces in Indonesia, but clove productivity in Maluku is higher which affects the amount of production. One of the causes of the high productivity of cloves in Maluku is the high diversity of superior clove germplasm and is supported by the suitability of the agro-climate.

The main product of cloves is flowers, which are widely used as raw materials for the cigarette and spice industry. Another product is in the form of essential oils which can be distilled from flowers, flower stalks, and clove leaves [5–9]. Most of the clove production in Indonesia is used to meet domestic needs as raw material for cigarettes and less than 10% is used for other needs such as cooking spices, medicinal ingredients, food preservatives and others [2,9–17].

Clove product development in the future is directed at diversifying cloves into a variety of products that can provide added value economically. Although clove productivity in Maluku is quite high, its management into a commercial product is still very limited. Clove sales so far have been limited to dry flower buds, while processing in the form of essential oils is also still on a small scale and is still traditional [5,6,8,18–24]. The development of clove derivative products into various commercial products is also almost never done so that it has an impact on the welfare of clove farmers in Maluku, especially in Ambon City. The availability of clove raw materials for product diversification will provide opportunities to increase clove farmers' income in the future [25–31]. Therefore, this research needs to be carried out to evaluate the potential of clove processing diversification as a source of increasing clove business income in the city of Ambon, Maluku, Indonesia.

2. Methods

Diversification research on clove processing was carried out in several clove business units in Ambon City, Maluku, Indonesia. The location was chosen purposively because Ambon City is one of the clove sales centres in Maluku. The types of data used in this study are primary data and secondary data. Primary data was obtained through direct observation, recording, and direct interviews with 25 clove business actors to determine the diversification process of clove products. Secondary data include data on clove production, product prices, costs incurred during the diversification process, the amount of production obtained during the production cycle period and other supporting data. In addition, secondary data were obtained from the Central Statistics Agency (BPS), the Ministry of Agriculture, and relevant literature.

Data collection methods were carried out by means of observation, interviews, and discussions as well as filling out questionnaires to the respondents, namely 25 farmers who are members of the clove business group in Ambon City. The respondent sampling process was carried out by purposive sampling method, with the consideration that the respondents who were used as samples were considered to have the information needed by the researcher.

Data analysis conducted in this study consisted of qualitative analysis and quantitative analysis. Qualitative analysis was carried out through a descriptive approach. Qualitative analysis was conducted to find out the general description of the business unit, production process, and management applied by the business unit. Meanwhile, quantitative analysis was conducted to analyze the production of business unit diversification activities.

3. Results and discussion

3.1. An overview of the clove industry in Ambon City

Maluku Province is the largest clove producer in Indonesia from 2015 with an average contribution of 15.37% (Ditjenbun 2019). Although the area of development in this province is not as wide as in other provinces, it has high productivity which affects the amount of production (Table 1). One of the factors that determine the amount of production is the large number of superior clove varieties in the Maluku Province.

Ambon City as one of the centers of clove production in Maluku, has clove production reaching 404 tons of the total clove production in Maluku. Clove production in Maluku also comes from other regions, namely Eastern Seram Regency (4 763 tons), Central Maluku Regency (9 215 tons), South Buru Regency (2 226 tons), Western Seram Regency (2 879 tons), Buru Regency (412 tons), Southwest Maluku Regency (49 tons), Southeast Maluku Regency (4 tons), and Tual (9 tons).

Table 1. The area and Production of Smallholder Cloves in Maluku by Regencies and Crop Condition in 2018

No.	Regency/City	TBM *Immature*	TM *Mature*	TTM/TR *Damaged*	*Total*	Product (ton)	Productivity (kg/ha)	No of Farmer (KK)
1.	Southeast west Moluccas	-	-	-	-	-	-	-
2.	Southeast Moluccas	7	10	4	20	4	445	81
3.	Centre Moluccas	2 837	14 550	1 360	18 746	9 215	633	33 035
4.	Buru island	230	952	45	1 226	412	433	6 935
5.	Aru Island	0	-	-	0	-	-	3
6.	Western Seram	1 385	5 230	501	7 116	2 879	551	13 819
7.	Eastern Seram	1 985	6 802	1 196	9 983	4 763	700	10.275

8.	Southwest Moluccas	70	130	18	218	49	380	367
9.	South Buru	1 735	3 737	123	5 596	2 266	606	5 541
10.	Ambon city	381	899	221	1 501	404	449	2 517
11.	Tual city	11	17	0	28	9	521	195

Source: Ditjenbun (2019)

3.2. Product and market diversification
Based on the results of the study, the main product of cloves in Ambon City has so far only been used in the form of selling dry flower buds. Processing in the form of essential oils is still limited and traditional [8,32,33]. Sales in the form of other derivative products are also almost non-existent. Clove flower production purchased from farmers is usually sold immediately after being dried without going through a diversification process to become other derivative products. Clove flowers are usually purchased by large collectors and then sent to large companies in Surabaya via the Yos Sudarso sea port, Ambon.

The selling price of aromatic clove flowers in Ambon City in 2018 was in the range of Rp. 85 000 per kilogram, where from the sale of 1 clove tree, only Rp. 800 000-1 000 000 per tree. If the diversification of the clove flower into essential oil, an even greater profit can be obtained, namely Rp. 1 500 000 – 2 000 000 per tree [6–8]. Profits from the production of other processed cloves are expected to increase again if diversification is carried out into various other commercial products.

3.3. Description of the main actors in the clove value chain in Ambon City
3.3.1. Farmers
Clove farmers in Ambon City carry out activities on cultivated land (on farm), starting from planting, maintaining to harvesting. The average land ownership of farmers is 2-3 hectares. Planting new plants still uses seedlings that grow under the stands and rarely through seedlings first. The maintenance of the garden is only limited to cleaning the nuisance plants that grow in the farmers' garden area. Garden sanitation is usually carried out simultaneously with the initiation of clove flowers.

Clove harvesting is done by climbing up by the nuclear family (father and son) or hired labor to reap the results. Female family members usually collect the harvest under the tree while helping to separate (remove) the clove flowers from the stems [34–39]. Clove threshing usually continues at night with the help of other relatives. Farmers whose plantations are located far away and need money urgently usually sell their harvested raw cloves directly to collectors in the village who visit the farmers' gardens.

3.3.2. Collectors at village level
Collectors at the village level buy clove flowers from farmers that are still on the tree or that have been knocked down in the form of wet flowers or dried flowers. For the purchase of clove flowers that are still on the tree, traders usually hire workers to do the picking and threshing. The results of the purchase of clove flowers by farmers are then sold to large collectors in the center of Ambon city.

3.3.3. Big traders in the center of Ambon City
Large traders in the center of Ambon City buy cloves which are sold by farmers and by collectors in the village. The collectors bought cloves on a large scale. The results of the cloves are then processed through cleanliness checks, weighing, packaging, storage and shipping through the Yos Sudarso sea port, Ambon to companies that use clove raw materials in Surabaya.

3.3.4. Supporting actor
The success of strengthening the clove commodity value chain in Ambon City is determined by the existence of access to information or knowledge, technology and finance as well as other supporting services. Conditions of supporting actors in the clove value chain in Ambon City are as follows:

➢ *Financial Aspect*

Farmers do not need large capital for clove cultivation. Farmers' access to sources of financing in Ambon City is wide open, but the majority of access to banking by farmers is only used for educational and consumptive funding needs.

➢ *Information Aspect*

Farmers' information regarding the selling price of cloves in the market is very limited and not much is known by farmers. Traders so far have more control over clove price information. Market needs (quantity and quality) are also unknown to farmers. This condition causes farmers to have no power over the price of cloves set by collecting traders.

➢ *Business development Services*

Clove Business Development Service (BDS) in Ambon city so far has not been available and is only carried out by local governments through community economic development programs.

➢ *Research Institutions*

Research institutions that specialize in handling spices in Ambon City have been established such as the Maluku Spice Council, however, special studies on cloves are still very limited.

3.3.5. Market Chain

The results of the identification of clove farmers in Ambon City, most of them stated that they had no problems in marketing their clove harvest. Farmers' harvests are usually bought directly in the garden by collectors in the village or sold directly to large collectors at the center in Ambon City. The thing that farmers feel the most in the clove marketing system is that the selling price is completely determined by the collecting traders. Farmers have a very weak position in the clove marketing chain. Price formation rests entirely with middlemen in the village and wholesalers in the center of Ambon City.

The clove commodity marketing chain system in Ambon City has 2 chains. The first chain is from farmers to collectors in the village and then to large traders in the center of Ambon City. The second marketing chain is from farmers to large collectors in the center of Ambon City. From the large collectors in the two chains, the cloves are then sold to companies (manufactures) that use cloves as raw materials located in Surabaya. The two links are presented in Figure 1.

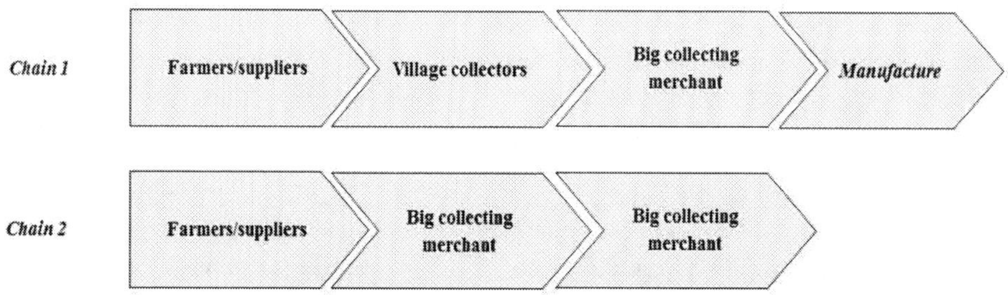

Figure 1. Clove Market Chain in Ambon City

4. Conclusion

Clove product diversification in business units in Ambon City is still very limited, where clove sales are still limited in the form of dry flower buds. Clove processing into products with economic value has the potential to increase the income of clove business actors in Ambon City.

References

[1] Milind P and Deepa K 2011 Clove: A champion spice *International Journal of Research in Ayurveda & Pharmacy* **2** 47–54

[2] Alfian A, Mahulette A S, Zainal M, Hardin and Bahrun A H 2019 Morphological character of raja clove (Syzygium aromaticum L . Merr & Perry .) native from Ambon Island *The First International Conference of Interdisiplinary Research on Green Environment Approach for Sustainable Development (ICROEST 2019)* (Makassar: IOP Conference Series: Earth and Environmental Science) pp 1–4

[3] Ditjenbun 2020 *Statistik Perkebunan Indonesia 2018-2020 (Cengkeh)* (Jakarta (ID): Kementerian Pertanian. 70 hlm.)

[4] BPS Maluku 2018 *Provinsi Maluku Dalam Angka* (BPS Provinsi Maluku. 745 hlm)

[5] Mahulette A S, Hariyadi, Yahya S and Wachjar A 2020 Physico-chemical properties of clove oil from three forest clove accession groups in Maluku *1st International Conference on Sustainable Plantation (1st ICSP 2019)* (Bogor: IOP Conf. Series: Earth and Environmental Science 418 (2020) 012028 IOP) pp 1–8

[6] Hariyadi, Mahulette A S, Yahya S and Wachjar A 2020 Morphological characters and essential oil constituents extracted of two clove varieties (Syzygium aromaticum (L.) Merr. & L. M. Perry.) from Ambon Island, Indonesia *Plant Archives* **20** 2208–14

[7] Mahulette A S, Yahya S and Wachjar A 2019 The physicochemical components and characteristic from essential oils of forest cloves Syzygium aromaticum (Myrtaceae) in Maluku Province, Indonesia *Plant Archives* **19** 466–72

[8] Hariyadi, Mahulette A S, Yahya S and Wachjar A 2020 Agro-morphologies and physicochemical properties of flower bud , stem and leaf oils in two clove varieties (Syzygium aromaticum L . Merr . and Perry .) originated from Ambon island *Chiang Mai University Journal of Natural Sciences* **19** 516–30

[9] Towaha J 2012 The benefits of cloves eugenol in various industries in Indonesia *Jurnal Perspektif* **11** 79–90

[10] Syukur C, Sukarman and Nurliani B 2016 Pengelompokan pohon induk cengkih terpilih di Kabupaten Sumedang berdasarkan karakter morfologi. *Jurnal Littri* **22** 29–36

[11] Hariyadi, Mahulette A S, Yahya S and Wachjar A 2019 Measuring the potential of biomass, carbon storage, and carbon sink of forest cloves *The 1st International Seminar on Natural Resources and Environmental Management (ISenREM 2019)* (Bogor: IOP Conf. Series: Earth and Environmental Science 399 (2019) 012063) pp 1–9

[12] Mahulette A S, Hariyadi, Yahya S, Wachjar A and Alfian A 2019 Morphological traits of Maluku native forest clove (Syzygium aromaticum L. Merr & Perry.) *Journal of Tropical Crop Science* **6** 105–11

[13] Mahulette A S, Hariyadi, Yahya S, Wachjar A and Marzuki I 2019 Morpho-agronomical diversity of forest clove in Moluccas, Indonesia *Hayati Journal of Biosciences* **26** 156–62

[14] Mahulette A S, Alfian A, Zainal M, Nendissa J I, Wattimena A Y, Tanasale V L, Makaruku M H and Laisina J K J 2020 Growth of forest clove seedlings at different concentrations of paclobutrazol *The 2nd International Conference of Interdisciplinary Research on Green Environmental Approach for Sustainable Development* (Makassar: IOP Conf. Series: Earth and Environmental Science 575 (2020) 012081) pp 1–7

[15] Gibriel A Y, Abdeldaiem M H and Ali H G M 2017 Antibacterial activity of clove (Syzigium aromaticum L.) essential oil and gamma irradiation against some food-borne pathogens in minced chicken meat *Arab Journal of Nuclear Sciences and Applications* **50** 179–93

[16] Uddin M A, Shahinuzzaman M, Sohel Rana M and Yaakob Z 2017 Study of chemical composition and medicinal properties of volatile oil from clove buds (Eugenia caryophyllus) *International Journal of Pharmaceutical Sciences and Research* **8** 895–9

[17] Hamed S F, Sadek Z and Edris A 2012 Antioxidant and antimicrobial activities of clove bud essential oil and eugenol nanoparticles in alcohol-free microemulsion *Journal of Oleo Science* **61** 641–8

[18] ossain M A, Al-hashmi R A, Weli A M, Al-riyami Q and Al- J N 2012 Constituents of the essential oil from different brands of Syzigium caryophyllatum L by gas chromatography-mass spectrometry *Asian Pacific Journal of Tropical Biomedicine* **2** 1446–9

[19] Bustaman S 2011 Potensi pengembangan minyak daun cengkih sebagai komoditas ekspor Maluku *Jurnal Litbang Pertanian* **30** 132–9

[20] Osman S M E, El-Obeid H A and Ayoub S M H 2012 Physical and chemical analysis of some imported essential oils in the Sudanese market *African Journal of Pharmacy and Pharmacology* **6** 1477–81

[21] Kaufman T S 2015 The multiple faces of Eugenol. A versatile starting material and building block for organic and bio-organic synthesis and a convenient precursor toward bio-based fine chemicals *Journal of the Brazilian Chemical Society* **26** 1055–85

[22] Gupta M, Singh D, Gularia P and Gupta S 2015 GCMS analysis and identifications of chemical constituents of Syzygium aromaticum , Brassica compestris and cow ghee *Journal of Chemical and Pharmaceutical Research* **7** 568–72
[23] Kamatou G P, Vermaak I and Viljoen A M 2012 Eugenol - From the remote Maluku islands to the international market place: A review of a remarkable and versatile molecule *Molecules* **17** 6953–81
[24] Tresniawati C and Randrian E 2011 Uji kekerabatan aksesi cengkeh di bebun Percobaan Sukapura *Buletin Plasma Nutfah* **17** 40–5
[25] Mardiana H and Kembauw E 2021 The Role of Diffusion of Innovation in Agricultural to Compete in Asean Community The Role of Diffusion of Innovation in Agricultural to Compete in Asean Community *IOP Conf. Series: Earth and Environmental Science 755 (2021) 012074* pp 1–13
[26] Pudjiastuti A Q and Kembauw E 2017 Sugar price policy and Indonesia's trade balance agnes *Journal of Advanced Research in Law and Economics DOI:* **VIII** 2540–8
[27] Putra P, Kembauw E, Sebayang A, Mukhlis H and Fauzi 2020 State owned enterprise for the creation of prosperity for all Indonesian *Journal of Critical Reviews* **7** 2032–6
[28] Ridjayanti R, Pattiasina M and Kembauw E 2020 Analisis produksi dan pendapatan petani pala (Myristica fragrans) di negeri Hila, Kecamatan Leihitu, Kabupaten Maluku Tengah *AGRILAN : Jurnal Agribisnis Kepulauan* **8** 244–60
[29] Umanailo M C B, Kembauw E, Mujib F, FaisAssagaf S S, Umanailo R, Meifilina A, Hentihu I, Kasmawati K, Hamiru H and Yusuf N 2018 Moral economics and survival strategies of the upland rice farmers *International Journal of Innovative Science and Research Technology* **3** 616–22
[30] Paley W B, Kembauw E and Tuhumury M T F 2021 Strategi pengembangan agroindustri kelapa ud wootay coconut di Kecamatan Teon Nila Serua Kabupaten Maluku Tengah *AGRILAN : Jurnal Agribisnis Kepulauan Pendahuluan* **9** 21–31
[31] Soegijono S P and Kembauw E 2021 Sustainable development of papalele's household from environmental issue *IOP Conf. Series: Earth and Environmental Science 755 (2021) 012026* pp 1–6
[32] Elim H I, Kembauw E, Siahainenia R H, Lamerkabel J S A, Sahusilawane A M, Djoko S W and Setha B 2019 Heavenly small islands simple technology & its cultural economy impacts in Maluku, Indonesia: a new proposed Multitasking Philosophy in Diversity (MPD) *Science Nature* **2** 192–207
[33] Sinay L J and Kembauw E 2021 Monthly rainfall components in Ambon City: evidence from the serious time analysis *IOP Conf. Series: Earth and Environmental Science 755 (2021) 012079* pp 1–8
[34] Sahusilawane A M and Kembauw E 2019 Gender role in farmer's household case study in Huaulu Village, North Seram Sub District, Central Maluku District, Maluku Province *Proceedings of the 1st Annual Conference on Education and Social Sciences (ACCESS 2019)* vol 465 pp 421–5
[35] Kembauw E, Sahusilawane A M and Sinay L J 2015 Sektor pertanian merupakan sektor unggulan terhadap pembangunan ekonomi Provinsi Maluku *Agriekonomika* **4** 210–20
[36] Kembauw E, Anindita R, Mustadjab M M and Muhaimin A W 2015 Agricultural sector investment impact of economy Province Maluku *Journal of Economics and Sustainable Development* **6** 31–7
[37] Kembauw E, Soekiman J F X S, Lydia E L, Shankar K and Huda M 2019 Benefits of corporate mentoring for business organization *Journal of Critical Reviews* **6** 101–6
[38] Miagina A, Biso H and Kembauw E 2021 Sustainable development through the One Village One Product (OVOP) approach for local commodities *IOP Conf. Series: Earth and Environmental Science 755 (2021) 012071* pp 1–7
[39] Lionardo A, Bugis M, Umanailo M C B, Kembauw E and Kurniawan R 2021 Labor social capital and human resource management for eucalyptus oil supply chain *Psychology and Education* **58** 1951–8

AUTHOR INDEX

Adam, F P ... 237
Adli, D N ... 79, 111, 417
Agisimanto, D ... 116
Aisah, A R ... 364
Aisyah ... 138
Akhdiya, A ... 44
Albasri ... 464
Alfian, A ... 9
Alvioliana, D ... 93
Anggraeni, I ... 50
Anggraeni, W ... 133
Anty, K ... 610
Apituley, R ... 642
Aponno, H S E S ... 642
Arif, A ... 464
Arini, N ... 93
Arisah, H ... 116
Asmi, Y ... 102
Asnur, P ... 138
Asyiah, I N ... 133
Augustyn, G H ... 169
Azmi, T K K ... 138
Barus, S ... 16
Basrudin ... 464
Bety, Y A ... 144
Boonraeng, S ... 622
Breemer, R ... 451, 588
Dalimunthe, M ... 355
Damanik, I P N ... 237
Darawati, M ... 50
Despal ... 68
Desrihastuti ... 106
Destiarni, R P ... 485
Dewi, K ... 526
Djoko, S W ... 650
Doloksaribu, T H ... 50
Dumais, J N K ... 27
Ega, La ... 642
Erawati, B T R ... 364
Ernita, M ... 281
Faisal, M ... 642
Fauzan, M ... 501
Febrianti ... 106
Ferdinandus, S J ... 642
Fitriatin, B N ... 133
Fransz, J J ... 121
Fredriksz, S ... 603
Gaffar, A ... 292
Gaswanto, R ... 408, 471
Girsang, W ... 385
Goo, N ... 102
Gusfarina, D S ... 244
Hadijah, M ... 507
Hadiwiyono ... 598
Haitami, A ... 106
Handoko, S ... 582
Harimurti, R ... 526
Harni, R ... 133
Hasinu, J V ... 177, 270
Hasudungan, P ... 428
Hehanussa, M L ... 183
Hehanussa, S Ch H ... 603
Hendrayani, H ... 535
Herawati, N ... 364
Herdiyantoro, D ... 370
Hidayah, B N ... 364
Hindersah, R ... 133, 478
Husna ... 464
Imlabla, W N ... 515
Indarwati, L D ... 336
Indradewa, D ... 229
Indrasti, R ... 292, 582
Irham, I ... 428
Irham ... 244
Iskar ... 22
Ismiasih ... 276
Jamhari ... 276
Jamil, A S ... 485
Jamilah ... 281
Jamlean, M ... 478
Jayanegara, A ... 111, 459
Jesajas, H ... 213
Kakisina, A P ... 653
Kakisina, L O ... 260, 588
Kalay, A M ... 1, 183, 478
Kamardiani, D R ... 501
Kanny, P I ... 138
Kapelle, I B D ... 650
Kaplale, E ... 588
Karjadi, A K ... 74
Karo, B ... 16
Kastanya, A ... 22, 560
Kaunang, R ... 27
Kembauw, E ... 9, 213, 317, 322, 642, 650, 653
Kembauw, M G I ... 653
Kesaulya, H ... 1, 9, 177, 183

Kesaulya, Henry 270
Khamdi, A 35
Kilkoda, A K 266
Kindangen, J G 292
Komul, Y D 56, 569
Kristiandi, K 50
Kunu, P J 438
Kurniasih, B 93, 153, 229, 336
Kurniasih, R 138
Kustanti, A 492
Kusumaningrum, S 526
Laimeheriwa, S 438
Laterissa, S 588
Latuamury, B 515
Latupapua, L 163, 629
Latupapua, Y Th 163, 629
Latupeirissa, Ch C E 603
Lawalata, I J 9
Lawalata, M 345
Lawalata, V N 169, 618
Lawalatta, I J 183
Leatemia, E D 345
Lestari, E S 128
Liur, I J 213, 255
Luhukay, J M 345, 588
Lumingkewas, J 27
Maail, R S 121
Mahardhika, B P 79, 86, 111
Mahulette, A S 9, 102, 106, 303, 317, 322, 653
Makaruku, M H 9, 106, 266, 317, 322
Marasabessy, D A 102
Marasabessy, H 507
Mardian, I 364
Mariati, W D 102, 106
Marpaung, A E 16
Matatula, E 183
Matatula, M J 213
Matinahoru, J M 577
Menap, M 535
Michala 102
Mukson 35
Mukti, R C 552
Murti, R H 153
Muspida 642
Mustafa, I 464
Mutia, R 79, 86
Nahrowi 459
Nasir, M A 276
Nendissa, D M 395
Nendissa, J I 106, 266
Nendissa, S J 183, 395
Ngabalin, I A 478
Nisa, R I 93
Nurfadilah, F S 370
Nurhayati, A 244
Nurjanah, S 203
Nurjanani 582
Nurlaeny, N 370
Osok, R M 438
Osok, R 1
Palijama, S 451, 588
Pangestuti, R 144, 153
Papilaya, B J 298
Pasambuna, M 50
Pasaribu, I A 355
Pattimahu, D V 22, 560
Pattinasarany, C K 163, 629
Pattiselanno, A E 260
Pelupessy, L. 56
Pelupessy, L 569
Perkasa, A Y 138
Permana, I G 68
Permanasari, I 229
Pesulima, Y 438
Picauly, P 421
Polnaya, F J 451, 588
Poromarto, Sh 598
Prasetya, R D D 459
Pratama, S N 552
Pratiwi, Arby'In 106
Pratiwi, W N 370
Prayantini, D C 128
Punyoyai, N 622
Purnamaningsih, R 544
Purwantoro, A 128
Putra, R M 370
Rahayu, A 408
Rahayu, S 544
Rahayu, W P 203
Rahmadani, M 459
Rajab 298
Ralahalu, T N 603
Ramdan, E P 138
Rapialdi 281
Ratuliami, N 370
Rawung, J B M 292, 582
Regar, E 642
Riansyah, K S 501
Ridla, M 79, 86
Rifqah, R A 102
Riry, J 9, 303
Riry, W A 303
Risnawati 138
Rizwan, M 355

Roessali, W ... 35
Rori, Y ... 27
Rosliani, R ... 408
Rosmalia, A ... 68
Ruhimat, I S ... 219
Sahupala, A ... 56, 569
Sahureka, M ... 507
Sahusilawane, A M ... 313
Sahusilawane, F ... 515
Sahusilawane, J F ... 313
Salamena, J F ... 298
Saleh, Khaerul ... 327
Sanduan, A ... 163, 629
Sangadji, M ... 642
Sanjaya, R A ... 44
Saptadi, D ... 116
Sarfan, R ... 298
Sastrawan, S ... 535
Satriawan, H ... 355
Seipala, B B ... 56, 569
Septya, F ... 485
Siahaya, L. ... 56
Siahaya, L ... 569
Siahaya, T E ... 56, 569
Sigalingging, R ... 401
Simanjorang, T M ... 379
Sinay, L J ... 642
Sipahelut, S G ... 169
Siregar, R ... 610
Sjofjan, O ... 111, 417
Sondang, Y ... 610
Sopamena, J F ... 260
Soselisa, F ... 163, 629
Sudolar, N R ... 582
Suhendy, C C V ... 22
Suherman ... 327
Suhesti, S ... 544
Sukmadjaja, D ... 544
Sulaiman, L ... 535
Sulastriningsih ... 128
Sulistyaningsih, E ... 128, 153, 229, 336, 526
Supriyadi ... 598
Supyani ... 598
Syahputra, R H ... 417
Tagueha, A D ... 213
Tahitu, M E ... 237
Talahaturuson, A ... 1, 183, 478
Talakua, S ... 438
Talaohu, M ... 515
Tanasale, V L ... 102, 266
Tapotubun, A M ... 303
Taribuka, J ... 266
Tetelay, F F ... 163, 629
Tetelepta, G ... 421
Thenu, S F W ... 593
Thenu, V J ... 451
Timisela, N R ... 345, 385, 588
Titarsole, J ... 121
Tiven, N C ... 379
Tuheteru, F D ... 464
Tuhumury, G N C ... 177, 270
Tuhumury, H C D ... 195
Tuhumury, M T F ... 653
Tupamahu, L ... 385
Turukay, M ... 237, 345
Umanailo, M C B ... 653
Utami, A W ... 428
Uyara, L ... 438
Veerman, M ... 255
Vinolina, N S ... 401
Wahyudin, C I ... 102, 106
Wahyuni, H D ... 203
Waluyo, N ... 74, 408
Wartono ... 44
Wattimena, A Y ... 9, 317, 322
Wenno, M ... 642
Widada, A W ... 244
Widiyanto, A ... 219
Wikan, T ... 408
Wulandari, R A ... 93
Yunianto, A E ... 50
Zahera, R ... 68